HEINEMANN PHYSICS 12

5TH EDITION

COORDINATING AUTHOR
Daniela Nardelli

AUTHORS
Liz Black
Amber Dommel
Norbert Dommel
Tracey Fisher
Kaitlin Jobson
Geoff Lewis
Greg Moran
Gregory White

VCE UNITS 3 AND 4 • 2024–2027

Pearson Australia
(a division of Pearson Australia Group Pty Ltd)
459–471 Church Street,
Level 1, Building B
Richmond, Victoria 3121
www.pearson.com.au

VCE Heinemann Project Leads: Fiona Cooke, Bryonie Scott, Misal Belvedere, Malcolm Parsons
Physics Content and Learning Specialists: Zoe Hamilton, Sam Trafford
Development Editor: Lucy Bates
Schools Programme Manager: Michelle Thomas
Production Editor: Chris Woods
Series Designer: Anne Donald
Rights & Permissions Editor: Amirah Fatin Binte Mohamed Sapi'ee
Publishing Services Analyst: Jit-Pin Chong
Design Analyst: Jennifer Johnston
Illustrators: Bruce Rankin, DiacriTech and QBS Learning
Editor: Geoffrey Marnell
Proofreader: Jane Fitzpatrick
Indexer: Max McMaster

Printed in Singapore by Markono/02 (2025)

National Library of Australia Cataloguing-in-Publication entry

A catalogue record for this book is available from the National Library of Australia

ISBN: 978 0 6557 0010 4

Pearson Australia Group Pty Ltd ABN 40 004 245 943

Disclaimer
The selection of internet addresses (URLs) provided for this book was valid at the time of publication and was chosen as being appropriate for use as a secondary education research tool. However, due to the dynamic nature of the internet, some addresses may have changed, may have ceased to exist since publication, or may inadvertently link to sites with content that could be considered offensive or inappropriate. While the authors and publisher regret any inconvenience this may cause readers, no responsibility for any such changes or unforeseeable errors can be accepted by either the authors or the publisher.

Indigenous Australians
We respectfully acknowledge the traditional custodians of the lands upon which the many schools throughout Australia are located. We acknowledge traditional Indigenous Knowledge systems are founded upon a logic that developed from Indigenous experience with the natural environment over thousands of years. This is in contrast to Western science, where the production of knowledge often takes place within specific disciplines. As a result, there are many cases where such disciplinary knowledge does not reflect Indigenous worldviews, and can be considered offensive to Indigenous peoples.

Some of the images used in *Heinemann Physics 12* 5th edition might have associations with deceased Indigenous Australians. Please be aware that these images might cause sadness or distress in Aboriginal or Torres Strait Islander communities.

Practical activities
All practical activities, including the illustrations, are provided as a guide only and the accuracy of such information cannot be guaranteed. Teachers must assess the appropriateness of an activity and take into account the experience of their students and facilities available. Additionally, all practical activities should be trialled before they are attempted with students and a risk assessment must be completed. All care should be taken and appropriate personal protective clothing and equipment should be worn when carrying out any practical activity. Although all practical activities have been written with safety in mind, Pearson Australia and the authors do not accept any responsibility for the information contained in or relating to the practical activities, and are not liable for loss and/or injury arising from or sustained as a result of conducting any of the practical activities described in this book.

HEINEMANN PHYSICS 12

5TH EDITION

Writing and development team

We are grateful to the following people for their time and expertise in contributing to the **Heinemann Physics 12** project.

Daniela Nardelli
Learning specialist
Coordinating author and reviewer

Doug Bail
Education consultant
SPARKlab developer

Stephanie Bernard
PhD candidate
Science communicator
Answer checker and digital question author

Liz Black
Teacher
Author and digital question author

Amber Dommel
Teacher and science coordinator
Author

Norbert Dommel
Lecturer
Author

Tracey Fisher
Lecturer
Author

Richard Hecker
Tutor
Digital question author

Kaitlin Jobson
Teacher
Author and digital question author

Allan Knight
Science education consultant
Reviewer

Geoff Lewis
Retired teacher
Author

Svetlana Marchouba
Laboratory technician
Teacher support author

Veronika Miles
Tutor
Answer checker

Greg Moran
Teacher
Author and digital question author

Michael Ojaimi
Student
Answer checker

Bryonie Scott
Science communicator
Digital question author

Gregory White
Lecturer
Author

The Publisher wishes to thank and acknowledge Doug Bail, Keith Burrows, Robert Chapman, Carmel Fry and Mark Hamilton for their contribution in creating the original works of the series and longstanding dedicated work with Pearson and Heinemann.

Contents

How to use this book

Heinemann Physics 12 5th edition

Heinemann Physics 12 5th edition has been written to the new VCE Physics Study Design 2024–2027. The book covers Units 3 and 4. Explore how to use this book below.

Case study

Case studies place physics in an applied situation or relevant context. Text and artwork refer to the nature and practice of physics, applications of physics and associated issues, and the historical development of physics concepts and ideas.

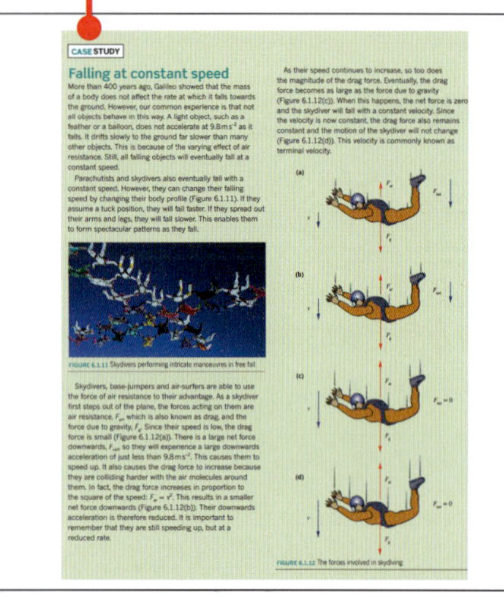

Case study: analysis

These case studies include real-world data that can be analysed and evaluated.

CHAPTER 06 Application of field concepts

The gravitational, magnetic and electric fields introduced in previous chapters have important applications. This chapter considers some of those applications. In particular, you will use the knowledge you gained about fields to understand the motion of satellites (natural and artificial), to see how electric motors operate and to predict how charged particles will behave in electric and magnetic fields.

Key knowledge

- model natural and artificial satellite motion as uniform circular motion **6.1**
- analyse the use of an electric field to accelerate a charge, including:
  - potential energy changes in a uniform electric field: $W = qV$, $E = \frac{V}{d}$ **6.3**
  - the magnitude of the force on a charged particle due to a uniform electric field: $F = qE$ **6.3**
- analyse the use of a magnetic field to change the path of a charged particle, including:
  - the magnitude and direction of the force applied to an electron beam by a magnetic field: $F = qvB$, in cases where the directions of v and B are perpendicular or parallel **6.3**
  - the radius of the path followed by an electron in a magnetic field: $qvB = \frac{mv^2}{r}$, where $v \ll c$ **6.3**
- apply the concepts of force due to gravity and normal force including in relation to satellites in orbit where the orbits are assumed to be uniform and circular **6.1**
- model satellite motion (artificial, Moon, planet) as uniform circular orbital motion: $a = \frac{v^2}{r} = \frac{4\pi^2 r}{T^2}$ **6.1**
- investigate and analyse theoretically and practically the force on a current carrying conductor due to an external magnetic field, $F = nIlB$, where the directions of I and B are either perpendicular or parallel to each other **6.2**
- investigate and analyse theoretically and practically the operation of simple DC motors consisting of one coil, containing a number of loops of wire, which is free to rotate about an axis in a uniform magnetic field and including the use of a split ring commutator **6.2**
- investigate, qualitatively, the effect of current, external magnetic field and the number of loops of wire on the torque of a simple motor **6.2**
- model the acceleration of particles in a particle accelerator (including synchrotrons) as uniform circular motion (limited to linear acceleration by a uniform electric field and direction change by a uniform magnetic field) **6.3**

VCE Physics Study Design extracts © VCAA (2022); reproduced by permission

Chapter opener

Chapter opening pages link the study design to the chapter content. Key knowledge addressed in the chapter is clearly listed. To help you find where each outcome is covered in the chapter, the relevant section numbers are written in bold.

Formula

Formula boxes focus on key formulas students that will need to apply in section and chapter reviews.

Worked example

Worked examples are set out in steps that show thinking and working. Each Worked example is followed by a Worked example: Try yourself that allows students to test their understanding.

Highlight

Highlight boxes focus on important information such as key definitions and summary points.

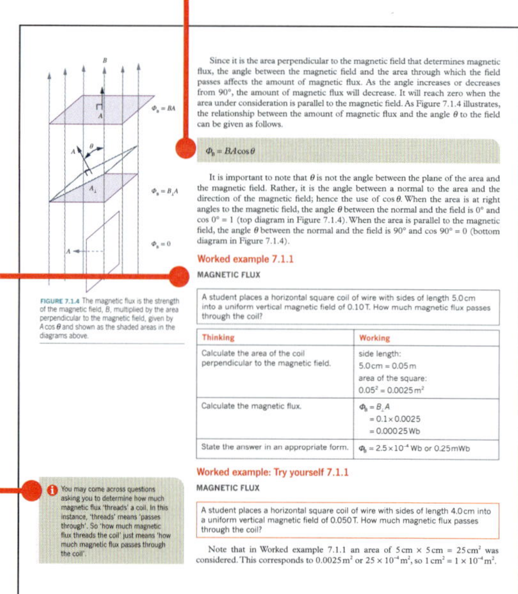

Since it is the area perpendicular to the magnetic field that determines magnetic flux, the angle between the magnetic field and the area through which the field passes affects the amount of magnetic flux. As the angle increases or decreases from 90°, the amount of magnetic flux will decrease. It will reach zero when the area under consideration is parallel to the magnetic field. As Figure 7.1.4 illustrates, the relationship between the amount of magnetic flux and the angle θ to the field can be given as follows.

$$\Phi_B = BA\cos\theta$$

It is important to note that θ is not the angle between the plane of the area and the magnetic field. Rather, it is the angle between a normal to the area and the direction of the magnetic field; hence the use of $\cos\theta$. When the area is at right angles to the magnetic field, the angle θ between the normal and the field is 0° and $\cos 0° = 1$ (top diagram in Figure 7.1.4). When the area is parallel to the magnetic field, the angle θ between the normal and the field is 90° and $\cos 90° = 0$ (bottom diagram in Figure 7.1.4).

FIGURE 7.1.4 The magnetic flux is the strength of the magnetic field, B, multiplied by the area perpendicular to the magnetic field, given by $A\cos\theta$ and shown as the shaded areas in the diagrams above.

Worked example 7.1.1

MAGNETIC FLUX

A student places a horizontal square coil of wire with sides of length 5.0 cm into a uniform vertical magnetic field of 0.10 T. How much magnetic flux passes through the coil?

Thinking	Working
Calculate the area of the coil perpendicular to the magnetic field.	side length: 5.0 cm = 0.05 m area of the square: $0.05^2 = 0.0025\,m^2$
Calculate the magnetic flux.	$\Phi_B = B_\perp A = 0.1 \times 0.0025 = 0.00025\,Wb$
State the answer in an appropriate form.	$\Phi_B = 2.5 \times 10^{-4}\,Wb$ or 0.25 mWb

Worked example: Try yourself 7.1.1

MAGNETIC FLUX

A student places a horizontal square coil of wire with sides of length 4.0 cm into a uniform vertical magnetic field of 0.050 T. How much magnetic flux passes through the coil?

You may come across questions asking you to determine how much magnetic flux 'threads' a coil. In this instance, 'threads' means 'passes through'. So 'how much magnetic flux threads the coil' just means 'how much magnetic flux passes through the coil'.

Note that in Worked example 7.1.1 an area of 5 cm × 5 cm = 25 cm² was considered. This corresponds to 0.0025 m² or 25 × 10⁻⁴ m², so 1 cm² = 1 × 10⁻⁴ m².

Icons

This icon is used to alert you to engage with auto-corrected questions through Pearson Places.

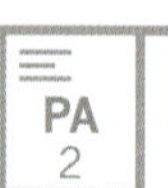

These icons indicate when it is the best time to engage with a worksheet (WS), a practical activity (PA) or exam questions (EQ) in the *Heinemann Physics 12 Skills and Assessment* book.

Chapter review

Each chapter concludes with a list of key terms and questions that test your understanding of the key knowledge covered in the chapter.

Section summary

Each section includes a summary to help you consolidate key points and concepts.

Section review

Each section concludes with questions that test your ability to recall, explain and apply key concepts.

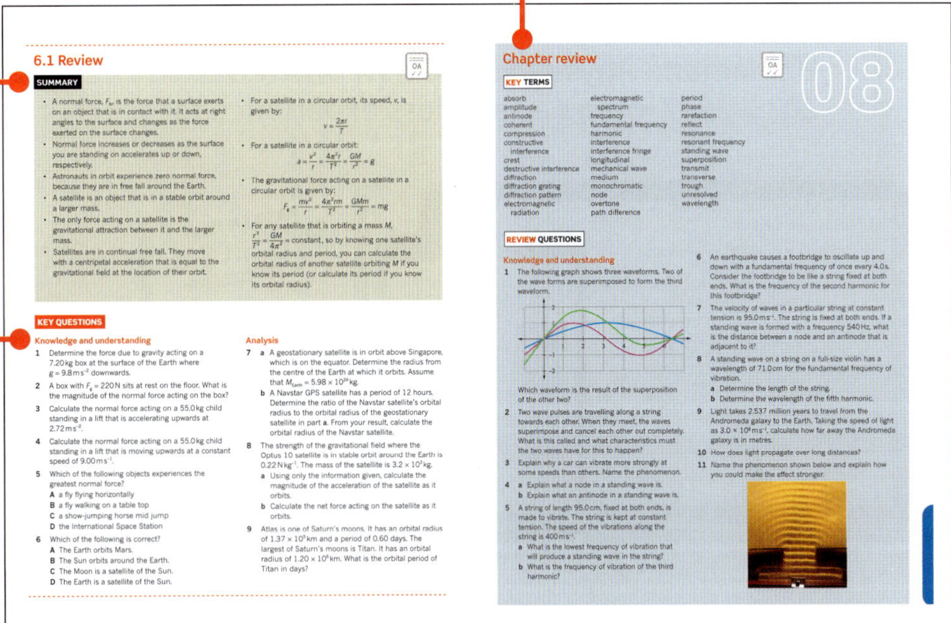

Area of Study review

Each area of study concludes with a comprehensive set of exam-style questions, including multiple-choice and short-answer, to support you in your exam preparation.

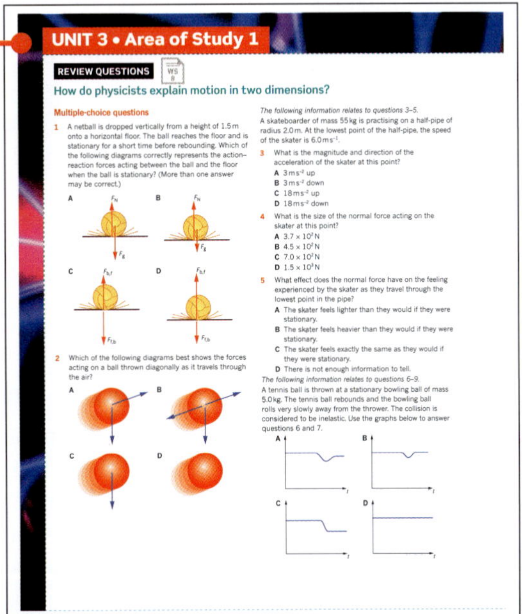

PhysicsFile

PhysicsFiles include interesting information and real-world examples.

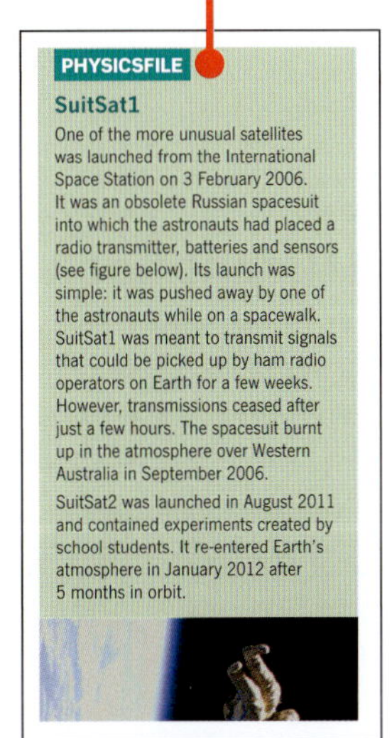

Answers

Comprehensive answers for all section review, chapter review and Area of Study review questions are provided via the *Heinemann Physics 12 5th edition eBook + Assessment.*

Glossary

Key terms are shown in **bold** throughout and listed at the end of each chapter. A comprehensive glossary at the end of the book defines all key terms.

Revision

Revision sections contain vital information from Year 11.

CHAPTER

01 Scientific investigation

Gaining a sound understanding of key science skills in a variety of contexts is essential preparation for undertaking scientific investigations and evaluating the research of others. The development of such skills is a core component of VCE Physics, and these skills apply across all units and areas of study.

Chapter 1 describes some of the most fundamental science skills. The chapter can be read as a whole or referred to as you work through other chapters. For example, it may provide you with a useful refresher on the scientific method of conducting investigations, or on what is typically included in a scientific report, at the time when you are to undertake an investigation as part of your studies.

Key science skills

Develop aims and questions, formulate hypotheses and make predictions

- identify, research and construct aims and questions for investigation **1.1**
- identify independent, dependent and controlled variables in experiments **1.1**
- formulate hypotheses to focus investigations **1.1**
- predict possible outcomes of investigations **1.1**

Plan and conduct investigations

- determine appropriate investigation methodology: case study; classification and identification; experiment; fieldwork; literature review; modelling; product, process or system development; simulation **1.1**
- design and conduct investigations: select and use methods appropriate to the selected investigation methodology, including consideration of equipment and procedures, taking into account potential sources of error and causes of uncertainty; determine the type and amount of qualitative and/or quantitative data to be generated or collated **1.2**
- work independently and collaboratively as appropriate and within identified research constraints, adapting or extending processes as required and recording such modifications in a logbook **1.2**

Comply with safety and ethical guidelines

- demonstrate safe laboratory practices when planning and conducting investigations by using risk assessments that are informed by safety data sheets (SDS), and accounting for risks **1.2**
- apply relevant occupational health and safety guidelines while undertaking practical investigations **1.2**
- demonstrate ethical conduct when undertaking and reporting investigations **1.2**

Generate, collate and record data

- systematically generate and record primary data, and collate secondary data, appropriate to the investigation, including use of databases and reputable online data sources **1.3**
- record and summarise both qualitative and quantitative data, including use of a logbook as an authentication of generated or collated data **1.2, 1.3**
- organise and present data in useful and meaningful ways, including tables and graphs **1.3**

Analyse and evaluate data and investigation methods

- process quantitative data using appropriate mathematical relationships and units **1.4**
- use appropriate numbers of significant figures in calculations **1.3, 1.4**
- construct graphs that show the relationship between variables **1.4**
- extrapolate to determine graph intercepts of significance **1.4**
- construct linearised graphs and identify the significance of the gradient (using relationships relevant to the key knowledge outlined in the areas of study) **1.4**
- identify and analyse experimental data qualitatively, handling, where appropriate, concepts of: accuracy, precision, repeatability, reproducibility, resolution and validity of measurements; and errors (random and systematic) **1.2, 1.3, 1.4**
- identify outliers, and contradictory, provisional or incomplete data **1.4**
- repeat experiments to evaluate the precision of data **1.4**
- evaluate investigation methods and possible causes of error and uncertainty, and suggest how precision can be improved, and how uncertainty can be reduced **1.4, 1.5**

Construct evidence-based arguments and draw conclusions

- distinguish between opinion and evidence, and between scientific and non-scientific ideas **1.5**
- evaluate data to determine the degree to which the evidence supports the aim of the investigation, and make recommendations, as appropriate, for modifying or extending the investigation **1.5**
- evaluate data to determine the degree to which the evidence supports or refutes the initial prediction or hypothesis **1.5**
- use reasoning to construct scientific arguments, and to draw and justify conclusions consistent with evidence and relevant to the question under investigation **1.5**
- identify, describe and explain the limitations of conclusions, including identification of further evidence required **1.5**
- discuss the implications of research findings **1.5**

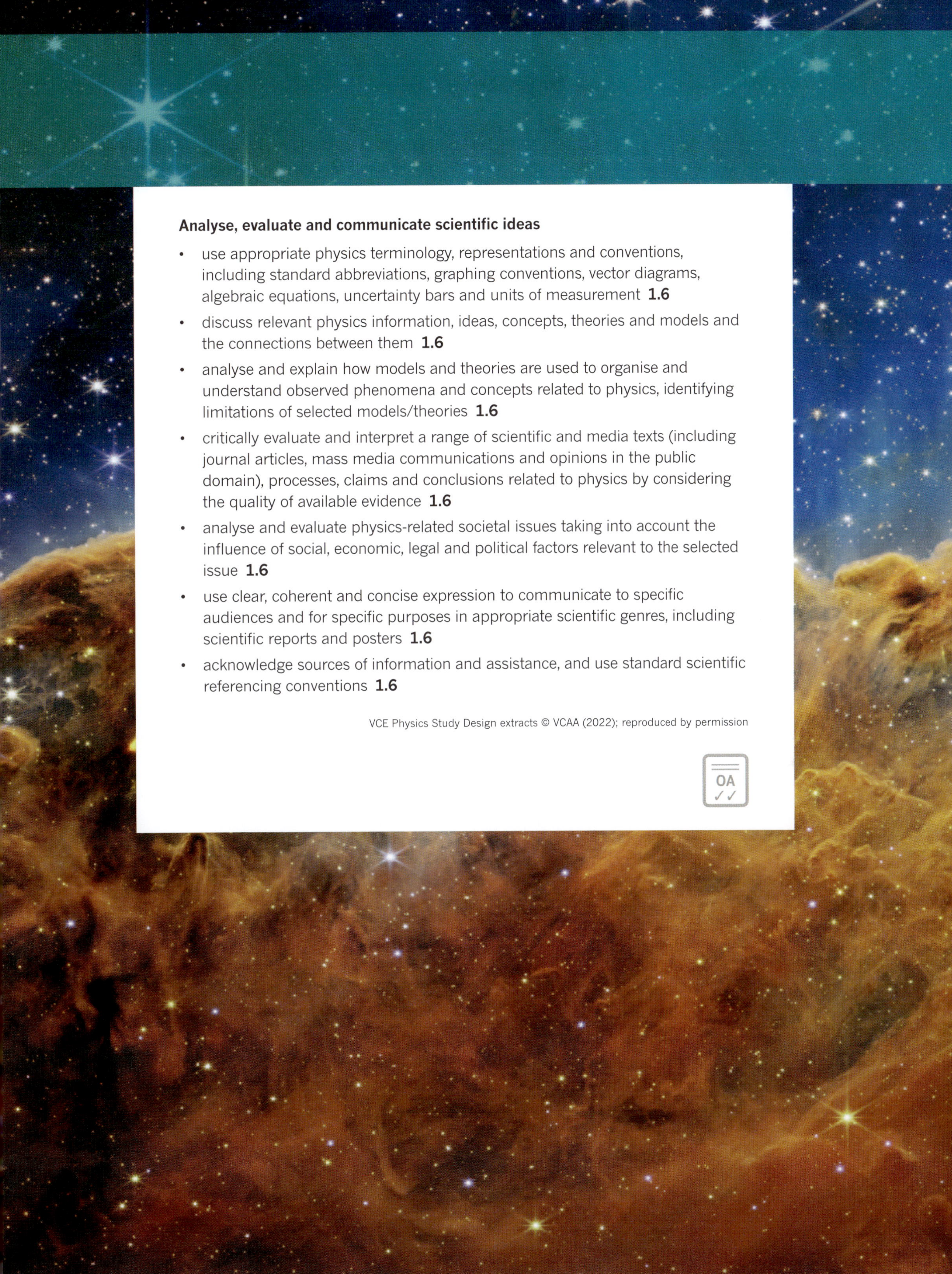

Analyse, evaluate and communicate scientific ideas

- use appropriate physics terminology, representations and conventions, including standard abbreviations, graphing conventions, vector diagrams, algebraic equations, uncertainty bars and units of measurement **1.6**
- discuss relevant physics information, ideas, concepts, theories and models and the connections between them **1.6**
- analyse and explain how models and theories are used to organise and understand observed phenomena and concepts related to physics, identifying limitations of selected models/theories **1.6**
- critically evaluate and interpret a range of scientific and media texts (including journal articles, mass media communications and opinions in the public domain), processes, claims and conclusions related to physics by considering the quality of available evidence **1.6**
- analyse and evaluate physics-related societal issues taking into account the influence of social, economic, legal and political factors relevant to the selected issue **1.6**
- use clear, coherent and concise expression to communicate to specific audiences and for specific purposes in appropriate scientific genres, including scientific reports and posters **1.6**
- acknowledge sources of information and assistance, and use standard scientific referencing conventions **1.6**

OA
✓✓

1.1 Planning scientific investigations

FIGURE 1.1.1 Students conduct an experiment investigating the wave nature of light.

Physics is the study of motion, force and energy. It deals with the laws, theories and principles governing motion, force and energy across all imaginable scales throughout the universe. Its scope ranges from the interactions between particles that make up individual protons and neutrons inside every atomic nucleus to the motions of galaxies that comprise the universe.

As scientists, physicists extend their understanding using the scientific method, which involves investigations that are carefully designed, conducted and reported (Figure 1.1.1). Well-designed research is based on a sound knowledge of what is already understood about a subject, as well as careful preparation and observation.

Taking the time to carefully plan and design a scientific investigation before beginning will help you keep focused throughout. Preparation is essential. You should ensure that you understand the theory behind your investigation and prepare a detailed plan regarding the practical aspects of the investigation. This section is a guide to some of the key steps that should be taken when planning and designing a scientific investigation.

OBSERVATION

Scientific investigations start with careful **observation**. Observation involves using all your senses: sight, sound, smell, taste and touch. For example, an observation may involve seeing a change in colour in an object or noticing the smell from a chemical reaction. It also involves using instruments and laboratory techniques that may allow observations you cannot make otherwise, or that are more detailed or more safely obtained (for example, using sensors to detect the radioactivity of a source).

The idea for a primary investigation of a complex problem arises from prior learning and observations that have raised further questions. How observations are interpreted depends on past experiences and knowledge. But to enquiring minds, observations will usually provoke further questions. Some examples are given below.

- What velocity must a satellite be moving in order to travel in a geosynchronous orbit?
- How does the motion of charged particles in an electric field compare to the movement of an object with mass in a gravitational field?
- What causes interference in signals travelling along a wire?
- What modifications could be made to improve the efficiency of a transformer?
- What applications are there of special relativity?
- What information does the spectrum from a star reveal?

Many of these questions cannot be answered by observation alone, but they can be answered through scientific investigation.

Many great discoveries have been made when a scientist is busy investigating some other problem. Good scientists have acute powers of observation and enquiring minds, and they make the most of these chance opportunities.

THE SCIENTIFIC METHOD

Before conducting an investigation, a scientist develops a clear and specific research question to explore. They state an aim that describes the purpose of their planned investigation. They then state a hypothesis, that is, a prediction based on scientific reasoning that can be tested experimentally. This is the basis of the **scientific method** (Figure 1.1.2).

- A **research question** describes the idea to be investigated. For example: What is the relationship between voltage and current?
- An **aim** is a statement describing in detail the purpose of the investigation. For example: The aim of the investigation is to investigate the relationship between voltage and current in a circuit of constant resistance.

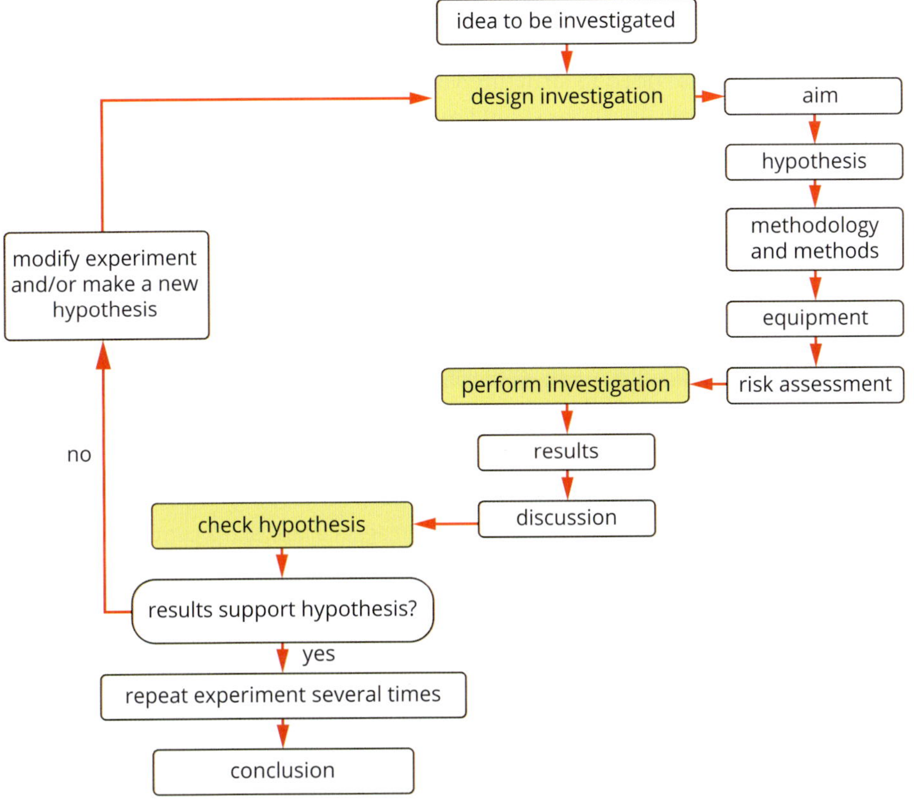

FIGURE 1.1.2 A flow chart illustrating the scientific method

- A **hypothesis** is a prediction that proposes an answer to a research question. It is a prediction that can be tested by observation or experimentation. For example: For the 5 Ω resistor used in this investigation, the current will increase as the voltage is increased from 3 V to 12 V.

Carefully designed scientific investigations are conducted to determine whether predictions are accurate or not. If the results of an investigation do not fall within an acceptable range, the hypothesis is rejected. If the predictions are found to be accurate, the hypothesis is supported. If, after many investigations, one hypothesis is supported by all the results obtained so far, then it is given the status of a theory or principle.

There is nothing mysterious about the scientific method. You might use the same process to find out how an unfamiliar machine works if you had no instructions. Careful observation is usually the first step.

Formulating a research question

The research question at the centre of a scientific investigation directs the inquiry. Its primary purpose is to clearly set the boundaries of the investigation, specify the direction of the research, and guide all stages of inquiry, analysis, interpretation and evaluation.

A research question should:

- clearly identify the topic of the experiment
- be specific enough to ensure a clear and unambiguous approach
- specify the scope or conditions of the inquiry
- propose to find trends, patterns or relationships between two measurable **variables**.

Relevant background research can help refine the question. This could include:

- information about the variables to be explored
- correlations between these variables
- ideas for refining the question.

PHYSICSFILE

The scientific method and gravitational waves

In the early part of the twentieth century, Einstein formulated the general theory of relativity. This theory predicted gravitational waves. Although hypothesised, scientists of the day were unsure how to detect them. Two American scientists, Kip Thorne from California Institute of Technology (Caltech) and Rainer Weiss from Massachusetts Institute of Technology (MIT), started to collaborate on possible ways to detect these waves. In 1984 Caltech and MIT set up a joint project called LIGO (Laser Interferometer Gravitational Wave Observatory).

There are over 1000 scientists working on the LIGO project. They are busy interpreting readings from their detection equipment, improving that equipment and, where necessary, modifying their hypotheses.

The LIGO team finally observed gravitational waves in 2015, almost 100 years after Einstein predicted them. It is the nature of scientific research and discovery that no scientist works alone. They work with others and build on the research done by their colleagues and predecessors.

To build on the discoveries of the LIGO team, the OzGrav program was started. OzGrav is a project funded by the Australian Government through the Australian Research Council, several Australian universities and other collaborating organisations in Australia and overseas. The aim of OzGrav is to understand the physics of black holes, and to inspire future Australian scientists and engineers to continue this field of study.

The structure of a research question

Table 1.1.1 provides examples of different types of research questions. You can use the guiding words provided to structure your research question.

If possible, research questions should make reference to the independent and the dependent variables that are to be explored. That is, the question should reference the variables the researcher will be manipulating (independent variables) and those that are expected to change in response (dependent variables). Each question should ask if, how or why the independent variable affects the dependent variable.

TABLE 1.1.1 Examples of how research questions can be constructed

Guiding word	Example research question
What	**What** effect does temperature have on the induced voltage produced by a generator?
Will	**Will** a motor rotate at a higher rate if the current through it is increased?
How	**How** does the angle between the magnetic field and a current-carrying wire affect the force on the wire?
Why	**Why** does the angle of release affect the range of a projectile?
Is/are	**Is** the speed of light the same for each colour of the rainbow? **Are** Kepler's laws valid for planets outside our solar system?
Can	**Can** standing waves form in air?
Do/does	**Does** the strength of the Earth's gravitational field depend on the height above sea level?

Before formulating a research question, it is good practice to:

1. conduct a literature review of the topic to be investigated
2. become familiar with the relevant scientific concepts and key terms
3. write down questions or correlations as they arise
4. compile a list of possible ideas.

Avoid rejecting ideas that initially seem impossible. Use these ideas to generate questions that are answerable.

Selecting and evaluating your question

When selecting a topic for your investigation, it is helpful to choose something that you already have some knowledge of and that you find interesting.

Ask yourself what the data you intend to collect might look like. If you cannot determine measurable variables, either pick another topic or modify your question to allow you to take measurements.

You should choose a question that can be explored within the conditions, time and equipment you have available to you.

Once a research question has been chosen, stop and evaluate it. Check if it requires further refinement or even further investigation before it is suitable as the basis for an achievable and worthwhile investigation. The checklist below will assist you in evaluating your research question.

- ☐ Relevance—make sure your research question is related to your chosen topic.
- ☐ Clarity and measurability—make sure your question can be framed as a clear hypothesis, otherwise it is going to be difficult to complete your research.
- ☐ Time frame—make sure your question can be answered within a reasonable period of time. This may not be possible if your question is too broad.
- ☐ Knowledge and skills—make sure you have the level of knowledge and the laboratory, fieldwork and technical skills that will enable you to explore the question within the specified time frame. Keep the question simple and the outcomes of the research achievable.

- ☐ Practicality—make sure that the resources that you will need, such as laboratory equipment and materials, are readily available. Keep things simple. Avoid investigations that require sophisticated or rare equipment. Instead, choose an investigation that requires only common laboratory equipment.
- ☐ Safety and ethics—consider the safety and ethical issues associated with the research question you will be investigating. If there are issues, determine if these need to be addressed before you begin your investigation. (Safety and ethics is discussed in more detail in Section 1.2.)
- ☐ Advice—seek advice from your teacher on your question. Their experience may lead you to consider aspects of the question that you have not thought about.

Finally, confirm the suitability of your research question with your teacher.

Defining the aim of the investigation

The aim sets out the purpose of the scientific investigation. It should directly mention the variables involved in the investigation and describe, in general terms, how each will be measured. An aim does not need to include the details of the proposed method.

An example for a laboratory experiment:

- Aim: The aim of the experiment is to investigate the relationship between force, mass and acceleration.
- In the first stage of this experiment, mass will be the independent variable (a number of different masses will be selected) and the force kept constant. The resulting acceleration (dependent variable) will be measured.
- In the second stage of the experiment, force will be the independent variable (a number of different forces will be tested) and the mass will be kept constant. The resulting acceleration (dependent variable) will be measured.
- Using the data collected from both stages of the experiment, the relationship between the three variables can be determined.

Formulating a hypothesis

Once the research question has been finalised and the aim of the research is clear, formulating a hypothesis is the next step. A hypothesis is the prediction that is to be tested by the evidence you will collect during your investigation. A hypothesis proposes a relationship between two or more variables. It should predict that a relationship exists or does not exist.

When writing the hypothesis, first identify the variables in your research question. There are different ways of constructing your hypothesis, but you may want to write the statement in the form of 'If ... is done, then ... will occur.' Ensure that the independent and dependent variables are included in this statement.

For example: If force (the independent variable) is increased, then acceleration (the dependent variable) will increase in proportion to the force.

A hypothesis does not need to include 'if' and 'then' in its wording. For example, the previous hypothesis could also be stated in the following way: As force increases, acceleration increases proportionally.

A good hypothesis should:

- be a statement, not a question
- be based on information contained in the research question and the aim
- be worded so that it can be tested in the experiment you are planning
- include an independent and a dependent variable
- include only variables that are measurable.

The hypothesis should also be falsifiable. This means that a negative outcome would disprove it. For example, the hypothesis that all apples are round cannot be proved beyond doubt, but it can be disproved—in other words, it is falsifiable. In fact, only one square apple is needed to disprove this hypothesis. Unfalsifiable hypotheses cannot be proved by science. These include hypotheses on ethical, moral and other subjective judgements. For instance, you could hypothesise that plagiarising your scientific report is wrong, but the results of this are a question of ethics not science.

PHYSICSFILE

Hypothetical neutrino

In 1934 Enrico Fermi proposed the nuclear electron hypothesis. He explained that the beta particles observed from radioactive processes were from the radioactive decay of a neutron. Theory suggested that a neutron would decay into a positively charged proton, a negatively charged electron, gamma radiation and a then-theoretical particle called the neutrino. This would meet the requirements of the conservation of energy and momentum. Experimental proof of the idea took another 20 years.

Variables

A good scientific hypothesis can be tested, that is, supported or refuted through investigation. To be a testable hypothesis, it must be possible to measure both what is changed and what will happen as a result. Thus a scientific investigation seeks to determine the relationship between variables.

There are three categories of variable:

- The **independent variable** is the variable that is manipulated by the researcher (that is, the one that is changed to see if there is an effect on the dependent variable).
- The **dependent variable** is the variable that may change in response to a change in the independent variable. This is the variable that will be observed.
- **Controlled variables** are the variables that must be kept constant during the investigation.

You should only test one variable at a time, otherwise you cannot be sure which variable has influenced a change in the dependent variable.

Completing a table like the one in Table 1.1.2 will assist you in evaluating the research question and formulating the hypothesis.

TABLE 1.1.2 Extract the relevant variables from your research question and formulate a hypothesis

Part of investigation	Example
research question	How does the angle of release of an arrow undergoing projectile motion affect the range of flight?
independent variable	the angle of release of the arrow
dependent variable	range of flight
controlled variables	• mass of the arrow • design of the arrow • extension of the bow (elastic potential energy stored in the system) • height of release of the arrow • height at which the arrow lands • release velocity of the arrow
hypothesis	If the release angle of a projectile is 45°, then the range of flight will be greatest.

Qualitative and quantitative variables

Variables are either qualitative or quantitative, with further subsets in each category.

- **Qualitative variables** can be observed but not measured. The data collected is known as **qualitative data**. They can only be sorted into groups or categories, such as brightness, type of material of construction or type of device. There are two main types:
  - Nominal variables are those in which the order is not important, for example, the type of material or type of device.
  - Ordinal variables are those in which order is important and values are therefore ranked, for example, brightness (Figure 1.1.3).
- **Quantitative variables** can be measured. The data collected is known as **quantitative data**. Length, area, weight, temperature and cost are examples of quantitative data. There are two main types:
  - **Discrete variables** can only take particular values, for example, the number of pins in a packet, the number of springs connected together or the energy levels in atoms.
  - **Continuous variables** allow for any numerical value within a given **range**, for example, the measurement of temperature, length, mass and frequency.

FIGURE 1.1.3 When recording qualitative data, describe in detail how each variable will be defined. For example, if recording the brightness of light globes, pictures are a good way of clearly defining what each assigned term represents.

Methodology and methods

When planning your investigation you will need to decide on the methodology and methods.

The **methodology** section in a research plan is a brief description of the general approach taken to investigate the research question or hypothesis and the reasons why this approach is taken. Examples of methodologies are a controlled experiment, fieldwork, a literature review, modelling and a simulation.

The **method** (also known as the procedure) is the set of specific steps that are to be taken to collect data during the investigation. The type of methodology and the methods selected will depend on the research question, the aim of the investigation and the equipment available to you.

For some investigations, setting up an experiment may require equipment that is not readily available to you in the school setting. This may mean you need to consider a computer simulation to model the outcomes of the investigation. Other approaches could include a literature review of other studies that considered a similar research question. The different approaches that you could use are outlined in Table 1.1.3.

TABLE 1.1.3 Scientific investigation methodologies

Type of methodology	Explanation	Example
case study	an investigation of a real or hypothetical situation, such as an activity, event, problem or behaviour, often involving analysis of data within a real-world context	determining the wavelength at which radiated power per unit area is maximum at any given temperature
classification and identification	arranging objects or events into manageable groups by identifying shared or similar features	classifying different stars in the sky
experiment	an experimental investigation that involves formulating a hypothesis and testing the effect of an independent variable on the dependent variable, while controlling all other variables in the experiment	investigating the effect of an object's mass on its momentum
fieldwork	collecting data outside the laboratory, such as from an excursion or by engaging with community experts	establishing the path of the Moon in the night sky
literature review	a critical analysis of what has already been investigated and published, using secondary data from other people's investigations or from experimental research to explain events or propose new ideas or relationships	analysing data looking at the impact of electromagnetic radiation on cells published in a variety of research papers to support, refute or develop a new hypothesis
modelling	using models as representations of objects, systems or processes to aid understanding or make predictions	constructing models of the atom
product, process or system development	using scientific understanding and advances in technology to design a new tool, method or process to meet the demands or needs of society	developing a new cochlear implant to restore hearing
simulation	using mathematical models or simulations to test hypotheses, conduct virtual experiments or model the complexity of a system or process	performing a computer simulation of the spectrum of light emitted by a blackbody as a function of wavelength

SOURCING INFORMATION

When you are sourcing information for your scientific investigation, consider whether the information is from primary or secondary sources. You should also consider the advantages and disadvantages of using such resources as books or the internet.

Primary and secondary sources

Primary sources of information are those created by a person directly involved in an investigation. An example is a peer-reviewed scientific article. **Secondary sources** of information are syntheses, reviews or interpretations of primary sources. Examples are textbooks, newspaper articles and websites.

Secondary sources of information may have a bias, so you need to determine if they are reliable sources of information. You will learn about assessing the accuracy, reliability and validity of data in Section 1.2.

Table 1.1.4 compares common primary and secondary sources.

TABLE 1.1.4 Summary of common primary and secondary sources

	Primary sources	Secondary sources
Characteristics	• first-hand records of events or experiences • written at the time the event happened • original documents	• interpretation of primary sources • written by people who did not see or experience the event • use information from original documents but rework it
Examples	• results of experiments • articles in scientific journals or magazines • reports of scientific discoveries • photographs, specimens, maps and artefacts • interviews with experts • websites (if they meet the criteria above)	• textbooks • biographies • newspaper articles • magazine articles • radio and television documentaries • websites that interpret the scientific work of others • podcasts

1.1 Review

SUMMARY

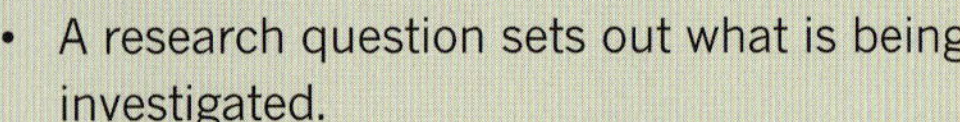

- A research question sets out what is being investigated.
- An aim is a statement describing in detail what will be investigated.
- A hypothesis is a prediction about the results of the investigation.
- Once a research question has been chosen, evaluate the question before proceeding.
- There are three main categories of variables:
  - The independent variable is the variable that is manipulated by the researcher (that is, the one that is changed to see if there is an effect on the dependent variable).
  - The dependent variable is the variable that may change in response to a change in the independent variable.
  - Controlled variables are all the variables that must be kept constant during the investigation.
- Qualitative data is unmeasurable and is comprised of descriptions.
- Qualitative data can be characterised as either:
  - nominal, when the order of the data is not important
  - ordinal, when the order of the data is important.
- Quantitative data is measurable and comprised of values, either directly observed or reported by instruments.
- Quantitative data can be characterised as either:
  - discrete, when the data can only be particular numerical values
  - continuous, when the data is not restricted to particular numerical values.
- The methodology describes the general approach or style of the investigation.
- The method (or procedure) is the set of specific steps taken to collect the data.
- Primary sources are those created by a person directly involved in the investigation.
- Secondary sources are syntheses, reviews or interpretations of primary sources.

KEY QUESTIONS

Knowledge and understanding

1 Define the term 'hypothesis'.

2 State the meaning of the term 'variable'.

3 Describe the differences between quantitative data and qualitative data.

4 The following steps in the scientific method are out of order. Rewrite them in your notebook in the correct order.
- form a hypothesis
- collect the results
- plan the experiment and equipment
- draw conclusions
- question whether the results support the hypothesis
- state the research question to be investigated
- perform the experiment

5 A student wants to investigate whether you can hit a ball harder with a two-handed grip of a bat instead of a one-handed grip. What would be the independent variable in the experiment?

6 Consider the hypothesis below. What are the dependent, independent and controlled variables?
Hypothesis: 'Releasing an arrow in archery at an angle greater or smaller than 45° will result in a shorter flight displacement (range) compared with a release angle of 45°'.

7 Consider the following four options.

A How does the diffraction pattern of light depend on the wavelength?

B If the wavelength of light is increased, then the spread of the diffraction pattern also increases.

C Diffraction and refraction have opposite trends with respect to wavelength.

D To investigate the effect of changing wavelength on the amount light that is diffracted.

a Which option (**A** to **D**) is a research question?

b Which option (**A** to **D**) is an aim?

c Which option (**A** to **D**) is a hypothesis?

8 For each of the following hypotheses, state the independent and dependent variables.

a If water at 90°C is allowed to cool to room temperature in different-shaped containers, the container with the largest surface area will reach room temperature in the shortest time.

b If you increase the thickness of foam bumpers attached to the front of a cart travelling at $1\,m\,s^{-1}$, then the force experienced during a head-on collision decreases.

c If the number of coils in the motor is increased, the force and therefore the torque will also increase.

d If you push an object of fixed mass (for example, a shot put) with a larger force, then the acceleration of that object will be greater.

e A springboard diver rotates faster when in a tucked position than when in a stretched position.

9 In an experiment, a student uses the following range of values to describe the brightness of a light: dazzling, bright, glowing, dim, off. What type of variable is brightness?

10 Consider the following research question: 'Is the radial acceleration experienced on the Luna Park roller-coaster proportional to the radius of curvature of the track?'
Which of the following is the independent variable, which is the dependent variable and which is a controlled variable?

a radial acceleration

b weight of the cart

c radius of track curvature

Analysis

11 Which is the most specific research question of the three options below. Explain your choice.

A Will a singly ionised atom with a higher atomic number undergo a smaller deflection angle when fired perpendicularly at the same velocity into a constant external magnetic field?

B Will shiny metals produce more photoelectrons in the photoelectric effect than dull metals?

C Will the diffraction pattern of green laser light become clearer when a narrower slit is used?

12 Select the best hypothesis from the options below. Give reasons for your choice.

A Take-off angular momentum and inertia affect angular (rotational) velocity.

B A body's position during angular airborne motion affects its inertia.

C A springboard diver's angular (rotational) velocity is slower when they hold a stretched (layout) position than when they are in a tuck position, if they take off with the same angular momentum.

1.2 Conducting investigations

Once you have written your research question, clarified the aim, stated the hypothesis and chosen a suitable methodology, you will need to develop a method to conduct the investigation. In this section you will learn about designing and selecting methods to use in scientific investigations in the laboratory and in field work. You will be introduced to different techniques and understand how selecting appropriate equipment and methods will allow you to obtain accurate and precise measurements. Risks and safety precautions will also be discussed.

WRITING A METHOD

The method is a step-by-step procedure that you will follow when conducting your investigation. The method must be detailed enough so that someone else can conduct the investigation using the same steps. It should be recorded in your logbook. For example, the step 'Place a sheet of semiconductive paper flat, printed side up, on the lab bench,' ensures that whenever the method is followed, the paper will be placed the same way up. Number your steps sequentially, covering only one action per step.

You must also ensure that the proposed method describes a way to collect data that is valid, repeatable and reproducible.

Validity

An experiment or investigation has **validity** only if it is actually testing the set research question or hypothesis.

Factors influencing validity include:

- whether your experiment measures what it claims to measure; in other words, your experiment should test your research question or hypothesis
- the certainty that something observed in your experiment was the result of your experimental conditions and not some other cause that you did not consider; in other words, whether the independent variable influenced the dependent variable in the way you have concluded and was not influenced by other variables that should have been kept constant.

Make sure you have identified the independent variable, the dependent variable and the variables you will control in your experiment (and how you will control them). This information should be included in your method. In the example step discussed above, the way the paper is placed (printed side up) is a controlled variable.

You should also be clear about what raw data you will collect (quantitative or qualitative). If necessary, re-read the relevant text about variables in Section 1.1.

Repeatability and reproducibility

Repeatability is the ability to obtain the same results if an investigation is repeated under the exact same set of conditions. (This is sometimes called reliability.) Several steps can be taken to help improve the repeatability of an experiment by reducing the influence of natural variation, random error, uncalibrated instruments or instrument error, and the influence of unforeseen variables. These steps include:

- selecting an appropriate sample size to reduce natural variation, errors and uncertainty
- selecting the appropriate equipment to take the measurements you seek
- taking several measurements under the same conditions (that is, repeat readings of each trial)
- specifying the materials and methods in detail.

Repetition can minimise random errors, but it will not minimise systematic errors. (Errors are discussed later in this section.) Repeating the investigation and averaging the results will generate data that is more reliable. To ensure that all variables are being tested under the same conditions, modifications to your method may need to be considered before repeating the investigation. The goal is to ensure that every measurement can be repeated and the same result obtained (within a reasonable margin of experimental error, such as less than 5%).

Reproducibility is the ability to obtain the same results if an experiment is repeated under different conditions. Different conditions might include a different researcher conducting the experiment, the use of different equipment or instruments, or conducting the experiment at a different time or location. It is important to write a clear and detailed method so that the experiment can be reproduced successfully.

Sample size is extremely important in scientific experiments. A larger sample size:

- reduces the effects of natural variation, errors and uncertainty
- provides more evidence to support the experimental results and conclusions
- improves repeatability (i.e. reliability).

Measurements or observations could vary, so the greater the sample size, the more reliable the data.

EQUIPMENT

When conducting your scientific investigation, it is important to choose the correct equipment. This will ensure that your measurements are accurate (by minimising error) and that your results are reproducible and repeatable.

It is important to understand how to use the equipment correctly and how your choice of equipment will affect the accuracy and precision of the results you collect. You should ensure that the equipment is suitable for the measurement required. For example, a ticker timer (Figure 1.2.1(a)) and a motion detector (Figure 1.2.1(b)) would be suitable choices for measuring the velocity of an object. It is also important to ensure that the equipment is properly calibrated.

COLLECTING AND RECORDING DATA

For an investigation to be scientific, it must be objective and systematic. When working, keep asking questions. Is the work biased in any way? If changes are made, how will they affect the study? Will the investigation still be valid given the aim and hypothesis?

It is essential that you record the following information in your logbook during your investigation:

- all quantitative and qualitative data collected
- the methods used to collect the data
- any incident, feature or unexpected event that may have affected the quality or validity of the data.

The data recorded in a logbook is the **raw data**. Usually this data needs to be processed in some manner before it is presented. If an error occurs in the processing, or you decide to present the data in an alternative format, the recorded raw data will still be available for you to refer back to. How to collect and process your raw data is covered in Section 1.3.

IDENTIFYING ERRORS

When an instrument is used to measure a physical quantity and obtain a numerical value, the aim is to determine the true value. The **true value** is the value, or range of values, that would be obtained if the variable could be measured perfectly. However, for a number of reasons the measured value is often not the true value.

Most practical investigations will have some errors in the data collected. Errors can occur for a variety of reasons. Being aware of potential errors will help you to avoid or minimise them.

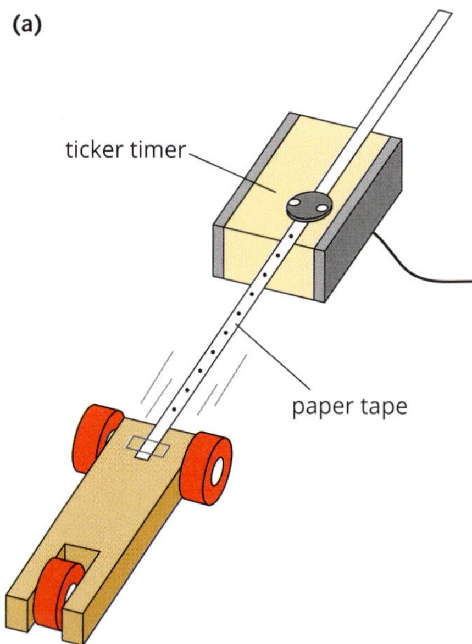

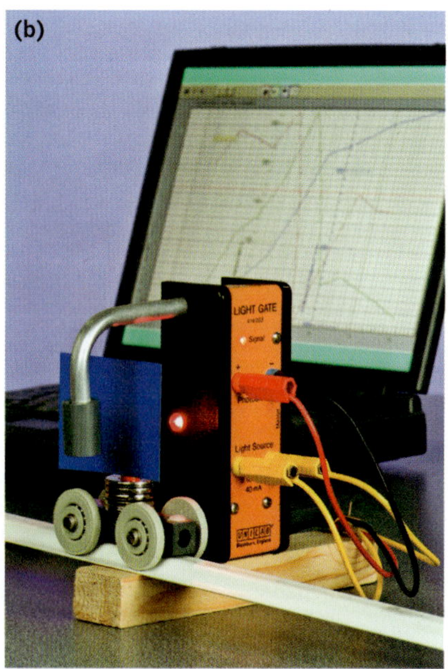

FIGURE 1.2.1 Some useful items of equipment in scientific investigations are (a) a ticker timer and (b) a motion sensor.

For an investigation to be valid, it is important to identify and record any errors. There are three types of error (Figure 1.2.2):

- systematic errors
- random errors
- mistakes.

Systematic errors

A **systematic error** is an error that is consistent and will occur again if the investigation is repeated in the same way.

Systematic errors are usually a result of instruments not being calibrated correctly, methods that are flawed or environmental factors (such as noise sources).

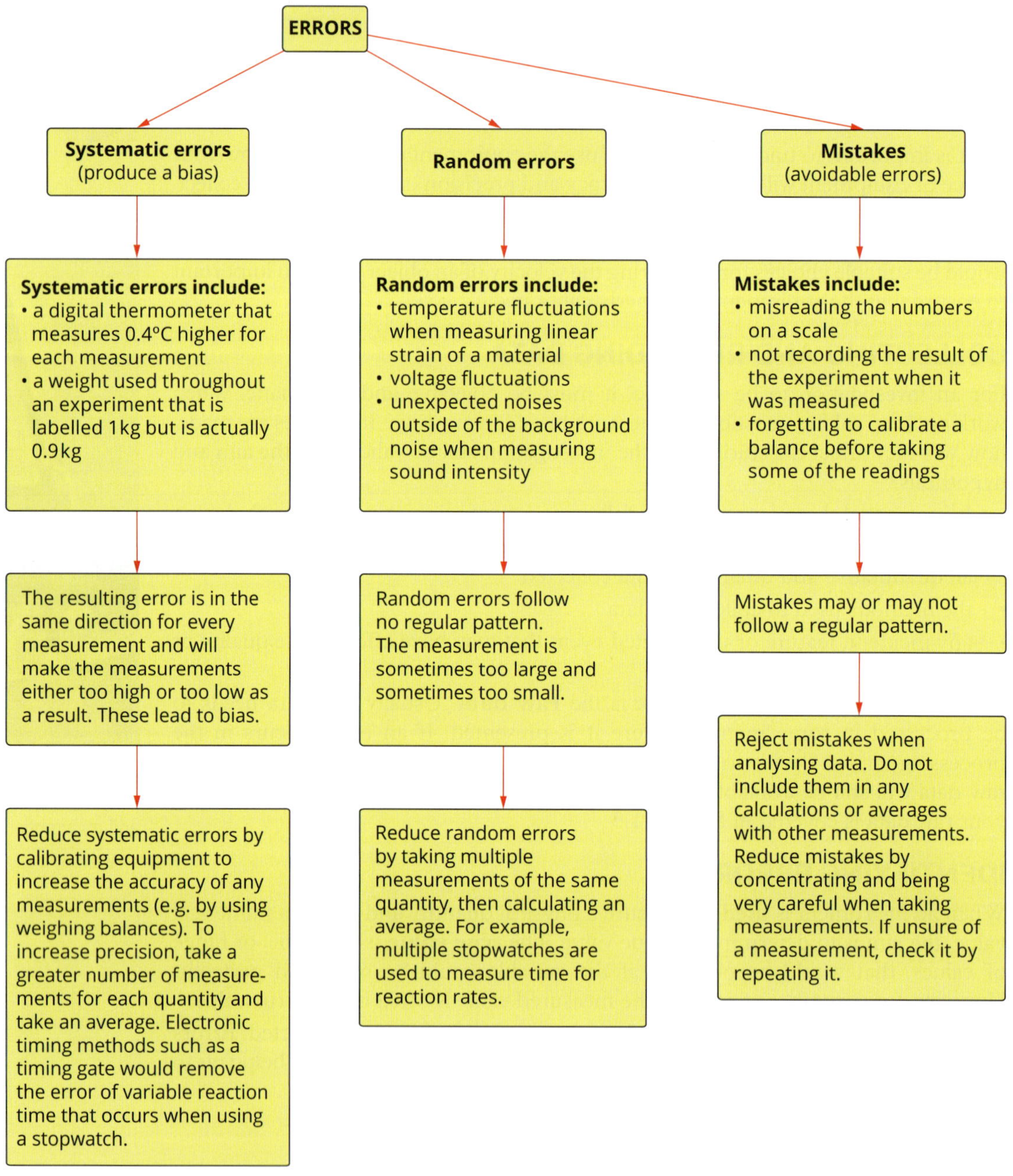

FIGURE 1.2.2 Types of error

An example of a systematic error is if a ruler mark indicating 5 cm from 0 cm was actually only 4.9 cm from 0 cm, perhaps due to a manufacturing error or shrinkage of the wood. Another example is if a researcher repeatedly used a piece of equipment incorrectly throughout the entire investigation.

Make sure you choose appropriate equipment that is in good working order before conducting your investigation. This will reduce the possibility of introducing systematic errors.

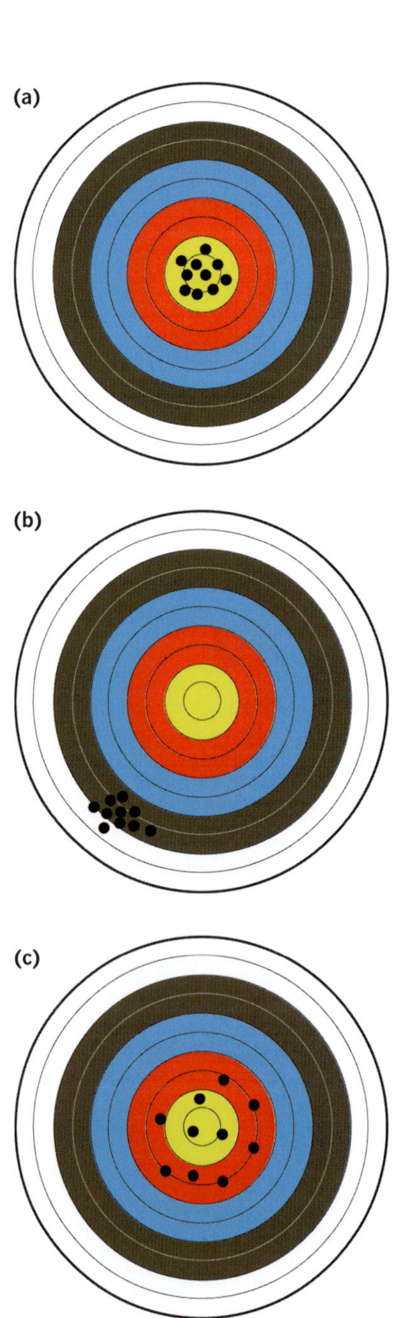

Random errors

Random errors occur in an unpredictable manner and are generally small. A random error could, for example, result from a researcher reading the same output value correctly one time and incorrectly another time. Another example is if an instrument's readings fluctuated during periods of low battery power. To reduce the impact of random errors, make sure to build repetition into your method.

Mistakes

Mistakes are avoidable errors. Examples include:

- misreading the numbers on a scale
- not identifying or labelling data points adequately.

A measurement that involves a mistake must not be included in any calculation or averaged with other measurements of the same quantity. Mistakes are often not referred to as errors because they are caused by the experimenter rather than the experiment or the experimental method. Sometimes they are referred to as personal errors.

Accuracy and precision

Two very important aspects of any measurement are accuracy and precision.

- **Accuracy** is how close a measurement is to the true value, including during repeated trials of the experiment. To obtain accurate results you must minimise systematic errors.
- **Precision** is how closely a set of measurements agree with each other. A set of precise measurements will have values very close to the mean value of the measurements. Precision is different from accuracy in that it does not indicate how close the measurements are to the true value. To obtain precise results, you must minimise random errors.

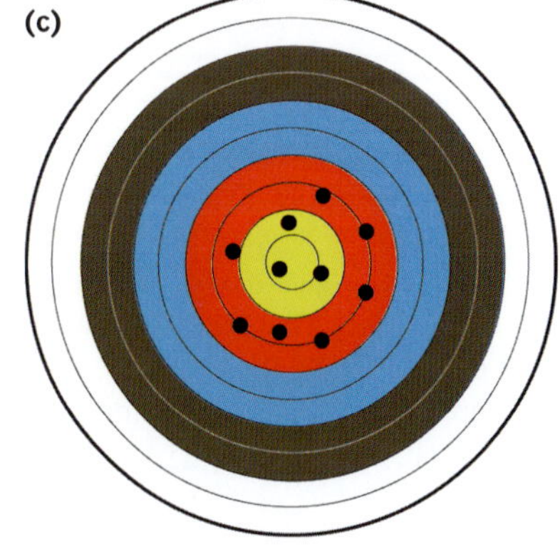

Instruments are accurate if they truly report the quantity being measured. For example, if a tape measure is correctly manufactured it can be used to measure lengths accurately to the nearest centimetre.

Instruments are precise if they can differentiate between slightly different quantities. Precision refers to the fineness of the scale being used.

To understand more clearly the difference between accuracy and precision, think about firing arrows at an archery target (Figure 1.2.3). Accuracy is being able to hit the bullseye. Precision is being able to hit the same spot every time you shoot. If you hit the bullseye every time you shoot, you are both accurate and precise (Figure 1.2.3(a)). If you hit the same area of the target every time but not the bullseye, you are precise but not accurate (Figure 1.2.3(b)). If you hit the area around the bullseye each time but don't always hit the bullseye, you are accurate but not precise (Figure 1.2.3(c)). If you hit a different part of the target every time you shoot, you are neither accurate nor precise (Figure 1.2.3(d)).

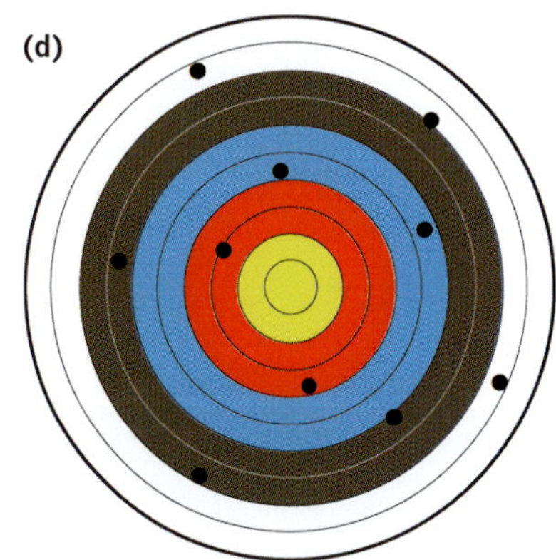

FIGURE 1.2.3 Examples of accuracy and precision: (a) both accurate and precise, (b) precise but not accurate, (c) accurate but not precise and (d) neither accurate nor precise

Consider a metre rule, a tape measure and a measuring wheel used to mark out a sports field. All three will measure distance and all three can be accurate. However, the metre rule is the most precise of the three. This is because it measures to the nearest millimetre while the tape measures only to the nearest centimetre and the measuring wheel measures only to the nearest metre.

To see that the tape measure is a more precise instrument than the measuring wheel, suppose two distances of 2673 mm and 2691 mm are being measured with these two instruments. Each distance would be measured as 3 m, to the nearest metre, by the wheel. They would be measured by the tape measure as 2.67 m and 2.69 m to the nearest centimetre. The tape measure is more precise because it has a finer scale. You might also say that it has greater resolution. The measuring wheel has such low precision that it cannot be used to measure which of the two distances is greater. Measuring instruments that have less precision give measurements that are less certain. The uncertainty in the measurement is due to the coarser scale. The measuring wheel gives less-certain measurements than the tape measure even though both instruments may be equally accurate.

PHYSICSFILE

Uncertainty

The uncertainty in an instrument's measurements is half the smallest scale of division in the measurements offered by the instrument. This is written using a plus or minus symbol (±). This indicates that the true value of the measurement can be anywhere between the measured value less the uncertainty and the measured value plus the uncertainty. For example, a rule may have gradations that are 1 mm apart. This makes the uncertainty in the rule ±0.5 mm. If you measured the distance of a gap in a diffraction experiment using this rule and found it to be 10 mm, then the true value is between 9.5 mm (10 mm − 0.5 mm) and 10.5 mm (10 mm + 0.5 mm).

The **uncertainty** of a measurement is due to the limited precision of the instrument that does the measuring. All measurements have some uncertainty. The uncertainty is generally one half of the finest scale division on the measuring instrument. This means that the actual measurement could be anywhere from half of the smallest graduation too big to half of the smallest graduation too small. The measuring wheel in the example above has an uncertainty of 0.5 m (Figure 1.2.4). The tape measure has an uncertainty of 0.5 cm. The metre rule has an uncertainty of 0.5 mm.

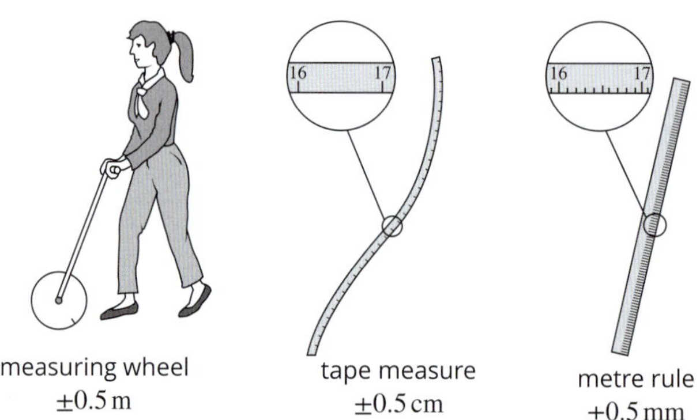

FIGURE 1.2.4 The measuring wheel has low precision and only measures to the nearest metre. It has an uncertainty of 0.5 m. The tape measure has more precision and has an uncertainty of 0.5 cm or 0.005 m. The metre rule has an uncertainty of 0.5 mm or 0.0005 m.

Sometimes this uncertainty is referred to as error. It is not an error, or something that has gone wrong. All measuring instruments have limited precision.

When conducting experiments, check that the instruments to be used are sensitive enough. Build some testing into your investigation to confirm the accuracy and precision of those instruments.

Calibrated equipment

Some equipment is sensitive to the conditions in which it operates, such as temperature and humidity. An example is the motion sensor. The accuracy of this equipment should be tested before each use to account for this. This is called calibration. Before carrying out the investigation, make sure the equipment is properly calibrated and functioning correctly. For example, measure the temperature and, if necessary, apply a correction to the speed of sound to calibrate a motion sensor.

Correct use of equipment

Always use equipment properly. Complete whatever training in the use of the equipment is necessary and practise using the equipment before beginning your investigation. Improper use of equipment can result in inaccurate and imprecise data with large errors, which weakens the validity of the data.

RECORDING INFORMATION FROM PRIMARY AND SECONDARY SOURCES

You may recall from Section 1.1 that research results can come from secondary sources as well as primary sources. As you conduct your investigation, it is important to note any information you use that has come from secondary sources. This must be acknowledged in your final written report. Categorising the information and evidence you find while you are researching will make it easier to locate the information later when you are writing the report. For example, you could use the following categories for identifying the sources of your information:

- research methods—the steps and methods to conduct an experiment
- key findings—key information and facts related to the experiment
- research relevance—how relevant the source of information is to the experiment being conducted.

Record this information in your logbook.

MODIFYING THE METHOD

The method may need modifying as the investigation proceeds. The following actions will help you identify any issues and decide how to address them.

- Record everything in your logbook.
- Be prepared to make changes to your approach.
- Note any difficulties encountered and the ways they were overcome. What were the failures and successes? Every test carried out can contribute to an understanding of the investigation as a whole, no matter how much of a disaster it may first appear.
- Don't panic. Go over the theory again and talk to your teacher and other students. A different perspective can lead to a solution.

If the expected data is not obtained, don't worry. As long as it can be critically and objectively evaluated, the limitations of the investigation identified and further investigations proposed, your work will have been worthwhile.

ETHICAL CONSIDERATIONS

When you are planning a scientific investigation, identify all possible ethical considerations that might arise and consider how you could reduce or eliminate them. Asking questions such as those below may help you uncover ethical implications.

- How could this affect the wider society?
- Does one individual or group benefit over another? Is it fair?
- Who will have access to the data and results?
- Does it prevent anyone from meeting their basic needs?

SAFETY AND RISK ASSESSMENTS

When planning your investigation, you need to be aware of any risks or safety concerns and consider ways to mitigate them. Always use safe procedures and common sense. For example, all equipment and instruments should be used at the back of the bench so that students walking by do not knock them or trip over them and cause an accident. Place a sign on the lab bench warning other students and staff not to touch the equipment.

You must follow the safety and **risk assessment** guidelines of your teacher and your school. Completing a risk assessment may mean filling out a form or working through an online process.

While conducting an investigation, it is important for your own safety and the safety of others that all potential risks are considered. Risk assessments are undertaken to identify, assess and control potential hazards. A risk assessment should be undertaken for any situation—in the laboratory or outside in the field. Always identify the risks and control them to keep everyone safe.

PHYSICSFILE

Connecting the world

Have you ever wondered how the internet sends data around the world? Most people believe the world is connected using satellites that transfer information from device to device across countries separated by the oceans. But this is not the case. Most data is transferred along fibre optic cables that lie across the vast ocean floors. There are more than 440 underwater cables keeping the world connected.

Laying these fibre optic cables can be complex. The likely impact on coral reefs, aquatic animals and other aquatic systems must be considered when cables are being laid. Underwater topography, volcanic activity and ocean currents also influence where the cables can be laid. Where the cables cross international borders, negotiations between countries may lead to routes that are less than optimal. And when something goes wrong in such a complex network, maintenance and repairs might need to be done in countries far away. So the next time your internet connection is down, the solution may be far more complicated than simply resetting your modem.

Underwater cables form the backbone of the internet.

FIGURE 1.2.5 Avoid electrical hazards by identifying, assessing and controlling them during the planning phase

For example, conduct voltage–current experiments only with low voltages (less than 6.0V DC or 4 × 1.5V batteries) coupled to resistors so that the currents in the circuits are of the order of milliamps. At all times avoid direct exposure to 240V AC household voltages (Figure 1.2.5).

To identify risks think about:

- the activity that will be carried out
- where it will be carried out
- the equipment that will be used.

The hierarchy of risk controls in Figure 1.2.6 is organised from the least effective (at the bottom) to the most effective (at the top).

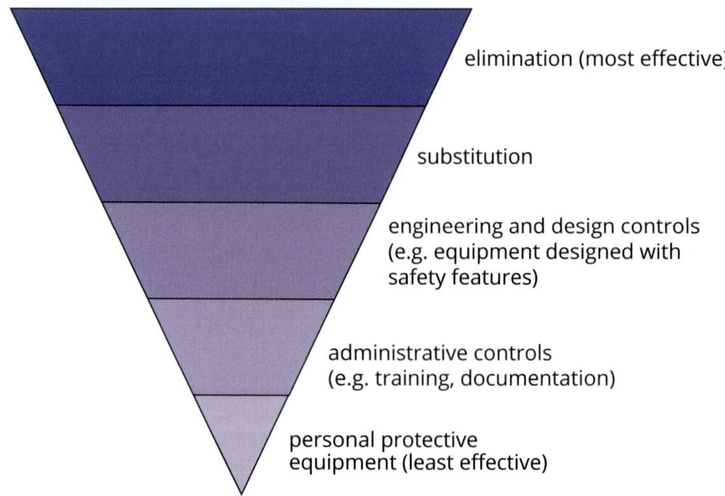

FIGURE 1.2.6 A hierarchy of measures for controlling risk, with the least effective at the bottom and the most effective at the top.

FIGURE 1.2.7 Examples of PPE shown are protective eyewear, lab coats and gloves.

Personal protective equipment

Everyone who works in a laboratory should wear **personal protective equipment** (PPE) to help keep them safe. Consult your teacher or laboratory technician, and each relevant safety data sheet (SDS), to see what PPE you are likely to need. Examples include:

- safety glasses
- lab coats
- shoes with covered tops
- disposable gloves.

Examples of PPE are shown in Figure 1.2.7.

Chemical codes

In January 2017 Australia adopted the Globally Harmonized System of Classification and Labelling of Chemicals (GHS). This system is used for labelling containers and in safety data sheets. The chemicals at school or at a hardware shop have a warning symbol on the label. These symbols—mandated by the GHS—indicate the nature of the contents (Table 1.2.1). You will need to analyse them when you are planning and conducting scientific investigations involving chemicals. You will perform a risk assessment in which these symbols will be provided. After analysing them you may need to modify your experimental plan so that safety is improved.

TABLE 1.2.1 GHS symbols used as warnings on chemical labels

GHS symbol	Use	GHS symbol	Use	GHS symbol	Use
	flammable liquids, solids and gases; including self-heating and self-igniting substances		oxidising liquids, solids and gases; may cause or intensify fire		explosion, blast or projection hazard
	corrosive chemicals; may cause severe skin and eye damage and may be corrosive to metals		gases under pressure		fatal or toxic if swallowed, inhaled or in contact with skin
	low level toxicity; this includes respiratory, skin and eye irritation, skin sensitisers and chemicals harmful if swallowed, inhaled or in contact with skin		hazardous to aquatic life and the environment		chronic health hazards; this includes aspiratory and respiratory hazards, carcinogenicity, mutagenicity and reproductive toxicity

Safety data sheets

Every chemical substance has an associated document called a **safety data sheet (SDS)**. The SDS contains important safety, environmental and first aid information about the chemical, including how the chemical should be handled and stored (Figure 1.2.8). For example, if an SDS states that the products of a reaction with the chemical are toxic to the environment, you must pour your waste into a special container, not down the sink.

An SDS provides employers, workers, and health and safety representatives with the necessary information to safely manage the risk of exposure to a hazardous substance.

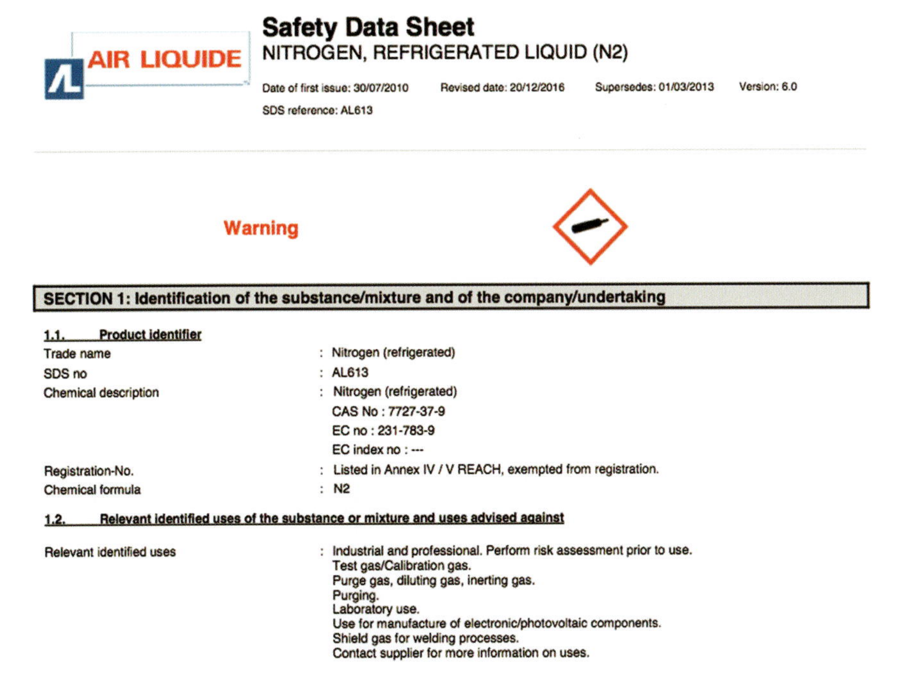

AIR LIQUIDE

Safety Data Sheet

NITROGEN, REFRIGERATED LIQUID (N2)

Date of first issue: 30/07/2010 Revised date: 20/12/2016 Supersedes: 01/03/2013 Version: 6.0

SDS reference: AL613

Warning

SECTION 1: Identification of the substance/mixture and of the company/undertaking

1.1. Product identifier

Trade name	: Nitrogen (refrigerated)
SDS no	: AL613
Chemical description	: Nitrogen (refrigerated) CAS No : 7727-37-9 EC no : 231-783-9 EC index no : ---
Registration-No.	: Listed in Annex IV / V REACH, exempted from registration.
Chemical formula	: N2

1.2. Relevant identified uses of the substance or mixture and uses advised against

Relevant identified uses	: Industrial and professional. Perform risk assessment prior to use. Test gas/Calibration gas. Purge gas, diluting gas, inerting gas. Purging. Laboratory use. Use for manufacture of electronic/photovoltaic components. Shield gas for welding processes. Contact supplier for more information on uses.

FIGURE 1.2.8 An example of part of an SDS for liquid nitrogen showing the symbol for a compressed gas to alert the reader to potential hazards when using the substance. The SDS also includes measures to reduce the risk of harm.

Science outdoors

Sometimes investigations and experiments will be carried out outdoors. Working outdoors has its own set of potential risks and it is equally important to consider ways of eliminating or reducing those risks.

Table 1.2.2 gives examples of risks associated with working outdoors.

TABLE 1.2.2 Risks associated with working outdoors

Risks	Control measures
sunburn	wear sunscreen, a hat and sunglasses
hot or cold weather	wear clothing to protect against heat or cold
projectile launch	create barriers so that people know not to enter the area
trip hazards	minimise the use of cables (electrical, computer etc.) and cover them with matting
landscape	be aware of tree roots, rocks, roads etc.

First aid measures

Minimising the risk of injury reduces the chance of requiring first aid assistance. However, it is still important to have someone with first aid training present during a practical investigation.

Always tell your teacher or laboratory technician if an injury or accident happens.

1.2 Review

OA ✓✓

SUMMARY

- It is essential that during an investigation, the following is recorded in your logbook:
  - all quantitative and qualitative data collected
  - the methods used to collect the data
  - any incident, feature or unexpected event that may have affected the quality or validity of the data
  - all sources of information, both primary and secondary.
- An experiment or investigation is valid only if it is testing the set research question or hypothesis.
- Repeatability (or reliability) means that if an experiment is repeated many times, the results obtained are consistent.
- Repeatability is improved by:
  - replication (having multiple samples within an experiment)
  - repeat trials (repeating the experimental test).
- The accuracy of a measurement is how close it is to the true value.
- The precision of an instrument describes the smallest value it can measure.
- The uncertainty of a measurement due to the limited precision of the instrument is plus or minus half of the finest scale division on the instrument.
- A systematic error is an error that is consistent. It will occur again if the investigation is repeated in the same way. Systematic errors are usually a result of instruments that are not calibrated correctly or methods that are flawed.
- Random errors are unpredictable and are generally insignificant. A random error could arise when a researcher misreads one of a number of measurements.
- Mistakes are avoidable. Do not include in the analysis of your results any measurement that resulted from a mistake.
- Identify and address any ethical issues while you plan your investigation.
- Risk assessments identify and assess hazards, and propose controls to minimise their occurrence or effect.
- GHS symbols identify the types of hazards associated with substances.

KEY QUESTIONS

Knowledge and understanding

1 The masses of numerous cubes of potato, each 1 cm^3 in volume, were recorded and the cubes placed in distilled water. After 60 minutes the cubes were weighed again and the difference in mass was calculated. What type of error occurs:
   - a if the electronic scales are calibrated incorrectly
   - b if the electronic scales were briefly affected by a power surge?

2 Explain the terms 'accuracy' and 'precision'.

3 State why it is important to choose appropriate equipment and instruments when conducting an experiment.

4 What is one way to ensure that an experiment is valid, and one way to ensure that it is repeatable?

5 Identify whether the following are systematic errors, random errors or mistakes.
   - a A student spills some water from the container while measuring the temperature increase.
   - b Some reported measurements are above the true value and some below the true value.
   - c A 25.00 g mass measures 25.5 g on a scale.
   - d A scale that should have measured a mass as 25.00 ± 0.03 g actually measured 24.92 ± 0.03 g.
   - e A student misreads the value on the slotted mass during a Hooke's law investigation.
   - f The mass of a ball was taken three times, with the following results: 147.9 g, 174.8 g and 147.6 g.

6 List the following types of hazard control from the most effective to the least effective.
substitution, personal protective equipment, administrative controls, elimination, engineering controls

7 Explain the reasons for having an SDS for each chemical used in the laboratory.

8 Describe the appropriate action to take if you came into contact with a chemical substance with the following label on the container.

Analysis

9 A student included the following diagram of the experimental setup in her method. Use the diagram to write clear instructions for the method. The aim of the experiment is to investigate the impact of launch velocity on the range of a projectile.

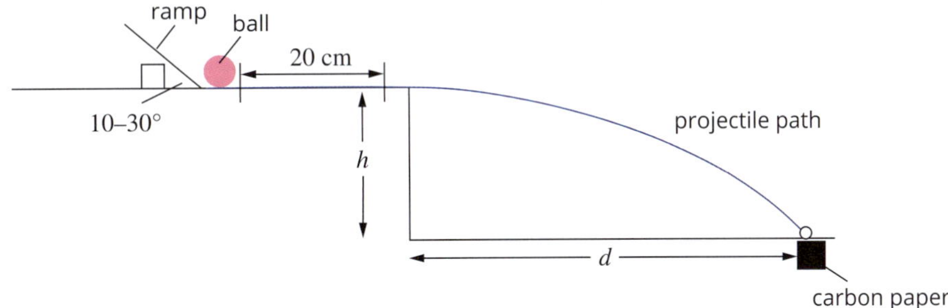

1.3 Data collection and quality

Measurements and observations made during a scientific investigation together form a picture of what occurred during the investigation. This is the raw data (Figure 1.3.1). The raw data needs to be analysed and then represented using tables, graphs, schematics or diagrams in accordance with correct mathematical and scientific conventions.

Choosing not to record certain measurements or observations in your raw data makes the experiment invalid, may show bias and is scientifically fraudulent. Unusual and unexpected measurements and observations may be due to valid relationships between variables that are unknown to science. This cannot be determined until the raw data is processed, analysed and interpreted. Results that are determined to be mistakes can be ignored only once proper analysis has occurred.

In this section you will learn about recording quantitative data. You will also learn about the various factors that contribute to data quality and the importance of controlled experiments in producing valid results.

FIGURE 1.3.1 All measurements and data collected during an experiment constitute the raw data.

RECORDING QUANTITATIVE DATA

The measurements or observations that you make during your investigation are your **primary data**. When planning your investigation, you may have decided to use **secondary data** as well (that is, data you have not collected yourself).

The logbook

Careful and accurate record keeping is essential to good science. It is important that the methods you use in your investigations, observations, analysis of data and conclusions are recorded. This ensures that the way your investigation was carried out, as well as your observations, analysis and conclusions, can be shared with others who can verify your work.

One way of recording investigations is with logbooks. Throughout Units 3 and 4, and during your practical investigation for Unit 4 Area of Study 2, you must keep a logbook that includes every detail of your research.

The following checklist will help you remember what to include in your logbook.

- ☐ your ideas when planning the research
- ☐ clear protocols for each stage of your investigation (for example, what standard procedures you will use and follow exactly each time)
- ☐ instructions, noting exactly what needs to be recorded
- ☐ tables ready for data entry
- ☐ records of all materials, methods, experiments and raw data
- ☐ all notes, sketches, photographs and results
- ☐ records of any incidents or errors that may influence the results

The data you record in your logbook is raw data. This data needs to be processed or analysed before it can be presented. **Processed data** is raw data that has been organised, altered or analysed to produce meaningful information. If an error occurs in processing the data, or you decide to present the data in a different format, you will always have the recorded raw data to refer back to.

Raw data is unlikely to be used directly to validate a hypothesis. However, raw data is essential to the investigation, and plans for collecting the raw data should be made carefully. Consider the formulas or graphs that will be used to analyse the data at the end of the investigation. This will help you to determine the type of raw data that needs to be collected.

For example, to calculate take-off velocity for a vertical jump, three sets of raw data will need to be collected using a force platform: the athlete's standing body mass, the ground reaction force and the time during the vertical jump. The data can then be processed to obtain the take-off velocity.

The data that you collect must relate directly to the variables in your experiment and be relevant to the proposed relationship set out in your research question and hypothesis. It must also be sufficient to provide accuracy and precision, otherwise the analysis and interpretation of the data will not be valid in relation to the research question or hypothesis.

Collecting sufficient data

You need to collect enough data to substantiate whether or not a relationship exists between the variables you are studying. This includes collecting an appropriate number of individual samples (also known as observational or collection points) and an appropriate number of replicates. It is also important to collect data around interesting points in your range, such as where a curve representing the data reaches a maximum or minimum.

Together, the number of individual samples and replicates determines the sample size. A sufficient sample size is essential for your interpretation to be considered supported.

Collecting relevant data

The variables you measure (that is, the data collected) must be directly related to the independent–dependent variable relationship specified in your hypothesis. Additional variables can be measured that are indirectly related to your hypothesis if your background research shows it could be beneficial in the analysis or interpretation of the relationship specified in your hypothesis.

MEASUREMENT AND UNITS

Every science needs a system of units in order to fully describe the measurements that are made. In physics, measurements are described using the International System of Units (known as SI).

Using unit symbols

The correct use of unit symbols removes ambiguity, as symbols are recognised internationally. The symbols for units are not abbreviations and should not be followed by a full stop unless they are at the end of a sentence.

Seven fundamental units are specified in the SI. They are the metre, kilogram, second, kelvin, ampere, mole and candela.

The majority of other units used in physics are a mathematical combination involving at least one of the seven fundamental units. These are called derived units.

The names and symbols for units are treated differently. Upper case letters are not used for the names of any physics unit names when written in full. For example, we write 'newton' for the unit of force, while we write 'Newton' if referring to someone with that name. Upper case letters are only used for the symbols of the units that are named after people. For example, the unit of length is metre and its symbol is m. This is lower case because metre is not named after anyone. However, the unit for force is newton and its symbol is N; the unit for energy is joule and its symbol is J. N is upper case because the newton is named after Isaac Newton and J is upper case because the joule is named after James Joule (who is famous for his studies of energy conversion). The exception to this lower case – upper case rule is L for litre. Litre is not named after anyone, but we capitalise L because, in many fonts, a lower case 'l' looks very similar to the numeral 1, a similarity that could lead to ambiguity and confusion.

The product of units is shown by separating the symbol for each unit with a dot or a space. Most teachers prefer a space but a dot is equally correct. The division or ratio of two or more units can be shown in fraction form, by using a slash or by using negative indices.

Prefixes—such as the k in kg—should not be separated by a space.

Table 1.3.1 gives some examples of the correct symbols and format for SI derived units.

TABLE 1.3.1 Examples of the use of symbols for derived units

Majority preference	Also correct	Wrong
m s^{-2}	m.s^{-2} m/s^{2}	ms^{-2}
kW h	kW.h	kWh k Wh
kg m^{-3}	kg.m^{-3} kg/m^{3}	kgm^{-3}
μm		μ m
N m	N.m	Nm

Units take the plural form by adding an 's' when used with numbers greater than one. Never do this with the unit symbols. Hence it is acceptable to write 'two newtons' but wrong to write '2 Ns'. It is also acceptable to say 'two newton'.

Numbers should always be used if a unit is abbreviated. For example, twenty metres, 20 m and 20 metres are all correct but twenty m is incorrect.

Significant figures

Significant figures are the numbers that convey meaning and precision. The number of significant figures used depends on the scale of the instrument you are using. It is important to record data to the number of significant figures available with the instrument. Using either a greater or smaller number of significant figures can be misleading.

The following examples indicate how the number of significant figures is determined.

- Non-zero numbers are always significant: 15 has two significant figures; 3.5 has two significant figures.
- Trailing zeroes to the right of the decimal point are significant: 3.50 has three significant figures.
- Leading zeros are not significant: 0.037 has two significant figures.
- Zeros between non-zero digits are significant: 1401 has four significant figures.

Some numbers may have an ambiguous number of significant figures. For example, 100 could have three or one significant figure. In VCE Physics, for simplicity, trailing zeros are significant, thus 100 is taken as having three significant figures.

The number of significant figures in the result of a calculation should never exceed the minimum significant figures in any component of the calculation. For example:

Calculate gravitational potential energy (E_g) using the formula $E_g = mg\Delta h$ when $g = 9.8\,\text{m s}^{-2}$, $m = 7.50\,\text{kg}$ and $h = 0.64\,\text{m}$.

The calculation is:

$$E_g = 7.50 \times 9.8 \times 0.64 = 47.04\,\text{J}$$

However, you should only quote the answer to the least number of significant figures in the component data. In this case, quote the answer to two significant figures: $E_g = 47\,\text{J}$.

Although digital scales can measure to many more than two figures, and calculators can give 12 figures, you should follow the significant-figures rule.

> In VCE Physics, for simplicity, trailing zeros are significant, thus 100 is taken as having three significant figures.

Scientific notation

For clarity, quantities are often written in scientific notation. A number between one and ten (but excluding ten) is written and then multiplied by an appropriate power of ten. Note that 'scientific notation', 'standard notation' and 'standard form' all have the same meaning.

Examples of some measurements rewritten in scientific notation are:

0.054 m becomes $5.4 \times 10^{-2}\,\text{m}$

245.7 J becomes $2.457 \times 10^{2}\,\text{J}$

2080 N becomes $2.080 \times 10^{3}\,\text{N}$

You should routinely be using scientific notation to express very large or very small numbers. This also involves learning to use your calculator intelligently. Scientific and graphing calculators can be put into a mode where all numbers are displayed in scientific notation. It is useful when doing calculations to use this mode rather than converting to scientific notation by counting digits on the calculator display. It is quite acceptable to write all numbers in scientific notation, although most people prefer not to use scientific notation when writing numbers between 0.1 and 1000.

An important reason for using scientific notation is that it removes ambiguity about the precision of some measurements. For example, a measurement recorded as 240 m could be a measurement to the nearest metre, that is, somewhere between 239.5 m and 240.5 m. It could also be a measurement to the nearest ten metres, that is, somewhere between 235 m and 245 m. Writing the measurement as 240 m does not indicate which level of precision is the case. If the measurement was taken to the nearest metre, it would be written in scientific notation as $2.40 \times 10^{2}\,\text{m}$. If it was taken only to the nearest ten metres, it would be written as $2.4 \times 10^{2}\,\text{m}$.

Prefixes and conversion factors

Conversion factors should be used carefully. You should be familiar with the prefixes and conversion factors in Table 1.3.2 on page 26. The most common mistake made with conversion factors is multiplying rather than dividing. Some simple strategies can help you avoid this problem. Note that the table gives all conversions as a multiplying factor.

It is important to give the symbol the correct case (upper or lower case). There is a big difference between 1 mm and 1 Mm.

There is no space between prefixes and unit symbols. For example, one-thousandth of an ampere is given the symbol mA. Writing it as m A is incorrect. The space between m and A would mean that the symbol is for a derived unit—a metre ampere.

TABLE 1.3.2 Prefixes and conversion factors

Multiplying factor	Index form	Prefix	Symbol
1 000 000 000 000	10^{12}	tera	T
1 000 000 000	10^{9}	giga	G
1 000 000	10^{6}	mega	M
1000	10^{3}	kilo	k
0.01	10^{-2}	centi	c
0.001	10^{-3}	milli	m
0.000 001	10^{-6}	micro	μ
0.000 000 001	10^{-9}	nano	n
0.000 000 000 001	10^{-12}	pico	p

PRESENTING DATA

The raw data you have obtained should be presented in a way that is clear, concise and accurate.

There are a number of ways of presenting data, including tables, graphs, flow charts and diagrams. The best way of visualising the data depends on its nature. To create the best possible presentation, try several formats before making a final decision.

Presenting raw and processed data in tables

Tables organise data into rows and columns and can vary in complexity according to the nature of the data. Tables can be used to organise raw data and processed data. They can also be used to summarise results.

The simplest form of a table is one with two columns. In a two-column table, the first column should contain the independent variable (the one being changed) and the second column should contain the dependent variable (the one that may change in response to a change in the independent variable).

Tables should have the following features:

- a descriptive title (preceded by 'Table *n*' where *n* is the table number)
- column headings (including the units)
- scientific notation used in the column header for very small or very large numbers
- an indication of the precision of the data
- consistent use of significant figures
- figures that align on the decimal points
- the independent variable placed in the left column
- the dependent variable placed in the next column to the right
- replicate measurements
- calculated averages as required.

The table in Figure 1.3.2 has been used to organise raw and processed data about the effect of current on voltage.

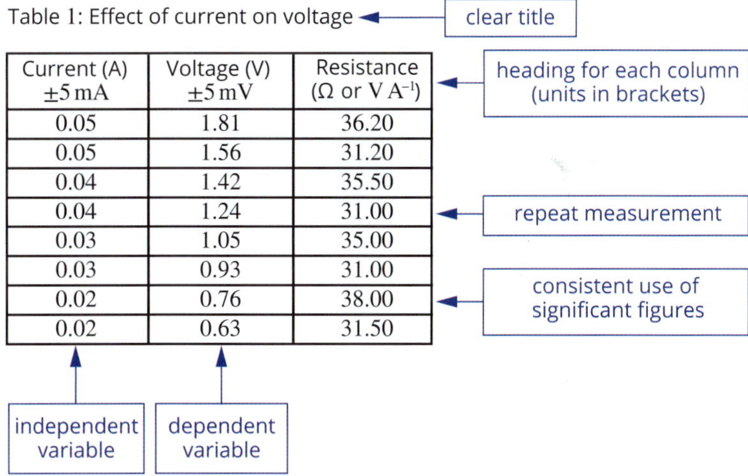

Table 1: Effect of current on voltage

Current (A) $\pm 5\,\text{mA}$	Voltage (V) $\pm 5\,\text{mV}$	Resistance (Ω or V A^{-1})
0.05	1.81	36.20
0.05	1.56	31.20
0.04	1.42	35.50
0.04	1.24	31.00
0.03	1.05	35.00
0.03	0.93	31.00
0.02	0.76	38.00
0.02	0.63	31.50

FIGURE 1.3.2 A simple table listing the raw data obtained in the first and second columns and processed data in the third column

Several statistical measures are used that help describe data accurately. They include the mean, median, mode and uncertainty.

The **mean** is the average of the data, and can be obtained by adding all the measurements and dividing by the total number of measurements.

The **median** is the middle value of an ordered list of values (that is, there are as many values less than the median as there are greater than it). For example, the median of the values 5, 5, 5, 8, 9, 10, 20 is 8. The median is preferred when the data range is spread, especially when the data includes unusual results (also known as outliers). In this situation the mean is unreliable.

The **mode** is the value that appears most often in a data set. This measure is useful to describe qualitative or discrete data. For example, the mode of the values 5, 5, 5, 8, 9, 10, 20 is 5.

The measurement uncertainty is an indicator of the precision of the equipment used (as discussed in Section 1.2) and should be included in the heading of each relevant column, as shown in Figure 1.3.2.

It is good practice to repeat experiments and collect data for several trials. In this case, the table of processed data usually presents the mean of the replicates. However, the mean on its own does not provide an accurate picture of the results. To report processed data more accurately, the overall uncertainty should be presented as well. More detail about uncertainty is discussed in Section 1.4.

1.3 Review

OA ✓✓

SUMMARY

- Record primary and secondary data in a logbook.
- Make sure you record your data with the correct units and to an appropriate number of significant figures.
- Collect enough data to substantiate whether or not a relationship exists between the variables you are studying.
- Make sure the data you collect is directly related to the independent–dependent variable relationship specified in your hypothesis.
- Physics uses the International System of Units (SI).
- There are seven fundamental units: metre, second, kilogram, kelvin, mole, ampere and candela. Most other units in physics are derived from these units.
- The SI prefixes are symbols that go before a unit and indicate multiplication of the unit by a power of 10.
- Units with a prefix (such as km or mA) should be converted into scientific notation before they are used in a physics formula.
- Significant figures should be considered in calculations using your data. Quote the results of calculations to the least number of significant figures of your data.
- Descriptive statistics includes three measures of central tendency:
  - mean, which is the sum of the values divided by the number of values
  - median, which is the middle value in an ordered list of values
  - mode, which is the value that occurs most often in a list of values.

KEY QUESTIONS

Knowledge and understanding

1 Explain the difference between raw and processed data.

2 For the data set 21, 28, 19, 19, 25, 24, 20, determine:
  a the mean
  b the mode
  c the median.

3 In a practical investigation a student changes the voltage by adding or subtracting batteries in series to the circuit.
  a How could the voltage be a discrete value?
  b How could it be continuous?

4 Convert 2.5 mm (millimetres) into μm (micrometres).

5 If using the quantities mass = 7.50 kg and speed = 1.4 m s^{-1} in a calculation, what would be the appropriate number of significant figures in the answer?

6 a Write 255 000 in scientific notation.
  b Write 0.000 000 432 in scientific notation.
  c Explain why scientific notation is used.

7 Write the following measurements in scientific notation to the stated number of significant figures.
  a 6.626×10^{-34} J s to 3 significant figures
  b 0.001 783 30 MeV to 3 significant figures

Analysis

8 Clancey recorded the data below to test the following hypothesis: the length of the wire wound into a solenoid affects the magnetic field inside the solenoid. Assess whether or not sufficient and relevant data was collected and suggest improvements in the way the data is presented in the table.

Length of wire (m)	Thickness of wire (mm)	Current through wire (A)	Length of solenoid (cm)	Diameter of solenoid (cm)	Type of wire	Magnetic field inside solenoid (T)	Brand of wire
0.50	0.31	0.02	12	3.4	copper	40.0	Ruby and Macey's House of Wire
1.00	0.31	0.02	12	3.4	copper	20.0	P & L
1.50	0.31	0.02	12	3.4	copper	13.3	P & L

1.4 Data analysis and presentation

A major problem with making a calculation from just one set of measurements is that a single incorrect measurement can significantly affect the result. Scientists like to collect a large amount of data and observe the trends in that data. This gives more precise measurements and allows scientists to recognise and eliminate problematic data.

Physicists commonly use tables and graphical techniques to analyse large sets of data (Figure 1.4.1). Trends in data are often easier to observe in graphs than in tables. In this section, the basic graphing techniques physicists use will be outlined and a general method for fitting a mathematical relationship to a set of data will be explored.

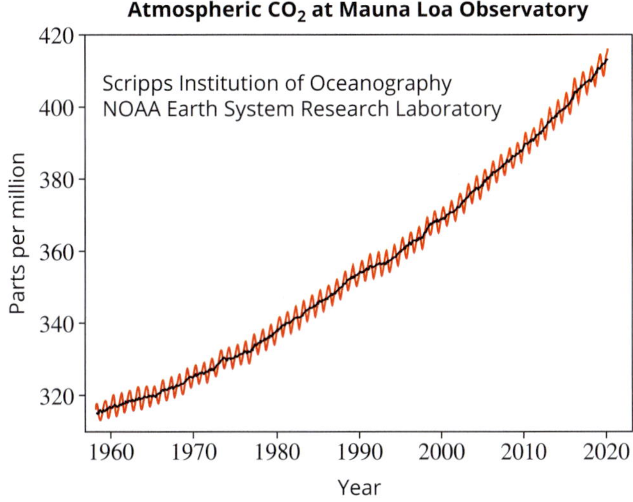

FIGURE 1.4.1 Graphical techniques are often used to find trends in data.

GRAPHICAL ANALYSIS OF DATA

There are several types of graphs that can be used, including line graphs, bar graphs and pie charts. The best one to use will depend on the nature of the data. Table 1.4.1 on page 30 lists some suitable graphs for quantitative discrete data (data from measuring discrete variables) and continuous data (data from measuring continuous variables).

General rules to follow when plotting a graph are listed below. Figure 1.4.2 on page 31 illustrates these rules. (Note that these rules do not apply to all types of graph.)

- Keep the graph simple and uncluttered.
- Use a descriptive title.
- Represent the independent variable on the x-axis and the dependent variable on the y-axis.
- Make axes proportionate to the data.
- Clearly label axes with both the variable and the unit in which it is measured.
- Use scientific notation where appropriate.

You can extrapolate (extend the trend line beyond the obtained data) to predict other values—for example, to determine where the graph intersects the axes. Take care, however, because the relationship between variables may not hold beyond the measured data.

TABLE 1.4.1 Suitable graph types for quantitative data

Type of data	Appropriate type of graph	Examples
discrete	bar graph histogram pie chart	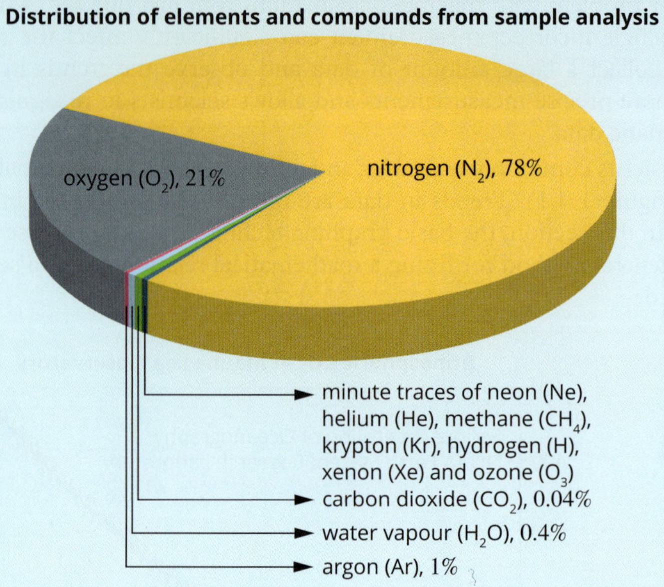 Distribution of elements and compounds from sample analysis oxygen (O_2), 21% nitrogen (N_2), 78% minute traces of neon (Ne), helium (He), methane (CH_4), krypton (Kr), hydrogen (H), xenon (Xe) and ozone (O_3) carbon dioxide (CO_2), 0.04% water vapour (H_2O), 0.4% argon (Ar), 1% Pie chart showing distribution of elements and compounds from an analysis of a sample
continuous	line graph or scatterplot, including a trend line if appropriate	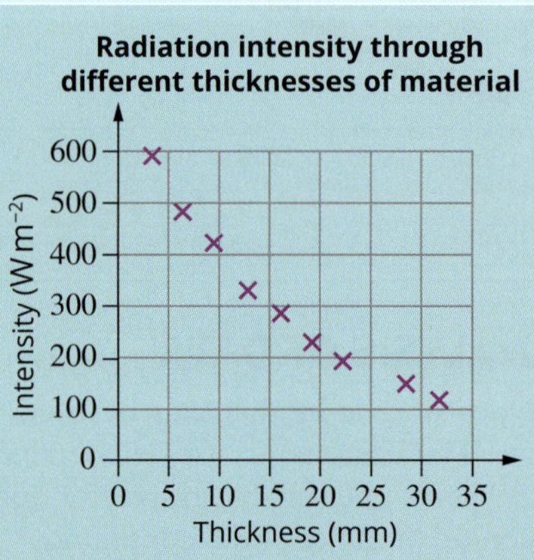 Radiation intensity through different thicknesses of material Intensity (W m^{-2}) 600 500 400 300 200 100 0 0 5 10 15 20 25 30 35 Thickness (mm) Scatterplot of the radiation intensity through different thicknesses of material 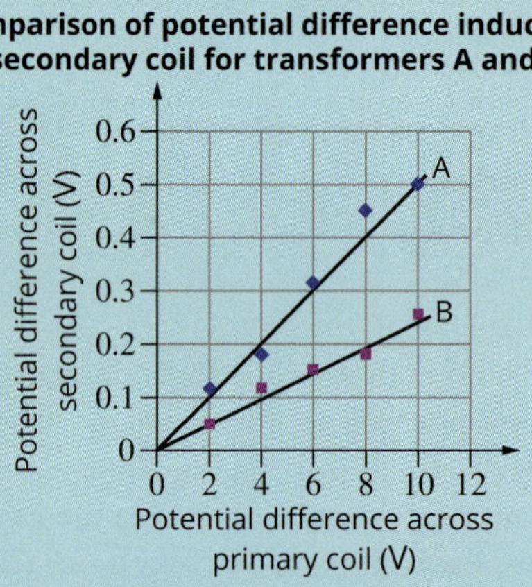 Comparison of potential difference induced in secondary coil for transformers A and B Potential difference across secondary coil (V) 0.6 0.5 0.4 0.3 0.2 0.1 0 A B 0 2 4 6 8 10 12 Potential difference across primary coil (V) Graph, including a trend line, showing results of an investigation into two different transformers to see which is better to use with a model motor

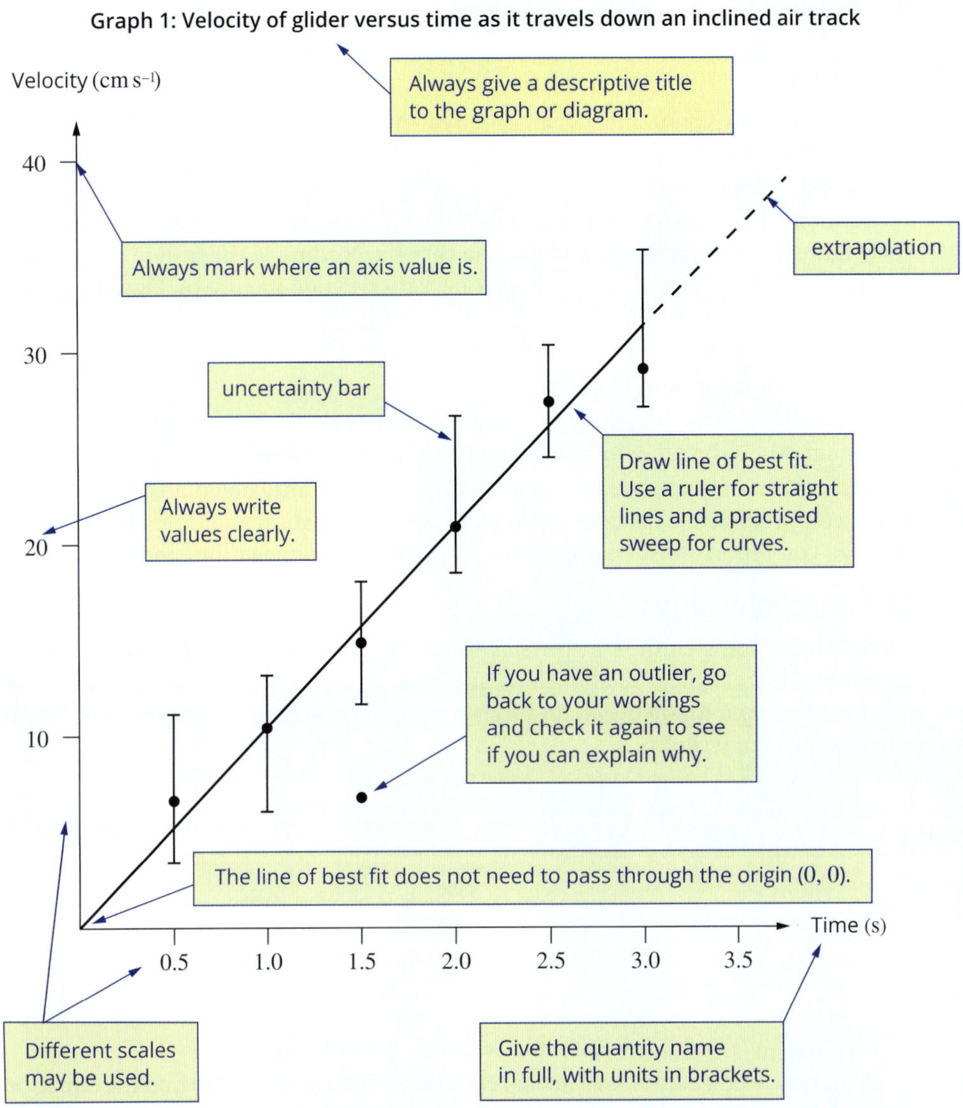

Graph 1: Velocity of glider versus time as it travels down an inclined air track
Velocity (cm s⁻¹)
Always give a descriptive title to the graph or diagram.
40
30
20
10
Always mark where an axis value is.
extrapolation
uncertainty bar
Draw line of best fit. Use a ruler for straight lines and a practised sweep for curves.
Always write values clearly.
If you have an outlier, go back to your workings and check it again to see if you can explain why.
The line of best fit does not need to pass through the origin (0, 0).
Time (s)
0.5
1.0
1.5
2.0
2.5
3.0
3.5
Different scales may be used.
Give the quantity name in full, with units in brackets.

FIGURE 1.4.2 A graph showing the relationship between two variables

Line graphs

Line graphs are a good way of representing continuous quantitative data. In a line graph, the values are plotted as a series of points on the graph. There are two ways of joining these points.

- A line can be ruled from each point to the next (Figure 1.4.3(a)). This may show the overall trend but it is not meant to predict the value of the points between the plotted data.
- The points can be joined with a single smooth line, straight or curved (Figure 1.4.3(b)). This creates a **line of best fit**, also known as a trend line. The line of best fit does not have to pass through every point, but should go close to as many points as possible. It is used when there is an obvious relationship between the variables.

Outliers

Sometimes when the data is collected, there may be data points that do not fit with the trend and are clearly a mistake or a random error. These points are called **outliers**. An outlier is often caused by a mistake made in measuring or recording data, or from a random error that occurred during the investigation. If there is an outlier, include it on the graph, but ignore it when adding a line of best fit. In Figure 1.4.2 the point (1.5, 6) is an outlier.

(a)
Height of river at bridge flood gauge
Height of river (m)
3
2
1
0
J F M A M J J A S O N D
Months 2010
(b)
Stopping distance for car
Stopping distance (m)
50
40
30
20
10
0
0
20
40
60
80
Speed of car (km h⁻¹)

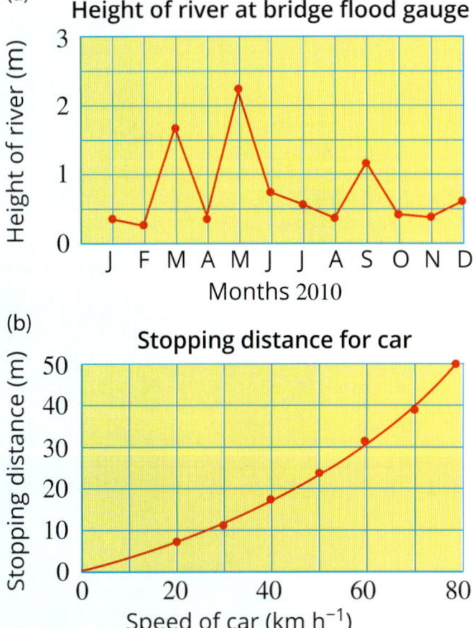

FIGURE 1.4.3 (a) Data in the graph is joined from point to point. (b) Data in the graph is joined with a line of best fit, which shows the general trend.

Describing trends in line graphs

Graphs are drawn to show the relationship, or trend, between two variables (Figure 1.4.4).

- Variables that change in linear or direct proportion to each other produce a straight, sloping trend line that shows a **linear relationship**, illustrated in Figures 1.4.4(a) and (b).
- Variables that produce curved trends include inversely proportional variables (Figure 1.4.4(c)); variables where one varies according to the square of the other (Figure 1.4.4(d)); variables where one varies exponentially or logarithmically with respect to the other (Figure 1.4.4(e) and (f)); and many other relationships.
- Variables that have a periodic relationship produce an oscillating relationship (Figures 1.4.4(g) and (h)).
- When there is no relationship between two variables, one variable does not change when the other variable changes (Figure 1.4.4(i)).

Remember that your results may be unexpected and not match the type of graph you predicted. This does not make your investigation a failure. However, the findings you report must be related to the hypothesis, aims and method.

Linear relationships

Some relationships studied in physics are linear, that is, they can be represented by a straight line. Linear relationships and their graphs are fully specified with just two numbers: the gradient, m, and the vertical axis intercept, c. In general, linear relationships are written:

$$y = mx + c$$

Linear relationships are usually written as:
$y = mx + c$
where m is the gradient
c is the y-intercept.

The gradient, m, can be calculated from the coordinates of two points on the line:

$$m = \frac{\text{rise}}{\text{run}} = \frac{y_2 - y_1}{x_2 - x_1}$$

where (x_1, y_1) and (x_2, y_2) are any two points on the line.

When analysing data from a linear relationship, you may need to find the equation for the line that best fits the data. This line of best fit is also called a regression line. The entire process can be done on paper, but it may be more convenient to use a computer spreadsheet, calculator or some other computer-based application. The benefit of using a digital tool is that the straight line that is produced will be the very best fit for the data. If you plot a line of best fit by eye, some subjectivity will be introduced.

(a)

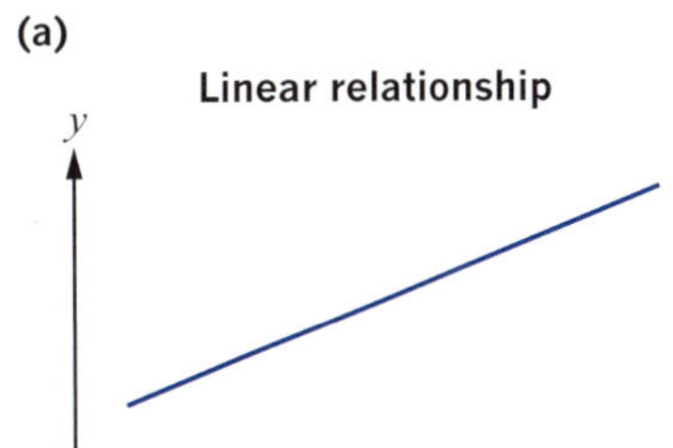

- Positive relationship—as x increases, y increases.

General equation:
$y = mx + c$
m = gradient
c = y-intercept

(b)

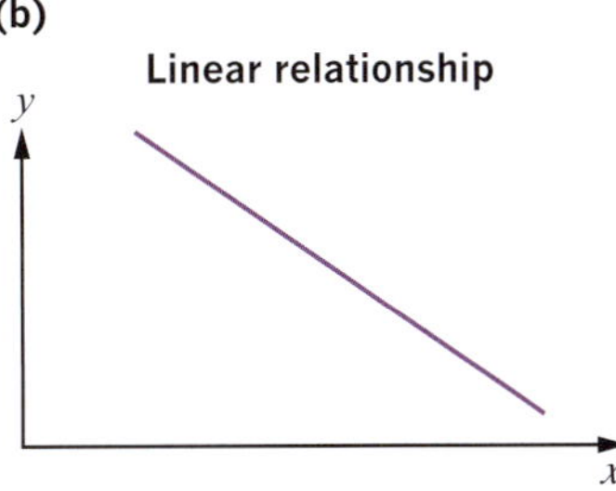

- Negative relationship—as x increases, y decreases.

General equation:
$y = mx + c$
m = gradient
c = y-intercept

(c)

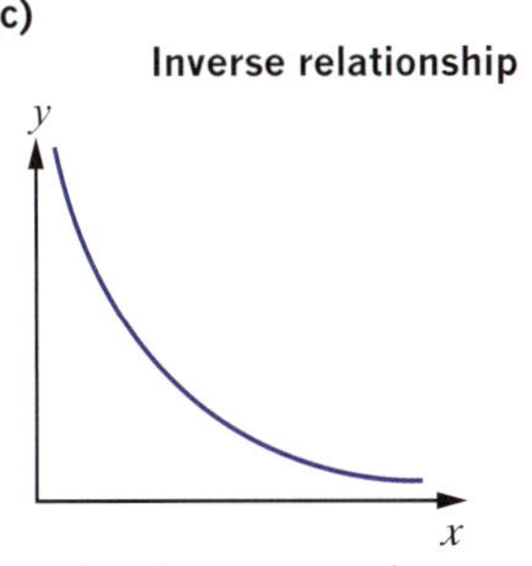

- As x increases, y decreases rapidly, then more slowly, approaching a minimum y value.

General equation:
$y = \frac{k}{x}$
k = constant

(d)

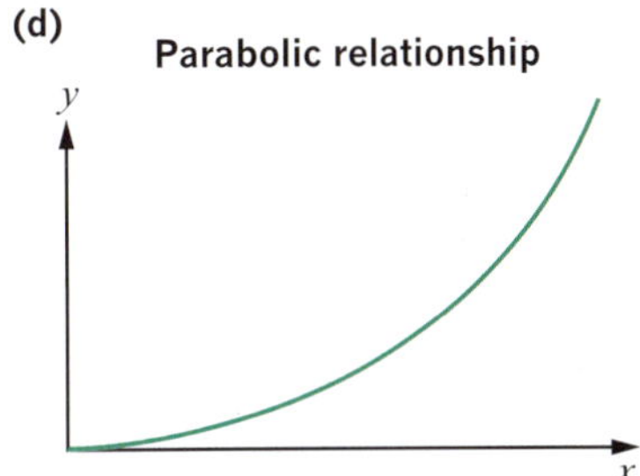

- As x increases, y increases slowly, then more rapidly.

General equation:
$y = kx^2$
k = constant

(e)

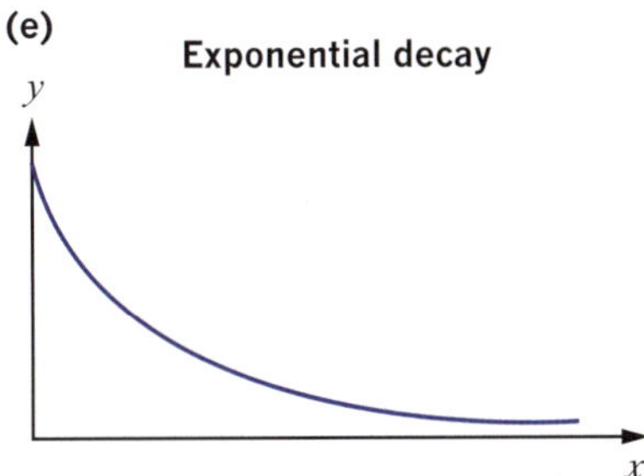

- As x increases, y decreases rapidly, then more slowly, until a minimum y-value is reached.

General equation:
$y = Ae^{-kx}$
k = constant
e = mathematical constant 2.718...

(f)

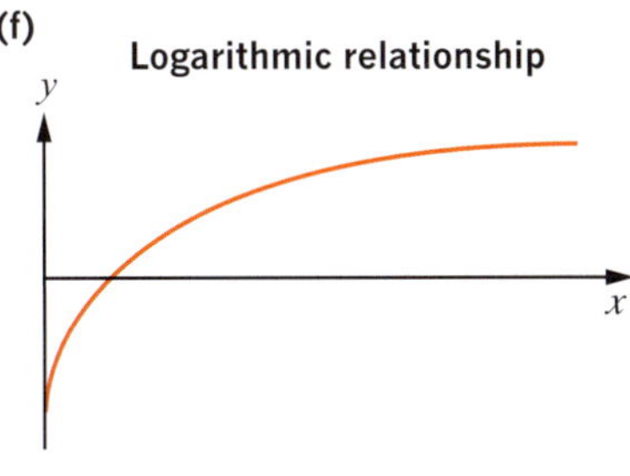

General equation:
$y = \log(x)$

(g)

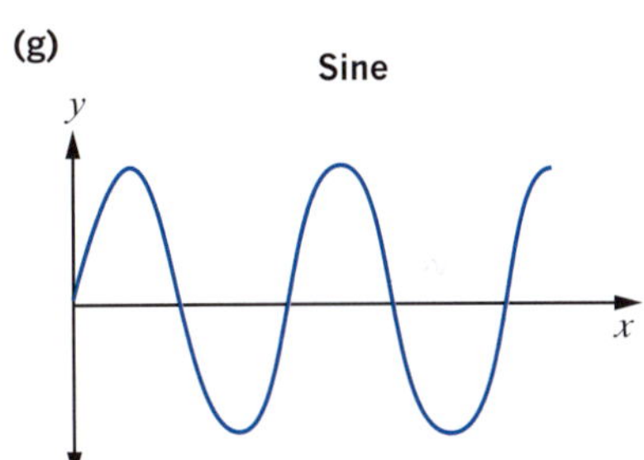

- Periodic relationship—oscillates between a maximum and minimum.

General equation:
$y = A\sin(\omega x)$
A = amplitude
ω = angular frequency of the motion

(h)

Cosine

y

x

- Periodic relationship—oscillates between a maximum and minimum.

General equation:
$y = A\cos(\omega x)$
A = amplitude
ω = angular frequency of the motion

(i)

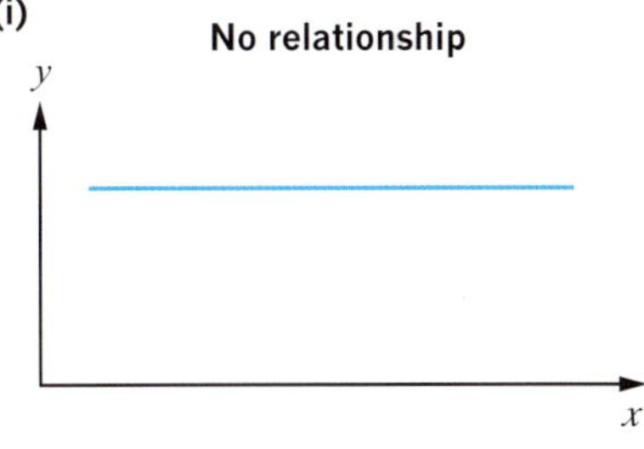

- As x increases, y remains the same.

FIGURE 1.4.4 Various relationships can exist between two variables.

If you are plotting your graph manually on paper and fitting a regression line by eye, proceed as follows:

1 Plot each data point on clearly labelled, unbroken axes.
2 Label, but otherwise ignore, any suspect data points (outliers).
3 Draw by eye the line that best fits the points. The points should be evenly scattered either side of the line.
4 Locate the vertical axis intercept and record its value as c.
5 Choose any two points on the line of best fit and calculate the gradient. These points can be from the original data you plotted, or just any two points, so long as they lie on the line. The gradient is the value m.
6 Write $y = mx + c$, replacing x and y with appropriate symbols, and use this equation for any further analysis.

If you are using a spreadsheet or calculator, proceed as follows:

1 Create a table of the raw data.
2 Plot a graph of the raw data.
3 Identify outliers and create another data table without them.
4 Plot a graph of the data without the outliers. Keep both graphs as you should not discard suspect data, but you can eliminate it from your analysis.
5 Plot the line of best fit, that is, the regression line.
6 Compute the equation of the line of best fit that will give you values for m and c. (Many software programs and calculators will calculate these values for you.)
7 Write $y = mx + c$, replacing x and y with appropriate symbols, and substituting in the values of m and c that were calculated. Use this equation for any further analysis.

Make sure you consider the units on your x and y axes when you state the values for m and c or use them for further calculations. For example, if you have plotted time in ms on the x-axis and speed in $m\,s^{-1}$ on the y-axis, your value for m, the acceleration, will not be in $m\,s^{-2}$.

Worked example 1.4.1

FINDING A LINEAR RELATIONSHIP FROM DATA

A group of students used a computer with an ultrasonic detector to obtain the speed–time data for a falling tennis ball. They wished to measure the acceleration of the ball as it fell. Their hypothesis is that the acceleration would be nearly constant and that the relevant relationship is $v = u + at$, where v is the speed of the ball at any given time, u is the speed when the measurements began, a is the acceleration of the ball and t is the time since the measurements began.

Their computer returned the following data:

Time (s)	Speed ($m\,s^{-1}$)
0.0	1.3673
0.1	2.4973
0.2	3.4804
0.3	4.5991
0.4	5.7969
0.5	6.8139
0.6	7.8874

Using a spreadsheet or calculator, find their experimental value for acceleration.

Thinking	Working
Decide which axes each variable should be placed on.	The dependent variable is the speed, so it will go on the *y*-axis. The independent variable is the time, so it will go on the *x*-axis.
Graph the data as a scatterplot and generate the line of best fit through the points.	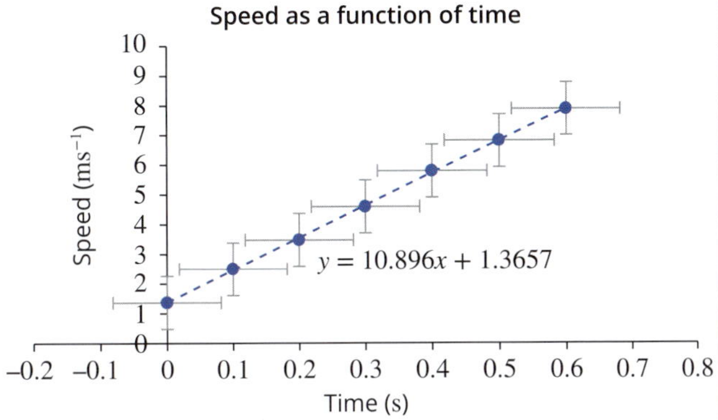 Speed as a function of time Speed (ms^{-1}) Time (s) $y = 10.896x + 1.3657$
Using the functionality of the spreadsheet or calculator, find the equation for the line of best fit. Instead of *y* and *x*, you should use *v* and *t*.	$v = 10.896t + 1.3657$
State the linear relationship in the form required. Express all numbers to 2 significant figures.	$v = 10.896t + 1.3657$ $v = 1.4 + 11t$ Note that linear relationships can be written in the form $y = mx + c$ or $y = c + mx$.
State the answer.	The acceleration is $11\,m\,s^{-2}$.

Worked example: Try yourself 1.4.1

FINDING A LINEAR RELATIONSHIP FROM DATA

A student conducted an experiment to calculate the acceleration due to gravity, *g*. The force due to gravity, F_g, is known to be equal to *mg*.

The downward force was measured for a variety of different masses. It was measured using a spring scale with precision to 2 decimal places. The mass was measured with electronic scales to the nearest 10 g.

Mass (kg)	Force (N)
0.25	2.49
0.50	4.89
0.75	7.48
1.00	9.79
1.25	12.41

Using a spreadsheet or calculator, find the experimental value for the acceleration due to gravity, *g*, to 2 significant figures.

Non-linear relationships

Suppose you were examining the relationship between two quantities, B and d, and had good reason to believe that the relationship can be expressed as:

$$B = \frac{k}{d}$$

where k is some constant. This is similar to $y = \frac{1}{x}$. If you draw this graph using your calculator, you will see that this relationship is non-linear. Therefore a graph of B against d will not be a straight line. However, the relationship can be restated by making the following substitutions:

$$B = k\frac{1}{d}$$

$$\uparrow \quad \uparrow\uparrow$$

$$y = m\ x + c$$

A graph of B (on the vertical axis) against $\frac{1}{d}$ (on the horizontal axis) will be linear. The gradient of the line will be k and the vertical intercept, c, will be zero. The line of best fit would be expected to go through the origin because, in this case, there is no constant added and so c is zero.

In this example, a graph of the raw data would simply show that as d increases, B decreases. It would be impossible to determine this relationship just by looking at a graph of the raw data.

Although a graph of raw data will not reveal the mathematical relationship between the variables, it can give some clues. The shape of the graph might suggest a possible relationship. Several relationships can be tried and then the best chosen. This does not *prove* any relationship, but it could provide strong evidence of a particular relationship.

When an experiment suggests that the relationship being explored is non-linear, the following procedure—called linearising the data—is followed.

1 Plot a graph of the original raw data.

2 Choose a possible relationship based on the shape of the initial graph and your knowledge of various graphical forms. Refer to Figure 1.4.4 on page 33 for guidance.

3 Restate the relationship so that it mimics the form $y = mx + c$.

4 Make a new table of the data using the linear relationship.

5 Derive the line of best fit.

It may be necessary to try several relationships to find the one that best fits the data.

Worked example 1.4.2

FINDING A NON-LINEAR RELATIONSHIP FROM DATA

A group of students investigated the relationship between current and resistance for a new solid-state electronic device. They obtained the data shown in the following table. The current was measured using an ammeter with precision to 1 decimal place and the resistance was measured to the nearest ohm using a multimeter.

Current, I (A)	Resistance, R (Ω)
1.5	22
1.7	39
2.2	46
2.6	70
3.1	110
3.4	145
3.9	212
4.2	236

According to the theory they had researched, the students believed that the relationship between I and R is

$R = dI^3 + g$

where d and g are constants.

By linearising, manipulating the data accordingly and graphing, find the experimental values for d and g. Use a spreadsheet or calculator to assist with finding the line of best fit.

Thinking	Working
Plot a graph of the raw data.	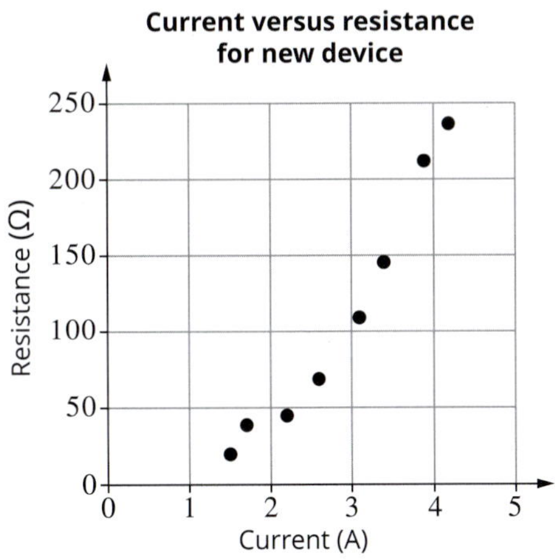 The second data point was considered an outlier and the students chose to ignore it.
Linearise the relationship; that is, restate it to mimic the form $y = mx + c$. A graph of R on the vertical axis and I^3 on the horizontal axis has a gradient of d and a vertical axis intercept of g.	$R = d\ I^3 + g$ ↑ ↑↑ ↑ $y = m\ x + c$

<table>
<tr><td>Make a new table of data manipulated according to the linear relationship. The data is manipulated by finding the cube of each of the values for current, that is, I^3.</td><td>
<table>
<tr><th>Current cubed, I^3 (A^3)</th><th>Resistance, R (Ω)</th></tr>
<tr><td>3.38</td><td>22</td></tr>
<tr><td>10.65</td><td>46</td></tr>
<tr><td>17.58</td><td>70</td></tr>
<tr><td>29.79</td><td>110</td></tr>
<tr><td>39.30</td><td>145</td></tr>
<tr><td>59.32</td><td>212</td></tr>
<tr><td>74.09</td><td>236</td></tr>
</table>
</td></tr>
<tr><td>Plot the graph using the manipulated data.</td><td> 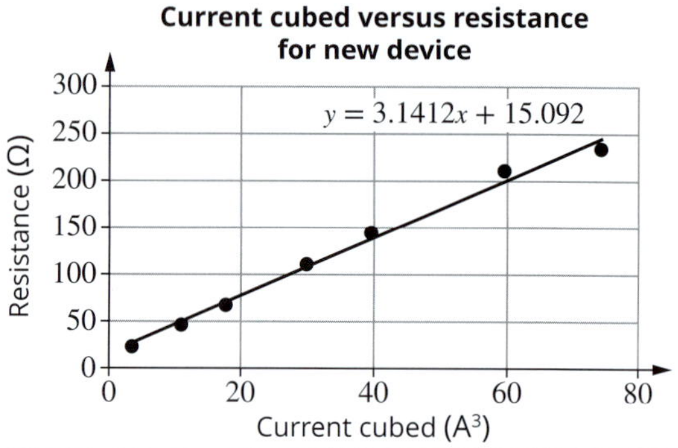
Current cubed versus resistance for new device
$y = 3.1412x + 15.092$
Resistance (Ω): 0, 50, 100, 150, 200, 250, 300
Current cubed (A^3): 0, 20, 40, 60, 80
</td></tr>
<tr><td>Using the functionality of a spreadsheet or calculator, find the equation for the line of best fit.</td><td>$y = 3.1x + 15.1$</td></tr>
<tr><td>Restate the equation using I and R.</td><td>$R = 3.1I^3 + 15.1$</td></tr>
<tr><td>Give the values for d and g. Remember to include the correct units.</td><td>$d = 3.1\ \Omega\ A^{-3}$
$g = 15.1\ \Omega$</td></tr>
</table>

Worked example: Try yourself 1.4.2

FINDING A NON-LINEAR RELATIONSHIP FROM DATA

A group of students investigated the relationship between the distance from a source and the intensity of sound emanating from that source. They obtained the data shown in the following table. Distance was measured using a metre rule and intensity was measured using an app that displays the intensity to 3 decimal places.

Distance, r (m)	Intensity, I ($W\ m^{-2}$)
1	0.040
2	0.010
3	0.005
4	0.003
5	0.002

According to a theory they had researched, the students believed that the relationship between I and r is $I = \frac{P}{r^2}$, where P is a constant.

By appropriate manipulation and graphical techniques, find the students' experimental value for P. Use a spreadsheet or calculator to assist with finding the line of best fit.

EVALUATING THE QUALITY OF DATA

Raw data can be processed in numerous ways. Processing data is usually done to reveal any trends, patterns, uncertainties, mistakes, outliers and results of significance that may exist in the data, including relationships between any variables (for example, the dependent and independent variables). It is important to discuss the limitations of your method of investigation and any effect these limitations may have had on the data collected. Specifically, you should look for anything that may have affected the validity, accuracy, precision or repeatability of the data. Sources of errors and uncertainty must also be stated in the discussion section of the final report on your investigation.

When analysing data, it is important not to select processes that demonstrate only what you want to see. Bias will result from using analysis tools (such as statistics) inappropriately and this may lead to invalid conclusions. It might also be a case of academic fraud. Quality scientific analysis processes raw data as it is and is open to any result.

Bias

In Section 1.3 you learnt about recording data during an experiment. It is important that bias is not introduced during the data-recording process. **Bias** is a form of systematic error resulting from the researcher's personal preferences or motivations. There are many types of bias, including:

- poor definitions of concepts or variables (e.g. classifying cricket pitches as slow or fast according to their interaction with a cricket ball without defining 'slow' and 'fast')
- incorrect assumptions (e.g. that footwear type, model and manufacturer does not affect ground reaction forces and, as a result, failing to control for these variables during an investigation into slip risk on different indoor and outdoor surfaces)
- errors in the design or methodology of the investigation (e.g. testing car-safety features without taking into account different heights of people).

Bias may occur in any part of an investigation, including sampling and measurement.

Some biases cannot be eliminated, but they should be noted in the discussion.

Analysing precision

Section 1.2 highlighted the importance of designing an investigation that will minimise errors and ensure accuracy, precision and validity. Understanding uncertainty and precision is also vital in any analysis of data. In physics there is always variation in measurements. In your experiment you should determine whether the variation is caused by systematic or random errors, in other words, how much variation in the collected data is due to the instruments and how much is due to nature.

The precision and uncertainty of instruments must always be displayed as a range of data next to the results (measurement ± uncertainty). If calculations are performed with the results, then corresponding calculations must also be done with the uncertainties. When the total uncertainty is known, then it can be established whether variation in the data is due to the instrument or to the variables being tested.

If the measurements from repeated trials fall within the uncertainty range of the instrument, then the variation in results could simply be due to the instrument. If the difference between the measured results is greater than the uncertainty range, then the variation in the results is not due to the instrument and must therefore be due to other variables.

It is important to understand the accuracy and precision of the instruments used in an investigation, because they affect the interpretation of the results. There are a few ways to analyse precision, such as by considering:

- measurement uncertainty, which is the precision of the instrument and explains instrumental variation in the measured results
- range, which is the difference between the smallest and largest measurement
- tendency (e.g. the mean), which is the potential variation in instrumental measurements due to the instrument's design or increments.

Analysing validity and theoretical relationships

In processing the data and results you will look for trends, patterns or differences. A common process for analysing data is to make statistical calculations to determine likely true values, uncertainties, errors and the significance of the measurements. Once the quality of the data is understood, then the validity can be analysed in relation to established theoretical concepts.

Analysis can also find anomalies and outliers in data that are not valid measurements. During the experiment, your record of observations may provide a reason for any outlier in your data. Based on this reason, you may be able to suggest improvements in the methodology that could eliminate outliers.

ESTIMATING THE UNCERTAINTY IN A RESULT

Uncertainty has been considered in this chapter in relation to instrument precision. You should be aware that scientists also routinely estimate uncertainty in calculations due to the variation in results. Uncertainty due to natural variation can be treated through statistical analysis of the data.

You will see a combination of the measurement uncertainty and the uncertainty due to variation indicated in processed data. The measurement uncertainty will likely be shown in the column title of a table as ± uncertainty, as shown in Figure 1.3.2 on page 27. An indication of the overall uncertainty will also appear in graphical data with the use of uncertainty bars, as shown in Figure 1.4.2 on page 31 and in Figure 1.4.5 below. You are not expected to calculate the overall uncertainty for the purposes of VCE Physics, but when you interpret results you should recognise that uncertainty indicates a limitation of the data.

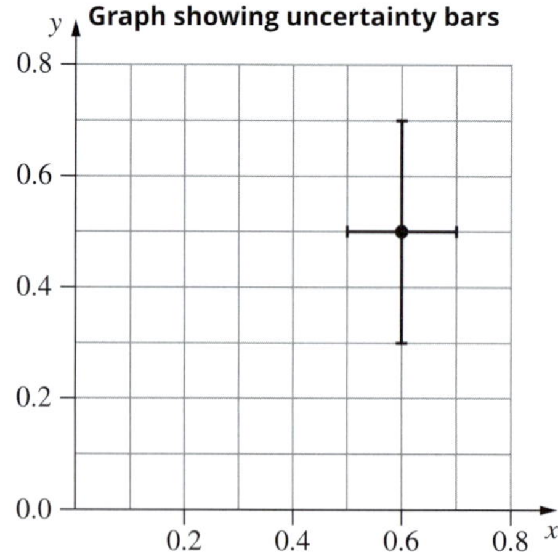

FIGURE 1.4.5 The horizontal uncertainty bar indicates an x-value of 0.6 ± 0.1. The vertical uncertainty bar indicates a y-value of 0.5 ± 0.2.

There are several ways to estimate the overall uncertainty in your data and your teacher will explain the method that best suits your needs. Some of these methods track the effect of instrument precision on calculations, others analyse the variation of the data from the mean.

Uncertainty bars

Uncertainty bars (often called error bars) indicate the absolute uncertainties in the independent values or dependent values. They are drawn as a horizontal line centred on the data point and with a length indicating the uncertainty of the independent variable, and as a vertical line centred on the data point with a length indicating the uncertainty of the dependent variable (Figure 1.4.5). The uncertainty bars can be viewed as forming an uncertainty rectangle, with the true measurement falling somewhere within that rectangle.

> Uncertainty bars (or error bars) provide a graphical representation of the uncertainties in the independent and dependent variables.

The shorter the uncertainty bars are for a given point, the more precise is the measurement.

When data is represented by a straight line of best fit, uncertainty bars can be used to determine the uncertainty in the gradient and the y-intercept. Vertical uncertainty bars extend from the minimum to maximum gradient.

- The maximum gradient line is found by drawing a straight line that passes through all the uncertainty rectangles at the steepest possible gradient.
- The minimum gradient line is found by drawing a straight line that passes through all uncertainty rectangles at the shallowest possible gradient.

Note that this can be tedious to do by hand. Software is available that will do it automatically.

Figure 1.4.6 shows an example of a linear line of best fit with maximum and minimum gradient lines.

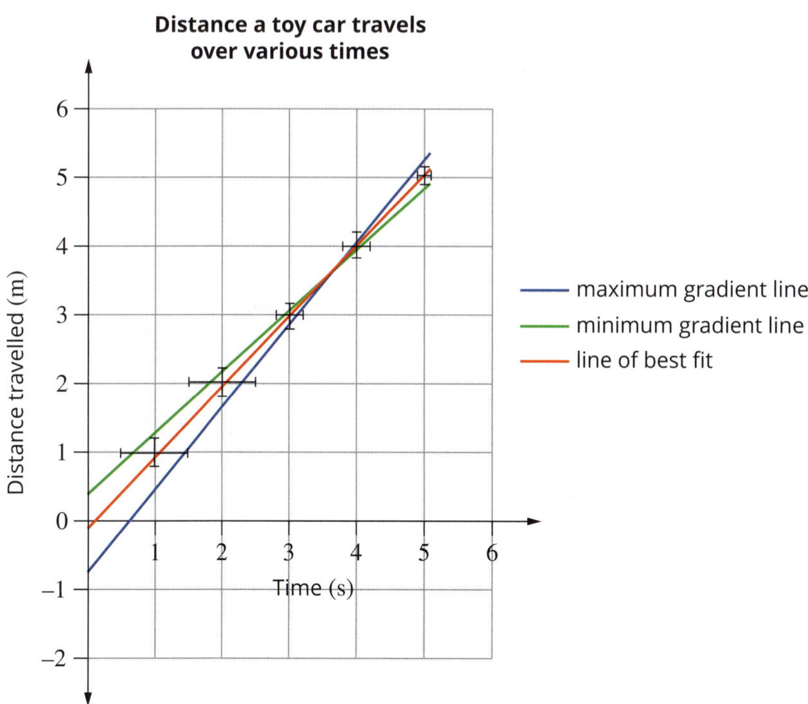

FIGURE 1.4.6 An example of a linear graph showing the line of best fit, uncertainty bars, and maximum and minimum gradient lines. The minimum and maximum gradient lines are used to determine the uncertainty in the gradient and y-intercept.

CASE STUDY **ANALYSIS**

The Hubble constant

In 1924 the Russian scientist Alexander Friedmann used Einstein's general theory of relativity to develop a mathematical proof that the universe was either expanding or contracting. For astronomers to decide if the universe was expanding or contracting, or that Friedmann was wrong and the universe was in fact static, they needed some evidence.

At the time, astronomers were unsure about the size of the universe. For example, they had not yet determined that spiral nebulae are actually galaxies like our Milky Way, but at vast distances away. It was Edwin Hubble who pioneered a technique to find the distance to these other galaxies. He was able to calculate these distances from the brightness of stars known as cepheid variables. Cepheid variables pulsate at regular intervals, and there is a clear relationship between their intensity and the period of this pulsation.

Hubble also examined the spectra from distant stars and galaxies. He found that light from stars in the galaxies he was analysing showed the familiar spectra of the elements that we know on the Earth, such as hydrogen and helium. However, all the spectral lines were redshifted, that is, they had moved towards the red end of the spectrum. Astronomers interpreted this redshift as indicating that the galaxies were receding from us, and at huge speeds. This allowed them to calculate a velocity of recession from the redshift.

This discovery provided evidence that the universe is expanding. These galaxies are not moving through the universe relative to us. They appear to be moving away from us only because the universe between us and them is getting bigger.

Combining his research on the distance to galaxies and their redshift, Hubble produced a graph which made it clear that there was a relationship between certain properties of the galaxies he had measured (Figure 1.4.7). He found that the further away a galaxy, the faster it seemed to be moving; that is, the speed of recession was proportional to distance. This has become known as Hubble's law and is expressed by the equation:

$$v = H_0 d$$

where v is the speed of recession ($km\,s^{-1}$)

H_0 is the Hubble constant ($km\,s^{-1}\,Mpc^{-1}$)

d is distance (Mpc, the abbreviation for the astronomical distance unit megaparsec)

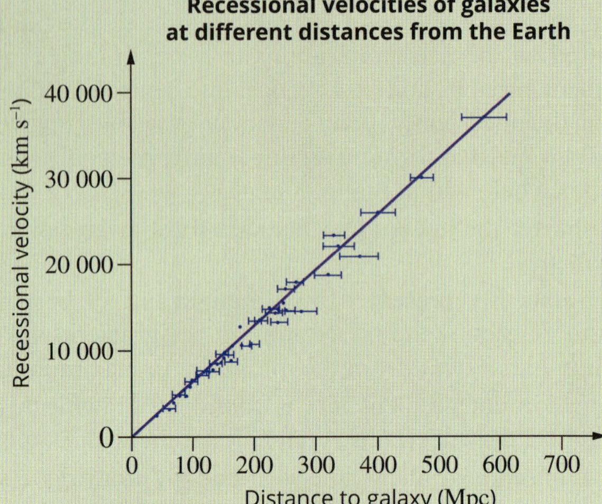

FIGURE 1.4.7 Hubble discovered that the galaxies were receding from Earth and that their recessional velocities were proportional to their distance from the Earth.

Analysis

Astronomers have found the following data for the recessional velocity of distant galaxies:

Distance (Mpc)	Recessional velocity ($km\,s^{-1}$)
10 ± 5	730
31 ± 10	2200
80 ± 20	5600
84 ± 22	6000
207 ± 20	15 000

1. Construct a graph of the distance versus the velocity. Include the uncertainty bars.
2. Create a line of best fit. Also create the maximum and minimum gradient lines.
3. Determine the Hubble constant from your graph.
4. Using the relationship
   $$\text{uncertainty} = \frac{\text{max. gradient} - \text{min. gradient}}{2}$$
   determine the uncertainty in the experimental value of the Hubble constant.

1.4 Review

OA ✓✓

SUMMARY

- Graphs can show a wealth of information about how two or more variables are related, including their uncertainties.
- Graphs can be used to spot outliers in data.
- Graphs in physics should display:
  - a title
  - labels, scales and units on each axis
  - plotted data
  - uncertainty bars in the *x*- and *y*-directions
  - a line of best fit.
- Common graphs in physics show linear, inverse, parabolic, exponential, logarithmic, sine and cosine relationships.
- An outlier is often caused by a mistake made in measuring or recording data, or from a random error in the measuring equipment. If there is an outlier, include it on the graph but ignore it when adding the line of best fit.
- A linear graph is easiest to analyse because it is straightforward to calculate the gradient and *y*-intercept. Such graphs can be used in further processing the data.
- Any non-linear graph can be linearised to produce a linear graph with modified *x*- and *y*-axes.

KEY QUESTIONS

Knowledge and understanding

1 Identify the type of relationship between the *x* and *y* variables that is suggested in each of the following graphs.

a

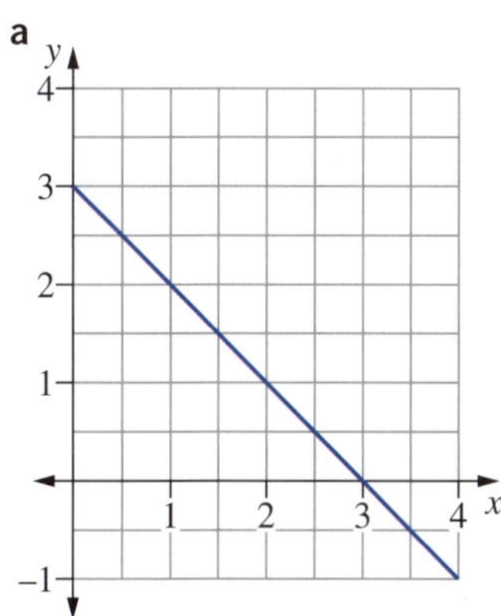

b

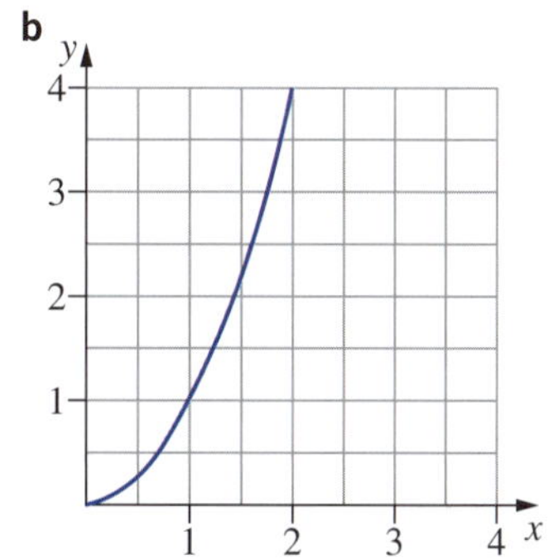

c

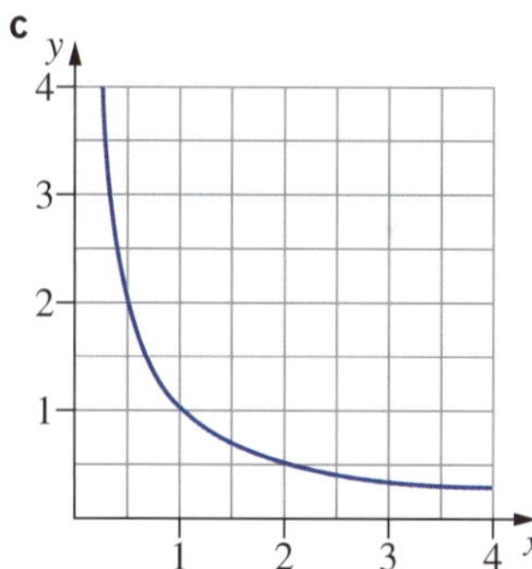

2 If your hypothesis is that impact force is directly proportional to drop height, what would you expect a graph of the data to look like?

3 Describe what bias is in an investigation. Provide an example.

4 Define the term 'outlier'.

5 Plot the following data set, assigning each variable to the appropriate axis.

Current (A)	Voltage (V)
0.06	2.07
0.05	1.56
0.04	1.24
0.03	0.93
0.02	0.63

continued over page

1.4 Review *continued*

6 Determine the experimental equation from the following graph.

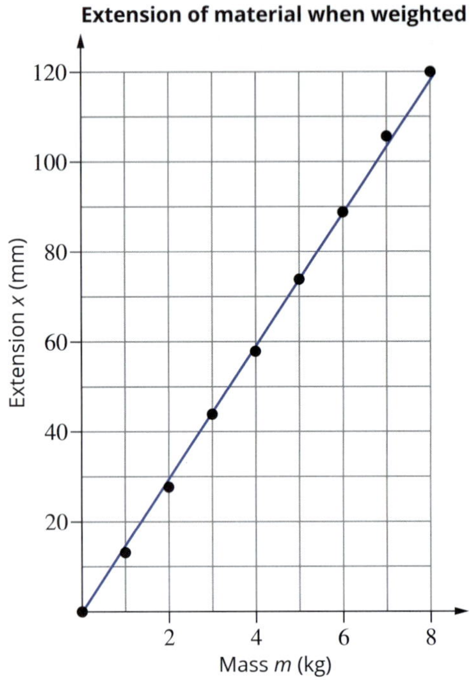

Analysis

7 Consider the graph shown below.

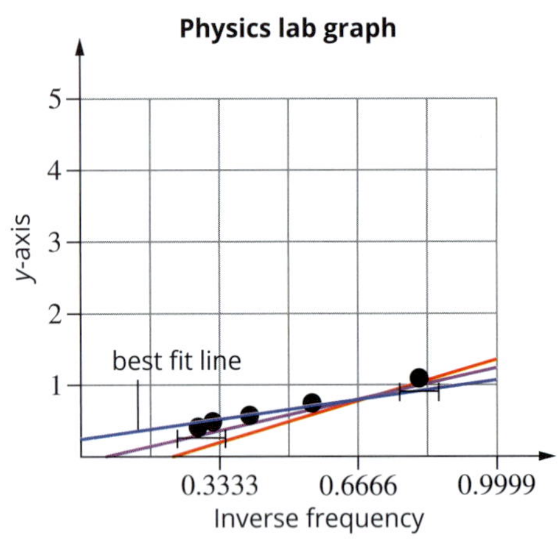

a List the poor graphing techniques displayed in this graph.

b For each poor technique you noted in part **a**, indicate how the graph could be improved.

8 Below are two formulas used in physics. A description of the shape of a graph when related raw data is plotted is given in brackets. Assume that the symbols *A* and *r* are constants.

Formula 1: $P = A\sigma T^4$ (*P* on the *y*-axis and *T* on the *x*-axis produces a rapidly rising curve.)

Formula 2: $a = \frac{4\pi r}{T^2}$ (*a* on the *y*-axis and *T* on the *x*-axis produces a curve that is symmetric about the *y*-axis, and that rapidly decreases as *T* moves away from zero.)

a Describe what modified variables, if any, need to be plotted on each axis to produce a linear graph.

b Identify what the gradient and *y*-intercept become in each case.

9 Zoe researches the period, *T*, and average orbital radius, *r*, of several planets in our solar system. Her results are shown in the table below.

Planet	Period *T* ($\times 10^6 \pm 10^6$ s)	Average orbital radius *r* ($\times 10^7 \pm 10^7$ km)
Mercury	7.6	5.8
Venus	19.4	10.8
Earth	31.5	15.0
Mars	59.4	22.8

The relationship between *T* and *r* is given in:

$$\frac{T^2}{r^3} = \frac{4\pi^2}{GM}$$

where *G* is the universal gravitational constant, $6.67 \times 10^{-11}\,\text{N}\,\text{m}^2\,\text{kg}^{-2}$, and *M* is the mass of the Sun in kg (when all other units are expressed as SI units).

a Plot a graph of the data shown in the table, with *r* as the independent variable (in m) and *T* as the dependent variable (in s). Add a line or curve of best fit as well as uncertainty bars.

b Linearise the data by creating new variables for the *x*- and *y*-axes.

c Draw a line of best fit for the new variables.

d Determine the gradient of the line of best fit.

e Use your answer to part **d** to calculate the mass of the Sun.

1.5 Conclusion and evaluation

After your chosen topic has been thoroughly researched, your investigation conducted and the data collected and analysed, it is time to draw it all together. The final part of the investigation is to summarise your findings in an objective, clear and concise manner in a scientific report. Ways of presenting your report, and report formats, are discussed in Section 1.6.

EXPLAINING RESULTS IN THE DISCUSSION

The discussion is the part of a report where the method is explained and evaluated. It is also where the results are interpreted.

The key sections of the discussion are:

- an evaluation of the investigative method
- an analysis and evaluation of the data
- an explanation of how the findings relate to established concepts in physics.

When writing the discussion section, consider the message to be conveyed and your expected audience. Statements need to be clear and concise. At the conclusion of the discussion, the audience must have a clear idea of the context, results and implications of the investigation.

Evaluating the investigative method

Your discussion should evaluate your investigative methodology and methods and identify any issues that could have affected the validity, repeatability, accuracy or precision of the data. Any possible source of error in your experiment should be stated. Remember that controls are essential to the repeatability and validity of your investigation, so if you have overlooked, or were unable to control, a variable that should have been controlled, this may explain unexpected results.

The discussion should also make recommendations for modifying or extending the investigation. If there were sources of error in any of the methods or steps, provide suggestions for how they could be improved so that future researchers can benefit from your experience.

It is also important to acknowledge contradictions in data and information. Do any of the results not match the predictions? If so, is this a result of a limitation of the experimental design or methods? In your discussion, acknowledge these sorts of issues and make suggestions for further experiments to address them.

Some experimental findings may lead you to formulate new research questions and develop new hypotheses. An extension of the experiment may be to make an alteration that will enable further investigation. For example, if the effect of temperature has been investigated, further understanding of temperature could be determined by using a different temperature range in a modification of the original method.

Analysing and evaluating data

In the discussion, the findings of the investigation need to be analysed and interpreted. A number of things need to be considered.

- State whether a pattern, trend or relationship was observed, and whether it was between the independent and dependent variables. Describe what kind of pattern it is and specify the conditions under which it was observed. Section 1.4 provided guidance about how to analyse trends in data.
- Acknowledge and explain any discrepancies, deviations or anomalies in the data.
- Identify any limitations in the data you collected. For example, you might think that a larger sample or further variations in the independent variable would have led to a more valid and reliable conclusion.

Relating your investigation to relevant physics concepts

To make your investigation more useful—and more interesting to other physicists or physics students—your discussion should explain how your investigation is related to established ideas, concepts, theories or models in physics. In particular, you should explain why you considered your hypothesis to be a worthwhile idea to explore.

For example, if studying the impact of temperature on the linear strain of a material (e.g. a rubber band), some of the information relevant to physics that might be included in the discussion is:

- the functions of linear strain
- the factors known to affect linear strain
- existing knowledge on the role of temperature on linear strain
- the range of temperatures investigated and the reason they were chosen
- the materials studied and the reasons for this choice
- methods of measuring the linear strain of a material.

FRAMING YOUR DISCUSSION

By relating your investigation to relevant concepts in physics, you will have created a framework for discussing whether the data you collected supports or refutes your hypothesis. Ask the following questions:

- Was the hypothesis supported?
- Has the research question been fully answered? (If not, give an explanation of why that is and suggest what could be done to either improve or complement the investigation.)
- Do the results contradict the hypothesis? If so, why? (The explanation must be plausible and must be based on the results and on previous evidence.)

After identifying the major findings of your investigation, compare your results with existing relevant research and knowledge. Consider questions such as:

- How does the data fit with the literature?
- Does the data contradict the literature?
- Do the findings fill a gap in the literature?
- Do the findings lead to further questions?
- Can the findings be extended to other situations?

Be sure to discuss the broader implications of your findings. Ask such questions as:

- Do the findings contribute to our current knowledge of the topic?
- Do the findings suggest any practical applications?

Use the points in Figure 1.5.1 to help frame your discussion.

DRAWING EVIDENCE-BASED CONCLUSIONS

The **conclusion** to a scientific report or paper links the collected evidence to the hypothesis and provides a justified response to the research question.

Indicate whether the hypothesis was supported or refuted, and the evidence on which this is based (i.e. your results). Do not provide irrelevant information. Refer only to the specifics of the hypothesis and the research question and do not make generalisations.

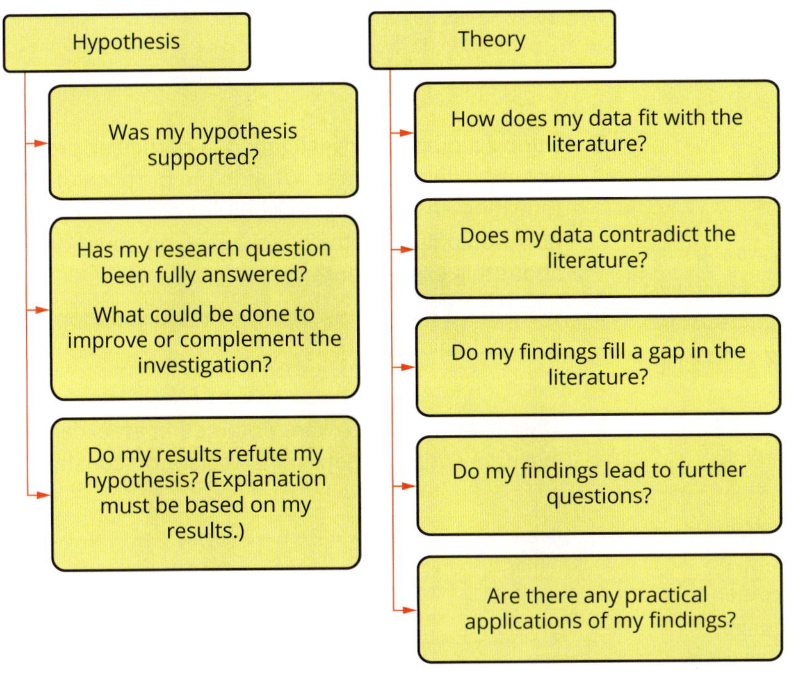

FIGURE 1.5.1 Points to help frame your discussion

The examples of poor and better conclusions in Table 1.5.1 may be of assistance.

TABLE 1.5.1 Examples of poor and better conclusions

Hypothesis/Research question	Poor conclusion	Better conclusion
An increase in temperature will cause an increase in linear deformation (change in length) before failure.	Linear deformation has value y_1 at temperature 1 and value y_2 at temperature 2.	An increase in temperature from 1 to 2 produced an increase in linear deformation in the rubber band.
Does temperature affect the maximum linear deformation the material can withstand?	The results show that temperature does affect the maximum deformation of a material.	Analysis of the results of the effect of an increase in temperature from 1 to 2 on the rubber band supports our current knowledge that an increase in temperature increases the maximum linear deformation.

1.5 Review

OA ✓✓

SUMMARY

- The discussion evaluates and explains the investigation methods and the results you obtained.
- Analyse and interpret your results in the discussion by:
  - identifying and describing any patterns, trends or relationships in the data. Specify the conditions under which they were observed.
  - acknowledging and explaining any discrepancies, deviations or anomalies in the data
  - identifying any limitations in the data collected.
- Explain the hypothesis and investigation within the context of current thinking in physics.
- Use evidence from the data to conclude whether the hypothesis was supported or refuted.

continued over page

1.5 Review *continued*

KEY QUESTIONS

Knowledge and understanding

1 Consider this hypothesis: *An increase in the current passing through a single resistor in an electric circuit will cause an increase in the voltage drop across the resistor.* Given that hypothesis, improve the following conclusion:

When the current was 0.03 A, the voltage was 0.93 V and when the current was 0.05 A, the voltage was 1.81 V.

2 Which of the following would not support a strong conclusion to a report?

A The conclusion is relevant and provides evidence.
B The conclusion is written in emotive language.
C The conclusion makes reference to the limitations of the research.
D The conclusion includes suggestions for further avenues of research.

3 Before beginning an investigation, a student proposes the following hypothesis: *According to Newton's second law, for a constant force, if the mass is increased the acceleration is decreased.*

The table shows the results obtained during the investigation.

Mass (kg)	Acceleration ($m s^{-2}$)
1.0	3.0
2.0	2.0
3.0	1.0

Do the results support or refute the hypothesis? Justify your answer.

4 You conduct an investigation to test the following hypothesis: *If two objects are simultaneously dropped from the same height, they will both land at the same time.*

What is one conclusion you could draw if your results showed the following times were recorded for objects dropped from a height of 1 m?

Object	Time (s)
feather	2
tennis ball	0.5
bowling ball	0.4

5 During a practical investigation, a particular procedure was repeated 30 times. What is the purpose of repeating an experiment multiple times like this, and how would it affect the conclusion you could draw about this experiment?

6 What are the key subsections of the discussion section of a scientific report?

Analysis

7 The following results were obtained in an experiment testing a spring to determine the spring constant.

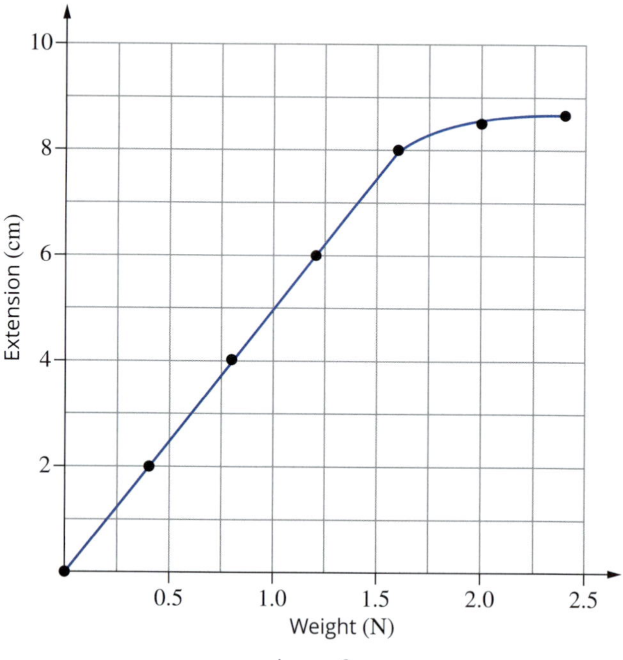

$$k = \frac{\text{rise}}{\text{run}} = \frac{8}{1.6} = 5 \text{ cm N}^{-1}$$

Which sentence below is the best concluding statement? Give a reason why the other options are not appropriate.

A All springs have a stiffness of 5 cm N^{-1} when a force is applied between 0 N and 1.6 N.
B The stiffness is always measured in cm N^{-1}.
C The stiffness of the spring tested was 5 cm N^{-1} for forces between 0 N and 2.5 N.
D For forces between 0 N and 1.6 N, the spring tested had a stiffness of 5 cm N^{-1}.

1.6 Reporting investigations

Scientists report their findings in a number of ways: as written peer-reviewed journal articles, on web pages, and at scientific conferences with short oral presentations or scientific posters (Figure 1.6.1). In this section you will learn how to present your findings effectively.

FIGURE 1.6.1 Posters at a scientific conference are one way of presenting the findings of your research.

PRESENTATION FORMATS

There are numerous ways to present the results of a scientific investigation, each with varying emphasis on visual and textual components. Table 1.6.1 provides some guidelines for different presentation formats.

TABLE 1.6.1 Main formats for presenting research work

Format	Characteristics	General guidelines for the presentation format
scientific poster	• concise visual display of information • suitable for presenting information to many people • summary of ideas	• title that attracts attention • large headings that stand out • subheadings of a smaller size • attractive presentation • combination of written material and visual material (such as diagrams, photographs, tables and graphs) • writing large enough to be read from a distance
written practical report	• presents clear and detailed information on a topic • suitable for providing detailed and more comprehensive background information	• appropriate writing style for a report that provides sections for an introduction, methodology and methods, results, discussion, conclusion and references • subheadings for organising sections • text should be supported by tables, graphs, diagrams and/or photographs
oral communication with supporting slides and/or handouts	• easy-to-follow format • good for presenting to a large audience • supporting slides can be printed as notes and given to the audience • opportunity to answer questions from the audience	• brief oral descriptions • clear visuals that complement what is spoken • minimal text on each slide • consistent format on all slides—background, colours and text • images, diagrams and graphs are clear and large
online presentation, e.g. website, blog	• accessible to a worldwide audience • easy to follow • easy to update with new information	• hyperlinks to related information • multimedia, such as video and audio, if appropriate • same format throughout—font, background, colours • clear headings • your name, credentials and date of publication

STRUCTURING A WRITTEN REPORT

To write a scientific report you need to follow some general conventions. Even though there are many ways to present a report, the report must broadly follow the structure set out below to meet the requirements of the syllabus.

Headings are an essential feature of a scientific report. There is no single convention for headings in scientific report writing and what is required is often specific to a particular journal. Figure 1.6.2 lists the main headings that are commonly used in scientific reports and describes the information that is usually provided under each heading. Sections can be broken down further into subsections. Although some subheadings may be suitable for more than one section, a scientific report will only use each heading and subheading once. It is best to ask your teacher about which headings are preferred to meet the requirements of the Physics VCE Study Design.

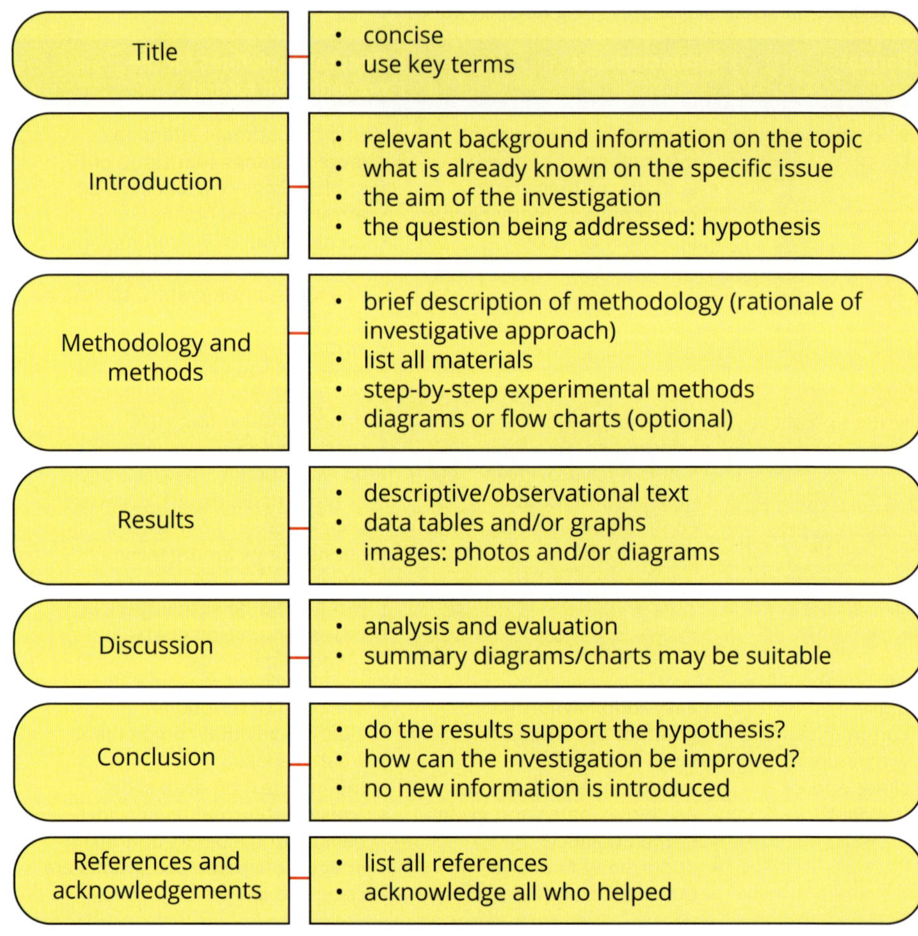

FIGURE 1.6.2 Elements of a scientific report or presentation

Title

The title should include key terms and give a clear idea of what your investigation and the report is about, without being too long.

Introduction

The introduction sets the context of your report. It should outline relevant physics ideas, concepts, theories and models, and how they relate to your research question and hypothesis. It introduces the key terms, the specific question to be addressed, and states your hypothesis and aim. Any references used in the introduction should be correctly cited. This section should also identify the independent, dependent and controlled variables.

Methodology and methods

The methodology and methods section outlines the reason why you adopted your particular investigative approach and describes in detail all the steps that were undertaken during the investigation. It also includes a list of the materials used. Use step-by-step lists, diagrams of specific methods, and/or flow charts of the overall experimental design.

There should be enough detail in this section for someone else to be able to reproduce your experiment. Therefore your method needs to be in the correct sequence and include how you observed, measured and recorded your results.

Results

The results section is a record of your observations. It is where you present your data using graphs, diagrams, tables or photographs. In Section 1.4 you learnt tips on using graphs and tables appropriately.

In general, tables provide more detailed data than graphs. However, it is easier to observe trends and patterns in graphs, making them a very useful tool for presenting evidence. Pie charts illustrate percentages well, while scatterplots illustrate relationships between variables. Bar charts are best used for qualitative data and discrete quantitative data. Scatterplots are best used for continuous quantitative data.

Discussion

In the discussion section you interpret your results and discuss how they relate to your research question and hypothesis, and to the research of others. You should link your discussion to the key concepts discussed in the introduction.

It is also important to evaluate the methods used and the impact of any errors on the results and on the conclusions drawn.

Conclusion

Your conclusion should reference the evidence (that is, your data) that supports or refutes your hypothesis. It should provide a carefully considered response to your research question based on your results and discussion. You should clearly state whether your hypothesis was supported or not. Draw your conclusions by identifying trends, patterns and relationships in the data.

It is important to recognise the limitations of both your data and the scientific method. Be careful not to overstate your conclusion. Your results will support or refute the hypothesis. They will not prove that something is true, as you can only ever provide evidence that indicates the probability of something being true.

Do not provide irrelevant information, or introduce new information, in your conclusion. Refer to the specifics of your hypothesis and research question, and do not make generalisations.

The conclusion section should be a short, succinct paragraph.

References

You must cite the source of any information you used in your report wherever it is used, and also provide a list of these references at the end of your report. This demonstrates that you are aware of previous work in the area and allows readers to locate sources of information if they want to study them further.

Acknowledgements

It is good practice to acknowledge anyone who helped you during your investigation, such as your teacher, lab technician or anyone else who provided you with guidance. This only needs to be a sentence or two. An example is 'Thanks to my teacher for useful discussions about the feasibility of my experiment, to our lab technician for help setting up the motion sensor correctly and to my fellow students for repeatedly running along the sports track so I could time them'.

WRITING FOR SCIENCE

A scientific report is written for a scientific audience, so it is important to ensure that the report uses appropriate scientific language and follows the expected conventions. Scientific language and its conventions are different from everyday English writing. For example, scientific reports should be written using:

- past tense—the experiment was conducted in the past, so the report should be in the past tense
- third person, passive voice with impersonal verbs (Table 1.6.2)
- scientific language—the terms used are specific to concepts, models and theories
- objective, unbiased language—avoid subjective and emotional or persuasive writing (Table 1.6.3).
- concise language—avoid unnecessary repetition and express ideas succinctly. Scientific language allows more details, knowledge and understanding to be communicated in fewer words. Use short sentences (Table 1.6.4).

Scientific language must be used without error so that the reader understands the meaning of the information easily. For the report to be concise, there should not be any repetition. The report must remain within the required word limit. Being precise and concise will help you stay within that limit.

TABLE 1.6.2 Examples of first-person and third-person writing

First person	Third person
I first tied a rubber stopper of known mass onto one end of a piece of fishing line, and a brass cradle of 200 g to the other end.	First, a rubber stopper of known mass was tied to one end of a piece of fishing line. A brass cradle of mass 200 g was tied to the other end.
After the current was switched on, I found that...	After the current was turned on, the results showed...
My colleagues and I found...	Researchers found...

TABLE 1.6.3 Persuasive writing versus scientific writing styles

Persuasive writing examples	Equivalent scientific writing examples
Use of biased and subjective language: • The results are extremely bad, atrocious, wonderful etc. • This is terrible because...	Use of unbiased and objective language: • The results showed... • The implications of these results are... • The results imply...
Use of exaggeration: • The object weighed a colossal amount, like an elephant. • Safety crisis...	Use of non-emotive language: • The object weighed 256 kg. • Safety issue...
Use of everyday or colloquial language: • The experiment didn't work because we didn't know what we were doing. • We did not obtain the results we expected because we were not sure how to use the equipment, which led to significant errors in measurement. • The results don't... • The researchers had a sneaking suspicion that...	Use of formal language: • Further research is needed to fully determine why the results of the experiment were not as expected. • The results do not... • The researchers predicted (or hypothesised or theorised) that...

TABLE 1.6.4 Examples of wordy language written more concisely

Wordy language	Concise language
Due to the fact that...	Because...
Anog and Walsh (2022) undertook an investigation into...	Anog and Walsh (2022) investigated...
It is possible that the cause could be...	The cause may be...
End result...	Result...
In the event that...	If...
Shorter in length...	Shorter...

Paragraphs in scientific writing

In a scientific report each paragraph should explain only one topic. The first sentence of a paragraph (called the topic sentence) introduces the topic. The sentences that follow the topic sentence provide details about the topic, with the final sentence concluding the discussion of the one topic that the paragraph is about.

Each sentence in a paragraph should refer to only one subject (two only if necessary), and each sentence should flow on to the next. Readers should be able to see how each sentence relates to the previous one.

EDITING YOUR REPORT

Editing your report is an important part of the process. After editing your report, save new drafts with a different file name and always back up your files in a second location. Once you have completed a draft, it is good practice to read your work a day or two after you have completed it.

When reading your own work, do not read it as you intend it to be read by others. Instead, carefully read your work, following the punctuation, grammar and spelling as it appears on the page. This is more easily achieved if you read the report aloud. When editing, look for content that:

- is ambiguous or unclear
- is repetitive
- is awkwardly phrased
- is too lengthy
- is not relevant to your research question
- is poorly structured
- lacks evidence
- lacks a reference (if it is another researcher's work)
- contains spelling mistakes.

ACKNOWLEDGING SOURCES

The source of all quotations used in your report must be listed in the references section. You should also acknowledge the ideas of others that have been instrumental in forming the idea for your research or the interpretation you have made of your results. References and acknowledgements also give credibility to your study and allow the audience to locate information sources for further study.

Plagiarism is using other people's work without acknowledging them as the author or creator. To avoid plagiarism, include a reference every time you report the work of others, placing it at the end of a sentence or following a diagram. If you use a direct quotation from a source, enclose it in quotation marks.

Referencing

Each time you write about the findings of other people or organisations, you need to provide an in-text citation. You also need to provide the full details of the source in the reference list at the back of the report. Numerous referencing systems are in use, but the American Psychological Association (APA) system is common in the sciences. Check with your teacher which system you should use for your report.

In most referencing systems, sources in a reference list are listed in alphabetical order (by the author's last name or the organisation's name). Compile your references in a separate document as you conduct your practical investigation. This will save you time later.

A bibliography is a list of all the sources used during your research that helped you develop an understanding of your research topic even if such a source is not cited in your final report. A reference list only lists those sources that you cite. Your Unit 4 Area of Study 2 practical investigation report does not require a bibliography. A reference list is sufficient.

The following examples show the use of in-text citation and the corresponding reference list entry for an article in a journal in APA (seventh edition) style. The format varies slightly according to the type of document or source you are referencing. Guides can be found online, including on many university websites.

Examples of citing a reference in the text	Formats for listing a reference in the reference list (with examples)
Research article or review article in a scientific journal A single atom of the rare-earth metal holmium has been made into the world's smallest, stable magnet. This was then used to make an atomic hard drive, in which each holmium atom stored one bit of information (Natterer et al., 2017).	Author, initials. (year). Title of article. Journal title, volume number(issue number), page numbers. Digital object identifier (doi) or URL Natterer, F., Yang, K., Paul, W., Willke, P., Choi, T., Greber, T., Heinrich, A., & Lutz, C. (2017). Reading and writing single-atom magnets. *Nature*, 543, 226–228.
Book Hawking (1988) discusses the theories of general relativity and quantum mechanics and seeks to describe a unifying theory that combines these.	Author, initials. (year). Title of book (edition, if not first). Publisher Hawking, S. (1988). *A brief history of time: From the big bang to black holes.* Bantam Books.
Online article or page A layered crystal (created with hafnium oxide and zirconium oxide) reduces the required voltage by around 30% (Perfetto, 2017).	Author, initials/name of organisation. (year, month day). Title of webpage or web document. URL Perfetto, I. (2022, April 1). New crystal could help transistors run on less power. Cosmosmagazine.com. https://cosmosmagazine.com/technology/computing/crystal-for-transistors/

STRUCTURING A POSTER

Scientific posters are used at conferences to grab people's attention and quickly convey a summary your research. A poster should follow much the same format as a scientific report, outlined earlier in this section. A poster is meant to be more succinct and direct in its approach to presenting your findings. You should carefully pick the information that you want to present so that the impact of your investigation is best communicated. For instance, in analysing your data you may have produced both a table and a graph. In the poster, it may be best to simply show the graph, which will more clearly represent any trends in your data.

A scientific poster should be understood by both technical and non-technical audiences. Make sure to keep the language concise and as free of jargon as possible.

Figure 1.6.3 is an example of a poster from a student's investigation of the research question 'Is the period of a pendulum affected by the mass or length of the pendulum?'

Is the period of a pendulum affected by the mass or length of the pendulum?

Yarran Jones
Heinemann Physics College
Unit 3–4 Physics 2022

Introduction

A simple pendulum consists of a mass or 'bob' attached to a pivot point. As the bob is lifted and released, the pendulum sweeps back and forth. One full movement, from left to right and back again, is a 'period' (Figure 1).

When a pendulum is lifted, it acquires gravitational potential energy (E_g). When the pendulum is released, the bob's mass causes it to swing back and forth. Its E_g is transformed into kinetic energy (E_k). When the pendulum passes through its equilibrium, the E_g it has 'lost' has been transformed into E_k. As it continues to swing and move upwards on the next arc, the E_k is transformed back into E_g.

(Museum of Science and Industry, Chicago, n.d.)

AIM To determine if the mass or length of a pendulum affects its period.

HYPOTHESIS The period of a pendulum is independent of its mass and depends on the square root of its length.

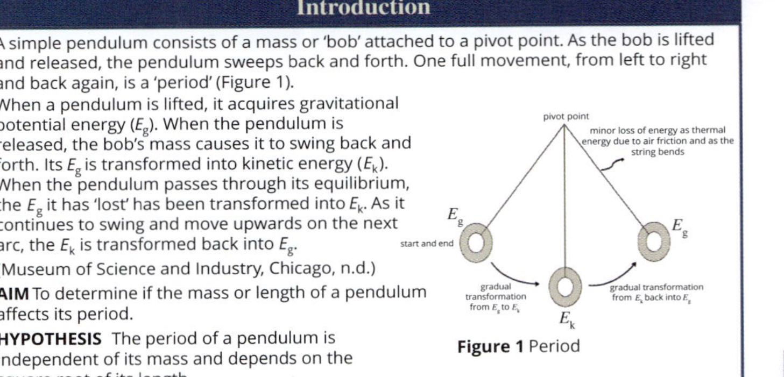

Figure 1 Period

Methodology and methods

METHODOLOGY

Controlled experiment

METHODS

Set up

1. Fold A4 paper into a 'ramp' approximately 11 cm × 2 cm with a V-notch in one end.
2. Mark a line at 45° on vertical desk edge.
3. Extend ramp over desk edge, with V lined up with line on desk edge.
4. Tie a washer to a 0.45 m piece of string.
5. Measure from the bottom of the washer along the string. Put coloured dots at 0.10 m, 0.15 m, 0.20 m, 0.25 m, 0.30 m and 0.35 m.
6. Make two more pendulums of two and three washers each and mark their lengths.

Testing the pendulums

1. Hang pendulum over ramp with the 0.30 m mark on the string with one washer attached.
2. Pull washer so string lines up with 45° line.
3. Release pendulum, start timing. Stop after three complete swings.
4. Repeat twice more, calculate averages.

Part A: Change the mass

1. Change mass of the pendulum by adding a washer.
2. Repeat test procedure, record three times for this pendulum.
3. Change to three-washer pendulum and repeat.

Part B: Change the length

1. Return to one washer on the pendulum.
2. Change length of the pendulum to 0.35 m by adjusting the string length.
3. Use different lengths for one-washer pendulum (0.10 m, 0.15 m, 0.20 m and 0.25 m).
4. Calculate averages.

Communication statement

The period of a pendulum is independent of its mass. The period does depend on the square root of its length (i.e. $T \propto \sqrt{L}$).

Results

Table 1 Effect of mass on period

No. of washers	Length (m)	Release angle (°)	Time for three swings (s)				Period T (s)
			Trial 1	Trial 2	Trial 3	Average	
1	0.30	45	3.39	3.36	3.30	3.35	1.12
2	0.30	45	3.24	3.49	3.42	3.38	1.13
3	0.30	45	3.32	3.31	3.36	3.33	1.11

Table 2 Effect of length on period

No. of washers	Length (m)	Release angle (°)	Time for three swings (s)				Period T (s)
			Trial 1	Trial 2	Trial 3	Average	
1	0						0
1	0.10	45	1.60	1.69	1.63	1.64	0.55
1	0.15	45	2.27	2.26	2.15	2.23	0.74
1	0.20	45	2.69	2.76	2.68	2.71	0.90
1	0.25	45	3.02	3.00	3.01	3.01	1.00
1	0.30	45	3.30	3.31	3.43	3.35	1.12
1	0.5	45	3.62	3.60	3.61	3.61	1.20

Discussion

EVALUATION

Graph 1 Period squared v length

Period squared (s²)
1.60 1.40 1.20 1.00 0.80 0.60 0.40 0.20 0.00
$y = 4.05x$
0 0.05 0.1 0.15 0.2 0.25 0.3 0.35 0.4
Length (m)

Table 1 shows no significant variation in the three periods when the mass changed. The period of a pendulum is not, therefore, dependent on the mass.

According to the formula $T = 2\pi\sqrt{\frac{L}{g}}$ (Bunn, 1990), the relationship between T and L is a square root relationship. This is evident in Table 2 and confirmed by the linear relationship of Graph 1.

LIMITATIONS

- measurements of length limited by use of a ruler
- error bars not included as graph is too small to show up precisely
- V supported string and held it away from desk edge by a few mm to reduce friction while pendulum swinging.

IMPROVEMENTS

- better method of removing friction at pivot point
- more accurate measure of length of the pendulum
- measure masses of washers more accurately.

Conclusion

The hypothesis was supported by the results showing that the period of a pendulum is independent of the mass and dependent on the square root of the length.

Further experiments could be conducted:

- with a larger range of masses to further confirm the independence of the period on mass
- to determine if the angle of release (other than 45°) has an effect on the period
- to determine if different types of strings have an effect on the period
- to determine the acceleration due to gravity.

References and acknowledgements

1. Bunn, D.J. (1990). *Physics for a modern world*, (p. 237). John Wiley & Sons Australia.
2. Museum of Science and Industry, Chicago. (n.d.). *Back and Forth*. https://www.msichicago.org/fileadmin/assets/educators/learning_labs/documents/Back_and_Forth.pdf

Acknowledgements: with thanks to lab technicians Jarrah and Ivan and my lab partner Michelle Chan.

FIGURE 1.6.3 Example of a scientific poster

1.6 Review

OA ✓✓

SUMMARY

- A scientific report generally includes the following headings:
  - Title
  - Introduction
  - Methodology and methods
  - Results
  - Discussion
  - Conclusion
  - References
  - Acknowledgements
- Your scientific report must meet the requirements of the syllabus. Check with your teacher about the specific requirements for your task.
- A scientific poster follows a similar structure to that of a scientific report.
- Scientific reports must be written in scientific language: past tense, third person, passive voice, objective and concise.
- Where the work of another scientist is used or referred to in a scientific report, that scientist must be acknowledged, both in the text and in the reference list.
- A scientific poster should be more succinct than a report and written for both technical and non-technical audiences.

KEY QUESTIONS

Knowledge and understanding

1 Which of the following statements is written in scientific style?
  - **A** The results were fantastic ...
  - **B** The data in Table 2 indicates ...
  - **C** The researchers felt ...
  - **D** The smell was awful ...

2 Which of the following statements is written in the first-person?
  - **A** The researchers reported ...
  - **B** Samples were analysed using ...
  - **C** The experiment was repeated three times ...
  - **D** I reported ...

3 Recall in which section of a scientific report you would find processed data.

4 What is the purpose of referencing and acknowledging documents, ideas, images and quotations in your investigation?

5 Explain the difference between a bibliography and a reference list.

Analysis

6 A scientist conducted an experiment to test the following hypothesis: *Increasing the temperature of a wire would result in a decrease in the resistance of the wire*. The discussion section of the scientist's report included comments to support the repeatability (reliability), validity, accuracy and precision of the investigation. Determine whether the following sentences indicate repeatability, validity, accuracy or precision.
  - **a** Five different wires of the same length were tested at each temperature. The resistance was measured three times and averaged.
  - **b** The temperature and resistance of the wires were initially recorded using data-logging equipment. However, the resistance of some of the wires was measured using an analogue voltmeter and ammeter.
  - **c** The data-logging equipment was calibrated for temperature before use.
  - **d** The voltage/current sensor (data logger) measured voltage to the nearest 0.1 V and the current to the nearest 0.1 mA. An analogue voltmeter measured the voltage to the nearest 1 V, and the ammeter measured to the nearest 10 mA.

Chapter review

01

KEY TERMS

accuracy
aim
bias
conclusion
continuous variable
controlled variable
dependent variable
discrete variable
hypothesis
independent variable
line of best fit
linear relationship
mean
median
method
methodology
mistake
mode
observation
outlier
personal protective equipment (PPE)
precision
primary data
primary source
processed data
qualitative data
qualitative variable
quantitative data
quantitative variable
random error
range
raw data
repeatability
reproducibility
research question
risk assessment
safety data sheet (SDS)
scientific method
secondary data
secondary source
significant figures
systematic error
true value
uncertainty
uncertainty bars
validity
variable

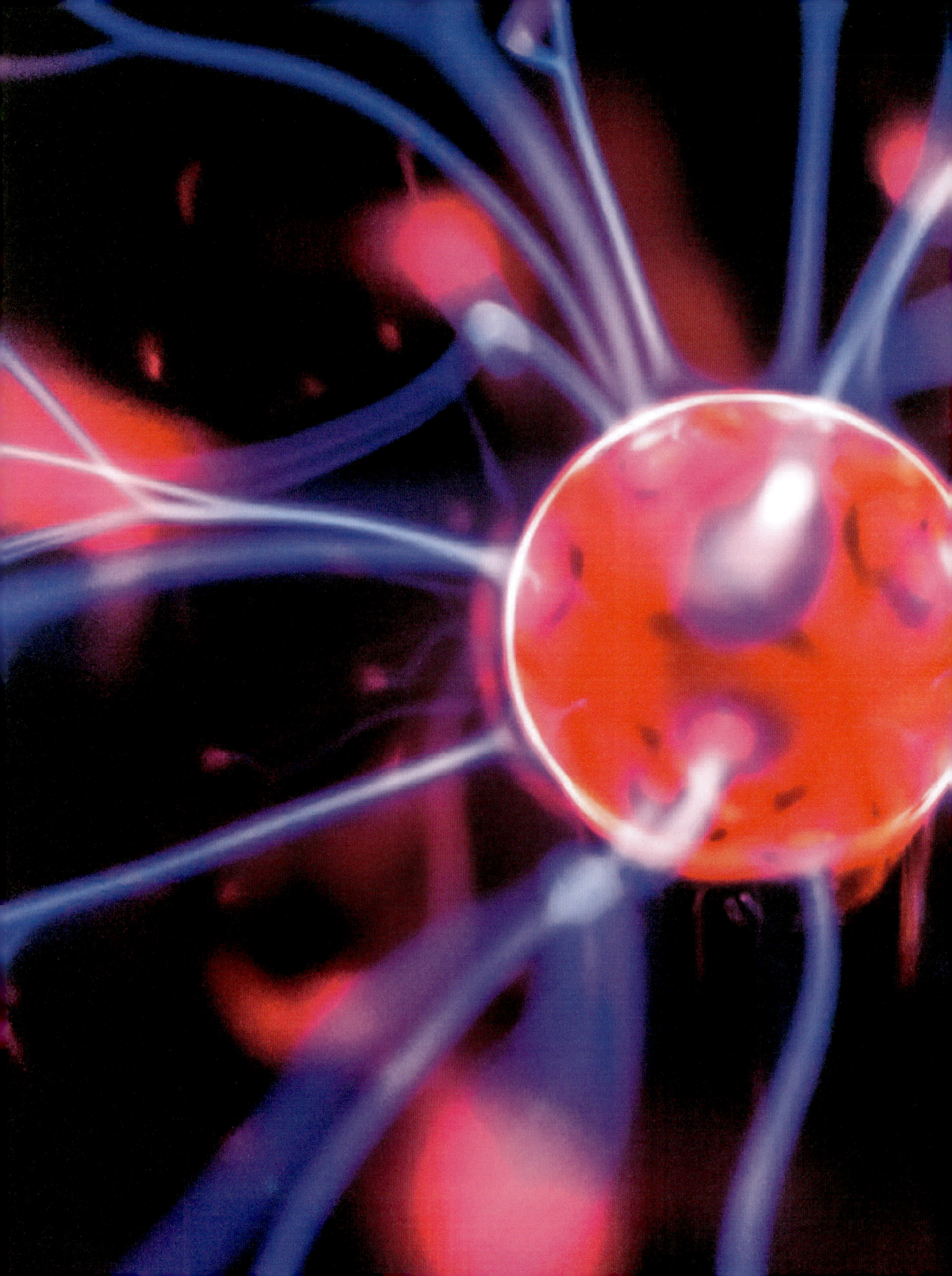

UNIT 3 How do fields explain motion and electricity?

To achieve the outcomes in Unit 3, you will draw on key knowledge outlined in each area of study and the related key science skills on pages 11 and 12 of the study design. The key science skills are discussed in Chapter 1 of this book.

AREA OF STUDY 1

How do physicists explain motion in two dimensions?

Outcome 1: On completion of this unit the student should be able to investigate motion and related energy transformations experimentally, and analyse motion using Newton's laws of motion in one and two dimensions.

AREA OF STUDY 2

How do things move without contact?

Outcome 2: On completion of this unit the student should be able to analyse gravitational, electric and magnetic fields, and apply these to explain the operation of motors and particle accelerators, and the orbits of satellites.

AREA OF STUDY 3

How are fields used in electricity generation?

Outcome 3: On completion of this unit the student should be able to analyse and evaluate an electricity generation and distribution system.

LIZZIE ARMANTO
BIRD HOUSE

CHAPTER

02 Newtonian theories of motion

An understanding of forces and fields has allowed humans to land on the Moon and explore the outer reaches of the solar system. Satellites in orbit around the Earth have changed the way people live. These advances have been achieved using Newton's laws of motion, which were published in the seventeenth century. Newton suggested that it should be possible to put satellites in orbit around the Earth almost 300 years before it became technically possible. While relativistic corrections introduced by Einstein are important in a limited number of contexts, Newton's description of gravitation and the laws governing motion are accurate enough for most practical purposes.

In this chapter Newton's laws will be used to analyse motion when two or more forces act on a body and how projectiles travel in the Earth's gravitational field. The chapter also covers how forces keep objects travelling in a circular path.

Key knowledge

- investigate and apply theoretically and practically Newton's three laws of motion in situations where two or more coplanar forces act along a straight line and in two dimensions **2.1**
- investigate and analyse theoretically and practically the uniform circular motion of an object moving in a horizontal plane: ($F_{net} = \frac{mv^2}{r}$), including:
  - a vehicle moving around a circular road **2.2**
  - a vehicle moving around a banked track **2.3**
  - an object on the end of a string **2.2**
- investigate and apply theoretically Newton's second law to circular motion in a vertical plane (forces at the highest and lowest positions only) **2.4**
- investigate and analyse theoretically and practically the motion of projectiles near Earth's surface, including a qualitative description of the effects of air resistance **2.5, 2.6**

2.1 Newton's laws of motion

REVISION

Equations of motion

The equations of motion can be used in situations where there is a constant acceleration a (in $m s^{-2}$). These equations allow you to model the motion of objects and predict values for the initial velocity u (in $m s^{-1}$), final velocity v (in $m s^{-1}$), displacement s (in m) and time t (in s). A direction convention should also be followed when using the equations of motion.

The equations of motion for uniform acceleration are:

$$v = u + at$$

$$s = \frac{1}{2}(u + v)t$$

$$s = ut + \frac{1}{2}at^2$$

$$v^2 = u^2 + 2as$$

These equations will be used in this chapter in addition to a number of new equations.

On 14 July 2015, NASA's New Horizons spacecraft (Figure 2.1.1) sped past Pluto and sent back images to the Earth that appeared on news broadcasts across the world. The principles of physics on which this mission depended were published by Isaac Newton in 1687 in a set of laws that radically challenged the understanding of his time.

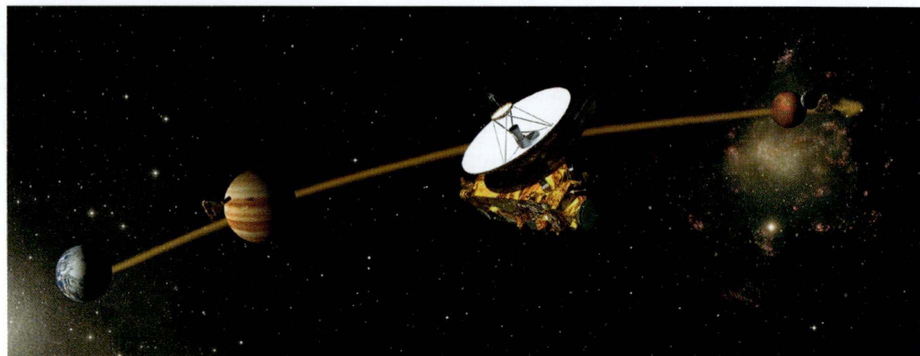

FIGURE 2.1.1 An artist's impression of New Horizons flying past Jupiter on its way to Pluto

Newton's laws are, in fact, only an approximation and have been superseded by Einstein's relativistic theories. In situations involving extremely high speeds (greater than 10% of the speed of light) or strong gravitational fields, Newton's laws are imprecise, and Einstein's theories must be used instead. However, Newton's laws are not obsolete. In most cases, Newton's laws remain invaluable for describing the motion of objects as diverse as planets and ping-pong balls.

NEWTON'S THREE LAWS OF MOTION

Newton's three laws of motion describe how forces affect the motion of bodies. The first law describes what happens to a body when there is no net force on it. The second law explains motion when there is an unbalanced force acting on a body. The third law states that all forces act in action–reaction pairs (that is, for every action there is an equal but opposite reaction).

Newton's first law

Newton's first law states that every object continues to be at rest, or continues with constant velocity, unless it experiences an unbalanced force. This is also called the law of inertia. An object that is moving at constant velocity will keep moving. This is seldom observed in everyday life due to the presence of forces such as friction and **air resistance**, which eventually slow the motion of the object. To maintain constant motion, frictional forces must be balanced with some other force. For example, an object can keep moving at a constant velocity if it is driven by a motor.

An object that is stationary will remain stationary while the forces acting on it are balanced. For example, an object will fall due to the force of gravity, but it will remain at rest when this force is balanced by the **normal force** applied by a table on which the object comes to rest. A normal force is one that exists between a surface and an object, and it always acts at right angles to the surface.

Newton's second law

Newton's second law states that the acceleration of a body experiencing an unbalanced force is directly proportional to the net force acting on it and inversely proportional to the mass of the body, i.e. $a = \frac{F_{net}}{m}$. This is often written as follows.

$F_{net} = ma$
where F_{net} is the net or resultant force acting (N)
m is the mass of the object (kg)
a is the acceleration of the object ($m\,s^{-2}$)

In other words, an object will accelerate at a greater rate when the force acting on it is increased. Heavy objects are harder to accelerate than lighter ones, so the rate of acceleration decreases as mass increases.

Newton's third law

Newton's third law states that when one body exerts a force on another body (an action force), the second body exerts an equal force on the first body but in the opposite direction (a reaction force):

$$F_{\text{on A by B}} = -F_{\text{on B by A}}$$

To simplify the notation, this text will use the following convention:

$$F_{AB} = F_{\text{on A by B}}$$

In this convention, the first subscript always indicates the body experiencing the force.

The forces in an action–reaction pair:

- are the same **magnitude**
- act in opposite directions and
- are exerted on two different objects.

It is important to note that action–reaction pairs can never be added together. This is because they act on different bodies. This is explained in Figure 2.1.2. In Figure 2.1.2(b) the pair of forces shown, F_g and F_N, are not an action–reaction pair because both forces act on the same object (the basketball).

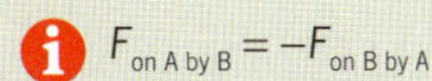

$F_{\text{on A by B}} = -F_{\text{on B by A}}$

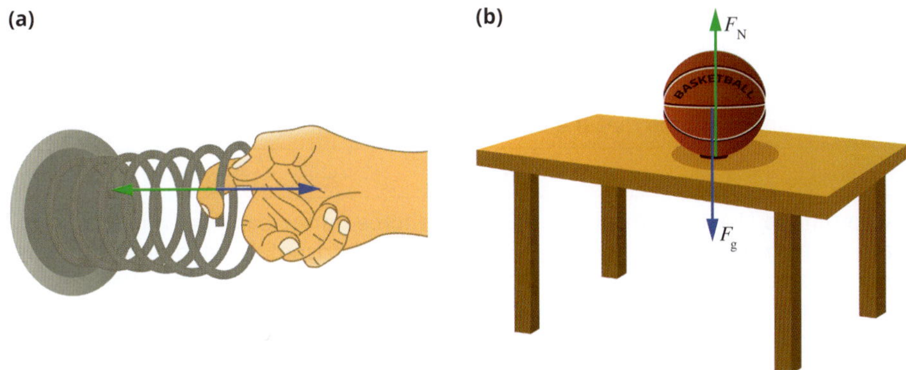

FIGURE 2.1.2 (a) An action–reaction pair: the hand pulls on the spring and the spring pulls back on the hand with an equal and opposite force. Figure (b) does not show an action-reaction pair. This is because the force due to gravity and the normal force both act on the same object, the basketball.

While the force is the same size on both objects, the resulting acceleration may not be. That is because the rate of acceleration depends on the mass of the objects concerned (from Newton's second law). Sometimes, when the objects have very different masses, the effect of one force in an action–reaction pair is much more noticeable. For example, if you stub your toe on a large heavy rock, the force exerted on your toe by the rock causes your foot to decelerate significantly. The equal and opposite force exerted by your toe on the rock does not cause any significant acceleration of the rock. This is because of its much greater mass.

PHYSICSFILE

Tethered spacewalks

When stationed on the International Space Station (ISS), astronauts are often required to conduct spacewalks, that is, they need to complete tasks outside their spacecraft. During spacewalks, astronauts are tethered (i.e. attached) to their spacecraft. If they weren't, they would float off into space (remember Newton's first law of motion!). All the astronaut's tools are attached to their spacesuits, otherwise they too would float off into space. If an astronaut were to become accidentally untethered, it could be a disaster. Without a surface to push against, the astronaut would float off into space and be unable to return to the spacecraft. As a safety precaution, every astronaut is fitted with a small jet pack they can use to manoeuvre themselves back to their spacecraft. The jet pack propels the astronaut forward when it is fired backwards (remember Newton's third law of motion).

Worked example 2.1.1

APPLICATION OF NEWTON'S FIRST AND THIRD LAWS

A toddler drags a 4.5 kg cart of blocks across a floor at a constant speed of $0.75\,m\,s^{-1}$. It is being dragged by a handle which is at an angle of 35° above the horizontal. The force of friction between the cart and the floor is 5.0 N.

a Calculate the net force on the cart.	
Thinking	**Working**
The cart has constant speed, i.e. no acceleration. According to Newton's first law, the net force acting on the cart is zero.	$F_{net} = 0\,N$

b Calculate the force that the toddler exerts on the cart.	
Thinking	**Working**
Draw a force diagram.	
If the net force is zero then the horizontal forces must be balanced. Therefore the horizontal component of the force on the cart by the toddler, F_{CT_x}, is equal to the magnitude of the frictional force, F_{CF}.	$F_{CT_x} = F_{CT} \cos 35° = F_{CF}$ $F_{CT} \cos 35° = 5.0\,N$ $F_{CT} = \dfrac{5.0}{\cos 35°} = 6.1\,N$ 35° above the horizontal

c Determine the force that the cart exerts on the toddler.	
Thinking	**Working**
According to Newton's third law, the force on the cart by the toddler is equal and opposite to the force on the toddler by the cart: $F_{CT} = -F_{TC}$	Since the force on the cart is at an angle of 35° above the horizontal, the force of the cart on the toddler is 6.1 N at an angle of 35° below the horizontal.

Worked example: Try yourself 2.1.1

APPLICATION OF NEWTON'S FIRST AND THIRD LAWS

The toddler adds extra blocks to the cart and drags it across the floor more slowly. The 5.5 kg cart travels at a constant speed of $0.65\,m\,s^{-1}$. The force of friction between the cart and the floor is 5.2 N and the handle is now at an angle of 30° above the horizontal.

a Calculate the net force on the cart.

b Calculate the force that the toddler exerts on the cart.

c Determine the force that the cart exerts on the toddler.

Applying Newton's first or second laws

When solving motion problems, a key strategy is to determine whether Newton's first law or second law should be applied. In the following examples, the objects in the questions are accelerating. Hence the second law should be used and the net force is proportional to the acceleration. In problems involving connected bodies, both the whole system and each component of the system have the same acceleration.

Worked example 2.1.2

APPLICATION OF NEWTON'S LAWS

A vehicle towing a caravan accelerates at 1.8 m s^{-2} to overtake the car in front. The vehicle's mass is 2700 kg and the caravan's mass is 2000 kg. The drag force on the vehicle is 1100 N and the drag force on the caravan is 1500 N.

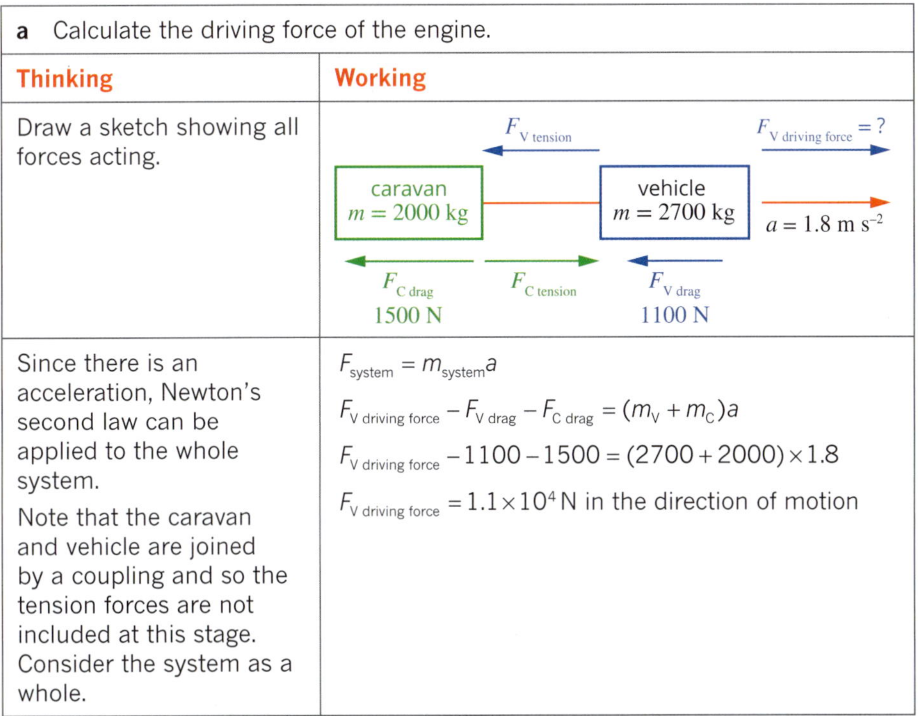

a Calculate the driving force of the engine.	
Thinking	**Working**
Draw a sketch showing all forces acting.	
Since there is an acceleration, Newton's second law can be applied to the whole system. Note that the caravan and vehicle are joined by a coupling and so the tension forces are not included at this stage. Consider the system as a whole.	$F_{\text{system}} = m_{\text{system}}a$ $F_{\text{V driving force}} - F_{\text{V drag}} - F_{\text{C drag}} = (m_V + m_C)a$ $F_{\text{V driving force}} - 1100 - 1500 = (2700 + 2000) \times 1.8$ $F_{\text{V driving force}} = 1.1 \times 10^4$ N in the direction of motion

b Calculate the magnitude of the tension in the coupling.	
Thinking	**Working**
Consider only one part of the system, for example the caravan, and once again apply Newton's second law.	$F_{\text{C net}} = m_C a$ $F_{\text{C tension}} - F_{\text{C drag}} = m_C a$ $F_{\text{C tension}} = 2000 \times 1.8 + 1500$ $= 5.1 \times 10^3$ N

Worked example: Try yourself 2.1.2

APPLICATION OF NEWTON'S LAWS

A vehicle towing a trailer accelerates at 2.8 m s^{-2} to overtake a car in front. The vehicle's mass is 2700 kg and the trailer's mass is 600 kg. The drag force on the vehicle is 1100 N and the drag force on the trailer is 500 N.

a Calculate the driving force of the engine.

b Calculate the magnitude of the tension in the coupling.

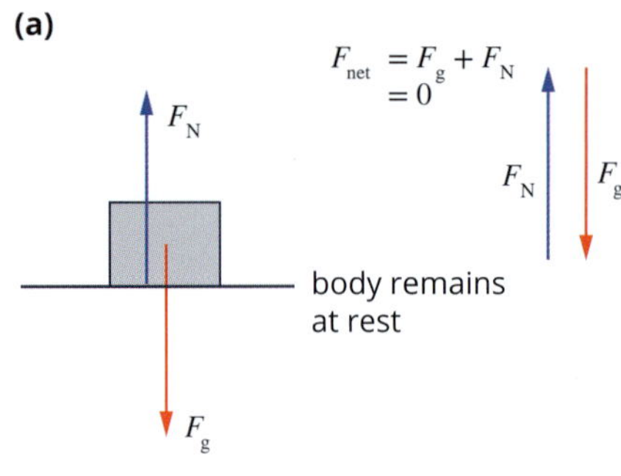

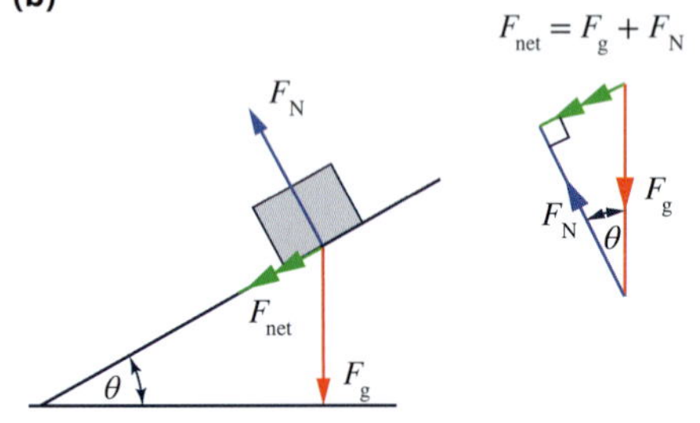

FIGURE 2.1.3 (a) A block on a level surface experiences a net force of zero, as F_N and F_g balance each other. (b) With the block on an incline, $F_N = F_g \cos \theta$, and the net force is given by $F_{net} = F_g + F_N$ added as vectors.

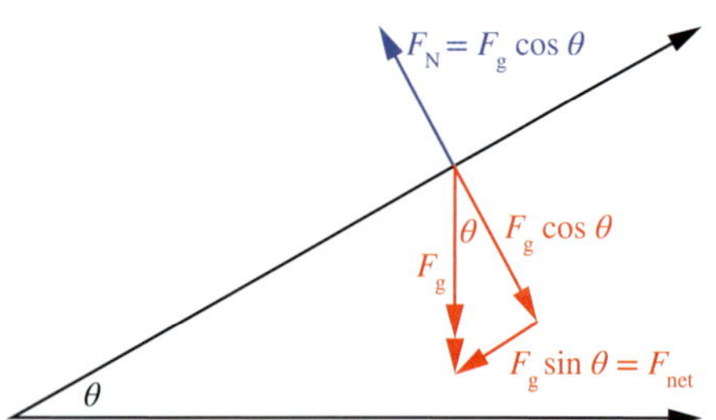

FIGURE 2.1.4 For a block on an incline, the force due to gravity can be resolved into a force perpendicular to the surface and a force parallel to the surface.

THE NORMAL FORCE AND INCLINED PLANES

One reaction force deserves a special mention. When an object exerts a force on a surface, the surface exerts a force back on the object that is at right angles (i.e. normal) to the surface. For example, the block in Figure 2.1.3(a) exerts a force on the level surface and the surface exerts a normal force back on the block. The force the block exerts on the surface is equal in size to the force due to gravity, F_g. Thus F_g is balanced by F_N, as shown in the figure. As there is no net force on the block, the object remains stationary.

Consider an **inclined plane** (Figure 2.1.3(b)). The normal force is still at right angles to the surface. However, as the surface is not horizontal, F_N will be at an angle to F_g. There is a net force down the slope and the block accelerates, as predicted by Newton's second law.

Another way of viewing the forces along an inclined plane is to resolve the vector of the force due to gravity, F_g, into two components: one perpendicular to the slope and one parallel to the slope (Figure 2.1.4). The component perpendicular to the surface is balanced by the normal force F_N. The component of the force due to gravity that is parallel to the slope is the force that actually causes the acceleration.

Worked example 2.1.3

INCLINED PLANES

A skier of mass 50 kg is skiing down an icy slope that is inclined at 20° to the horizontal. Assume that friction is negligible and that the acceleration due to gravity is 9.8 m s^{-2}.

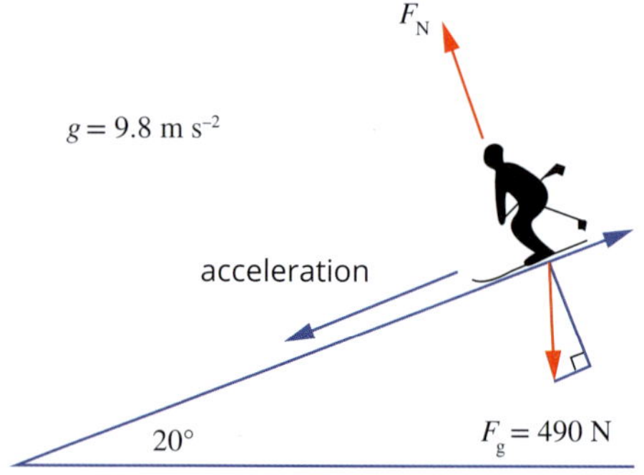

a Determine the components of the force due to gravity on the skier perpendicular to the slope and parallel to the slope.	
Thinking	**Working**
Draw a sketch and include the values provided.	F_g cos 20° F_g 20° 20° F_g sin 20°
Resolve the force due to gravity into the component perpendicular to the slope.	The perpendicular component is: $F_\perp = F_g \cos 20°$ $= 490 \cos 20°$ $= 460$ $= 4.6 \times 10^2$ N

Resolve the force due to gravity into the component parallel to the slope.	The parallel component is: $F_{\parallel} = F_g \sin 20°$ $= 490 \sin 20°$ $= 168$ $= 1.7 \times 10^2$ N

b Determine the normal force that acts on the skier.	
Thinking	**Working**
The normal force is equal in magnitude to the perpendicular component of the force due to gravity.	$F_N = 4.6 \times 10^2$ N

c Calculate the acceleration of the skier down the slope.	
Thinking	**Working**
Apply Newton's second law, rearranged to make *a* the subject. The net force along the slope is the component of the force due to gravity parallel to the slope.	$a = \frac{F_{net}}{m}$ $= \frac{168}{50}$ $= 3.4$ m s^{-2} down the slope

Worked example: Try yourself 2.1.3

INCLINED PLANES

A skier of mass 85 kg travels down the same icy slope inclined at 20° to the horizontal. Assume that friction is negligible and that the acceleration due to gravity is 9.8 m s^{-2}.

a Determine the components of the force due to gravity on the skier perpendicular to the slope and parallel to the slope.

b Determine the normal force that acts on the skier.

c Calculate the acceleration of the skier down the slope.

Aside from rounding differences, the acceleration calculated in the Worked example and Try yourself questions above are equal. This is because acceleration is independent of the mass of the object (if we ignore friction forces). Mathematically, the relationship can be written as follows.

$$a = \frac{F_{net}}{m} = \frac{mg \sin \theta}{m} = g \sin \theta$$

where a is the acceleration of the object (m s^{-2})

g is the acceleration due to gravity (m s^{-2})

θ is the angle of the inclined plane from the horizontal

STRATEGIES FOR SOLVING FORCE AND MOTION PROBLEMS

Where forces on a body are given, Newton's laws can be applied. Two questions should be asked:

1. Is the object stationary or travelling at constant velocity? In these cases, $F_{net} = 0$.
2. Is the object accelerating? In this case, $F_{net} = ma$.

When dealing with connected bodies, consider the whole system first, and then consider the separate parts of the system.

For coplanar forces that are not aligned (for example, on an inclined plane), resolve forces into their components.

Newton's second law can be used to find the acceleration of an object. It can then be used with the other equations of motion to find such quantities as displacement and final velocity.

2.1 Review

OA ✓✓

SUMMARY

- Newton's first law states that every object continues to be at rest, or continues with constant velocity, unless it experiences an unbalanced force. This is also called the law of inertia.
- Newton's second law states that the acceleration of a body experiencing an unbalanced force is directly proportional to the net force on the body and inversely proportional to the mass of the body: $F_{net} = ma$.
- Newton's third law states that when one body exerts a force on another body (an action force), the second body exerts an equal force on the first body but in the opposite direction (a reaction force): $F_{AB} = -F_{BA}$.
- The forces in an action–reaction pair are of the same magnitude, act in opposite directions and are exerted on two different objects.
- A normal force, F_N, acts between an object and a surface at right angles to the surface.
  - On a horizontal surface, $F_N = F_g$ and the object is stationary.
  - On an inclined surface, F_N is equal and opposite to the component of the force due to gravity acting perpendicular to the plane: $F_N = F_g \cos \theta$.
- The net force (F_{net}) acting on an object on a plane inclined at an angle θ is $F_g \sin \theta$ (assuming that friction is negligible).

KEY QUESTIONS

Knowledge and understanding

1. Phil is standing inside a tram when it starts off suddenly. Lisa, who is sitting down, comments that Phil was thrown backwards as the tram started moving. Is this a correct statement? Explain your answer in terms of Newton's laws.
2. Consider an object of mass 5.3 kg sliding across a frictionless surface. What force is required to accelerate it at a rate of 2.2 m s^{-2}?
3. On each of the following force diagrams, draw the reaction force that is the partner of the action force that is shown. For each force you draw, state what the force is acting on and what is providing the force.

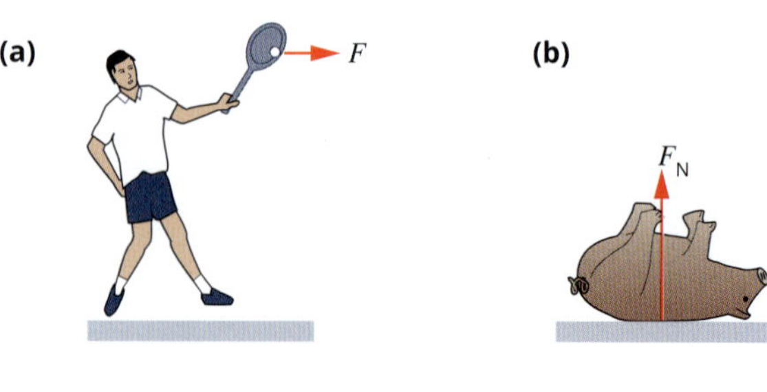

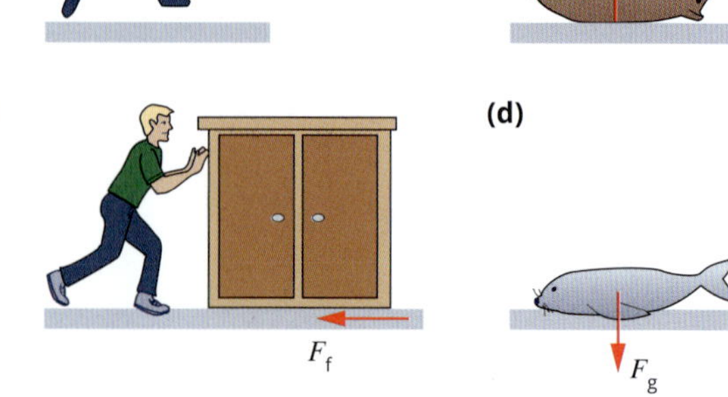

4 A table-tennis ball of mass 10 g is falling towards the ground at a constant speed of 8.2 m s^{-1}. Calculate the magnitude and direction of the force due to air resistance acting on the ball.

5 Ishtar is riding a motorised scooter along a level path. The combined mass of Ishtar and her scooter is 80.0 kg. The frictional and drag forces that are acting total 45.0 N. Determine the magnitude of the driving force provided by the motor under the following conditions.

- **a** Ishtar is moving at a constant speed of 10 m s^{-1}.
- **b** Ishtar is accelerating at 1.50 m s^{-2}.

6 A cyclist and his bike have a combined mass of 80 kg. When starting off from traffic lights, the cyclist accelerates uniformly and reaches a speed of 7.5 m s^{-1} in 5.0 s.

- **a** What is the acceleration during this time?
- **b** Calculate the driving force being provided by the cyclist's legs as he starts off. Assume that drag forces are negligible during this time.
- **c** The cyclist now rides at a constant speed of 15 m s^{-1}. If the force being provided by his legs is now 60 N, determine the magnitude of the drag forces that are acting.

7 During preseason football training, Matt was required to run dragging behind him a bag of sand of mass 50 kg. The bag was attached to a rope which made an angle of 25° to the horizontal. When Matt ran with a constant speed of 4.0 m s^{-1}, a frictional force of 60 N was acting on the bag.

- **a** What was the net force acting on the bag?
- **b** Calculate the size of the tension force that was acting in the rope.
- **c** What was the magnitude of the force the rope exerted on Matt as he ran?

Analysis

8 A block on a table is accelerated by a falling mass, as shown below. Calculate the acceleration of the blocks and the tension in the cord if the block on the table experiences a frictional force of 2.0 N as it slides along.

9 A 950 kg car is used to tow a small trailer of mass 100 kg. The car and trailer have an acceleration of 0.800 m s^{-2}. The resistive forces acting on the car total 500 N. An additional 500 N of resistive forces act on the trailer.

- **a** Calculate the driving force required by the car's engine.
- **b** What tension exists in the tow rope between the car and trailer?

10 Kirsty is riding in a bobsled that is sliding down a snow-covered hill with a slope of 30° to the horizontal. The total mass of the sled and Kirsty is 100 kg. Initially the brakes are on and the sled moves down the hill with a constant velocity.

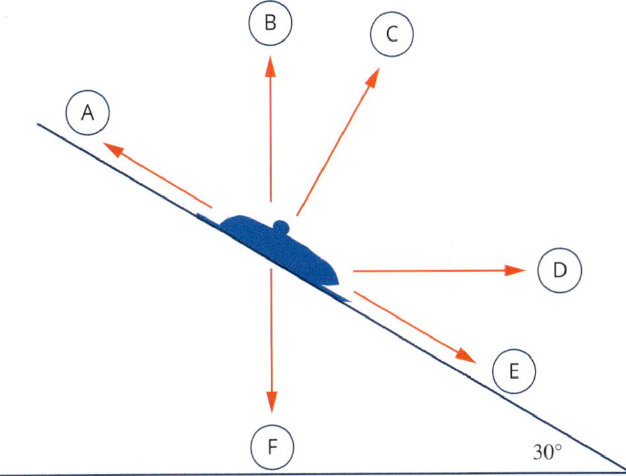

- **a** Which one of the arrows (A–F) best represents the direction of the frictional force acting on the sled?
- **b** Which one of the arrows (A–F) best represents the direction of the normal force acting on the sled?
- **c** Calculate the net frictional force acting on the sled.
- **d** Kirsty releases the brakes and the sled accelerates. What is the magnitude of her initial acceleration?
- **e** Kirsty returns to the top of the hill. A friend now joins her in the bobsled, taking the total mass to 140 kg. The bobsled takes off down the same slope and with the brakes off (thus friction can be ignored). How will the extra mass affect the acceleration of the bobsled?

2.2 Circular motion in a horizontal plane

Circular motion is common throughout the universe. Children on a fairground ride (Figure 2.2.1) move in a circular path, and so do those in a car as it travels around a roundabout. In athletics, hammer throwers swing the hammer in a circular path before releasing it. On a much larger scale, the planets orbit the Sun in paths that are approximately circular. On an even grander scale, stars can travel in circular paths around the centres of their galaxies. This section explains the nature of circular motion in a horizontal plane and applies Newton's first and second laws to problems involving circular motion.

FIGURE 2.2.1 The people on this ride are travelling in a circular path.

UNIFORM CIRCULAR MOTION

In Figure 2.2.2, an athlete in a hammer-throw event swings a hammer—which is usually a steel ball—in a horizontal circle with a constant speed of $25\,\text{m}\,\text{s}^{-1}$. Although its speed is constant, its velocity is continually changing. This means that it is accelerating.

Remember that velocity is a vector. Since the direction of the hammer is changing, so too is its velocity, even though its speed is not changing. The velocity of the hammer at any instant is **tangential** (i.e. at a tangent) to its path. At one instant, the hammer is travelling at $25\,\text{m}\,\text{s}^{-1}$ north. An instant later it is travelling at $25\,\text{m}\,\text{s}^{-1}$ west, and then $25\,\text{m}\,\text{s}^{-1}$ south, and so on.

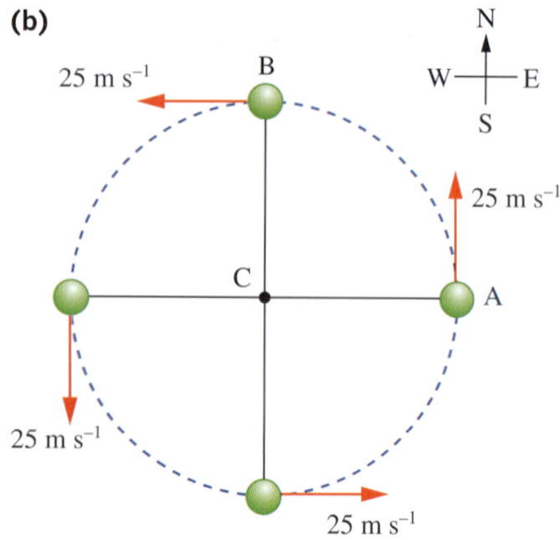

FIGURE 2.2.2 (a) An athlete in a hammer-throw event. (b) The velocity of the hammer at any instant is tangential to its path and is continually changing, even though it has constant speed. Because its velocity is changing, the hammer is accelerating.

PERIOD AND FREQUENCY

Imagine that an object is moving in a circular path of radius r metres with a constant speed of v and takes T seconds to complete one revolution. The time taken to travel once around a circle is called the **period**, T, of the motion. The number of rotations each second is the **frequency**, f.

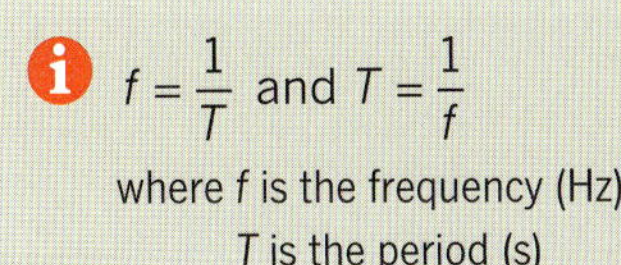

$f = \frac{1}{T}$ and $T = \frac{1}{f}$

where f is the frequency (Hz)

T is the period (s)

SPEED

An object that travels in a circle will travel a distance equal to the circumference of the circle, $2\pi r$, with each revolution (Figure 2.2.3). Given that the time for each revolution is the period, T, the average speed of the object is:

$$\text{speed} = \frac{\text{distance}}{\text{time}} = \frac{\text{circumference}}{\text{period}} = \frac{2\pi r}{T}$$

The average speed of an object moving in a circular path is:

$$v = \frac{2\pi r}{T}$$

where v is the speed of the object (m s^{-1})

r is the radius of the circle (m)

T is the period of motion (s)

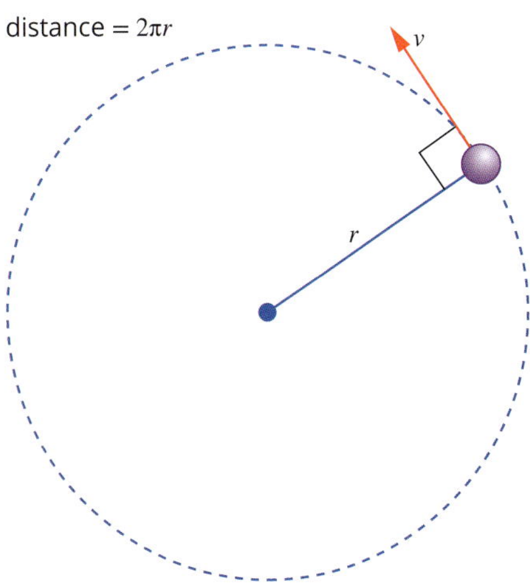

FIGURE 2.2.3 The average speed of an object moving in a circular path is given by the distance travelled in one revolution (the circumference, $2\pi r$) divided by the time taken (the period, T).

PHYSICSFILE

Wind generators

The wind generators in the figure below are part of a wind farm at Macarthur in south-west Victoria. The wind farm has 140 turbines and the towers are 85 m high. Each blade is 55 m long and can rotate at a maximum rate of 20 revolutions per minute. Although the blades are moving in a vertical, not horizontal, plane, their motion can be described using the same equation for circular motion given above. From the information given, you should be able to calculate that the tip of each blade is travelling at around $400\ \text{km h}^{-1}$.

The tips of these wind-generator blades travel in a circular path and can reach speeds of approximately $400\ \text{km h}^{-1}$.

Worked example 2.2.1

CALCULATING SPEED

A wind turbine has blades 55.0 m in length that rotate at a frequency of 20 revolutions per minute. At what speed do the tips of the blades travel? Express your answer in $km\,h^{-1}$.

Thinking	Working
Calculate the period, T. Remember to express frequency in the correct units. Alternatively, recognise that 20 revolutions in 60 seconds means that each revolution takes 3 seconds.	20 revolutions per minute $= \frac{20}{60} = 0.333\,Hz$ $T = \frac{1}{f}$ $= \frac{1}{0.333} = 3.0\,s$
Substitute r and T into the appropriate formula for speed and solve for v.	$v = \frac{2\pi r}{T}$ $= \frac{2 \times \pi \times 55.0}{3}$ $= 115.2\,m\,s^{-1}$ $= 1.15 \times 10^2\,m\,s^{-1}$
Convert $m\,s^{-1}$ into $km\,h^{-1}$ by multiplying by 3.6.	$115.2 \times 3.6 = 4.2 \times 10^2\,km\,h^{-1}$

Worked example: Try yourself 2.2.1

CALCULATING SPEED

A water wheel has blades 2.0 m in length that rotate at a frequency of 10 revolutions per minute. At what speed do the tips of the blades travel? Express your answer in $km\,h^{-1}$.

CENTRIPETAL ACCELERATION

Since the velocity of an object travelling in a horizontal circle is changing, it is accelerating even though its speed is not changing. The object is continually deviating inwards from a direction tangential to the circle and so has an acceleration towards the centre. This acceleration is known as **centripetal acceleration**. (The word 'centripetal' means to move towards a centre.)

In Figure 2.2.4, the velocity vector of an object travelling in a circular path is shown with an arrow labelled v, at a tangent to the circular path. The centripetal acceleration, a, is towards the centre of the circular path.

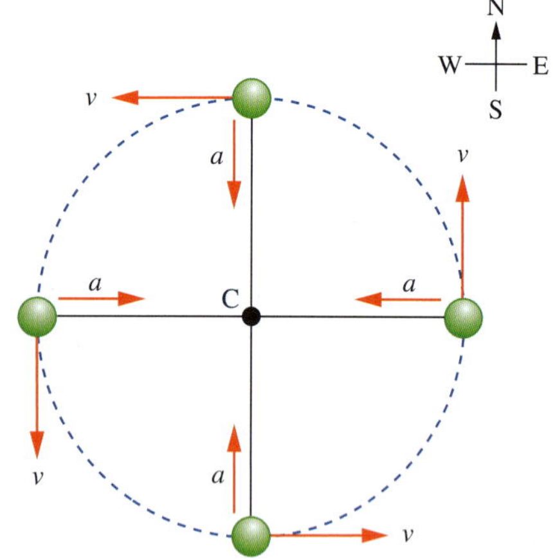

FIGURE 2.2.4 A body moving in a circular path has an acceleration towards the centre of the circle. This is known as centripetal acceleration.

However, as Figure 2.2.5 shows, even though the object is accelerating towards the centre of the circle, it never gets any closer to the centre.

The centripetal acceleration, a, of an object moving in a circular path of radius r with a velocity v can be found from the relationship:

$$a = \frac{v^2}{r}$$

A substitution can be made in this equation for the speed of the object, which was found earlier to be:

$$v = \frac{2\pi r}{T}$$

Thus:

$$\begin{aligned} a &= \frac{v^2}{r} \\ &= \left(\frac{2\pi r}{T}\right)^2 \times \frac{1}{r} \\ &= \frac{4\pi^2 r}{T^2} \end{aligned}$$

Centripetal acceleration is always directed towards the centre of the circular path and is given by:

$$a = \frac{v^2}{r} = \frac{4\pi^2 r}{T^2}$$

where a is the centripetal acceleration (m s^{-2})
v is the speed (m s^{-1})
r is the radius of the circle (m)
T is the period of motion (s)

FORCES THAT CAUSE CIRCULAR MOTION

As with all forms of motion, an analysis of the forces that are acting is needed to understand why circular motion occurs. In the hammer throw described earlier in this section, the hammer is continually accelerating. It follows from Newton's second law that there must be a net unbalanced force continuously acting on it. The net unbalanced force that gives the ball its acceleration towards the centre of the circle is known as a **centripetal force**.

In every case of circular motion, a real force is necessary to provide the centripetal force. The force acts in the same direction as the acceleration, that is, towards the centre of the circle. This centripetal force can be provided in a number of ways. For the hammer in Figure 2.2.5(a), the centripetal force is the tension force in the cable. Three other examples of centripetal force are also shown in Figure 2.2.5.

Consider the consequences if the unbalanced force ceases to act. In the example of the hammer thrower, if the tension in the cable became zero—as happens when the thrower releases the hammer—there is no longer a force causing the hammer to change direction. The result is that the hammer moves in a straight line tangential to its circular path, as would be expected from Newton's first law.

Centripetal force is given by:

$$F_{net} = ma = \frac{mv^2}{r} = \frac{4\pi^2 rm}{T^2}$$

where F_{net} is the net or centripetal force on the object (N)
m is the mass (kg)
a is the acceleration (m s^{-2})
v is the speed (m s^{-1})
r is the radius of the circle (m)
T is the period of motion (s)

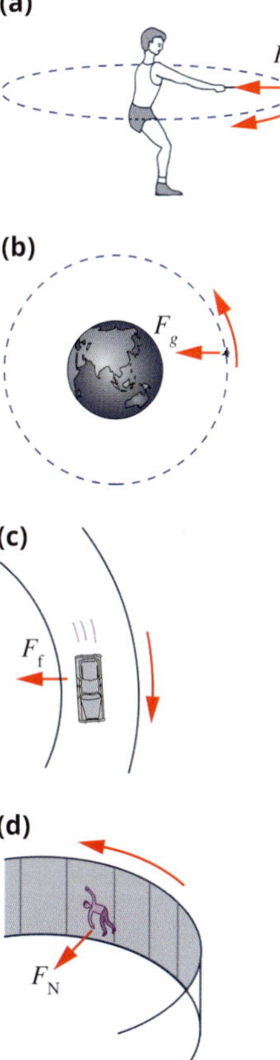

FIGURE 2.2.5 (a) In a hammer throw, tension in the cable provides the centripetal force. (b) For planets and satellites, the gravitational attraction towards the central body provides the centripetal force. (c) For a car on a curved road, the friction between the tyres and the road provides the centripetal force. (d) For a person in a Gravitron ride, it is the normal force from the wall that provides the centripetal force.

Worked example 2.2.2

CENTRIPETAL FORCES

An athlete in a hammer-throw event is swinging a metal ball of mass 7.0 kg in a horizontal circular path of radius 60 m. The ball is moving at 20.0 m s^{-1}.

a Calculate the magnitude of the acceleration of the ball.	
Thinking	**Working**
As the object is moving in a circular path, the centripetal acceleration is towards the centre of the circle. To find the magnitude of this acceleration, consider the variables that are given.	$v = 20.0\ \text{m s}^{-1}$ $r = 1.60\ \text{m}$ $a = ?$
Select the equation for centripetal acceleration that fits the values you have, substitute the values and solve the equation.	$a = \frac{v^2}{r}$ $= \frac{20.0^2}{1.60}$ $= 250\ \text{m s}^{-2}$
State the magnitude only, as no direction is required.	The acceleration of the ball is 250 m s^{-2}.

b Calculate the magnitude of the tensile force (tension) acting in the wire used to swing the ball.	
Thinking	**Working**
Identify the unbalanced force that is causing the object to move in a circular path. Write down the relevant variables that you have.	$m = 7.0\ \text{kg}$ $a = 250\ \text{m s}^{-2}$ $F_{net} = ?$
Select the appropriate equation for centripetal force, substitute the relevant variables and solve the equation.	$F_{net} = ma$ $= 7.0 \times 250$ $= 1750$ $= 1.8 \times 10^3\ \text{N}$
Only the magnitude is required, so no direction is needed in the answer.	The force of tension in the wire is the unbalanced force that is causing the ball to accelerate. Tensile force $F_t = 1.8 \times 10^3\ \text{N}$

Worked example: Try yourself 2.2.2

CENTRIPETAL FORCES

An athlete in a hammer-throw event is swinging a ball of mass 7.0 kg in a horizontal circular path of radius 1.20 m. The ball is moving at 25.0 m s^{-1}.

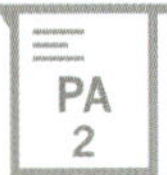

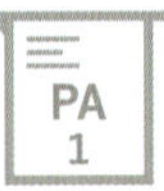

a Calculate the magnitude of the acceleration of the ball.

b Calculate the magnitude of the tensile force (tension) acting in the wire.

CASE STUDY

The Gravitron

When a car turns sharply to the left, the passengers in the car seem to sway to the right. Many mistakenly think that a force to the right is acting. In fact, the passengers are simply maintaining their motion in the original direction of the car, as described by Newton's first law; that is, they are experiencing inertia. If the passengers are (unwisely) not wearing seatbelts, they may be squashed against the right-hand door as the car turns. This will exert a large force to the left on them, which causes them to move to the left.

People moving rapidly in circular paths might also mistakenly think that there is an outward force acting on them. In order for the physics to be explained, it is necessary to use different frames of reference. For example, riders on the Gravitron (also known as the Vortex or Rotor), like those in Figure 2.2.6, will feel a force pushing them into the wall. This outwards force is commonly known as a centrifugal force. (Centrifugal means 'centre-fleeing'.) This force does not actually exist in their frame of reference. The riders think that it does because they are in a rotating frame of reference. From outside the Gravitron, it is evident that there is an inwards force (the normal force) that is holding them in a circular path. If the walls disintegrated and this normal force ceased to act, they would not fly outwards, but move at a tangent to their circle.

A Gravitron can rotate at 24 rpm with a radius of 7 m. The centripetal acceleration can be over 40 m s^{-2}. This is caused by a very large centripetal force from the wall i.e. the normal force, F_N. In the vertical direction, F_g is balanced by an upwards frictional force, F_f, so the riders experience no vertical motion even if the floor then drops away. It is important to remember that there is no force acting outwards. In fact, as you can see in Figure 2.2.7, the forces are unbalanced and the net force is equal in size and direction to the normal force towards the centre of the circle.

FIGURE 2.2.6 There is a large inwards force from the wall (a normal force) that causes these Gravitron riders to travel in a circular path.

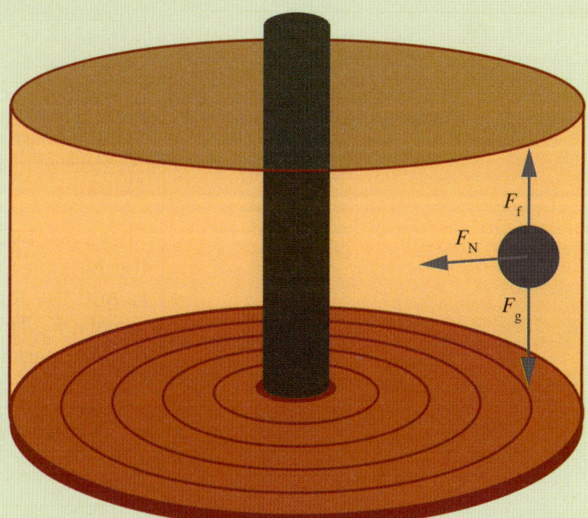

FIGURE 2.2.7 There are three forces acting on a rider in a Gravitron. Vertically, the forces are balanced and so no motion occurs in this direction. The remaining force, F_N, provides the net (centripetal) force to the centre of the ride.

BALL ON A STRING

You may have played totem tennis. This is a game where a ball is attached to a pole by a string and can travel in a horizontal circle, although the string itself is not horizontal (Figure 2.2.8).

If the ball at the end of the string is swinging slowly, the string swings down at an angle closer to the pole. If the ball was swung faster, the string would become closer to being horizontal. In fact, it is not possible for the string to be absolutely horizontal, although as the speed increases, the closer to horizontal it becomes. This system is known as a conical pendulum.

If the angle of the conical pendulum is known, trigonometry can be used to find the radius of the circle and the forces involved.

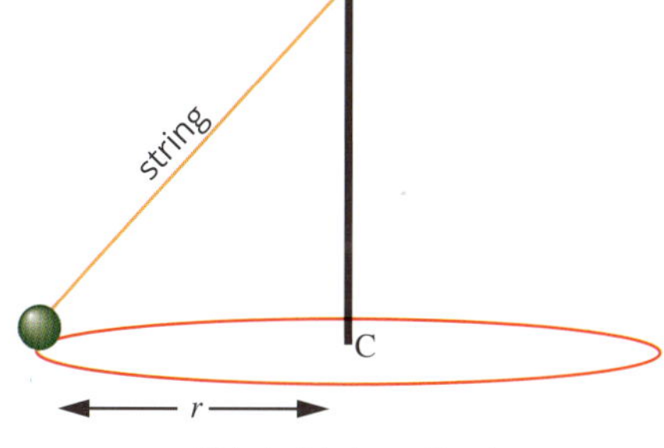

FIGURE 2.2.8 This ball is travelling in a horizontal circular path of radius r. The centre of its circular motion is at C.

Worked example 2.2.3

OBJECT ROTATING ON THE END OF A STRING

During a game of totem tennis, a ball of mass 150 g is swinging freely in a horizontal circular path. The cord is 1.50 m long and at an angle of 60.0° to the vertical.

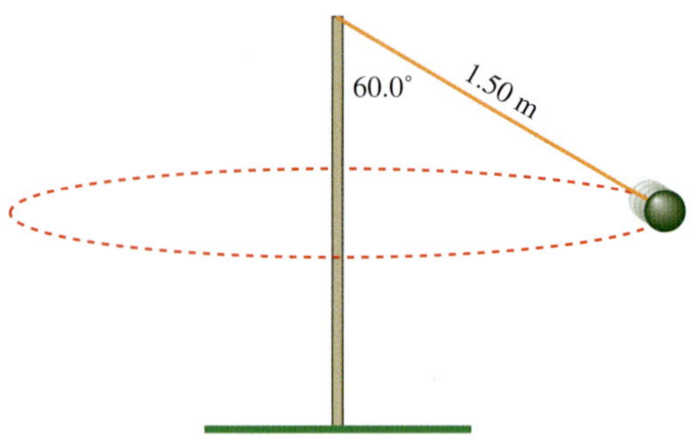

a Calculate the radius of the ball's circular path.	
Thinking	**Working**
The radius of the circular path and the pole form a right angle. Use trigonometry to find the radius.	$r = 1.50 \sin 60.0° = 1.30\,\text{m}$

b Draw and label the forces that are acting on the ball at the instant shown in the diagram.	
Thinking	**Working**
There are two forces acting: the tension in the cord, F_t, and the force due to gravity, F_g. These forces are unbalanced.	F_t F_g

c Determine the net force that is acting on the ball at this time.	
Thinking	**Working**
First calculate the force due to gravity, F_g.	$F_g = mg$ $= 0.150 \times 9.8$ $= 1.47\,\text{N}$
The ball has an acceleration that is towards the centre of its circular path. This is horizontal and acts towards the left at this instant. The net force will also be acting in this direction. A force triangle and trigonometry can be used to determine the net force.	60.0° $F_t = ?$ $F_g = 1.47\,\text{N}$ $F_{net} = ?$ $F_{net} = 1.47 \tan 60.0°$ $= 2.55\,\text{N}$ towards the centre of the circular path

d Calculate the size of the tensile force in the cord.	
Thinking	**Working**
Use trigonometry to find F_t.	$F_t = \dfrac{1.47}{\cos 60.0°}$ $= 2.94\,\text{N}$

Worked example: Try yourself 2.2.3

OBJECT ROTATING ON THE END OF A STRING

During a game of totem tennis, a ball of mass 200 g is swinging freely in a horizontal circular path. The cord is 2.00 m long and at an angle of 50.0° to the vertical.

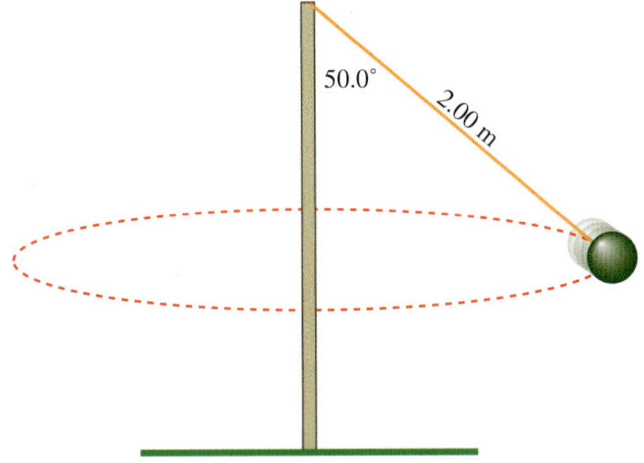

a Calculate the radius of the ball's circular path.

b Draw and label the forces that are acting on the ball at the instant shown in the diagram.

c Determine the net force that is acting on the ball at this time.

d Calculate the size of the tensile force in the cord.

2.2 Review

OA ✓✓

SUMMARY

- Frequency, f, is the number of revolutions each second and is measured in hertz (Hz).
- Period, T, is the time for one revolution and is measured in seconds.
- The relationship between T and f is:

$$f = \frac{1}{T} \text{ and } T = \frac{1}{f}$$

- An object moving with a uniform speed in a circular path of radius r and with period T has an average speed given by:

$$v = \frac{2\pi r}{T}$$

- The velocity of an object moving with a constant speed in a circular path is continuously changing. The velocity vector is always directed at a tangent to the circular path.
- An object moving in a circular path with a constant speed has an acceleration due to its circular motion. This acceleration is directed towards the centre of the circular path and is called centripetal acceleration, a, where:

$$a = \frac{v^2}{r} = \frac{4\pi^2 r}{T^2}$$

- Centripetal acceleration is a consequence of a centripetal force acting to make an object move in a circular path.
- A centripetal force is directed towards the centre of the circle and its magnitude can be calculated using Newton's second law:

$$F_{net} = \frac{mv^2}{r} = \frac{4\pi^2 rm}{T^2}$$

- A centripetal force is always supplied by a real force the nature of which depends on the situation. The real force is commonly friction, gravitation or the tension in a string or cable.

KEY QUESTIONS

Knowledge and understanding

The following information relates to questions 1–5.

A car of mass 1200 kg is travelling in a roundabout in a circular path of radius 9.2 m. The car moves with a constant speed of 8.0 m s^{-1}. The direction of the car is clockwise around the roundabout when viewed from above.

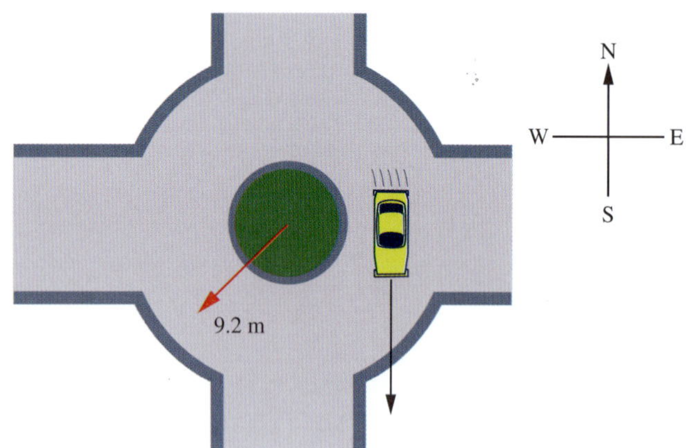

1 Which two of the following statements correctly describe the motion of the car as it travels around the roundabout?

A It has a constant speed.
B It has a constant velocity.
C It has zero acceleration.
D It has an acceleration that is directed towards the centre of the roundabout.

2 When the car is in the position shown in the diagram, what is the:

a speed of the car
b velocity of the car
c magnitude and direction of the acceleration of the car?

3 Calculate the magnitude and direction of the net force acting on the car at the position shown.

4 When the car has travelled halfway around the roundabout, what is the:

a velocity of the car at this point
b direction of its acceleration at this point?

5 If the driver of the car kept speeding up, what would eventually happen to the car as it travelled around the roundabout? Explain your answer.

6 An ice skater of mass 75 kg is skating in a horizontal circle of radius 2.5 m at a constant speed of 1.5 m s^{-1}.

a Determine the magnitude of the skater's acceleration.

b Are the forces acting on the skater balanced or unbalanced? Explain your answer.

c Calculate the magnitude of the centripetal force acting on the skater.

7 A 1.5 kg ball is made to swing in a horizontal circle of radius 1.2 m at 2.5 revolutions per second.

a What is the period of rotation of the ball?

b What is the orbital speed of the ball?

c What is the magnitude of the acceleration of the ball?

d What is the magnitude of the net force acting on the ball?

8 A child of mass 30.0 kg is playing on a maypole swing in a playground. The rope is 2.40 m long and at an angle of 60.0° to the horizontal as she swings freely in a circular path. In answering the following questions, ignore the mass of the rope in your calculations.

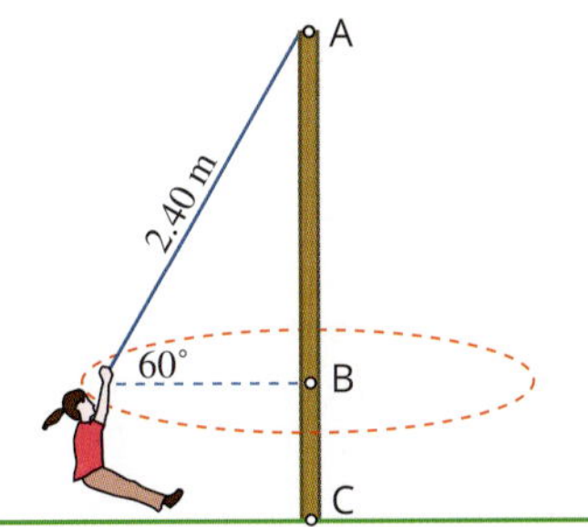

a Calculate the radius of the child's circular path.

b Identify the forces that are acting on her as she swings freely.

c What is the direction of her acceleration when she is at the position shown in the diagram?

d Calculate the net force acting on the girl.

e What is her speed as she swings?

Analysis

9 A car with a mass of 1500 kg is travelling around a circular curve of radius 30 m at a constant speed of 25 m s^{-1}.

a Calculate the centripetal force required for it to round the curve.

b What provides the centripetal force?

c Why do the passengers in the car slide towards the outside of the car when the car rounds the curve?

d If there is any ice or oil on the road, the friction between the car tyres and the road is reduced. If the car is travelling too fast for the turn, what path will the car take?

2.3 Circular motion on banked tracks

FIGURE 2.3.1 Australia's Paige Greco, a Paralympic cyclist who won a gold medal at the 2020 Tokyo Paralympics in the 3000 m Individual Pursuit C1–3

The previous section considered relatively simple situations involving uniform circular motion in a horizontal plane. However, there are more complex situations involving circular motion. On many road bends, the road is not horizontal, but at a small angle to the horizontal. This enables vehicles to maintain their speed without skidding. A similar situation is at a cycling velodrome (Figure 2.3.1). The velodrome at the Darebin International Sports Centre in Thornbury has banked or inclined corners that peak at 43°. This enables cyclists to travel at much higher speeds than if the track were flat. This section examines the principles underlying banked cornering and applies Newton's laws to problems involving circular motion on banked tracks.

BANKED TRACKS

Cars and bikes can travel much faster around corners when the road or track surface is inclined or banked at an angle to the horizontal. Banked tracks are used at cycling velodromes and certain motor sport events, such as NASCAR races. Road engineers design roads to be banked in places where there are sharp corners, such as exit ramps from freeways.

When cars travel in circular paths on horizontal roads, they are relying on the force of friction between the tyres and the road to provide the sideways force that keeps the car in its circular path.

Consider a car travelling clockwise around a horizontal roundabout at a constant speed, v (Figure 2.3.2). The car has an acceleration towards C (the centre of the circle) and so the net force is also sideways on the car towards C. The vertical forces (gravity and the normal force) are balanced (Figure 2.3.3). The only horizontal force is the sideways force that the road exerts on the car tyres. This is a force of friction, F_f. It is unbalanced and so must equal the net force, F_{net}.

If the car drove over an icy patch, there would be no friction and the car would not be able to turn. It would skid in a straight line at a tangent to the circular path.

Creating a **banked track** by angling the road reduces the need for a sideways frictional force and allows cars to travel faster without skidding off the road and away from the circular path. Consider the same car travelling around a circular, banked road at constant speed, v (Figure 2.3.4). It is possible for the car to travel at a speed so that there is no need for a sideways frictional force. This is called the **design speed** and it is dependent on the angle, θ, at which the road is banked. At this speed, the car exhibits no tendency to drift higher or lower on the road.

The car still has an acceleration towards the centre of the circle, C, and so there must be an unbalanced force in this direction. Due to the banking, there are now only two forces acting on the car: its force due to gravity, F_g, and the normal force, F_N, from the road. As can be seen in Figure 2.3.4(b), these forces are unbalanced. They add together to give a net force that is horizontal and directed towards C (Figure 2.3.4(c)).

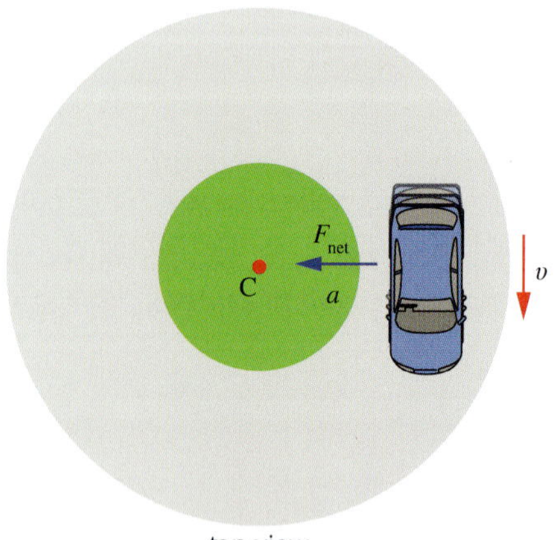

FIGURE 2.3.2 A car travelling in a circular path on a horizontal track

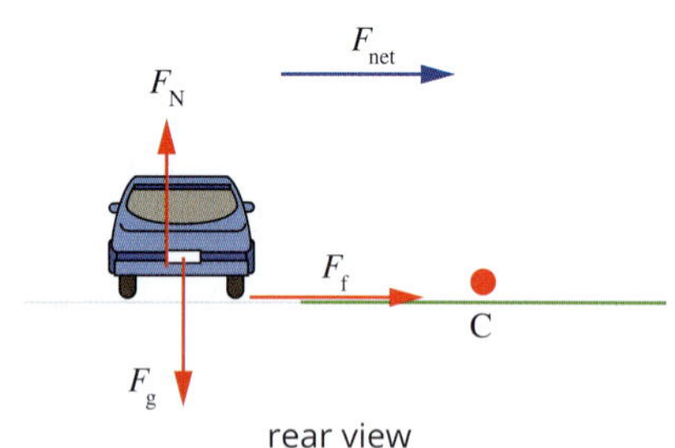

FIGURE 2.3.3 The vertical forces are in balance. It is friction between the tyres and the road that enables the car to turn.

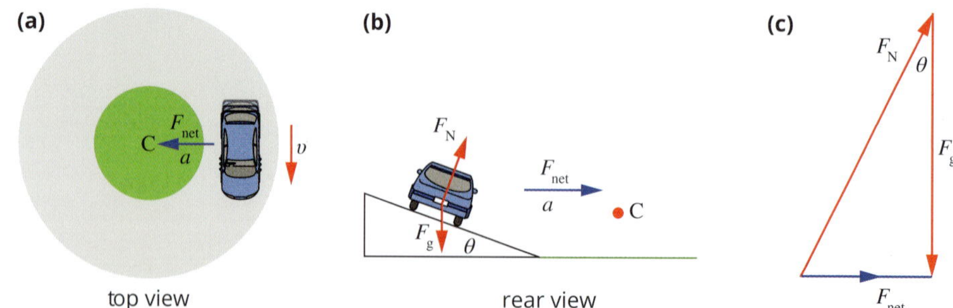

FIGURE 2.3.4 (a) A car is travelling in a circular path on a banked road. (b) The acceleration and net force are towards the centre of the path, C. The banked road means that the normal force (F_N) has an inwards component. This is what enables the car to turn the corner. (c) Vector addition gives the net force (F_{net}) acting horizontally towards the centre.

Banking angle

From Figure 2.3.4(c), it can be seen that the banking angle, θ, at the design speed of a road or track can be found by trigonometry:

$$\tan\theta = \frac{F_{net}}{F_g}$$

where F_{net} is the force acting towards the centre of the circle (N)
F_g is the force due to gravity on the object (N).

Substituting $F_{net} = \frac{mv^2}{r}$ and $F_g = mg$ into the equation and simplifying gives the following equation.

$$\tan\theta = \frac{v^2}{rg}$$

$$\therefore \theta = \tan^{-1}\left(\frac{v^2}{rg}\right)$$

where v is the speed of the vehicle (m s^{-1})
r is the radius of the track (m)
θ is the banking angle (degrees)
g is the acceleration due to gravity ($9.8\,\text{m s}^{-2}$ near the surface of the Earth)

Thus if the banking angle is known, trigonometry can be used to calculate the design speed. Rearranging $\tan\theta = \frac{v^2}{rg}$ gives the following equation for the design speed, v.

$$v^2 = rg\tan\theta$$

$$v = \sqrt{rg\tan\theta}$$

Note that the normal force on an object will be larger when it travels on a banked track than when it travels on a flat track. For example, the cyclist in Figure 2.3.5 would feel a larger force acting from the road when she is on a banked track than when she is cycling on a flat track.

FIGURE 2.3.5 Australian cyclist Anna Meares on a banked velodrome track is cornering at speeds far higher than she could on a flat track. Cyclists on a velodrome will have a greater normal force from the track than cyclists on a flat track.

PHYSICSFILE

Inclined planes vs banked tracks

It is easy to confuse problems involving inclined planes with those involving banked tracks. Inclined-plane problems involve static objects overcoming the coefficient of static friction to slide down the plane. Banked-track problems involve an object moving in uniform circular motion on an inclined plane.

For banked-track problems, it is helpful to consider the vector components of the normal force, F_N, when completing calculations. For inclined-plane problems, components of the force due to gravity, F_g, are used.

Worked example 2.3.1

BANKED TRACKS

A curved section of track on an Olympic velodrome has a radius of 50 m and is banked at an angle of 42° to the horizontal. A cyclist of mass 75 kg is riding on this section of track at the design speed. Assume that g is $9.8\,\text{m s}^{-2}$.

a Calculate the net force acting on the cyclist at this instant.

Thinking	Working
Draw a force diagram and include all forces acting on the cyclist. These forces are the force due to gravity and the normal force from the track, and these are unbalanced. The net force is horizontal and towards the centre of the circular track, as shown in diagram (a) and the force triangle of diagram (b).	
Calculate the force due to gravity, F_g.	$F_g = mg$ $= 75 \times 9.8$ $= 735\,\text{N}$
Use the force triangle and trigonometry to calculate the net force, F_{net}.	$\tan\theta = \frac{F_{net}}{F_g}$ $\tan 42° = \frac{F_{net}}{735}$ $F_{net} = 0.90 \times 735$ $= 662\,\text{N}$
As force is a vector, a direction is needed in the answer.	The net force is 6.6×10^2 N horizontally towards the centre of the circle.

b Calculate the design speed for this section of the track.

Thinking	Working
Note the relevant values.	$g = 9.8\,\text{m s}^{-2}$ $r = 50\,\text{m}$ $\theta = 42°$ $v = ?$
Use the design speed formula.	$v = \sqrt{rg\tan\theta}$ $= \sqrt{50 \times 9.8 \times \tan 42°}$ $= 21\,\text{m s}^{-1}$

Worked example: Try yourself 2.3.1

BANKED TRACKS

A curved section of track on an Olympic velodrome has radius of 40 m and is banked at an angle of 37° to the horizontal. A cyclist of mass 80 kg is riding on this section of track at the design speed. Assume that g is $9.8\,\text{m s}^{-2}$.

a Calculate the net force acting on the cyclist at this instant.

b Calculate the design speed for this section of the track.

Worked example 2.3.2

FINDING THE BANKING ANGLE

The curved portion of a highway needs to be banked to prevent cars from skidding off it. Assume that the banked track of the highway is designed for a top vehicle speed of $80\,km\,h^{-1}$ (i.e. the maximum speed limit for this portion of the highway is $80\,km\,h^{-1}$). The banked track portion of the highway has a radius of 500 m.

What is the value of the banking angle, θ, such that the forces acting on a car keep it on the highway without the need for friction? Assume that g is $9.8\,m\,s^{-2}$.

Thinking	Working
Recall the formula for finding the banking angle.	$\theta = \tan^{-1}\left(\frac{v^2}{rg}\right)$
Convert the design speed from $km\,h^{-1}$ to $m\,s^{-1}$.	$v = \frac{80\,km\,h^{-1}}{3.6}$ $= 22.2\,m\,s^{-1}$
Calculate the angle.	$\theta = \tan^{-1}\left(\frac{v^2}{rg}\right)$ $= \tan^{-1}\left(\frac{22.2^2}{500 \times 9.8}\right)$ $= 5.7°$

Worked example: Try yourself 2.3.2

WS 2

FINDING THE BANKING ANGLE

The curved portion of a highway needs to be banked to prevent cars from skidding off it. Assume that the banked track of the highway is designed for a top vehicle speed of $110\,km\,h^{-1}$. The banked track portion of the highway has a radius of 750 m.

What is the value of the banking angle, θ, such that the forces keep the car on the highway without the need for friction? Assume that g is $9.8\,m\,s^{-2}$.

PHYSICSFILE

Wall of Death

In some amusement parks around the world, there is a ride known menacingly as the Wall of Death (see the figure below). It consists of a cylindrical enclosure with vertical walls. People on bicycles and motorbikes ride into the enclosure and around the vertical walls, so the angle of banking is 90°! The riders need to keep moving and are depending on friction to hold them up. By travelling fast, the centripetal force (the normal force from the wall) is large and this increases the size of the grip (i.e. friction) between the wall and tyres. If the rider slammed on the brakes and stopped, they would simply plummet to the ground.

For a rider to successfully conquer the Wall of Death, they need to travel fast and there must be a good grip between the tyres and the track. The rider is relying on friction to maintain their motion along the wall.

2.3 Review

OA ✓✓

SUMMARY

- A banked track is one where the track is inclined at an angle to the horizontal. This enables vehicles to travel at higher speeds when cornering than when travelling around a flat curved path.
- Banking a track eliminates the need for a sideways frictional force to make a turn. When the speed and angle are such that there is no sideways frictional force, the speed is known as the design speed.
- The forces acting on a vehicle travelling at the design speed on a banked track are gravity and the normal force from the track. These forces are unbalanced and add to give a net force directed towards the centre of the circular motion.
- At the design speed, the banking angle of the track is given by:

$$\theta = \tan^{-1}\left(\frac{v^2}{rg}\right)$$

- For a given banking angle and curve radius, the design speed is given by:

$$v = \sqrt{rg\tan\theta}$$

KEY QUESTIONS

Knowledge and understanding

1 A cyclist is riding along a circular section of a velodrome where the radius is 30 m and the track is inclined at 30° to the horizontal. The cyclist is riding at the design speed and maintains a constant speed. Describe the direction of the acceleration on the cyclist.

2 Copy the following diagram and then draw on it the normal force, the force due to gravity and the net force acting on the bicycle. Label each force.

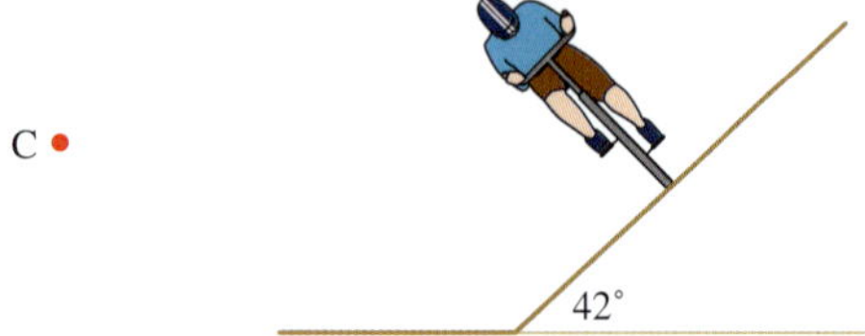

3 The net force acting on an 80 kg bike racing around a banked track is 780 N. What is the banked angle of the track, given that the bike is racing at the design speed?

4 A cycling velodrome has a turn that is banked at 25° to the horizontal. The radius of the track at this point is 35 m.

a Determine the speed (in km h^{-1}) at which a cyclist of mass 75 kg would experience no sideways force as they ride on this section of the track.

b Calculate the size of the normal force acting on the cyclist.

c How would this compare with the normal force if they were riding on a flat track?

5 A car-racing track is banked so that when the cars corner at 55 m s^{-1}, they experience no sideways frictional forces. The track is circular with a radius of 275 m. Calculate the angle to the horizontal at which the track is banked.

6 A curved portion of a highway with a speed limit of 90 km h^{-1} needs to be banked to prevent cars from skidding off it. The curved portion has a radius of 450 m. What is the value of the banking angle that will keep a car travelling at the speed limit on that portion of the highway without the need for friction?

Analysis

7 An architect is designing a velodrome and the original plans have semi-circular sections of radius 15 m and a banking angle of 30°. The architect is asked to make changes to the plans that will increase the design speed for the velodrome. What two design elements could the architect change in order to meet this requirement?

2.4 Circular motion in a vertical plane

Just as a body moving with constant speed in a horizontal circular path has an acceleration that is directed towards the centre of the circle, so does a body moving in a vertical circular path.

If you have been on a rollercoaster ride you will have travelled over humps and down dips at high speeds and, at times, in circular arcs. Some rides even have 360° circular tracks that are entirely vertical (Figure 2.4.1). During these rides, your body may experience forces that you find unpleasant.

When you travel on a rollercoaster, you can experience quite strong forces pushing you down into the seat as you fly through the dips. Then, as you travel over the humps, you tend to lift off your seat. These forces will be discussed in this section. As in the previous sections, Newton's laws are used to solve problems involving this type of circular motion.

FIGURE 2.4.1 This rollercoaster has a circular path in a vertical plane.

MOVING IN VERTICAL CIRCLES

A body moving with constant speed in a horizontal circular path has an acceleration that is directed towards the centre of the circle. The same applies for vertical circular paths. However, circular motion in a vertical plane in real life is often more complex, because it does not usually involve constant speeds.

An example is illustrated in Figure 2.4.2(a). The speed of the skateboarder practising in a half-pipe will increase on the way down as gravitational potential energy is converted into kinetic energy. This means that the skater will experience linear acceleration, a_l, as well as centripetal acceleration, a_c. The resultant acceleration is not directed towards the centre of the circular path.

At the bottom of the half-pipe, the skateboarder will be neither slowing down nor speeding up, so the acceleration is entirely centripetal at this point (Figure 2.4.2(b)). The same applies at the very top of a circular path. For this reason, motion at these points is easier to analyse.

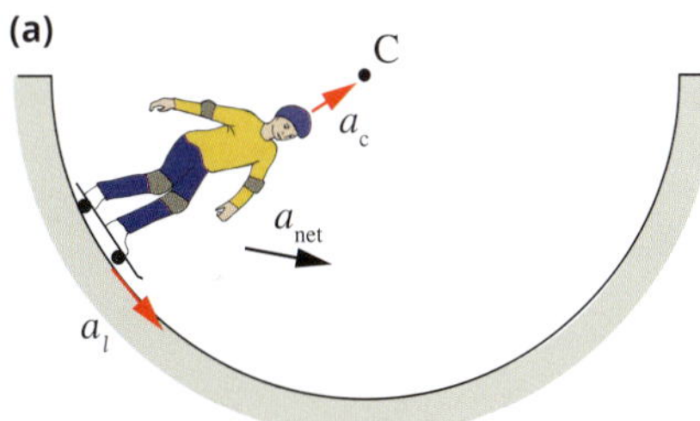

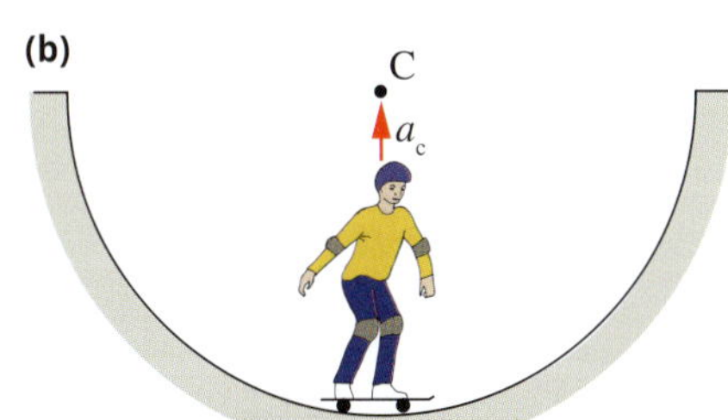

FIGURE 2.4.2 (a) When coming down the sides of a half-pipe, the skateboarder speeds up, and so has both a linear and a centripetal acceleration. The net acceleration, a_{net}, is not towards the centre, C. (b) At the lowest point, the velocity of the skateboarder is momentarily constant, so there is no linear acceleration. The acceleration is supplied completely by the centripetal acceleration, a_c, which is acting towards C.

Uniform horizontal motion

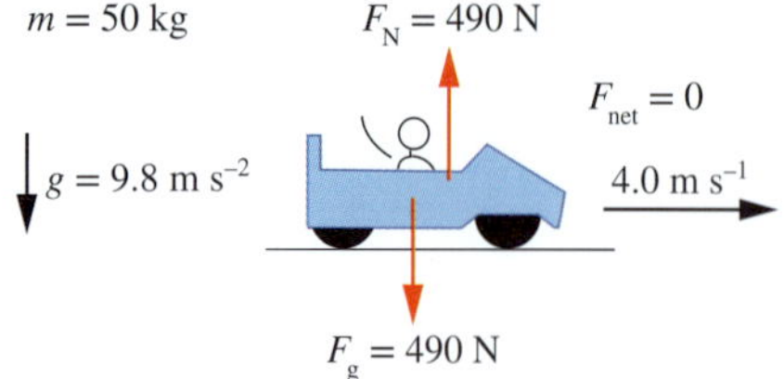

FIGURE 2.4.3 The vertical forces are in balance in this situation, i.e. $F_N = F_g$.

Theme park rides make you appreciate that the forces you experience throughout a ride can vary greatly. First, consider the case of a person in a rollercoaster cart, like that shown in Figure 2.4.3, travelling horizontally at $4.0\,m\,s^{-1}$. If the person's mass is 50 kg and the gravitational field strength is $9.8\,m\,s^{-2}$, the forces acting on the person can be easily calculated. These forces are the gravitational force, F_g, and the normal force, F_N, from the seat.

The person is moving in a straight line with a constant speed, so there are no unbalanced forces acting. The force due to gravity balances the normal force from the seat. The normal force is therefore 490 N up, which is what usually acts upwards on the person when moving horizontally. Hence they would feel the same as their usual force due to gravity.

Circular motion: travelling through dips

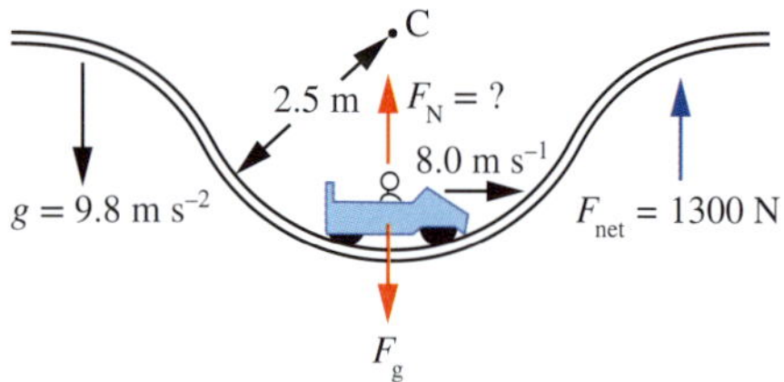

FIGURE 2.4.4 The person has a centripetal acceleration that is directed upwards towards the centre of the circle, and so the net force is also upwards. In this case, the magnitude of the normal force, F_N, is greater than the force due to gravity, F_g, and produces a situation where the rider feels heavier than usual.

Now consider the forces that act on the person as the cart reaches the bottom of a circular dip. Suppose that the dip has a radius of 2.5 m and the cart is moving at $8.0\,m\,s^{-1}$ (Figure 2.4.4).

The person will have a centripetal acceleration due to the circular path. This centripetal acceleration is directed towards the centre, C, of the circular path—in this case, vertically upwards. The person's centripetal acceleration, a, is:

$$\begin{aligned} a &= \frac{v^2}{r} \\ &= \frac{8.0^2}{2.5} \\ &= 26\,\text{m}\,\text{s}^{-2} \text{ upwards towards C} \end{aligned}$$

The net centripetal force acting on the person is given by:

$$\begin{aligned} F_{net} &= ma \\ &= 50 \times 26 \\ &= 1300\,\text{N upwards} \end{aligned}$$

The normal force, F_N, and the force due to gravity, F_g, are no longer in balance. They add together to give an upwards force of 1300 N. This indicates that the normal force must be greater than the force due to gravity by 1300 N. In other words, the normal force is 490 N + 1300 N = 1790 N up. This is more than three times greater than the normal force of 490 N that usually acts on a person of mass 50 kg. That is the reason why, when you are at this point in a ride, you feel the seat pushing up against you much more strongly and you feel much heavier than usual.

Circular motion: travelling over humps

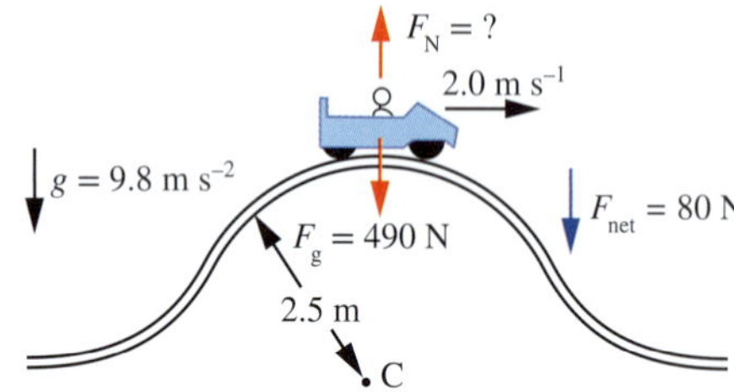

FIGURE 2.4.5 The centripetal acceleration is downwards towards the centre of the circle, and so the net force is also in that direction. At this point, the magnitude of the normal force, F_N, is less than the force on the person due to gravity, F_g.

Now consider the situation as the cart moves over the top of a hump of radius 2.5 m with a lower speed of $2.0\,m\,s^{-1}$ (Figure 2.4.5).

The person now has a centripetal acceleration that is directed vertically downwards towards the centre of the circle, C. Therefore the net force acting at this point is directed vertically downwards. The centripetal acceleration is:

$$\begin{aligned} a &= \frac{v^2}{r} \\ &= \frac{2.0^2}{2.5} \\ &= 1.6\,\text{m}\,\text{s}^{-2} \text{ downwards towards C} \end{aligned}$$

The net centripetal force is:

$$\begin{aligned} F_{net} &= ma \\ &= 50 \times 1.6 \\ &= 80\,\text{N downwards} \end{aligned}$$

As in the dip, the force due to gravity and the normal force are not in balance. They add to give a net force of 80 N down. The force due to gravity, F_g, must therefore be 80 N greater than the normal force, F_N. This tells us that the normal force is 490 N − 80 N = 410 N up. This explains why you feel lighter when travelling over a hump.

Circular motion: travelling through loops

You might have been on a rollercoaster like the one in Figure 2.4.6 where you were upside down at times during the ride. The speed of these rides and the radius of their circular path is what prevents riders from falling out. In theory, the safety harness worn by a rider is not needed to hold them in their seats.

FIGURE 2.4.6 The thrill seekers on this rollercoaster ride don't fall out when upside down because the centripetal acceleration of their cart is greater than $9.8\,m\,s^{-2}$ down.

The reason people don't fall out of the rollercoaster is that their centripetal acceleration is greater than the acceleration due to gravity ($9.8\,m\,s^{-2}$). To illustrate this, try the following activity. Extend a hand palm up, place an eraser on your palm then turn your hand over and move it rapidly towards the floor. You should find that it is possible to keep the eraser in contact with your hand as you move your hand down. The eraser is under your hand but it is not falling out of your hand. Your hand must be, for a short time, moving downwards with an acceleration in excess of $9.8\,m\,s^{-2}$ and continually exerting a normal force on the eraser. If your hand had an acceleration less than $9.8\,m\,s^{-2}$, the eraser would fall away from your hand to the floor.

A similar principle holds with rollercoaster rides. The people on the ride don't fall out at the top because the motion of the rollercoaster gives them a centripetal acceleration that is greater than $9.8\,m\,s^{-2}$ down. The engineers who designed the ride would have ensured that the rollercoaster can move with sufficient speed and in a circle of an appropriate radius so that this happens.

To explore this further, consider a rollercoaster ride that has a section of radius 15 m in a vertical circle (Figure 2.4.7).

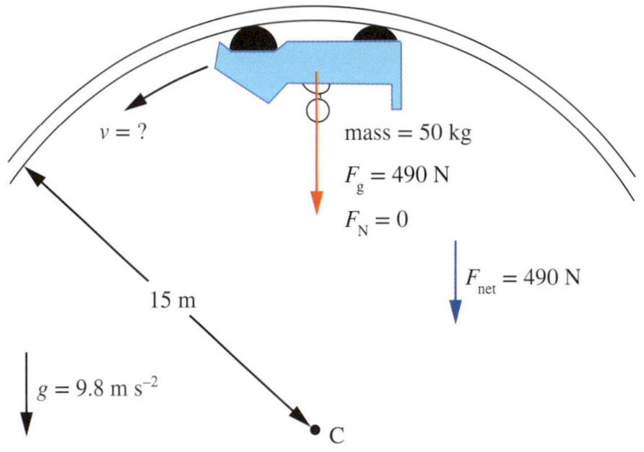

FIGURE 2.4.7 A rollercoaster cart travelling upside down through a loop. At the critical point where the cart just stays in contact with the track, the normal force can be considered to be zero.

PHYSICSFILE

Fighter pilots

A fighter pilot in a vertical loop manoeuvre can safely experience centripetal accelerations of up to around 9 g, or 88 m s^{-2}. In a loop where the g-force is greater than this, the pilot may pass out. The centripetal acceleration of the plane will push her into her seat and make the blood flow away from her head. The resulting reduction of blood in the brain may cause her to experience visual disturbance (a 'grey out') or even lose consciousness (a 'black out'). This type of force is called a positive g-force. Fighter pilots wear g-suits which pressurise the legs and limit the blood flowing to them. This helps them to maintain consciousness.

On the other hand, if the pilot's head is on the outside of the loop, the centripetal acceleration will pull the pilot onto their harness and the additional blood flow to the head can make the whites of the eyes turn red. The excess blood flow in the head may cause 'red out'. This type of force is called a negative g-force.

It is possible to calculate the speed that would ensure that a rider does not fall out. At the critical speed (i.e. the minimum speed), the normal force, F_N, on the person will be zero. In other words, the seat will exert no force on them at this speed. The critical speed is independent of the mass of the person. Assuming that $g = 9.8\,\text{m s}^{-2}$, the centripetal force, F_{net}, is:

$$F_{net} = F_g + F_N$$

but:

$$F_N = 0,$$

so:

$$F_{net} = F_g$$

Therefore:

$$\frac{mv^2}{r} = mg$$

$$v^2 = \frac{mgr}{m}$$

$$= gr$$

$$v = \sqrt{gr}$$

$$= \sqrt{9.8 \times 15}$$

$$= 12\,\text{m s}^{-1}$$

This speed is approximately equal to 43 km h^{-1} and is the minimum speed needed to prevent riders—whatever their mass—from falling out of their cart in a loop of that particular radius. In practice, the rollercoaster would move with a speed much greater than this to ensure that there was a significant force between the riders and their seats, rather than zero normal force as calculated for the critical speed. Corkscrew rollercoasters can travel at up to 110 km h^{-1} and the riders can experience accelerations of up to 50 m s^{-2} (or 5 g). So safety harnesses are really only needed when the speed is below the critical value. Their primary function is to prevent people from moving around while on the ride.

How the normal force varies during the ride

It is interesting to compare the normal force that acts on the 50 kg rollercoaster rider in the three situations explored, that is, when they are travelling horizontally with uniform motion, when they are at the bottom of a dip and when they are at the top of a loop.

- The normal force when travelling horizontally is 490 N upwards.
- At the bottom of a dip, the normal force is 1790 N upwards. In other words, in the dip the seat pushes into the rider with a greater force than usual. As the rider experiences a normal force of 1790 N, they feel much heavier than normal. If the rider had been sitting on weighing scales at this time, the scales would have shown a higher-than-usual reading.
- At the top of a hump, the normal force is 410 N upwards. In other words, over the hump the seat pushes into the rider with a smaller force than usual. As the rider experiences a normal force of 410 N, this gives them the sensation of feeling lighter.

The force on the rider due to gravity has not changed throughout the ride: F_g remains at 490 N. It is the normal force acting on them that varies. It is the normal force that makes the rider feel heavier and lighter as they travel through the dips and humps respectively.

Worked example 2.4.1

VERTICAL CIRCULAR MOTION

A student arranges a toy car track with a vertical loop of radius 20.0 cm.

A toy car of mass 150 g is released from rest at a height of 1.00 m (point X). The car rolls down the track and travels inside the loop. Assume that g is $9.8\,\text{m s}^{-2}$ and ignore friction.

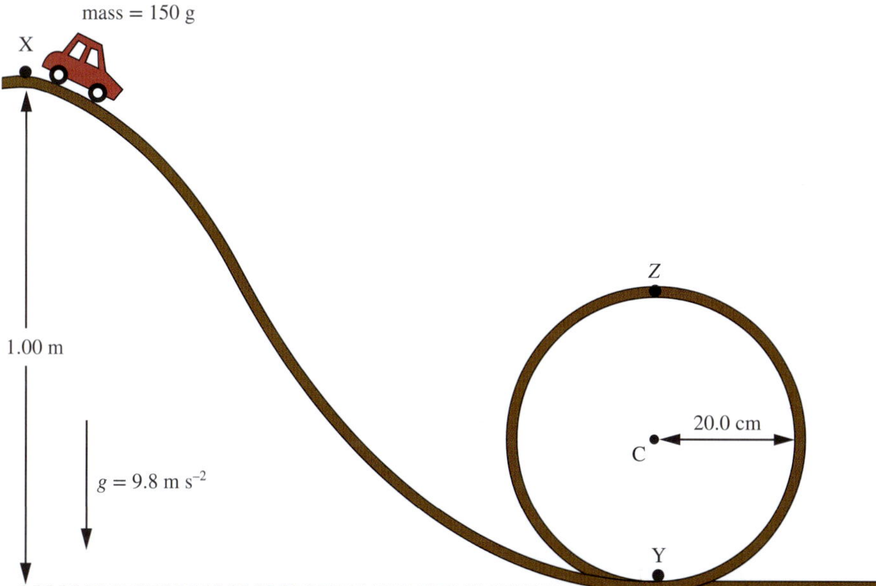

a Calculate the speed of the car as it reaches point Y at the bottom of the loop.	
Thinking	**Working**
Note all the variables given in the question.	At X: $m = 150\,\text{g} = 0.150\,\text{kg}$ $\Delta h = 1.00\,\text{m}$ $v = 0$ $g = 9.8\,\text{m s}^{-2}$
Approach the problem by considering that energy is conserved during the car's motion. Calculate the total mechanical energy first. Note that the initial speed is zero, so E_k at X is zero.	The total mechanical energy, E_m, at X is: $E_m = E_k + E_g$ $= \frac{1}{2}mv^2 + mg\Delta h$ $= 0 + (0.150 \times 9.8 \times 1.00)$ $= 1.47\,\text{J}$
Use conservation of energy ($E_m = E_k + E_g$) to determine the velocity at point Y. As the car rolls down the track, it loses its gravitational potential energy and gains kinetic energy. At the bottom of the loop (Y), the car has zero potential energy.	At Y: $E_m = 1.47\,\text{J}$ $\Delta h = 0$ $E_g = 0$ $E_m = E_k + E_g$ $E_m = \frac{1}{2}mv^2 + mg\Delta h$ $1.47 = \frac{1}{2} \times 0.150v^2 + 0$ $v^2 = \frac{1.47}{0.0750}$ $v = \sqrt{19.6}$ $= 4.4\,\text{m s}^{-1}$

b Calculate the normal force from the track at point Y.	
Thinking	**Working**
To solve for F_N, start by working out the net, or centripetal, force. At Y the car has a centripetal acceleration towards C (i.e. upwards), so the net centripetal force must also be vertically upwards at this point.	$F_{net} = \frac{mv^2}{r}$ $= \frac{0.150 \times 4.43^2}{0.200}$ $= 14.7\text{ N up}$
Calculate the force due to gravity, F_g, and add it to a force diagram.	$F_g = mg$ $= 0.150 \times 9.8$ $= 1.47\text{ N down}$ At point Y $F_N = ?$ $F_{net} = 14.7\text{ N}$ $F_g = 1.47\text{ N}$
Work out the normal force using vectors. Note up as positive and down as negative in your calculations. The forces acting are unbalanced, as the car has a centripetal acceleration upwards (towards C). The upwards (normal) force must be larger than the downwards force.	$F_{net} = F_g + F_N$ $+14.7 = -1.47 + F_N$ $F_N = +14.7 + 1.47$ $= 16\text{ N up}$ Note that the force the track exerts on the car is much greater (by about ten times) than the force due to gravity. If the car were travelling horizontally on a flat surface, the normal force would be just 1.47 N up.

c What is the speed of the car as it reaches point Z?	
Thinking	**Working**
Calculate the velocity from the values you have, using $E_m = E_k + E_g$.	At Z: $m = 0.150\text{ kg}$ $\Delta h = 2 \times 0.200 = 0.400\text{ m}$ Mechanical energy is conserved, so $E_m = 1.47\text{ J}$ (from part **a**). At Z: $E_m = E_k + E_g$ $= \frac{1}{2}mv^2 + mg\Delta h$ $1.47 = (0.5 \times 0.150 \times v^2) + (0.150 \times 9.8 \times 0.400)$ $1.47 = 0.075 \times v^2 + 0.588$ $v^2 = 11.76$ $v = 3.4\text{ m s}^{-1}$

d What is the normal force acting on the car at point Z?	
Thinking	**Working**
To find F_N, start by working out the net, or centripetal, force. At Z the car has a centripetal acceleration towards C (i.e. downwards), so the net centripetal force must also be vertically downwards at this point.	$F_{net} = \frac{mv^2}{r}$ $= \frac{0.150 \times 3.43^2}{0.200}$ $= 8.82\text{N down}$
Work out the normal force using vectors. Note up as positive and down as negative in your calculations.	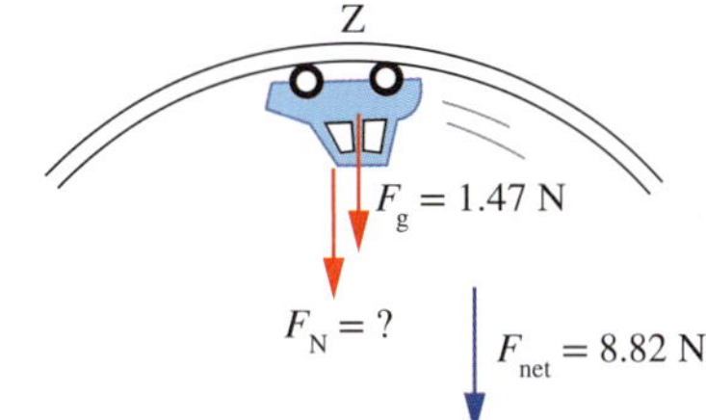 $F_{net} = F_g + F_N$ $-8.82 = -1.47 + F_N$ $F_N = -8.82 + 1.47$ $= -7.35$ $= 7.4\text{N down}$ Note that there is still strong contact between the car and the track—as given by the normal force—but that it is only about half the size compared to when the car was at the bottom of the track. If the car had progressively lower speeds, the normal force at Z would decrease and eventually drop to zero. At this point, the car would lose contact with the track, fall off the track and its acceleration would be equal to g.

Worked example: Try yourself 2.4.1

VERTICAL CIRCULAR MOTION

A student arranges a toy car track with a vertical loop of radius 25.0 cm, as shown.

A toy car of mass 150g is released from rest at a height of 1.20m (point X). The car rolls down the track and travels around the loop. Assume that g is 9.8 m s^{-2} and ignore friction.

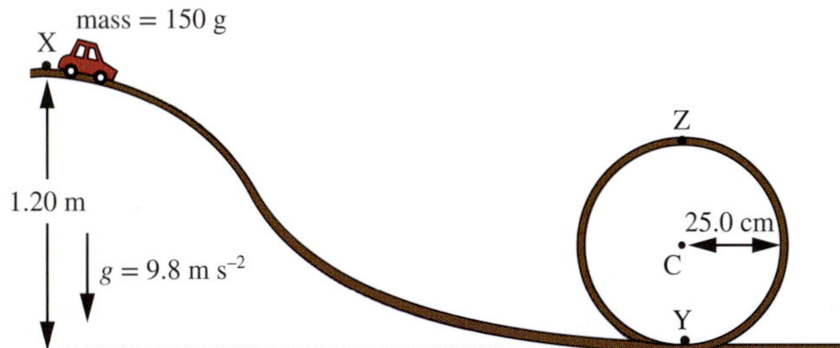

a Calculate the speed of the car as it reaches point Y at the bottom of the loop.

b Calculate the normal force from the track at point Y.

c What is the speed of the car as it reaches point Z?

d What is the normal force acting on the car at point Z?

2.4 Review

OA ✓✓

SUMMARY

- The gravitational force, F_g, and normal force, F_N, must be considered when analysing the motion of an object moving in a vertical circle.
- If the normal force is greater than the gravitational force ($F_N > F_g$) the passenger or rider will feel heavier than they really are.
- If the normal force is less than the gravitational force ($F_N < F_g$) the passenger or rider will feel lighter than they really are.
- In vertical circular motion, the gravitational force always acts vertically downwards regardless of the position of the rider or passenger around the circle, the net force always acts towards the centre of the circle, and the normal force always acts between the seat and the passenger or rider.
- The normal force and the gravitational force are added together as vectors in a force diagram to give the resultant as the net force.
- At the point where a moving object falls from its circular path, the normal force is zero. The object will be moving with a centripetal acceleration equal to that due to gravity (9.8 m s^{-2} down).
- Problems relating to motion in vertical circles can often be solved by noting that energy is conserved at all points in the motion:

$$E_m = E_k + E_g = \frac{1}{2}mv^2 + mg\Delta h$$

KEY QUESTIONS

Knowledge and understanding

In answering the following questions, assume that $g = 9.8$ m s^{-2} and ignore the effects of air resistance.

1 A yo-yo is swung with a constant speed in a vertical circle.
  - a Describe the magnitude of the acceleration of the yo-yo along its path.
  - b At which point in the circular path is there the most tension in the string?
  - c At which point in the circular path is there the least tension in the string?
  - d At which point is the string most likely to break?
  - e If the yo-yo has a mass of 100 g and the radius of the circle is 1.25 m, find the minimum speed that the yo-yo must have at the top of the circle so that the string does not slacken.

2 A car of mass 800 kg encounters a speed hump of radius 10 m. The car drives over the hump at a constant speed of 14.4 km h^{-1}.
  - a Name all the vertical forces acting on the car when it is at the top of the hump.
  - b Calculate the resultant force acting on the car when it is at the top of the hump.
  - c After travelling over the hump, the driver remarked to a passenger that she felt lighter as the car moved over the top of the hump. Is this possible? Explain your answer.
  - d What is the maximum speed (in km h^{-1}) that this car can have at the top of the hump and still have its wheels in contact with the road?

3 A student is designing an amusement park ride in which a cart descends a steep incline at point X, enters a circular loop of radius 17 m at point Y and makes one complete revolution of the loop. The cart has a mass of 700 kg, and it carries passengers at a speed of 1.75 m s^{-1} before it begins its descent from point X, which is 70 m higher than the bottom of the loop.

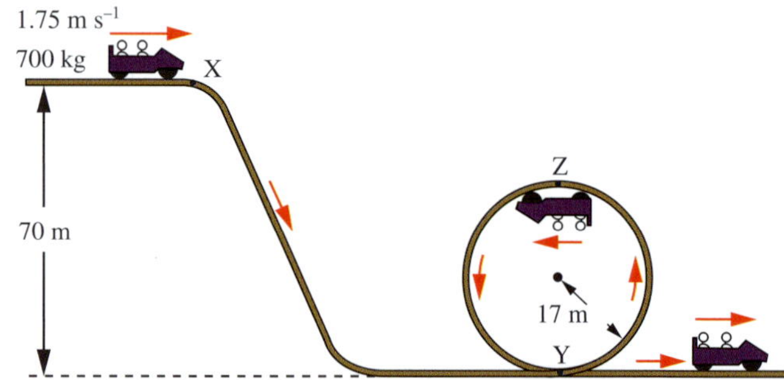

  - a Calculate the speed of the cart at point Y.
  - b What is the speed of the cart at point Z?
  - c Calculate the normal force acting on the cart at point Z.
  - d What is the minimum speed that the cart can have at point Z and still stay in contact with the track?

4 A stunt pilot appearing at an air show decides to perform a vertical loop so that she is upside down at the top of the loop. During the stunt she maintains a constant speed of 35.0 m s^{-1} and the loop has a radius of 100 m.

Calculate the normal force acting on the 80.0 kg pilot when she is at the top of the loop.

5 A skateboarder of mass 72 kg is practising on a half-pipe of radius 3.0 m. At the lowest point of the half-pipe, the speed of the skater is 7.0 m s^{-1}.

a What is the acceleration of the skater at this point? Specify both the magnitude and direction.

b Calculate the size of the normal force acting on the skater at this point.

Analysis

6 The maximum value of acceleration that the human body can safely tolerate for a short time is nine times that due to gravity. Calculate the maximum speed with which a pilot could safely pull out of a circular dive of radius 400 m.

7 A student rolls a toy car of mass 50 g along a smooth track in the shape of a loop-the-loop. They try to give the car a launch speed at point A so that the car just maintains contact with the track as it passes through point B.

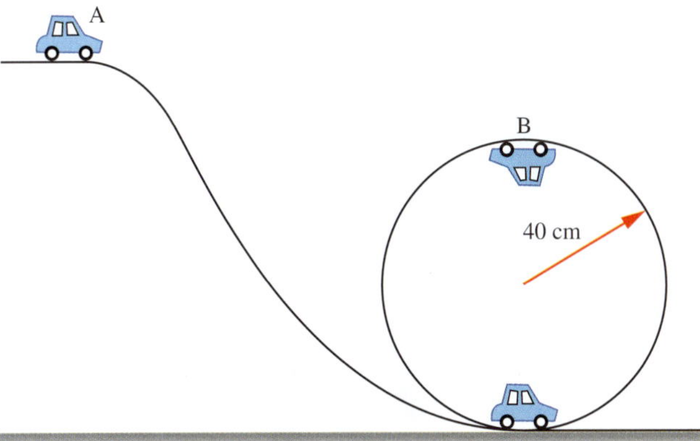

a Determine the acceleration of the toy car as it passes point B.

b How fast is the toy car travelling at point B?

8 A car of mass 1500 kg slows to travel over an old stone bridge of radius 10 m. The car's speed at the top of the bridge is 8.0 m s^{-1}.

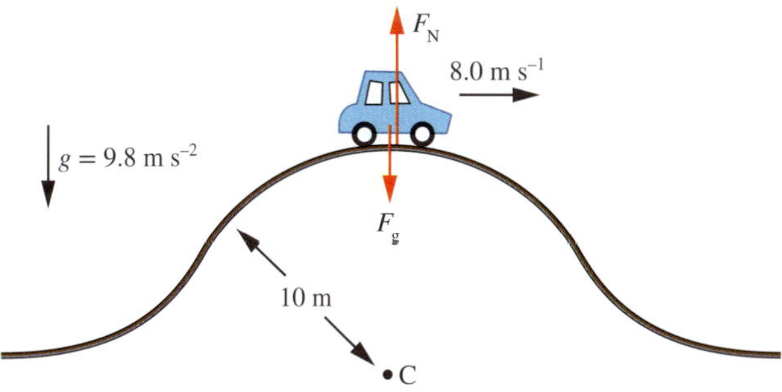

a Calculate the magnitude and direction of the resultant force acting on the car when it is at the top of the bridge.

b Calculate the magnitude and direction of the normal force acting on the car when it is at the top of the bridge.

c What is the maximum speed that the car can have at the top of the bridge and still have its wheels in contact with the road?

2.5 Projectiles launched horizontally

FIGURE 2.5.1 A multi-flash photograph of a golf ball that has been bounced on a hard surface. The ball moves in a series of parabolic paths.

A **projectile** is any object that is thrown or projected into the air and is moving freely, that is, it has no power source (such as a rocket engine or propeller) driving it. A netball as it is passed, a cricket ball that is hit for six and an aerial skier flying through the air are examples of projectiles. People have long argued about the path that projectiles follow, with some thinking that they were circular or had straight sections. It is now known that if projectiles are not launched vertically, and if air resistance is ignored, they move in smooth parabolic paths (Figure 2.5.1). This section considers projectiles that are launched horizontally and shows how Newton's laws can be used to solve problems involving projectile motion.

PROJECTILE MOTION

It is a very common misconception that when a projectile travels forwards through the air, it has a forwards force acting on it. This is incorrect. There may have been some forwards force acting as the projectile was launched, but once the projectile is released, this forwards force is no longer acting.

In fact, if air resistance is ignored, the only force acting on a projectile during its flight is the force due to gravity, F_g. This force is constant and always directed vertically downwards. This causes the projectile to continually deviate from a straight-line path and follow a parabolic path (Figure 2.5.2).

Projectile motion is quite complex compared to straight-line motion. It must be analysed by considering the different components—horizontal and vertical—of the actual motion. The vertical and horizontal components are independent of each other and must be treated separately.

FIGURE 2.5.2 The motorcycle and rider travel in a parabolic path as they fly through the air.

Given that the only force acting on a projectile is the force due to gravity, F_g, it follows that the projectile must have a vertical acceleration of $9.8\,\mathrm{m\,s^{-2}}$ downwards throughout its motion.

> In the vertical direction, a projectile accelerates due to the force of gravity, that is, at a rate of $9.8\,\mathrm{m\,s^{-2}}$ downwards.
> In the horizontal direction, a projectile has a uniform velocity. This is because there are no forces acting on it in this direction (if air resistance is ignored). Thus the horizontal acceleration is zero.

PROJECTILES LAUNCHED HORIZONTALLY

Projectiles can be launched at any angle. The launch velocity needs to be resolved into vertical and horizontal components and trigonometry used to solve most problems involving projectile motion. For projectiles launched horizontally, calculating the vector components of the launch velocity is straightforward: the initial vertical velocity is zero (although it increases during the flight) and the horizontal velocity is constant (equal to its launch velocity). This can be verified using trigonometric ratios and a launch angle of 0°.

Tips for solving projectile motion problems

1 Construct a diagram showing the projectile's motion. Write down the information supplied for the horizontal and vertical components.

2 In the horizontal direction, the velocity, v, is constant, and the only formula needed to calculate it is $v_{av} = \frac{s}{t}$.

3 In the vertical direction, the projectile is moving with a constant acceleration ($9.8\,\mathrm{m\,s^{-2}}$ downwards), so the equations of motion for uniform acceleration can be used. These include:

$$v = u + at$$

$$s = ut + \frac{1}{2}at^2$$

$$v^2 = u^2 + 2as$$

4 In the vertical direction, it is important to clearly specify whether up or down is the positive or negative direction. Either choice will work just as effectively. However, the same convention needs to be used consistently throughout each problem.

5 If a projectile is launched horizontally, its horizontal velocity throughout the flight is the same as its initial velocity.

6 Pythagoras's theorem can be used to determine the actual speed of the projectile at any point.

7 If the velocity of the projectile is required, it is necessary to provide a direction with respect to the horizontal plane as well as its speed.

PHYSICSFILE

Cartoon physics

It is easy to get the wrong idea about projectile motion from watching cartoon characters running or driving off cliffs. In many cartoons, the character leaves the cliff and travels horizontally outwards, stopping in mid-air (see figure at right). Once they realise where they are, they immediately fall vertically downwards. Clearly, this is not what happens in reality! The character should start falling in a smooth parabolic arc as soon as they leave the cliff-top.

Many misconceptions can arise from what is shown in cartoons. In real life, this car would start falling as soon as it leaves the cliff top and travel in a parabolic arc.

Worked example 2.5.1

PROJECTILE LAUNCHED HORIZONTALLY

A golf ball of mass 150 g is hit horizontally with a speed of 25.0 m s^{-1} from the top of a cliff 40.0 m high. Assume that $g = 9.8$ m s^{-2} and ignore air resistance.

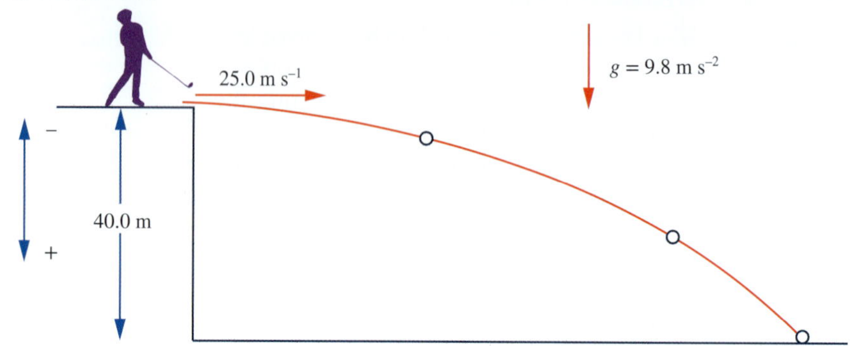

a Calculate the time the ball takes to land.	
Thinking	**Working**
Let the downwards direction be positive. Write down the information relevant to the vertical component of the motion. Note that the instant the ball is hit, it is travelling only horizontally, so its initial vertical velocity is zero.	Down is positive. Vertically: $u = 0$ m s^{-1} $s = 40.0$ m $a = 9.8$ m s^{-2} $t = ?$
In the vertical direction, the ball has constant acceleration, so use an equation for uniform acceleration. Select the equation that best fits the information you have.	$s = ut + \frac{1}{2}at^2$
Substitute values in the equation, rearrange it and solve for t.	$40.0 = 0 + \frac{1}{2} \times 9.8t^2$ $t = \sqrt{\frac{40.0}{4.90}}$ $= 2.86$ $= 2.9$ s

b Calculate the distance the ball travels from the base of the cliff, i.e. the range of the ball.	
Thinking	**Working**
Write down the information relevant to the horizontal component of the motion. As the ball is hit horizontally, the initial speed gives the horizontal component of the velocity throughout the flight.	Horizontally: $u = 25.0$ m s^{-1} $t = 2.86$ s from part **a** $s = ?$
Select the equation that best fits the information you have.	As the horizontal speed is constant (i.e. $u = v$), you can use $v_{av} = \frac{s}{t}$.
Substitute values in the equation, rearrange it and solve for s.	$25.0 = \frac{s}{2.86}$ $s = 25.0 \times 2.86$ $= 72$ m Note that the mass of the ball does not affect its motion, as is the case with all objects in projectile motion or in free fall.

c Calculate the velocity of the ball as it lands.

Thinking	Working
Find the horizontal and vertical components of the ball's speed as it lands. Write down the information relevant to both the vertical and horizontal components.	Horizontally: $u = v = 25.0\,\text{m s}^{-1}$ Vertically, with downwards as positive: $u = 0$ $a = 9.8\,\text{m s}^{-2}$ $s = 40.0\,\text{m}$ $t = 2.86\,\text{s}$ $v = ?$
To find the final vertical speed, use the equation for uniform acceleration that best fits the information you have.	$v = u + at$
Substitute values in the equation and solve for the variable you are looking for: in this case, v.	Vertically: $v = 0 + 9.8 \times 2.86$ $= 28\,\text{m s}^{-1}$ down
Add the components as vectors.	25.0 m s^{-1} θ v 28.0 m s^{-1}
Use Pythagoras's theorem to calculate the actual speed, v, of the ball.	$v = \sqrt{v_h^2 + v_v^2}$ $= \sqrt{25.0^2 + 28.0^2}$ $= \sqrt{1409}$ $= 38\,\text{m s}^{-1}$
Use trigonometry to find the angle, θ.	$\theta = \tan^{-1}\left(\frac{28.0}{25.0}\right)$ $= 48.2°$
Specify the velocity with a magnitude and a direction relative to the horizontal. Express the answer to 2 significant figures.	The final velocity of the ball is $38\,\text{m s}^{-1}$ at 48° below the horizontal.

Worked example: Try yourself 2.5.1

PROJECTILE LAUNCHED HORIZONTALLY

A golf ball of mass 100 g is hit horizontally with a speed of $20.0\,\text{m s}^{-1}$ from the top of a 30.0 m high cliff. Assume that $g = 9.8\,\text{m s}^{-2}$ and ignore air resistance.

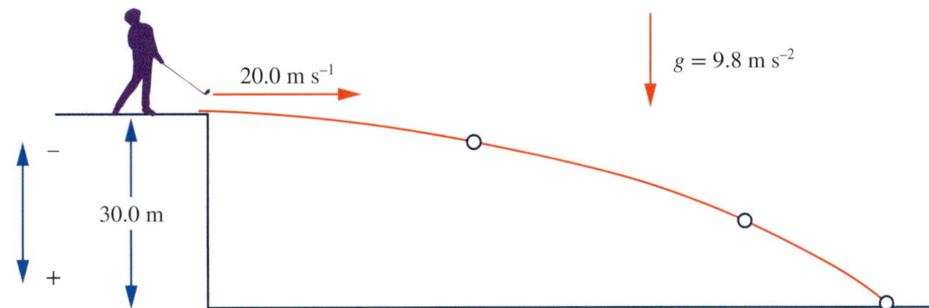

a Calculate the time the ball takes to land.

b Calculate the distance the ball travels from the base of the cliff, i.e. the range of the ball.

c Calculate the velocity of the ball as it lands.

PHYSICSFILE

Aerodynamic design

In the track-and-field event of javelin, the aerodynamic shape of the javelin once used proved to be too successful. The javelin had been progressively streamlined to reduce the drag force so that the athletes could throw it further. This was not a problem until the 1980s—the javelin could now be thrown so far that runners competing in nearby track events were endangered. The design of the javelin was changed again. It was made more snub-nosed to increase drag and reduce the distance it could be thrown (see figure below). In 1983, the world record was 104.8 m. In 1986, with the modified design, the world record dropped to 85.7 m.

Australian Kelsey-Lee Barber winning a bronze medal in the women's javelin final at the 2020 Tokyo Olympics. Note the javelin's snub-nosed end.

THE EFFECTS OF AIR RESISTANCE

The interaction between a projectile and the air can have a significant effect on its motion, particularly if the projectile has a large surface area and a relatively low mass. If you try throwing an inflated party balloon, it will not travel very far compared to throwing a marble at the same speed.

The size of the air resistance (i.e. the drag force) that acts on an object as it moves depends on such factors as:

- the speed of the object: the faster an object moves, the greater the drag force becomes
- the cross-sectional area of the object in its direction of motion: a greater area means greater drag
- the aerodynamic shape of the object: a more streamlined shape experiences less drag
- the density of the air: higher air density means greater drag.

When a pilot drops a supply parcel from a plane, the drag force from the air acts in the opposite direction to the parcel's velocity. If the parcel were dropped on the Moon, where there is no air and hence no air resistance, the parcel would continue its horizontal motion and would remain directly below the plane as it fell.

Figure 2.5.3 shows a supply parcel being dropped from a plane moving at a constant velocity. If air resistance is ignored, the parcel falls in the parabolic arc shown by the darker blue curved line in diagram (a). It would continue moving horizontally at the same rate as the plane, that is, as the parcel falls it would stay directly beneath the plane until it hits the ground. The effect of air resistance is shown by the light-blue curved path. Air resistance (i.e. drag) is a retarding force and it acts in a direction that is opposite to the motion of the projectile. Air resistance makes the parcel fall more slowly and over a shorter path. If air resistance is taken into account, there are now two forces acting, as shown in diagram (b): the force due to gravity, F_g, and air resistance, F_a. Therefore the resultant force, F_{net}, that acts on the projectile is not vertically down and nor is its acceleration.

(a)

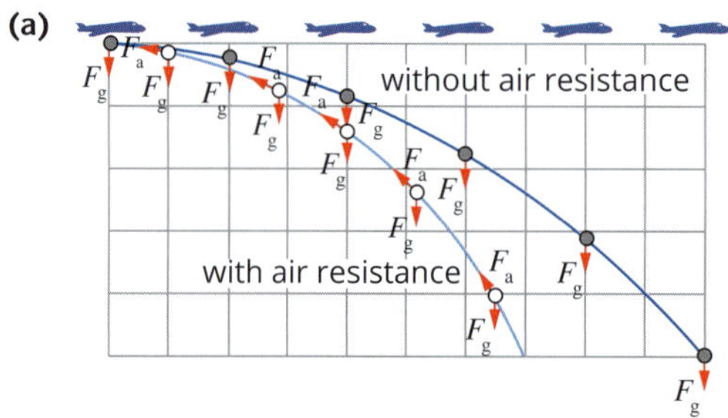

(b)

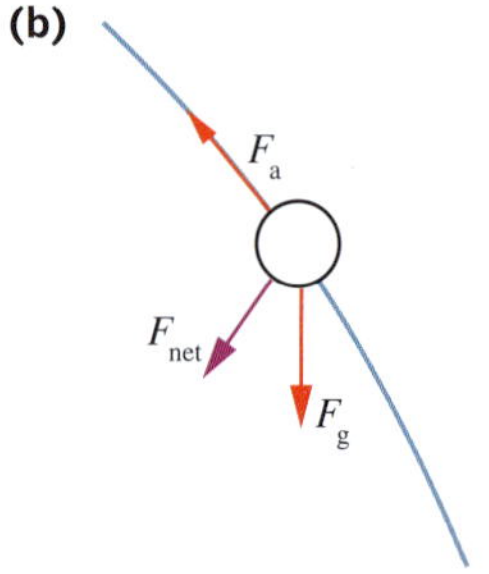

FIGURE 2.5.3 (a) The paths of a supply parcel dropped from a plane with and without air resistance. (b) When air resistance is acting, the net force on the parcel is not vertically down.

2.5 Review

OA ✓✓

SUMMARY

- If air resistance is ignored, the only force acting on a projectile is the force due to gravity, F_g. This results in the projectile having a vertical acceleration of 9.8 m s^{-2} downwards during its flight.
- Projectiles move in parabolic paths that can be analysed by considering the horizontal and vertical components of their motion.
- The following equations of motion for uniform acceleration can be used to determine the vertical component of the motion:

$$v = u + at$$

$$s = ut + \frac{1}{2}at^2$$

$$v^2 = u^2 + 2as$$

- The horizontal velocity of a projectile remains constant throughout its flight if air resistance is ignored. Therefore the following equation for average velocity can be used for the horizontal component of its motion:

$$v_{av} = \frac{s}{t}$$

- Pythagoras's theorem can be used with the vertical and horizontal components of velocity to work out the speed, v, of the projectile.
- If the velocity of the projectile is required, trigonometry can be used to find the angle to the horizontal at which the projectile is travelling.

KEY QUESTIONS

Knowledge and understanding

For the following questions, assume that the acceleration due to gravity is 9.8 m s^{-2} and ignore the effects of air resistance unless otherwise stated.

1 A skateboard travelling at 5.5 m s^{-1} rolls off a horizontal bench that is 1.7 m high.
- **a** How long does the board take to hit the ground?
- **b** How far does the board land from the base of the bench?
- **c** What is the magnitude and direction of the board's acceleration just before it lands?

2 Two identical tennis balls, X and Y, are hit horizontally from a point 2.5 m above the ground with different initial speeds: ball X has an initial speed of 7.5 m s^{-1} and ball Y has an initial speed of 12 m s^{-1}.
- **a** Calculate the time it takes for ball X to strike the ground.
- **b** Calculate the time it takes for ball Y to strike the ground.
- **c** How much further than ball X does ball Y travel in the horizontal direction before bouncing?

3 An archer stands on top of a platform that is 45 m high and fires an arrow horizontally at 70 m s^{-1}.
- **a** What is the speed of the arrow as it reaches the ground?
- **b** At what angle relative to the horizontal is the arrow travelling as it reaches the ground?

4 A bowling ball of mass 9.5 kg travelling at 6.5 m s^{-1} rolls off a horizontal table 1.0 m high.
- **a** Calculate the ball's horizontal velocity just as it strikes the floor.
- **b** What is the vertical velocity of the ball as it strikes the floor?
- **c** Calculate the velocity of the ball as it strikes the floor.
- **d** What time interval has elapsed between the ball leaving the table and striking the floor?
- **e** Calculate the horizontal distance travelled by the ball as it falls.
- **f** Draw a diagram showing the forces acting on the ball as it falls towards the floor.

Analysis

5 A golf ball of mass 175 g is hit horizontally from the top of a cliff 75.0 m high. The golf ball lands 100 m from the base of the cliff. Calculate the horizontal speed at which the golf ball left the cliff top.

2.6 Projectiles launched obliquely

The previous section looked at projectiles that were launched horizontally. Another common situation is when projectiles are launched obliquely (i.e. at an angle) by being thrown forwards and upwards at the same time. An example of an oblique launch is shooting for a goal in basketball (Figure 2.6.1). Once the ball is released, the only forces acting on it are gravity (pulling it down to the Earth) and air resistance (which slightly retards the ball's motion).

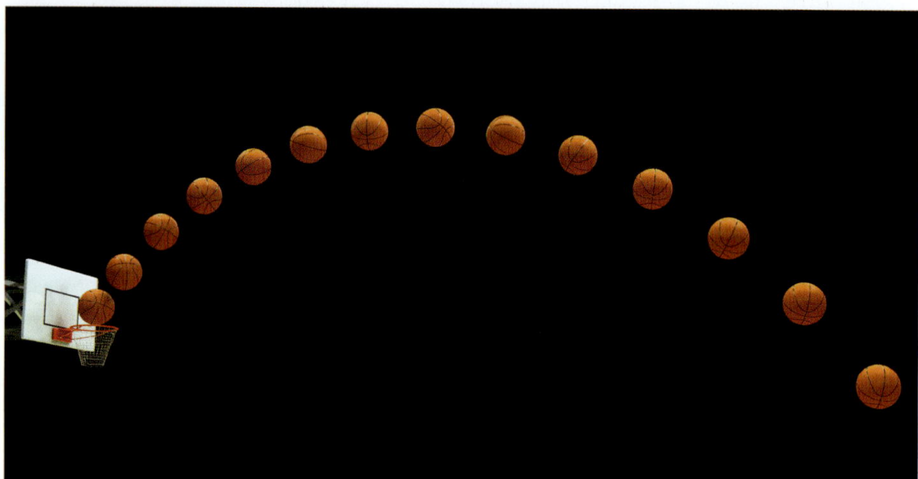

FIGURE 2.6.1 A basketball thrown up towards the ring travels in a smooth parabolic path.

PROJECTILES LAUNCHED AT AN ANGLE

If drag forces are ignored, the only force acting on a projectile that is launched at an angle to the horizontal is gravity, F_g. This is the same as with projectiles launched horizontally.

Gravity acts vertically downwards and so it has no effect on the projectile's horizontal motion. The vertical and horizontal components of the motion are independent of each other and once again must be treated separately.

In the vertical direction, a projectile accelerates due to the force of gravity, that is, at a rate of $9.8\,\text{m}\,\text{s}^{-2}$ downwards. Thus the vertical component of the projectile's velocity decreases as the projectile rises. It is momentarily zero at the top of the flight and then it increases again as the projectile descends.

In the horizontal direction, the projectile has a uniform velocity since there are no forces acting in this direction (if air resistance is ignored).

Tips for problems involving projectile motion

General rules for solving problems involving projectile motion were given in the previous section—see page 95 for a reminder.

If a projectile is launched at an angle to the horizontal, trigonometry can be used to find the components of the initial horizontal and vertical velocity. Pythagoras's theorem can then be used to determine the actual velocity of the projectile at any point as well as its direction with respect to the horizontal plane. Worked example 2.6.1 demonstrates how this is done.

Worked example 2.6.1

PROJECTILE LAUNCHED AT AN ANGLE

A 65 kg athlete in a long-jump event leaps with a velocity of 7.50 m s^{-1} at an angle of 30.0° to the horizontal.

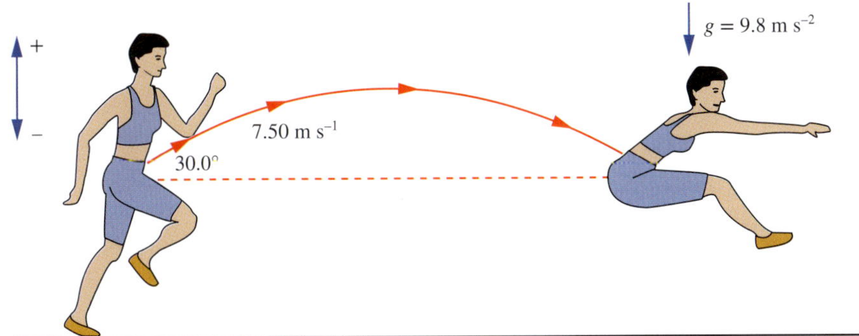

In answering the following questions, treat the athlete as a point mass, ignore air resistance and assume that $g = 9.8$ m s^{-2}.

a What is the athlete's velocity at the highest point in the jump?	
Thinking	**Working**
First find the horizontal and vertical components of the initial speed.	7.50 m s^{-1}, 30.0°, u_V, u_H Using trigonometry: $u_H = 7.50 \cos 30.0°$ $= 6.50$ m s^{-1} $u_V = 7.50 \sin 30.0°$ $= 3.75$ m s^{-1}
Projectiles that are launched obliquely move only horizontally at their highest point. The vertical component of the velocity at this point is zero. Thus the actual velocity is given by the horizontal component of the velocity.	At maximum height, $v = 6.50$ m s^{-1} horizontally to the right.

b What is the maximum height gained by the athlete's centre of mass during the jump?	
Thinking	**Working**
To find the maximum height you must work with the vertical component of the velocity. Recall that the vertical component of velocity at the highest point is zero.	Vertically, taking up as positive: $u = 3.75$ m s^{-1} $a = -9.8$ m s^{-2} $v = 0$ $s = ?$
Substitute these values into an appropriate equation for uniform acceleration.	$v^2 = u^2 + 2as$ $0 = 3.75^2 + 2 \times -9.8 \times s$
Rearrange the equation and solve for s.	$s = \frac{3.75^2}{19.6}$ $= 0.72$ m

PHYSICSFILE

Motorcycle jumping

A motorcycle jumping is an example of projectile motion at an angle. The distance the motorcycle can jump is a function of its approach velocity to the ramp, the ramp angle, and the mass of the motorcycle and rider. Other factors also play a significant role, such as frictional forces, air resistance, wind speed, etc.

Robert 'Robbie' Maddison is an Australian motorbike stunt performer. He has broken several world records for the distance jumped and has successfully jumped distances over 105 m and made a perfect landing. Some of his feats include jumping from one side to the other of the Tower Bridge in London, with a backflip, while the drawbridge was open.

A motorcycle jumping on a dirt track is an example of a projectile launched at an angle

c Assuming a return to the original height, what is the total time the athlete is in the air?	
Thinking	**Working**
As the motion is symmetrical, the time required to complete it will be double the time taken to reach the maximum height. First, the time it takes to reach the maximum height must be found.	Vertically, taking up as positive: $u = 3.75\,\text{m s}^{-1}$ $a = -9.8\,\text{m s}^{-2}$ $v = 0$ $t = ?$
Substitute the relevant values into an appropriate equation for uniform acceleration.	$v = u + at$ $0 = 3.75 - 9.8t$
Rearrange the formula and solve for t, the time needed to reach the maximum height.	$t = \frac{3.75}{9.8}$ $= 0.38\,\text{s}$
The time to complete the jump is double the time it takes to reach the maximum height.	Total time $= 2 \times 0.383$ $= 0.77\,\text{s}$

Worked example: Try yourself 2.6.1

PROJECTILE LAUNCHED AT AN ANGLE

A 50 kg athlete in a long-jump event leaps with a velocity of $6.50\,\text{m s}^{-1}$ at 20.0° to the horizontal.

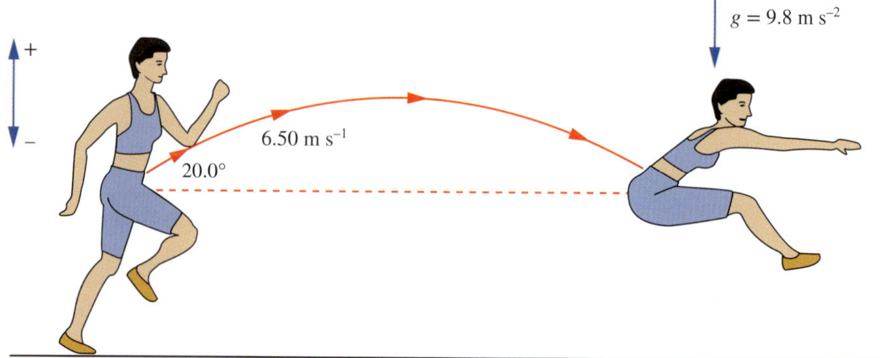

In answering the following questions, treat the athlete as a point mass, ignore air resistance and use $g = 9.8\,\text{m s}^{-2}$.

a What is the athlete's velocity at the highest point in the jump?

b What is the maximum height gained by the athlete's centre of mass during the jump?

c Assuming a return to the original height, what is the total time the athlete is in the air?

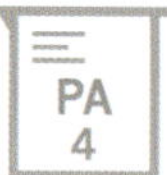

CASE STUDY ANALYSIS

The physics of shot putting

In shot-put competitions there is an advantage in being tall. It means that the release height of the shot will be higher than that of a competitor who is not as tall. It also means the distance travelled by the shot will be greater.

At the 2020 Tokyo Olympic Games the men's event was won by Ryan Crouser of the United State of America, with a distance of 23.30 m. Crouser is 201 cm tall. The gold medal for women was won by Gong Lijiao of China (175 cm tall), with a distance of 20.58 m.

When a projectile is launched at an angle to the horizontal, the theoretical launch angle that gives the maximum range is 45°. This applies only where a projectile lands at the same height as it was launched. A projectile could land at a point lower than its launch height. For example, shot putters launch their shot from above the ground. The theoretical launch angle for maximum range in this case is approximately 43°, depending on the actual release height. In reality, however, shot putters never release at this angle. This is because the speed at which they can launch the shot decreases as the angle gets further from the horizontal. Figure 2.6.2 shows the relationship between launch speed and launch angle.

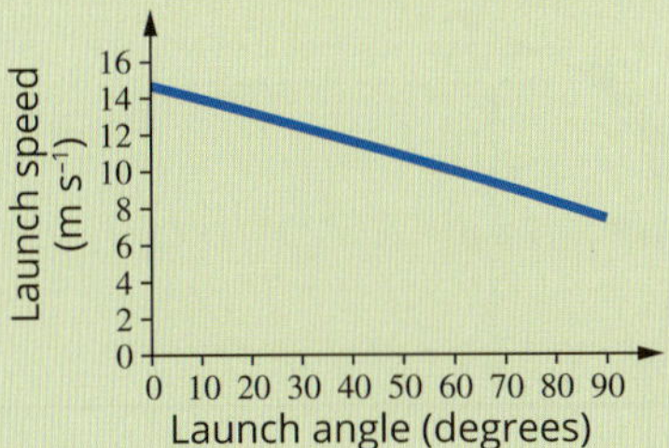

FIGURE 2.6.2 A graph showing how launch speed is greatest with a horizontal launch and decreases as the launch angle increases

The decrease in launch speed with an increase in launch angle is caused by two factors:

- When throwing with a high launch angle, the shot putter must expend a greater effort during the delivery phase to overcome the force of gravity. This reduces the launch speed.
- The structure of the shoulder and arm favours the production of putting force in the horizontal direction more than in the vertical direction.

The optimum launch angle for an athlete is obtained by combining the speed–angle relation for the athlete with the equation for the range of a projectile in free flight. For these reasons, the optimum launch angle for shot putters is actually around 34°.

Analysis

In a shot-put event a 3.0 kg shot is launched with an initial velocity of 7.5 $m\ s^{-1}$ from a height of 1.6 m and at an angle of 30° to the horizontal.

1. What is the initial horizontal speed of the shot?
2. What is the initial vertical speed of the shot?
3. How long does it take the shot to reach its maximum height?
4. What is the maximum height from the ground reached by the shot?
5. What is the speed of the shot when it reaches its maximum height?
6. What distance does the shot put travel?

2.6 Review

SUMMARY

- Projectiles move in parabolic paths that can be analysed by considering the horizontal and vertical components of their motion.
- If air resistance is ignored, the only force acting on a projectile is the force due to gravity, F_g. This results in the projectile having a vertical acceleration of $9.8\,m\,s^{-2}$ downwards during its flight.
- The equations for uniform acceleration can be used to determine the vertical component. These equations are:

$$v = u + at$$
$$s = ut + \frac{1}{2}at^2$$
$$v^2 = u^2 + 2as$$

- If air resistance is ignored, the horizontal velocity of a projectile remains constant throughout its flight. The appropriate equation is:

$$v_{av} = \frac{s}{t}$$

- For objects initially launched at an angle to the horizontal, it is useful to calculate the initial horizontal and vertical velocities using trigonometry.
- At its highest point, the projectile is moving horizontally. Its velocity at this point is given by the horizontal component of its launch velocity. The vertical component of the velocity is zero at this point.

KEY QUESTIONS

Knowledge and understanding

In answering the following questions, assume that the acceleration due to gravity is $9.8\,m\,s^{-2}$ and ignore the effects of air resistance unless otherwise stated.

1 A javelin thrower launches her javelin at 35° above the horizontal. Describe how the horizontal velocity of the javelin changes during its flight.

2 James and Ollie are using a garden hose to water some raised veggie beds that are approximately waist height. The hose is quite short, and the water does not reach the farthest plants. James thinks that if they hold the hose horizontally, all of the velocity of the water will be in the horizontal direction, so the water will reach the furthest. Ollie thinks the water will reach furthest if they hold the hose at 45°. Who is correct, and why?

3 A rugby player kicks for a goal by taking a place kick with the ball at rest on the ground. The ball is kicked at 40° to the horizontal at $25\,m\,s^{-1}$. At its highest point, what is the speed of the ball?

4 A basketballer shoots for a goal by launching the ball at $25\,m\,s^{-1}$ at 30° to the horizontal.

- **a** Calculate the initial horizontal speed of the ball.
- **b** What is the initial vertical speed of the ball?
- **c** What is the magnitude and direction of the acceleration of the ball when it is at its maximum height?
- **d** What is the speed of the ball when it is at its maximum height?

Analysis

5 In a shot-put event a 3.5 kg shot is launched from a height of 1.8 m at an angle of 25° to the horizontal. Its initial velocity is $12\,m\,s^{-1}$.

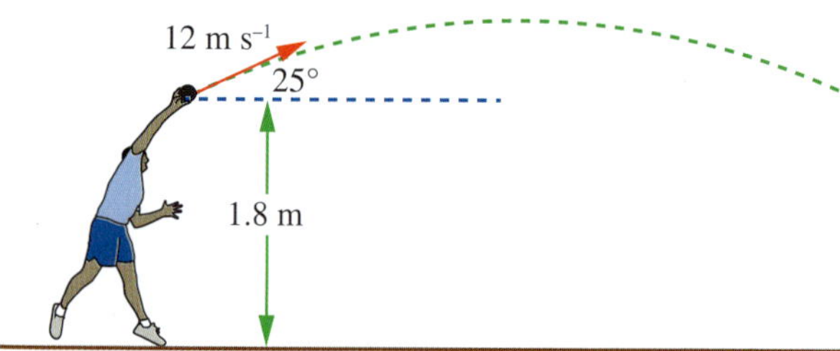

- **a** What is the initial horizontal speed of the shot?
- **b** What is the initial vertical speed of the shot?
- **c** How long does it take the shot to reach its maximum height?
- **d** What is the maximum height from the ground reached by the shot?
- **e** What is the speed of the shot when it reaches its maximum height?
- **f** What distance does the shot put travel?

6 A tennis player is using a tennis ball machine to practise her forehand. She sets the machine to launch tennis balls with an initial velocity of 22.0 $m\,s^{-1}$ at an angle of 10.0° above the horizontal. The balls are launched from approximately the same height as her racquet. Give your answers to 3 significant figures.

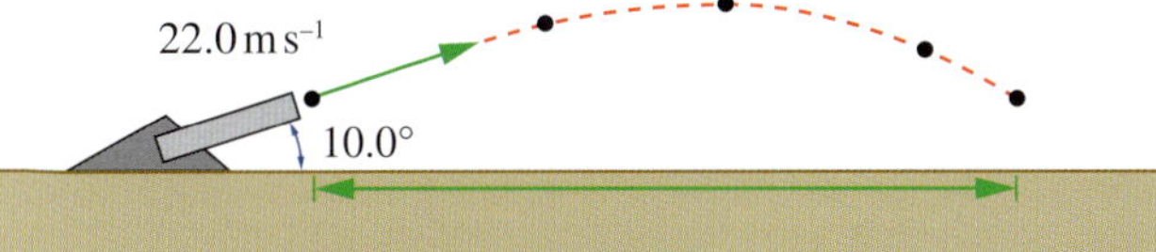

- **a** Calculate the horizontal component of the velocity of the ball:
  - **i** initially
  - **ii** after 0.25 s
  - **iii** after 0.50 s
- **b** Calculate the vertical component of the velocity of the ball:
  - **i** initially
  - **ii** after 0.25 s
  - **iii** after 0.50 s
- **c** What is the speed of the tennis ball after 0.50 s?
- **d** What is the speed of the ball as it hits her racquet?
- **e** What horizontal distance does the ball travel before it hits her racquet—that is, what is its range?
- **f** Describe what effect air resistance has on the ball.

Chapter review

OA ✓✓

02

KEY TERMS

air resistance
banked track
centripetal acceleration
centripetal force
design speed
frequency
inclined plane
magnitude
normal force
period
projectile
tangential

REVIEW QUESTIONS

Knowledge and understanding

1 A bowling ball is rolling down a smooth, straight ramp. Describe the speed and acceleration of the ball.

2 A cyclist is riding at high speed around a steeply banked section of a velodrome. Choose the option below that best describes the magnitude of the normal force acting on the cyclist.

   A zero

   B greater than the force of gravity on the cyclist

   C equal to the force of gravity on the cyclist

   D less than the force of gravity on the cyclist

3 A bowling ball is rolling down a smooth track that is inclined at 45° to the horizontal.

   a What is the magnitude of the acceleration of the ball?

   b How does the magnitude of the normal force acting on the ball compare to the force due to gravity?

4 Two students are discussing a demonstration in which a bucket half filled with water was swung at high speed in a vertical circle. No water spilled out of the bucket as it passed the overhead position. Dave states that at the top of the circle, a centrifugal force acts outwards against the force due to gravity and so holds the water in the bucket. Emma states that inertia keeps the water in the bucket and the normal force from the bucket keeps the water travelling in a circular path. Who is correct, and why?

5 Two blocks are connected by a string that passes over a smooth pulley. One of the blocks is placed on a frictionless table and the other is free to move up or down. The block on the table has a mass of 5.0 kg and the connected block has a mass of 10.0 kg.

   a At what rate do the blocks accelerate?

   b What is the magnitude of the tension in the string?

6 A skier of mass 90 kg is skiing down an icy slope that is inclined at 15° to the horizontal. In answering the following questions, assume that friction is negligible and that the acceleration due to gravity is $9.8\,m\,s^{-2}$.

   a Determine the components of the force on the skier due to gravity perpendicular to the slope and parallel to the slope.

   b Determine the normal force that acts on the skier.

   c Calculate the acceleration of the skier down the slope.

7 A locomotive of mass 7500 kg is pulling two carriages. Carriage A has a mass of 10 000 kg and carriage B has a mass of 5000 kg. The locomotive accelerates at $2.0\,m\,s^{-2}$. The drag force on carriage A is 2000 N, on carriage B it is 1000 N and on the locomotive it is 1500 N.

   a Calculate the driving force of the locomotive engine.

   b Calculate the magnitude of the tension in the coupling between the two carriages.

8 A 1000 kg car tows a 200 kg trailer along a level surface at an acceleration of $2.5\,m\,s^{-2}$. The frictional drag on the car is 800 N and the frictional drag on the trailer is 700 N. Calculate the driving force provided by the car engine to give this acceleration.

9 A marble is rolled from rest down a smooth slide that is 3.5 m long. The slide is inclined at an angle of 40° to the horizontal.

   a Calculate the acceleration of the marble.

   b What is the speed of the marble as it reaches the end of the slide?

10 Marshall has a mass of 57 kg and is riding his 3.0 kg skateboard down a 5.0 m long ramp that is inclined at an angle of 65° to the horizontal. Ignore friction when answering questions **a** to **d**.

   a Calculate the magnitude of the normal force acting on Marshall and his skateboard.

   b What is the acceleration of Marshall as he travels down the ramp?

   c What is the net force acting on Marshall and his board when no friction acts?

   d If Marshall's initial speed is zero at the top of the ramp, calculate his final speed as he reaches the bottom of the ramp.

   e Marshall now stands halfway up the incline while holding his board in his hands. Friction now acts on him. Calculate the frictional force acting on Marshall while he is standing stationary on the ramp.

11 During a high-school physics experiment, a copper ball of mass 25.0 g is attached to a very light piece of steel wire 0.920 m long and whirled in a circle at 30.0° to the horizontal, as shown in diagram (a). The ball moves in a circular path of radius 0.800 m with a period of 1.36 s. The top view of the ball's motion is shown in diagram (b).

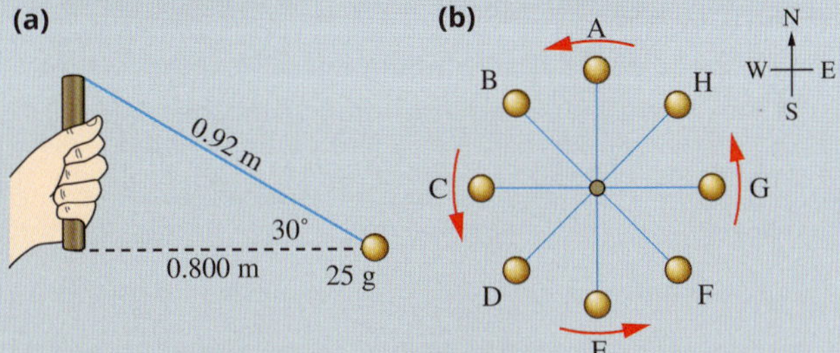

a Calculate the orbital speed of the ball.

b What is the centripetal acceleration of the ball?

c What is the magnitude of the centripetal force acting on the ball?

d Draw a diagram similar to diagram (a) showing all the forces acting on the ball.

e What is the magnitude of the tension in the wire?

12 A toy car is travelling in a circular path of radius 15 m at a constant speed of 7.5 m s^{-1}.

a What is the acceleration of the toy car?

b What force is keeping the toy car moving in its circular path?

13 A cycling track has a turn that is banked at 40° to the horizontal. The radius of the track at this point is 30 m. Determine the speed at which a cyclist of mass 60 kg would experience no sideways force as they ride along this section of track.

14 The Ferris wheel at an amusement park has an arm of 10 m radius and its compartments move with a constant speed of 5.0 m s^{-1}.

a Calculate the normal force that a 50 kg boy would experience from the seat when at the:

i top of the ride

ii bottom of the ride.

b After getting off the ride, the boy remarks to a friend that he felt lighter than usual at the top of the ride. Which option explains why he might feel lighter at the top of the ride?

A He lost weight during the ride.

B The strength of the gravitational field was weaker at the top of the ride.

C The normal force there was larger than the gravitational force.

D The normal force there was smaller than the gravitational force.

15 A toy car is moving at 3.75 m s^{-1} as it rolls off a horizontal table. The car takes 1.5 s to reach the floor.

a How far does the car land from the table?

b What is the magnitude and direction of acceleration when the car is halfway to the floor?

16 A bowling ball of mass 8.75 kg travelling at 15.0 m s^{-1} rolls off a horizontal table that is 1.27 m high.

a What is the horizontal speed of the ball as it strikes the floor?

b What is the vertical speed of the ball as it strikes the floor?

c Calculate the speed of the ball as it reaches the floor.

Application and analysis

17 In a tennis match, a tennis ball is hit from a height of 1.70 m with an initial velocity of 18.5 m s^{-1} at an angle of 46.0° to the horizontal. Ignore drag forces in answering the following questions.

a What is the initial horizontal speed of the ball?

b What is the initial vertical speed of the ball?

c What is the maximum height that the ball reaches above the court surface?

18 An experiment examined the relationship between the distance an object travels and the time it takes to travel that distance as it falls from a height of 20 m. The following data show the distance the object has fallen and the time when the object was at that distance. For example, 1.212 s after the object was dropped it had fallen 7.2 m.

Distance (m)	Time (s)
0.0	0.000
1.0	0.452
2.5	0.714
3.0	0.782
5.0	1.010
7.2	1.212
9.1	1.363
11.5	1.532
13.2	1.641

a What is the rate at which the object drops?

b Plot the time to fall (*s*) as a function of distance the object has fallen (*m*).

c Predict the time taken to fall 15 m.

19 An experiment is conducted to examine the relationship between the horizontal distance an object travels and the vertical distance it falls when launched at an angle of 25° from the horizontal, and from a vertical height of 15 m. An electronic device recorded the horizontal distance travelled and the vertical height after each 0.1 second.

Horizontal distance travelled (m)	Time (s)	Height (m)
0.10	0.1	15.0
0.20	0.2	15.2
0.30	0.3	15.6
0.40	0.4	16.2
0.50	0.5	17.0
0.60	0.6	18.0
0.70	0.7	19.2
0.80	0.8	18.0
0.90	0.9	17.0
1.00	1.0	16.2
1.10	1.1	15.6
1.20	1.2	15.2
1.30	1.3	15.0
1.40	1.4	14.3
1.45	1.5	13.6
1.50	1.6	12.8
1.55	1.7	12.2
1.60	1.8	11.5
1.65	1.9	10.9
1.70	2.0	10.1
1.75	2.1	9.5
1.80	2.2	8.7
1.85	2.3	8.2
1.90	2.4	7.6
1.95	2.5	6.5
2.00	2.6	5.8
2.05	2.7	5.1
2.10	2.8	4.3
2.15	2.9	3.7
2.20	3.0	2.9
2.25	3.1	2.0

a Plot the horizontal distance travelled by the object as a function of time.

b Using the data, calculate the horizontal velocity of the object.

c Plot the horizontal velocity as a function of time. What observations can you make from the plot?

d Using the data, calculate the horizontal acceleration of the object.

e Plot the horizontal acceleration as a function of time. What observations can you make from the plot?

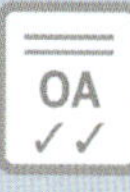

CHAPTER

03 The relationship between force, energy and mass

In 1675 Isaac Newton said, 'If I have seen further it is by standing on the shoulders of giants'. What he meant by this is that he had relied on the work of brilliant minds that preceded him, in particular Galileo Galilei (1564–1642). Newton refined Galileo's work on motion and gravity when he published *Philosophiae Naturalis Principia Mathematica* in 1687, a work in which he outlined the connection between the force and motion of bodies with mass. In this chapter we will investigate the connection between force and energy, and their effect on mass.

Key knowledge

- investigate and apply theoretically and practically the laws of energy and momentum conservation in isolated systems in one dimension **3.1, 3.6**
- investigate and analyse theoretically and practically impulse in an isolated system for collisions between objects moving in a straight line: $F\Delta t = m\Delta v$ **3.2**
- investigate and apply theoretically and practically the concept of work done by a force using:
  - work done = force × displacement **3.3**
  - work done = area under force vs distance graph (one dimensional only) **3.3**
- analyse transformations of energy between kinetic energy, elastic potential energy, gravitational potential energy and energy dissipated to the environment (considered as a combination of heat, sound and deformation of material):
  - kinetic energy at low speeds: $E_k = \frac{1}{2}mv^2$; elastic and inelastic collisions with reference to conservation of kinetic energy **3.5**
  - elastic potential energy: area under force-distance graph including ideal springs obeying Hooke's Law: $E_s = \frac{1}{2}kx^2$ **3.4**
  - gravitational potential energy: $E_g = mg\Delta h$ or from area under a force-distance graph and area under a field-distance graph multiplied by mass **3.5**

OA ✓✓

3.1 Conservation of momentum

FIGURE 3.1.1 Newton's cradle is a popular illustration of almost perfect conservation of momentum. As the raised sphere collides with the other spheres, the sphere's momentum is passed on until the final sphere continues with almost the same momentum as the original.

Where there are moving objects, there are bound to be collisions. These can range from the interaction of sub-atomic particles to events on a galactic scale. Newton's cradle (Figure 3.1.1) provides another example of a collision. It also provides a demonstration of the law of conservation of momentum, a powerful tool with which to analyse collisions.

THE LAW OF CONSERVATION OF MOMENTUM

The product of the mass of an object and its velocity is called its **momentum**, and is given by the following equation.

> $p = mv$
> where p is momentum (kg m s^{-1})
> m is the mass of the object (kg)
> v is the velocity of the object (m s^{-1})

Given that velocity is a vector quantity and momentum is calculated from velocity, it follows that momentum is also a vector quantity.

The **law of conservation of momentum** states that in any collision or interaction between two or more objects in an isolated system, the total momentum of the system will be **conserved** (that is, it will remain constant). The total initial momentum is equal to the total final momentum:

sum of the initial momentum ($\Sigma p_{\text{initial}}$) = sum of the final momentum (Σp_{final})

where Σ is the mathematical symbol representing the addition of each factor. Hence $\Sigma p_{\text{initial}}$ is the sum of the initial momentum of each object in the system and Σp_{final} is the sum of the final momentum of each object in the system.

> $\Sigma p_{\text{initial}} = \Sigma p_{\text{final}}$

Another way of putting this is that the total change in momentum of the system is zero. This is often found by adding up the change in momentum of all the parts of the system:

$$\Sigma \Delta p = 0$$

In physics, a collision refers to a situation where two objects interact and exert action–reaction forces on each other. They do not have to make physical contact. For instance, two identically charged particles could approach and repel one another, moving off in opposite directions without ever making physical contact. This is still considered a collision and the law of conservation of momentum still applies.

Note that the law refers to objects in an **isolated system**. For a system to be isolated, there are only internal forces acting between the objects, with no interaction with any objects outside the system. In reality, perfectly isolated systems cannot exist on the Earth because of friction and gravity. There are, however, many situations where treating a system as isolated is a useful approximation.

In the rear-end collision between a car and a bus examined in Worked example 3.1.1, friction is relatively small compared to the forces exerted by the vehicles on one another. Therefore the vehicles can be treated as an isolated system.

PHYSICSFILE

Discovering the neutrino

Conservation of momentum helped scientists discover the neutrino. In the 1920s, it was observed that in beta decay, a nucleus emitted a beta particle. However, when the nucleus recoiled, it was not in the exact opposite direction to the emitted electron. Thus the momentum of these particles did not appear to comply with the law of conservation of momentum. In 1930, Wolfgang Pauli proposed that another particle must have also been emitted in order to conserve the total momentum of the system. This particle, the neutrino, was not detected experimentally until 1956. As you read this, billions of neutrinos originating from the Sun are passing through your body and the Earth.

Worked example 3.1.1

CONSERVATION OF MOMENTUM

In a head-on collision on a demolition derby track, a car of mass 1000 kg travelling east at 20.0 m s^{-1} crashes into a mini-bus of mass 5000 kg travelling west at 8.00 m s^{-1}. Assume that the car and mini-bus lock together on impact and ignore the effect of friction.

a Calculate the final common velocity of the vehicles.	
Thinking	**Working**
First assign a direction that will be considered positive. Note: as long as directions are assigned positive or negative consistently in the same problem, it does not matter which direction is assigned positive.	In this case we will consider vectors directed eastwards to be positive. $m_c = 1000\,\text{kg}$ $u_c = 20.0\,\text{m s}^{-1}$ $m_b = 5000\,\text{kg}$ $u_b = -8.00\,\text{m s}^{-1}$
Apply the law of conservation of momentum.	$\Sigma p_{\text{initial}} = \Sigma p_{\text{final}}$ $m_c u_c + m_b u_b = (m_c + m_b)v$ $(1000)(20.0) + (5000)(-8.00) = (1000 + 5000)v$ $(-20000) = (6000)v$ $v = \dfrac{(-20000)}{(6000)}$ $= -3.33333$ $= 3.33\,\text{m s}^{-1}$ west

b Calculate the change in momentum of the car.	
Thinking	**Working**
The change in momentum of the car is its final momentum minus its initial momentum.	$\Delta p_c = p_{\text{final}} - p_{\text{initial}}$ $= m_c(v - u)$ $= 1000(-3.33333 - 20.0)$ $= -23333.3$ $= 2.33 \times 10^4\,\text{kg m s}^{-1}$ west

c Calculate the change in momentum of the bus.	
Thinking	**Working**
The change in momentum of the bus is its final momentum minus its initial momentum.	$\Delta p_b = p_{\text{final}} - p_{\text{initial}}$ $= m_b(v - u)$ $= 5000(-3.33333 - (-8.00))$ $= 23333.3$ $= 2.33 \times 10^4\,\text{kg m s}^{-1}$ east

d Verify that the momentum of the system is constant.	
Thinking	**Working**
The total change in the momentum of a system is the vector sum of the change of momentum of its parts. This should be zero, from the conservation of momentum.	$\Delta p_c + \Delta p_b = (-2.33 \times 10^4) + (2.33 \times 10^4) = 0$ Therefore the momentum of the system is constant (i.e. conserved) as expected.

Worked example: Try yourself 3.1.1

CONSERVATION OF MOMENTUM

In a safety-rating test of head-on collisions, a car of mass 1200 kg travelling east at 22.0 m s^{-1} crashes into a bus of mass 7000 kg travelling west at 15.0 m s^{-1}. Assume that the car and bus lock together on impact. You can ignore the effect of friction.

a Calculate the final common velocity of the vehicles.

b Calculate the change in momentum of the car.

c Calculate the change in momentum of the bus.

d Verify that the momentum of the system is constant.

CONSERVATION OF MOMENTUM FROM NEWTON'S LAWS

The principle of conservation of momentum follows directly from Newton's second and third laws. This can be shown in the following way.

Consider a bowling ball of mass m_b moving with an initial velocity of u_b. It collides with a stationary pin of mass m_p. The velocities of both the ball and pin change. The pin's final velocity is represented by v_p in Figure 3.1.2.

When the ball and pin collide, they exert action–reaction forces on each other and, according to Newton's third law:

$$F_{bp} = -F_{pb}$$

The forces cause the ball to decelerate and the pin to accelerate. Thus from Newton's second law ($F = ma$):

$$m_b a_b = -m_p a_p$$

The ball and pin are in contact for time Δt. Thus we can rewrite acceleration in terms of velocity:

$$m_b \frac{(v_b - u_b)}{\Delta t} = -m_p \frac{(v_p - u_p)}{\Delta t}$$

The times are the same and so they cancel out, leaving:

$$m_b(v_b - u_b) = -m_p(v_p - u_p)$$

Expanding and rearranging gives:

$$m_b u_b + m_p u_p = m_b v_b + m_p v_p$$

The left-hand side of this equation describes the initial momentum of the system and the right-hand side represents the final momentum of the system. Thus an application of Newton's second and third laws has produced the equation for the conservation of momentum.

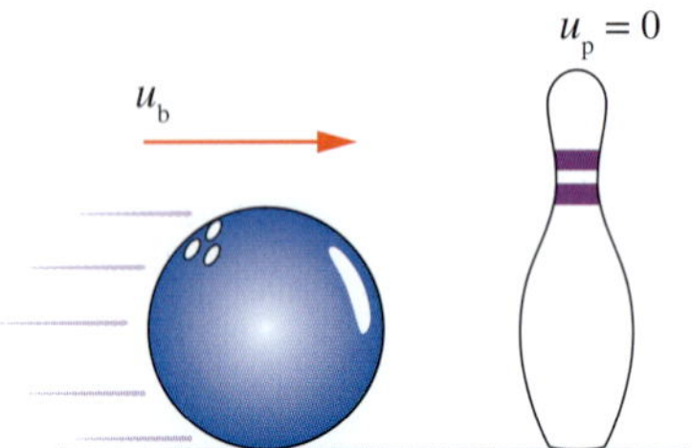

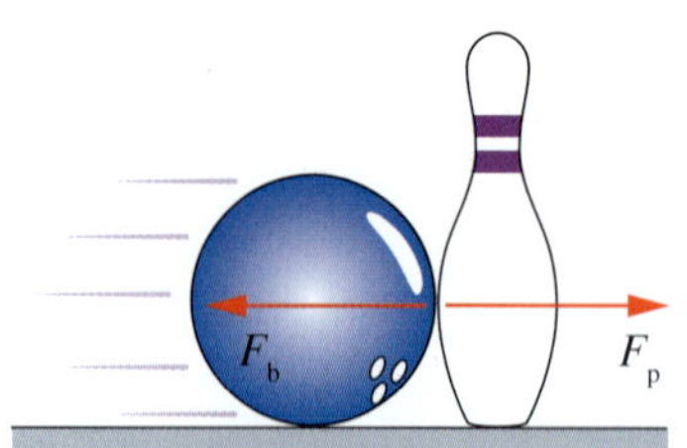

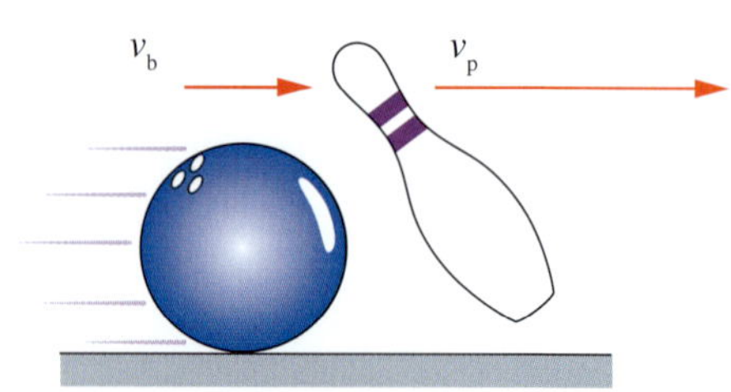

FIGURE 3.1.2 When a bowling ball collides with a pin, they exert equal but opposite forces on each other. These forces cause the ball to lose some momentum and the pin to gain an equal amount of momentum.

Worked example 3.1.2

REBOUNDING

In a football game, player A of mass 80.0 kg travelling towards the goal at 9.00 m s^{-1} collides with an opposition player (player B) of mass 72.0 kg travelling away from the goal at 6.50 m s^{-1}. After the collision, player A is travelling towards the goal at 0.50 m s^{-1}. Assume that the two players rebound off each other on impact and ignore the effects of friction.

a Calculate the sum of the momentum of the two players before the collision.	
Thinking	**Working**
First assign a direction that will be considered positive.	In this case we will consider vectors directed towards the goal to be positive. $m_A = 80.0\,\text{kg}$ $u_A = 9.00\,\text{m s}^{-1}$ $v_A = 0.50\,\text{m s}^{-1}$ $m_B = 72.0\,\text{kg}$ $u_B = -6.50\,\text{m s}^{-1}$
Use the equation of momentum for each player and substitute the values.	$\Sigma p_{\text{initial}} = p_A + p_B$ $= m_A u_A + m_B u_B$ $= (80.0)(9.00) + (72.0)(-6.50)$ $= (720) + (-468)$ $= 252\,\text{kg m s}^{-1}$ towards the goal

b Calculate the final velocity of the opposition player (player B).	
Thinking	**Working**
The sum of the momentum after the collision is equal to the sum of the momentum before the collision.	$\Sigma p_{\text{initial}} = \Sigma p_{\text{final}}$ $\Sigma p_{\text{initial}} = m_A v_A + m_B v_B$ $252 = (80.0)(0.50) + (72.0)v_B$ $v_B = \dfrac{252 - 40.0}{72.0}$ $= 2.944\,44$ $= 2.94\,\text{m s}^{-1}$ towards the goal

c Calculate the change in momentum of player A.	
Thinking	**Working**
The change in momentum of player A is their final momentum minus their initial momentum.	$\Delta p_A = p_{\text{final}} - p_{\text{initial}}$ $= m_A(v_A - u_A)$ $= (80.0)(0.50 - 9.00)$ $= -680\,\text{kg m s}^{-1}$ $= 680\,\text{kg m s}^{-1}$ away from the goal

d Calculate the change in momentum of player B.	
Thinking	**Working**
The change in momentum of player B is their final momentum minus their initial momentum.	$\Delta p_B = p_{final} = p_{initial}$ $= m_B(v_B - u_B)$ $= (72.0)(2.944\,44 - (-6.50))$ $= 680\,kg\,m\,s^{-1}$ towards the goal

Worked example: Try yourself 3.1.2

REBOUNDING

In a child's toy, a blue marble rolls along a track and collides with a red marble rolling the other way. The blue marble has a mass of 0.003 20 kg and is travelling south at 0.800 m s^{-1} as it hits the red marble. The red marble has a mass of 0.001 50 kg and is travelling north at 1.00 m s^{-1} when it hits the blue marble. After the collision the blue marble begins travelling towards the south at 0.450 m s^{-1}. Assume that the two marbles bounce off each other on impact and ignore the effect of friction.

a Calculate the sum of the momentum of the two marbles before they hit.

b Calculate the final velocity of the red marble.

c Calculate the change in momentum of the blue marble.

d Calculate the change in momentum of the red marble.

Worked example 3.1.3

EXPLOSIVE MOMENTUM

Two friends are standing on their stationary skateboards facing each other with their hands gripped together. They place their feet together and push with their legs as they release their hands. After releasing their grip, skater A, of mass 50.0 kg, travels towards the east at 3.50 m s^{-1}. Skater B, of mass 34.0 kg, travels in the opposite direction. You can ignore the effect of friction.

a Calculate the sum of the momentum of the two skaters before they release their grip.	
Thinking	**Working**
Assign a direction that will be considered positive.	In this case we will consider vectors directed towards the east to be positive. $m_A = 50.0\,kg$ $u_A = 0$ $v_A = 3.50\,m\,s^{-1}$ $m_B = 34.0\,kg$ $u_B = 0\,m\,s^{-1}$
Use the equation of momentum for the combined mass of the skaters.	$\Sigma p_{initial} = p_A + p_B$ $= (m_A + m_B)u$ $= (50.0 + 34.0)(0)$ $= (84.0)(0)$ $= 0\,kg\,m\,s^{-1}$

b Calculate the final velocity of skater B.	
Thinking	**Working**
The sum of the momentum after the skaters release their grip is equal to the sum of the momentum before the release.	$\Sigma p_{initial} = \Sigma p_{final}$ $\Sigma p_{initial} = m_A v_A + m_B v_B$ $(0) = (50.0)(3.50) + (34.0)v_B$ $(34.0)v_B = (-175)$ $v_B = -5.14706$ $= 5.15\,m\,s^{-1}$ west

c Calculate the change in momentum of skater A.	
Thinking	**Working**
The change in momentum of skater A is their final momentum minus their initial momentum.	$\Delta p_A = p_{final} - p_{initial}$ $= m_A(v_A - u_A)$ $= (50.0)(3.50 - 0)$ $= 175$ $= 175\,kg\,m\,s^{-1}$ east

d Calculate the change in momentum of skater B.	
Thinking	**Working**
The change in momentum of skater B is their final momentum minus their initial momentum.	$\Delta p_B = p_{final} - p_{initial}$ $= m_B(v_B - u_B)$ $= (34.0)(-5.14706 - 0)$ $= -175$ $= 175\,kg\,m\,s^{-1}$ west

Worked example: Try yourself 3.1.3

EXPLOSIVE MOMENTUM

Two ice dancers are standing still in the centre of an ice rink facing each other with their palms together. They then begin their routine by pushing with their hands. After pushing away, ice dancer A, of mass 62.0 kg, travels towards the north at 2.20 m s^{-1}. Ice dancer B, of mass 98.0 kg, travels towards the south. You can ignore the effect of friction.

a Calculate the sum of the momentum of the two ice dancers before they push away.

b Calculate the final velocity of the dancer B.

c Calculate the change in momentum of ice dancer A.

d Calculate the change in momentum of ice dancer B.

PHYSICSFILE

Not so strongman

Traditionally, circus strongmen would often perform a feat where they place a large rock on their chest and then invite another person to smash the rock with a sledgehammer. This might seem at first to be an act of extreme strength and daring. However, a quick analysis using the principle of conservation of momentum will show otherwise. Assume that the rock has a mass of 27 kg and that a sledgehammer of mass 3.0 kg strikes it at 5.0 m s^{-1}. From the law of conservation of momentum, we can show that the rock and sledgehammer will move together at just 0.50 m s^{-1} after impact:

$$m_1u_1 + m_2u_2 = (m_1 + m_2)v$$
$$(3.0 \times 5.0) + (27 \times 0) = (3.0 + 27)v$$
$$15 = 30v$$
$$v = 0.50\,m\,s^{-1}$$

This is a very low speed. The large mass of the rock means that the final common speed is too low to hurt the strongman. A more dangerous feat would be to use the sledgehammer to smash a pebble.

3.1 Review

OA ✓✓

SUMMARY

- The momentum of an object is the product of its mass and its velocity: $p = mv$. Momentum is measured in $kg\,m\,s^{-1}$.
- The total momentum of an isolated system is conserved; that is, the sum of the momentum of the parts of a system before a collision is equal to the sum of their momentum after the collision:

$$\Sigma p_{initial} = \Sigma p_{final}.$$

- In a simple collision between two objects of mass m_1 and m_2, this equation becomes:

$$m_1u_1 + m_2u_2 = m_1v_1 + m_2v_2$$

- A collision between two objects can be described as:
  - combined, where the objects lock together on impact
  - rebound, where the objects bounce off each other and remain separate
  - explosive, where the objects were combined before moving away from each other.

KEY QUESTIONS

Knowledge and understanding

1 Using the concepts investigated in this section, explain why you are more likely to end up in the water when attempting to step onto a dock from a stand-up paddle board than when attempting to step onto a dock from a river ferry.

2 Two toy cars have Velcro attached such that they will stick together on contact. A child makes the cars collide and both cars come to rest. Explain how momentum is conserved in this situation.

Analysis

3 In an experiment, a student hangs a green sphere (25.0 kg) from a long rope and an orange sphere (50.0 kg) from another long rope. Each sphere has a magnet attached. The spheres are pushed towards each other. Just before they collide, the green sphere is moving east at $3.50\,m\,s^{-1}$ and the orange sphere is moving west at $6.00\,m\,s^{-1}$. They collide in mid-air and remain locked together due to the magnets. Calculate the final velocity of the combined spheres.

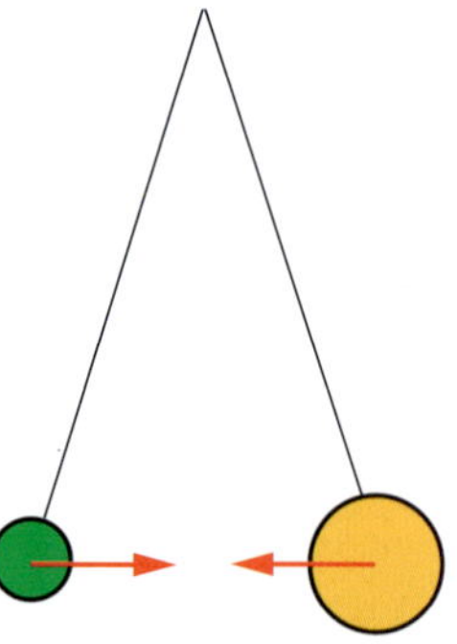

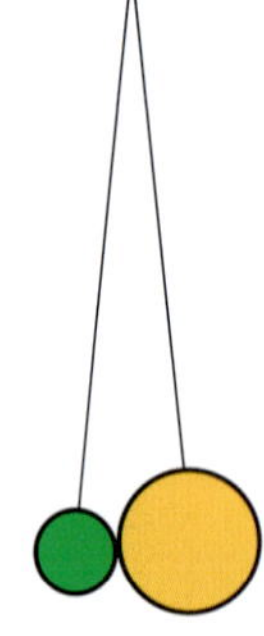

4 While shunting empty railway carriages, an 11.0 t passenger carriage travelling at $7.50\,m\,s^{-1}$ north collides with a diner carriage of 16.0 t that is also travelling north but at $3.50\,m\,s^{-1}$. The two carriages lock together after the collision. Ignoring friction, find their combined velocity.

5 A sports car of mass 1000 kg travelling east at $36.0\,km\,h^{-1}$ approaches a station wagon of mass 2000 kg moving west at $18.0\,km\,h^{-1}$.

a **i** Calculate the momentum of the sports car.

ii Calculate the momentum of the station wagon.

iii Determine the sum of the momentum of these vehicles.

b The two vehicles now collide head-on on an icy stretch of road where there is negligible friction. The vehicles remain locked together after the collision.

i Calculate the common velocity of the two vehicles after the collision.

ii Where has the initial momentum of the vehicles gone?

iii Determine the change in momentum of the sports car.

iv Determine the change in momentum of the station wagon.

6 A 155 g pink snooker ball travelling with initial velocity $5.00\,m\,s^{-1}$ to the right collides with a stationary green ball of mass 132 g. The two balls rebound off each other. If the final velocity of the pink ball is $3.00\,m\,s^{-1}$ to the left, calculate the velocity of the green ball after the collision.

7 Two students at a fun fair run at each other with large fitness balls held in front of them. One student has a yellow ball and is running at 4.20 m s^{-1} east with a total mass of 71.0 kg. The other student has an orange ball and is running at 5.30 m s^{-1} west with a total mass of 65.0 kg. After they collide and bounce off each other, the final velocity of the student with the orange ball is 1.40 m s^{-1} east. Calculate the final velocity of the student with the yellow ball.

8 In a training exercise a group of astronauts investigate what would happen in the event of a failed coupling of a cargo ship with the International Space Station (ISS). In a large hangar a model of the ISS, with a mass of 4.20×10^5 kg, is suspended from the ceiling and is stationary. A 3.20×10^4 kg model of a supply ship is nearby and also suspended. The model of the supply ship is pushed towards the model of the ISS. If it is moving at 5.00 m s^{-1} south as it strikes the model of the ISS and rebounds at 5.00 m s^{-1} north, with what velocity does the model of the ISS move after the collision?

9 A stationary 1000 kg cannon mounted on wheels fires a 10.0 kg shell east with a horizontal speed of 505 m s^{-1}. Assuming that friction is negligible, calculate the recoil velocity of the cannon.

10 An astronaut is floating in deep space while holding a toolbox. The total mass of the astronaut, including their suit, is 235 kg and the mass of the toolbox is 46.0 kg. The combined astronaut and toolbox are drifting away from the spaceship at 0.750 m s^{-1}. With no other way to get back to the spaceship the astronaut decides to sacrifice the toolbox and throw it as fast as possible in a direction away from the spaceship. With what speed should the astronaut throw the toolbox if they hope to move towards the spaceship at 0.300 m s^{-1} after the throw?

3.2 Impulse

In a car crash, it is not only how fast the car travels that determines the damage, but how quickly it stops. This is a direct consequence of Newton's second law of motion. From $F = ma$, if a is small you can conclude that the force required to bring the car to a stop will also be relatively small. On the other hand, if the car is brought to a halt very rapidly (Figure 3.2.1), there will be a large deceleration requiring a large force. The force determines the damage. Ignoring the likelihood of injury caused by a large force that acts for a short time, such as in a car crash, a small force acting for a longer time has the same effect: suddenly applying the brakes and gradually applying the brakes both bring the car to rest. One way to quantify the similarity between these situations is to describe the impulse in a collision, which considers both the force and the time over which the force acts.

FIGURE 3.2.1 Rapid deceleration requires a large force and often results in damage and injury.

CHANGE IN MOMENTUM

Newton's original formulation of his second law was not expressed in terms of acceleration. Rather, he spoke of the 'motion' of a body that would be altered when a force acted on that body over a time interval. This is very close to saying that the momentum of the body changes when a resultant force acts on it. This is equivalent to the more familiar $F = ma$ formulation of Newton's second law, as will now be shown.

Consider a body of mass m with a resultant force F acting on it for time Δt. The mass will accelerate as described by Newton's second law:

$$F = ma$$

$\therefore F = \frac{m\Delta v}{\Delta t}$ after substituting the definition of acceleration.

By rearranging this equation we can write:

$$F\Delta t = m\Delta v = \Delta p$$

The term $m\Delta v$ is the change in the momentum of the body. It is also called the **impulse**. The force involved in a collision can change in value during the collision, so the average force is used. The average force acting on the body for a time Δt causes a change in the momentum.

The term $F\Delta t$ is called the impulse of the resultant force and is equal to the change in momentum of the object. That is:

impulse = $F\Delta t = \Delta p$

where F is the average force acting on the object (N)

Δt is the time over which the force acts (s)

Δp is the change in momentum of the object (kg m s^{-1}).

Impulse is measured in newton seconds (N s)

It is important to note that impulse is a vector quantity. Its direction is the same as that of the average force or of the change in momentum (or velocity).

Momentum units

Since impulse can be expressed in terms of a momentum change, the units for momentum (kg m s^{-1}) and impulse (N s) must be equivalent. This can be shown using Newton's second law.

Given that 1 N = 1 kg m s^{-2} (from $F = ma$), it follows that 1 N s = 1 kg m s^{-2} × s.

That is, 1 N s = 1 kg m s^{-1}.

Even though the units are equivalent, they should be used with the appropriate quantities as a reminder of the quantity that is being considered: momentum or impulse. The newton second (N s) is the product of a force and a time interval and so should be used with impulse. The kilogram metre per second (kg m s^{-1}) is the product of a mass and a velocity and so should be used with momentum. Even so, it is not uncommon to see newton seconds used as the unit of momentum.

Worked example 3.2.1

CALCULATING THE IMPULSE

Calculate the impulse of a tree on a 1480 kg sports car if the vehicle is travelling at 93.0 km h^{-1} north when the driver loses control of the vehicle on an icy road. The car comes to rest against the tree.

Thinking	Working
Convert the speed to m s^{-1}.	$93.0\ \text{km h}^{-1} = \frac{93.0}{3.6}\ \text{m s}^{-1}$ $= 25.8333\ \text{m s}^{-1}$
Calculate the change in momentum. The negative sign indicates that the change in momentum, and therefore the impulse, is in a direction opposite to the initial momentum (with north as positive).	$\Delta p = m(v - u)$ $= (1480)(0 - 25.8333)$ $= -3.82333 \times 10^4$ $= 3.82 \times 10^4$ N s south
The impulse is equal to the change in momentum.	impulse = 3.82×10^4 N s south

Worked example: Try yourself 3.2.1

CALCULATING THE IMPULSE

Calculate the impulse of the braking system on the 1480 kg sports car if the vehicle was travelling at 95.5 km h^{-1} north-east before coming to an abrupt halt.

FORCE VERSUS TIME GRAPHS

In many situations the force applied to an object is not constant. For example, when a tennis player hits a ball, the initial force exerted by the racquet is relatively small. As the strings stretch and the ball deforms, this force builds up to a maximum before decreasing again as the ball rebounds from the racquet. Where the force changes over time, the relationship can be represented graphically (Figure 3.2.2).

The impulse of the ball, or the change in momentum, can be found from the product $F\Delta t$. This is simply the area under the force vs time graph.

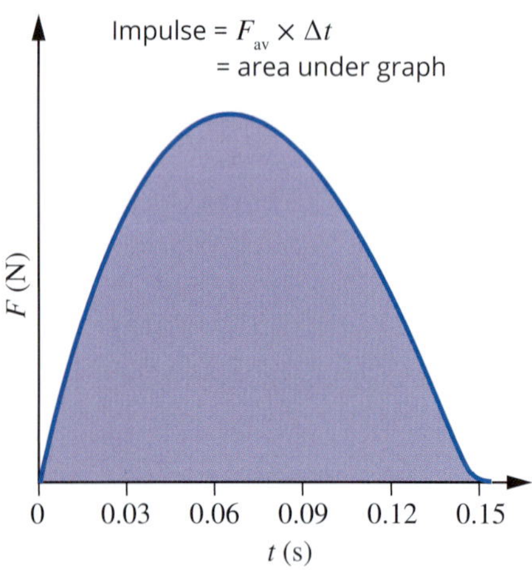

FIGURE 3.2.2 The force changes with time as the racquet strikes the ball.

Worked example 3.2.2

IMPULSE OF RUNNING SHOES

A running-shoe company plots the following force vs time graph for a running shoe. Use the data to calculate the magnitude of the impulse.

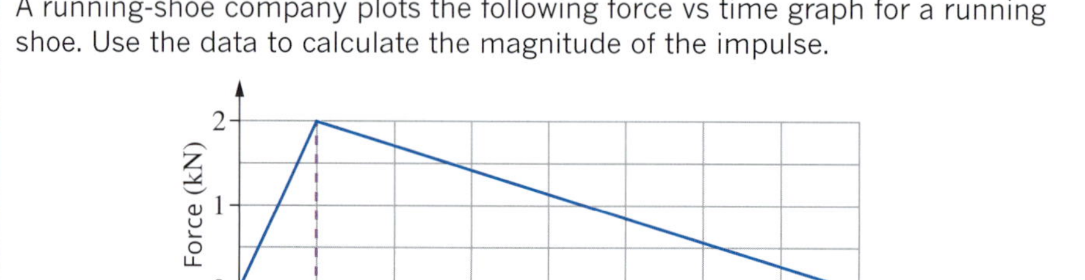

Thinking	Working
Recall that impulse = $F\Delta t$. This is the area under the force vs time graph.	$\text{impulse} = \frac{1}{2} \times \text{base} \times \text{height}$ $= \frac{1}{2} \times 160 \times 10^{-3} \times 2 \times 10^{3}$ $= 160\,\text{N s}$

Worked example: Try yourself 3.2.2

IMPULSE OF RUNNING SHOES

A running-shoe company plots the following force vs time graph for an alternative design intended to reduce the peak force on the heel. Calculate the magnitude of the impulse.

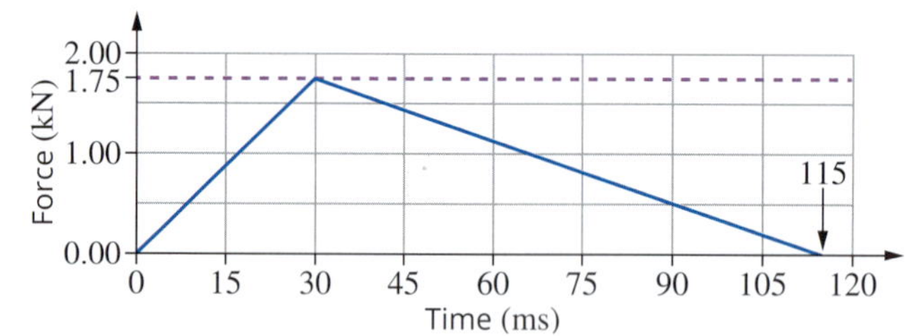

APPLICATIONS OF IMPULSE

The relationship between impulse, force and collision duration is useful in analysing collisions. When a vehicle collides with another object and comes to rest, the vehicle and occupants undergo a rapid deceleration. The impulse depends on the initial speed of the vehicle and on its mass.

Since impulse = $\Delta p = F\Delta t$, a large force is exerted to bring the vehicle to rest in a very short time. Extending the time taken for a vehicle to stop reduces the force exerted. Examples of increased stopping times in different activities are shown in Figure 3.2.3.

FIGURE 3.2.3 (a) The landing mat extends the time over which the athlete comes to rest, reducing the size of the stopping force. If the high jumper missed the mat and landed on the ground, the force would be larger, but their momentum change would be the same. (b) Thick padding around the goal post extends the time over which a player who collides with it comes to rest, thereby reducing the size of the stopping force. (c) Wicketkeepers follow the ball's final trajectory with their gloves when keeping. This extends the ball's stopping time, reduces the stopping force and softens the blow on the gloves.

Worked example 3.2.3

BRAKING FORCE

A 2520 kg truck is travelling at 30.0 m s^{-1} before the brakes are applied. Calculate the magnitude of the average force exerted by the brakes to bring the vehicle to rest in 12.0 s.

Thinking	Working
Calculate the change in momentum. The negative sign indicates that the change in momentum, and therefore the braking force, is in the direction opposite to the initial momentum.	$\Delta p = m(v - u)$ $= 2520(0 - 30.0)$ $= -75600$ $= -7.56 \times 10^4\ \text{kg m s}^{-1}$
Transpose $\Delta p = F\Delta t$ to find the force. The sign of the momentum can be ignored, since you are only finding the magnitude of the average force.	$F = \dfrac{\Delta p}{\Delta t}$ $= \dfrac{75600}{12.0}$ $= 6.30 \times 10^3\ \text{N}$

Worked example: Try yourself 3.2.3

BRAKING FORCE

The same 2520 kg truck travelling at 30.0 m s^{-1} needs to stop in 1.50 s because a vehicle in front has suddenly stopped. Calculate the magnitude of the average braking force required to stop the truck in that time.

Safety features in cars—such as crumple zones and airbags (Figure 3.2.4)—are designed to extend Δt. This reduces the force on the occupants of the vehicle, potentially saving lives and preventing injuries.

FIGURE 3.2.4 Airbags reduce the force on passengers by extending the time it takes for them to stop in the event of a collision.

Figure 3.2.5(b) shows a force vs time graph for a collision where an airbag is inflated compared with one where there is no airbag. The change in momentum, or impulse, of the passenger is the same in each case. Thus the area under each curve should be equal. Note, however, that both the peak force and the average force are significantly higher where there is no airbag. The broader peak for the airbag indicates that the passenger is losing their momentum over a longer time and thus experiencing a lower force.

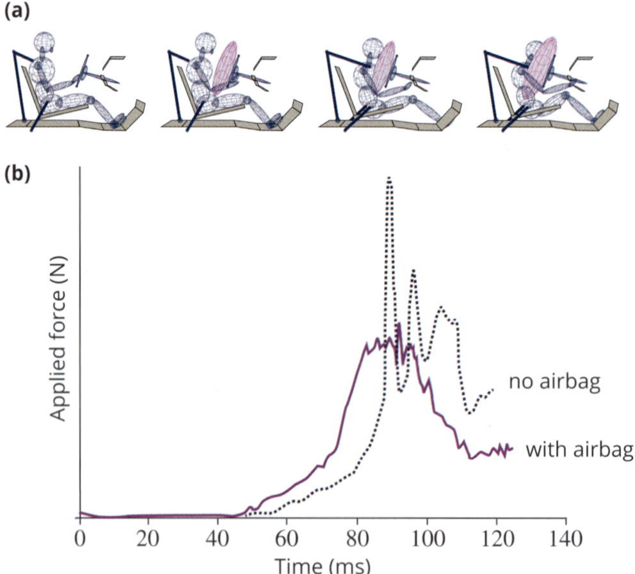

FIGURE 3.2.5 (a) Diagram of an airbag being inflated in a collision. (b) Graph illustrating the difference in the force on a passenger over time when an airbag is inflated in a collision (solid line), and when no airbag is present (dotted line).

CASE STUDY ANALYSIS

Car safety and crumple zones

Worldwide, car accidents are responsible for millions of deaths each year. Many times this number of people are injured. One way of reducing the road toll is to design safer vehicles. Modern cars employ a variety of safety features that help to improve the occupants' chances of surviving an accident. Some of these safety features are the antilock braking system (ABS), electronic stability control (ESC), inertia reel seatbelts, variable-ratio-response steering systems, collapsible steering columns, head rests, shatterproof windscreen glass, padded dashboards, front and side airbags, front and rear crumple zones and a rigid passenger compartment.

Some cars today are equipped with collision avoidance systems. These have radar, laser or infrared sensors that advise the driver of a hazardous situation. They may even take control of the car when an accident appears likely.

The purpose of such safety features as inertia reel seatbelts, collapsible steering columns, padded dashboards, airbags and crumple zones is not to reduce the size of the impulse, but to reduce the size of the forces that act to bring the car to a stop. Automotive engineers strive to achieve this by extending the time over which the driver loses momentum.

Crumple zones

A popular misconception is that cars would be much safer if they were sturdier and more rigid. Drivers often complain that cars seem to collapse too easily during collisions, and that it would be better if cars were structurally stronger—more like an army tank. In fact, cars are specifically designed to crumple to some extent (Figure 3.2.6). This makes them safer and actually reduces the seriousness of injuries suffered in car accidents.

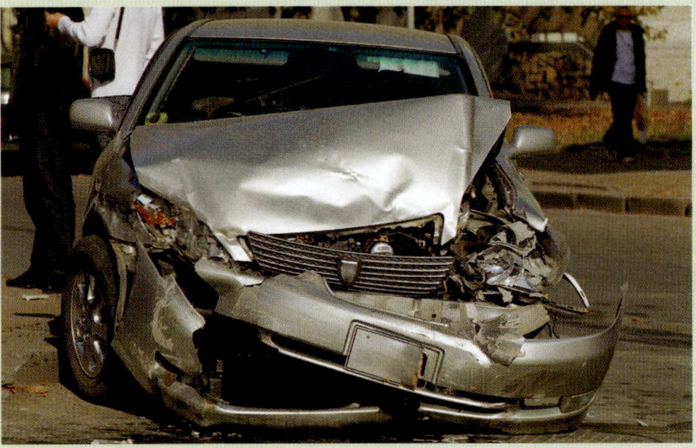

FIGURE 3.2.6 Cars are designed with weak points in their chassis that enable them to crumple in the event of a collision. This extends the time over which they come to rest and so reduces the size of the forces acting on the occupants.

Army tanks are designed to be extremely sturdy and rigid vehicles. They are able to withstand the effect of collisions without suffering serious structural damage. If a tank travelling at 60.0 km h^{-1} crashed into a solid obstacle, the tank would be relatively undamaged. However, its occupants would very likely be killed. This is because the tank has no give in its structure and so the tank and its occupants would stop in an extremely short time interval. The occupants would lose all their momentum in an instant, which means that the forces acting on them would necessarily be very large. These large forces would cause the occupants of the tank to sustain very serious injuries, even if they were wearing seatbelts.

Cars today have strong and rigid passenger compartments; however, they are also designed with non-rigid sections—such as bonnets and boots—that crumple if the car is struck from the front or rear (Figure 3.2.7). The chassis contains parts that have grooves or beads cast into them. In a collision, these grooves or beads act as weak points, causing the chassis to crumple in a concertina shape.

FIGURE 3.2.7 The Australian New Car Assessment Program (ANCAP) assesses the crashworthiness of new cars. This car has just crashed at 50 km h^{-1} into a 5 t concrete block. The crumpling effect can be seen.

By crumpling the front or rear of the car, the time interval over which the car and its occupants come to a stop is extended. This stopping time is typically longer than 0.1 s in a 50 km h^{-1} crash. Because the occupants' momentum is lost more gradually, the peak forces that act on them are smaller and so the chances of injury are reduced.

continued over page

CASE STUDY ANALYSIS *continued*

Analysis

1 Consider the driver of a car that crashes into a tree while driving north at 60.0 km h^{-1}. If the driver has a mass of 90.0 kg, calculate the momentum of the driver just before the collision.

2 If the driver comes to a complete stop as a result of the collision, calculate their change in momentum (i.e. the impulse).

3 How would the impulse change if the collision occurred over a longer period of time?

4 Compare the impulse experienced by a 90.0 kg driver of a car to the impulse experienced by a 90.0 kg driver of a tank if both were to crash and come to a stop from 60.0 km h^{-1}.

5 Calculate the force on the 90.0 kg car driver if the impulse experienced by the driver occurred over a period of 985 ms.

6 Calculate the force on the 90.0 kg tank driver if the impulse experienced by the driver occurred over a period of 81.5 ms. Assume that the tank, like the car, is travelling north before the impact.

7 Given that the conditions of a collision are identical except for the period of time over which the collision occurs, how would you describe the relationship between the force experienced and the period of time?

3.2 Review

OA ✓✓

SUMMARY

- When a force is exerted on an object over a time interval, Δt, it brings about a change in momentum, Δp, by changing the velocity of the object:

$$F\Delta t = m\Delta v = \Delta p$$

- Impulse is the change in the momentum of an object.
- The unit of impulse is the newton second (N s) and the unit of change in momentum is kg m s^{-1}. These units are equivalent.
- Impulse can be calculated from the area under a force vs time graph.
- When designing for safety during collisions, measures are taken to increase the time of an interaction in order to reduce the maximum force experienced during that interaction.

KEY QUESTIONS

Knowledge and understanding

1 A 165 g cricket ball flies past the wicket at 155 km h^{-1} and is stopped by the wicket keeper. Calculate the magnitude of the impulse delivered by the ball to the wicket keeper.

2 When a tennis player serves, she hits a 57.0 g tennis ball at the top of its flight when the ball is momentarily stationary. It then leaves the racquet at 144 km h^{-1}. If the ball and racquet are in contact for 0.0600 s, calculate the magnitude of the average force exerted by the racquet on the ball.

3 A basketball of mass 0.625 kg is bounced against the court at a speed of 32.0 m s^{-1}. It rebounds at 24.5 m s^{-1}. Calculate the average force exerted by the court on the ball if the interaction lasts 16.5 ms.

4 Consider a 100 t train travelling at 50.0 km h^{-1}.

a Calculate the momentum of the train.

b Calculate the magnitude of the impulse if the train were to collide with a 5.00 t truck at a level crossing and push the truck for 15.0 m before coming to rest.

Analysis

5 Three balls of identical mass are thrown against a surface at the same speed.
Ball A stops on impact.
Ball B rebounds with 75% of its initial speed.
Ball C rebounds with 50% of its initial speed.
Order the balls in terms of their change in momentum, from least to most.

6 A child wearing a backpack jumps from a tree and lands on her feet. Describe at least three factors that will influence the force on her knees and ankles when she lands. In your response, you may wish to refer to the child's footwear, her landing technique and the surface on which she lands.

7 Two crash-test cars of identical mass are travelling at 22.0 m s^{-1} towards a solid concrete block. Car A is designed with crumple zones built into the front of the chassis and car B is built with a rigid chassis. The passenger compartment of car A comes to rest in 0.0896 s after hitting the concrete block, while the passenger compartment in car B comes to rest in 0.004 00 s. By what factor does the average force on car B compare to car A?

8 The graph below represents the force exerted by an athlete's foot over the 200 ms that his foot is in contact with the ground.

a Calculate the magnitude of the impulse of the athlete on the ground.

b Calculate the magnitude of the average force exerted by his foot over the duration of the contact.

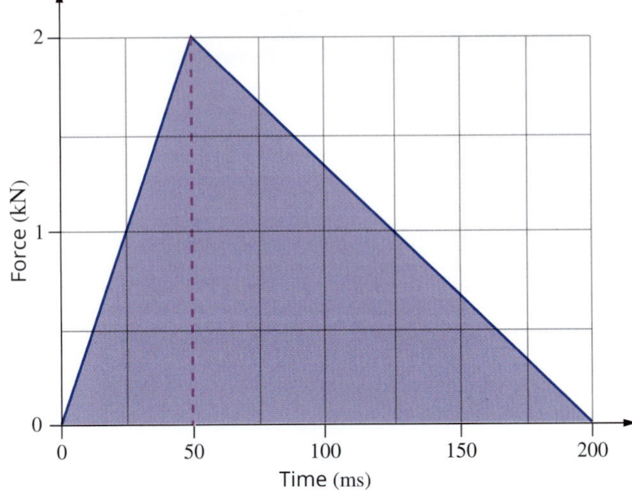

9 A tennis ball of mass 57.5 g is tested for compliance with tennis regulations by being dropped from a height of 251 cm onto concrete. A bounce height of 146 cm is deemed acceptable. Find the magnitude of the average force on a ball that just reaches the acceptable height if it is in contact with the concrete for 0.0550 s.

3.3 Work done

In everyday language, the concept of work is associated with effort and putting energy into something, whether it be your studies, sports or a part time job. Although the word **work** has a much more specific meaning in physics, it is still connected with energy.

When an unbalanced force acts on an object over a time interval, the object accelerates and its momentum changes. When the force causes a displacement in the direction of the force, the energy of the object changes, and we say that work has been done. The weightlifter in Figure 3.3.1 does work by exerting a force and causing the barbell to undergo a displacement. The **gravitational potential energy** of the barbell is increased and the store of chemical energy in the muscles of the weightlifter is decreased.

FIGURE 3.3.1 Suamili Nanai broke the Australian clean-and-jerk record in July 2021 by lifting 201 kg above his head in two movements.

CALCULATING WORK

Work is the transfer of energy from one object to another and/or the transformation of energy from one form to another. A force does work on an object when it acts on that object and causes a displacement in the direction of the force. Where the force is constant, the work done by the force is given by the following equation.

$W = Fs$

where W is the work done by the force (J)

F is the magnitude of the constant force (N)

s is the displacement (m)

If the force is applied at an angle to the displacement, only the component of the force in the direction of the displacement contributes to the work done. If the force and displacement vectors are at an angle θ to each other, then $F\cos\theta$ is the component of the force that does the work.

$W = Fs\cos\theta$

where W is the work done by the force (J)

F is the magnitude of the constant force (N)

s is the displacement (m)

θ is the angle between the force and displacement vectors

While both force and displacement are vectors, work and energy are scalar quantities and are measured in joules (J).

To find the work done on an object, it is the net force that needs to be used. For instance, if a person pushes a heavy couch across a carpeted floor, the work done on the couch depends on the force applied by the person less the frictional force that opposes the motion:

$$W = \Delta E = F_{net}s$$

In this section we will assume that any force that acts to do work on an object is the net force.

Worked example 3.3.1

FORCE APPLIED AT AN ANGLE TO THE DISPLACEMENT

A rope that is 30.0° to the horizontal is used to pull a 10.0 kg crate across a rough floor. The crate is initially at rest and is dragged 4.00 m along the floor. The tension, F_t, in the rope is 50.0 N and the frictional force, F_f, opposing the motion is 20.0 N.

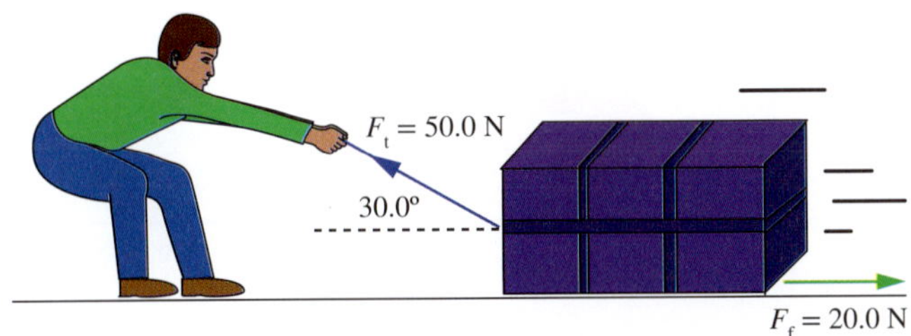

a Determine the work done by the person.	
Thinking	**Working**
Draw a diagram of the forces in action.	F_t F F_f = 20.0 N
Find the component of the tension in the rope that is in the direction of the displacement (shown by the red arrow).	$F = F_t \cos 30.0°$ $= 50.0 \times \cos 30.0°$ $= 43.3013$ N
Find the work done by the person.	$W = Fs$ $= 43.3013 \times 4.00$ $= 1.73205 \times 10^2$ $= 1.73 \times 10^2$ J

PHYSICSFILE

How much work does it take to break a record?

In July 2021, weightlifter Suamili Nanai (Figure 3.3.1) became the male Australian record holder for the clean-and-jerk when he lifted 201 kg from the ground to above his head. While it is difficult to determine how much time and effort had gone into Nanai's preparations for breaking the record, we can calculate the work he did against gravity to lift the record mass. Assuming that he lifted the mass from the ground to 205 cm above the ground, we can calculate the work done as:

$$\begin{aligned} W &= Fs \\ &= mgs \\ &= 201 \times 9.8 \times 2.05 \\ &= 4038.1 \\ &= 4040\text{ J} \end{aligned}$$

b Calculate the work done on the crate.	
Thinking	**Working**
The work done on the crate is the net force acting on it multiplied by the displacement. (This is also the increase in the kinetic energy of the crate.)	$W = Fs$ $= (F - F_f)s$ $= (43.3013 - 20.0) \times 4.00$ $= 93.20$ $= 93.2\text{ J}$

c Calculate the energy transformed into heat and sound due to the frictional force.	
Thinking	**Working**
The energy transformed into heat and sound due to the frictional force is the difference between the work done by the person and the energy gained by the crate.	$E = 173.2 - 93.20$ $= 79.99$ $= 80.0\text{ J}$
This is equal to the work done against friction, which can also be calculated from the frictional force.	$W_f = F_f s$ $= 20.0 \times 4.00$ $= 80.0\text{ J}$

Worked example: Try yourself 3.3.1

FORCE APPLIED AT AN ANGLE TO THE DISPLACEMENT

A boy moves a toy car by pulling on a cord that is attached to the car at 45.0° to the horizontal. The boy applies a force of 15.0 N and pulls the car for 10.0 m along a path against a frictional force of 6.00 N.

a Determine the work done by the boy pulling on the cord.

b Calculate the work done on the toy car.

c Calculate the energy transformed into heat and sound due to the frictional force.

When a force performs no work

It is important to remember that work is only done when a force, or a component of a force, is applied in the direction of displacement. Hence it is possible to exert a force and feel very tired without doing work. This would mean no energy has been transferred. For example, if you hold a heavy object in outstretched arms you will get tired very quickly, but you are not doing any work on the object.

Similarly, an object moving in a circular path in a horizontal plane is constantly accelerated by a centripetal force. Because this force is perpendicular to the displacement at each instant, the force does no work, and no energy is transferred to the object. It does not get faster or slower; it only changes direction (Figure 3.3.2).

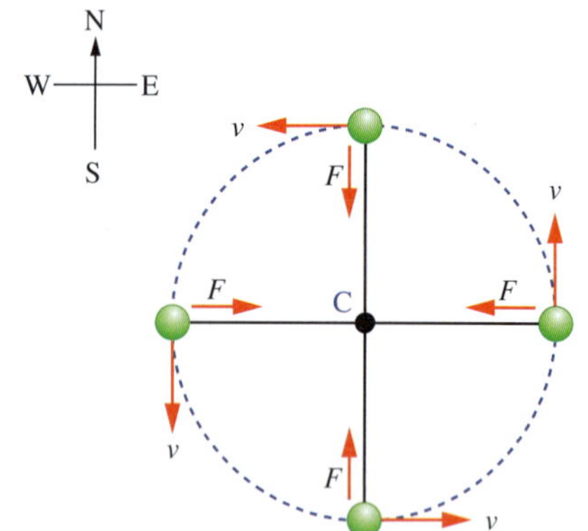

FIGURE 3.3.2 A body moving in a circular path has a force directed towards the centre of the path. The displacement is in the direction of the velocity. There is therefore no force in the direction of the displacement and thus no work is done.

FORCE VS DISTANCE GRAPHS

When the force is constant, the work done is easily calculated. However, in many situations the net force is changing. In these situations, a graph can be used to calculate the work done. Where the force vs distance relationship is represented graphically, the work done is the area under the graph. This principle is very similar to the way in which impulse can be calculated from the area under a force vs time graph. However, it is important not to confuse these two quantities.

When graphed, the relationship between the force and the distance stretched of an **elastic** object, such as a spring, offers a way to calculate the work done in stretching the material: the work done is the area under the force vs distance stretched graph.

If the force vs distance graph, or force vs distance stretched graph, is not linear, the area can be calculated by counting squares. It is important to take careful note of the units in order to calculate the work represented by each square.

Worked example 3.3.2

CALCULATING WORK DONE FROM A GRAPH

The force required to stretch a piece of bungee cord is represented in the graph below. Calculate the work done when a 60 N force is applied to the cord.

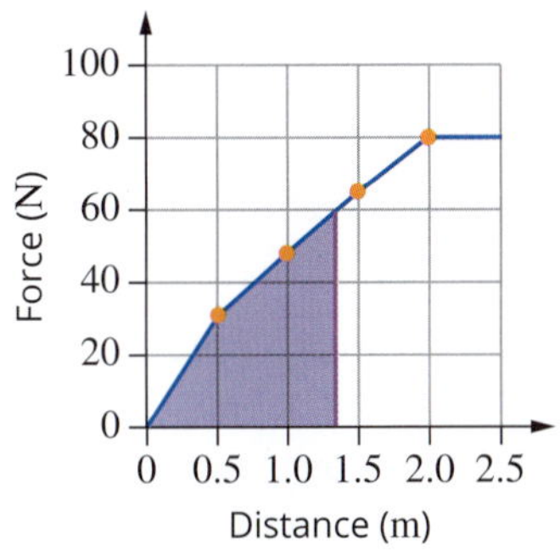

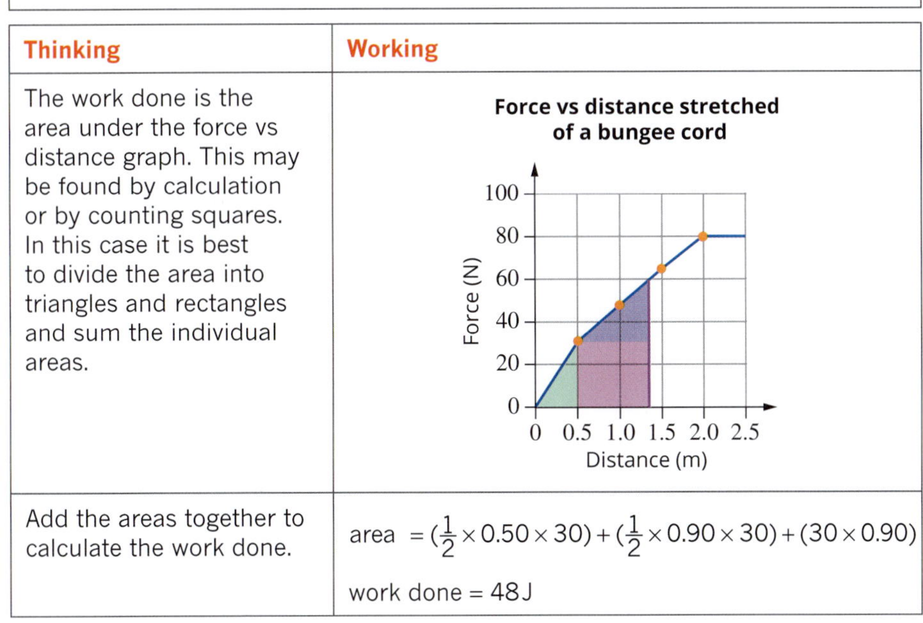

Thinking	Working
The work done is the area under the force vs distance graph. This may be found by calculation or by counting squares. In this case it is best to divide the area into triangles and rectangles and sum the individual areas.	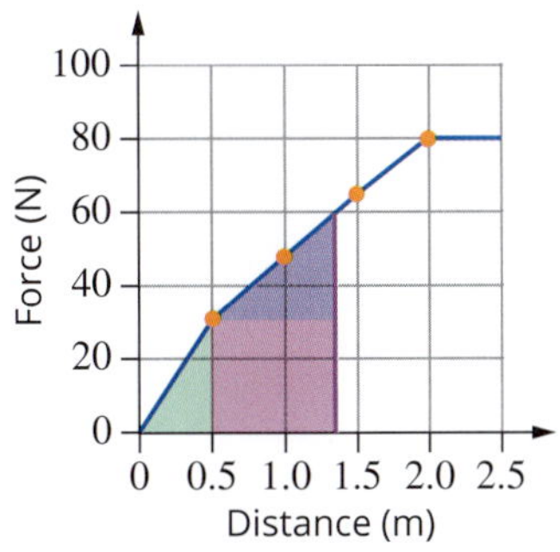
Add the areas together to calculate the work done.	area $= (\frac{1}{2} \times 0.50 \times 30) + (\frac{1}{2} \times 0.90 \times 30) + (30 \times 0.90)$ work done = 48 J

Worked example: Try yourself 3.3.2

CALCULATING WORK DONE FROM A GRAPH

The force required to elongate a piece of rubber tubing is represented in the graph below. Calculate the work done when the tubing is stretched by 2.0 m.

Force vs distance stretched of rubber tubing

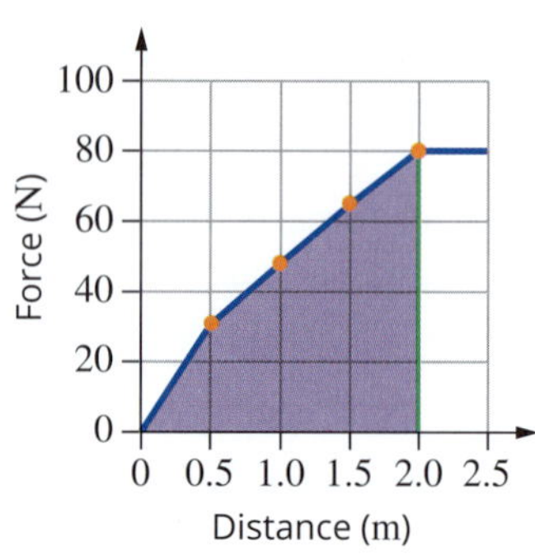

3.3 Review

SUMMARY

- When a force does work on an object, there is a change in the displacement and energy of the object.
- Work, W, is a scalar and is measured in joules (J).
- The work done on an object is the net force on the object multiplied by the displacement moved in the direction of the force: $W = Fs$.
- When the force is applied to an object at an angle to the displacement, work is only done by the component of the force in the direction of the displacement: $W = Fs\cos\theta$.
- A centripetal force does no work on an orbiting object, as the force and displacement are perpendicular.
- The work done by a varying force is the area under the force vs distance graph.

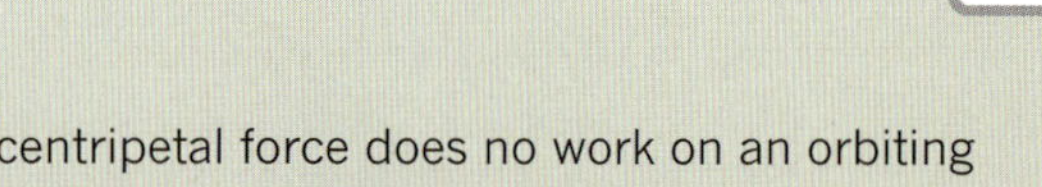

KEY QUESTIONS

Knowledge and understanding

1 Describe a scenario in which a force is applied but no work is done.

2 If we consider the Earth to orbit the Sun in a circular path with a constant gravitational force of attraction, justify the statement that the Sun does no work to move the Earth.

3 A child uses a string to drag a 2.00 kg toy across a floor. The string is held at an angle of 60.0° to the horizontal and the child applies a force of 30.0 N on the toy, which is initially at rest. A constant frictional force of 10.0 N acts on the toy as it is dragged 2.40 m along the floor.

a Calculate the work done by the horizontal component of the 30.0 N force.

b Calculate the work that the child does in overcoming friction.

c Calculate the kinetic energy gained by the toy.

4 The graph below shows the force vs distance graph as a sports shoe is compressed during the stride of an athlete. Calculate the work done in compressing the shoe by 7 mm.

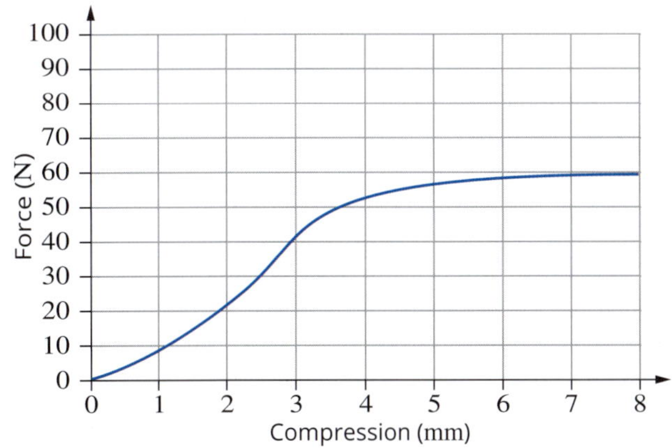

5 A weightlifter raises a 155 kg barbell to a height of 1.20 m at constant speed. Calculate the work done by the weightlifter.

6 Krisha pushes a lawnmower at constant speed across 15.0 m of lawn. She applies a force of 68.0 N at an angle of 60.0° to the horizontal. Calculate the work she does against friction.

Analysis

7 An engineer is testing a new material for its elastic properties. By applying various forces on a sample and measuring the corresponding distance stretched, the following data were obtained.

Force (kN)	Distance stretched (mm)
0.0	0.00
5.0	2.00
10.0	6.00
15.0	8.00
20.0	7.50
25.0	5.50
30.0	3.00
35.0	1.00
40.0	0.00

a Construct a graph of the data with force on the *y*-axis and distance stretched on the *x*-axis. Ensure that you draw a smooth curve of best fit.

b Use the graph to calculate the work done up to the point of maximum distance stretched.

8 An 806 g javelin is released at an angle of 45.0° from a height of 1.90 m and at a speed of 108 km h^{-1}. Calculate the work done by the gravitational force on the javelin from its release to the point where it lands on the ground.

3.4 Elastic potential energy

In everyday life you frequently encounter situations in which work is done to stretch or compress materials. Think of bungee jumping, pole vaulting (Figure 3.4.1), trampolining and tennis, where the elastic properties of materials are harnessed to generate thrills for spectators and participants. Computer keyboards have tiny springs in the keys, and wind-up toys, old-fashioned watches, door-closing mechanisms and car suspensions are some of the other devices that use elastic springs.

Elastic potential energy is the energy stored in a material when it is stretched or compressed. If the material is elastic, this energy can be returned to the system, but in inelastic materials, permanent change occurs.

PHYSICSFILE

Recovery straps and tow ropes

Recovery straps are used to pull bogged cars out of their predicament using the energy stored in the elasticised straps. When the recovery vehicle moves forward, the kinetic energy of the recovery vehicle causes the strap to stretch. The energy is stored in the recovery strap and is then transferred to the bogged vehicle over an extended period of time, which pulls it out of the sand or mud. Using a tow rope, which has no elasticity, would cause a greater force over a shorter period, which could cause damage to the recovery vehicle, the bogged vehicle, or the tow rope itself.

Conversely, you would not use a recovery strap to tow a broken down vehicle, as the energy stored in the strap would cause the vehicle in tow to 'bounce' forwards and backwards, which would make it difficult to control. Using a tow rope with less stretch means that the kinetic energy of the towed vehicle would not fluctuate and so its velocity could be maintained safely.

FIGURE 3.4.1 The elastic potential energy stored in the pole is what allows the pole vault competitor to propel herself over the bar.

HOOKE'S LAW

It is relatively easy to start stretching a spring, but more and more force is required for each incremental amount of extension (distance stretched). This is expressed in Hooke's law.

$F = -kx$

where F is the force exerted by the spring (N)

k is the spring constant (N m^{-1})

x is the displacement (the extension or compression) of the spring (m)

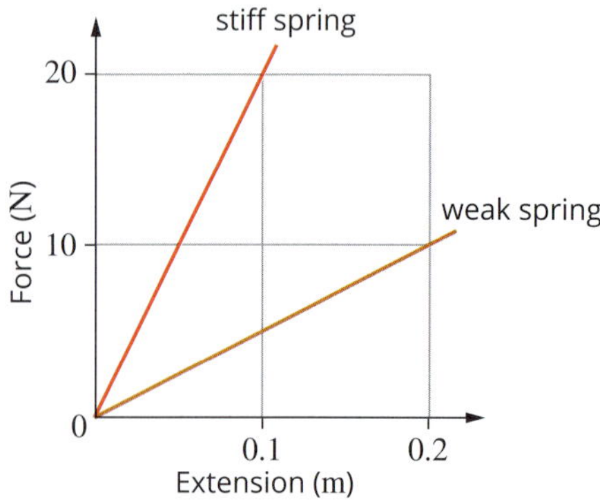

FIGURE 3.4.2 Both springs represented in this graph are ideal (i.e. they obey Hooke's law). The springs obey Hooke's law because they both have linear graphs, but they have different degrees of stiffness. The stiff spring has a spring constant of 200 N m^{-1}. The spring constant of the other more elastic spring is just 50 N m^{-1}. The stiffer spring has the higher gradient (i.e. a steeper line) on the graph.

Hooke's law describes how the force exerted by a spring is directly proportional to, but opposite in direction to, the distance that the spring is extended or compressed. The **spring constant** k is a measure of the stiffness of the spring. The behaviour of a spring under force is often illustrated graphically by plotting the force applied versus the extension achieved (Figure 3.4.2). The spring constant is represented by the gradient of the graph. Notice that a stiffer spring has a greater gradient and thus a larger spring constant.

When considering the work done in deforming a spring, the force applied is in the direction of the displacement and hence the negative sign in $F = -kx$ can be ignored. The applied force is directly proportional to displacement and, as discussed in the previous section, when force is not constant, the work done by the force can be calculated (or estimated) by determining the area under the force vs extension graph.

An expression for the work done in extending or compressing a spring, and the elastic potential energy which is then stored in the spring, can now be derived. Consider the graph in Figure 3.4.3. The elastic potential energy when the spring is extended by 5 metres is represented by the area of the shaded triangle.

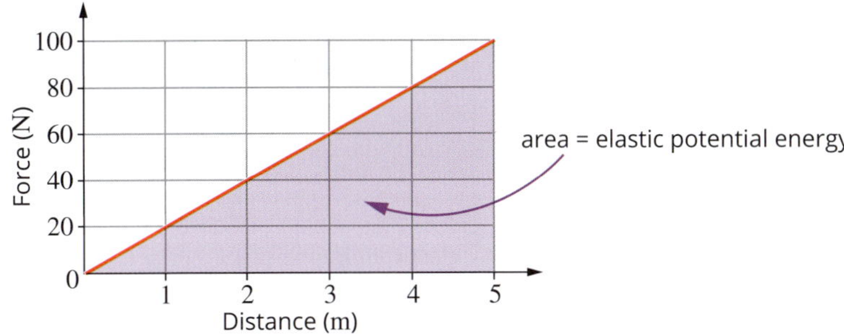

FIGURE 3.4.3 The elastic potential energy is calculated by the area under the force vs distance graph.

The elastic potential energy, E_s, is calculated as follows:

$$E_s = \frac{1}{2} \times \text{ base } \times \text{ height}$$
$$= \frac{1}{2} \times x \times F$$
$$= \frac{1}{2} \times x \times kx$$
$$= \frac{1}{2} kx^2$$

$$E_s = \frac{1}{2} kx^2$$

where k is the spring constant (N m^{-1})

x is the distance the spring is extended, also called extension (m)

Elastic potential energy is the energy stored in any elastic medium—such as a rope, spring or rubber band—due to forces stretching or compressing the bonds between atoms.

We call the directly proportional relationship between force and extension 'elastic behaviour'. Elastic behaviour obeys Hooke's law. Springs that exhibit elastic behaviour will be able to do work with the elastic potential energy when the applied force is removed.

It is possible to exceed the **elastic limit** of a spring or other elastic material. At this point permanent **deformation** occurs, that is, the spring no longer returns to its initial shape. If the force is increased further, the **breaking point** is reached, at which point the material fails or breaks down (Figure 3.4.4).

While the work done in permanently deforming a spring can still be calculated from the area under the force vs distance curve, the energy stored may not all be recoverable, as work has been done to permanently change the material.

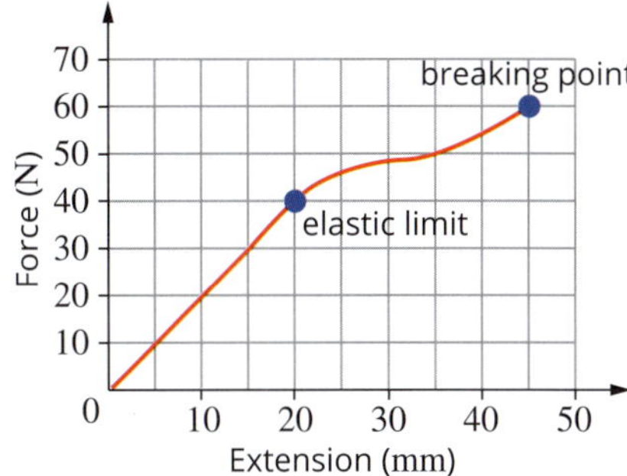

FIGURE 3.4.4 The point at which the force vs distance curve first deviates from linear behaviour is the elastic limit, i.e. the point where permanent damage is done to the spring.

Worked example 3.4.1

CALCULATING THE SPRING CONSTANT, ELASTIC POTENTIAL ENERGY AND WORK

A fine steel wire has the force and extension properties shown in the graph below.

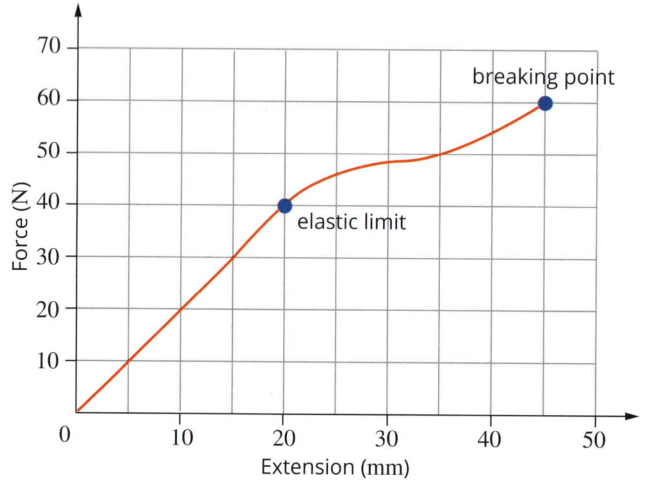

a Calculate the spring constant, *k*, for the wire.	
Thinking	**Working**
The spring constant is the gradient of the first linear section of the graph (in units $N\,m^{-1}$).	$k = \frac{\Delta F}{\Delta x}$ $= \frac{40}{0.020}$ $= 2000\,N\,m^{-1}$

b Calculate the elastic potential energy that the wire can store before permanent deformation begins.	
Thinking	**Working**
The elastic potential energy is the area under the curve up to the elastic limit.	$E_s = \frac{1}{2} \times \text{height} \times \text{base}$ $= \frac{1}{2} \times 40 \times 0.020$ $= 0.40\,J$
This value can also be obtained using the formula for elastic potential energy.	$E_s = \frac{1}{2}kx^2$ $= \frac{1}{2} \times 2000 \times (0.020)^2$ $= 0.40\,J$

c Calculate the work done to break the wire.	
Thinking	**Working**
Add up the number of squares under the curve up to the breaking point.	number of squares = 33 (approx.)
Calculate the energy per square. This is given by the area of a square. Remember to convert mm to m.	energy for one square = 10×0.005 $= 0.050\,\text{J}$
Multiply the energy per square by the number of squares.	work = energy per square × number of squares $= 0.050 \times 33$ $= 1.650$ $= 1.7\,\text{J}$ (approx.)

Worked example: Try yourself 3.4.1

CALCULATING THE SPRING CONSTANT, ELASTIC POTENTIAL ENERGY AND WORK

An alloy sample is tested under tension, giving the force vs extension graph shown below. X indicates the elastic limit and Y indicates the breaking point.

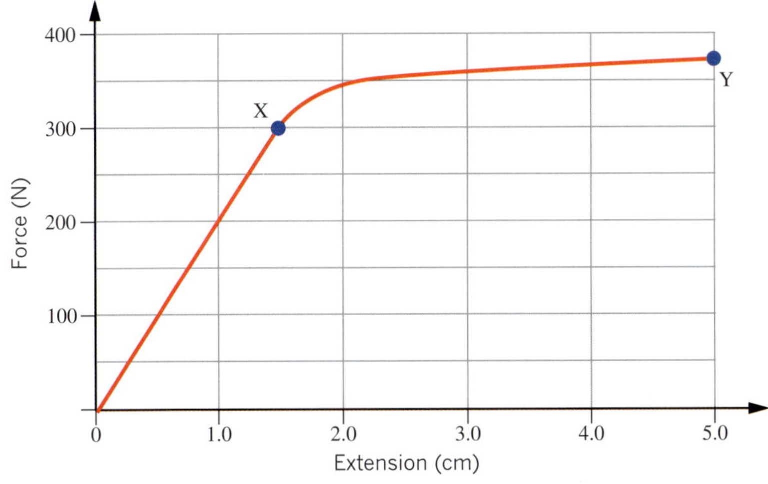

a Calculate the spring constant, *k*, for the sample.

b Calculate the elastic potential energy that the alloy can store before permanent deformation begins.

c Calculate the work done to break the sample.

3.4 Review

OA ✓✓

SUMMARY

- Hooke's law states that the force exerted by a spring is $F = -kx$. The negative sign indicates that the force opposes the displacement.
- k is the spring constant and is measured in N m^{-1}. This can be calculated (or estimated) from the gradient of the linear section (or the first linear section if there is more than one) of a force–displacement graph.
- The work done to a spring is equal to the elastic potential energy stored in the spring:

$$E_s = \frac{1}{2}kx^2$$

- The elastic potential energy (E_s) is measured in J or N m. This can be calculated (or estimated) from the area under a force–displacement graph.
- When a material displays elastic behaviour, it obeys Hooke's law, and the elastic potential energy stored is returned when the force is removed.
- When a material exceeds its elastic limit, permanent deformation occurs and not all the elastic potential energy is returned when the force is removed.

KEY QUESTIONS

Knowledge and understanding

1 Rank the springs below in order of increasing stiffness.

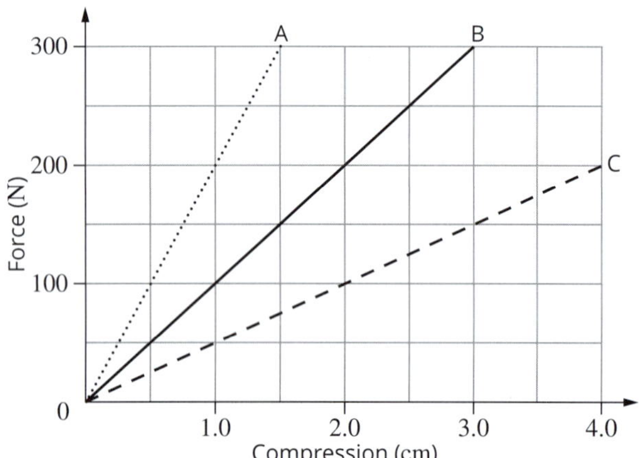

2 Consider the following tasks and decide whether you would prefer a rope with a high, medium or low spring constant.

a lowering a prefabricated concrete panel into place on a high-rise building site

b towing a bogged car out of a muddy track

c making a cargo net to secure various loads on a trailer

3 The graph of the stretching force versus extension for two springs is shown below.

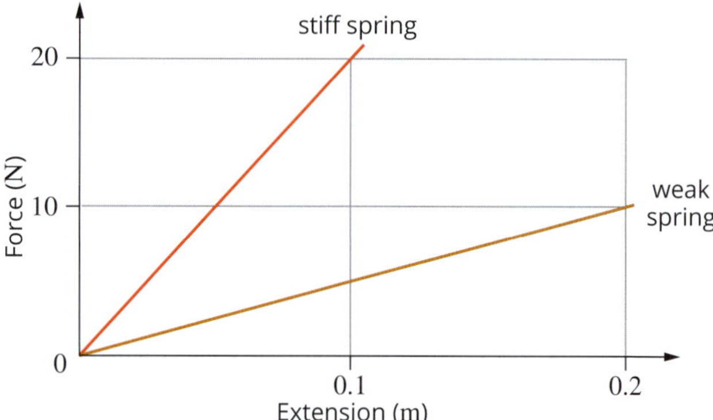

a Calculate the spring constant of each spring.

b Find the difference between the elastic potential energy stored when each spring is extended by 20 cm. Assume the elastic limit has not been reached.

4 A 1.00 m piece of rubber has a spring constant of 50.0 N m^{-1}. Calculate how much the rubber will stretch if a force of 4.00 N is exerted on it.

5 A stretched rubber band is used to launch a toy plane into the air. The rubber band is stretched by 25.0 cm and has a spring constant of 128 N m^{-1}. Assume that the rubber band follows Hooke's law and ignore its mass.

a Calculate the magnitude of the force applied to the rubber band to stretch it by 25.0 cm.

b Calculate the elastic potential energy stored in the stretched rubber band.

Analysis

6 An archer purchased a new bow for the Olympics. The table below shows the force required to pull the string back by various distances (the distance between X and Y in the diagrams).

Force (N)	Distance between bow and string (m)
0.0	0.100
30.0	0.150
40.0	0.200
45.0	0.250
50.0	0.300

The illustrations below show the bow and its string when (a) no force is applied and (b) some force is applied.

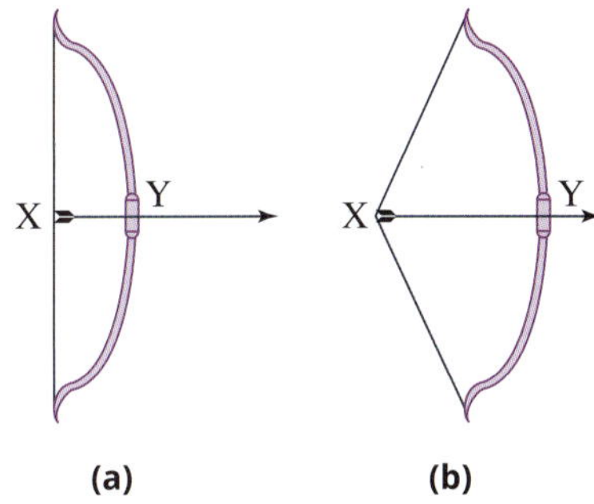

(a) (b)

Answer the following questions in the case where the archer has drawn the string back so that the distance between the bow and the string (XY) is 30.0 cm.

a Construct a graph of the force (N) vs XY distance (m).

b Use the graph to calculate the elastic potential energy stored in the stretched string.

c Calculate the work done by the archer.

d Does the string obey Hooke's law as it is drawn back until the distance between X and Y is 30.0 cm? Justify your answer.

e Where on the graph is the elastic limit of the string?

3.5 Kinetic and potential energy

FIGURE 3.5.1 The bungee jumper is in free fall until the cord starts to take up some of the kinetic energy and convert it to potential energy.

A bungee jumper stakes their life on the principle of the conservation of energy (Figure 3.5.1). The gravitational potential energy they lose as they begin their jump is rapidly converted to kinetic energy. As the bungee jumper approaches the ground, the kinetic energy is converted to elastic potential energy in the bungee cord. The jumper is then jerked back upwards (no doubt relishing the adrenalin rush) as the elastic potential energy is converted back to kinetic and potential energy. The calculations that ensure their safety are the subject of this section.

KINETIC ENERGY

Kinetic energy (E_k) is the energy of motion of a body. For low speeds, it is calculated using the following equation.

$$E_k = \frac{1}{2}mv^2$$

where E_k is the kinetic energy of the object (J)
m is the mass of the object (kg)
v is the velocity of the object (m s^{-1})

This equation can be derived from the definition of work. Recall that if a force, F, acts on a body of mass m and causes a horizontal displacement of s, the work done is given by the formula $W = Fs$, which is equivalent to $W = mas$.

Start by rearranging the equation $v^2 = u^2 + 2as$ to make s the subject:

$$s = \frac{v^2 - u^2}{2a}$$

Substitute this into the second equation for work given above: $W = mas$. This yields:

$$\begin{aligned} W &= ma\left(\frac{v^2 - u^2}{2a}\right) \\ &= m\left(\frac{v^2 - u^2}{2}\right) \\ &= \frac{1}{2}mv^2 - \frac{1}{2}mu^2 \end{aligned}$$

As the work is done to change the kinetic energy, then:

$$\Delta E_k = \frac{1}{2}mv^2 - \frac{1}{2}mu^2$$

For a particular speed, the equation can be simplified to:

$$E_k = \frac{1}{2}mv^2$$

Kinetic energy in collisions

In perfectly **elastic collisions**, kinetic energy is transferred between objects and no energy is transformed into heat, sound or deformation. In these cases the following relationship holds:

$$E_k \text{ (before)} = E_k \text{ (after)}$$

In Section 3.1 you saw that, in a closed system, momentum is always conserved in a collision. The total energy is also conserved in a closed system. However, in general, kinetic energy is not conserved in collisions. These collisions are called **inelastic collisions**.

Perfectly elastic collisions do not exist in everyday situations, but they do exist in the interactions between atoms and subatomic particles. A collision between two billiard balls or the spheres in Newton's cradle is almost perfectly elastic, because very little of their kinetic energy is transformed into heat and sound energy.

Collisions such as a bouncing basketball, a gymnast bouncing on a trampoline or a tennis ball being hit are moderately elastic, with about half the kinetic energy of the system being retained. Perfectly inelastic collisions are those in which the colliding bodies stick together after impact with no kinetic energy. Some car crashes, a collision between a meteorite and the Moon, and a collision involving two balls of plasticine could all be perfectly inelastic. In these collisions, most—and sometimes all—of the initial kinetic energy of the system is transformed into other forms of energy.

Worked example 3.5.1

ELASTIC OR INELASTIC COLLISION?

A car of mass 1000 kg travelling west at 20.0 m s^{-1} crashes into the rear of a stationary bus of mass 5000 kg. The vehicles lock together on impact. Using appropriate calculations, show whether the collision is elastic or inelastic.

Thinking	Working
Use conservation of momentum to find the final velocity of the wreck.	$\Sigma p_{\text{initial}} = \Sigma p_{\text{final}}$ $p_{\text{initial c}} + p_{\text{initial b}} = p_{\text{final(c+b)}}$ $m_c u_c + m_b u_b = m_{c+b} u_{c+b}$ $1000 \times 20.0 + 5000 \times 0 = (1000 + 5000)v$ $20\,000 = 6000v$ $v = 3.33 \text{ m s}^{-1}$
Calculate the total initial kinetic energy before the collision.	Initially: $E_k = \frac{1}{2}mu^2$ $= \frac{1}{2} \times 1000 \times 20.0^2$ $= 2.00 \times 10^5 \text{ J}$
Calculate the total final kinetic energy of the joined vehicles.	Finally: $E_k = \frac{1}{2}mv^2$ $= \frac{1}{2} \times (1000 + 5000) \times 3.33^2$ $= 33266.7$ $= 3.33 \times 10^4 \text{ J}$
Compare the kinetic energy before and after the collision to determine whether the collision is elastic or inelastic.	The kinetic energy after the collision is less than the kinetic energy before it. Therefore the collision is inelastic. The missing energy has been transformed into heat, sound and deformation of the vehicles.

Worked example: Try yourself 3.5.1

ELASTIC OR INELASTIC COLLISION?

A 209 g softball with initial velocity 9.00 m s^{-1} to the right collides with a stationary baseball of mass 112 g. After the collision, both balls move to the right and the softball has a speed of 3.00 m s^{-1}. Using appropriate calculations, show whether the collision is elastic or inelastic.

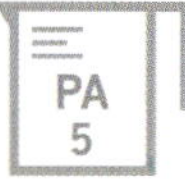

POTENTIAL ENERGY

The gravitational potential energy of an object, E_g, is the energy stored in it due to its position in a gravitational field above a reference point. It is directly proportional to the mass of the object, m, its height above the reference point, Δh, and the strength of the gravitational field, g. This is combined in the following equation.

$$E_g = mg\Delta h$$

where E_g is the gravitational potential energy (J)
m is the mass of the object (kg)
g is gravitational field strength ($N\,kg^{-1}$)
Δh is the height above the reference point (m)

This equation is derived from the fact that, in order to lift an object of mass m through a distance Δh, work needs to be done against the force of gravity. Close to the surface of the Earth, this force is simply $F = mg$ (where $g = 9.8\,N\,kg^{-1}$) and the distance travelled, s, is Δh. Thus the work done is $W = Fs$, which is equal to the potential energy gained.

Calculating changes in gravitational potential energy from a force graph

When the gravitational force acting on an object varies, the gravitational potential energy can be calculated using a graph (in the same way that you calculated the work done by a varying force in sections 3.3 and 3.4). If the force is plotted as a function of distance, a graph like the one in Figure 3.5.2 is obtained.

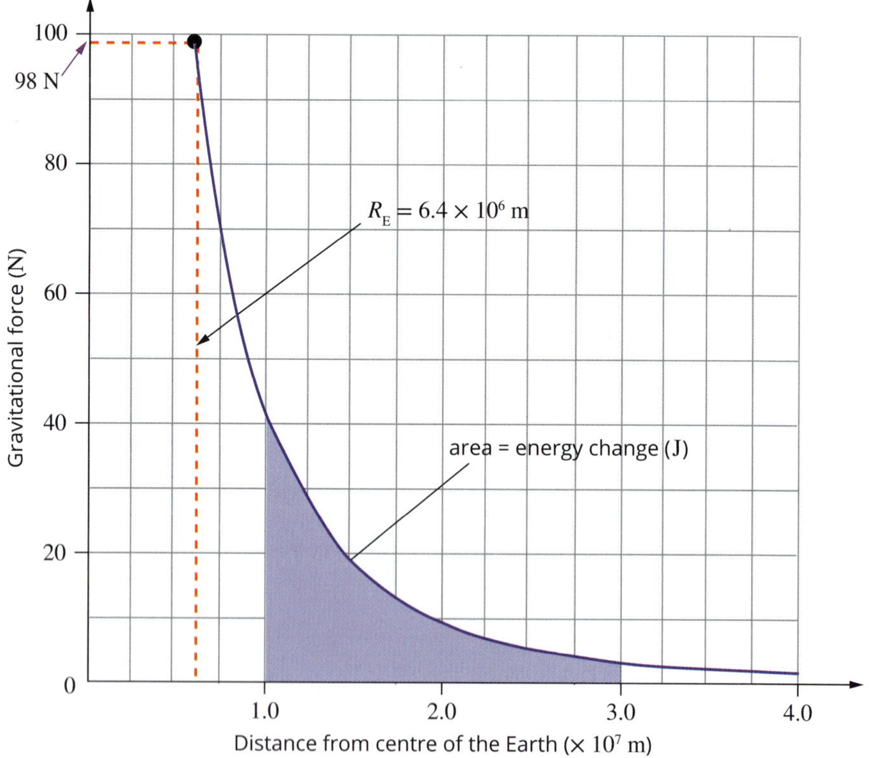

FIGURE 3.5.2 Plot of the gravitational force acting on a 10 kg body as a function of the distance from the centre of the Earth. The shaded area represents the work done in moving the body from 1.0×10^7 m to 3.0×10^7 m above the centre of the Earth.

Worked example 3.5.2

DETERMINING CHANGES IN GRAVITATIONAL POTENTIAL ENERGY USING A FORCE VS DISTANCE GRAPH

Using the graph in Figure 3.5.2, estimate the work done against the gravitational force in moving the 10 kg object from an orbital radius of 1.0×10^7 m to 3.0×10^7 m, and hence find the gravitational potential energy gained.

Thinking	Working
Find the energy represented per square in the graph.	One square represents: $10 \times 0.25 \times 10^7 = 2.5 \times 10^7$ J
Identify the two values of distance that are relevant to the question.	The object starts at 1.0×10^7 m and finishes at 3.0×10^7 m.
Count the squares under the curve between the two distances and multiply the number by the energy per square.	Work done: 10.5 squares (approx.) $\times 2.5 \times 10^7$ $= 2.6 \times 10^8$ J (approx.)
Potential energy gained = work done	2.6×10^8 J (approx.)

Worked example: Try yourself 3.5.2

DETERMINING CHANGES IN GRAVITATIONAL POTENTIAL ENERGY USING A FORCE VS DISTANCE GRAPH

Using the graph in Figure 3.5.2, calculate the gravitational potential energy gained if the 10 kg object is moved from the surface of the Earth to 2.0×10^7 m above the centre of the Earth.

The disadvantage of the graph in Figure 3.5.2 is that it is specific to the mass of the object under consideration. Further, to construct the graph, the force on the 10 kg object has to be calculated at each distance.

Sometimes it is more useful to create a graph of the force exerted per unit mass. Recall Newton's law of universal gravitation:

$$F_g = \frac{GMm}{r^2}$$

This can be rearranged as follows, to give an expression for the force per unit mass, i.e. the gravitational field strength, g:

$$g = \frac{F_g}{m} = \frac{GM}{r^2}$$

This is often called the gravitational field strength equation and it is dependent only on the mass that is generating the gravitational field. Such a graph can be used to calculate the work done on any mass in the field.

Worked example 3.5.3

CHANGES IN GRAVITATIONAL POTENTIAL ENERGY USING A GRAVITATIONAL FIELD STRENGTH GRAPH

A decommissioned satellite of mass 1000 kg has an elliptical orbit around the Earth. At its closest approach (its perigee), it is 600 km above the Earth's surface. At its furthest point (its apogee) it is 2000 km from the Earth's surface. The Earth has a radius of 6.4×10^6 m. The gravitational field strength of the Earth is shown in the graph.

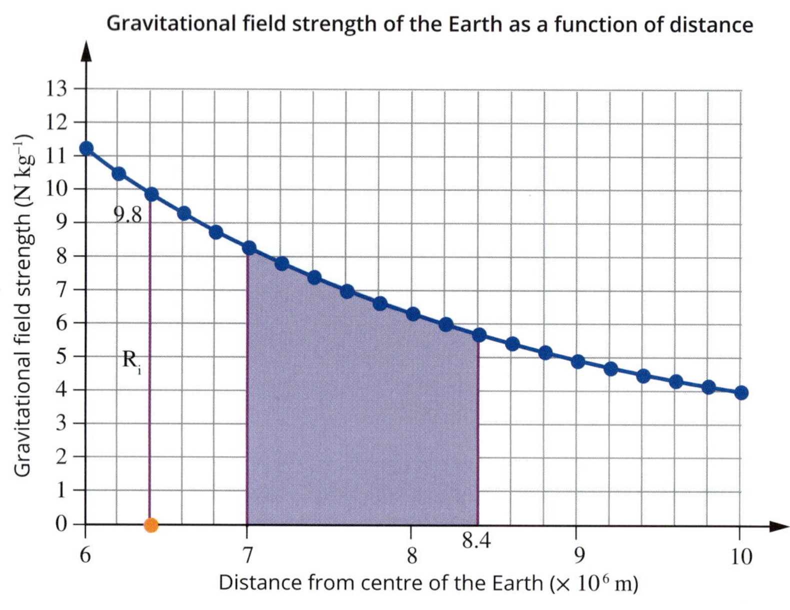

a Calculate the change in potential energy of the satellite as it moves from its perigee to its apogee.	
Thinking	**Working**
Find the energy represented by each square in the graph.	One square represents $1.0 \times 0.20 \times 10^6 = 2.0 \times 10^5$ J kg^{-1}
Count the squares under the curve for the relevant area, and multiply the total by the energy per kg represented by each square.	49 squares (approx.) $\times 2.0 \times 10^5$ $= 9.8 \times 10^6$ J kg^{-1}
Calculate the potential energy gained by the satellite by multiplying the work done by the mass of the satellite.	Energy gained: $E_g = 9.8 \times 10^6 \times 1000$ $= 9.8 \times 10^9$ J (approx.)

b The satellite is moving with a speed of 15 km s^{-1} at its perigee. How fast is it travelling at its apogee?	
Thinking	**Working**
First calculate the satellite's kinetic energy at its perigee.	$E_{kp} = \frac{1}{2}mv_p^2$ $= \frac{1}{2} \times 1000 \times (15 \times 10^3)^2$ $= 1.125 \times 10^{11}$ J

The gain in gravitational potential energy at its apogee is at the expense of kinetic energy. Calculate the kinetic energy of the satellite at its apogee.	$E_{ka} = E_{kp} - E_g$ $= 1.125 \times 10^{11} - 9.8 \times 10^{9}$ $= 1.0 \times 10^{11}$ J
Calculate the speed of the satellite at its apogee.	$E_{ka} = \frac{1}{2}mv_a^2$ $1.0 \times 10^{11} = \frac{1}{2} \times 1000 \times v_a^2$ $v_a = \sqrt{\frac{2 \times 1.0 \times 10^{11}}{1000}}$ $= 14142.1$ m s^{-1} $= 14$ km s^{-1}

Worked example: Try yourself 3.5.3

CHANGES IN GRAVITATIONAL POTENTIAL ENERGY USING A GRAVITATIONAL FIELD STRENGTH GRAPH

A satellite of mass 1100 kg is in an elliptical orbit around the Earth. At its closest approach (perigee), it is just 600 km above the Earth's surface. Its furthest point (apogee) is 2600 km from the Earth's surface. The Earth has a radius of 6.4×10^6 m. The gravitational field strength of the Earth is shown in the graph.

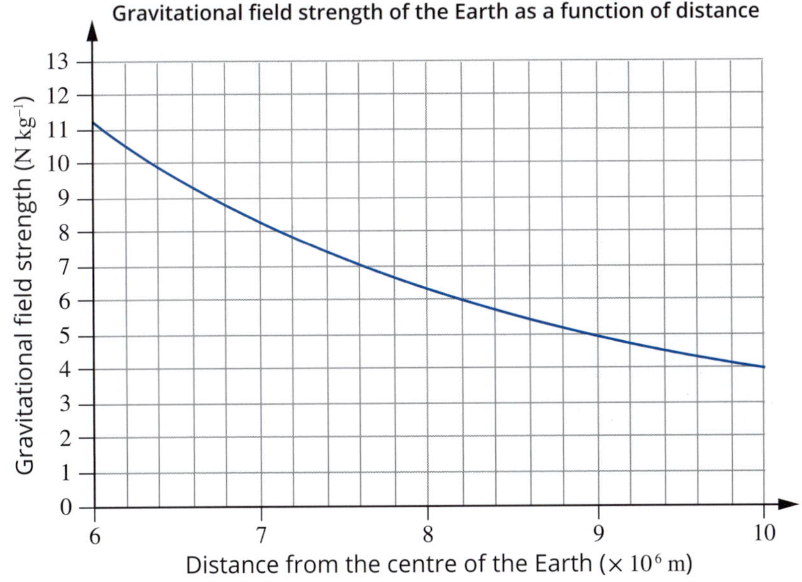

a	Calculate the change in potential energy of the satellite as it moves from its perigee to its apogee.

b	The satellite is moving with a speed of 8.0 km s^{-1} at its perigee. How fast will it be travelling at its apogee?

WORK AND ENERGY

Both work and energy are scalar quantities and thus have only magnitude. It is important, however, that you keep account of whether kinetic energy is being gained or lost by an object or whether gravitational potential energy is being gained or lost by the gravitational field. If work is being done by a body, it could lose kinetic energy as it slows down, or, if work is being done by the gravitational field, the field loses gravitational potential energy as the object falls. Conversely, if work is done on the body by an external force, the body would gain kinetic energy as it speeds up, or the gravitational field would gain gravitational potential energy as the object rises.

A weightlifter loses chemical potential energy as they exert a force on a barbell to lift the bar. If they lift the bar at constant speed, the bar does not gain kinetic energy, but the gravitational field gains gravitational potential energy. In drawing back an arrow, an archer does work on the bow, and this elastic potential energy is transformed to the kinetic energy of the arrow when the string does work on the arrow as it is released (Figure 3.5.3).

FIGURE 3.5.3 The archer does work on the bow and elastic potential energy is stored. This is later transformed into the kinetic energy of the arrow.

3.5 Review

OA ✓✓

SUMMARY

- Kinetic energy is the energy of motion of a body:

$$E_k = \frac{1}{2}mv^2$$

- For perfectly elastic collisions, the kinetic energy before the collision is equal to the kinetic energy after the collision.
- Close to the surface of the Earth, where the force of gravity can be taken as constant, the change in gravitational potential energy of an object of mass m is $E_g = mg\Delta h$, where the height changes by Δh.
- For a non-constant gravitational force, the gravitational potential energy can be calculated from the area under a graph of force versus distance.
- For convenience, force–distance graphs are often plotted as force per unit mass (for example, gravitational field strength) versus distance. This enables the same graph to be used for any mass. In this case the area under the graph is the potential energy per unit mass.

KEY QUESTIONS

Knowledge and understanding

1 The figure below shows a meteor plunging towards the Earth, partially burning up in the atmosphere on its way.
Choose which statements are correct. More than one correct answer is possible.

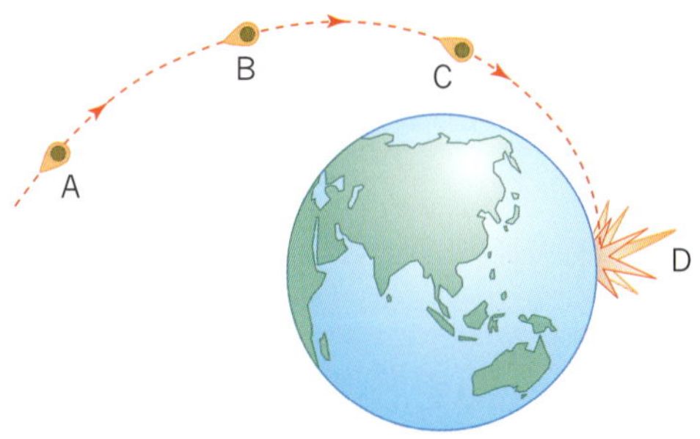

A The kinetic energy of the meteor increases as it travels from A to D.
B The gravitational potential energy of the meteor relative to the surface of the Earth increases as it travels from A to D.
C The total energy of the meteor increases as it travels from A to D.
D The total mechanical energy of the meteor remains constant.
E The gravitational potential energy of the meteor relative to the surface of the Earth decreases as it travels from A to D.

2 In a cable car system, two cars of the same mass are attached to a moving cable that is powered by a motor at one end. As car A is pulled upwards, car B descends, both at the same speed. Select the statements that are correct. More than one correct answer is possible.
A Car A and car B each have constant kinetic energy.
B Car A and car B each have constant gravitational potential energy.
C As the gravitational potential energy of car A increases, that of car B decreases.
D The motor does work on the cable.

3 Calculate the gravitational potential energy of a 115 kg climber standing at the top of Mount Kosciuszko 2228 m above sea level.

4 A 283 g volleyball is hit into the opposition court with a velocity of 9.50 m s^{-1}. Calculate the kinetic energy of the volleyball as it leaves the player's hand.

5 Calculate the gravitational potential energy that a 3.00 kg watermelon has when it has travelled 45.0 m up into the air after having been fired from a slingshot.

continued over page

3.5 Review *continued*

Analysis

6 The 11 t Hubble Space Telescope is in a circular orbit at an altitude of approximately 600 km above the surface of the Earth. A geosynchronous weather satellite of the same mass is in an orbit at an altitude of approximately 3600 km. Select the statements that are correct. More than one correct answer is possible.

- **A** The gravitational potential energy of the geosynchronous satellite is six times that of the Hubble telescope, relative to the surface of the Earth.
- **B** The Hubble telescope's orbital speed is greater than that of the weather satellite.
- **C** The kinetic energy of the weather satellite is greater than that of the Hubble telescope.
- **D** The weather satellite has more gravitational potential energy than the Hubble telescope, relative to the surface of the Earth.

7 A 500 kg lump of space junk is plummeting towards the Moon. Its speed when it is 2.7×10^6 m from the centre of the Moon is 250 m s^{-1}. The Moon has a radius of 1.7×10^6 m.

The gravitational force–distance graph for the space junk is shown below.

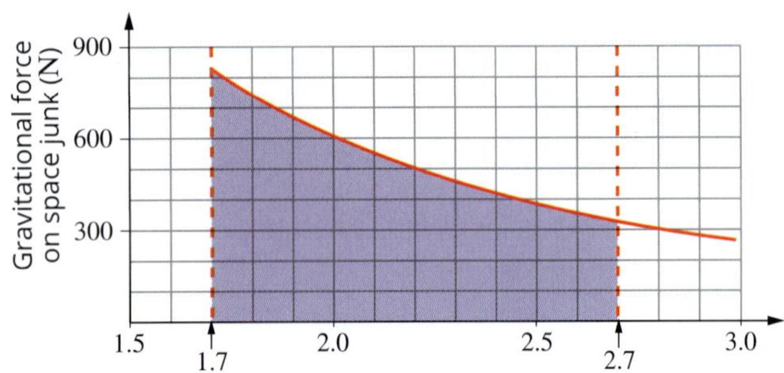

- **a** Calculate the kinetic energy of the junk when it is travelling at 250 m s^{-1}.
- **b** Calculate the increase in kinetic energy of the junk as it falls from 2.7×10^6 m from the Moon's centre to 1.7×10^6 m from the Moon's centre.
- **c** Calculate the speed of the junk as it crashes into the Moon.

8 A 20 t piece of space junk is in orbit at an altitude of 600 km above the surface of the Earth. In order to move it from the path of an oncoming satellite, it is shifted into an orbit of 2600 km above the surface of the Earth. Calculate the work done in moving the space junk into the higher orbit. The surface of the Earth is 6.4×10^6 m from the centre of the Earth.

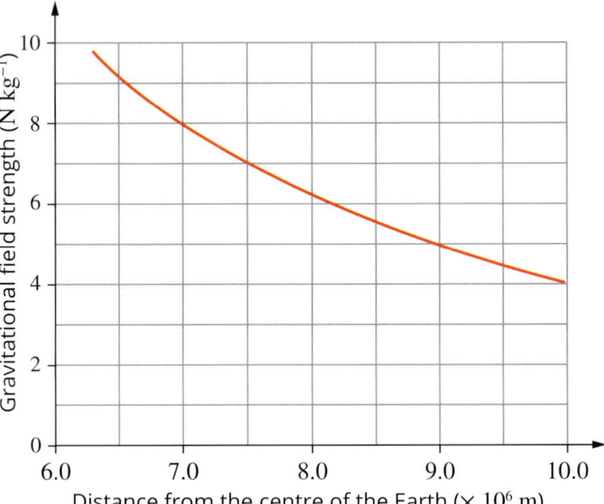

3.6 Conservation of energy

We can classify everything we know about the universe as either matter or energy. In his famous equation $E = mc^2$, Einstein showed that matter is actually a store of energy, so everything in the universe is really just energy. This section explores the law of conservation of energy, a fundamental principle that can be applied to all interactions between objects.

THE LAW OF CONSERVATION OF ENERGY

Energy comes in many forms, such as heat, light, sound, chemical and electrical. It is a scalar quantity and is measured in joules (J). Energy is also associated with the motion and position of an object. Collectively this energy is called the **mechanical energy** of the object. In the motion problems explored in this chapter, moving objects are described as having kinetic energy. An object can also have access to stored or gravitational potential energy because of its position in a gravitational field. For instance, a building crane lifting a steel beam several stories is doing work against the gravitational field, giving the beam access to the gravitational potential energy stored in the gravitational field. If the lifting chain were to break, the field will then do work on the beam and increase its kinetic energy as it accelerates under the influence of gravity.

The transformation of gravitational potential energy to kinetic energy is an illustration of the **law of conservation of energy**, a fundamental principle of nature. This law states that energy is neither created nor destroyed. However, it can change from one form to another, or in other words, **transform**. As the gravitational potential energy available to a falling object decreases, its kinetic energy increases. The total amount of mechanical energy remains constant, that is, it is conserved.

While energy is never destroyed, it can be transformed into other energies that are not easily recoverable. For instance, the kinetic energy of a vehicle is reduced as it encounters friction, with the energy transformed into heat in the tyres. It could also be transformed into heat in the brakes as the vehicle stops. The mechanical energy before and after an event is only the same under ideal conditions, but in many cases, this equality is a useful approximation.

Problems combining gravitational potential and kinetic energy

Energy is a scalar quantity and hence easier to work with than a vector quantity. Therefore it is worth analysing a problem to see if calculations involving energy are possible without resorting to techniques involving forces and other vectors.

The sum of the potential and kinetic energy of an object is its mechanical energy, and this is constant unless work is done by an external force:

$$E_m = E_k + E_g = \frac{1}{2}mv^2 + mg\Delta h$$

Energy is frequently transformed from potential energy to kinetic energy and vice versa. But in any transformation, total mechanical energy is conserved. To illustrate this, consider a 60 g tennis ball dropped from a height of 1.0 m (Figure 3.6.1 on page 148). Before it is released, its kinetic energy is 0 J and its gravitational potential energy (assuming that $g = 9.8\,m\,s^{-2}$) is:

$$E_g = mg\Delta h = (0.060)(9.8)(1.0) = 0.59\,J$$

Thus the ball's mechanical energy is 0 + 0.59 = 0.59 J.

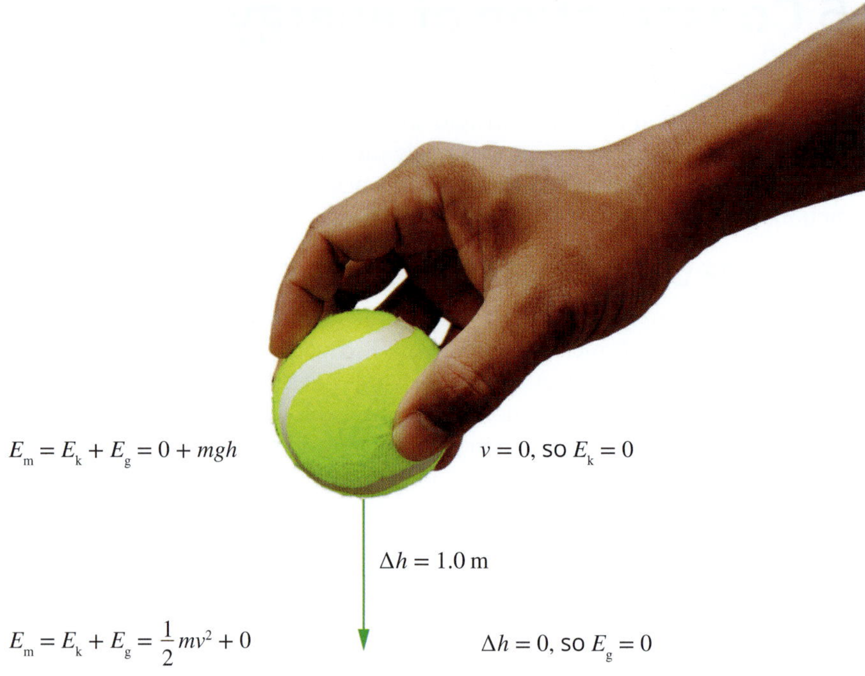

FIGURE 3.6.1 A falling tennis ball provides an illustration of the conservation of mechanical energy.

At the instant the ball hits the ground, the total mechanical energy consists of the gravitational potential energy available to it, which would be 0 J, and the kinetic energy it has just prior to hitting the ground. To calculate its kinetic energy, its velocity just before it hits the ground needs to be calculated using an appropriate equation of motion. Knowing that $s = -1.0\,m$, $a = -9.8\,m\,s^{-2}$ and $u = 0\,m\,s^{-1}$, the final velocity can be calculated as:

$$\begin{aligned} v^2 &= u^2 + 2as \\ &= (0)^2 + 2(-9.8)(-1.0) \\ v &= \sqrt{19.6} = 4.43\,m\,s^{-1} \end{aligned}$$

Therefore the kinetic energy of the ball just before it hits the ground is:

$$E_k = \frac{1}{2}mv^2 = \frac{1}{2}(0.0600)(4.43)^2 = 0.59\,J$$

Notice that the total mechanical energy prior to the ball's release is the same as its total mechanical energy as it hits the ground. Before release:

$$E_m = E_k + E_g = 0 + 0.59 = 0.59\,J$$

On hitting the ground:

$$E_m = E_k + E_g = 0.59 + 0 = 0.59\,J$$

In fact, mechanical energy is constant throughout the drop. To see this, consider the tennis ball when it has fallen halfway to the ground. At this point, $h = 0.50\,m$, $v = 3.13\,m\,s^{-1}$ and its mechanical energy is:

$$\begin{aligned} E_m &= E_k + E_g \\ &= \frac{1}{2}(0.060)(3.13)^2 + (0.060)(9.8)(0.50) \\ &= 0.294 + 0.294 \\ &= 0.59\,J \end{aligned}$$

Note that at the halfway point, the mechanical energy is evenly split between kinetic energy (0.294 J) and gravitational potential energy (also 0.294 J).

In reality, as a ball drops through the air, a small amount of its energy is transformed into heat and sound, and the ball slows down slightly. This means that mechanical energy is not entirely conserved. However, this small effect can be considered negligible for many falling objects.

Using conservation of mechanical energy to calculate velocity

The speed of a falling object does not depend on its mass. This can be demonstrated by applying the law of conservation of energy to mechanical energy.

Consider an object of mass m dropped from a height of h. At the moment it is dropped, its initial kinetic energy is zero. At the moment before it hits the ground, its final gravitational potential energy is zero. From the conservation of mechanical energy it follows that:

$$E_{\text{m initial}} = E_{\text{m final}}$$

$$E_{\text{k initial}} + E_{\text{g initial}} = E_{\text{k final}} + E_{\text{g final}}$$

$$0 + mgh = \frac{1}{2}mv^2 + 0$$

$$mgh = \frac{1}{2}mv^2$$

$$gh = \frac{1}{2}v^2$$

$$v^2 = 2gh$$

$$v = \sqrt{2gh}$$

This equation can be used to find the final velocity of a falling object. Note that the equation does not mention the mass of the falling object. Thus if air resistance is negligible, any object will have the same final velocity when it is dropped from the same height, whatever its mass.

Conservation of mechanical energy in complex situations

Knowing that mechanical energy is conserved allows us to determine outcomes in non-linear situations where equations of motion cannot be used. For example, consider a pendulum with a bob displaced from its mean position such that its height has increased by 20 cm (Figure 3.6.2).

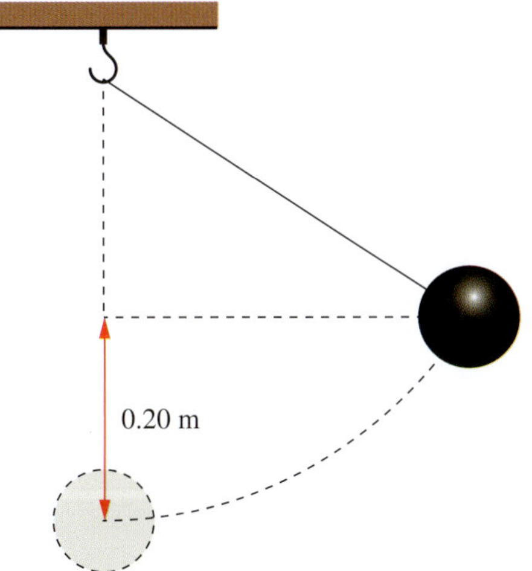

FIGURE 3.6.2 A falling pendulum is an example of conservation of mechanical energy.

Since a pendulum involves transforming gravitational potential energy into kinetic energy, the conservation of mechanical energy applies to the situation. Therefore the formula developed earlier for the velocity of a falling object can be used to find the velocity of the pendulum bob at its lowest point:

$$v = \sqrt{2gh} = \sqrt{2(9.8)(0.20)} = 2.0\,\text{m}\,\text{s}^{-1}$$

The speed of the pendulum bob will be approximately $2.0\,\text{m}\,\text{s}^{-1}$. However, unlike a falling object, in this case the direction of the bob's motion will be horizontal instead of vertical at its lowest point.

The equations of motion relate to linear motion and cannot be applied to the motion of the pendulum.

Worked example 3.6.1

APPLYING THE LAW OF CONSERVATION OF ENERGY

Consider a rollercoaster with a lift hill of height 25 m and a loop height of 18 m. At the top of the lift hill, a rollercoaster car has zero velocity just before it begins to roll down the hill. Calculate the speed of the car at point P on the loop when the car is 6.0 m above the ground. Assume that friction is negligible and that $g = 9.8\,\text{m}\,\text{s}^{-2}$.

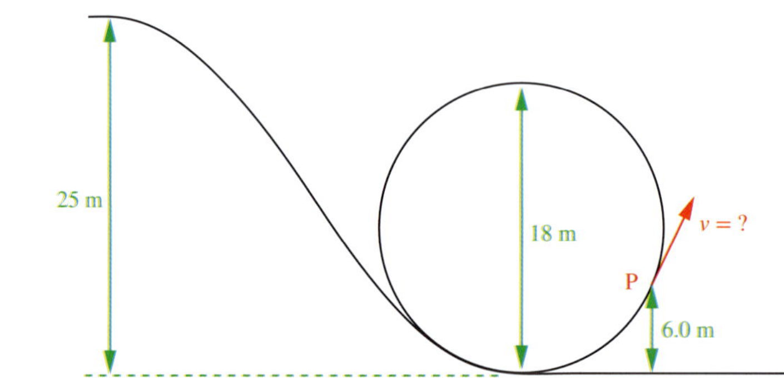

Thinking	Working
Because of the law of conservation of mechanical energy, the total mechanical energy, E_m, of the car before rolling down the hill can be equated with the total mechanical energy at point P.	$E_{\text{m before}} = E_{\text{m at P}}$
Expand the equation and cancel m from both sides.	$\frac{1}{2}mu^2 + mg\Delta h = \frac{1}{2}mv^2 + mg\Delta h$ $\frac{1}{2}u^2 + g\Delta h = \frac{1}{2}v^2 + g\Delta h$
Substitute the given values into the equation.	$\frac{1}{2}(0)^2 + (9.8)(25.0) = \frac{1}{2}v^2 + (9.8)(6.0)$
Rearrange the equation and solve for v.	$(0) + (245.0) = \frac{1}{2}v^2 + (58.8)$ $v = \sqrt{2(245.0 - 58.8)}$ $= \sqrt{372.4}$
Present your answer with the correct number of significant figures and the correct units.	$v = 19\,\text{m}\text{s}^{-1}$

Worked example: Try yourself 3.6.1

APPLYING THE LAW OF CONSERVATION OF ENERGY

Use the law of conservation of energy to determine the height of the lift hill required to ensure that the speed of a rollercoaster car at the top of the 18 m loop is 25 m s^{-1}. Assume that the velocity of the car at the top of the hill is zero just before it begins to roll down the hill, friction is negligible and that $g = 9.8$ m s^{-2}.

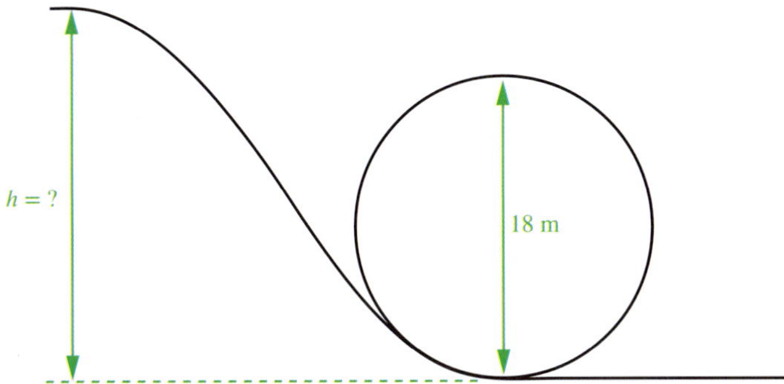

Worked Example 3.6.2

USING THE CONSERVATION OF ENERGY TO ANALYSE PROJECTILE MOTION

A cricket ball of mass 142 g is thrown upwards at a speed of 15 m s^{-1}. Calculate the speed of the ball when it has reached a height of 8.0 m. Assume that the ball is thrown from a height of 1.5 m above the ground and that $g = 9.8$ m s^{-2}.

Thinking	Working
Equate the total mechanical energy, E_m, of the cricket ball as it is released with the total mechanical energy at a height of 8.0 m.	$E_{\text{m before}} = E_{\text{m at 8.0 m}}$
Expand the equation and then cancel *m* from both sides.	$\frac{1}{2}mu^2 + mg\Delta h = \frac{1}{2}mv^2 + mg\Delta h$ $\frac{1}{2}u^2 + g\Delta h = \frac{1}{2}v^2 + g\Delta h$
Substitute the given values into the equation.	$\frac{1}{2}(15)^2 + (9.8)(1.5) = \frac{1}{2}v^2 + (9.8)(8.0)$
Rearrange the equation and solve for *v*.	$(112.5) + (14.7) = \frac{1}{2}v^2 + (78.4)$ $v = \sqrt{2(112.5 + 14.7 - 78.4)}$ $= \sqrt{97.6}$
Present your answer with the correct number of significant figures and the correct units.	$v = 9.9$ m s^{-1}

Worked Example: Try yourself 3.6.2

USING THE CONSERVATION OF ENERGY TO ANALYSE PROJECTILE MOTION

An arrow of mass of 35 g is fired into the air at 80 m s^{-1} from a height of 1.4 m above the ground. Calculate the speed of the arrow when it is 30 m above the ground. Assume that $g = 9.8$ m s^{-2}.

Loss of mechanical energy

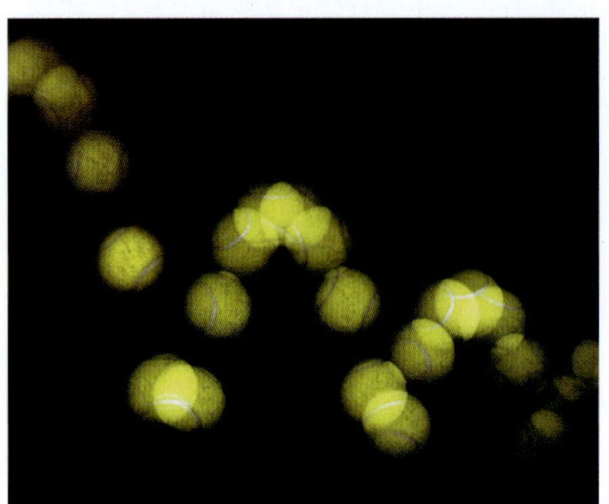

FIGURE 3.6.3 Mechanical energy is lost with each bounce of a tennis ball.

Mechanical energy is not conserved in every situation. This can be seen in the fact that when a tennis ball bounces a number of times, each bounce is lower than the bounce before it (Figure 3.6.3).

While mechanical energy is largely conserved as the ball moves through the air, a significant amount of kinetic energy is transformed into heat and sound when the ball compresses and decompress each time it bounces. This means that the ball does not have as much kinetic energy when it leaves the ground as it did when it landed. Therefore the gravitational potential energy that can be stored on the second bounce will be less than the gravitational potential energy that was stored initially, and so the second bounce is lower.

CASE STUDY

Coefficient of restitution

The bounce of the ball is an important factor in many sports. Physicists describe the bounciness of balls using a concept known as the coefficient of restitution (e). This coefficient is the ratio of the speed of a ball directly after a bounce to its speed before that bounce:

$$e = \frac{v_2}{v_1}$$

where v_1 is the speed before the bounce and v_2 is the speed after the bounce.

Since the coefficient of restitution is defined in terms of speed, v, and kinetic energy is proportional to v^2, it follows that:

$$e = \frac{v_2}{v_1} = \sqrt{\frac{(E_k)_2}{(E_k)_1}}$$

If we consider a ball dropped from height H and rebounding to height h, then, according to conservation of mechanical energy, the kinetic energy as the ball hits the ground is the same as the gravitational potential energy at the top of the bounce. Therefore:

$$e = \sqrt{\frac{(E_k)_2}{(E_k)_1}} = \sqrt{\frac{(E_g)_2}{(E_g)_1}} = \sqrt{\frac{mgh}{mgH}} = \sqrt{\frac{h}{H}}$$

So the coefficient of restitution (CoR) can be calculated from the initial drop height and the height of the first bounce. This is how many sports bodies specify the acceptability of balls used in playing the sport.

For example, according to the rules of the International Table Tennis Federation, a table tennis ball must bounce between 24 and 26 cm when dropped from a height of 30.5 cm onto a steel block. This corresponds to a CoR between 0.89 and 0.92. Similarly, a basketball must have a CoR between 0.81 and 0.85 before it can be used in competition. Likewise a tennis ball must have a CoR between 0.73 and 0.76.

The CoR depends on both the ball and the surface it is bouncing on. A tennis ball bouncing on grass has a different CoR to one bouncing on clay. This is one reason why different tennis players prefer to play on some surfaces rather than others.

3.6 Review

SUMMARY

- Energy is a scalar quantity and is measured in joules (J).
- Energy is not created or destroyed, but merely transformed. This is called the law of conservation of energy.
- When work is done on a body it gains mechanical energy.
- When the body does work, energy is dissipated to the environment—as, for example, heat, sound or deformation—and the body loses mechanical energy.
- The sum of the kinetic and potential energy (i.e. the total mechanical energy) of an isolated system is always conserved.
- Because it is simpler to work with scalars, it is often helpful to solve motion problems by considering the energy involved.

OA ✓✓

KEY QUESTIONS

Knowledge and understanding

1 Choose the best alternative to complete the following sentence. In physics, the law of conservation of energy entails that:

- **A** All energy must be converted from one form into only one other form.
- **B** When energy is converted from one form to another, any missing energy must have been destroyed.
- **C** When energy is converted from one form to another, any extra energy gained must have come from gravitational potential energy.
- **D** No energy is gained or lost when one form of energy is converted into another form.

2 A student drops two sticks—one brown, one green—from a bridge into the water below to see which one emerges first at the other side of the bridge. The brown stick is twice the mass of the green stick. In answering the following questions, ignore any air resistance and friction that might be involved.

- **a** Which stick would hit the water first if they were dropped at the same time?
- **b** Which stick would have access to the greatest amount of gravitational potential energy at the top of the bridge?
- **c** Which stick would hit the water with the greatest speed?
- **d** Which stick would have the greatest kinetic energy just before it hit the water?

3 A group of people decide to film themselves throwing various objects off tall places. In one video they drop a bowling ball from the top of a dam wall. The ball hits the ground with a speed of 45.5 m s^{-1}. Calculate the height of the dam wall.

4 A high-diver steps off a 10.0 m high platform and plunges into the pool below. Calculate the speed at which the diver hits the water.

5 If a high-jumper with a mass of 63.0 kg just clears a height of 2.10 m, what was the high-jumper's speed as they left the ground?

6 A girl throws a 198.4 g softball directly up into the air. It leaves her hand at a speed of 21.7 m s^{-1}.

- **a** Calculate the kinetic energy of the softball as it leaves the girl's hand.
- **b** If air resistance is ignored, what gain in gravitational potential energy occurs as the softball reaches the top of its flight?
- **c** If air resistance is ignored, calculate the height the softball reaches above the girl's hand.

Analysis

7 A box slides down a frictionless plane that is inclined at 35.0° to the horizontal.

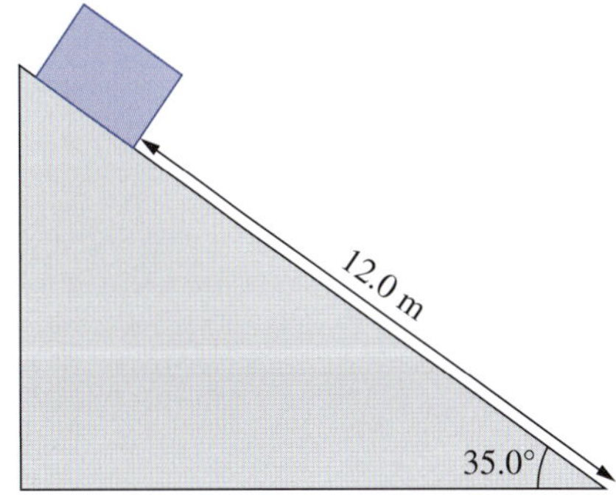

Use the law of conservation of energy to calculate the speed of the box after it has travelled 12.0 m down the plane.

continued over page

3.6 Review *continued*

8 A group of students sets up a pendulum with a thin 20.0 cm chain holding a heavy metal ball.

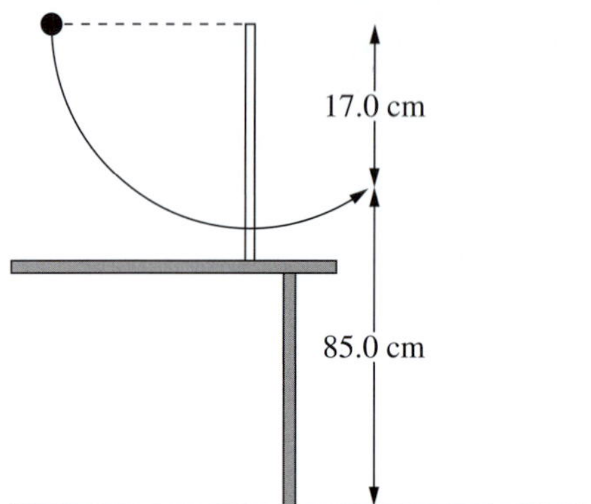

When the students tried to swing the pendulum for the first time, the chain broke just as the ball was 85.0 cm from the floor and 17.0 cm below the point at which it started to swing.

a Calculate the speed of the ball at the point at which the chain breaks.

b The students were not expecting the chain to break. Use the law of conservation of energy and the answer from part **a** to calculate the maximum height above the ground the ball would have achieved if the chain didn't break.

c Use the law of conservation of energy and the answer from part **b** to calculate the speed at which the ball strikes the ground.

d Use the law of conservation of energy and the starting point of the ball to calculate the speed at which the ball strikes the ground.

9 A 75.0 kg student swings out over a river on a rope attached to a tree on the riverbank. The student's final speed when they hit the water is 6.27 m s^{-1}.

a Calculate the kinetic energy of the student the moment before they hit the water.

b Determine the gravitational potential energy available to the student at the top of the riverbank before they began their swing.

c Using your answer to part **b**, calculate the height of the riverbank above the water level at the point where the student began their swing.

Chapter review

KEY TERMS

breaking point
conserved
deformation
elastic
elastic collision
elastic limit
elastic potential energy
gravitational potential energy
impulse
inelastic collision
isolated system
kinetic energy
law of conservation of energy
law of conservation of momentum
mechanical energy
momentum
spring constant
transform
work

REVIEW QUESTIONS

Knowledge and understanding

1 Arrange the following objects in order of decreasing magnitude of momentum.
  A 10.0 kg dog running west at 5.00 m s^{-1}
  B 42.0 kg child jogging south at 2.00 m s^{-1}
  C 25.0 kg fish swimming east at 3 m s^{-1}
  D 1250 kg car stationary at the traffic lights

2 Which alternative from the list below shows the unit for momentum that is equivalent to kg m s^{-1}?
  A J m
  B N s^{-1}
  C N s
  D J s^{-1}

3 In an explosive collision, a combined mass separates into two masses. If one of the masses has momentum of 345 kg m s^{-1} south, what is the momentum of the other mass?

4 Which of the following statements correctly describes impulse? More than one correct answer is possible.
  A Impulse is the rate of change of momentum.
  B Impulse is the final momentum minus the initial momentum.
  C Impulse is a scalar.
  D Impulse can be calculated from the force and the time over which the force acts.

5 Use the concept of impulse to explain how airbags can help reduce injury during a car crash.

6 In the case of a person pushing against a solid brick wall, explain why no work is being done.

7 Contrast the meanings of the words 'energy' and 'work'.

8 A group of students has conducted an investigation into the properties of an elastic band. They collected data by hanging different masses on the elastic band and measuring the extension from its original length. Unfortunately the students cannot agree on what the gradient of the graph represents and what the area under the graph represents. Explain how you could resolve their confusion using the equations for gradient and area, and the units for force and extension.

9 A squash ball that is repeatedly hit against a wall during a game becomes hot. Which of the following options explains this best?
  A The racquet gives the ball kinetic energy.
  B The impulse is positive.
  C The collisions are perfectly elastic.
  D Kinetic energy is not conserved in the collisions.

10 A student carries a fully loaded backpack along a horizontal footpath for 450 m on their way home from school. What work was done by the student on the backpack during this journey if they walked at a constant pace all the way?

11 An apple falls to the ground from a tree and strikes the ground with 45.5 J of kinetic energy. Ignoring any air resistance, how much gravitational potential energy did the apple have when it was on its branch?

12 A tennis ball is hit with the frame of a racquet and goes straight upwards. While it is travelling upwards it is slowing down until it reaches its maximum height, where its speed is zero. Where has all of the kinetic energy gone? In your answer ignore any air resistance.

13 A 70.0 kg rower steps out of a stationary boat with a velocity of 2.50 m s^{-1} onto a riverbank. The boat has a mass of 495 kg. With what velocity does the boat begin to move as the rower steps out?

continued over page

14 A spacecraft of mass 1.00×10^4 kg that is initially at rest burns 5.00 kg of fuel to produce an equal mass of exhaust gases. The gases are ejected at a velocity of 6.00×10^3 m s^{-1}. Calculate the velocity of the spacecraft after this burn.

15 A batter blocks a 165 g cricket ball travelling towards him at 104 km h^{-1}. The ball leaves the bat at 20.0 km h^{-1}. Calculate the magnitude of the change in momentum of the ball.

16 Calculate the magnitude of the average force required to be applied by the brakes of a 15.0 kg bicycle carrying a 65.0 kg rider if the bike and rider are travelling at 12.0 m s^{-1} and come to rest in 2.00 s.

17 A polar research worker uses a tractor to drag a sled with supplies across a glacier. The harness is held at an angle of 60.0° to the horizontal and applies a force of 316 N on the sled, which is initially at rest. A constant frictional force of 105 N acts on the sled as it is dragged for a distance of 245 m.

- **a** For this distance, calculate the work done by the tractor on the 152 kg sled.
- **b** Find the speed of the sled after travelling 245 m.

18 A student wanting to increase their upper-body strength decides to stretch a piece of bungee cord 150 times each morning before school. From the force vs extension graph for the cord given below, estimate how much energy is expended in the workout if the student stretches the cord from 0.5 m to 1.0 m each time.

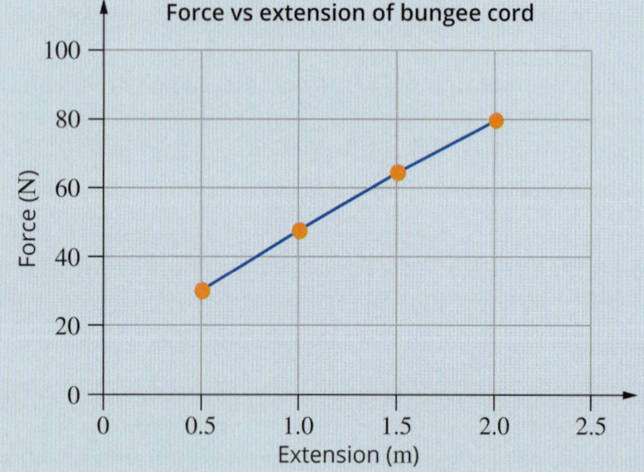

19 A steel cable 1.50 m long is stretched by fixing it at one end and applying a force to the other end. The graph of the force applied and the extension achieved is shown below.

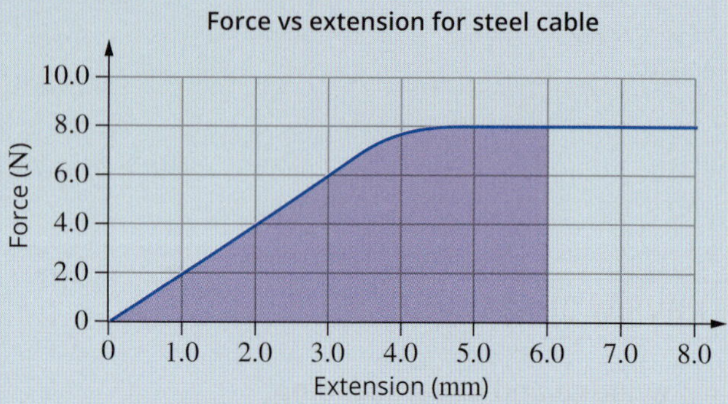

Estimate the elastic potential energy stored in the cable when stretched by a distance of 6.0 mm.

20 The mass of a motorbike and its rider is 232 kg. If they are travelling at 80.0 km h^{-1}, calculate their combined kinetic energy.

21 A car of mass 1540 kg is travelling at 17.0 m s^{-1}. How much work would its engine need to do to accelerate the car to 28.0 m s^{-1}?

22 A 57.0 g tennis ball is thrown 8.20 m into the air.

- **a** Calculate the gravitational potential energy of the ball at the top of its flight.
- **b** Calculate the gravitational potential energy of the ball when it has fallen halfway back to the ground.

23 When climbing Mount Everest (h = 8848 m), a mountain climber stops to rest at North Base Camp (h = 5150 m). If the climber has a mass of 65.0 kg, how much gravitational potential energy will she gain in the her climb from North Base Camp to the summit? For simplicity, assume that g is 9.80 m s^{-2} for the whole climb.

Application and analysis

24 Two identical bowling balls, each of mass 4.00 kg, move towards each other across a frictionless horizontal surface with equal speeds of 3.00 m s^{-1}. During the collision 20.0 J of kinetic energy is transformed into heat and sound. After the collision the balls move in opposite directions.

- **a** Is momentum conserved in this collision?
- **b** Is this an elastic or inelastic collision? Explain your answer.
- **c** Calculate the speed of each ball after the collision.

25 An 80.0 kg student jumps from a bridge at the end of a bungee rope. When the student drops the full length of the 134 m bungee rope it then stretches by 10.0% as the student comes to a stop. Calculate the spring constant of the rope.

26 A student throws a basketball upwards with an initial speed of $u\,\text{m s}^{-1}$ and notes that it reaches a maximum height above their hand of h m. If the student then throws the ball with an initial speed of $2u\,\text{m s}^{-1}$, how high will the ball go? Give your answer in terms of h.

27 Two children are standing on a bridge throwing stones into the river below. Susan throws a stone upwards, and Peter throws a stone downwards at the same speed. Select the correct answer from the following options and justify your choice.

A Both stones will hit the water at the same speed.

B The stone that is thrown downwards by Peter will hit the water at a greater speed than Susan's stone, which was thrown upwards.

C Susan's stone will hit the water at a greater speed than Peter's stone.

D More information is required to determine which stone hits the water at the greatest speed.

28 An 11.0 t satellite is in orbit at an altitude of 1130 km above the surface of the Earth. A booster rocket is fired, putting the satellite into an orbit of altitude 2130 km. The graph below shows how the gravitational field changes as the distance from the centre of the Earth varies. The radius of the Earth is 6.37×10^6 m and the mass of the Earth is 5.98×10^{24} kg.

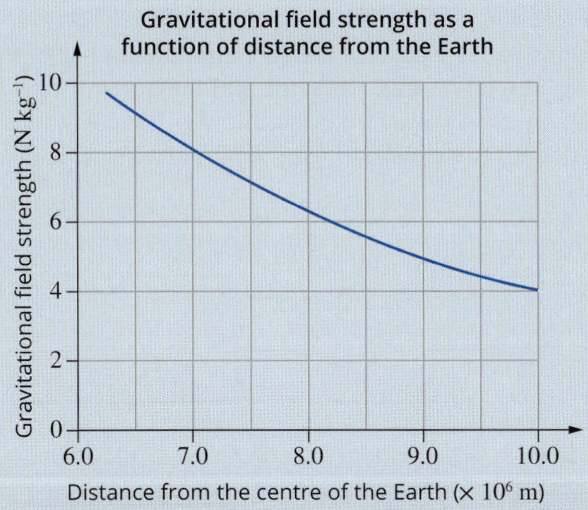

a Using the graph above, estimate the work done by the booster rocket in increasing the potential energy of the satellite.

b Calculate the kinetic energy of the satellite in its final orbit.

29 A new space telescope is 631 km above the surface of the Earth and in a circular orbit. Its mass is 1.10×10^7 kg. Use the graph below to estimate its gravitational potential energy relative to the surface of the Earth.

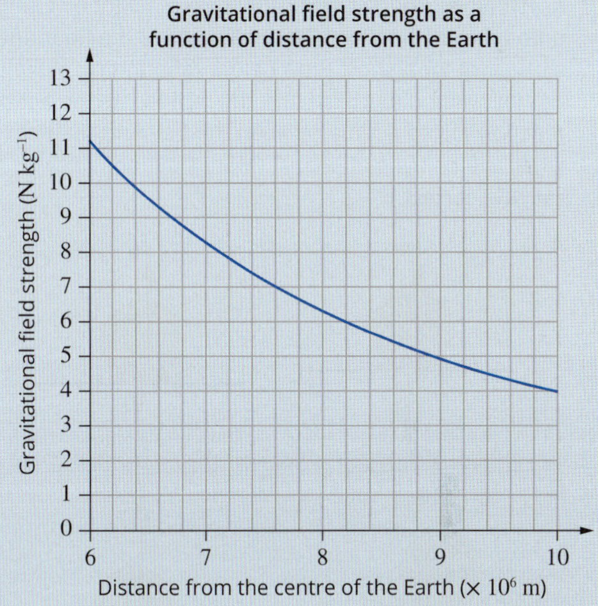

30 A 264 g toy truck with a springy bumper is travelling at $0.300\,\text{m s}^{-1}$. It collides with a 112 g toy car travelling in the same direction at $0.200\,\text{m s}^{-1}$. The car moves forwards at a speed of $0.300\,\text{m s}^{-1}$.

a Calculate the speed of the truck after the collision.

b Calculate the total kinetic energy of the system before the collision.

c Calculate the total kinetic energy of the system after the collision.

d Complete the following statements by selecting the appropriate option from those in bold.

i The total kinetic energy before the collision is **more than/less than/equal to** the total kinetic energy after the collision.

ii The kinetic energy of the system of toys **is/is not** conserved.

iii The total energy of the system of toys **is/is not** conserved.

iv The total momentum of the system of toys **is/is not** conserved.

v The collision **is/is not** perfectly elastic because **kinetic energy/total energy/momentum** is not conserved.

UNIT 3 • Area of Study 1

REVIEW QUESTIONS

How do physicists explain motion in two dimensions?

Multiple-choice questions

1 A netball is dropped vertically from a height of 1.5 m onto a horizontal floor. The ball reaches the floor and is stationary for a short time before rebounding. Which of the following diagrams correctly represents the action–reaction forces acting between the ball and the floor when the ball is stationary? (More than one answer may be correct.)

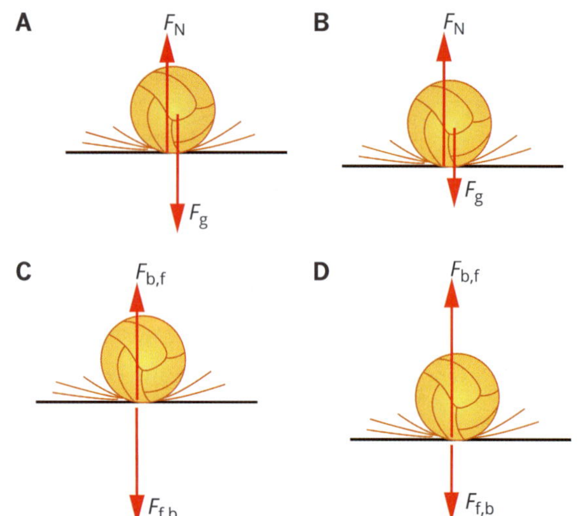

2 Which of the following diagrams best shows the forces acting on a ball thrown diagonally as it travels through the air?

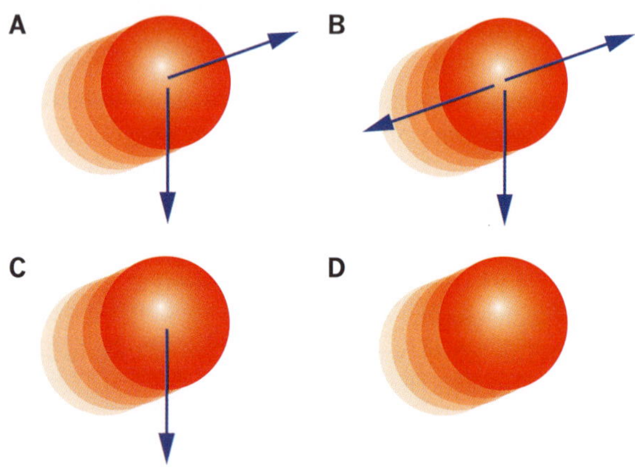

The following information relates to questions 3–5.

A skateboarder of mass 55 kg is practising on a half-pipe of radius 2.0 m. At the lowest point of the half-pipe, the speed of the skater is 6.0 m s^{-1}.

3 What is the magnitude and direction of the acceleration of the skater at this point?

A 3 m s^{-2} up
B 3 m s^{-2} down
C 18 m s^{-2} up
D 18 m s^{-2} down

4 What is the size of the normal force acting on the skater at this point?

A 3.7×10^2 N
B 4.5×10^2 N
C 7.0×10^2 N
D 1.5×10^3 N

5 What effect does the normal force have on the feeling experienced by the skater as they travel through the lowest point in the pipe?

A The skater feels lighter than they would if they were stationary.
B The skater feels heavier than they would if they were stationary.
C The skater feels exactly the same as they would if they were stationary.
D There is not enough information to tell.

The following information relates to questions 6–9.

A tennis ball is thrown at a stationary bowling ball of mass 5.0 kg. The tennis ball rebounds and the bowling ball rolls very slowly away from the thrower. The collision is considered to be inelastic. Use the graphs below to answer questions 6 and 7.

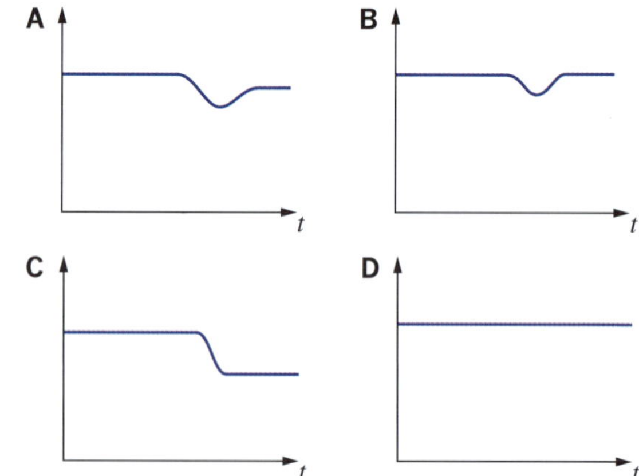

6 Which graph best shows the total kinetic energy of the system before, during and after the collision?

7 Which graph best shows the total momentum of the system before, during and after the collision?

8 How does the change in momentum of the tennis ball compare with the change in momentum of the bowling ball?

A They are equal.

B The tennis ball experiences a greater change of momentum.

C The bowling ball experiences a greater change of momentum.

D They are equal in magnitude and opposite in direction.

9 How do the forces that the two balls exert on each other during the collision compare?

A The forces are equal.

B The tennis ball exerts the greater force.

C The bowling ball exerts the greater force.

D The forces are equal in magnitude and opposite in direction.

The following information relates to questions 10–12.

As part of an investigation into energy conservation, students attach a 2.0 kg mass to a spring with a spring constant of 100 N m^{-1}. The mass is initially supported at position A so that there is no spring extension. When the mass is released, the students observe it falling through position B until it momentarily comes to rest at position C. The mass then passes through position D and returns to near its original position, oscillating back and forth for some time before coming to rest.

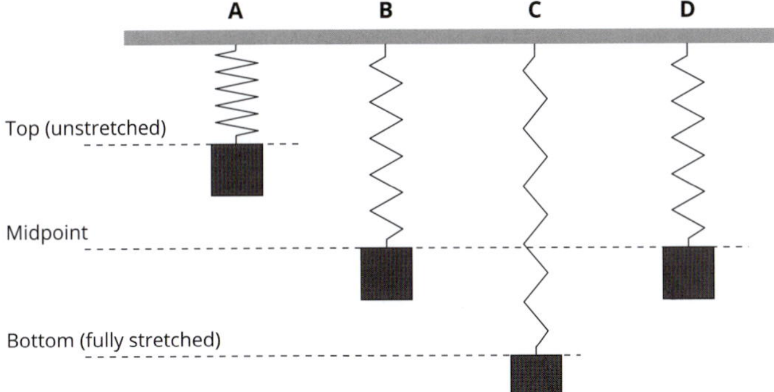

10 In the diagram below left, at which position (A, B, C or D, and more than one may be correct) is there:

a the greatest elastic potential energy?

b the greatest gravitational potential energy?

c the greatest kinetic energy?

11 How far has the spring extended when the mass is at position C?

A 0.10 cm

B 0.39 cm

C 10 cm

D 39 cm

12 How far has the spring extended when the mass stops oscillating?

A 0.10 cm

B 0.20 cm

C 10 cm

D 20 cm

Short-answer questions

13 In the Gravitron ride, patrons enter a cylindrical chamber which rotates rapidly, causing them to be pinned to the walls as the floor drops away. A particular Gravitron has a radius of 5.00 m and rotates with a period of 2.50 s. Jodie, of mass 60.0 kg, is on the ride.

a Choose the correct responses in the following statement from the options given in bold:
As the Gravitron spins at a uniform rate and Jodie is pinned to the wall, the horizontal forces acting on her are **balanced/unbalanced** and the vertical forces are **balanced/unbalanced**.

b Calculate Jodie's speed as she revolves on the ride.

c What is the magnitude of her centripetal acceleration?

d Calculate the magnitude of the normal force that acts on Jodie from the wall of the Gravitron.

14 The diagram below shows a car travelling at $40\,m\,s^{-1}$ on a banked racing track. The track is banked so that when a car corners at $40\,m\,s^{-1}$, it experiences no sideways frictional forces.
The track is circular with a radius of 150 m.

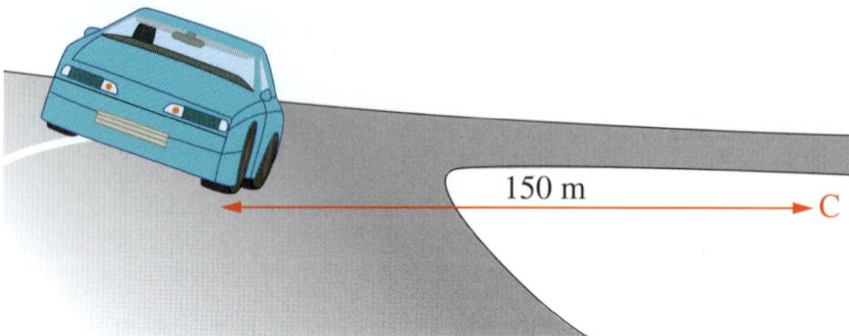

a Draw and identify the forces that are acting on the car at this instant.

b Calculate the angle to the horizontal at which the track is banked.

15 Two friends, Elvis and Kurt, are having a game of catch. Elvis throws a baseball to Kurt who is standing 8.0 m away. Kurt catches the ball 2.0 s later at the same height, 2.0 m, from which it is thrown. The mass of the baseball is 250 g. Ignore the effects of air resistance.

a What is the maximum height of the ball during its flight?

b What is the acceleration of the ball at its maximum height?

c Calculate the speed at which the ball was thrown.

16 A ball bearing of mass 25 g is rolled along a smooth track a section of which is in the shape of a loop. The ball bearing is given a launch speed at point A so that it just maintains contact with the track as it passes through point C. Ignore drag forces when answering the following questions.

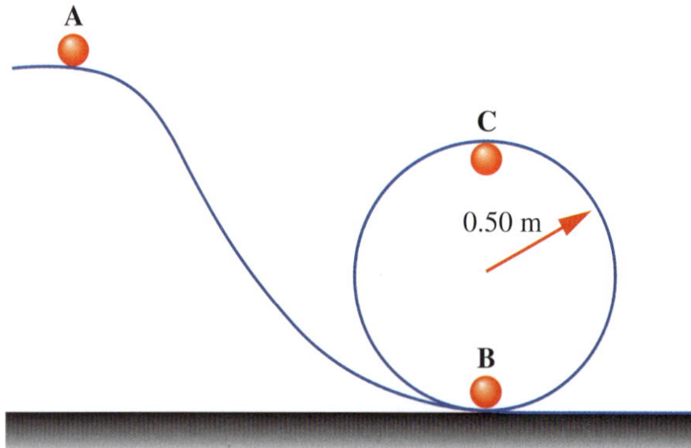

a Determine the magnitude of the acceleration of the ball bearing as it passes point C.

b How fast is the ball bearing travelling at point C?

c How fast is the ball bearing travelling at point B?

17 A small-time gold prospector sets up a cable-pulley system that allows him to move a trolley full of ore of total mass 200 kg from rest a distance of 20 m along a level section of rail track, as shown in the following diagram.

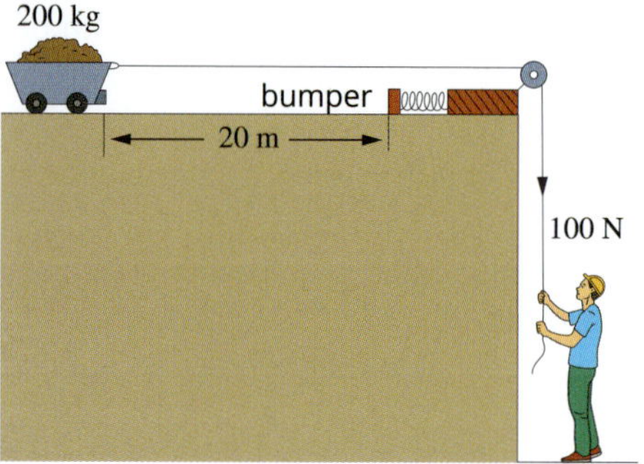

When the load reaches the end of the track, it is momentarily brought to rest by a powerful spring-bumper system, which can be assumed to have negligible mass. A constant frictional force of 30 N acts on the wheels of the trolley. The prospector applies a constant force of 100 N to the cable as the trolley moves along the track. Assume that there is negligible friction between the pulley and the cable.

a How much work is done on the trolley as it moves along the track?

b Calculate the change in kinetic energy of the load as it moves along the track.

c What is the speed of the load when it reaches the end of the track?

d The spring bumper has a force constant of $1500\,N\,m^{-1}$. How much kinetic energy does the trolley lose as the spring is compressed by 18 cm?

18 At football training, players are throwing themselves at a large tackle bag of mass 45 kg. A ruckman of mass 120 kg running at $6.0\,m\,s^{-1}$ crashes into the stationary bag and carries it forward.

a At the moment of impact, what is the combined speed of the bag and the ruckman?

b **i** How much momentum does the ruckman lose?

ii How much momentum does the tackle bag gain?

c What is the force experienced by the ruckman if the collision with the tackle bag occurs over 120 ms?

d Is the collision of the ruckman and the tackle bag elastic or inelastic? Use calculations to justify your answer.

19 A physics student decides to study the properties of a bungee rope by recording the extension produced by various masses attached to the end of the rope. The results of the experiment are shown in the following table.

Mass (kg)	Extension (m)
0.5	0.24
1.0	0.52
1.5	0.73
2.0	0.95
2.5	1.20
3.0	1.48
3.5	1.70

a Draw the force versus extension graph for the bungee rope.

b Estimate the value of the spring constant of the bungee rope.

During an investigation, a student stretches the rope horizontally by 15 m.

c Assuming that the bungee rope behaves ideally, determine the potential energy stored in the rope at this point.

The student stands on a skateboard and allows the rope, stretched by 15 m, to drag her across the smooth floor of the school gymnasium. The combined mass of the student and her board is 60 kg.

d Calculate the maximum speed that the student reaches as she is pulled by the bungee cord.

20 Aristotle suggested that the natural state of motion of any object is rest. Galileo introduced the principle of inertia and claimed that the natural state of motion of any object is constant velocity (zero velocity being just one example). Explain why Aristotle's view was so hard to overcome and why, if we had spent time as an astronaut on a space station, Galileo's principle would have been much easier to accept.

WS 9

EQ U301

CHAPTER

04 Gravity

Gravity is, quite literally, the force that drives the universe. It was gravity that first caused particles to coalesce into atoms, and atoms to congregate into nebulas, planets and stars. It holds the planets in orbit around the Sun, and the Moon in orbit around the Earth. An understanding of gravity is fundamental to understanding the universe.

This chapter considers Newton's law of universal gravitation. This law can be used to predict the size of the force experienced by an object at various locations on the Earth and on other planets. It will also be used to develop the idea of a gravitational field. Since the concept of a field is also used to describe other basic forces—such as electromagnetism and the strong and weak nuclear forces—this chapter provides an important foundation for further study in physics.

Key knowledge

- describe gravitation, magnetism and electricity using a field model **4.2**
- investigate and compare theoretically and practically gravitational, magnetic and electric fields, including directions and shapes of fields, attractive and repulsive effects, and the existence of dipoles and monopoles **4.2**
- investigate and compare theoretically and practically gravitational fields and electrical fields about a point mass or charge (positive or negative) with reference to:
  - the direction of the field **4.2**
  - the shape of the field **4.2**
  - the use of the inverse square law to determine the magnitude of the field **4.1, 4.2**
  - potential energy changes (qualitative) associated with a point mass or charge moving in the field **4.3**
- analyse the use of gravitational fields to accelerate mass, including
  - gravitational field and gravitational force concepts: $g = G\frac{M}{r^2}$ and $F_g = G\frac{m_1 m_2}{r^2}$ **4.1, 4.2**
  - potential energy changes in a uniform gravitational field: $E_g = mg\Delta h$ **4.3**
- analyse the change in gravitational potential energy from area under a force vs distance graph and area under a field vs distance graph multiplied by mass **4.3**
- identify fields as static or changing, and as uniform or non-uniform **4.2**

OA
✓✓

4.1 Newton's law of universal gravitation

FIGURE 4.1.1 Sir Isaac Newton is one of the most influential physicists who ever lived.

In 1687 Sir Isaac Newton (Figure 4.1.1) published a book that changed the world. Entitled *Philosophiæ Naturalis Principia Mathematica* (Mathematical Principles of Natural Philosophy), Newton's book (Figure 4.1.2) used a new form of mathematics now known as calculus and outlined his famous laws of motion.

The *Principia* also introduced **Newton's law of universal gravitation**. This was particularly significant because, for the first time in history, it scientifically explained the motion of the planets. This led to a change in humanity's understanding of its place in the universe.

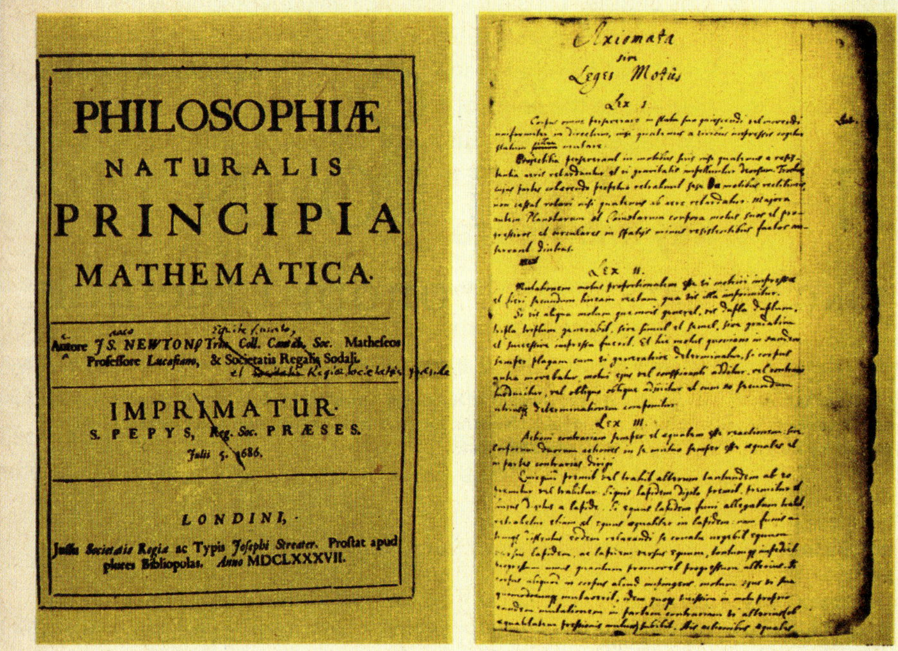
PHILOSOPHIÆ
NATURALIS
PRINCIPIA
MATHEMATICA.

Autore J S. NEWTON, Trin. Coll. Cantab. Soc. Matheseos Professore Lucasiano, & Societatis Regalis Sodali.

IMPRIMATUR.
S. PEPYS, Reg. Soc. PRÆSES.
Julii 5. 1686.

LONDINI,
Jussu Societatis Regiæ ac Typis Josephi Streater. Prostat apud plures Bibliopolas. Anno MDCLXXXVII.

FIGURE 4.1.2 The *Principia* is one of the most influential books in the history of science.

UNIVERSAL GRAVITATION

Newton's law of universal gravitation states that any two bodies in the universe attract each other with a **gravitational force** that is directly proportional to the product of their masses and inversely proportional to the square of the distance between them.

Mathematically, Newton's law of universal gravitation can be expressed as follows.

$$F_g = G\frac{m_1 m_2}{r^2}$$

where F_g is the gravitational force (N)

m_1 is the mass of object 1 (kg)

m_2 is the mass of object 2 (kg)

r is the distance between the centres of m_1 and m_2 (m)

G is the **gravitational constant**, 6.67×10^{-11} N m^2 kg^{-2}

CASE STUDY ANALYSIS

Measuring the gravitational constant, G

The gravitational constant, G, was first accurately measured by the British scientist Henry Cavendish in 1798, more than a century after Newton's death. Cavendish used a torsion balance (Figure 4.1.3), a device that can measure very small twisting forces, to conduct his experiment. Cavendish's experiment could measure forces smaller than 1 μN (i.e. 10^{-6} N). He used this balance to measure the force of attraction between two lead balls held a small distance apart. Once the size of the force was known for a given combination of masses at a known separation distance, a value for G could be determined.

Analysis

This case study describes Henry Cavendish's 1798 experiment. The experiment used two pairs of lead balls: a pair each of mass 158 kg and a pair each of mass 0.730 kg. The different sized balls were held 230 mm apart.

1 Using $G = 6.67 \times 10^{-11}$ N m^2 kg^{-2}, calculate the magnitude of the force of gravitational attraction between one of the large and one of the small lead balls.

The value of the gravitational constant, G, used in this course is 6.67×10^{-11} N m^2 kg^{-2}. However, Henry Cavendish calculated it to be 6.75×10^{-11} N m^2 kg^{-2}.

2 What is the percentage error in Henry Cavendish's result compared with the value we use in this course?

FIGURE 4.1.3 A model of the torsion balance that Henry Cavendish used to measure the small twisting force created by the gravitational attraction of lead balls.

The fact that r appears in the denominator of Newton's law of universal gravitation indicates an inverse relationship. Since r is also squared, this relationship is known as the **inverse square law**. The implication is that as r increases, F_g will decrease dramatically. This type of law will appear again later in the chapter.

As its name suggests, the law of universal gravitation predicts that any two objects that have mass will attract each other. However, because the value of G is so small, the gravitational force between two everyday objects is too small to be noticed. This is demonstrated in Worked example 4.1.1, where the result is an attractive force that is negligible.

PHYSICSFILE

Universal laws

The use of the term 'universal' in Newton's law assumes that this law applies throughout the universe. While many measurements have supported this law, the universe is so large that our measurements can only sample a tiny fraction of the universe. We may learn one day that there are regions of the universe where this law doesn't apply.

Worked example 4.1.1

GRAVITATIONAL ATTRACTION BETWEEN SMALL OBJECTS

Two bodies with masses 90 kg and 75 kg are at a distance of 80 cm between their centres. Calculate the force of gravitational attraction between them.

Thinking	Working
Recall the equation for Newton's law of universal gravitation.	$F_g = G\frac{m_1 m_2}{r^2}$
Note the information provided and convert values into appropriate units where necessary.	$G = 6.67 \times 10^{-11}\,\text{N m}^2\,\text{kg}^{-2}$ $m_1 = 90\,\text{kg}$ $m_2 = 75\,\text{kg}$ $r = 80\,\text{cm} = 0.80\,\text{m}$
Substitute the values into Newton's equation.	$F_g = 6.67 \times 10^{-11} \times \frac{90 \times 75}{0.80^2}$
Solve the equation.	$F_g = 7.0 \times 10^{-7}\,\text{N}$

Worked example: Try yourself 4.1.1

GRAVITATIONAL ATTRACTION BETWEEN SMALL OBJECTS

Two bowling balls are sitting next to each other on a shelf. The centres of the balls are 60 cm apart. Ball 1 has a mass of 7.0 kg and ball 2 has a mass of 5.5 kg. Calculate the force of gravitational attraction between them.

GRAVITATIONAL ATTRACTION BETWEEN MASSIVE OBJECTS

Gravitational forces between everyday objects are so small (as seen in Worked example 4.1.1) that they are hard to detect without special equipment and can usually be considered negligible.

For the gravitational force between objects to become significant, at least one of the objects must have a very large mass, such as the mass of a planet (Figure 4.1.4).

FIGURE 4.1.4 Gravitational forces between objects only become significant when at least one of the objects has a large mass, e.g. the Earth and the Moon.

Worked example 4.1.2

GRAVITATIONAL ATTRACTION BETWEEN MASSIVE OBJECTS

Calculate the force of gravitational attraction between the Earth and the Moon given the following data:

$m_{Earth} = 5.98 \times 10^{24}$ kg

$m_{Moon} = 7.35 \times 10^{22}$ kg

$r_{Moon-Earth} = 3.84 \times 10^{8}$ m

Thinking	Working
Recall the equation for Newton's law of universal gravitation.	$F_g = G\frac{m_1 m_2}{r^2}$
Note the information provided.	$G = 6.67 \times 10^{-11}$ N m² kg^{-2} $m_1 = 5.98 \times 10^{24}$ kg $m_2 = 7.35 \times 10^{22}$ kg $r = 3.84 \times 10^{8}$ m
Substitute the values into Newton's equation.	$F_g = 6.67 \times 10^{-11} \times \frac{5.98 \times 10^{24} \times 7.35 \times 10^{22}}{(3.84 \times 10^{8})^2}$
Solve the equation.	$F_g = 1.99 \times 10^{20}$ N

Worked example: Try yourself 4.1.2

GRAVITATIONAL ATTRACTION BETWEEN MASSIVE OBJECTS

Calculate the force of gravitational attraction between the Sun and the Earth given the following data:

$m_{Sun} = 1.99 \times 10^{30}$ kg

$m_{Earth} = 5.98 \times 10^{24}$ kg

$r_{Sun-Earth} = 1.50 \times 10^{11}$ m

The force calculated in Worked example 4.1.2 is much greater than the force calculated in Worked example 4.1.1, illustrating the difference in the size of the gravitational force when at least one of the objects has a very large mass.

EFFECT OF GRAVITY

According to Newton's third law of motion, forces occur in action–reaction pairs. An example of such a pair is shown in Figure 4.1.5. The Earth exerts a gravitational force on the Moon and, conversely, the Moon exerts an equal and opposite force on the Earth. Using Newton's second law of motion, you can see that the effect of the gravitational force of the Moon on the Earth will be much smaller than the corresponding effect of the Earth on the Moon. This is because of the Earth's larger mass.

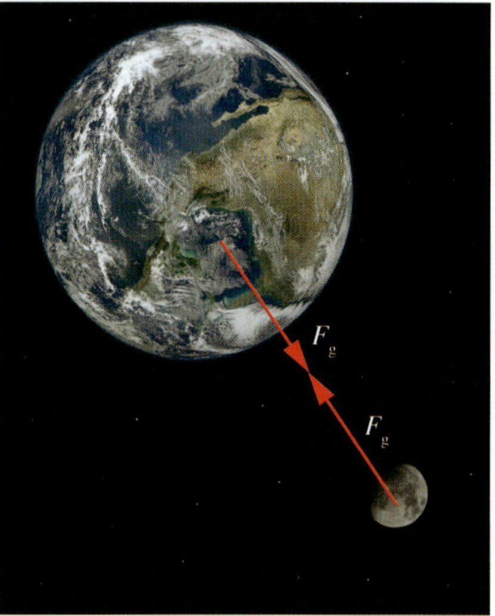

FIGURE 4.1.5 The Earth and the Moon exert gravitational forces on each other.

Worked example 4.1.3

ACCELERATION CAUSED BY A GRAVITATIONAL FORCE

The force of gravitational attraction between the Moon and the Earth is approximately 2.0×10^{20} N. Calculate the acceleration of the Earth and the Moon caused by this force. Compare these accelerations by calculating the ratio $\frac{a_{Moon}}{a_{Earth}}$.

Use the following data:

$m_{Earth} = 5.98 \times 10^{24}$ kg

$m_{Moon} = 7.35 \times 10^{22}$ kg

Thinking	Working
Recall the equation for Newton's second law of motion.	$F_{net} = ma$
Transpose the equation to make *a* the subject.	$a = \frac{F_{net}}{m}$
Substitute values into this equation to find the accelerations of the Moon and the Earth.	$a_{Earth} = \frac{2.0 \times 10^{20}}{5.98 \times 10^{24}} = 3.3 \times 10^{-5}\,\text{ms}^{-2}$ $a_{Moon} = \frac{2.0 \times 10^{20}}{7.35 \times 10^{22}} = 2.7 \times 10^{-3}\,\text{ms}^{-2}$
Compare the two accelerations.	$\frac{a_{Moon}}{a_{Earth}} = \frac{2.7 \times 10^{-3}}{3.3 \times 10^{-5}} = 82$ The acceleration of the Moon is approximately 82 times greater than the acceleration of the Earth.

Worked example: Try yourself 4.1.3

ACCELERATION CAUSED BY A GRAVITATIONAL FORCE

The force of gravitational attraction between the Sun and the Earth is approximately 3.5×10^{22} N. Calculate the acceleration of the Earth and the Sun caused by this force. Compare these accelerations by calculating the ratio $\frac{a_{Earth}}{a_{Sun}}$.

Use the following data:

$m_{Earth} = 5.98 \times 10^{24}$ kg

$m_{Sun} = 1.99 \times 10^{30}$ kg

Gravity in the solar system

Although the accelerations caused by gravitational forces in Worked example 4.1.3 are small, over billions of years they created the motion of the solar system.

In the Earth–Moon system, the acceleration of the Moon is many times greater than that of the Earth, which is why the Moon orbits the Earth. Although the Moon's gravitational force causes a much smaller acceleration of the Earth, it does have other significant effects, such as causing the tides.

Similarly, the Earth and other planets orbit the Sun because their masses are much smaller than the Sun's mass. However, the combined gravitational effect of the planets of the solar system (and Jupiter in particular) causes the Sun to wobble slightly as the planets orbit it.

CASE STUDY **ANALYSIS**

Extrasolar planets

Planets orbiting stars other than the Sun are known as extrasolar planets (or 'exoplanets'). One of the ways in which exoplanets can be detected is by the gravitational effect they have on their host star.

When a planet orbits a star, the gravitational force between them causes the star to wobble. Sometimes this wobble is big enough to be detected on the Earth. Exoplanets that have been detected in this way range in size from the mass of the Earth to many times the mass of Jupiter.

Consider Newton's law of universal gravitation: $F_g = G\frac{m_1m_2}{r^2}$. From this equation it follows that the greater the mass of a planet, the greater the gravitational force between that planet and its host star. Further, the greater the gravitational force between a planet and its host star, the greater is the host star's wobble. Similarly, the closer the planet's orbit is to the star (i.e. the smaller r is), the greater the gravitational force and therefore the greater the wobble. This implies that the easiest exoplanets to spot are those known as 'hot Jupiters'. These are planets with a mass at least that of Jupiter that orbit their star closer than Mercury orbits the Sun. They are the planets that exert the greatest forces on their host star.

Hot Jupiters were the first exoplanets to be discovered using the gravitational wobble method. In fact, before hot Jupiters were detected, astronomers didn't believe that planets that large could form so close to a star. In 1995 the very first exoplanet discovered orbiting a regular star, 51 Pegasi b, revolutionised our understanding of how solar systems form.

Since then, astronomers' instruments have improved and can now detect the gravitational tug of much smaller planets.

In 2016 it was discovered that Proxima Centauri, the star closest to our solar system, has at least one planet not much larger than the Earth tugging at the star. With such a small planet, the effect on the star's orbit is tiny—a change in velocity of only $1.4\,m\,s^{-1}$. The instrumentation used by astronomers is precise enough to detect such small changes in velocity.

Instruments at the observatory at Mt Kent in southern Queensland can detect even smaller gravitational wobbles in distant stars, changes as small as $1\,m\,s^{-1}$, or about the same as a very slow walking pace.

Queensland astronomer Jonti Horner (Figure 4.1.6) has discovered many exoplanets using gravitational wobble and other planet-hunting methods. But he is still impressed by the science behind these discoveries: 'Stars are trillions or quadrillions of kilometres away. Proxima Centauri, the nearest, is more than 40 trillion km distant. I find it astonishing we can measure something that is so distant and see a wobble that is comparable to the speed at which you or I would walk around the shops!'

FIGURE 4.1.6 Queensland astronomer Jonti Horner identifies distant exoplanets by the gravitational pull they have on their host star.

Analysis

1 Explain the term 'hot Jupiters' and how they are reasonably easy for astronomers to detect.

2 Astronomers can discover planets orbiting distant stars by observing the effect of the exoplanet's gravitational pull on its host star. The huge exoplanet Hypatia has a mass of 1.68×10^{28} kg (almost 10 times the mass of Jupiter).

   a Calculate the magnitude of the gravitational force that the planet exerts on its host star given that Hypatia orbits at an average distance of 195 000 000 km and the mass of its host star is 3.62×10^{30} kg.

   b Calculate the magnitude of the resulting acceleration of the planet's host star.

GRAVITATIONAL FORCE

In Unit 2 Physics, the force on an object due to gravity was calculated using the equation $F_g = mg$. Worked example 4.1.4 below shows that the equation $F_g = mg$ and Newton's law of universal gravitation give the same answer for the gravitational force acting on objects on the Earth's surface. It is important to note that the distance used in these calculations is the distance between the *centres* of the two objects, which, in Worked example 4.1.4, is effectively the radius of the Earth. This should not be confused with **altitude**, which is the height above the Earth's surface.

Worked example 4.1.4

GRAVITATIONAL FORCE

Compare the force due to gravity on a 40 kg person calculated using $F_g = mg$ with the gravitational force calculated using $F_g = G\frac{m_1m_2}{r^2}$.

Use the following data:

$g = 9.8\,\text{m}\,\text{s}^{-2}$

$m_{\text{Earth}} = 5.98 \times 10^{24}\,\text{kg}$

$r_{\text{Earth}} = 6.37 \times 10^{6}\,\text{m}$

Thinking	Working
Apply the equation $F_g = mg$.	$F_g = mg$ $= 40 \times 9.8$ $= 392$ $= 3.9 \times 10^2$ N to 2 significant figures
Apply Newton's law of universal gravitation.	$F_g = G\frac{m_1m_2}{r^2}$ $= 6.67 \times 10^{-11} \times \frac{5.98 \times 10^{24} \times 40}{(6.37 \times 10^6)^2}$ $= 393$ $= 3.9 \times 10^2$ N to 2 significant figures
Compare the two values.	The equations give the same result when rounded to 2 significant figures.

Worked example: Try yourself 4.1.4

GRAVITATIONAL FORCE

Compare the force due to gravity on a 1.0 kg mass on the Earth's surface calculated using the equations $F_g = mg$ and $F_g = G\frac{m_1m_2}{r^2}$.

Use the following data:

$g = 9.8\,\text{m}\,\text{s}^{-2}$

$m_{\text{Earth}} = 5.98 \times 10^{24}\,\text{kg}$

$r_{\text{Earth}} = 6.37 \times 10^{6}\,\text{m}$

Worked example 4.1.4 shows that the constant for the **acceleration due to gravity**, g, can be derived directly from the dimensions of the Earth. Given that:

- an object with mass m on the surface of the Earth is at a distance of 6.37×10^6 m from the centre of the Earth and
- the mass of the Earth is 5.98×10^{24} kg

the force due to gravity can be derived from mg and Newton's law of universal gravitation:

$$mg = G\frac{m_{\text{Earth}}m}{(r_{\text{Earth}})^2}$$

$$mg = mG\frac{m_{\text{Earth}}}{(r_{\text{Earth}})^2}$$

$$g = G\frac{m_{\text{Earth}}}{(r_{\text{Earth}})^2}$$

$$g = 6.67\times10^{-11}\times\frac{5.98\times10^{24}}{(6.37\times10^6)^2}$$

$$g = 9.8\,\text{m}\,\text{s}^{-2}$$

Thus the rate of acceleration of objects near the surface of the Earth is a result of the Earth's mass and radius. A planet with a different mass and/or different radius will therefore have a different value for g. Likewise, if an object is above the Earth's surface, the value of r will be greater and thus the value of g will be smaller (since r is in the denominator of the equation). This is why the strength of the Earth's gravity reduces as you travel away from the Earth.

FORCES DURING VERTICAL ACCELERATION

A person accelerating upwards or downwards in a lift will experience different forces acting on them compared to when they are standing on the ground. This is because the contact force between the person and the ground (the **normal force**) changes as they accelerate upwards or downwards.

In Worked example 4.1.4 you explored the force due to gravity. When a person is standing on the ground, the force they experience due to gravity (F_g) is equal to the normal force (F_N) of the ground pushing up against their feet. The result is that the person does not accelerate upwards or downwards. They remain stationary on the ground. However, when they are accelerating upwards or downwards in a lift, F_N is no longer equal to F_g. If F_N is greater than F_g (as when they are accelerating upwards), the person will feel heavier than when they are standing on the ground. If F_N is less than F_g (as when they are accelerating downwards), they will feel lighter than when they are standing on the ground.

PHYSICSFILE

g at the ISS

The International Space Station (ISS) orbits the Earth at an altitude of approximately 400 km. It is easy to think that because astronauts on the ISS are floating they must experience little or no gravitational forces. In fact the value of g at the ISS is only a little less than that on the Earth: it is approximately $8.7\,\text{N}\,\text{kg}^{-1}$. The astronauts float because they are in free fall. You will learn more about this in Chapter 6.

Worked example 4.1.5

FORCES DURING VERTICAL ACCELERATION

A 74 kg person is standing in a lift that is accelerating upwards at 1.5 m s^{-2}. Calculate the force on the person due to gravity and the normal force they experience. Describe that person's experience. Assume that $g = 9.8$ m s^{-2}.

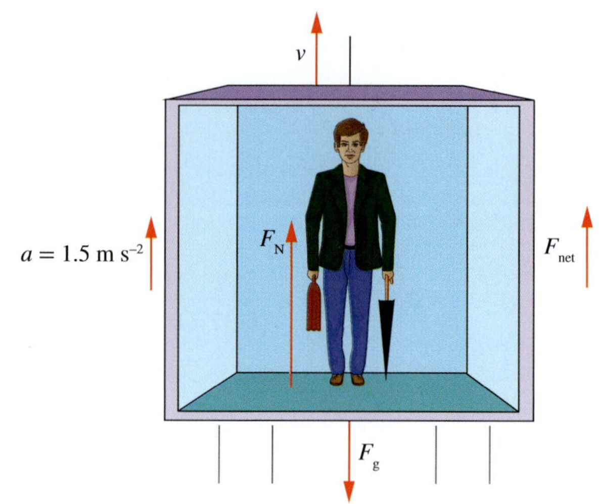

Thinking	Working
Calculate the force due to gravity on the person using $F_g = mg$.	$F_g = mg = 74 \times 9.8 = 725$ N
Calculate the force required to accelerate the person upwards at 1.5 m s^{-2}.	$F_{net} = ma = 74 \times 1.5 = 111$ N
The net force that causes the acceleration results from the normal force (upwards) and the force due to gravity (downwards). Since the lift is accelerating upwards, $F_N > F_g$.	$F_{net} = 111$ $F_N - F_g = 111$ $F_N - 725 = 111$ $F_N = 725 + 111$ $= 836$ $= 8.4 \times 10^2$ N The person will feel heavier than when they are standing on the ground.

Worked example: Try yourself 4.1.5

FORCES DURING VERTICAL ACCELERATION

Calculate the force due to gravity and the normal force acting on a 90 kg person in a lift that is accelerating downwards at 0.80 m s^{-2}. Describe that person's experience. Assume that $g = 9.8$ m s^{-2}.

4.1 Review

OA ✓✓

SUMMARY

- All objects with mass attract one another with a gravitational force. The gravitational force acts equally on each of the masses.
- The magnitude of the gravitational force is given by Newton's law of universal gravitation:

$$F_g = G\frac{m_1 m_2}{r^2}$$

- Gravitational forces are usually negligible unless at least one of the objects is massive, e.g. a planet.
- The force due to gravity on an object at the Earth's surface is due to the gravitational attraction of the Earth, F_g.
- The acceleration due to gravity of an object near the Earth's surface can be calculated using the dimensions of the Earth:

$$g = G\frac{m_{Earth}}{(r_{Earth})^2} = 9.8\,\text{m s}^{-2}$$

- If you experience a normal force (F_N) that is *greater than* the force due to gravity (F_g) you will feel heavier than when you are standing stationary on the ground. Similarly, you will feel lighter if the normal force you experience is *less than* the force due to gravity.

KEY QUESTIONS

Knowledge and understanding

1 Newton's law of universal gravitation tells us that a gravitational force exists between all objects. However, the magnitude of the force can be insignificant in many circumstances. Describe a situation where the gravitational force between objects is negligible.

2 What does the symbol r represent in Newton's law of universal gravitation? In what unit is it measured?

3 Describe what happens to the gravitational force acting between two masses m_1 and m_2 a distance r apart in each case below.

a The mass of m_1 is doubled.

b The distance r is doubled.

c The distance r is quadrupled.

4 The gravitational attraction between the Sun and Venus keeps the planet in orbit.

a Calculate the force of gravitational attraction between the Sun and Venus given the following data:

$m_{Sun} = 2.0 \times 10^{30}\,\text{kg}$

$m_{Venus} = 4.9 \times 10^{24}\,\text{kg}$

$r_{Sun\text{–}Venus} = 1.1 \times 10^{11}\,\text{m}$

b Calculate the acceleration of Venus towards the Sun.

5 Deimos is a small, rocky moon of Mars. The acceleration of Deimos caused by the gravitational force of Mars is much larger than the acceleration of Mars due to the gravitational force of Deimos. Explain why this is the case.

6 Two astronauts, each of mass 150 kg (including their suits), float in outer space 1.00 m apart.

a Calculate the gravitational force between the astronauts.

b Calculate the resulting acceleration of each astronaut.

7 Calculate the force due to gravity on an 80.0 kg astronaut standing on the surface of Mars. Mars has a mass of $6.40 \times 10^{23}\,\text{kg}$ and a radius of $3.40 \times 10^{6}\,\text{m}$.

8 Calculate the normal force acting on a 60.0 kg person in a lift under the following circumstances.

a accelerating upwards at $1.40\,\text{m s}^{-2}$

b moving upwards at a constant speed of $8.00\,\text{m s}^{-1}$

continued over page

4.1 Review *continued*

Analysis

9 In December 2020, Jupiter and Saturn coincided to form a great conjunction. The two planets, when viewed from the Earth, were about a tenth of a degree apart, so close that, to the naked eye, they looked like a single point. However, they were 734 million km apart.

Use the following data to answer the questions below:

$m_{Jupiter} = 1.90 \times 10^{27}$ kg

$m_{Saturn} = 5.68 \times 10^{26}$ kg

$m_{Sun} = 1.99 \times 10^{30}$ kg

a Calculate the gravitational force between Saturn and Jupiter during the great conjunction of December 2020.

b Calculate the force of the Sun on Jupiter when the distance between them is 778 million km.

c The distance between Jupiter and the Sun is greater than the distance between Jupiter and Saturn, yet the force of attraction between Jupiter and the Sun is greater. Explain why this is the case.

d Compare, quantitatively, the gravitational forces calculated in parts **a** and **b**. Also compare the masses of the Sun and Saturn. Use these values to confirm your response to part **c**.

10 Every few years the orbits of Mars and the Earth come closer than usual. On average, the centres of the planets are approximately 225 million km apart. But in August 2003 the planets had their closest encounter in nearly 60 000 years—the centres of the planets were just 56 million km apart. Use the following data to answer the questions below.

$m_{Earth} = 5.98 \times 10^{24}$ kg

$m_{Mars} = 6.39 \times 10^{23}$ kg

a Calculate the gravitational force between the Earth and Mars in August 2003.

b Calculate the gravitational force between the Earth and Mars when their centres are at their average separation distance (225 million km).

c Compare your answers to parts **a** and **b** by expressing the average force as a percentage of the force during the August 2003 approach.

11 Consider the force of gravity on Mercury, a small, rocky planet.

a State the equation you would use to calculate gravitational acceleration on the surface of Mercury.

b Mercury has a radius only a third of the radius of the Earth. Explain the effect the much smaller radius of Mercury has on gravitational acceleration. (You do not need to refer to the effect of mass in this instance.)

c The mass of Mercury is much smaller than that of the Earth. Explain the effect the much smaller mass has on gravitational acceleration.

d Calculate gravitational acceleration on the surface of Mercury given that Mercury has a mass of 3.29×10^{23} kg and a radius of 2440 km.

e Calculate the force due to gravity on a person standing on the surface of Mercury if this force measured 735 N on the surface of the Earth.

4.2 Gravitational fields

Newton's law of universal gravitation describes the force acting between two mutually attracting bodies. In reality, complex systems involve many more objects (such as the Sun and planets in our solar system, shown in Figure 4.2.1), with each object exerting an attractive force on each other at the same time.

To simplify the process of calculating the effect of simultaneous gravitational forces, scientists in the eighteenth century developed a mental construct known as the **gravitational field**. A **field** describes a region around an object where other objects experience a force due to a property of the field. In the centuries that followed, the idea of a field was also applied to other forces and it has become a very important concept in physics.

A gravitational field is a region in which a gravitational force is exerted on all matter within that region. Every physical object has an accompanying gravitational field. For example, the space around your body contains a gravitational field because any other object that comes near you will experience a force of gravitational attraction to your body (although it may be very small).

The gravitational field around a large object, such as a planet, is much more significant than that around a small object. For example, the Earth's gravitational field exerts a significant influence on objects on its surface and even up to thousands of kilometres into space.

FIGURE 4.2.1 The solar system is a complex gravitational system.

CASE STUDY

Discovery of Neptune

The planet Neptune (Figure 4.2.2) was discovered through its gravitational effect on other planets. Two astronomers, Urbain Le Verrier of France and John Couch Adams of England, each independently noted that the observed orbit of Uranus varied significantly from predictions made based on the gravitational effects of the Sun and the other known planets. Both suggested that this was due to the influence of a distant, undiscovered planet.

FIGURE 4.2.2 Neptune, our solar system's outermost planet, was discovered in 1846 after astronomers observed an irregularity in the orbit of Uranus that suggested the gravitational influence of another planet.

Le Verrier sent a prediction of the location of the new planet to Gottfried Galle at the Berlin Observatory and, on 23 September 1846, Neptune was discovered within 1° of Le Verrier's prediction (Figure 4.2.3).

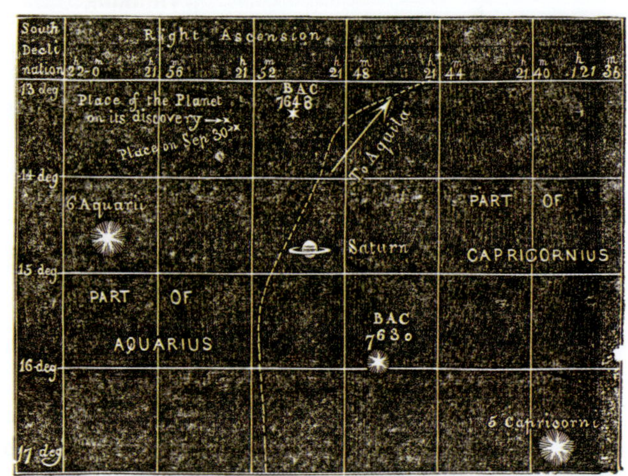

FIGURE 4.2.3 This star chart, published in 1846, shows the location of Neptune in the constellation Aquarius when it was discovered on 23 September, and its location one week later.

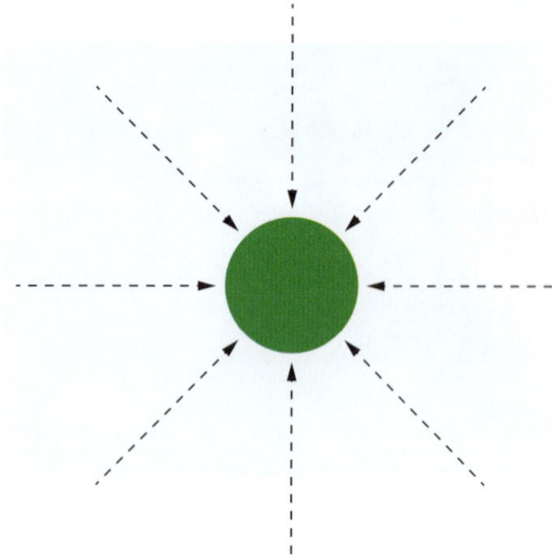

FIGURE 4.2.4 The arrows in this gravitational field diagram indicate that objects will be attracted towards the mass. The spacing of the lines shows that force will be strongest at the surface of the mass and weaker further away from it.

REPRESENTING GRAVITATIONAL FIELDS

Scientists have developed a way of representing fields by a series of arrows known as **field lines** (Figure 4.2.4). For gravitational fields, these are constructed as follows:

- the direction of the arrowhead indicates the direction of the gravitational force
- the space between the arrows indicates the relative magnitude of the field:
  - closely spaced arrows indicate a strong field
  - widely spaced arrows indicate a weaker field
  - parallel field lines indicate constant (or **uniform**) field strength
- the field lines never cross.

An infinite number of field lines could be drawn, so only a few are chosen to represent the rest. The size of the gravitational force acting on a mass in the region of a gravitational field is determined by the strength of the field, and the force acts in the direction of the field.

Worked example 4.2.1

INTERPRETING GRAVITATIONAL FIELD DIAGRAMS

The diagram below represents the gravitational field of a moon.

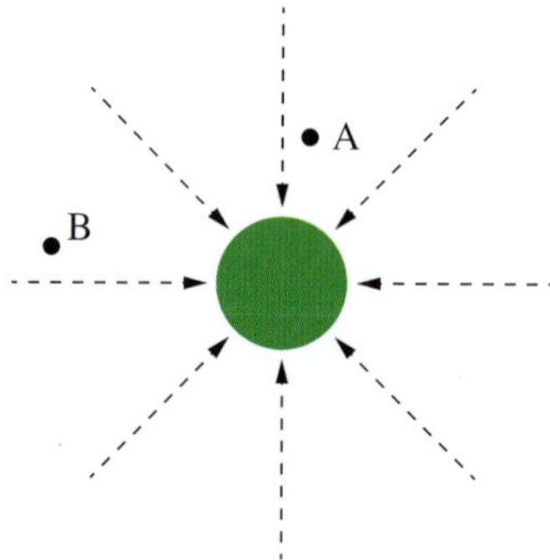

a Draw arrows to indicate the direction of the gravitational force acting at points A and B.	
Thinking	**Working**
The direction of the field arrows indicates the direction of the gravitational force, which is inwards, towards the centre of the moon.	A B

b State the relative strength of the gravitational field at each point.	
Thinking	**Working**
The closer the field lines, the stronger the force. The field lines are closer together at point A than they are at point B, because point A is closer to the moon.	The field is stronger at point A than at point B.

Worked example: Try yourself 4.2.1

INTERPRETING GRAVITATIONAL FIELD DIAGRAMS

The diagram below represents the gravitational field of a planet.

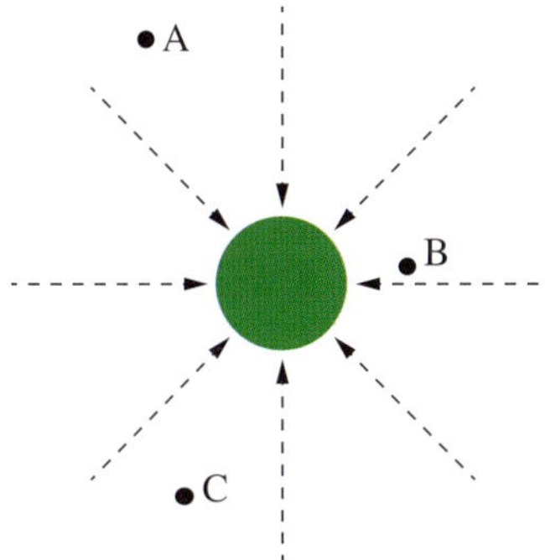

a Draw arrows to indicate the direction of the gravitational force acting at points A, B and C.

b State the relative strength of the gravitational field at each point.

GRAVITATIONAL FIELD STRENGTH

In theory, gravitational fields extend infinitely out into space. However, since the magnitude of the gravitational force decreases with the square of distance, eventually these fields become so weak as to become negligible.

In Section 4.1 it was shown that it is possible to calculate the acceleration due to gravity of objects near the Earth's surface using the dimensions of the Earth:

$$g = G\frac{m_{\text{Earth}}}{(r_{\text{Earth}})^2} = 9.8 \text{ m s}^{-2}$$

The constant g can also be used as a measure of the strength of the gravitational field. When understood in this way, the constant is written with the equivalent units of N kg^{-1} rather than m s^{-2}. This means that $g_{\text{Earth}} = 9.8 \text{ N kg}^{-1}$.

These units indicate that objects on the surface of the Earth experience 9.8 N of gravitational force for every kilogram of their mass.

It follows that the familiar equation $F_g = mg$ can be transposed so that the **gravitational field strength**, g, can be calculated as follows.

$$g = \frac{F_g}{m}$$

where g is gravitational field strength (N kg^{-1})
F_g is the force due to gravity (N)
m is the mass of an object in the field (kg)

PHYSICSFILE

$\text{N kg}^{-1} = \text{m s}^{-2}$

It is a simple matter to show that N kg^{-1} and m s^{-2} are equivalent units.

From Newton's second law, $F = ma$:

$1 \text{ N} = 1 \text{ kg m s}^{-2}$

Therefore $1 \text{ N} \times \text{kg}^{-1} = 1 \text{ kg m s}^{-2} \times \text{kg}^{-1}$

So $1 \text{ N kg}^{-1} = 1 \text{ m s}^{-2}$

Worked example 4.2.2

CALCULATING GRAVITATIONAL FIELD STRENGTH

When a student hangs a 5.0 kg mass from a spring balance, the balance measures a downwards force of 49 N.

According to this experiment, what is the gravitational field strength of the Earth at this location?

Thinking	Working
Recall the equation for gravitational field strength.	$g = \frac{F_g}{m}$
Substitute the appropriate values.	$g = \frac{49}{5.0}$
Solve the equation.	$g = 9.8\,\text{N kg}^{-1}$

Worked example: Try yourself 4.2.2

CALCULATING GRAVITATIONAL FIELD STRENGTH

A student uses a spring balance to measure the force due to gravity on a piece of wood as 2.5 N.

If the piece of wood has a mass of 260 g, calculate the gravitational field strength indicated by this experiment.

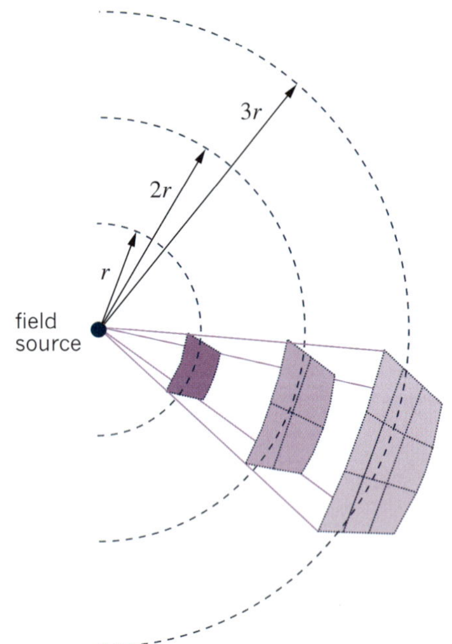

FIGURE 4.2.5 The gravitational field spreads out radially in three dimensions. As the distance from the source of a field increases, the field is spread over an area that increases with the square of the distance from the source, resulting in the strength of the field decreasing by the same ratio.

The equation for gravitational field strength, $g = \frac{F_g}{m}$, can be combined with Newton's law of universal gravitation, $F_g = G\frac{Mm}{r^2}$, to develop the equation for gravitational field strength:

$$g = \frac{F_g}{m} = \frac{\left(G\frac{Mm}{r^2}\right)}{m}$$

This can be simplified as follows.

$$g = G\frac{M}{r^2}$$

where g is the gravitational field strength (N kg^{-1})

G is the gravitational constant, 6.67×10^{-11} ($\text{N m}^2\,\text{kg}^{-2}$)

M is the mass of the central body (kg)

r is the radius of the central body (m)

Inverse square law

The field model is a very powerful tool for understanding forces that act at a distance. It has also been applied to other forces, such as the electrostatic force between charged objects and the magnetic force between two magnets.

The study of gravitational fields introduces the concept of the inverse square law. From the point source of a field, whether it is gravitational, electric or magnetic, the field will spread out radially in three dimensions. When the distance from the source is doubled, the field will be spread over four times the original area. This is shown in Figure 4.2.5. In going from r to $2r$, the area shown increases from one square to four squares (2^2). In going from r to $3r$, the area shown increases from one square to nine squares (3^2).

From the inverse nature of the inverse square law, at a distance $2r$ the strength of the field is a quarter of that at r. The force the field will exert at that distance will also be one quarter. At $3r$ from the source, the field will be reduced to one-ninth of that at the source, and so on.

One key difference between the gravitational force and other inverse square forces is that the gravitational force is always attractive, whereas like charges or like magnetic poles repel one another.

Inverse square laws are an important concept in physics, not only in the study of fields but also for other phenomena where energy is moving away from its source in three dimensions, such as with sound and other waves.

Variations in gravitational field strength of the Earth

The gravitational field strength of the Earth, g, is usually assigned a value of $9.81\,\text{N kg}^{-1}$. However, the field strength experienced by objects on the surface of the Earth can vary between $9.76\,\text{N kg}^{-1}$ and $9.83\,\text{N kg}^{-1}$, depending on the location. For example, the gravitational field is slightly stronger at the poles and slightly weaker at the equator.

In addition, certain geological formations can create differences in gravitational field strength. Geologists use a sensitive instrument known as a **gravimeter** (Figure 4.2.6) to detect small local variations in gravitational field strength. These variations indicate particular underground features. For example, rocks with above-average density, such as those containing mineral ores, create slightly stronger gravitational fields, whereas less-dense sedimentary rocks produce weaker fields.

In a gravitational field, the strength of the force varies inversely with the square of the distance between the objects:

$$F \propto \frac{1}{r^2}$$

where F is the force
r is the distance from the source of the gravitational field

This is referred to as the inverse square law.

FIGURE 4.2.6 A gravimeter can be used to measure the strength of the local gravitational field.

PHYSICSFILE

The shape of the Earth

The shape of the Earth is known as an oblate spheroid, that is, a partially flattened sphere. Mathematically, this is the shape made when an ellipse is rotated around its minor axis. The diameter of the Earth between the North Pole and South Pole is approximately 40 km shorter than its diameter at the equator. This means that the gravitational field is slightly stronger at the poles and slightly weaker at the equator.

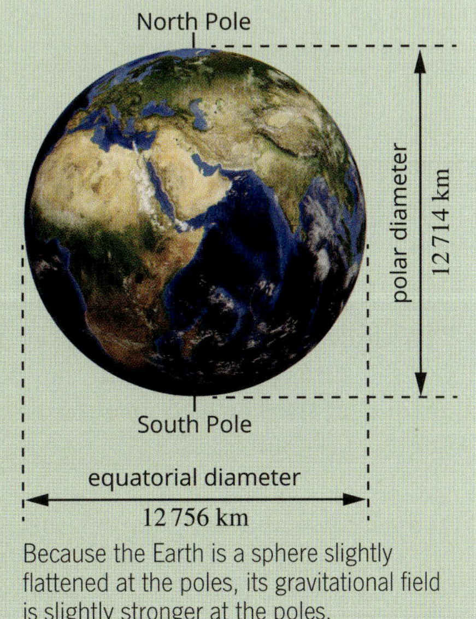

Because the Earth is a sphere slightly flattened at the poles, its gravitational field is slightly stronger at the poles.

PHYSICSFILE

Variations in gravitational field strength

The measured value for g is not constant everywhere on Earth. It is higher at the poles than at the equator due to the slightly squashed shape of the Earth. At the equator g is about $9.780\,\text{N kg}^{-1}$ and at the poles it is about $9.832\,\text{N kg}^{-1}$. The value of g also varies according to certain geological features, such as valleys, mountains and the density of nearby rocks. One of the lowest readings of g is on Mount Nevado Huascaran in Peru, while the highest is on the Arctic Ocean surface. Globally, the average value for g is $9.8\,\text{N kg}^{-1}$.

If the surface of the Earth is considered to be flat, as it appears in everyday life, then the gravitational field lines are approximately parallel, indicating a uniform field (Figure 4.2.7).

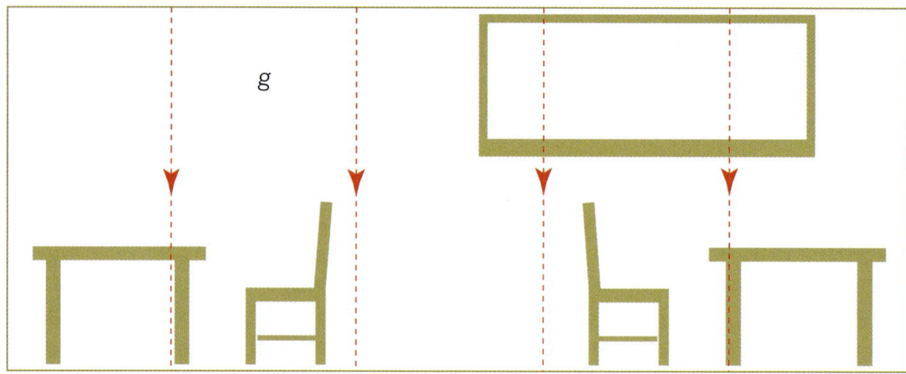

FIGURE 4.2.7 The uniform gravitational field, g, is represented by evenly spaced parallel lines in the direction of the force.

However, when the spherical shape of the Earth is viewed from a distance, it becomes clear that the Earth's gravitational field is not uniform at all (Figure 4.2.8). The increasing distance between the field lines as the distance from the Earth increases indicates that the field becomes progressively weaker. Values for g at different altitudes are shown in Figure 4.2.9.

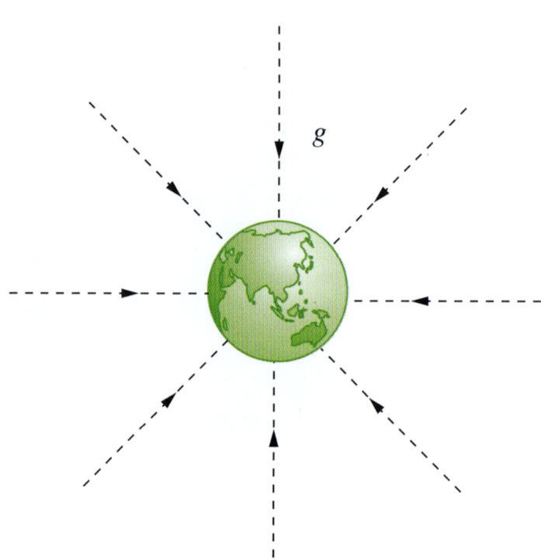

FIGURE 4.2.8 The Earth's gravitational field becomes progressively weaker as the distance from the Earth increases. This is evident in the greater distance between the field lines further from the Earth.

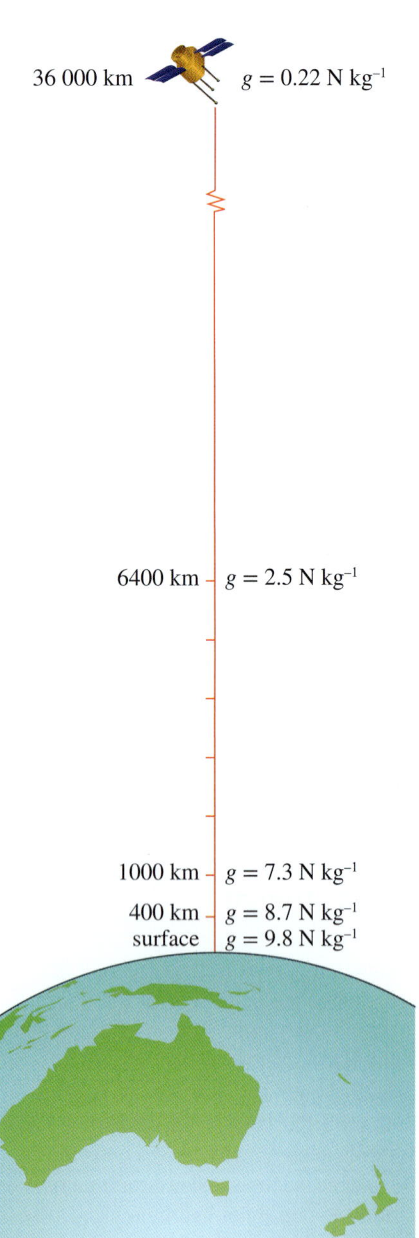

FIGURE 4.2.9 The Earth's gravitational field strength is weaker at higher altitudes.

This variation in the value of g is because gravitational field strength, like gravitational force, is governed by the inverse square law:

$$g = G\frac{M_{\text{Earth}}}{(r_{\text{Earth}})^2}$$

This is shown graphically in Figure 4.2.10 on page 181.

The gravitational field strength at different altitudes (that is, different heights above the surface of the Earth) can be calculated by first adding the altitude to the radius of the Earth to give the distance of the object from the Earth's centre (Figures 4.2.9 and 4.2.10).

At an altitude above the Earth's surface:

$$g = \frac{GM_{\text{Earth}}}{(r_{\text{Earth}} + \text{altitude})^2}$$

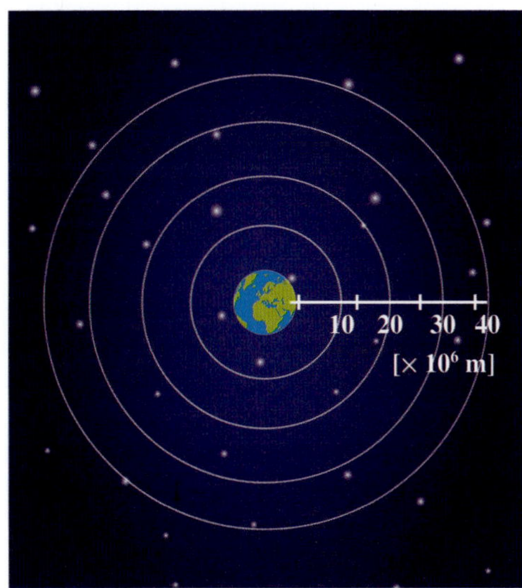

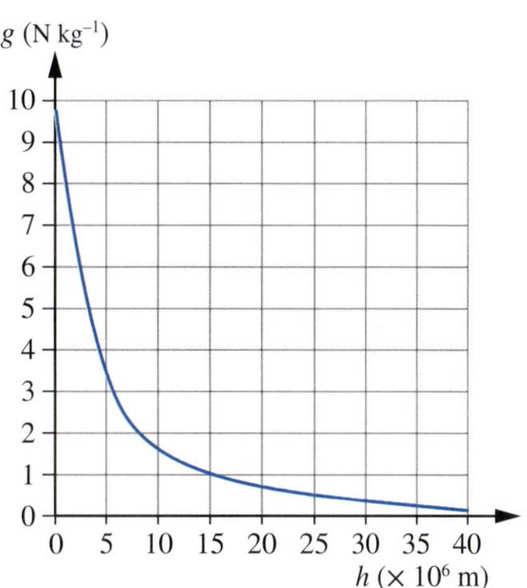

FIGURE 4.2.10 As the distance from the surface of the Earth increases, the value of g decreases rapidly from 9.8 N kg^{-1}, according to the inverse square law. The blue line on the graph gives the value of g at various altitudes (h).

Worked example 4.2.3

CALCULATING GRAVITATIONAL FIELD STRENGTH AT DIFFERENT ALTITUDES

The Concorde was a passenger aircraft capable of supersonic speeds. It flew at much higher altitudes than most other aircraft. Calculate the strength of the Earth's gravitational field at the cruising altitude of the Concorde using the following data:

$r_{Earth} = 6.37 \times 10^6$ m

$m_{Earth} = 5.98 \times 10^{24}$ kg

cruising altitude of the Concorde = 18 000 m

Thinking	Working
Recall the equation for gravitational field strength.	$g = G\frac{M}{r^2}$
Add the height of the cruising altitude to the radius of the Earth.	$r = 6.37 \times 10^6 + 18000$ $= 6.388 \times 10^6$ m
Substitute values into the equation and solve it.	$g = G\frac{M}{r^2}$ $= 6.67 \times 10^{-11} \times \frac{5.98 \times 10^{24}}{(6.388 \times 10^6)^2}$ $= 9.77$ N kg^{-1}

Worked example: Try yourself 4.2.3

CALCULATING GRAVITATIONAL FIELD STRENGTH AT DIFFERENT ALTITUDES

Commercial airlines typically fly at an altitude of 12 000 m. Calculate the gravitational field strength of the Earth at this height using the following data:

$r_{Earth} = 6.37 \times 10^6$ m

$m_{Earth} = 5.98 \times 10^{24}$ kg

FIGURE 4.2.11 The gravitational field strength on the surface of Mars is different to the gravitational field strength on the surface of the Earth and the Moon.

PHYSICSFILE

Moon walking

Walking is a process of repeatedly stopping yourself from falling over. When astronauts first tried to walk on the Moon, they found that they fell too slowly to walk easily. So they invented a kind of shuffling jump that was a much quicker way of moving around in the Moon's weak gravitational field.

Astronauts had to invent a new way of walking to deal with the Moon's weak gravitational field

Gravitational field strengths of other astronomical bodies

The equation $g = G\frac{M}{r^2}$ can be used to calculate the gravitational field strength on the surface of any astronomical body, such as Mars (Figure 4.2.11) or the Moon.

The gravitational field strength on the Moon is much less than on the Earth, at approximately $1.6\,\text{N kg}^{-1}$. This is because the Moon's mass is smaller than the Earth's.

Worked example 4.2.4

GRAVITATIONAL FIELD STRENGTH ON ANOTHER PLANET OR MOON

Calculate the strength of the gravitational field on the surface of the Moon given that the Moon's mass is 7.35×10^{22} kg and its radius is 1740 km. Compare your answer with the average gravitational field strength on the Earth ($9.8\,\text{N kg}^{-1}$).

Thinking	Working
Recall the equation for gravitational field strength.	$g = G\frac{M}{r^2}$
Convert the Moon's radius to metres.	$r = 1740\,\text{km}$ $= 1740 \times 1000$ $= 1.74 \times 10^6\,\text{m}$
Substitute values into the equation and solve it.	$g = G\frac{M}{r^2}$ $= 6.67 \times 10^{-11} \times \frac{7.35 \times 10^{22}}{(1.74 \times 10^6)^2}$ $= 1.62\,\text{N kg}^{-1}$
Compare the field strength with the Earth's average gravitational field strength by calculating the ratio $\frac{g_{Earth}}{g_{Moon}}$.	$\frac{g_{Earth}}{g_{Moon}} = \frac{9.8}{1.62}$ $= 6.0$ At their surfaces, the Earth's gravitational field strength is six times stronger than that of the Moon.

Worked example: Try yourself 4.2.4

GRAVITATIONAL FIELD STRENGTH ON ANOTHER PLANET OR MOON

Calculate the strength of the gravitational field on the surface of Mars.

$m_{Mars} = 6.42 \times 10^{23}$ kg

$r_{Mars} = 3390$ km

Compare your answer with the Earth's average gravitational field strength ($9.8\,\text{N kg}^{-1}$).

4.2 Review

OA ✓✓

SUMMARY

- A gravitational field is a region in which a gravitational force is exerted on all matter within that region.
- A gravitational field can be represented by a gravitational field diagram.
  - The arrowheads indicate the direction of the gravitational force.
  - The spacing of the lines indicates the relative strength of the field. The closer the line spacing, the stronger the field.
- The gravitational field strength on the surface of any astronomical body depends on the mass and radius of the body.
- The strength of a gravitational field can be calculated using the following equations:

$$g = \frac{F_g}{m} \text{ or } g = G\frac{M}{r^2}$$

- The gravitational field strength on the Earth's surface is approximately 9.8 N kg^{-1}. This varies from location to location and with altitude.

KEY QUESTIONS

Knowledge and understanding

1 Answer the following questions about the Earth's gravitational field.

a State the average value of the gravitational field strength at the Earth's surface. Express your answer with appropriate units. Explain why the average value for the gravitational field strength of the Earth is often used in calculations.

b The Earth's gravitational field points directly to the centre of the Earth. Explain, then, why the field close to the Earth's surface can be treated as a uniform field.

2 A spring balance is used to measure the force due to gravity on a 200 g set of slotted masses. The force measured is 1.5 N. What is the gravitational field strength at this location?

3 A gravitational field, g, is measured as 6.0 N kg^{-1} at a distance of 300 km from the centre of a planet. If the gravitational field were to be measured at a distance of 1200 km from the centre of this same planet, what would be the ratio of the new measurement to the original measurement?

4 The International Space Station orbits the Earth at an altitude of approximately 400 km above the Earth's surface. Calculate the strength of the Earth's gravitational field on the International Space Station. Use the following data:

r_{Earth} = 6370 km

m_{Earth} = 5.98 × 10^{24} kg

5 On 12 November 2014 the Rosetta spacecraft landed a probe on the comet 67P/Churyumov–Gerasimenko. Assuming that this comet is roughly spherical and has a mass of 1.0 × 10^{13} kg and a radius of 900 m, calculate the gravitational field strength 100 m above its surface.

6 The mass and radius of three of Saturn's moons are listed in the table below. Calculate the gravitational field strength, g, at the surface of each moon (or at the radius stated for moons without a solid surface).

	Moon	Mass (kg)	Radius (m)
a	Titan	1.35 × 10^{23}	2.57 × 10^{6}
b	Dione	1.10 × 10^{21}	5.61 × 10^{5}
c	Tethys	6.17 × 10^{20}	5.31 × 10^{5}

Analysis

7 When a star dies, its atomic structure may collapse to form a very small, very dense body known as a neutron star. A typical neutron star can have a mass of 3.0 × 10^{30} kg and a radius of just 10 km.

a Calculate the gravitational field strength at the surface of a typical neutron star.

b Compare the gravitational field strength of a typical neutron star to the strength of the gravitational field at the Earth's surface. (Use g_{Earth} = 9.8 m s^{-2}.)

c Calculate the distance from a typical neutron star where the gravitational field strength is the same as the gravitational field strength at the surface of the Earth. (Use g_{Earth} = 9.8 m s^{-2}.)

continued over page

4.2 Review *continued*

8 A hypothetical planet in a distant solar system is distinctly non-spherical in shape. Its polar radius is 5000 km and its equatorial radius is 6000 km. The gravitational field strength at the poles is 8.0 N kg^{-1}.

  a Calculate the ratio between the polar and equatorial radii and use this to find the gravitational field strength at the equator of the planet.

  b Confirm your answer to part **a** using the equation $g = G\frac{M}{r^2}$.

9 There is a point between the Earth and the Moon where the total gravitational field is zero. The significance of this is that lunar missions are able to return to the Earth under the influence of the Earth's gravitational field once they pass this point on their journey back. Given that the mass of the Earth is 5.98×10^{24} kg, the mass of the Moon is 7.3×10^{22} kg and the radius of the Moon's orbit is 3.8×10^{8} m, calculate the distance of this point of zero gravity from the centre of the Earth.

10 An astronaut travels away from the Earth to a region in space where the gravitational force due to the Earth is only 1.0% of that at the Earth's surface. What distance, in Earth radii, is the astronaut from the centre of the Earth?

4.3 Work in a gravitational field

The concept of **gravitational potential energy** will be familiar to you from Section 3.5. However, the nature of a gravitational field requires a more sophisticated understanding of gravitational potential energy when considering the motion of such objects as rockets and satellites (Figure 4.3.1).

FIGURE 4.3.1 Satellites in orbit have gravitational potential energy.

REVISITING WORK AND CONSERVATION OF ENERGY

The gravitational potential energy of an object, E_g, is the energy of an object due to its position in a gravitational field. It is directly proportional to the mass of the object, m, its height above a reference point, Δh, and the strength of the gravitational field, g.

$$E_g = mg\Delta h$$

where E_g is the gravitational potential energy of an object (J)
- m is the mass of the object (kg)
- g is the gravitational field strength ($N\,kg^{-1}$; $9.8\,N\,kg^{-1}$ near the surface of the Earth)
- Δh is the height of the object above a reference point (m)

The equation for gravitational potential energy is developed from the work–energy theorem, which assumes that work done against the force of gravity is converted into potential energy.

$$\Delta E = W = Fs$$

where ΔE is the change in gravitational potential energy (J)
- W is the work done (J)
- F is the force of gravity (N)
- s is the distance moved in the gravitational field (m)

Worked example 4.3.1

WORK DONE FOR A CHANGE IN GRAVITATIONAL POTENTIAL ENERGY

A mountaineer climbs from a height of 700 m above sea level to the top of Mount Everest, which is 8848 m above sea level.

The total mass of the mountaineer (with equipment) is 100 kg. Assuming that the gravitational field strength of the Earth (g) is 9.8 N kg^{-1}, calculate the amount of work done (in MJ) by the mountaineer in climbing to the summit of the mountain. Assume that g is 9.8 N kg^{-1} and give your answer to two significant figures.

Thinking	Working
Calculate the change in height.	$\Delta h = 8848 - 700$ $= 8148\,\text{m}$
Substitute appropriate values into $E_g = mg\Delta h$. Remember to give your answer in MJ to two significant figures.	$E_g = mg\Delta h$ $= 100 \times 9.8 \times 8148$ $= 7985040$ $= 8.0\,\text{MJ}$
The work done by the mountaineer is equal to the change in gravitational potential energy.	$W = \Delta E = 8.0\,\text{MJ}$

Worked example: Try yourself 4.3.1

WORK DONE FOR A CHANGE IN GRAVITATIONAL POTENTIAL ENERGY

Calculate the work done (in MJ) to lift a weather satellite of 3.2 tonnes from the Earth's surface to the limit of the atmosphere, which ends at the Karman line (exactly 100 km up from the surface of the Earth). Assume that $g = 9.8$ N kg^{-1}.

Gravitational, kinetic and mechanical energy

Gravitational potential energy calculations are important because, when combined with the concepts of kinetic energy and conservation of mechanical energy, they allow the speed of a falling object to be determined.

Recall that kinetic energy can be defined as follows.

$$E_k = \frac{1}{2}mv^2$$

where E_k is the kinetic energy of an object (J)

m is the mass of the object (kg)

v is the speed of the object (m s^{-1})

Worked example 4.3.2

SPEED OF A FALLING OBJECT

An astronaut standing on the Moon dropped a hammer of mass of 450 g from a height of 1.4 m. Calculate the speed of the hammer as it hit the Moon's surface, where $g = 1.6\,\text{N kg}^{-1}$.

Thinking	Working
Calculate the gravitational potential energy of the hammer on the Moon. Change the units of measurement where necessary.	$E_g = mg\Delta h$ $= 0.45 \times 1.6 \times 1.4$ $= 1.0\text{ J}$
Assume that when the hammer hit the surface of the Moon, all its gravitational potential energy had been converted into kinetic energy.	$E_k = E_g = 1.0\text{ J}$
Use the equation for kinetic energy to calculate the speed of the hammer as it hit the ground.	$E_k = \frac{1}{2}mv^2$ $1.0 = \frac{1}{2} \times 0.45 \times v^2$ $v^2 = \frac{1.0 \times 2}{0.45}$ $v = 2.1\text{ m s}^{-1}$

Worked example: Try yourself 4.3.2

SPEED OF A FALLING OBJECT

Calculate how fast a 450 g hammer is going as it hits the ground after being dropped from a height of 1.4 m on the Earth, where $g = 9.8\,\text{N kg}^{-1}$.

Work in a non-constant gravitational field

The equation $E_g = mg\Delta h$ assumes that work is done against a constant force of gravity: $\Delta E = W = Fs$. While this assumption holds true for objects close to the surface of a planet, it is not adequate for objects such as satellites at altitudes where the gravitational field of the planet is significantly reduced.

Newton's law of universal gravitation indicates that the strength of the Earth's gravitational field changes with altitude: the field is stronger closer to the ground and weaker at higher altitudes (Figure 4.3.3).

Consider the example of a 10 kg meteor falling towards the Earth from deep space (Figure 4.3.2). As the meteor gets closer to the Earth, it encounters an increasing gravitational field strength. So the gravitational force, F_g, on the meteor increases as it approaches the Earth. Since the force is not constant, the work done on the meteor (which corresponds to the change in its gravitational potential energy) cannot be found by simply multiplying the gravitational force by the distance travelled.

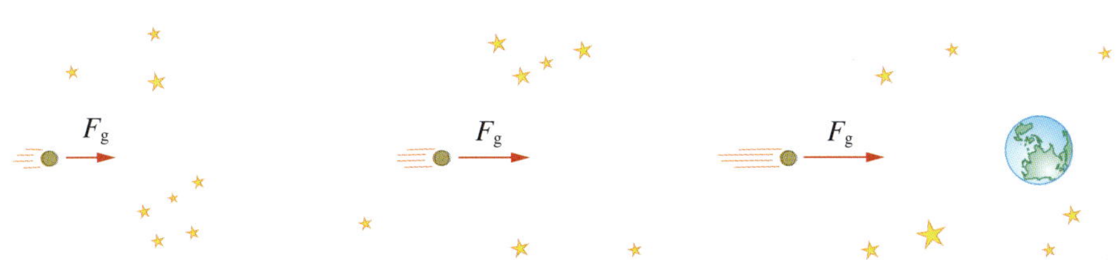

FIGURE 4.3.2 As a meteor approaches the Earth, it moves through a gravitational field of increasing strength and so is acted upon by a greater gravitational force.

PHYSICSFILE

Hammer drop

A variation of the hammer drop experiment outlined in Worked example 4.3.2 was used to show that a hammer and a feather would fall at the same speed and land simultaneously on the Moon's surface. The theory behind the experiment had been well understood since the time when Galileo concluded that all objects, no matter their mass, fall at the same rate. The only reason the hammer and feather fall at different rates on the Earth is because of the presence of resistive forces in the air. The Moon was the perfect place to test the theory due to the environment being essentially a vacuum.

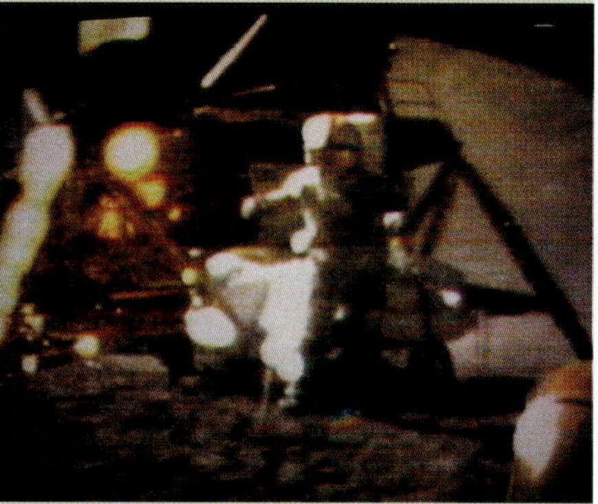

Apollo 15 astronaut David Scott simultaneously drops a hammer and a feather from the same height while standing on the Moon

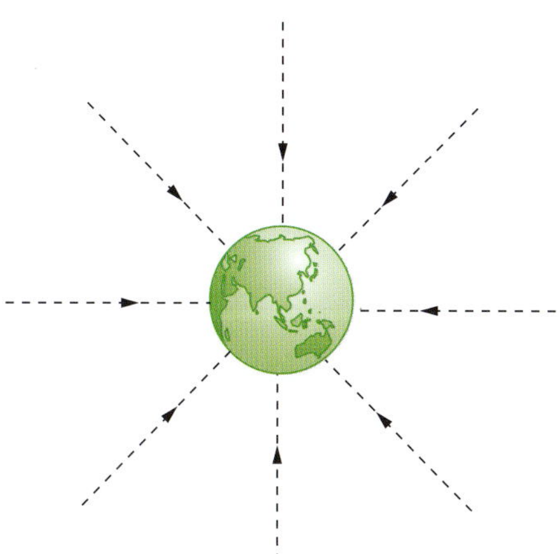

FIGURE 4.3.3 The Earth's gravitational field extends out into space, and the field is stronger closer to the Earth (where the field lines are closer together).

The area under a gravitational force vs distance graph gives the change in energy that an object will undergo as it moves through the gravitational field.

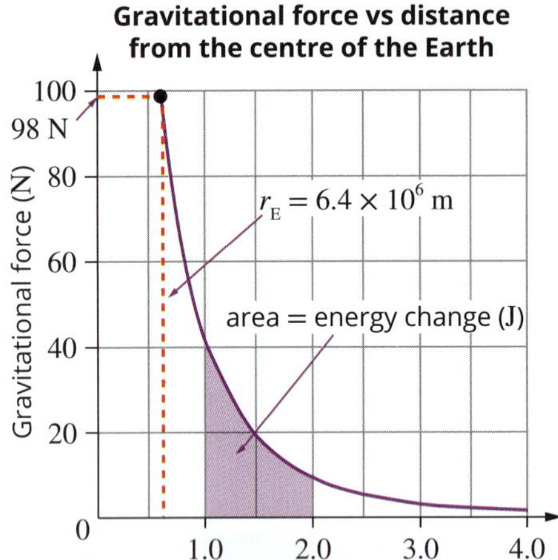

FIGURE 4.3.4 The gravitational force acting on a 10 kg meteor at different distances from the Earth. The shaded region represents the work done by the gravitational field as the body moves between 2.0×10^7 m and 1.0×10^7 m from the centre of the Earth.

Using the force vs distance graph

When a freefalling body, such as the meteor in Figure 4.3.2 on the previous page, is acted upon by a varying gravitational force, the body's change in energy can be analysed using a gravitational force vs distance graph. As with other force vs distance graphs, the area under the graph is equal to the work done, that is, the change in the energy of the body. The area under the graph has units of newton metres (N m), which are equivalent to joules (J).

The shaded area in Figure 4.3.4 represents the decrease in gravitational potential energy of the 10 kg meteor as it falls from a distance of 2.0×10^7 m to 1.0×10^7 m from the centre of the Earth. This area also represents the amount of kinetic energy that the meteor gains as it approaches the Earth.

Note that the energy change of the meteor will be the same regardless of whether the meteor falls directly towards the planet (Figure 4.3.5(a)) or follows an indirect path (Figure 4.3.5(b)).

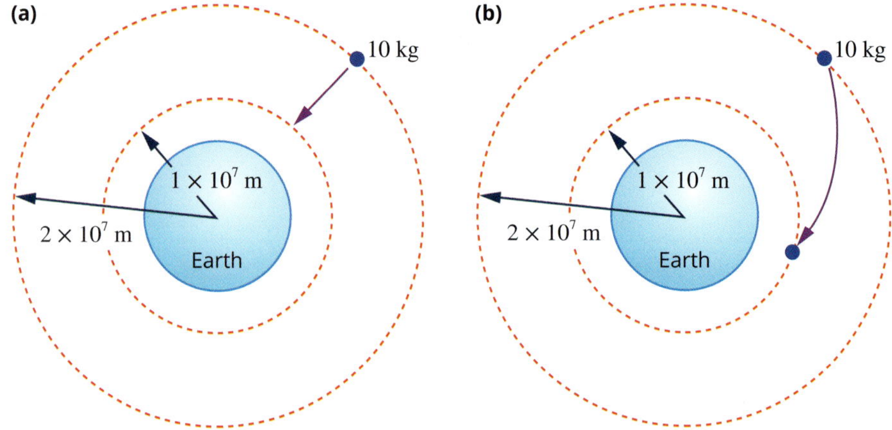

FIGURE 4.3.5 Two freefall situations: (a) direct and (b) indirect. The shaded region below the gravitational force vs distance graph in Figure 4.3.4 represents the change in energy in both situations.

Worked example 4.3.3 shows how a force vs distance graph can be used to determine the change in gravitational potential energy of a meteor.

Worked example 4.3.3

CHANGE IN GRAVITATIONAL POTENTIAL ENERGY USING A FORCE VS DISTANCE GRAPH

A 10 kg meteor falls from a distance of 2.0×10^7 m to 1.0×10^7 m as measured from the centre of the Earth. Use the graph below to determine the approximate change in gravitational potential energy of the meteor.

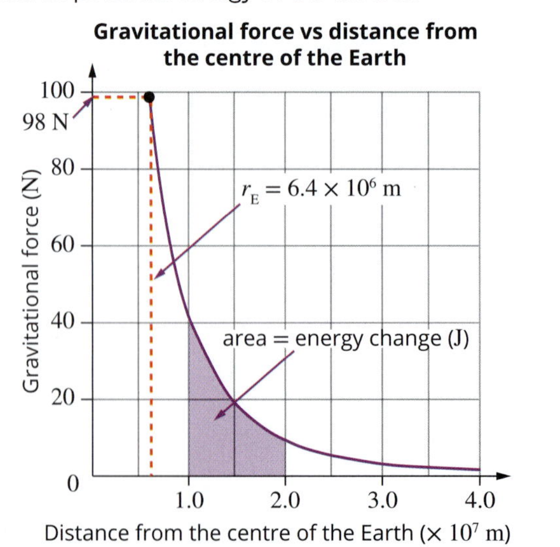

Thinking	Working
Count the number of shaded squares. (In this example, count the partially shaded squares as half squares.)	number of shaded squares = 2
Calculate the area (energy value) of each square.	$E_{square} = 0.5 \times 10^7 \times 20$ $= 1 \times 10^8$ J
To calculate the change in energy, multiply the number of shaded squares by the energy value of each square.	$\Delta E_g = 2 \times (1 \times 10^8)$ $= 2 \times 10^8$ J

Worked example: Try yourself 4.3.3

CHANGE IN GRAVITATIONAL POTENTIAL ENERGY USING A FORCE VS DISTANCE GRAPH

A 500 kg lump of space junk is plummeting towards the Moon. It falls a distance of 1.0×10^6 m and then strikes the surface of the Moon. Using the diagram and the force vs distance graph shown, determine the approximate decrease in gravitational potential energy of the space junk as it hits the Moon's surface.

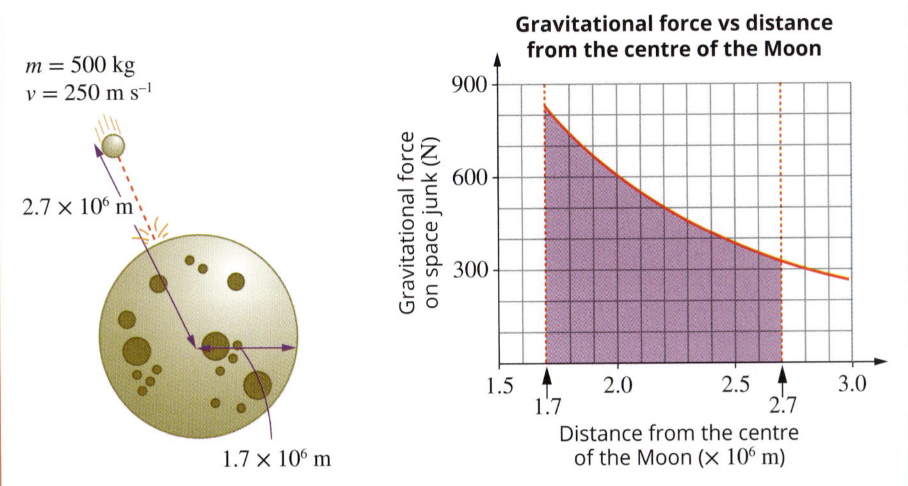

USING THE GRAVITATIONAL FIELD STRENGTH VS DISTANCE GRAPH

The change in gravitational potential energy of an object at different distances from the centre of a gravitational field can also be determined from a graph of the gravitational field strength (Figure 4.3.6).

The area under a gravitational field strength vs distance graph is a quantity with units $\text{N kg}^{-1} \times \text{m}$, which is equivalent to J kg^{-1}, so the area indicates the change in energy for each kilogram of the object's mass. To find the work done or energy change (J), the area (J kg^{-1}) must therefore be multiplied by the mass (kg) of the object.

Worked example 4.3.4 on the next page shows how to use a gravitational field strength vs distance graph to determine the change in gravitational potential energy of an object.

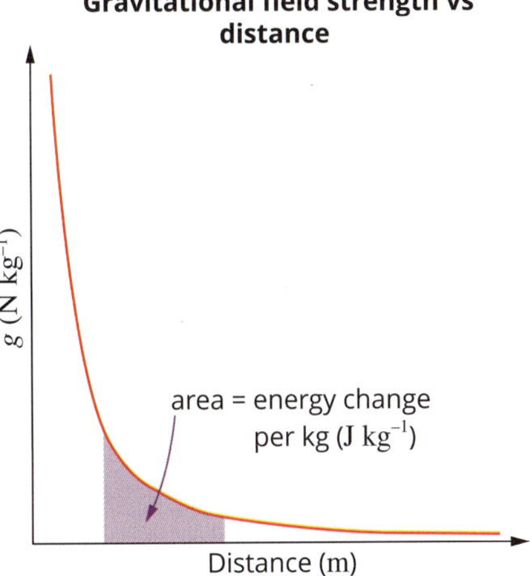

FIGURE 4.3.6 A gravitational field strength vs distance graph can also be used to determine the change in energy as a body moves through a gravitational field.

The area under a gravitational field strength vs distance graph gives the energy change per kilogram of mass. To find the change in energy, the area must be multiplied by the mass of the object in kg.

Worked example 4.3.4

CHANGE IN GRAVITATIONAL POTENTIAL ENERGY USING A GRAVITATIONAL FIELD STRENGTH VS DISTANCE GRAPH

A wayward satellite of mass 1500 kg has developed a highly elliptical orbit around the Earth. At its closest approach (perigee) it is just 500 km above the Earth's surface. Its furthest point (apogee) is 3000 km from the Earth's surface. Using the graph below, determine the approximate change in the gravitational potential energy of the satellite as it orbits. (Take the radius of the Earth to be 6400 km.)

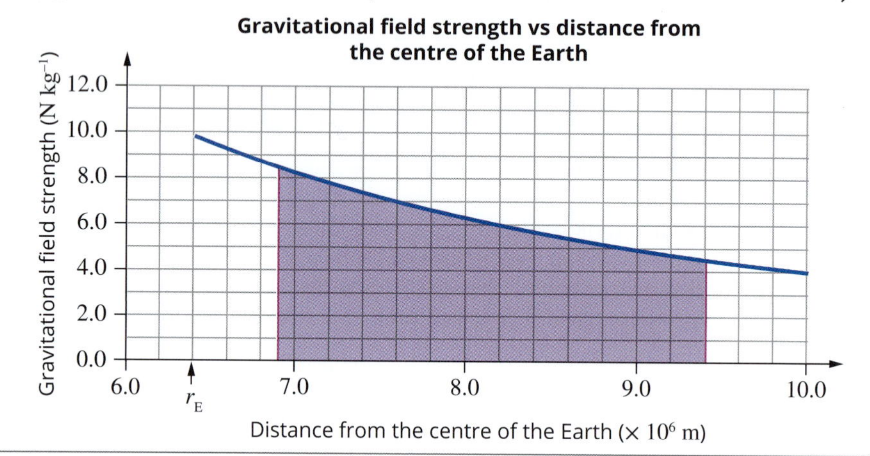

Thinking	Working
Count the number of shaded squares. Only count squares that are at least 50% shaded.	number of shaded squares = 82
Calculate the energy value of each square.	$E_{\text{square}} = 0.2 \times 10^6 \times 1 \text{N kg}^{-1}$ $= 2 \times 10^5 \text{ J kg}^{-1}$
To calculate the change in energy multiply the number of shaded squares by the energy value of each square and the mass of the satellite.	$\Delta E_g = 82 \times (2 \times 10^5) \times 1500$ $= 2.5 \times 10^{10} \text{ J}$

Worked example: Try yourself 4.3.4

CHANGE IN GRAVITATIONAL POTENTIAL ENERGY USING A GRAVITATIONAL FIELD STRENGTH VS DISTANCE GRAPH

A 3000 kg Soyuz rocket moves from an orbital height of 300 km above the Earth's surface to dock with the International Space Station at a height of 500 km. Using the graph below, determine the approximate change in the gravitational potential energy of the rocket.

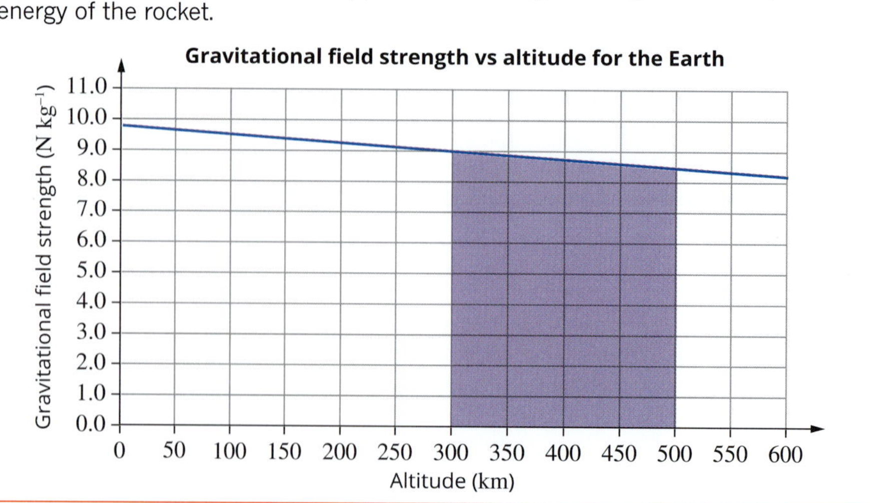

4.3 Review

OA ✓✓

SUMMARY

- The principles of work and the conservation of energy are useful for understanding gravitational potential energy. The following equations are particularly useful.

$$W = Fs$$
$$W = \Delta E$$
$$E_k = \frac{1}{2}mv^2$$

- The gravitational potential energy equation $E_g = mg\Delta h$ assumes that the Earth's gravitational field is constant. This is approximately true for objects that are within a few kilometres of the Earth's surface.
- The strength of the Earth's gravitational field decreases as altitude increases.
- The area under a gravitational force vs distance graph gives the change in kinetic energy or change in gravitational potential energy of a body. It also indicates the work done by the gravitational field.
- The area under a gravitational field strength vs distance graph gives the change in energy per kilogram ($J\,kg^{-1}$) of the object. To calculate the energy change, the area is multiplied by the mass of the object (kg).

KEY QUESTIONS

Knowledge and understanding

1 Each year, there is a stair-climb competition from the ground floor of the Eureka Tower in Melbourne to its eighty-eighth floor, which is 285 m above ground level. Calculate the work done against gravity by a 75 kg competitor as they race up to the eighty-eighth floor.

2 Consider a rocket of mass 2000 t launched from the Earth. After firing its stage 1 rockets it had reached an altitude of 70 km. Assuming that g is $9.8\,N\,kg^{-1}$ over this distance, how much work did the stage 1 rockets do in this time?

3 The gravitational field strength at the surface of Mars is $3.7\,N\,kg^{-1}$. Calculate the speed at which a 1.5 kg rock hits the surface of the planet if it falls from a height of 2.2 m.

4 The path of a meteor plunging towards the Earth is as shown. (Ignore air resistance when answering these questions.)

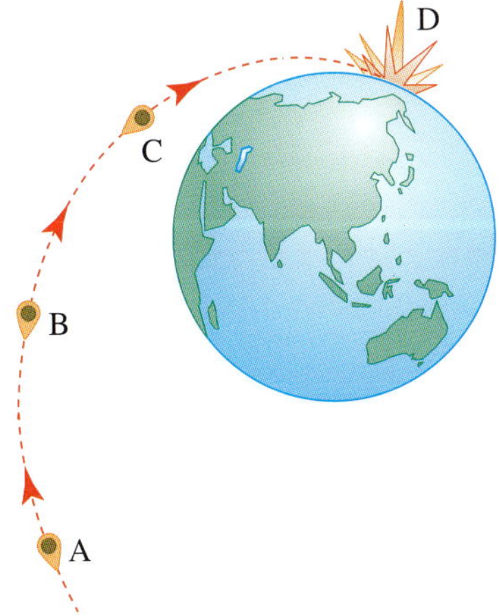

a Describe, qualitatively, how the gravitational field strength of the Earth changes from point A to point D.

b How will the acceleration of the meteor change as it travels along the path shown?

c For each of the following, decide how the energy described will change. Select the correct response to complete the sentence.

i The kinetic energy of the meteor as it travels from A to D will **increase/decrease/stay the same**.

ii The gravitational potential energy of the meteor as it travels from A to D will **increase/decrease/stay the same**.

iii The total energy of the meteor as it travels from A to D will **increase/decrease/stay the same**.

Analysis

5 The Valles Marineris on Mars is one of the most spectacular land formations in the solar system. It is a gigantic canyon 4000 km long, 200 km wide and 7 km deep.

a If a 600 g rock on the edge of the canyon drops to the canyon's floor 7000 m below, how fast would it be going when it hit the bottom? The gravitational field strength on Mars is $3.7\,N\,kg^{-1}$.

b If the same rock was to fall to the deepest point of the Grand Canyon in Arizona, USA, it would reach a speed of $189\,m\,s^{-1}$ (if air resistance is ignored). How deep is this point in the Grand Canyon? (Use $g = 9.8\,N\,kg^{-1}$.)

continued over page

4.3 Review *continued*

6 The graph shows the force on a mass of 1.0 kg as a function of its distance from the centre of the Earth. Assume that the radius of the Earth is 6.4×10^6 m.

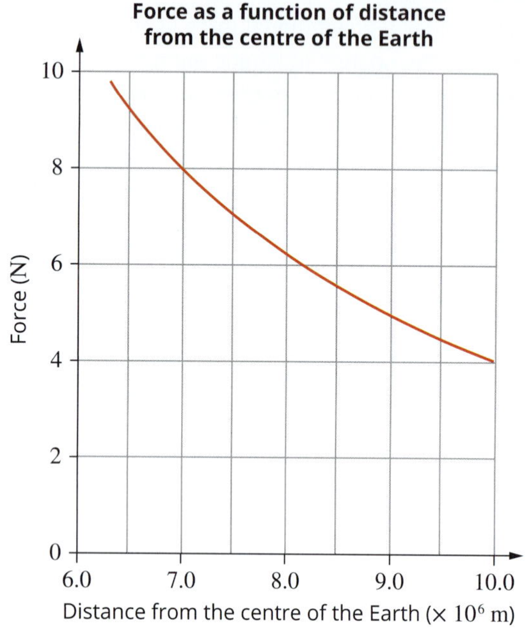

a Use the graph to determine the gravitational force between the Earth and a 1.0 kg mass 100 km above the Earth's surface.

b Use the graph to determine the height above the Earth's surface at which a 1.0 kg mass would experience a gravitational force of 5.0 N.

A 1.0 kg mass is speeding towards the Earth. When it is at a distance of 9.5×10^6 m from the centre of the Earth, its speed is 4.0 km s^{-1}.

c Determine the kinetic energy of the object when it is 9.5×10^6 m from the centre of the Earth.

d Calculate the increase in kinetic energy of the object as it moves from a distance of 9.5×10^6 m from the centre of the Earth to a point that is 6.5×10^6 m from the centre.

e Ignoring air resistance, what is the kinetic energy of the object when it is 6.5×10^6 m from the centre of the Earth?

f How fast is the object travelling when it is 6.5×10^6 m from the centre of the Earth?

7 A communications satellite of mass 240 kg is launched from a space shuttle that is in orbit 600 km above the Earth's surface. The satellite travels directly away from the Earth and reaches a maximum distance of 8000 km from the centre of the Earth before stopping due to the influence of the Earth's gravitational field.

a Use the graph in question **6** to estimate the kinetic energy of this satellite as it was launched.

b An identical satellite is launched from the Earth's surface. Rockets do 2.64×10^9 J of work to launch the satellite. To approximately what altitude was the satellite launched?

8 A 20 t remote-sensing satellite is in a circular orbit around the Earth at an altitude of 100 km. The satellite is moved to a new stable orbit with an altitude of 2600 km. Use the following graph to estimate the increase in the gravitational potential energy of the satellite as it moved from its lower orbit to its higher orbit.

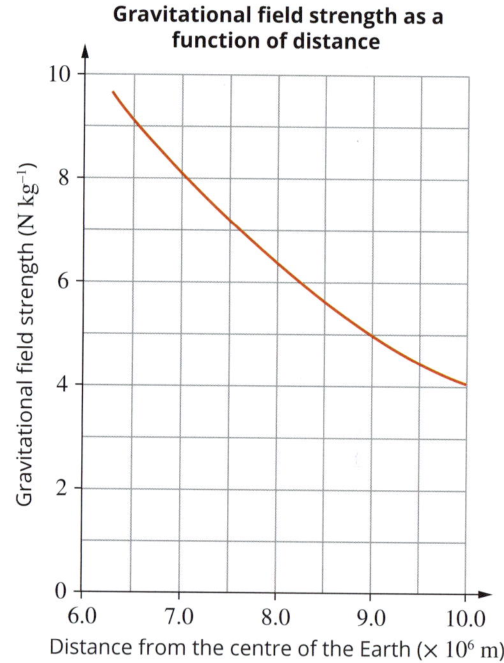

Chapter review

KEY TERMS

acceleration due to gravity
altitude
field
field lines
gravimeter
gravitational constant
gravitational field
gravitational field strength
gravitational force
gravitational potential energy
inverse square law
Newton's law of universal gravitation
normal force
uniform

REVIEW QUESTIONS

Knowledge and understanding

1 Use Newton's law of universal gravitation to calculate the gravitational force acting on a person with a mass of 100 kg.
$m_{Earth} = 5.98 \times 10^{24}$ kg
$r_{Earth} = 6370$ km

2 The gravitational force of attraction between Saturn and Dione, a moon of Saturn, is 2.79×10^{20} N. Calculate the orbital radius of Dione.
$m_{Dione} = 1.05 \times 10^{21}$ kg
$m_{Saturn} = 5.69 \times 10^{26}$ kg

3 Two bodies, one of mass 2.0×10^{4} kg and the other of mass 2.0×10^{3} kg, exert a gravitational force of 1.0×10^{-3} N on each other. How far apart are their centres?

4 Jupiter and the Sun exert gravitational forces on each other.
  a Compare, qualitatively, the force exerted on Jupiter by the Sun to the force exerted on the Sun by Jupiter.
  b Compare, qualitatively, the acceleration of Jupiter caused by the Sun to the acceleration of the Sun caused by Jupiter.

5 A planet in a distant solar system is known to exert a force of 2.1×10^{23} N on its closest star. Calculate the acceleration of the star due to this force given that $m_{star} = 1.0 \times 10^{30}$ kg.

6 A comet of mass 1000 kg is plummeting towards Jupiter. Jupiter has a mass of 1.90×10^{27} kg and a radius of 7.15×10^{7} m. If the comet is about to crash into Jupiter, calculate the:
  a magnitude of the gravitational force that Jupiter exerts on the comet
  b magnitude of the gravitational force that the comet exerts on Jupiter
  c acceleration of the comet towards Jupiter
  d acceleration of Jupiter towards the comet.

7 Calculate the acceleration due to gravity on the surface of Mars given that it has a mass of 6.4×10^{23} kg and a radius of 3400 km.

8 Calculate the normal force acting on a 50 kg person in a lift under the following circumstances.
  a The lift is accelerating downwards at 0.6 m s^{-2}.
  b The lift is moving downwards at a constant speed of 2.0 m s^{-1}.

9 During a space mission, an astronaut of mass 80 kg initially accelerates at 30 m s^{-2} upwards, then travels in a stable circular orbit at an altitude where the gravitational field strength is 8.2 N kg^{-1}.
  a What is the normal force acting on the astronaut during lift-off?
  b Calculate the normal force acting on the astronaut during the orbit phase.
  c What is the force due to gravity on the astronaut during the orbit phase?

10 The following is a diagram of the Moon.

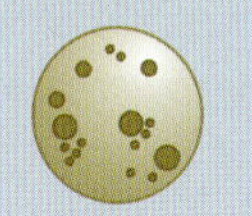

  a Redraw the diagram to indicate the direction of the Moon's gravitational field. Draw at least 8 field lines.
  b At which point, A or B, is the gravitational field strength strongest? Justify your answer.

11 An experiment is conducted with a spring balance on the surface of Mars. The experiment shows that an object experiences a force of 1.86 N due to gravity.

a What is the mass of the object if the gravitational field strength at the surface of Mars is 3.72 N kg^{-1}?

b What is the mass of the same object on the Earth?

12 A set of bathroom scales is calibrated so that when the person standing on it experiences a force of 700 N due to gravity, the scales read 71.4 kg. What gravitational field strength has been assumed in this setting? Give your answer to 2 decimal places.

13 A hypothetical planet is a flattened sphere. Its radius at the poles is 4750 km compared to 4800 km at the equator. The planet's mass is 4.00×10^{24} kg.

a Calculate the planet's gravitational field strength at the equator. Give your answer to 2 decimal places.

b Calculate how much stronger the gravitational field would be at the planet's poles compared with the equator. Give your answer as a percentage of the strength at the equator. Express your answer to 2 decimal places.

14 Neptune has a radius of 2.48×10^{7} m and a mass of 1.02×10^{26} kg.

a Calculate the gravitational field strength on the surface of Neptune.

b A 250 kg lump of ice is falling directly towards Neptune. What is its acceleration as it nears the surface of Neptune? Ignore any drag effects.

15 Calculate the distance from the Sun (which has a mass of 1.99×10^{30} kg) at which an astronaut would experience the same gravitational field strength as they experience at the Earth's surface (that is, 9.8 N kg^{-1}).

16 A person standing on the surface of the Earth experiences a gravitational force of 900 N. What gravitational force will this person experience at a height of two Earth radii above the Earth's surface?

17 A 20 kg rock is speeding towards Mercury. The following graph shows the force on the rock as a function of its distance from the centre of the planet. The radius of Mercury is 2.4×10^{6} m.

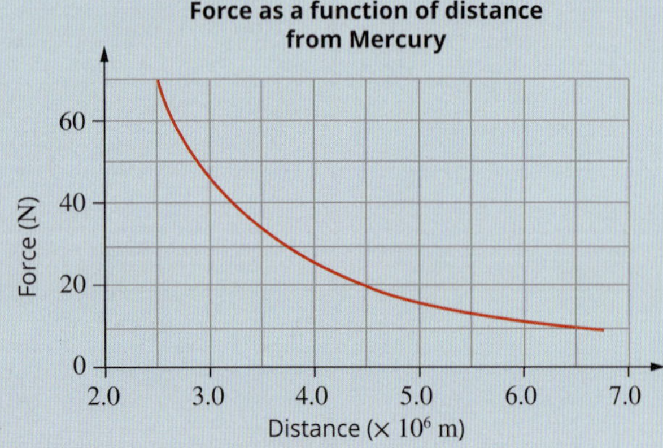

When the rock is 3.0×10^{6} m from the centre of the planet, its speed is estimated to be 1.0 km s^{-1}. Using the graph, estimate the:

a increase in the rock's kinetic energy as it moves to a point that is 2.5×10^{6} m from the centre of Mercury

b kinetic energy of the rock at this closer point

c speed of the rock at this point

d gravitational field strength at 2.5×10^{6} m from the centre of Mercury.

Application and analysis

18 According to the inverse square law, gravitational acceleration decreases as altitude above the Earth increases, as shown in the graph below. The blue line on the graph gives the value of gravitational acceleration (g) at various altitudes (h).

a Determine from the graph the approximate altitude at which the gravitational field strength will be 3.5 N kg^{-1}.

b Using the graph, compare the strength of the gravitational field at ground level with that at an altitude of 6500 km. Predict what the difference would be using the inverse square law. Is this prediction supported by the graph?

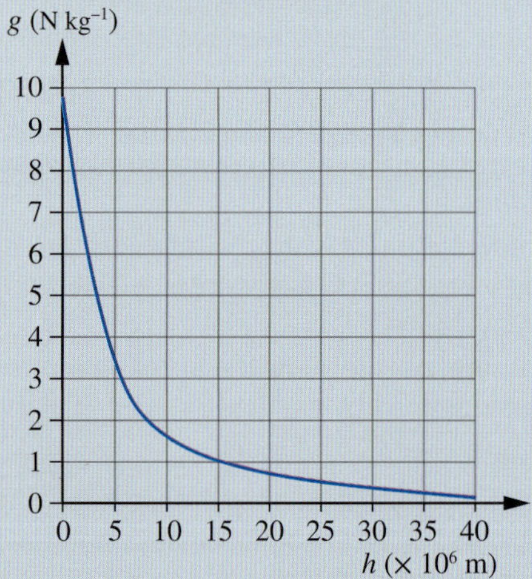

19 Two stars of masses M and m are in orbit around each other. As shown in the following diagram, they are a distance R apart. A spacecraft located at point X experiences zero net gravitational force from these stars. Calculate the value of the ratio $\frac{M}{m}$.

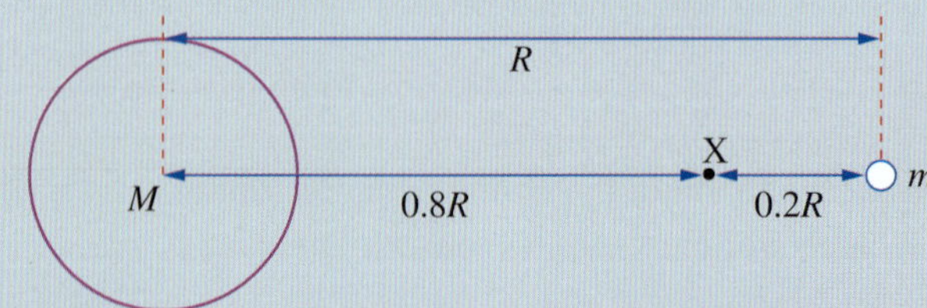

20 Two identical satellites orbit the Earth at altitudes r and $2r$, where r is the radius of the Earth.

a How does the force due to gravity acting on each of the satellites compare?

b How does the value of g at the site of each satellite compare?

21 The graph below shows the gravitational field at various distances above the Earth's surface. A wayward satellite of mass 2000 kg is drifting towards the Earth.

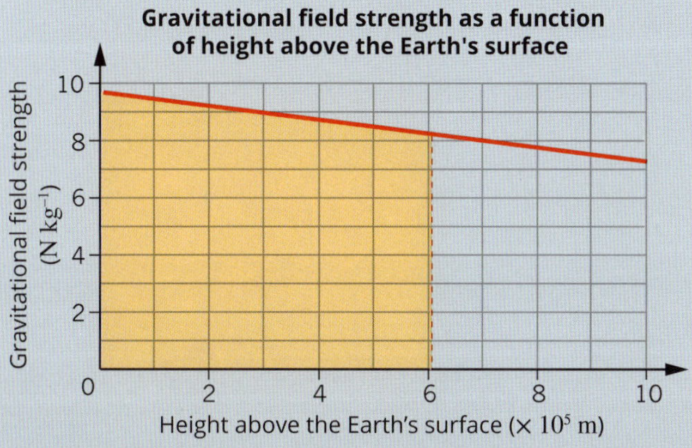

a What is the gravitational field strength at an altitude of 500 km?

b What quantities are represented by the shaded area on the graph?

c How much kinetic energy does the satellite gain as it travels from an altitude of 500 km back to the surface of the Earth?

d In reality, would the satellite gain the amount of kinetic energy that you have calculated in part **c**? Justify your answer.

CHAPTER

05 Electric and magnetic fields

In 1820, Hans Christian Oersted discovered that an electric current could produce a magnetic field. His work established the initial ideas behind electromagnetism. Since then, our understanding and application of electromagnetism has developed to such an extent that much of our modern way of living relies upon it.

In this chapter you will investigate electric and magnetic fields, the concepts that apply to each, and some of the interactions between these closely related phenomena.

Key knowledge

- describe gravitation, magnetism and electricity using a field model **5.1, 5.3**
- investigate and compare theoretically and practically gravitational, magnetic and electric fields, including directions and shapes of fields, attractive and repulsive effects, and the existence of dipoles and monopoles **5.1, 5.3, 5.5**
- investigate and compare theoretically and practically gravitational fields and electrical fields about a point mass or charge (positive or negative) with reference to:
  - the direction of the field **5.1, 5.5**
  - the shape of the field **5.1, 5.5**
  - the use of the inverse square law to determine the magnitude of the field **5.2, 5.5**
  - potential energy changes (qualitative) associated with a point mass or charge moving in the field **5.1, 5.5**
- investigate and apply theoretically and practically a field model to magnetic phenomena, including shapes and directions of fields produced by bar magnets, and by current-carrying wires, loops and solenoids **5.3**
- identify fields as static or changing, and as uniform or non-uniform **5.1, 5.3, 5.5**
- analyse the use of an electric field to accelerate a charge, including:
  - electric field and electric force concepts: $E = k\frac{Q}{r^2}$ and $F = k\frac{q_1q_2}{r^2}$ **5.2**
  - potential energy changes in a uniform electric field: $W = qV$, $E = \frac{V}{d}$ **5.1**
  - the magnitude of the force on a charged particle due to a uniform electric field: $F = qE$ **5.1**
- analyse the use of a magnetic field to change the path of a charged particle, including:
  - the magnitude and direction of the force applied to an electron beam by a magnetic field: $F = qvB$, in cases where the directions of v and B are perpendicular or parallel **5.4**

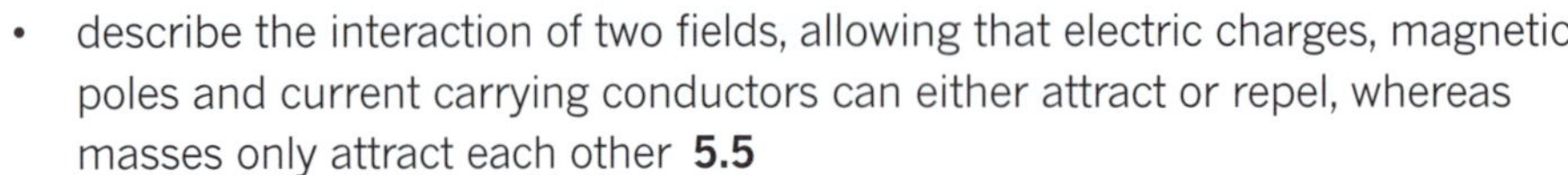

- describe the interaction of two fields, allowing that electric charges, magnetic poles and current carrying conductors can either attract or repel, whereas masses only attract each other **5.5**
- investigate and analyse theoretically and practically the force on a current carrying conductor due to an external magnetic field, $F = nIlB$, where the directions of I and B are either perpendicular or parallel to each other **5.4**

5.1 Electric fields

A **field** is a region of space where objects experience a force due to a property of that field. Gravity, electricity and magnetism can all be described by fields. In Chapter 4, the direction, shape and strength of gravitational fields around a mass were described. In this section the electric field will be explained.

An **electric field** surrounds positive and negative charges and exerts a force on other charges within the field. Just as a gravitational field can be represented by field lines, so can the electric field around a charged object (Figure 5.1.1).

FIGURE 5.1.1 Charged plasma follows lines of the electric field produced by a Van de Graaff generator.

There are four fundamental forces in nature that act at a distance; that is, they can exert a force on an object without making any physical contact with it. These are called non-contact forces. They are the strong nuclear force, the weak nuclear force, the electromagnetic force and the gravitational force.

In order to understand these forces, scientists use the idea of a field. A field is a region of space around an object that has certain properties, such as electric charge or mass. Another object with that same property in the field will experience a force without any contact between the two objects.

For example, there is a gravitational field around the Earth due to the mass of the Earth. Any object with mass that is within this gravitational field experiences a force of attraction towards the Earth. According to Newton's third law, there is also an equal and opposite force due to the gravitational field of the object. For example, the gravitational field of the Earth exerts a force on a skydiver (Figure 5.1.2) and the gravitational field of the skydiver exerts a force on the Earth.

FIGURE 5.1.2 As described by Newton's third law, the gravitational field of the Earth exerts a force on a skydiver and the gravitational field of the skydiver exerts a force on the Earth.

Similarly, any charged object has a region of space around it (an electric field) where another charged object will experience a force. However, unlike gravity, which only exerts an attractive force, electric fields can exert forces of attraction or repulsion.

ELECTRIC FIELD LINES

An electric field is a vector quantity: it has both direction and strength.

In order to visualise electric fields around charged objects you can use electric field lines. Some field lines are already visible—for example, the girl's hair in Figure 5.1.3 is tracing out the path of electric field lines. Diagrams of electric field lines can be constructed, just as they can for gravitational fields.

FIGURE 5.1.3 The girl's hair follows the lines of the electric field produced when she became charged while sliding down a plastic slide.

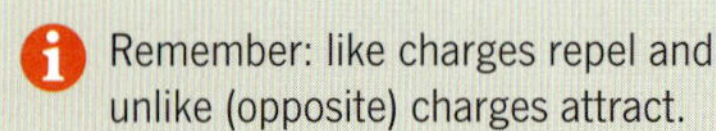

Remember: like charges repel and unlike (opposite) charges attract.

Field lines are drawn with arrowheads on them indicating the direction of the force that a small positive charge would experience if it were placed in the electric field. (This small charge is known as a positive *test charge*). Therefore field lines point away from positively charged objects and towards negatively charged objects. Usually only a few representative lines are drawn.

The density of field lines (that is, how close they are together) is an indication of the relative strength of the electric field. This is explained in more detail later in this section.

Rules for drawing electric field lines

When drawing electric field lines in two dimensions around a charged object, there are a few rules that need to be followed.

- Electric field lines go from positively charged objects to negatively charged objects.
- Electric field lines start and end at 90° to the charged surface, with no gap between the lines and the surface.
- Field lines can never cross; if they did it would indicate that the field is in two directions at that point, which can never happen.
- Around small charged spheres, called **point charges**, the field lines radiate like spokes on a wheel.
- Around point charges you should draw at least eight field lines: top, bottom, left, right and another field line in between each of these.
- Between two point charges, the direction of the field at any point is the resultant field vector determined by adding the field vectors due to each of the two point charges.
- Between two oppositely charged parallel plates, the field lines between the plates are evenly spaced and are drawn straight from the positive plate to the negative plate.

Note that field drawings are two-dimensional representations of a three-dimensional field.

Figure 5.1.4 provides some examples of how to draw electric field lines.

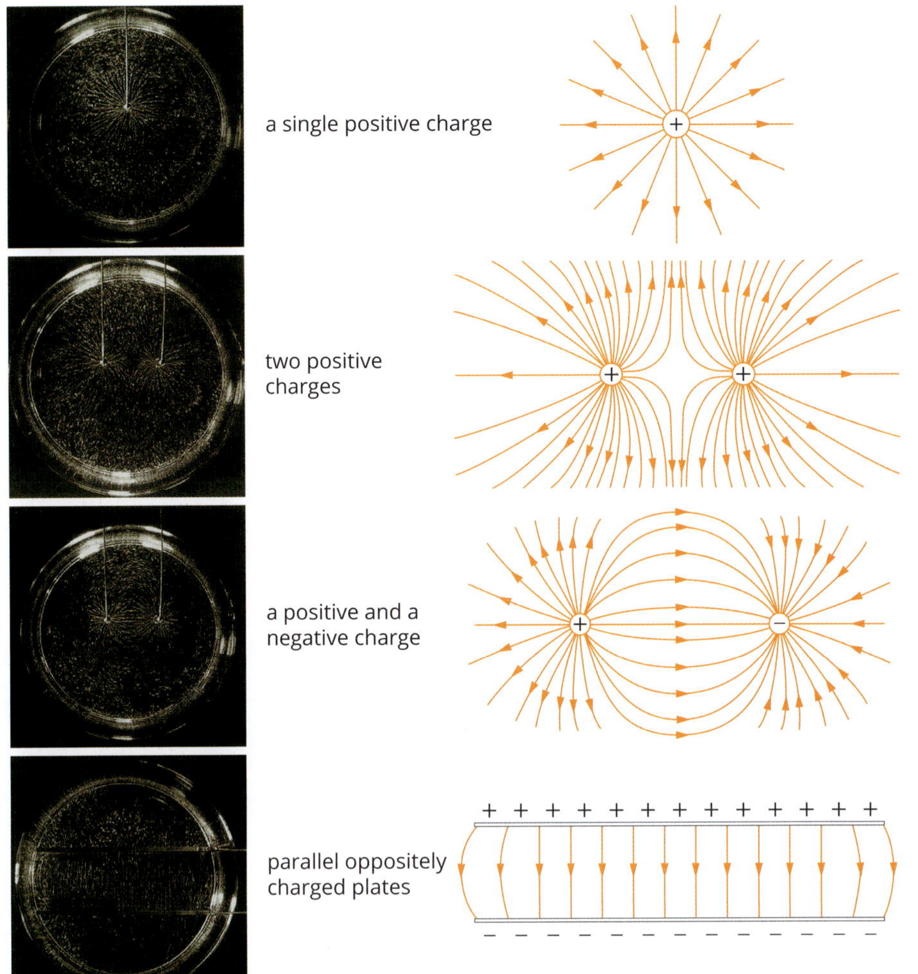

FIGURE 5.1.4 Grass seeds suspended in oil align themselves with an electric field. The diagram next to each photo shows lines representing the electric field.

Strength of the electric field

The distance between adjacent field lines indicates the strength of the field. Around a point charge the field lines are closer together near the charge and further apart the further you move away. You can see this in the field-line diagrams in Figure 5.1.4. Therefore the **electric field strength** decreases as the distance from a point charge increases.

A uniform electric field is established between two parallel metal plates that are oppositely charged. The field strength is constant at all points within a uniform electric field, so the field lines are parallel.

FORCES ON FREE CHARGES IN ELECTRIC FIELDS

If a charged particle, such as an electron, is placed in an electric field, it experiences a force. The direction of the field and the sign of the charge allow you to determine the direction of the force.

Figure 5.1.5 shows a positive test charge (a proton) and a negative test charge (an electron) in a uniform electric field. Recall that the direction of an electric field is defined as the direction of the force that a positive test charge would experience in that field. Thus an electron will experience a force in the opposite direction to the electric field, while a proton will experience a force in the same direction as the field.

PHYSICSFILE

Sharks and electric fields

Sharks have special sensory organs called electroreceptors. These electroreceptors help them sense electric fields in the water. Small electric fields are produced by muscle contractions in living creatures, including the creatures sharks prey on. Sharks may be able to detect prey by the electric field these muscle contractions create.

Sharks are able to hunt their prey using electroreceptors.

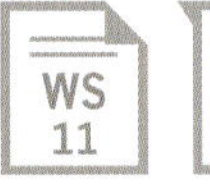

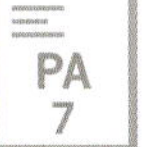

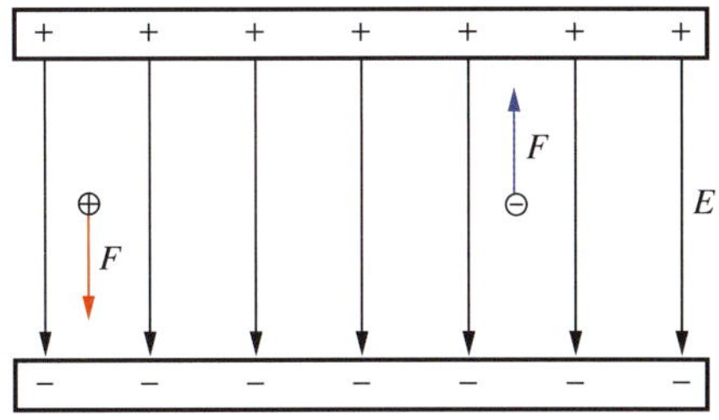

FIGURE 5.1.5 The direction of the electric field (E) indicates the direction in which a force would act on a positive charge. A negative charge would experience a force in the opposite direction to the field.

The magnitude of the force experienced by a charged particle due to an electric field can be determined using the following equation.

> $F = qE$
> where F is the force on the charged particle (N)
> q is the charge of the object experiencing the force (C)
> E is the strength of the electric field (N C^{-1})

This equation states that the force experienced by a charge is proportional to the strength of the electric field, E, and the size of the charge, q. The force on a charged particle will cause the particle to accelerate in the field. This means that the particle could increase or decrease its velocity while in the field. It could also change its direction.

To calculate the acceleration of a charged particle due to the force experienced, you can use the equation from Newton's second law: $F = ma$.

You may notice some similarities between the equations $F = qE$ and $F = mg$. Both equations describe the force on an object in a field. The first equation is for a charge in an electric field and the second is for a mass in a gravitational field. In the electric field equation, q is substituted for m and E for g.

Worked example 5.1.1

USING $F = qE$

Calculate the magnitude of the uniform electric field that would cause a force of 5.00×10^{-21} N on an electron. Assume that $q_e = -1.6 \times 10^{-19}$ C.

Thinking	Working
Rearrange the relevant equation to make electric field strength the subject.	$F = qE$ $E = \frac{F}{q}$
Substitute the values for F and q into the rearranged equation and solve for E. (As only the magnitude is required, q can be positive.)	$E = \frac{F}{q}$ $= \frac{5.00 \times 10^{-21}}{1.6 \times 10^{-19}}$ $= 3.1 \times 10^{-2}$ N C^{-1}

Worked example: Try yourself 5.1.1

USING $F = qE$

Calculate the magnitude of the uniform electric field that creates a force of 9.00×10^{-23} N on a proton. Assume that $q_p = +1.6 \times 10^{-19}$ C.

CASE STUDY **ANALYSIS**

Gravitational force and electric force

Oppositely charged parallel plates can be arranged one above the other. The electric field between them is vertical. The direction of the field can then be manipulated to create an upwards force on a charged particle in the field.

If the electric force created by the field on the charged object is equal to the gravitational force on the object, then these two forces will balance each other, resulting in a net force of zero. This means that the charged object will be suspended between the plates.

This phenomenon was used by Robert Millikan and his PhD student Harvey Fletcher in their oil drop experiment performed in 1909. The experiment determined the fundamental charge of an electron to within 1% of the currently accepted value.

Analysis

In his experiment, Millikan used known properties of the oil and drag forces to find the mass of an electron. He then used this to calculate the charge. Millikan determined that the charge on an electron was 1.6×10^{-19} C (to 2 significant figures). Assume that the oil drop was suspended by an electric field of strength 80×10^{3} N C^{-1}. Using these values, determine the value for the mass of an electron that Millikan must have used in his calculations. Provide your answer in grams.

WORK DONE IN UNIFORM ELECTRIC FIELDS

Electrical potential (V) is a form of energy that is stored in an electric field. Work is done on an electric field when a charged particle is forced to move in the field. Conversely, when the electric field moves a charged particle, work is done on the charged particle.

Electrical potential (V) is defined as the work required per unit charge to move a positive point charge from infinity to a place in the electric field. The electrical potential at infinity is defined as zero. This definition leads to the following equation.

$$V = \frac{W}{q}$$

$$\therefore W = qV$$

where W is the work done on a positive point charge or on the field (J)
q is the charge of the point charge (C)
V is the electrical potential (J C^{-1}) or volts (V)

This equation can also be derived using electric field strength. Electric field strength can be thought of as the force applied per coulomb of charge. It is expressed by the equation:

$$E = \frac{F}{q}$$

An alternative measure of the electric field strength is volts per metre, which is calculated using the equation:

$$E = \frac{V}{d}$$

where d is the distance in metres between points that are parallel to the field.

You can equate both expressions and rearrange them to find an expression for the work done (J) to make a charged particle move a distance against a **potential difference**:

$$\frac{F}{q} = \frac{V}{d}$$

$$Fd = qV \text{ and since } W = Fd$$

$$W = qV$$

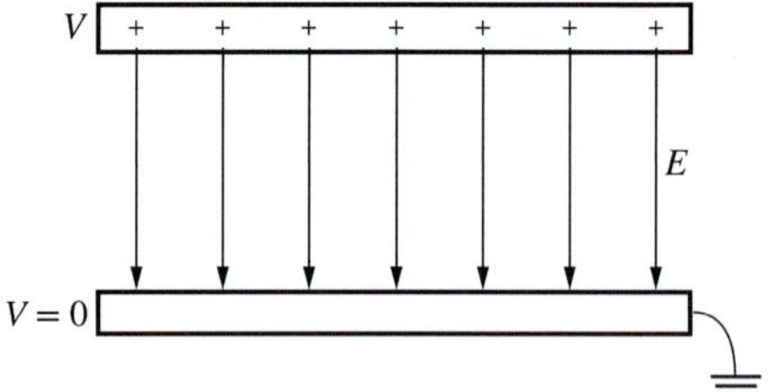

FIGURE 5.1.6 The potential of two plates when one has a positive potential and the other is earthed

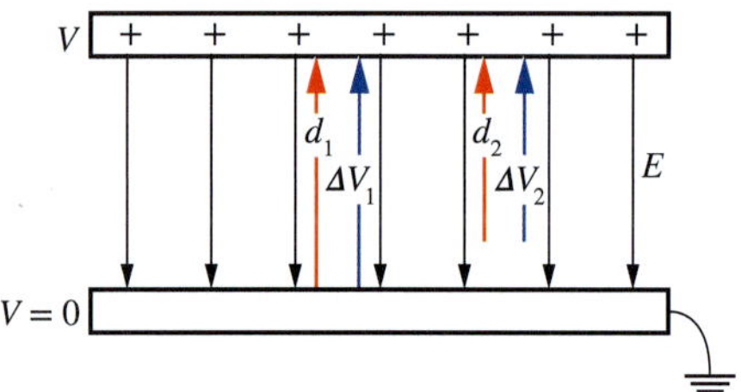

FIGURE 5.1.7 The potential difference between two points in a uniform electric field

Consider two parallel plates in which the positive plate is at a potential (V) and the other plate is earthed, which is defined as zero potential (Figure 5.1.6). The difference in potential between these two plates is called the electrical potential difference (V).

Between any two points in an electric field (E) separated by a distance (d) that is parallel to the field, the potential difference is defined as the change in the electrical potential between these two points (Figure 5.1.7).

$$E = \frac{V}{d}$$

$$V = Ed$$

where V is the difference in electrical potential (V)

E is the electrical field strength (V m^{-1})

d is the distance between points that are parallel to the field (m)

Sometimes the equation for the electric field strength is given in the form $E = -\frac{V}{d}$. In this version of the equation, the electric field is always from the higher potential (V_i) towards the lower potential (V_f), so that the change in potential difference ($V = V_f - V_i$) becomes a negative value. For example, if $V_i = 15\text{ V}$ and $V_f = 0\text{ V}$, $V = 0 - 15 = -15\text{ V}$. However, in this book, the potential, V, is given as the magnitude of the potential difference and so it is always a positive value. Thus the potential difference in the example just given would be 15 V.

CALCULATING WORK DONE

By combining two of the equations mentioned previously, you can derive an equation for the work done on a point charge to move it a distance across a potential difference.

$$W = qV \text{ and } V = Ed$$

$$\therefore W = qEd$$

where W is the work done on the point charge or on the field (J)

q is the charge of the point charge (C)

E is the electrical field strength (V m^{-1} or N C^{-1})

d is the distance the charged particle is forced to move in the electric field (m)

If the equation $W = qV$ is written in the form $W = qEd$, it is similar to the equation for gravitational potential energy ($E_g = mgh$), where the mass (m) is similar to the charge of the object experiencing the force (q), the strength of the electric field (E) is similar to the standard gravitational acceleration on the surface of the Earth (g) and the height (h) is similar to the distance (d).

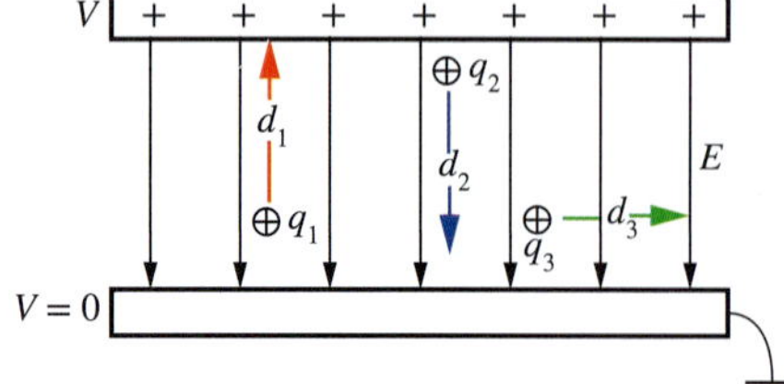

FIGURE 5.1.8 Work is being done on the field by moving q_1 and work is being done by the field on q_2. No work is done on q_3 since it is moving perpendicular to the field.

Work done by or on an electric field

When calculating work done, which changes the electrical potential energy, remember that work can be done either:

- by the electric field on a charged object or
- on the electric field by forcing the object to move.

You need to examine what is happening in a particular situation to know how the work is being done.

For example, if a charged object is moving in the direction it would naturally tend to go in an electric field, then work is done on the object by the field. So when a positive charge is moved in the direction of the electric field, the electric field has done work on the charge. (An example is q_2 in Figure 5.1.8.)

When work is done by a charged object on an electric field, the object is forced to move against the direction it would naturally go. Work has been done on the field by forcing the object to move. For example, if a force causes a positive charge to move towards the positive plate within a uniform electric field, work has been done on the electric field by forcing the object to move. (An example is q_1 in Figure 5.1.8.)

If a charge doesn't move any distance parallel to the field, then no work is done on or by the field. (An example is q_3 in Figure 5.1.8.)

Worked example 5.1.2

WORK DONE ON A CHARGE IN A UNIFORM ELECTRIC FIELD

A student sets up two parallel plates. One plate is at a potential of 12.0 V and the other plate, which is earthed, is positioned 0.50 m away. Calculate the work done to move a proton a distance of 10.0 cm towards the negative plate. Assume that $q_p = +1.6 \times 10^{-19}$ C.

In your answer identify what does the work and what the work is done on.

Thinking	Working
Identify the variables presented in the problem needed to calculate the electric field strength, *E*.	$V_2 = 12.0\text{ V}$ $V_1 = 0\text{ V}$ $d = 0.50\text{ m}$
Use the equation $E = \frac{V}{d}$ to determine the electric field strength.	$E = \frac{V}{d}$ $= \frac{12.0 - 0}{0.50}$ $= 24\text{ V m}^{-1}$
Use the equation $W = qEd$ to determine the work done. Note that *d* here is the distance that the proton moves.	$W = qEd$ $= 1.6 \times 10^{-19} \times 24 \times 0.10$ $= 3.8 \times 10^{-19}\text{ J}$
Determine if work is done on the charge by the field, or if work is done on the field by the charge.	As the positively charged proton is moving naturally towards the negative plate, work is done on the proton by the field.

Worked example: Try yourself 5.1.2

WORK DONE ON A CHARGE IN A UNIFORM ELECTRIC FIELD

A student sets up two parallel plates. One plate is at a potential of 36.0 V and the other plate, which is earthed, is positioned 2.00 m away. Calculate the work done to move an electron a distance of 75.0 cm towards the negative plate. Assume that $q_e = -1.6 \times 10^{-19}$ C.

In your answer identify what does the work and what the work is done on.

5.1 Review

OA ✓✓

SUMMARY

- An electric field is a region of space around a charged object in which another charged object will experience a force.
- Electric fields are represented by field lines.
- Electric field lines point in the direction of the force that a positive charge in the field would experience.
- A positive charge experiences a force in the direction of the electric field and a negative charge experiences a force in the opposite direction to the field.
- The spacing between field lines indicates the strength of the field. The closer the lines, the stronger the field.
- Electric field strength can be expressed as $E = \frac{F}{q}$ or $E = \frac{V}{d}$.
- Around point charges, the electric field radiates three-dimensionally in all directions.
- Between two oppositely charged parallel plates, the field is of uniform strength and so the field lines are parallel.
- When charges are in an electric field, they accelerate in the direction of the force acting on them.
- The force on a charged particle can be determined using the equation $F = qE$.
- Force can be related to the acceleration of a particle using the equation $F = ma$.
- Electrical potential energy is stored in an electric field.
- When a charged object is moved against the direction it would naturally move in an electric field, then work is done on the field.
- When a charged object is moved in the direction it would naturally tend to move in an electric field, then the field does work on the particle.
- The work done on or by an electric field can be calculated using the equations $W = qV$ or $W = qEd$.

KEY QUESTIONS

Knowledge and understanding

1 Draw the electric field lines between a positively charged object and a negatively charged object.

2 Draw the electric field lines between a negatively charged top plate and a positively charged bottom plate that is parallel to it.

3 Identify whether the rules below for drawing electric field lines are true or false.

- **a** Electric field lines start and end at 90° to the surface, with no gap between the lines and the surface.
- **b** Field lines can cross, which indicates that the field is in two directions at that point.
- **c** Electric fields go from negatively charged objects to positively charged objects.
- **d** Around small charged spheres called point charges you should draw at least eight field lines: top, bottom, left, right and in between each of these lines.
- **e** Around point charges the field lines radiate like the spokes of a wheel.
- **f** Between two point charges the direction of the field at any point is due to the closest of the two point charges.
- **g** Between two oppositely charged parallel plates, the field between the plates is evenly spaced and is represented by straight lines drawn from the negative plate to the positive plate.

4 For each of the following charged objects in a uniform electric field, determine if work is done on the field, by the field or if no work is done.

- **a** An electron moves towards a positive plate.
- **b** A positively charged point remains stationary.
- **c** A proton moves towards a positive plate.
- **d** A lithium ion (Li^+) moves parallel to the plates.
- **e** A positron moves away from a positive plate.

5 Calculate the force applied to a balloon carrying a charge of 5.00 mC in a uniform electric field of 2.50 N C^{-1}.

6 Calculate the charge on a plastic disk if it experiences a force of 0.0250 N in a uniform electric field of 18.0 N C^{-1}.

7 Calculate the acceleration of an electron in a uniform electric field of 3.25 N C^{-1}. The mass of an electron is 9.1×10^{-31} kg and its charge is -1.6×10^{-19} C.

Analysis

8 Calculate the potential difference between two points separated by 30.00 cm, parallel to the field lines and in an electric field of strength $4000\,V\,m^{-1}$.

9 An alpha particle is between two parallel plates, one of which is earthed. There is a uniform electric field of $34.0\,V\,m^{-1}$ between the plates.

a Calculate the work done to move the alpha particle a distance of 1.00 cm from the earthed plate to the plate with a positive potential. ($q_\alpha = +3.204 \times 10^{-19}\,C$)

b For the situation in part **a**, decide whether work was done on the field, by the field or if no work was done.

10 A researcher sees an oil drop with a mass of $1.161 \times 10^{-14}\,kg$ stationary between two horizontal parallel plates. There is an electric field of strength $3.55 \times 10^{4}\,N\,C^{-1}$ between the plates. The field is pointing vertically downwards. Calculate how many extra electrons are present on the oil drop. ($q_e = -1.6 \times 10^{-19}\,C$ and $g = 9.8\,N\,kg^{-1}$)

5.2 Coulomb's law

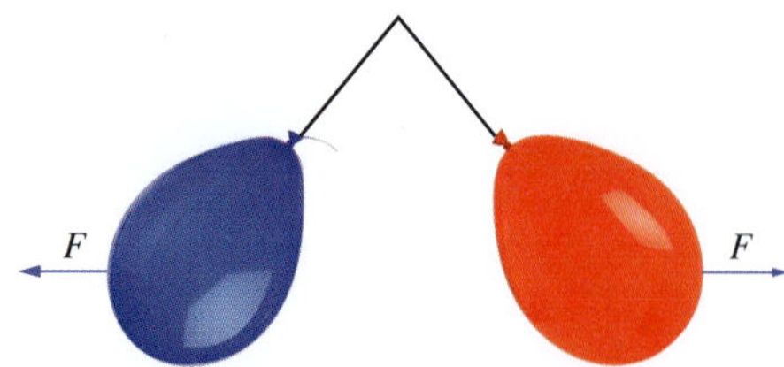

FIGURE 5.2.1 Two similarly charged balloons will repel each other by exerting a force on each other.

Electricity is one of nature's fundamental forces. It was Charles Coulomb who, in 1785, first published quantitative details of the force that acts between two electric charges. The force between any combination of electrical charges can be understood in terms of the force between two point charges separated by a certain distance (Figure 5.2.1). The effect of distance on the strength of an electric field from a single charge, and the force created by that field between charges, is explored in this section.

PHYSICSFILE

One coulomb of charge

One coulomb of charge is a huge amount. The amount of charge that can be placed on an ordinary object is a tiny fraction of a coulomb. Even a highly charged Van de Graaff generator will have only a few microcoulombs ($1\,\mu C = 10^{-6}\,C$) of excess charge. To understand how large one coulomb is, imagine two 1 C charges 1 metre apart. This configuration would produce a force of 10^{10} N—almost twice the weight of the Sydney Harbour Bridge!

THE FORCE BETWEEN CHARGED PARTICLES

Coulomb found that the force between two point charges (q_1 and q_2) separated by a distance (r) was proportional to the product of the two charges and inversely proportional to the square of the distance between them, r. This is another example of an inverse square law, discussed in Chapter 4.

Coulomb's law can be expressed as follows.

$$F = \frac{1}{4\pi\varepsilon_0}\frac{q_1 q_2}{r^2}$$

where F is the electrostatic force, i.e. the force on each charged object (N)

q_1 is the charge on one point (C)

q_2 is the charge on the other point (C)

r is the distance between the charged points (m)

ε_0 is a constant known as the permittivity of free space, equal to $8.8542 \times 10^{-12}\,C^2N^{-1}m^{-2}$ in air or a vacuum

By including the sign of each charge in the calculation, the force calculated could be positive or negative. A positive value indicates a repulsive force and a negative value indicates an attractive force.

For ease of calculations, Coulomb's law is simplified by combining the unvarying parts into a single constant known as Coulomb's constant (k) as follows:

$$k = \frac{1}{4\pi\varepsilon_0}$$

where $k = 8.99 \times 10^9\,N\,m^2\,C^{-2}$ in air or a vacuum, rounded to three significant figures.

The equation for Coulomb's law then can be written as follows.

$$F = k\frac{q_1 q_2}{r^2}$$

where $k = 8.99 \times 10^9\,N\,m^2\,C^{-2}$

The electric force between the two charges is similar to the gravitational force between two masses ($F_g = G\frac{m_1 m_2}{r^2}$), where the two masses (m_1 and m_2) are similar to the charges (q_1 and q_2) and the gravitational constant (G) is similar to Coulomb's constant (k). In both cases the force is inversely proportional to the square of the distance between the two objects.

Worked example 5.2.1

USING COULOMB'S LAW TO CALCULATE CHARGE

Two small positive point charges with equal charge are separated by 1.25 cm in air. Calculate the charge on each point charge if there is a repulsive force of 6.48 mN between them. Assume that $k = 8.99 \times 10^9\,\text{N m}^2\,\text{C}^{-2}$.

Thinking	Working
Convert the variables to SI units.	$F = 6.48 \times 10^{-3}\,\text{N}$ $r = 1.25 \times 10^{-2}\,\text{m}$
State Coulomb's law.	$F = k\dfrac{q_1q_2}{r^2}$
Substitute the values for F, r and k into the equation and calculate the answer. Remember: the question stated that the point charges are equal and positive.	$q_1q_2 = \dfrac{Fr^2}{k}$ $= \dfrac{6.48\times10^{-3}\times(1.25\times10^{-2})^2}{8.99\times10^9}$ $= 1.13\times10^{-16}$ Since $q_1 = q_2$: $q_1^{\,2} = 1.13\times10^{-16}$ $q_1 = \sqrt{1.13\times10^{-16}}$ $= +1.06\times10^{-8}\,\text{C}$ $q_2 = +1.06 \times 10^{-8}\,\text{C}$

Worked example: Try yourself 5.2.1

USING COULOMB'S LAW TO CALCULATE CHARGE

Two small point charges (q_1 and q_2) are charged by transferring a number of electrons from q_1 to q_2. The point charges are separated by 12.7 mm in air and their charges are equal and opposite. Calculate the charge on q_1 and q_2 if there is an attractive force of 22.5 μN between them. Assume that $k = 8.99 \times 10^9\,\text{N m}^2\,\text{C}^{-2}$.

Factors affecting the electric force

The force between two charged points is proportional to the product of their charges. If there is a force of 10 N between two charged points (Figure 5.2.2(a)) and either charge is doubled, then the force between them increases to 20 N (Figure 5.2.2(b)). Note that regardless of the charge on each point, the force on each point will be the same. If the force between two charges is 20 N, then both charges will experience the 20 N force regardless of their charge.

The force between two charged points is also inversely proportional to the square of the distance between them. This means that if the distance between q_1 and q_2 is doubled, the force on each point charge will decrease to one-quarter of the previous value.

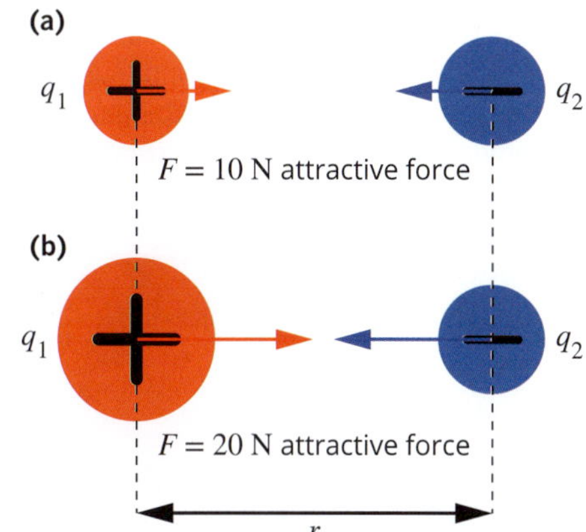

FIGURE 5.2.2 Forces acting between two point charges. (a) Two oppositely charged particles with a 10 N force of attraction. (b) Charge q_1 doubles, therefore the force of attraction between q_1 and q_2 also doubles, to 20 N.

PHYSICSFILE

Static electricity

Static electricity is a phenomenon you may have experienced when you felt an electric shock. The static shock occurs when a material or object that is negatively charged comes in contact with another material or object that is positively charged.

You may have experienced a static shock when walking across a carpet and then touching the handle of a door. Static electricity is generated on your shoes when the soles rub against the carpet. When you touch the door handle, the electrons are transferred to the door handle. It is this transfer that causes the static shock you feel.

Static electric shocks are often not harmful. There are cases, though, where static electricity and static electric shocks can be dangerous or cause damage. For example, static shock can be powerful enough to destroy electronic components. Anyone working on electronic components—during manufacturing, repair or testing—must use antistatic devices. These devices ensure that the wearer's body is electrically grounded and thus prevent any build-up of static charge. A commonly used device is an antistatic wrist strap (see figure below).

When building, testing or repairing electronic circuits, it is imperative to ensure that no static electricity is transferred to the electronic circuit.

THE ELECTRIC FIELD AT A DISTANCE FROM A CHARGE

In the previous section the electric field strength, E, was defined as the force applied per coulomb of charge, that is, $E = \frac{F}{q}$. In other words, E is proportional to the force exerted on a positive test charge and inversely proportional to the magnitude of that charge. It is measured in N C^{-1}. It is also useful to be able to determine the strength of the electric field at a distance from a single point charge.

The magnitude of the electric field at a distance from a single point charge is given by the following equation.

$$E = k\frac{Q}{r^2}$$

where E is the strength of the electric field around the point (N C^{-1})
Q is the charge on the point creating the field (C)
r is the distance from the charge (m)
k is Coulomb's constant ($8.99 \times 10^9 \text{ N m}^2 \text{C}^{-2}$)

The magnitude of E determined using this formula is independent of the value of the test charge and depends only on the charge producing the field (Q).

This equation can be derived from Coulomb's law. It is similar to the equation for gravitational field strength ($g = \frac{GM}{r^2}$). In both cases, the field strength (E or g) is inversely proportional to the square of the distance from the single point charge or mass.

Worked example 5.2.2

ELECTRIC FIELD OF A SINGLE POINT CHARGE

Calculate the magnitude and direction of the electric field at a point P that is 20 cm below a negative point charge, Q, of 2.0×10^{-6} C.

Thinking	Working
Convert the units to SI units as required.	$Q = -2.0 \times 10^{-6}$ C $r = 20$ cm $= 0.20$ m
Substitute the known values to find the magnitude of E.	$E = k\frac{Q}{r^2}$ $= 8.99 \times 10^9 \times \frac{2.0 \times 10^{-6}}{0.20^2}$ $= 4.5 \times 10^5 \text{ NC}^{-1}$
The direction of an electric field is defined as that acting on a positive test charge (see Section 5.1). Point P is below the charge.	Since the charge is negative, the direction will be towards the charge, i.e. upwards.

Worked example: Try yourself 5.2.2

ELECTRIC FIELD OF A SINGLE POINT CHARGE

Calculate the magnitude and direction of the electric field at a point P that is 15 cm to the right of a positive point charge, Q, of 2.0×10^{-6} C.

5.2 Review

OA ✓✓

SUMMARY

- Coulomb's law for the force between two charges, q_1 and q_2, separated by a distance of r is:

$$F = \frac{1}{4\pi\varepsilon_0}\frac{q_1 q_2}{r^2}$$

- By using Coulomb's constant (k) for the expression $\frac{1}{4\pi\varepsilon_0}$, Coulomb's law can be simplified to:

$$F = k\frac{q_1 q_2}{r^2} \text{ where } k = 8.99 \times 10^9\,\text{N m}^2\,\text{C}^{-2}$$

- The magnitude of the electric field, E, at a distance r from a single point charge, Q, is given by:

$$E = k\frac{Q}{r^2} \text{ where } k = 8.99 \times 10^9\,\text{N m}^2\,\text{C}^{-2}$$

KEY QUESTIONS

Knowledge and understanding

1 A charge of $+q$ is placed a distance r from another charge also of $+q$. A repulsive force of magnitude F is found to exist between them. Choose the correct answer from the options in bold to describe the changes, if any, that will occur to the force in the following situations.

  a If one of the charges is doubled to $+2q$, the force will **halve/double/quadruple/quarter** and the charges **repel/attract**.

  b If both charges are doubled to $+2q$, the force will **halve/double/quadruple/quarter** and the charges **repel/attract**.

  c If one of the charges is changed to $-2q$, the force will **halve/double/quadruple/quarter** and the charges **repel/attract**.

  d If the distance between the charges is halved to $0.5r$, the force will **halve/double/quadruple/quarter** and the charges **repel/attract**.

2 A point charge, Q, is moved from a position 30 cm away from a test charge to a position 15 cm from the same test charge. If the magnitude of the original electric field, E, was $6.0 \times 10^3\,\text{N C}^{-1}$, what is the magnitude of the electric field at the new position?

3 Calculate the magnitude of the force that would exist between two point charges each of 1.00 C separated by 1.00 km. Assume that $k = 8.99 \times 10^9\,\text{N m}^2\,\text{C}^{-2}$.

4 Calculate the repulsive force on each proton in a helium nucleus that is in a vacuum. The protons are 2.50 fm apart. Assume that $k = 8.99 \times 10^9\,\text{N m}^2\,\text{C}^{-2}$ and $q_p = +1.6 \times 10^{-19}\,\text{C}$. Note that 1 fm = 1×10^{-15} m.

Analysis

5 A point charge of 6.50 mC is suspended from a ceiling by an insulated rod. Calculate the distance from the point charge that a small sphere of mass 10.0 g and charge −3.45 nC will be located if it is suspended in air. Assume that $k = 8.99 \times 10^9\,\text{N m}^2\,\text{C}^{-2}$.

6 Two point charges 30.0 cm apart in air are charged by transferring electrons from one point to the other. Calculate how many electrons must be transferred to create an attractive force of 1.0 N. Assume that $k = 8.99 \times 10^9\,\text{N m}^2\,\text{C}^{-2}$ and $q_e = -1.6 \times 10^{-19}\,\text{C}$.

7 A photocopy machine is one application of the principles of Coulomb's law. Conduct some research and then explain the relevance of Coulomb's law to photocopy machines.

5.3 The magnetic field

Although naturally occurring magnets had been known for many centuries, it was still not known in the early 19th century how to create an artificial magnet. Then, in 1820, the Danish physicist Hans Christian Oersted (Figure 5.3.1) made a discovery that led him to develop a scientific explanation of the magnetic effect created by an electric current, an effect that led to the creation of artificial magnets.

Oersted was a keen believer in the unity of nature, that is, the concept that everything in the universe is somehow connected. He noticed that when he switched on a current from a voltaic pile (an early simple battery), a magnetic compass nearby moved. Intrigued by this observation, he conducted further experiments and demonstrated that it was the current from the voltaic pile that was affecting the compass. His experiments showed that the stronger the current, and the closer the compass was to it, the greater the effect on the compass. These observations led him to conclude that the electric current was creating a magnetic field. This connection between electric and magnetic fields is now used in ways that are fundamental to society.

FIGURE 5.3.2 A bar magnet attracting iron filings

FIGURE 5.3.1 In 1820, Hans Christian Oersted discovered that a magnetic effect is created by an electric current. Oersted is honoured by this statue in Oersted's Park, Copenhagen.

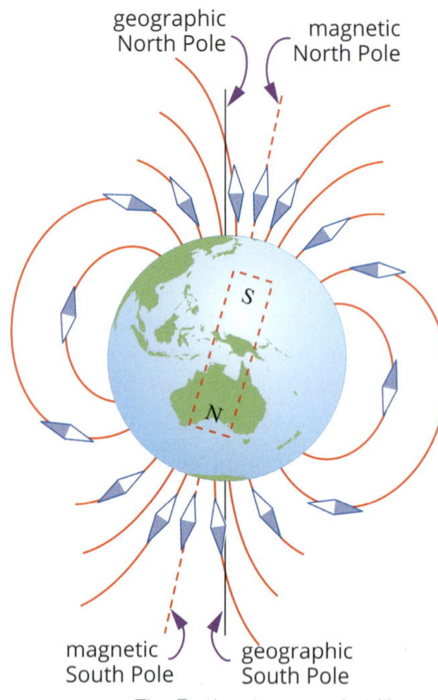

FIGURE 5.3.3 The Earth acts somewhat like a huge bar magnet. The south pole of this imaginary magnet (the magnetic North Pole) is near the geographic North Pole. This is the point to which the north pole of a compass appears to point.

Remember: Like magnetic poles repel each other; unlike magnetic poles attract each other.

MAGNETISM

Before looking further into the connection between electric current and magnetic fields, it is necessary to review some fundamentals of magnetism.

Magnetic materials include magnets, materials that can become magnetised and materials that feel a force of attraction to a magnet. The force of attraction is felt by a magnetic material in the **magnetic field** surrounding the magnet. Materials such as iron will be attracted to a magnet (Figure 5.3.2).

If you experiment with a magnet you will find that each end of it behaves differently, particularly when interacting with another magnet. One end will be attracted while the other is repelled. Each end of a magnet is referred to as a **magnetic pole**.

A suspended magnet that is free to move will always orientate itself in a north–south direction. That is basically what the needle of a compass is—a freely suspended, small magnet. If it is allowed to swing vertically as well, the magnet will tend to tilt vertically. The vertical direction (upwards–downwards) and the magnitude of the angle depend upon the direction of the magnet in relation to the Earth's poles.

The Earth has a huge magnetic field around it (Figure 5.3.3).

PHYSICSFILE

'Flipping' poles

The Earth's magnetic poles are not static like their geographic counterparts. For many years, the magnetic North Pole has been moving at approximately 9 km per year (see figure at right). In recent years this rate has increased, and is now averaging over 55 km per year. Once every few hundred thousand years the magnetic poles flip in a phenomenon called 'geomagnetic reversal', so after the next geomagnetic reversal, compasses will point south instead of north. The Earth is well overdue for the next flip, and recent measurements have shown that the Earth's magnetic field is starting to weaken faster than in the past, so the magnetic poles may be getting close to a flip. Although some studies have suggested such a flip is not instantaneous—it may take many hundreds if not a few thousands of years—some more recent studies have suggested that it could happen over a much shorter period.

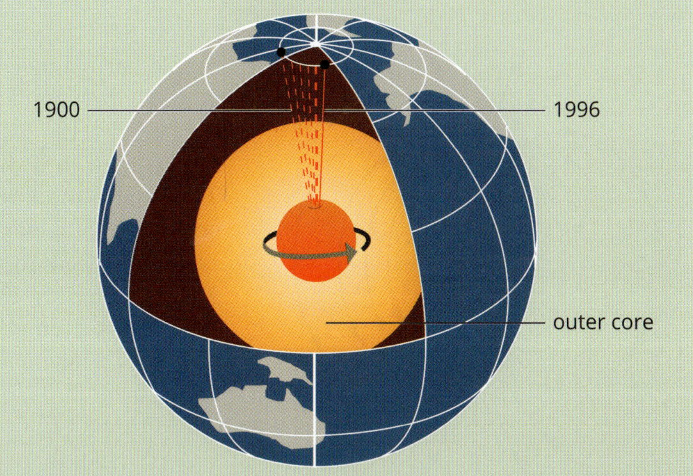

Diagram of the Earth's interior and the movement of the magnetic North Pole from 1900 to 1996. The Earth's outer core is believed to be the source of the geomagnetic field.

The names for the poles of a magnet derive from early observations of magnets orientating themselves with the Earth's geographic poles. The geographic poles are fixed points determined by the Earth's rotation.

Initially, the end of the magnet pointing toward the Earth's geographic north was denoted the North Pole, and compasses are thus marked with this end as north. However, it is now known that the geographic North Pole and the magnetic North Pole are not at the same point on Earth, and the same applies to the geographic South Pole and the magnetic South Pole.

Dipoles

If you break a magnet in half, you will get two smaller magnets, each with its own north and south poles. No matter how many times you break the magnet, and no matter how small the pieces are, each will be a separate magnet with two poles. Because magnets always have two poles, they are said to be **dipoles** (Figure 5.3.4). This is similar to electric charges: a positive charge and a negative charge in close proximity is said to form a dipole. A key difference is that you cannot have a **monopole** (a single magnetic pole) whether it be a south pole or a north pole; however, charges can exist on their own as either positive or negative.

FIGURE 5.3.4 Magnets are always dipolar, that is, if you break a magnet in half each part will still have a north pole and a south pole.

FIGURE 5.3.5 Iron fillings sprinkled around magnets trace the magnetic field lines. In (a) unlike poles are close together; in (b) like poles are close together. The patterns made by the filings show the attraction (a) and repulsion (b) between poles.

MAGNETIC FIELD

Earlier in this chapter you saw that point charges and charged objects produce an electric field in the space around them. As a result, charged bodies within the field experience a force. The direction of the force is determined by the direction of the electric field.

Magnets also create fields. If you place a pin near a magnet, you will observe that the pin is pulled towards the magnet. This shows that the space around the magnet is affected by the magnet.

If you sprinkle iron filings on a piece of material that is held over a magnet, the magnetic field will be clearly revealed (Figure 5.3.5). The iron filings line up with the field, tracing the field lines running from one end of the magnet to the other.

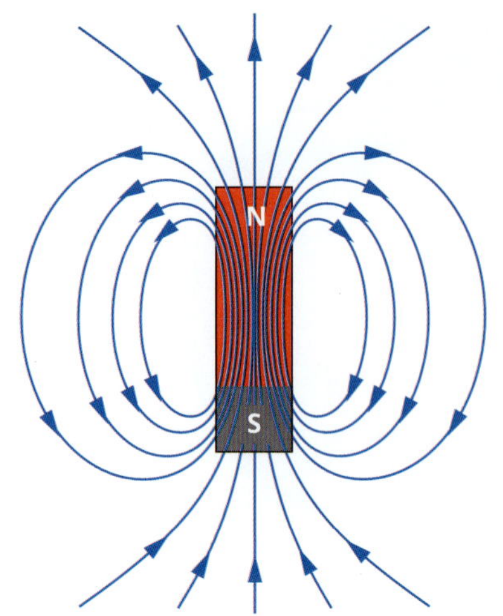

FIGURE 5.3.6 The field lines within and around a bar magnet. The lines show the direction of the force on a single north pole.

Vector field model for magnetic fields

Figure 5.3.6 is a representation of the magnetic field within and surrounding a simple bar magnet. The magnetic field can be defined in vector terms, specifying both the direction and magnitude.

The direction of the magnetic field at any point is the direction that a compass would point to if placed at that point—that is, towards the magnetic South Pole. This is also the direction of the force the magnetic field would exert on a single north pole (if it existed).

The strength of the magnetic field is shown by how close the lines are together. As the distance from the magnet increases, the magnetic field spreads over a greater area and its strength decreases. Because it has both a variable strength and variable direction, a magnetic field can be described as a vector quantity. The strength, or vector magnitude, of a magnetic field at a particular point is denoted by B and has units of tesla (T).

The fields between magnets are dependent on whether like or unlike poles are close together, the distance between the poles and the relative strength of the magnetic field of each magnet. Iron filings or small plotting compasses can be used to visualise the field between and around the magnets (Figure 5.3.7).

FIGURE 5.3.7 Small plotting compasses are placed around the magnets to help visualise the direction of the field lines.

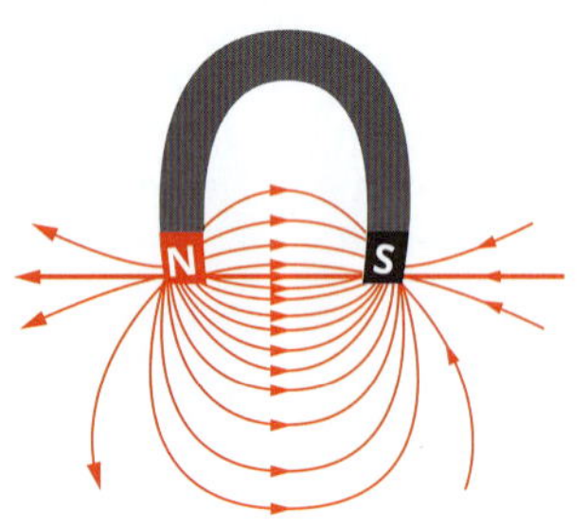

FIGURE 5.3.8 The horseshoe magnet has two unlike poles close to each other. This creates a very strong magnetic field.

Because the Earth has a giant magnetic field around it, you can predict how compasses on Earth will orient themselves—they will orient themselves along the magnetic field lines. Note from Figure 5.3.6 that the magnetic field lines close to the poles run almost vertically. Magnets placed near the Earth's magnetic poles will behave in the same way.

Magnets of different shapes produce fields of different shapes. Figure 5.3.8 shows the magnetic field around a horseshoe magnet and Figure 5.3.9(a) shows the magnetic field around a bar magnet.

The resultant direction of the magnetic field at a particular point will be the vector addition of each individual magnetic field acting at that point.

When two magnets are placed close together, two situations may arise. If the poles closest together are unlike, as in Figure 5.3.9(b), a magnetic field extending between the two poles is created and the poles experience attraction. If the poles closest together are alike, as in Figure 5.3.9(c), repulsion will occur. In this situation there will be a neutral point between the two poles where there is no magnetic field.

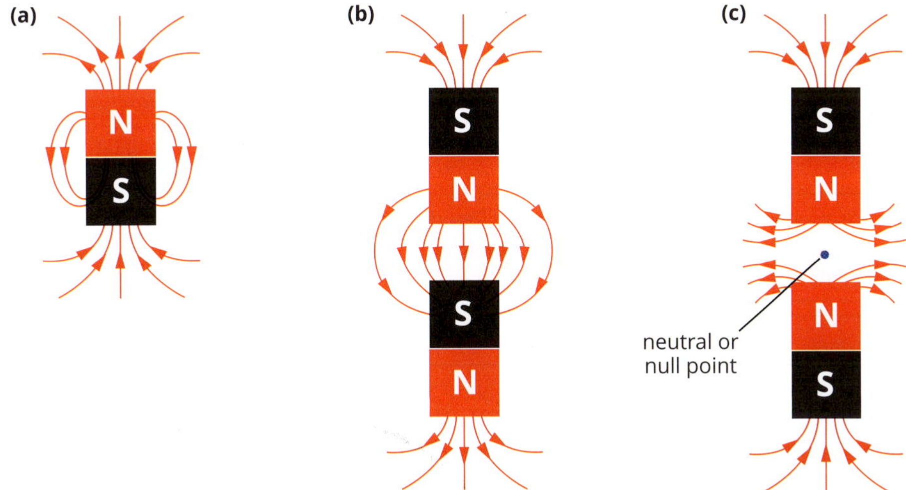

FIGURE 5.3.9 Magnetic field lines around (a) a single bar magnet; (b) around two magnets with unlike poles close together and (c) around two magnets with like poles close together

The bar magnets in Figure 5.3.9 have a fixed strength and position, thus the associated magnetic fields will be static.

MAGNETIC FIELDS AND CURRENT-CARRYING WIRES

The connection between electric current and magnetic fields was noted earlier. Oersted found that when he switched on the current from a voltaic pile, a nearby magnetic compass would move. It is believed that the Earth's magnetic field is created in a similar way—by the circulating electric currents in the Earth's molten metallic outer core.

A circular magnetic field is created around a wire that is carrying current (Figure 5.3.10). A compass aligns itself at a tangent to the concentric circles around the wire (i.e. to the magnetic field). The stronger the current, and the closer the compass is to the wire, the greater the effect.

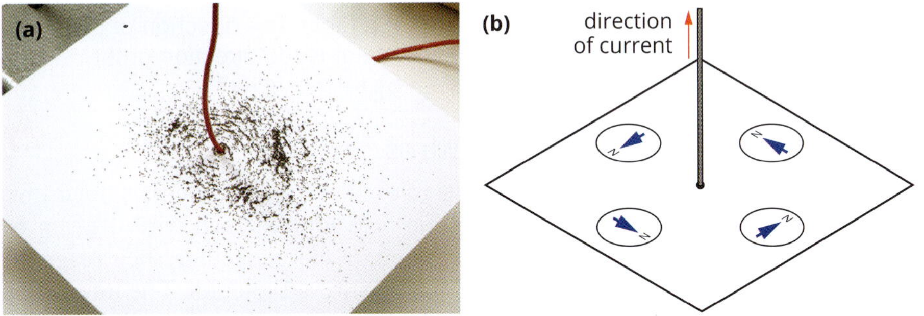

FIGURE 5.3.10 (a) The magnetic field around a current-carrying wire. The iron filings align with the field and show the circular nature of the magnetic field. (b) Compasses indicate the direction of the field.

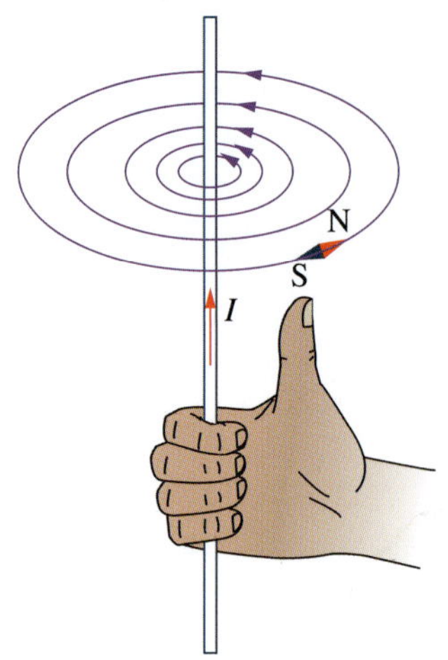

FIGURE 5.3.11 The right-hand grip rule can be used to find the direction of the magnetic field around a current-carrying wire, providing that the direction of the conventional current, *I*, is known.

The magnetic field is perpendicular to the current-carrying wire and the direction of the field depends on the direction of the current. There's a simple and easy way to determine the direction of the magnetic field, commonly referred to as the **right-hand grip rule**. Imagine grasping the conducting wire with your right hand and with your thumb pointing in the direction of the conventional electric current, *I* (i.e. positive to negative). Curl your fingers around the wire. The magnetic field will be perpendicular to the wire and in the direction your fingers are pointing (Figure 5.3.11).

Worked example 5.3.1

DIRECTION OF THE MAGNETIC FIELD

A horizontal wire is carrying current. The direction of the conventional current, *I*, is from left to right. What is the direction of the magnetic field created by the current?

Thinking	Working
Recall that the right-hand grip rule explains how to determine the direction of the magnetic field.	Hold your hand with your fingers aligned as if gripping the wire. Point your thumb to the right, i.e. in the direction of the current flow.
Describe the direction of the field in relation to the wire in simple terms so that the description can be easily understood.	The magnetic field direction is perpendicular to the wire and the field runs up behind the wire, over the top towards the front of the wire and continues in a loop.

Worked example: Try yourself 5.3.1

DIRECTION OF THE MAGNETIC FIELD

A current-carrying wire runs along the length of a table. The direction of the conventional current, *I*, is towards an observer. What is the direction of the magnetic field created by the current as seen by the observer?

Magnetic fields between parallel wires

Two current-carrying wires that are parallel to each other will each have their own magnetic field. The direction of the magnetic field around each wire is given by the right-hand grip rule. If the two wires are brought close together, their magnetic fields will interact, just as any two regular magnets would interact. The interaction could result in either an attraction or repulsion of the wires, depending on the direction of the conventional current and thus the direction of the magnetic fields between them. When the current runs in the same direction and the interacting magnetic fields are in opposite directions (Figure 5.3.12(a)), they act as opposite poles and the wires attract. When the current runs in opposite directions and the interacting magnetic fields are in the same direction (Figure 5.3.12(b)), they act as like poles and the wires repel.

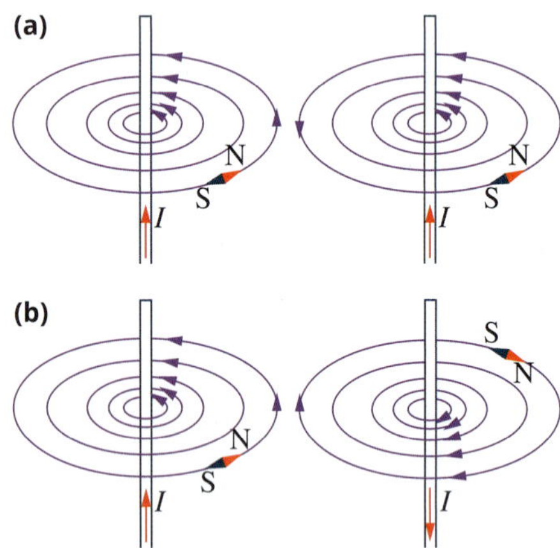

FIGURE 5.3.12 (a) Two current-carrying wires attract when current runs through them in the same direction. This is because at the point at which the magnetic fields interact between the wires, the direction of the fields is opposite. (b) Two current-carrying wires repel when the current passes through them in opposite directions. This is because the magnetic fields are in the same direction where they interact.

3-DIMENSIONAL FIELDS

Field lines can also be drawn for 3-dimensional fields, such as the field around the Earth or the fields around current-carrying loops and coils. In these cases the right-hand grip rule is still applicable (Figures 5.3.13 and 5.3.14).

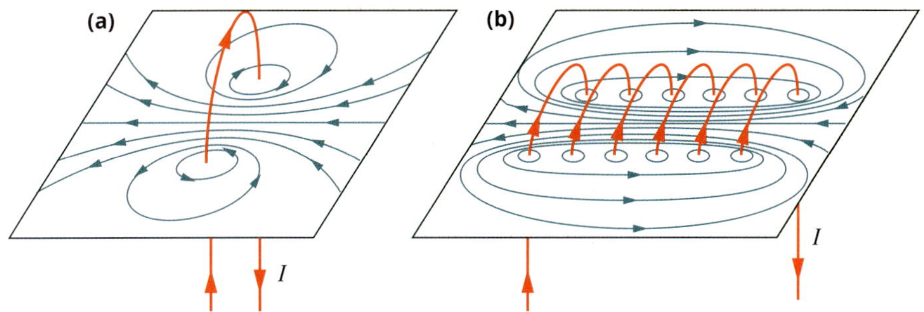

FIGURE 5.3.13 The magnetic field lines around (a) a single loop and (b) a series of loops. The blue arrows indicate the direction of the magnetic field. The more concentrated the lines are inside the loops, the stronger the magnetic field.

FIGURE 5.3.14 A 3-dimensional representation of the magnetic field around a loop of wire in the plane of the page. The blue arrows show the direction of the magnetic field. Notice that the magnetic field is a circular shape, with no field lines crossing.

Figure 5.3.14 shows a 3-dimensional representation of the magnetic field around a loop of wire. This same loop can also be represented in two dimensions using the following convention.

When a field is running directly into or out of the plane of a page:

- dots are used to indicate that the field is coming out of the page
- crosses are used to indicate that the field is going into the page.

This convention was adopted from the idea of viewing an arrow. The dot is the point of an arrow coming towards you and the cross is the tail's feathers as an arrow travels away from you.

Figure 5.3.15 shows the 2-dimensional representation of the same magnetic field around a simple loop of wire as that shown in Figure 5.3.14.

The strength of a field is indicated by the density of the lines, or the dots and crosses. Showing lines coming closer together indicates a strengthening of the field; showing lines moving apart indicates a weakening of the field. Similarly, a region where dots and crosses are relatively dense indicates a region where the field is stronger than a region where the dots and crosses are less dense. Lines that vary in their distance apart, or dots and crosses of varying densities, indicate a non-uniform field.

As the magnetic fields produced by current-carrying coils are dependent on the size of the current, the associated field may also be changing over time, either in magnitude or, if the current is reversed, in direction.

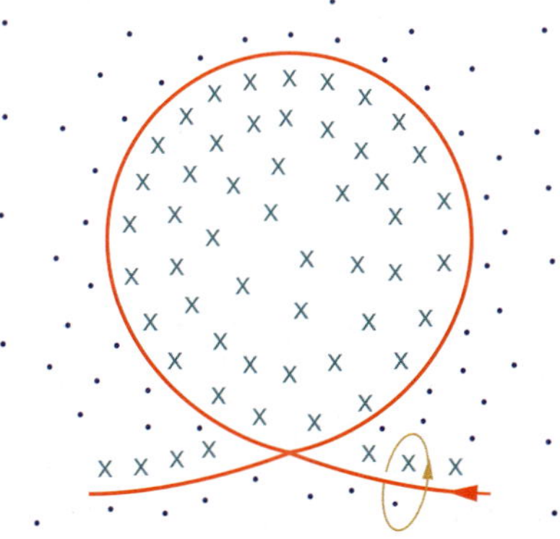

FIGURE 5.3.15 A 2-dimensional representation of the same current-carrying loop depicted in Figure 5.3.14. Areas where the magnetic field is stronger are shown with a greater density of dots or crosses.

The magnetic field around a solenoid

If many loops of wire are placed side by side and a current is passed along the wire, the magnetic field is much stronger than if there was only one loop. Winding many turns of wire into a coil forms what is called a **solenoid**. The magnetic field around a solenoid is like the field around a normal bar magnet. The direction of the field of the solenoid depends on the direction of the current in the wire making up the solenoid. This is explained in Figure 5.3.16.

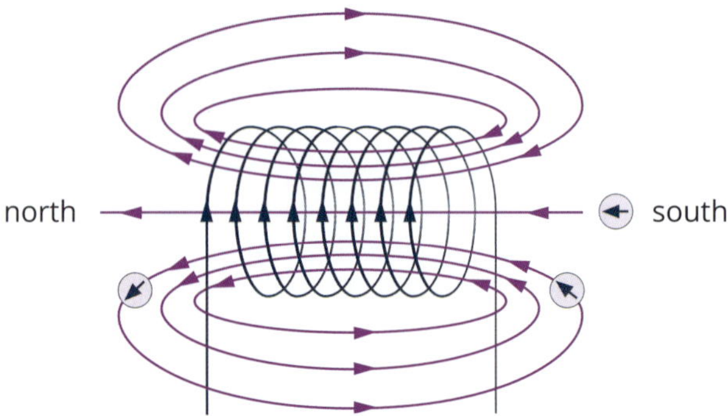

FIGURE 5.3.16 This solenoid has current running upwards at the front of the coil and downwards at the back. This arrangement creates an effective north end at the left and a south end at the right. The compass points in the direction of the field lines.

CREATING AN ELECTROMAGNET

The earliest magnets were all naturally occurring. If you wanted a magnet, you had to find one. They were regarded largely as curiosities. Hans Christian Oersted's discoveries made it possible to manufacture magnets, making the widescale use of magnets today possible.

An **electromagnet**, as the name infers, runs on electricity. It consists of a metallic core that is wrapped in a current-carrying conductor. It works because an electric current produces a magnetic field around a current-carrying wire. If the conductor is looped into a series of coils to make a solenoid, then the magnetic field can be concentrated within the coils. The more coils, the stronger the magnetic field and thus the stronger the electromagnet.

The magnetic field can be strengthened further by wrapping the coils around a core of magnetic material. A magnetic material is divided into magnetic domains. A magnetic domain is a region in the material where the magnetic field is aligned. Normally, the domains in materials such as iron point in random directions and the individual magnetic fields tend to cancel each other out (Figure 5.3.17(a)). However, the magnetic field produced by coils wrapped around an iron core can force the magnetic domains within the core to point in the same direction (Figure 5.3.17(b)). Their individual magnetic fields add together, creating a stronger magnetic field.

(a) (b)

FIGURE 5.3.17 (a) The magnetic fields in separate magnetic domains point in different directions. As a result, their magnetic fields cancel out. (b) The magnetic fields in separate magnetic domains point in the same direction, resulting in a strongly magnetised material.

The strength of an electromagnet can also be changed by varying the amount of electric current that flows through it.

The direction of the current creates poles in the electromagnet. The poles of an electromagnet can be reversed by reversing the direction of the electric current.

Today, electromagnets are used directly to lift heavy objects (Figure 5.3.18), as switches and relays, and as a way of creating new permanent magnets (by aligning the domains within magnetic materials).

FIGURE 5.3.18 A large electromagnet is being used to lift waste iron and steel at a scrapyard. Valuable metals such as these are separated and then recycled.

5.3 Review

SUMMARY

- Like magnetic poles repel; unlike magnetic poles attract.
- Magnets exist only as dipoles, i.e. they have both north and south poles. An object with a single magnetic pole (a monopole) is not known to exist.
- The direction of a magnetic field at a particular point is the same as that of the force on a (imaginary) single north pole.
- The resultant direction of interacting magnetic fields at any particular point will be the vector addition of each individual magnetic field acting at that point.
- The Earth has a dipolar magnetic field, with the south end (the magnetic North Pole) near the geographic North Pole.
- A uniform distribution of field lines represents a uniform magnetic field. A non-uniform field is shown by variations in the separation of the field lines.
- An electrical current produces a magnetic field that is circular around the conductor carrying the current. The direction of the field is determined by noting the direction of the conventional current and applying the right-hand grip rule.
- A magnetic field associated with a constant current is static.
- The direction of 3-dimensional fields can be determined by applying the right-hand grip rule to the loops or coils making up the current-carrying conductor.

KEY QUESTIONS

Knowledge and understanding

1 Describe what happens when a magnet is repeatedly cut in half.

2 If a magnet is suspended on a thin wire at its midpoint so that it is free to swing, approximately in which direction will the north end of the magnet point?

3 The following diagram shows two bar magnets separated by a distance *d*. At this separation the magnitude of the magnetic force between the poles is equal to *F*. What happens to *F* when *d* is increased?

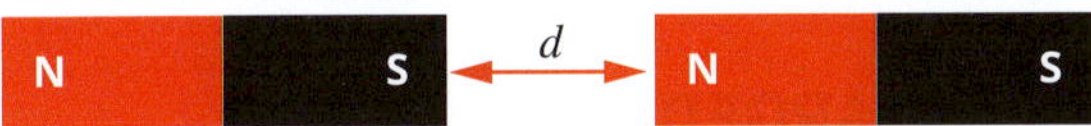

4 The following diagram represents a current-carrying solenoid.

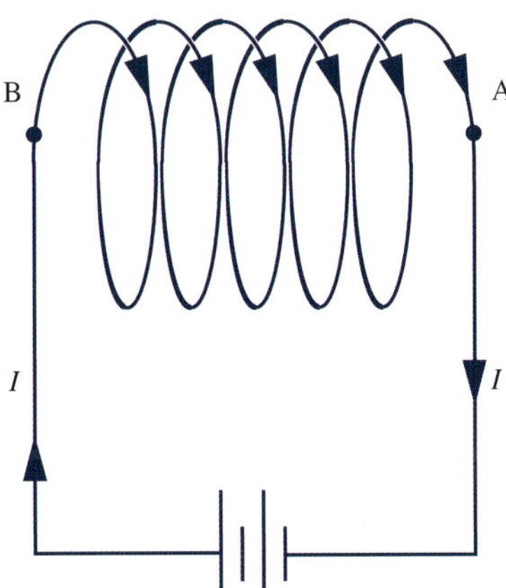

Which end (A or B) represents the south pole of the solenoid?

Analysis

5 Two strong bar magnets that produce magnetic fields of equal strength are arranged as shown.

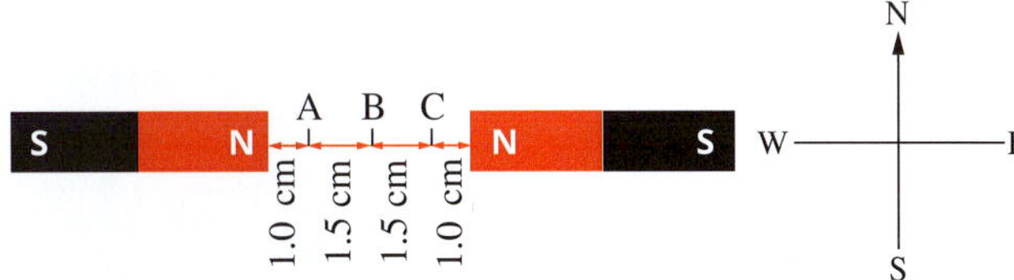

Ignoring the magnetic field of the Earth, determine:

a the approximate direction of the resulting magnetic field at point A and at point C

b the magnitude of the resulting magnetic field at point B.

6 The figure below shows a cross-sectional view of a long, straight, current-carrying conductor with its axis perpendicular to the plane of the page. The conductor carries an electric current into the page.

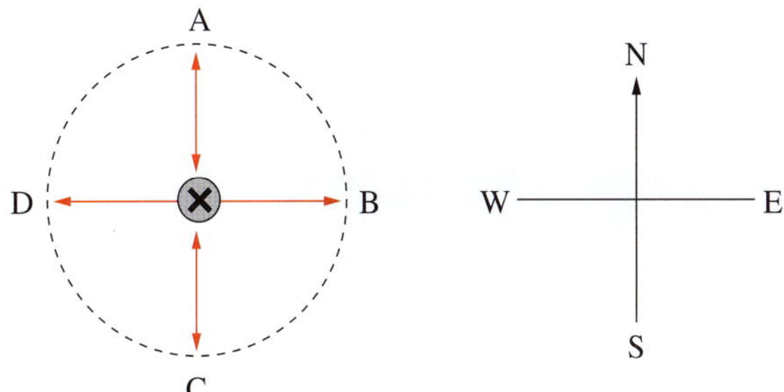

What is the direction of the magnetic field produced by this conductor at each of the points A, B, C and D?

5.4 Forces on charged objects due to magnetic fields

An electric current is a flow of electric charges. These may be electrons in a metal wire, electrons and mercury ions in a fluorescent tube, or cations and anions in an electrolytic cell. Whatever the nature of the flowing charge, a magnetic field is produced around the flow and a force is experienced within the field. The rate of flow of charge, i.e. the current, determines the field produced and the magnitude of the force.

Electrons rushing down the length of a **cathode ray tube** (CRT) were the basis on which old-style television sets worked (Figure 5.4.1). The electrons were deflected by the magnetic force they experienced as they passed through the yoke (coils of copper wire at the back of the tube that created a strong and variable magnetic field).

FIGURE 5.4.1 A cathode ray tube in an old-style television set

MAGNETIC FORCE ON CHARGED PARTICLES

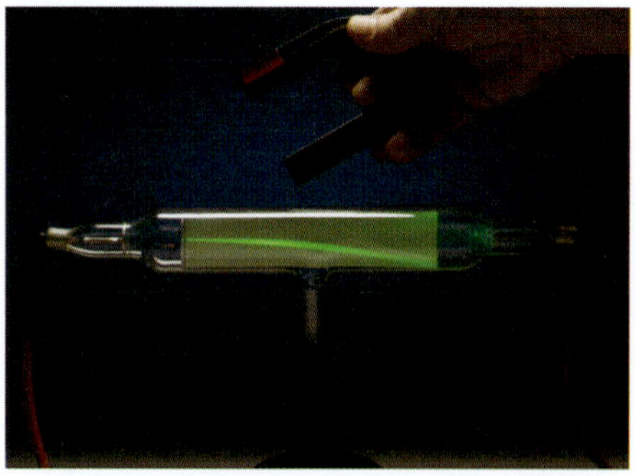

The principle behind a cathode ray tube (CRT) is that a charged particle moving within a magnetic field will experience a force. In Figure 5.4.2, a beam of electrons in a CRT is experiencing a force due to a magnetic field. The force causes the beam of electrons to bend. The magnitude of the force is proportional to the strength of the magnetic field, B, the component of the charge's velocity that is perpendicular to the field, v, and the charge on the particle (q).

FIGURE 5.4.2 An electron beam in a cathode ray tube being deflected by a magnet

When v and B are perpendicular to each other:

$$F = qvB$$

where F is the force in newtons (N)

- q is the electric charge on the particle in coulombs (C)
- v is the component of the velocity of the particle that is perpendicular to the magnetic field (m s^{-1})
- B is the strength of the magnetic field (T)

This force is referred to as the **Lorentz force**. The force is at a maximum when the charged particle is moving at right angles to the field. There is no force acting when the charged particles are travelling parallel to the field.

Worked example 5.4.1

MAGNITUDE OF FORCE ON A POSITIVELY CHARGED PARTICLE

A single positively charged particle with a charge of $+1.6 \times 10^{-19}$ C travels at a velocity of $10\,m\,s^{-1}$ perpendicular to a magnetic field of strength 4.0×10^{-5} T.

What is the magnitude of the force the particle will experience from the magnetic field?

Thinking	Working
Check the direction of the particle's velocity and determine whether a force will apply. Forces only apply on the component of the velocity perpendicular to the magnetic field.	The particle is moving perpendicular to the field, therefore a force will apply. The relevant equation is $F = qvB$.
Establish which quantities are known and which are required.	$F = ?$ $q = +1.6 \times 10^{-19}$ C $v = 10\,m\,s^{-1}$ $B = 4.0 \times 10^{-5}$ T
Substitute values into the force equation.	$F = qvB$ $= 1.6 \times 10^{-19} \times 10 \times 4.0 \times 10^{-5}$
Express the final answer in an appropriate form. Note that only the magnitude has been requested, so do not include the direction.	$F = 6.4 \times 10^{-23}$ N

Worked example: Try yourself 5.4.1

MAGNITUDE OF FORCE ON A POSITIVELY CHARGED PARTICLE

A single positively charged particle with a charge of $+1.6 \times 10^{-19}$ C travels at a velocity of $50\,m\,s^{-1}$ perpendicular to a magnetic field of strength 6.0×10^{-5} T.

What is the magnitude of the force the particle will experience from the magnetic field?

PHYSICSFILE

The tesla

The unit for the strength of a magnetic field, *B*, was given the name tesla (T) in honour of Nikola Tesla (1856–1943). Tesla was the first person to recommend the use of alternating current (AC) generators to supply power to towns. He was also a prolific inventor of electrical machines of all sorts, including the Tesla coil, which is a source of high-frequency, high-voltage electricity.

A magnetic field of 1 T is a very strong field. For this reason, a number of smaller units—such as the millitesla (mT, 10^{-3} T) and microtesla (μT, 10^{-6} T)—are in common use. Table 5.4.1 shows the strength of some magnets for comparison.

TABLE 5.4.1 Comparison of magnet strengths

Type of magnet	Strength of magnetic field
very strong electromagnets and super magnets	1 to 20 T
alnico and ferrite magnets	10^{-2} to 1 T
Earth's surface	5×10^{-5} T

Determining the direction of the force

The arrangement shown in Figure 5.4.3 can be used to determine the direction of the force on a charged particle moving in a magnetic field. With the fingers of your right hand stretched out in the direction of the magnetic field and the palm of your hand flat, point your thumb in the direction that a positive charge is moving (i.e., the direction of conventional current). The direction of the resulting force on the charge is the direction in which your palm is pointing. The force on a negatively charged particle will therefore be in the opposite direction to that on a positively charged particle. This is the **right-hand force rule**.

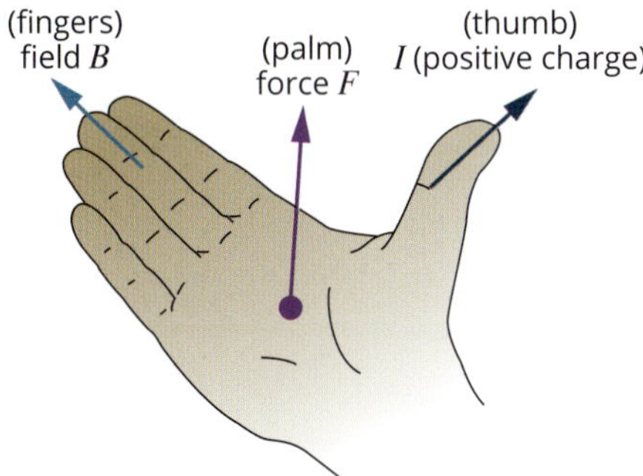

FIGURE 5.4.3 The right-hand force rule: Point the thumb of your right hand in the direction of the movement of a positive charge (conventional current direction) and your fingers in the direction of the magnetic field. The force on the charge will point out from your palm.

Worked example 5.4.2

DIRECTION OF FORCE ON A NEGATIVELY CHARGED PARTICLE

A single negatively charged particle with a charge of -1.6×10^{-19} C is travelling horizontally out of a computer screen and perpendicular to a magnetic field that runs horizontally from left to right across the screen. In what direction will the force experienced by the charge act?

Thinking	Working
(fingers) field B (palm) force F (thumb) I (positive charge) The right-hand force rule is used to determine the direction of the force on a positively charged particle.	Align your hand so that your fingers are pointing in the direction of the magnetic field, i.e. left to right and horizontal. If the negatively charged particle is travelling out of the screen, a positively charged particle would be moving in the opposite direction. Position your thumb so it is pointing into the screen, i.e. in the direction that a positive charge would travel. Your palm should be facing downwards. That is the direction of the force applied by the magnetic field on the negative charge coming out of the screen.

Worked example: Try yourself 5.4.2

DIRECTION OF FORCE ON A NEGATIVELY CHARGED PARTICLE

A single negatively charged particle with a charge of -1.6×10^{-19} C is travelling horizontally from left to right across a computer screen and perpendicular to a magnetic field that runs vertically down the screen. In what direction will the force experienced by the charge act?

OBJECTS MOVING AT AN ANGLE TO THE MAGNETIC FIELD

The force experienced by a charge moving in a magnetic field is a vector quantity. The force equation introduced earlier ($F = qvB$) applies only to that component of the velocity of the charge that is perpendicular to the magnetic field. To find the force acting on an object moving at an angle, θ, to the magnetic field, the equation to use is:

$$F = qvB\sin\theta$$

A charged particle travelling at a constant speed in a magnetic field experiences this force at an angle to its path and will be diverted.

This is the theory behind CRT screens. As the direction of a charged particle changes, so does the angle of the force acting on it. In a very large magnetic field, the charged particles will move in a circular path. Mass spectrometers and particle accelerators both work on this principle.

THE FORCE ON A CURRENT-CARRYING CONDUCTOR

Since a conducting wire is essentially a stream of charged particles flowing in one direction, it follows that a conductor carrying a stream of charges within a magnetic field will also experience a force. This is the theory behind the operation of electric motors, which will be explored in Chapter 6.

The current in a conductor is dependent on the rate at which charges are moving through the conductor. This is given by the following equation.

$$I = \frac{Q}{t}$$

where I is the current (A)
Q is the total charge (C)
t is the time taken (s)

For a 1 m length of conductor, the velocity of the charges through the conductor is:

$$v = \frac{s}{t} = \frac{1}{t}$$

Hence:

$$I = \frac{Q}{t} = Q \times \frac{1}{t} = Qv$$

As $F = qvB$ for a single charge, q, moving perpendicular to a magnetic field, then:

- for a one-metre conductor, $F = IB$
- for a conductor of any length l, $F = IlB$ and
- for a conductor made up of n loops of conductors of length l:

$$F = nIlB$$

where F is the force on the conductor perpendicular to the magnetic field, in newtons (N)
n is the number of loops
I is the current in the conductor in amperes (A)
l is the length of the conductor in metres (m)
B is the strength of the magnetic field in tesla (T)

Just as for a single charge moving in a magnetic field, the force on the conductor is maximum when the conductor is at right angles to the field. The force is zero when the conductor is parallel to the magnetic field. The right-hand force rule is used to determine the direction of the force.

CONDUCTORS AT AN ANGLE TO A MAGNETIC FIELD

The force experienced by a current-carrying conductor is a vector quantity. The force equations introduced in the previous section apply only to that component of the conductor that is perpendicular to the magnetic field. To find the force acting on any conductor, or part of a conductor, moving at an angle, θ, to the magnetic field, use the equation:

$$F = nIlB \sin \theta$$

This is particularly useful when considering electric motors.

CASE STUDY

Loudspeakers

A loudspeaker is a very common application of an electromagnet (Figure 5.4.4).

Inside the loudspeaker is a permanent magnet and a coil wrapped around a metal rod. The metal rod is connected to the diaphragm, or cone.

When the loudspeaker is connected to a music source (e.g. a stereo, iPhone, etc.), electric signals from the music source run through the coil inside the loudspeaker. This is labelled 'current from music source' in Figure 5.4.4. The electric signals are sent in the form of electric current, which creates an electromagnet. As the electric current flows back and forth in the cables, the magnetic field in the electromagnet either attracts or repels the permanent magnet. This causes the metal rod to move backwards and forwards, pushing or pulling the cone, which in turn causes sound pressure to change and you (the listener) to hear the sound.

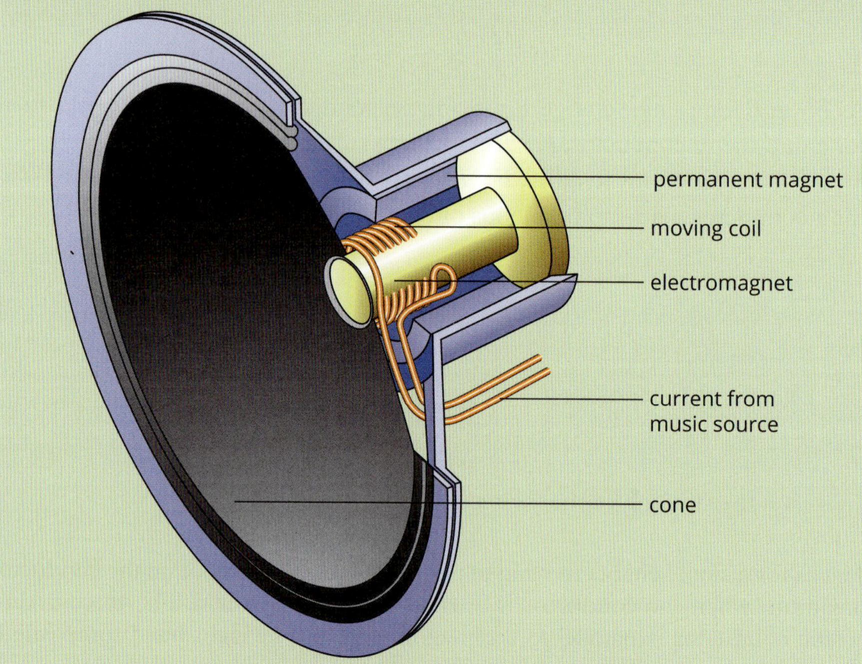

FIGURE 5.4.4 A diagram of a loudspeaker, showing the electromagnet inside

Worked example 5.4.3

MAGNITUDE OF THE FORCE ON A CURRENT-CARRYING WIRE

Determine the magnitude of the force per metre due to the Earth's magnetic field that acts on a suspended power line at the moment it carries a current of 100 A. The power lines are composed of two tightly coupled conductors that run in an east–west direction. Assume that the strength of the Earth's magnetic field at this point is 5.0×10^{-5} T.

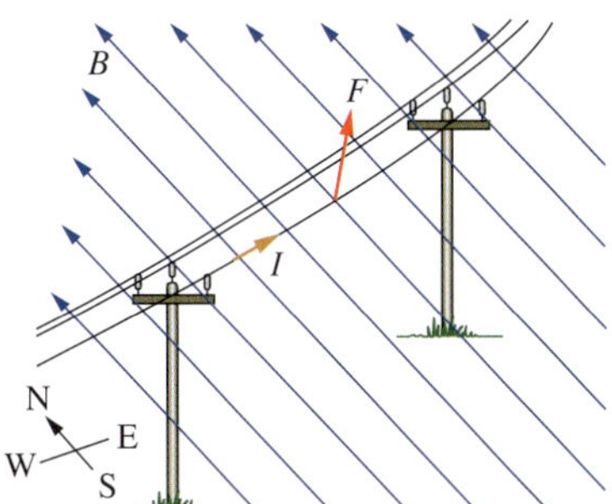

Thinking	Working
Check the direction of the conductor and determine whether a force will apply. Forces only apply to the component of the wire perpendicular to the magnetic field.	As the current is running east–west and the Earth's magnetic field runs south–north, the current and the field are at right angles and therefore a force will exist.
Establish what quantities are known and what are required. Since the length of the power line hasn't been supplied, consider the force per unit length (i.e. 1 m).	$F = ?$ $n = 2$ $I = 100$ A $l = 1.0$ m $B = 5.0 \times 10^{-5}$ T
Substitute values into the force equation and simplify.	$F = nIlB$ $= 2 \times 100 \times 1.0 \times 5.0 \times 10^{-5}$ $= 1.0 \times 10^{-2}$ N
Express the final answer in an appropriate form with a suitable number of significant figures. Note that only the magnitude has been requested, so do not include the direction.	$F = 1.0 \times 10^{-2}$ N per metre of power line

Worked example: Try yourself 5.4.3

MAGNITUDE OF THE FORCE ON A CURRENT-CARRYING WIRE

Determine the magnitude of the force per metre due to the Earth's magnetic field that acts on a single suspended power line running east–west at the moment it carries a current of 50 A. Assume that the strength of the Earth's magnetic field at this point is 5.0×10^{-5} T.

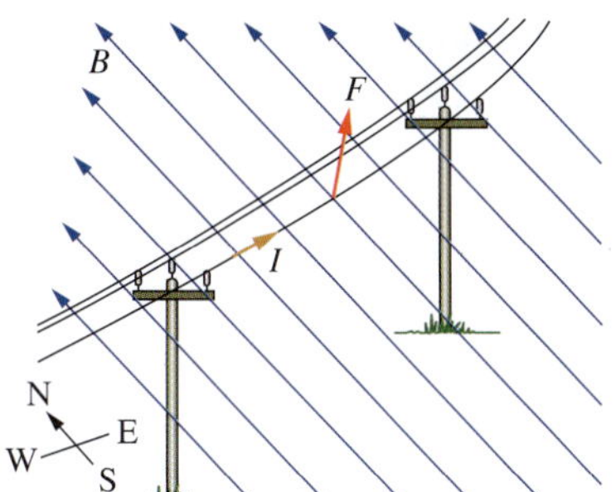

PHYSICSFILE

The current balance

A current balance can be used to determine the force on a conductor in a magnetic field.

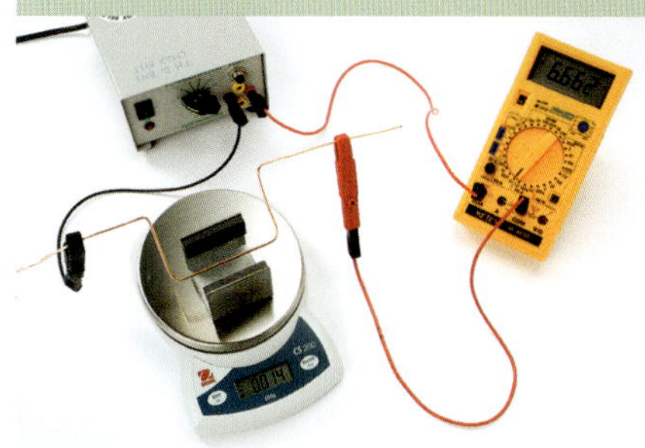

A current balance is used to measure the interaction between an electric conductor and a magnetic field. The relationship between force, current and conductor length can be shown.

Worked example 5.4.4

DIRECTION OF THE FORCE ON A CURRENT-CARRYING WIRE

A current balance is used to measure the force from a magnetic field on a wire 5.0 cm long running perpendicular to a magnetic field. The conventional current direction in the wire is from left to right. The magnetic field can be considered to be running away from you (the observer). What is the direction of the force on the wire?

Thinking	Working
(fingers) field B (palm) force F (thumb) I (positive charge) The right-hand force rule is used to determine the direction of the force.	Align your hand so that your fingers are pointing in the direction of the magnetic field, i.e. away from you. Align your thumb so it is pointing right, in the direction of the current. Your palm should be facing upwards. That is the direction of the force applied by the magnetic field on the wire.
State the direction in terms of the other directions given in the question. Make the answer as clear as possible to avoid any misunderstanding.	The force on the wire is acting vertically upwards.

Worked example: Try yourself 5.4.4

DIRECTION OF THE FORCE ON A CURRENT-CARRYING WIRE

A current balance is used to measure the force from a magnetic field on a wire of length 5.0 cm running perpendicular to the field. The conventional current direction in the wire is from left to right. The magnetic field can be considered running towards you (the observer). What is the direction of the force on the wire?

Worked example 5.4.5

FORCE AND DIRECTION ON A CURRENT-CARRYING WIRE

The Amundsen–Scott South Pole Station sits at the Earth's magnetic South Pole (which behaves like the north pole of a magnet).

Assuming the strength of the Earth's magnetic field at this point is 5.0×10^{-5} T, determine the magnitude and direction of the magnetic force on the following current-carrying wires.

a a 2.0 m length of wire carrying a conventional current of 10.0 A vertically up the exterior wall of one of the buildings of the station

Thinking	Working
Forces only apply to the components of the wire running perpendicular to the magnetic field. The direction of the magnetic field at the magnetic South Pole will be almost vertically upwards (that is, out of the ground).	The section of the wire running up the wall of the building will be parallel to the magnetic field. Hence no force will apply.
State your answer. A numeric value is required. Since there is no force, it is not necessary to state a direction.	$F = 0$ N

b a 2.0 m length of wire carrying a conventional current of 10.0 A running horizontally right to left across the back wall of one of the buildings	
Forces only apply to the components of the wire running perpendicular to the magnetic field. The direction of the magnetic field at the magnetic South Pole will be almost vertically upwards.	The section of the wire running horizontally across the building will be perpendicular to the magnetic field, *B*. A force, *F*, with a strength equivalent to $nIlB$ will apply.
Identify the known quantities.	$F = ?$ $n = 1$ $I = 10.0\text{ A}$ $l = 2.0\text{ m}$ $B = 5.0 \times 10^{-5}\text{ T}$
Substitute values into the appropriate equation and simplify.	$F = nIlB$ $= 1 \times 10.0 \times 2.0 \times 5.0 \times 10^{-5}$ $= 1.00 \times 10^{-3}\text{ N}$
(fingers) field B (palm) force F (thumb) I (positive charge) The direction of the magnetic force is also required to fully specify the vector quantity. Determine the direction of the magnetic force using the right-hand force rule.	Align your hand so that your fingers are pointing in the direction of the magnetic field, i.e. vertically up. Align your thumb so it is pointing left, in the direction of the current. Your palm should be facing inwards (towards the inside of the building). That is the direction of the force applied by the magnetic field on the wire.
State the magnetic force in an appropriate form with a suitable number of significant figures. Include the direction to fully specify the vector quantity.	$F = 1.0 \times 10^{-3}\text{ N}$ inwards

Worked example: Try yourself 5.4.5

FORCE AND DIRECTION ON A CURRENT-CARRYING WIRE

Santa's house sits at a point that can be considered the Earth's magnetic North Pole (which behaves like the south pole of a magnet).

Assuming that the Earth's magnetic field at this point is 5.0×10^{-5} T, calculate the magnetic force and its direction on the following current-carrying wires.

a a 2.0 m length of wire carrying a conventional current of 10.0 A vertically up the outside wall of Santa's house

b a 2.0 m length of wire carrying a conventional current of 10.0 A running horizontally right to left across the back wall of Santa's house

5.4 Review

OA ✓✓

SUMMARY

- The magnitude of the force on a charged object within a magnetic field is proportional to the strength of the magnetic field (B), the component of the velocity of the charge that is at right angles to the magnetic field (v) and the charge on the particle (q), i.e. $F = qvB$.
- This force is referred to as the Lorentz force.
- The force is at a maximum when the charged particle is moving at right angles to the magnetic field.
- The force is zero when the charged particle is travelling parallel to the magnetic field.
- The magnetic force on a current-carrying wire within a magnetic field is $F = nIlB$.
- The right-hand force rule is used to determine the direction of the force on a positive charge moving in a magnetic field. The direction of the force on a negatively charged particle is in the opposite direction.
- The right-hand force rule states that the force travels out of the palm of the hand once the thumb is pointing in the direction of the conventional current and the fingers are oriented in the direction of the magnetic field.

KEY QUESTIONS

Knowledge and understanding

1 The following diagram shows a particle, with initial velocity v, about to enter a uniform magnetic field directed out of the page.

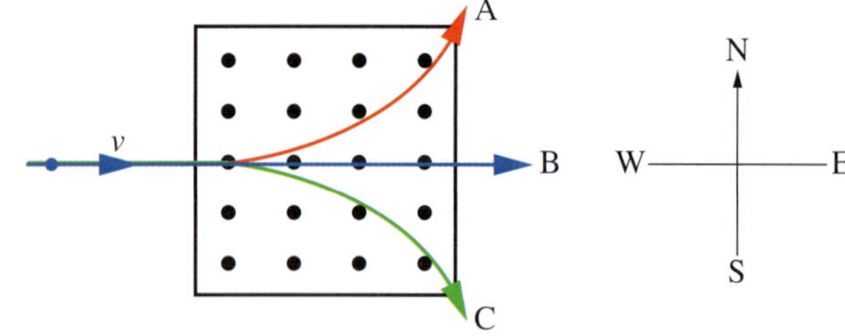

a If the charge on the particle is positive, what is the direction of the force on it just as it enters the field?

b What path (A, B or C) will the particle follow?

c Does the kinetic energy of the particle increase, decrease or remain constant?

d If this particle were negatively charged, what path would it follow?

e What kind of particle could follow path B?

2 A current balance is used to measure the force from a magnetic field on a wire of length 2 m running perpendicular to the magnetic field. The conventional current direction in the wire is from right to left. The magnetic field can be considered to be running towards the observer. What is the direction of the force on the wire?

3 Describe the effect of a force on a conductor if the number of individual wires is doubled or quadrupled.

4 A single negatively charged particle with a charge of -1.6×10^{-19} C travels at a velocity of 10.0 m s^{-1} from left to right parallel to a magnetic field of strength 12.0×10^{-5} T. What is the magnitude of the force the particle will experience from the magnetic field?

5 A rectangular loop of wire is carrying a current, I, in a magnetic field, B, as shown below. What is the direction of the force on the length of wire marked QR?

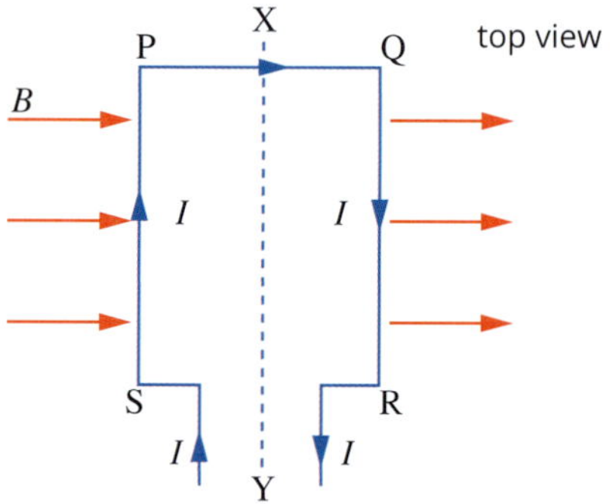

6 An east–west power line of length 200 m is suspended between two towers. Assume that the strength of the magnetic field of the Earth in this region is 5.0×10^{-5} T. Calculate the magnetic force on the power line (including its direction) at the moment it carries a current of 100 A from east to west.

7 An electron with a charge of -1.6×10^{-19} C is moving eastwards into a magnetic field of strength 1.5×10^{-5} T acting into the page, as shown below. If the magnitude of the initial velocity is $2\,m\,s^{-1}$, what is the magnitude and direction of the force the electron initially experiences as it enters the magnetic field?

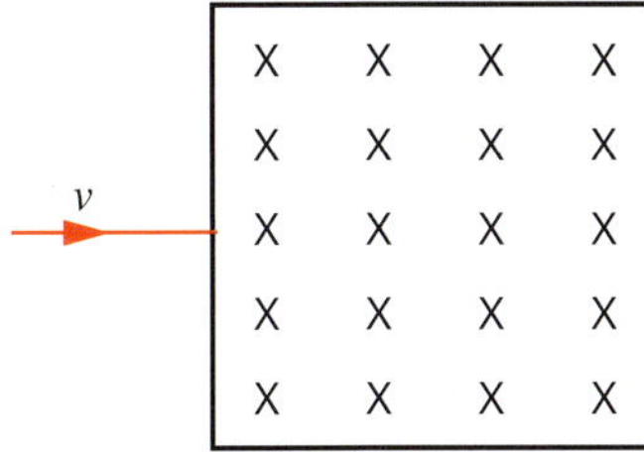

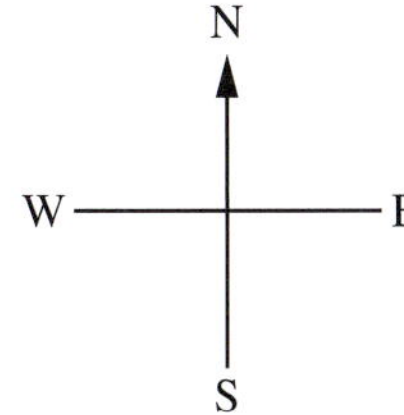

8 An alpha particle with a charge of $+3.2 \times 10^{-19}$ C is moving eastwards into a magnetic field acting into the page, as shown below. The force it experiences is *F*. If the particle's velocity, *v*, is doubled, what will be the magnitude and direction of the magnetic force it would experience in terms of *F*?

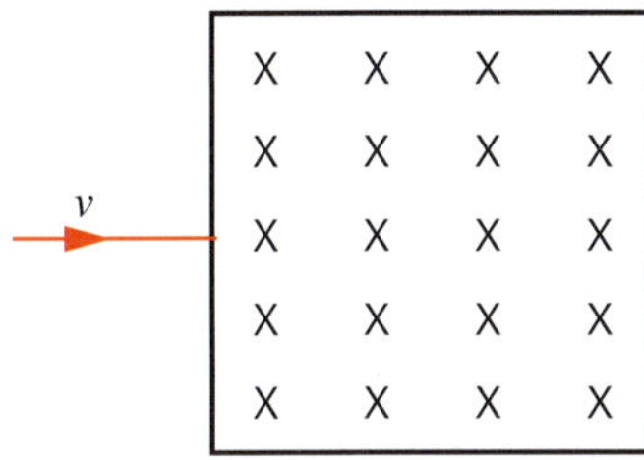

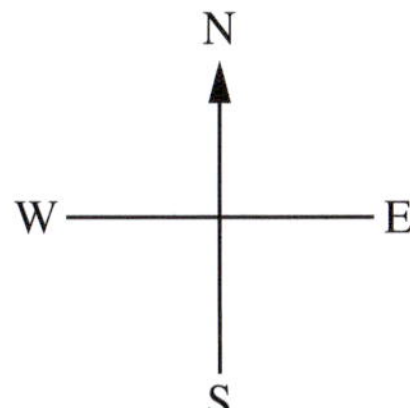

Analysis

9 The diagram below shows a cross-sectional view of a long, straight, current-carrying conductor positioned between the poles of a permanent magnet. The magnetic field, *B*, of the magnet and the current, *I*, are perpendicular to each other. Calculate the magnitude and direction of the magnetic force on a 5.0 cm section of the conductor when the current is 2.0 A into the page and *B* equals 2.0×10^{-3} T.

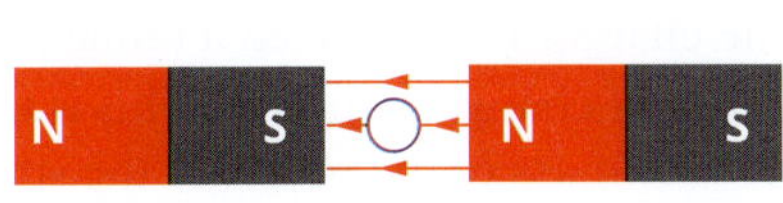

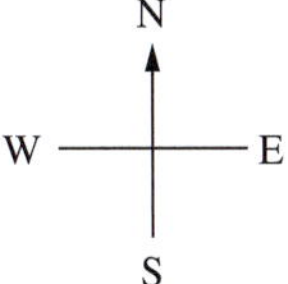

10 An east–west power line of length 80 m is suspended between two towers. Assume that the strength of the magnetic field of the Earth in this region is 4.5×10^{-5} T.

a Calculate the magnitude and direction of the magnetic force on the power line at the moment it carries a current of 50 A from east to west.

b Over time, the ground underneath the eastern tower subsides so that the power line is lower at that tower. Assuming that all other factors are the same, is the magnitude of the magnetic force on the power line greater than before, less than before or the same as before? Justify your answer.

5.5 Comparing fields—a summary

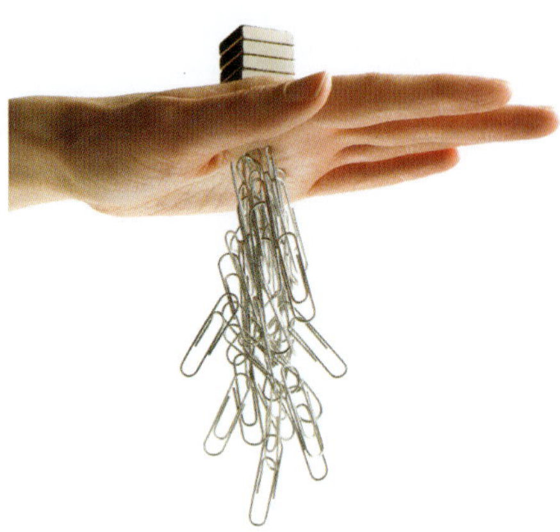

FIGURE 5.5.1 The magnet has an effect on the paper clips even though they are not in contact. This is because the paper clips are within the magnetic field produced by the magnet.

Many of the forces affecting us and the world around us can be described as contact forces. There is direct contact as you open a door, kick a ball or rest on a couch. By contrast, the forces of gravity, magnetism and electricity act over a distance without necessarily having any physical contact (Figure 5.5.1). This was a difficult idea for scientists to understand. Newton still had some misgivings even when publishing his ideas on universal gravitation. The concept of fields, used to explain how and why forces can act over a distance without contact, has proved to be a very powerful tool and one that has allowed us to better explain the fundamental forces of gravity and electromagnetism.

In this section key aspects of gravitational, electric and magnetic fields will be summarised and compared, giving an overview of the theoretical ideas covered so far in this book.

DIPOLES AND MONOPOLES

Gravitational fields consist essentially of monopoles. All objects with mass produce a gravitational field that can be considered as being towards the centre of the mass, i.e. towards a single pole. There is a concept of a gravitational dipole but it is a measure of how the mass of an object is distributed away from a particular centre in a particular direction, usually selected as the centre of mass.

Magnetic fields exist, in a practical sense, solely as dipoles; that is, they have opposite north and south poles. While a magnetic field is defined as having a direction that a north magnetic pole would move (i.e. towards a south pole), this is a theoretical single pole.

Electric fields can be either monopoles or dipoles. Single positive and negative charges are monopoles. Two equal point charges of opposite sign and at a distance constitute a dipole. These are common in physics and molecular biology.

PHYSICSFILE

Gravitational repulsive forces

A leading theory in the explanation of the expansion of the universe proposes the existence of dark energy. While little is understood about dark energy, it may be a source of a repulsive force of gravity originating from the interaction of matter and antimatter.

DIRECTION AND SHAPE OF FIELDS

Simple fields associated with a single monopole—whether gravitational or electric—look very similar since they are a representation of the spread of the field around a single point. Fields are vector quantities: they have both direction and size. Field lines are used to visualise the extent, shape and strength of the field, with arrows on the lines used to show the direction of the field.

A uniform field is indicated by lines that are evenly spaced throughout the region of the field. The electric field between two charged plates is uniform. Around a point charge, mass or pole, the field would not be uniform since the strength of the field decreases with the distance from the charge. This is called a radial field.

In a static field, the strength of the field doesn't change with time. This is true of most gravitational and magnetic fields, where the mass of the object or the strength of the magnet is unchanging. However, many electric fields are changing fields. Charges are moving or the amount of charge is changing regularly. Some electric fields can be static, with a fixed charge, just as there can be changing gravitational and magnetic fields. The magnetic field associated with a changing electric current is one example of a changing magnetic field.

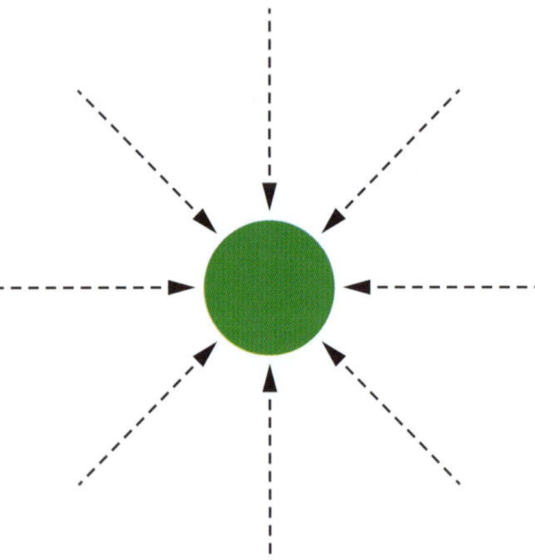

FIGURE 5.5.2 The arrows in this gravitational field diagram indicate that objects will be attracted towards the centre of the mass. The spacing of the lines shows that force is stronger as you approach the mass. Similar field diagrams apply to single electric point charges and magnetic poles.

A gravitational field is directed towards the point representing the centre of mass of the object and is always attractive (Figure 5.5.2). In the case of electric and magnetic fields, the field may be either attractive or repulsive, so a particular direction is defined as the positive direction. In the case of electric fields, this is the direction of the force on a positive test charge (i.e. positive to negative). For magnetic fields, the direction of the force is on a theoretical single north pole (i.e. north to south).

One other key difference between these fields is that, theoretically, a gravitational field around any mass extends an infinite distance from that mass. While the shape of the field will be influenced by the fields of other masses, there is no way of stopping the field. On the other hand, the extent of both electric and magnetic fields, while theoretically extending to infinity, can be constrained by external electric and magnetic influences.

The shape of electrical and magnetic fields around an object can be influenced by the shape of the object. An example is shown in Figure 5.5.3. The shape of the field around multiple masses, charges or poles becomes increasingly complex. However, the direction of a field at any point is always the resultant field vector determined by adding the individual field vectors due to each mass, charge or magnetic pole within the affected region.

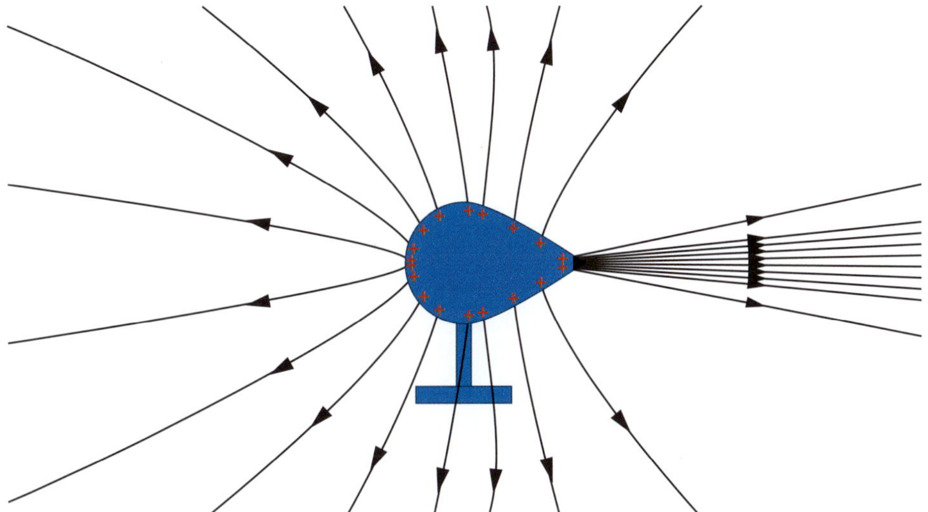

FIGURE 5.5.3 Field lines around a pear-shaped conductor. The uneven nature of their distribution is due to the contributions of each individual charge on the surface of the conductor and the greater density at more curved regions.

Whether a charge is positive or negative, or a magnetic pole is north or south, must also to be considered when determining the resultant field around multiple charges or magnetic poles (Figure 5.5.4).

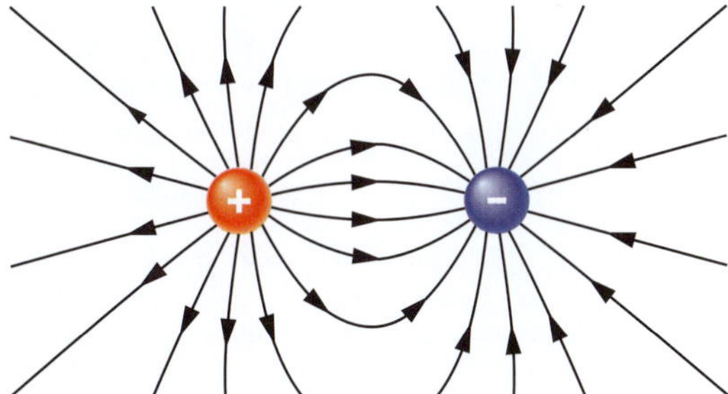

FIGURE 5.5.4 The electric field resulting when unlike charges are brought together. At any point the density and direction of the field lines represent the resultant field vector at that point.

PHYSICSFILE

Field strength around a dipole

While the theories of many particle physicists predict magnetic monopoles, in practical terms magnetic poles exist only as dipoles. The relevant relationships for dipoles are different to those for radial fields surrounding monopoles. For example, an inverse square law does not apply. In fact the strength of the field as it extends from a dipole decreases by the cube of the distance, that is, field strength $\propto \frac{1}{r^3}$.

5.5 Review

OA ✓✓

SUMMARY

- Gravitational, electric and magnetic fields are similar, but have some fundamental differences.
- The direction of a field at any point is always the resultant field vector determined by adding the individual field vectors due to each mass, charge or magnetic pole within the affected region.
- A uniform field is indicated by lines that are evenly spaced throughout the region of the field.
- In a static field, the strength of the field doesn't change with time.
- The field around a monopole is radial, static but not uniform. It varies with the distance from the point source.

Quantity or description	Gravitational fields	Electrical fields	Magnetic fields
type of poles	monopoles	monopoles/dipoles	dipoles
type of force	attractive	attractive/repulsive	attractive/repulsive
extent of the field	extends an infinite distance	can be constrained to a fixed distance	can be constrained to a fixed distance
effect of distance on field strength in a radial field	$g = G\frac{M}{r^2}$	$E = k\frac{Q}{r^2}$	not applicable
force between monopoles	$F_g = G\frac{m_1 m_2}{r^2}$	$F = k\frac{q_1 q_2}{r^2}$	not applicable
potential energy changes in a uniform field	$E_g = mg\Delta h$	$W = qV = qEd$	not applicable
force due to a uniform field	$F_g = mg$	$F = qE$	not applicable

KEY QUESTIONS

Knowledge and understanding

1 Determine which of the following statements is incorrect.
 A Gravitational fields are known only to be attractive.
 B In a static field, the strength of the field changes with time.
 C Gravitational fields consist essentially of monopoles.
 D Fields are vector quantities having both direction and size.

2 State the nature of the poles (monopoles, dipoles or both) for each of the following fields:
 a gravitational
 b electrical
 c magnetic.

3 Complete the following statement about the field around a monopole by selecting from the pairs of choices provided in bold.
 The field around a monopole is **linear/radial, static/dynamic** and **uniform/non-uniform**.

4 The diagram below shows the field between two point charges. The charge on the right is shown with no sign. What is the charge on the point charge on the right?

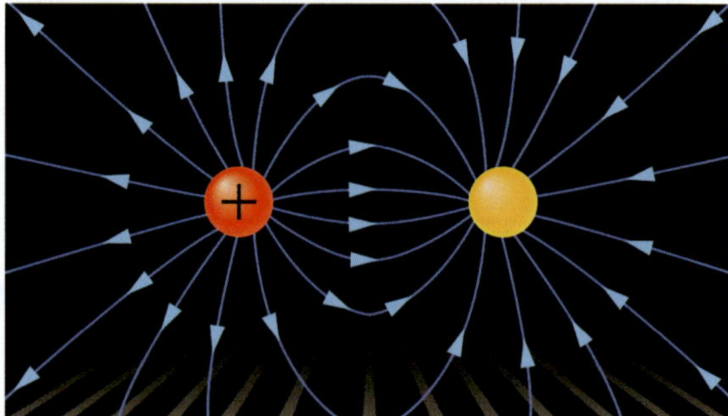

5 Which of the following statements explains why an inverse square law does not apply to the change in magnetic field strength with distance from the source.
 A Magnetic fields are considerably stronger than other field types.
 B Magnetic fields are uniform around each pole.
 C Magnetic fields are only associated with monopoles.
 D Magnetic fields are only associated with dipoles.

6 Complete the following statement about the direction of a field around a monopole by selecting from the choices provided in bold.

The direction of a field at any point is defined as the **maximum/resultant** field vector determined by adding the **total/individual** field vectors due to each mass, charge or magnetic pole within the field.

Analysis

7 The diagram below shows the electric field between two electrically charged parallel plates of opposite sign.

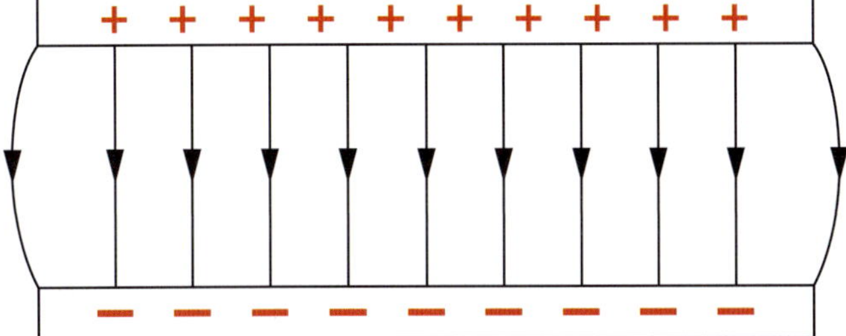

Describe why the electric field lines shown are bulging outwards at the ends of the plates. Hint: Consider the field between the plates and how this compares with the field outside the plates.

8 The gravitational force of attraction between two electrons is less than 8.0×10^{-37} N. At what minimum distance does this hold true for the two electrons? Assume that $G = 6.67 \times 10^{-11}\,m^3\,kg^{-1}\,s^{-2}$ and the mass of each electron is 9.1×10^{-31} kg.

9 The electron of a hydrogen atom orbits the atom's proton at an average distance of 0.53×10^{-10} m.

a The charges on both the electron and the proton are 1.6×10^{-19} C. What is the electrical force of attraction between the two particles? Assume that $k = 8.99 \times 10^9\,N\,m^2\,C^{-2}$.

b The mass of the electron is 9.1×10^{-31} kg and that of the proton is 1.67×10^{-27} kg. What is the gravitational force of attraction between the two particles? Assume that $G = 6.67 \times 10^{-11}\,m^3\,kg^{-1}\,s^{-2}$.

c Compare the answers to parts **a** and **b**, and comment on the difference between the gravitational force of attraction and the electrical force of attraction between the two particles.

Chapter review

KEY TERMS

cathode ray tube
dipole
electric field
electric field strength
electrical potential
electromagnet
field
Lorentz force
magnetic field
magnetic pole
monopole
point charge
potential difference
right-hand force rule
right-hand grip rule
solenoid

REVIEW QUESTIONS

Knowledge and understanding

1 Explain the difference between electrical potential and potential difference.

2 A positively charged particle moves across a potential difference, from a positive plate towards an earthed plate.

- **a** Does the electric field or the charged particle do the work?
- **b** Is the work done on the electrical field or on the charged particle?

3 Consider two parallel metal plates between which an electric field exists. Where is the electric field at a maximum between these two plates?

4 The diagram below shows a loop carrying a current, I, that produces a magnetic field of magnitude B in the centre of the loop. It is in a region where there is already a steady field of magnitude B (the same magnitude as that due to I) directed into the page. The resultant magnetic field in the centre of the loop has a magnitude of $2B$.

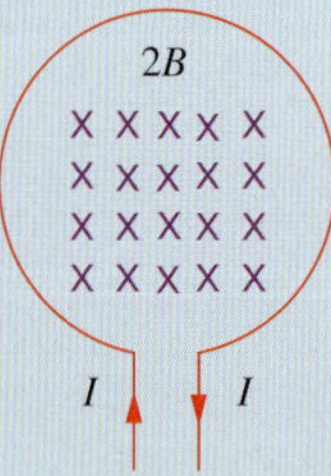

- **a** What would the magnitude and direction of the resultant field be at the centre of the loop if the current in the loop were switched off?
- **b** What would the magnitude and direction of the resultant field at the centre of the loop be if the current in the loop were increased by a factor of four?

5 A charge of $+q$ is placed a distance r from another charge also of $+q$. A repulsive force of magnitude F is found to exist between them.

- **a** The distance between the charges is increased by a factor of 3 (to $3r$). What effect will this have on the force between the charges?
- **b** The distance between the charges is reduced by a factor of 4 (to $0.25r$). What effect will this have on the force between the charges?

6 Explain what effect a magnetic field would have on a current-carrying conductor that is parallel to the field.

7 The right-hand force rule is used to determine the direction of the force on a current-carrying conductor perpendicular to a magnetic field. Explain what each part of the hand represents.

8 Calculate the force applied to an oil drop carrying a charge of 3.00 mC in a uniform electric field of $7.50\,\text{N}\,\text{C}^{-1}$.

9 A test charge is placed at a point, P, 30 cm directly above a charge, Q, of $+3 \times 10^{-5}$ C. What is the magnitude and direction of the electric field at point P?

- **A** $300\,\text{N}\,\text{C}^{-1}$ downwards
- **B** $300\,\text{N}\,\text{C}^{-1}$ upwards
- **C** $3 \times 10^{6}\,\text{N}\,\text{C}^{-1}$1 downwards
- **D** $3 \times 10^{6}\,\text{N}\,\text{C}^{-1}$ upwards

10 Calculate the potential difference between two points separated by 25 mm and parallel to the field lines in an electric field of strength $1000\,\text{V}\,\text{m}^{-1}$.

11 Calculate the work done to move a positively charged particle of 5.00×10^{-18} C a distance of 5.00 mm towards a positive plate in a uniform electric field of $650\,\text{N}\,\text{C}^{-1}$.

12 An electron gun accelerates an electron across a potential difference of 15 kV between a pair of parallel charged plates 12 cm apart. What is the magnitude of the force acting on the electron? Assume that $q_e = 1.6 \times 10^{-19}$ C.

13 Calculate the magnitude of the force that would exist between two point charges of 7.50 mC and 2.00 nC separated by 4.00 m. Assume that $k = 8.99 \times 10^{9}\,\text{N}\,\text{m}^{2}\,\text{C}^{-2}$.

14 How much current, I, must be flowing in a wire 3.7 m long if it is placed in a uniform magnetic field of 0.1100 T and the maximum force on it is 1.350 N?

15 Calculate the magnitude and direction of the magnetic force on conductors with the following characteristics.

a $B = 1.0\,\text{mT}$ left, $l = 5.0\,\text{mm}$, $I = 1.0\,\text{mA}$ upwards

b $B = 0.10\,\text{T}$ right, $l = 1.0\,\text{cm}$, $I = 2.0\,\text{A}$ upwards

16 Calculate the force exerted on an electron ($q = 1.6 \times 10^{-19}\,\text{C}$) travelling at a speed of $7.0 \times 10^{6}\,\text{m}\,\text{s}^{-1}$ at right angles to a uniform magnetic field of strength $8.6 \times 10^{-3}\,\text{T}$.

Application and analysis

17 Power lines carry an electric current in the Earth's magnetic field. Predict whether a power line running east–west or north–south would experience the greater magnetic force. Explain your answer.

18 A gold(III) ion is accelerated by the electric field created between two parallel plates separated by 0.020 m. The ion carries a charge of +5e and has a mass of $3.27 \times 10^{-25}\,\text{kg}$. A potential difference of 1000 V is applied across the plates. The work done to move the ion from one plate to the other results in an increase in the kinetic energy of the ion. If the ion starts from rest, calculate its final velocity. Assume that $q_e = -1.6 \times 10^{-19}\,\text{C}$.

19 A point charge of 3.50 mC is positioned on top of an insulated rod on a table. At what distance above the point charge would a sphere of mass 4.50 kg and charge of 5.01 mC be suspended? Assume that $k = 8.99 \times 10^{9}\,\text{N}\,\text{m}^2\,\text{C}^{-2}$.

$q_2 = 5.01$ mC
$m = 4.50$ kg

$q_1 = 3.50$ mC

20 A charged plastic ball of mass 7.50 g is placed in a uniform electric field pointing vertically upwards with a strength of $450.0\,\text{N}\,\text{C}^{-1}$. Calculate the magnitude and sign of the charge required on the ball in order to create an upwards force that exactly equals the force due to gravity acting on the ball.

21 **a** Two electrons are $5.4 \times 10^{-12}\,\text{m}$ apart. The charge on both is $-1.6 \times 10^{-19}\,\text{C}$. What is the electrical force of repulsion between the two electrons? Assume that $k = 8.99 \times 10^{9}\,\text{N}\,\text{m}^2\,\text{C}^{-2}$.

b Two electrons are $5.4 \times 10^{-12}\,\text{m}$ apart. The mass of each is $9.1 \times 10^{-31}\,\text{kg}$. What is the gravitational force of attraction between the two electrons? Assume that $G = 6.67 \times 10^{-11}\,\text{m}^3\,\text{kg}^{-1}\,\text{s}^{-2}$.

c Compare the answers to parts **a** and **b**, and comment on the difference between the gravitational force of attraction and the electrical force of attraction between the two particles.

OA
✓✓

HE-RBX 03
HE-RBX 02

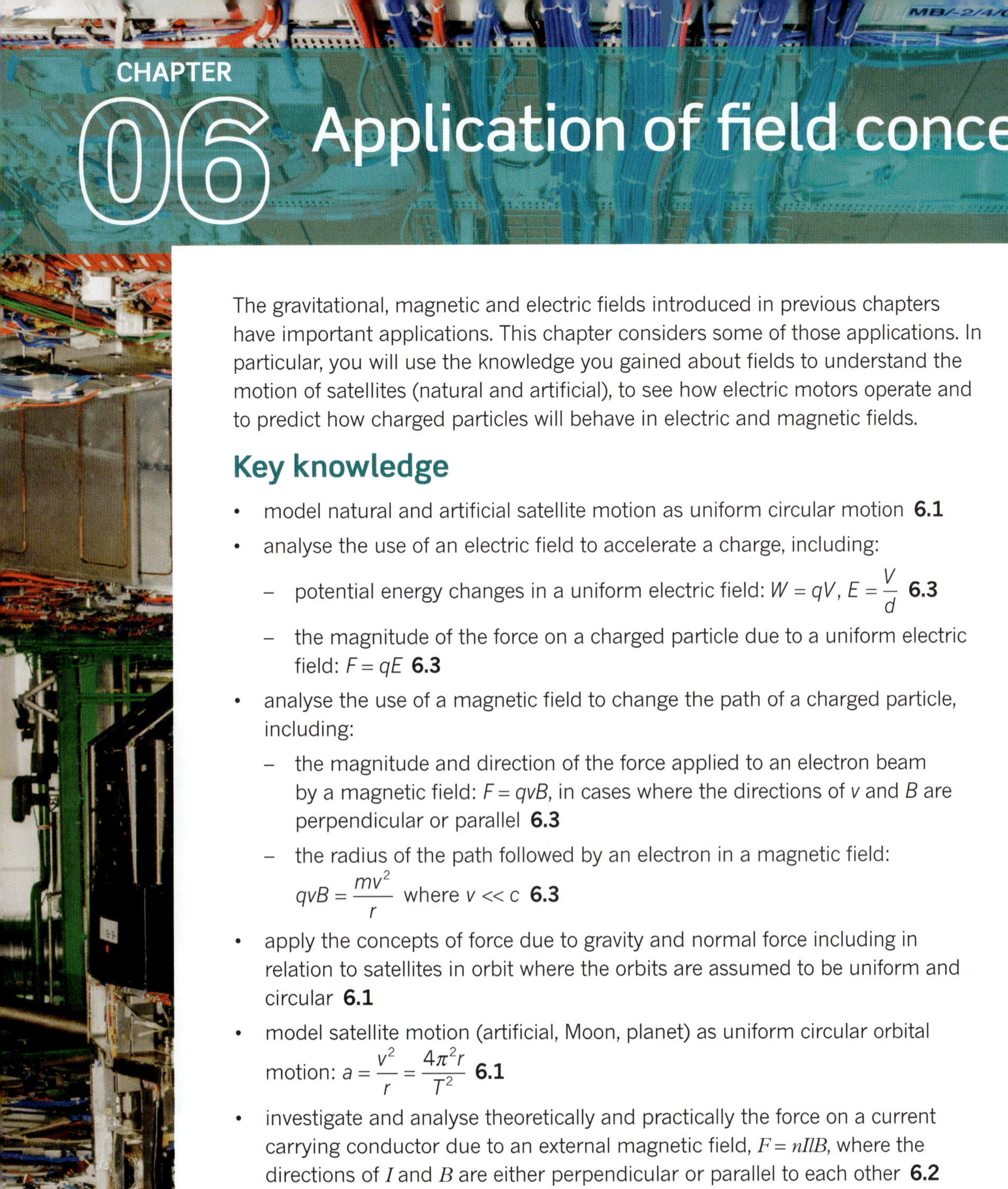

CHAPTER

06 Application of field concepts

The gravitational, magnetic and electric fields introduced in previous chapters have important applications. This chapter considers some of those applications. In particular, you will use the knowledge you gained about fields to understand the motion of satellites (natural and artificial), to see how electric motors operate and to predict how charged particles will behave in electric and magnetic fields.

Key knowledge

- model natural and artificial satellite motion as uniform circular motion **6.1**
- analyse the use of an electric field to accelerate a charge, including:
  - potential energy changes in a uniform electric field: $W = qV$, $E = \frac{V}{d}$ **6.3**
  - the magnitude of the force on a charged particle due to a uniform electric field: $F = qE$ **6.3**
- analyse the use of a magnetic field to change the path of a charged particle, including:
  - the magnitude and direction of the force applied to an electron beam by a magnetic field: $F = qvB$, in cases where the directions of v and B are perpendicular or parallel **6.3**
  - the radius of the path followed by an electron in a magnetic field: $qvB = \frac{mv^2}{r}$ where $v << c$ **6.3**
- apply the concepts of force due to gravity and normal force including in relation to satellites in orbit where the orbits are assumed to be uniform and circular **6.1**
- model satellite motion (artificial, Moon, planet) as uniform circular orbital motion: $a = \frac{v^2}{r} = \frac{4\pi^2 r}{T^2}$ **6.1**
- investigate and analyse theoretically and practically the force on a current carrying conductor due to an external magnetic field, $F = nIlB$, where the directions of I and B are either perpendicular or parallel to each other **6.2**
- investigate and analyse theoretically and practically the operation of simple DC motors consisting of one coil, containing a number of loops of wire, which is free to rotate about an axis in a uniform magnetic field and including the use of a split ring commutator **6.2**
- investigate, qualitatively, the effect of current, external magnetic field and the number of loops of wire on the torque of a simple motor **6.2**
- model the acceleration of particles in a particle accelerator (including synchrotrons) as uniform circular motion (limited to linear acceleration by a uniform electric field and direction change by a uniform magnetic field) **6.3**

OA
✓✓

6.1 Satellite motion

When Isaac Newton developed his law of universal gravitation (discussed in Chapter 4), he was building on work previously done by Nicolaus Copernicus, Johannes Kepler and Galileo Galilei. Copernicus had proposed a heliocentric (i.e. sun-centred) solar system, Galileo had developed laws relating to motion near the Earth's surface and Kepler had devised laws describing the motion of the planets. Kepler published his laws nearly 80 years before Newton published his law of universal gravitation.

In this section, you will see how Newton combined the work of Galileo and Kepler and proposed that the force that caused an apple to fall to Earth was the same force that kept the Moon in its orbit.

Newton was the first to propose that **satellites** could be placed in orbit around Earth—almost 300 years before it was technically possible to do so. Today thousands of artificial satellites are in orbit around Earth and are an essential part of modern life (Figure 6.1.1).

FIGURE 6.1.1 An artificial satellite in orbit around the Earth

FIGURE 6.1.2 Newton realised that the gravitational attraction of the Earth (F_g) was determining the motions of both the Moon and the apple.

NEWTON'S THOUGHT EXPERIMENT

A satellite is any object in a stable orbit around another object. Isaac Newton developed the notion of satellite motion while working on his theory of gravitation. He was comparing the motion of the Moon with the motion of a falling apple and realised that it was the gravitational force of attraction towards the Earth that determined the motion of both objects (Figure 6.1.2). He reasoned that if this force of gravity was not acting on the Moon, the Moon would move at constant speed in a straight line.

Newton proposed that the Moon, like the apple, was also falling. It was continuously falling towards the Earth without actually getting any closer to the Earth. He devised a thought experiment in which he compared the motion of the Moon with the motion of a cannonball fired horizontally from the top of a high mountain (Figure 6.1.3 on the following page). If the cannonball was fired at a low speed, it would not travel very far before gravity pulled it to the ground (indicated by the shortest dashed line in Figure 6.1.3(b)). If it was fired with a greater velocity, it would follow a less curved path and land a greater distance from the mountain (indicated by the next two dashed lines in Figure 6.1.3(b)).

Newton reasoned that, if air resistance was ignored and if the cannonball was fired fast enough, it could travel around the Earth and reach the place from where it had been launched (indicated by the solid circular line in Figure 6.1.3(b)). At this speed the cannonball would continue to circle the Earth indefinitely—even though it had no propulsion system.

In reality, satellites could not orbit the Earth at low altitudes. There would be too much air resistance. Nevertheless Newton had conceived of the idea of artificial satellites hundreds of years before one was actually launched. Any object with enough speed and placed at the right altitude would simply continue in its orbit.

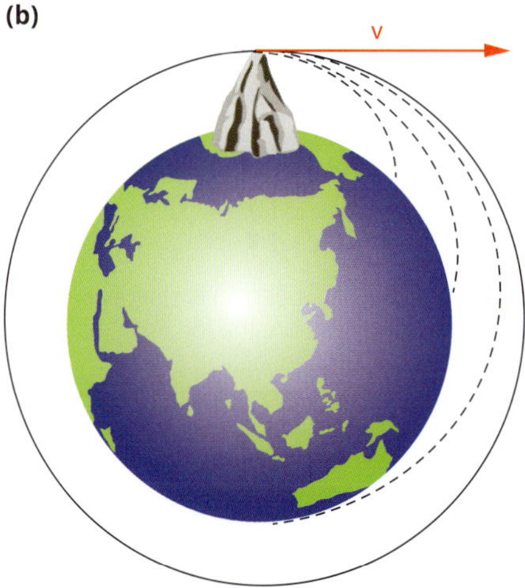

FIGURE 6.1.3 These diagrams illustrate how a projectile fired fast enough from a very high mountain would become a satellite of the Earth rather than fall to the Earth.

MASS AND FORCE DUE TO GRAVITY

In Unit 2 Physics, the concept of 'force due to gravity' was introduced. You will recall that this is a force on an object when that object is near the surface of a planet or other large body, such as the Earth or the Moon.

The force due to gravity (F_g) is equal to the mass (m) of the object multiplied by the acceleration due to gravity (g).

$$F_g = mg$$

where F_g is the force due to gravity (N)

m is the mass of the object (kg)

g is the gravitational field strength (N kg^{-1})

On the Earth the value of g is taken as $9.80665\,\text{m s}^{-2}$ or $9.80665\,\text{N kg}^{-1}$. This is often rounded to $9.8\,\text{m s}^{-2}$, or in some cases approximated to $10\,\text{m s}^{-2}$ for ease of calculations.

Like all forces, the force due to gravity is measured in newtons (N) and, because it is a vector quantity, it is not fully stated unless it is accompanied by a direction. The direction of the force due to gravity is always downwards, towards the centre of the planet or large body.

NORMAL FORCE

If you exert a force against a wall, Newton's third law says that the wall will exert an equal but opposite force on you (Figure 6.1.4(a)). If you push with a greater force, the wall will push back with a greater force (Figure 6.1.4(b)). In both cases in Figure 6.1.4, the force Joe exerts on the wall will be equal and opposite to the force the wall exerts on Joe. In other words, $F_{\text{on Joe by wall}} = -F_{\text{on wall by Joe}}$.

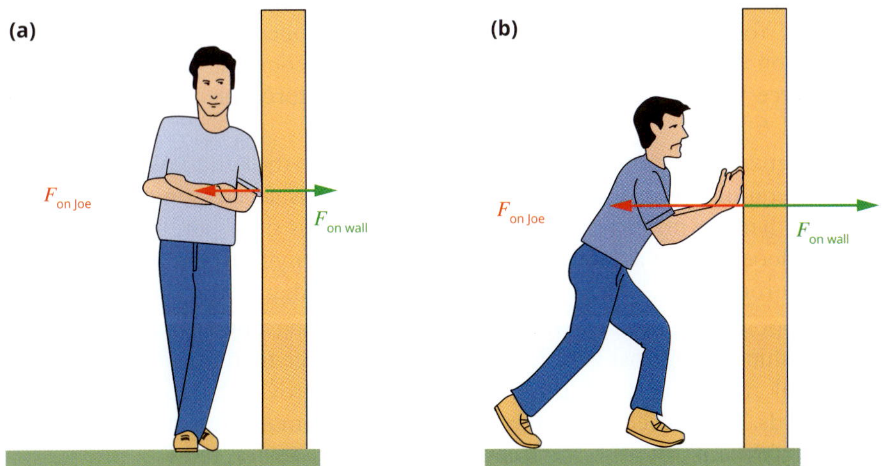

FIGURE 6.1.4 (a) If Joe exerts a small force on the wall, the wall exerts a small force on Joe. (b) When Joe pushes hard against the wall, the wall pushes back just as hard! In both (a) and (b), the red and green arrows are equal in size but opposite in direction.

FIGURE 6.1.6 When a person is stationary or in constant vertical motion, the forces that act on them, F_N and F_g, are equal.

The force from the wall acts at right angles to the surface, that is, it is normal to the surface and is thus called a **normal force**. The normal force is represented by F_N.

For an object at rest on the ground, the normal force is equal in size to the force due to gravity on the object.

During many interactions and collisions, the size of the normal force changes. For example, when a ball bounces, the forces that act on it during its contact with the floor are the gravitational force, F_g, and the normal force, F_N, from the floor. The diagrams in Figure 6.1.5 illustrate the changes in the magnitude of the normal force throughout the bounce of a ball.

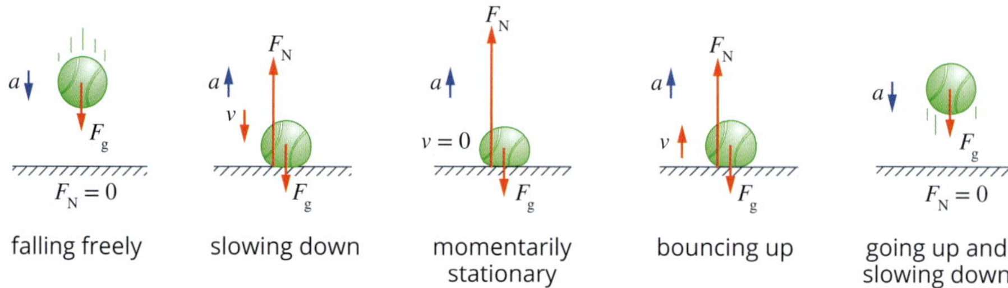

FIGURE 6.1.5 The forces acting on a bouncing ball before, during and after striking the floor

When contact has just been made, the ball is compressed, but only slightly, indicating that the force from the floor is minimal. This force becomes larger and larger, causing the ball to become more and more deformed. At the point of maximum compression, the normal force is at its maximum value and the bouncing ball is momentarily stationary.

The forces acting on a ball as it bounces (its gravitational force, F_g, and the normal force, F_N) are not an action–reaction pair. Both act on the same body whereas Newton's third law describes forces that bodies exert on each other. An action–reaction pair that does act during the bounce involves the upwards force, F_N, that the floor exerts on the ball and the downwards force that the ball exerts on the floor (not shown in Figure 6.1.5). This downwards force is equal in magnitude to the normal force, so it too varies during the bounce.

Experienced normal force

When you stand on a surface, you will feel the normal force acting upwards on your feet. It results from the force due to gravity pulling you downwards onto the floor and applying a force to the floor. You do not experience the force you apply to the floor. You only experience the forces that are applied on you. Typically, when you stand on a surface that is either stationary or in constant vertical motion, the normal force you feel is constant and equal to the force on you due to gravity (Figure 6.1.6).

The normal force that you experience changes when the surface you are standing on is accelerating upwards or downwards. If the floor is accelerating downwards at a rate less than $9.8\,\text{m}\,\text{s}^{-2}$, your feet will be pressing less firmly on the surface than when the floor was not accelerating. Therefore the normal force is less and you will feel a less strong force. That is, you would feel lighter than usual (Figure 6.1.7).

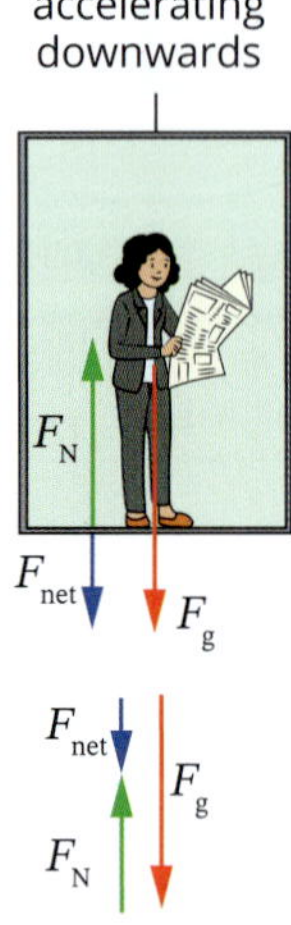

FIGURE 6.1.7 When accelerating downwards, the normal force acting on the person in the lift is less than usual ($F_N < F_g$) which causes her to feel lighter.

The opposite happens when the floor is accelerating upwards. In this case the floor is pushing up against your feet with a greater force than the normal force due to gravity alone. The upwards push of the floor must provide the force to accelerate you upwards. This accelerating force adds to the normal force, making it greater than it would be if you weren't accelerating. That is, you would feel heavier than usual (Figure 6.1.8).

FIGURE 6.1.8 When accelerating upwards, the normal force acting on the person in the lift is greater than usual ($F_N > F_g$) which causes her to feel heavier.

The normal force and the force due to gravity add as vectors to give the net force that causes the acceleration, as given in the following equation.

$F_{net} = F_N + F_g$

where F_N is the normal force that acts upwards on your feet

F_g is the force due to gravity (which never changes)

F_{net} is the net force causing the acceleration

Worked example 6.1.1

CALCULATING NORMAL FORCE

A 68.0 kg student rides a lift up to the top floor of an office block. During the journey, the lift accelerates upwards at $1.70\,m\,s^{-2}$ before travelling at a constant velocity of $6.78\,m\,s^{-1}$ and then finally decelerating at $2.53\,m\,s^{-2}$. Assume that $g = 9.8\,m\,s^{-2}$.

a Calculate the normal force acting on the student in the first part of the journey, i.e. while the lift was accelerating upwards at $1.70\,m\,s^{-2}$.

Thinking	Working
Ensure that the variables are in their standard units.	$m = 68.0\,kg$ $a = 1.70\,m\,s^{-2}$ up $g = 9.8\,m\,s^{-2}$ down
Apply the sign-and-direction convention for motion in one dimension: up is positive and down is negative.	$m = 68.0\,kg$ $a = +1.70\,m\,s^{-2}$ $g = -9.8\,m\,s^{-2}$
Apply the appropriate equations to calculate the normal force.	$F_{net} = F_N + F_g$ $F_N = F_{net} - F_g$ $= ma - mg$ $= (68.0 \times 1.70) - (68.0 \times -9.8)$ $= 115.6 + 666.4$ $= 782$ $= 7.8 \times 10^2\,N$

b Calculate the normal force acting on the student in the second part of the journey, i.e. while the lift is travelling upwards at a constant speed of $6.78\,m\,s^{-1}$.

Thinking	Working
Ensure that the variables are in their standard units.	$m = 68.0\,kg$ $a = 0\,m\,s^{-2}$ $g = 9.8\,m\,s^{-2}$ down
Apply the sign-and-direction convention for motion in one dimension: up is positive and down is negative.	$m = 68.0\,kg$ $a = 0\,m\,s^{-2}$ $g = -9.8\,m\,s^{-2}$
Apply the appropriate equations to calculate the normal force.	$F_{net} = F_N + F_g$ $F_N = F_{net} - F_g$ $= ma - mg$ $= (68.0 \times 0) - (68.0 \times -9.8)$ $= 0 + 666.4$ $= 6.7 \times 10^2\,N$

c Calculate the normal force acting on the student in the last part of the journey, i.e. while the lift is travelling upwards and decelerating at $2.53\,m\,s^{-2}$.	
Thinking	**Working**
Ensure that the variables are in their standard units. Also consider that deceleration is negative acceleration.	$m = 68.0\,kg$ $a = 2.53\,m\,s^{-2}$ down $g = 9.8\,m\,s^{-2}$ down
Apply the sign-and-direction convention for motion in one dimension: up is positive and down is negative.	$m = 68.0\,kg$ $a = -2.53\,m\,s^{-2}$ $g = -9.8\,m\,s^{-2}$
Apply the appropriate equations to calculate the normal force.	$F_{net} = F_N + F_g$ $F_N = F_{net} - F_g$ $= ma - mg$ $= (68.0 \times -2.53) - (68.0 \times -9.8)$ $= -172.0 + 666.4$ $= 494$ $= 4.9 \times 10^2\,N$

Worked example: Try yourself 6.1.1

CALCULATING NORMAL FORCE

A 68.0 kg student rides a lift down from the top floor of an office block to the ground floor. During the journey the lift accelerates downwards at $1.50\,m\,s^{-2}$ before travelling at a constant velocity of $3.08\,m\,s^{-1}$ and then finally decelerating at $3.80\,m\,s^{-2}$ until it reaches the ground floor.

a Calculate the normal force acting on the student in the first part of the journey, i.e. while accelerating downwards at $1.50\,m\,s^{-2}$.

b Calculate the normal force acting on the student in the second part of the journey, i.e. while travelling at a constant speed of $3.08\,m\,s^{-1}$.

c Calculate the normal force acting on the student in the last part of the journey, i.e. while travelling downwards and decelerating at $3.80\,m\,s^{-2}$.

From these worked examples, you can see that:

- when accelerating upwards, the student will feel heavier than normal: $F_N > F_g$. (Note: this is the same as decelerating while travelling downwards.)
- when accelerating downwards, the student will feel lighter than normal: $F_N < F_g$. (Note: this is the same as decelerating while travelling upwards.)
- when travelling upwards or downwards at a constant velocity, the student will feel their usual normal force, just as they would if the lift was stationary: $F_N = F_g$.

Free fall

The normal force is a contact reaction force that acts upwards on you from a surface because gravity is pulling you down on that surface. It follows that if you are not standing on a surface, then you will experience no force acting upwards. This experience is known as **free fall.** Only gravity is acting on you and you may feel as though you are floating. This could be what you feel the moment you step off the top platform of a diving tower at a swimming pool or skydive from a plane (although the air rushing passed you will probably deny you the sensation of floating when you skydive).

Felix Baumgartner experienced free fall as he jumped from his balloon 39 kilometres above the Earth (Figure 6.1.9). This height is almost equal to the widest part of Port Phillip Bay.

FIGURE 6.1.9 Felix Baumgartner experienced free fall on his return to Earth from 39 000 m.

Astronauts also experience the feeling of free fall, or zero normal force, in the International Space Station, which orbits about 370 kilometres above the surface of the Earth (about the distance from Melbourne to the town of Orbost).

Whenever you are in free fall you experience zero normal force. Conversely, whenever you experience zero normal force you must be in free fall. When astronauts orbiting the Earth experience this feeling, they are not floating in space without any gravitational force. They are actually in free fall. Astronauts and their spacecraft are both falling, but not directly towards the Earth like Baumgartner. The astronauts are actually moving horizontally (Figure 6.1.10). Baumgartner stayed approximately above the same place on the Earth from where he departed. However, astronauts are moving at a velocity relative to the Earth, so they are moving across the sky at the same time as they are falling. The combined effect is that they fall in a curved path that exactly mirrors the curvature of the Earth. So they fall, but continually miss the Earth because the surface of the Earth curves away from their path.

Importantly there is a significant difference between zero normal force and zero gravitational force. A zero force due to gravity only occurs when the gravitational field strength is zero, that is, when $F_g = 0$. This only occurs in deep space, far enough away from any bodies that their gravitational effect is zero. The lack of normal force, however, can occur when still under the influence of a gravitational field.

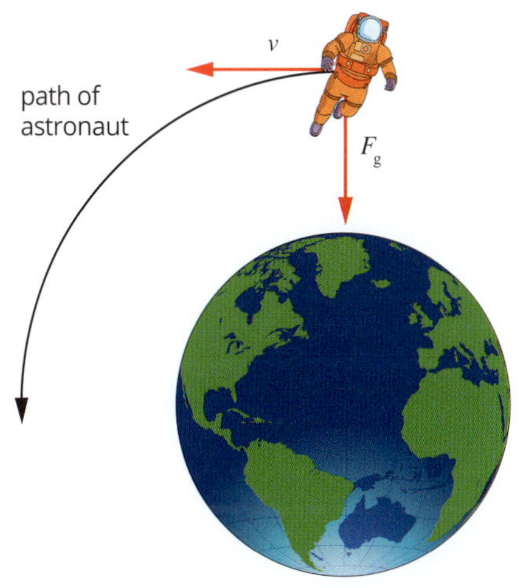

FIGURE 6.1.10 Astronauts are in free fall while orbiting the Earth.

CASE STUDY

Falling at constant speed

More than 400 years ago, Galileo showed that the mass of a body does not affect the rate at which it falls towards the ground. However, our common experience is that not all objects behave in this way. A light object, such as a feather or a balloon, does not accelerate at $9.8\,\mathrm{m\,s^{-2}}$ as it falls. It drifts slowly to the ground far slower than many other objects. This is because of the varying effect of air resistance. Still, all falling objects will eventually fall at a constant speed.

Parachutists and skydivers also eventually fall with a constant speed. However, they can change their falling speed by changing their body profile (Figure 6.1.11). If they assume a tuck position, they will fall faster. If they spread out their arms and legs, they will fall slower. This enables them to form spectacular patterns as they fall.

FIGURE 6.1.11 Skydivers performing intricate manoeuvres in free fall

Skydivers, base-jumpers and air-surfers are able to use the force of air resistance to their advantage. As a skydiver first steps out of the plane, the forces acting on them are air resistance, F_{ar}, which is also known as drag, and the force due to gravity, F_g. Since their speed is low, the drag force is small (Figure 6.1.12(a)). There is a large net force downwards, F_{net}, so they will experience a large downwards acceleration of just less than $9.8\,\mathrm{m\,s^{-2}}$. This causes them to speed up. It also causes the drag force to increase because they are colliding harder with the air molecules around them. In fact, the drag force increases in proportion to the square of the speed: $F_{ar} \propto v^2$. This results in a smaller net force downwards (Figure 6.1.12(b)). Their downwards acceleration is therefore reduced. It is important to remember that they are still speeding up, but at a reduced rate.

As their speed continues to increase, so too does the magnitude of the drag force. Eventually, the drag force becomes as large as the force due to gravity (Figure 6.1.12(c)). When this happens, the net force is zero and the skydiver will fall with a constant velocity. Since the velocity is now constant, the drag force also remains constant and the motion of the skydiver will not change (Figure 6.1.12(d)). This velocity is commonly known as terminal velocity.

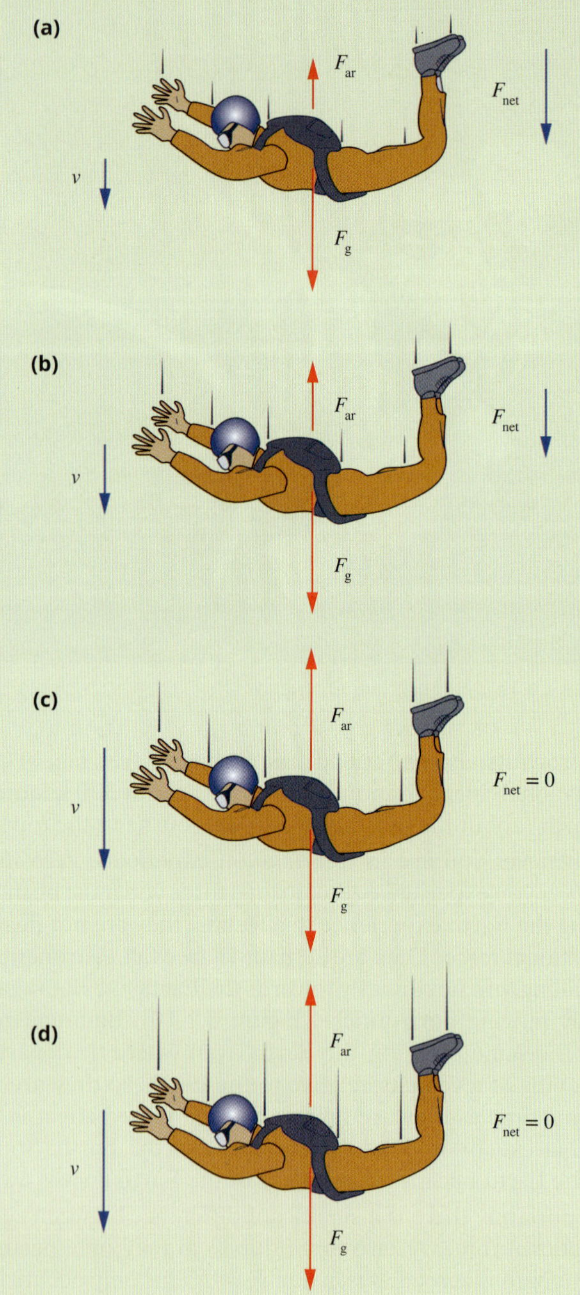

FIGURE 6.1.12 The forces involved in skydiving

NATURAL SATELLITES

A **natural satellite** is any body that orbits another body except orbiting bodies that were launched by humans. Natural satellites have existed throughout the universe for billions of years. The planets and asteroids of the solar system are natural satellites of the Sun (Figure 6.1.13).

The Earth has one natural satellite: the Moon. The largest planets—Jupiter and Saturn—have more than eighty natural satellites each in orbit around them. Most of the stars in the Milky Way galaxy have planets and more of these so-called exoplanets are being discovered each year.

FIGURE 6.1.13 The planets in the solar system are natural satellites of the Sun.

ARTIFICIAL SATELLITES

An **artificial satellite** is any body launched by humans that orbits another body. From the beginning of the Space Age in 1957 (with the launch of Sputnik) up until 2022, more than 14 000 artificial satellites have been launched into orbit around the Earth. As of 2022, there are approximately 9000 still in orbit, of which about 5500 are still active. This number is growing rapidly each year, owing to commercial space ventures.

Satellites in orbit around the Earth are classified as low-, medium- or high-orbit satellites.

- Low orbit: 180 km to 2000 km altitude. Most satellites orbit in this range. (An example is shown in Figure 6.1.14.). These include the Hubble Space Telescope, which is used by astronomers to view objects as far away as the edge of the universe.
- Medium orbit: 2000 km to 36 000 km altitude. The most common satellites in this region are the satellites used to run navigation systems, such as the Global Positioning System (GPS).
- High orbit: 36 000 km altitude or greater. An example is the Optus satellites that Australia uses for communications and the Japanese Himawari-8 satellite used to provide deep-space weather pictures. The satellites that sit at an altitude of approximately 36 000 km and orbit with a period of 24 hours are known as geostationary (or geosynchronous) satellites. Most communications satellites are geostationary.

FIGURE 6.1.14 A low-orbit satellite called the Soil Moisture and Ocean Salinity (SMOS) probe was launched in August 2014. Its role is to measure water movements and salinity levels on the Earth in order to monitor climate change. It was launched from northern Russia by the European Space Agency (ESA).

Earth satellites can have different orbital paths depending on their function. For example, a satellite can have:

- an equatorial orbit, where it always travels above the equator
- a polar or near-polar orbit, where it travels over or close to the North Pole and South Pole
- an inclined orbit, where it travels somewhere between an equatorial and a polar orbit.

Satellites are used for numerous purposes. Sixty per cent are used for communications. Many low-orbit satellites launched by the American National Oceanic and Atmospheric Administration (NOAA) have an inclination of 99° and an orbit that allows them to pass over each part of the Earth at the same time each day. These satellites are also known as Sun-synchronous satellites.

CASE STUDY ANALYSIS

Four satellites

Geostationary Meteorological Satellite: Himawari-8

The Japanese Himawari-8 is a weather satellite operated by the Japan Meteorological Agency. It provides important data for Australia's Bureau of Meteorology. It was launched in October 2014 and orbits directly over the equator. At its closest point to the Earth, known as the perigee, its altitude is 35 791 km. At its furthest point from the Earth, known as the apogee, its altitude is 35 795 km. Himawari-8 orbits at a longitude of 141°E, so it is just north of Cape York and ideally located for use by Australia's weather forecasters. It has a period of 24 hours, so is in a geostationary orbit.

Signals from Himawari-8 are transmitted every 10 minutes. They are received in Australia via a number of parallel sources, such as a dedicated fibre-optic line from Japan, a cloud-based internet service and satellite data reception sites in Melbourne, Darwin and Perth, as well as the Casey and Davis stations in Antarctica. Images are taken across a broad range of wavelengths, including infrared and visible colours. These images show temperature variations in the atmosphere, tropical cyclones and thunderstorms, fog and low cloud, and are invaluable in weather forecasting. Himawari-8 is box-like, with dimensions of approximately 5.2 m × 8.0 m × 5.3 m. It has a mass of 1300 kg and is powered by a single gallium arsenide solar panel.

Hubble Space Telescope (HST)

The Hubble Space Telescope is a cooperative venture between NASA and the European Space Agency (ESA). It was launched by the crew of the space shuttle Discovery on 24 April 1990. Hubble is a permanent unoccupied space-based observatory with a reflecting telescope 2.4 m in diameter, a number of spectrographs and a faint-object camera. It orbits above the Earth's atmosphere, producing images of distant stars and galaxies far clearer than those taken from ground-based observatories (Figure 6.1.15). The HST is in a low-Earth orbit and inclined at 28° to the equator. Its expected life span was 15 years, but service and repair missions have extended its life and it is still in use today.

FIGURE 6.1.15 In August 2014, astronomers used the Hubble Space Telescope to detect the blue companion star of a white dwarf in a distant galaxy. The white dwarf slowly siphoned fuel from its companion star, eventually igniting a runaway nuclear reaction in the star that eventually produced a supernova blast.

National Oceanic and Atmospheric Administration Satellite (NOAA-20)

Many of the US-owned-and-operated NOAA satellites are located in low altitudes and have near-polar orbits. This means that they pass close to the poles of the Earth as they orbit. NOAA-20 was launched in November 2017 and orbits at an inclination of 99° to the equator. Its low altitude means that it captures high-resolution pictures of small bands of the Earth. The data is used in local weather forecasting. It also provides information that helps in monitoring global warming and climate change.

James Webb Space Telescope (JWST)

The intended successor to the Hubble Space Telescope is the James Webb Space Telescope (JWST). This was a joint development between NASA, the ESA and the Canadian Space Agency (CSA). The launch suffered a series of long delays as the telescope went through numerous redevelopments. The JWST was finally launched on 25 December 2021. Like Hubble, the JWST is a reflecting telescope, but it has a much larger primary mirror (with a diameter of 6.5 m.) This extra size offers about 6.25 times more collecting area. It is also much more sensitive to infrared observations, which are necessary for collecting data about early-universe galaxies.

The JWST was designed for a much more ambitious route than Hubble. It orbits the Sun instead of the Earth! This enables the instruments on board to become extremely cold (<–225°C), which is necessary to clearly observe infrared signals. The telescope is equipped with a large sun shield to block out sources of heat and keep the electronics at their necessary operating temperatures.

The JWST was launched to a special point in orbit around the Sun called a Lagrange point. This is approximately 1.5 million kilometres from Earth—about four times further away than the Moon. At this point it is hidden behind the Earth, and always keeps the Sun, Moon, and Earth to one side.

TABLE 6.1.1 A comparison of three of the satellites discussed in this case study

Satellite	Orbit	Inclination	Perigee (km)	Apogee (km)	Period
Himawari-8	equatorial	0°	35 791	35 795	1 day
Hubble	inclined	28°	591	599	96.6 min
NOAA-20	near polar	99°	824	833	101.4 min

Analysis

1 Define the term 'geostationary orbit' and discuss why a weather satellite might be launched into this type of orbit.

2 For the apogee of each of the satellites listed in Table 6.1.1, compute the acceleration due to gravity, g, using the formula for gravitational field strength from Chapter 4.

3 For each of the satellites listed in Table 6.1.1, calculate the circumference, in metres, of the orbit at the average between its perigee and apogee.

4 For each of the satellites listed in Table 6.1.1, calculate the time it takes, in seconds, for a single orbit. Use the average orbital radius found in question **3** in your calculations.

5 Using the formula speed $= \frac{\text{distance}}{\text{time}}$, calculate the orbital speeds of each satellite listed in Table 6.1.1.

Artificial and natural satellites are not propelled by rockets or engines. They orbit in free fall. The only force acting on them is the gravitational attraction between themselves and the body they orbit. Thus the satellites have a centripetal acceleration equal to the gravitational field strength at their location (Figure 6.1.16). **Centripetal acceleration** describes the acceleration towards the centre of a circle when an object is in circular motion. It was covered in detail in Chapter 2.

Although not propelled by rockets or engines, artificial satellites are often equipped with tanks of propellant that are fired in the appropriate direction when the orbit of the satellite needs to be adjusted.

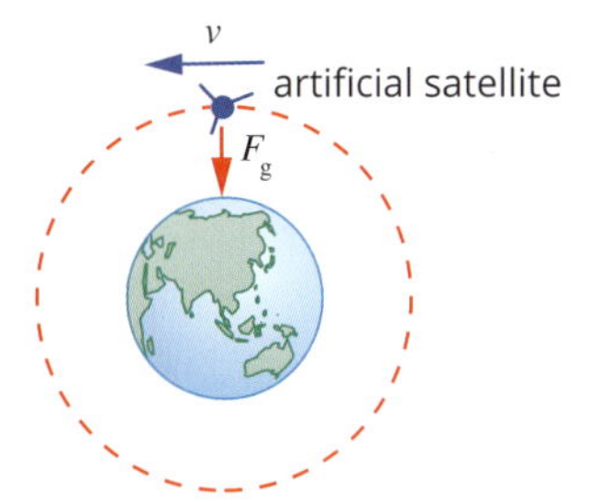

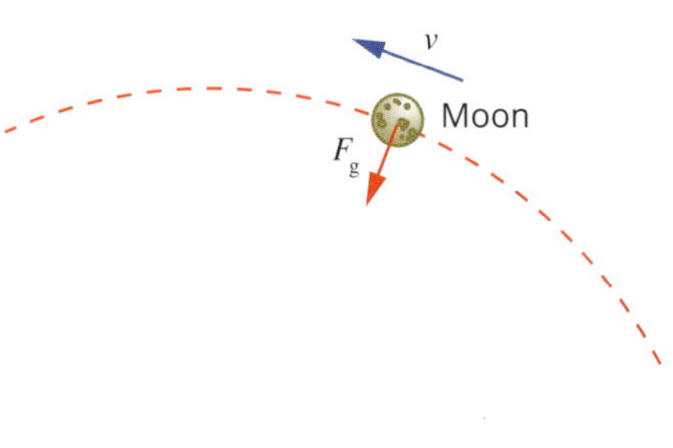

FIGURE 6.1.16 The only force acting on these artificial and natural satellites is the gravitational attraction of the Earth. Both orbit with a centripetal acceleration equal to the gravitational field strength at their location.

PHYSICSFILE

SuitSat1

One of the more unusual satellites was launched from the International Space Station on 3 February 2006. It was an obsolete Russian spacesuit into which the astronauts had placed a radio transmitter, batteries and sensors (see figure below). Its launch was simple: it was pushed away by one of the astronauts while on a spacewalk. SuitSat1 was meant to transmit signals that could be picked up by ham radio operators on Earth for a few weeks. However, transmissions ceased after just a few hours. The spacesuit burnt up in the atmosphere over Western Australia in September 2006.

SuitSat2 was launched in August 2011 and contained experiments created by school students. It re-entered Earth's atmosphere in January 2012 after 5 months in orbit.

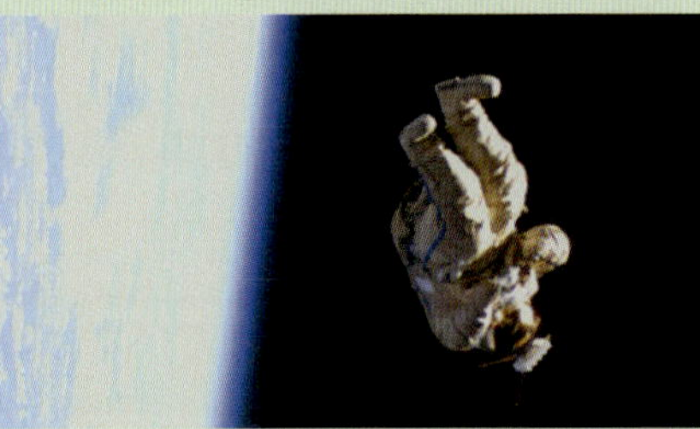

This photograph does not show an astronaut drifting off to certain death in space. This is SuitSat1, one of the strangest satellites ever launched, at the start of its short-lived mission.

KEPLER'S LAWS

In 1609 the German astronomer Johannes Kepler (Figure 6.1.17) published three famous laws regarding the motion of planets. This was nearly 80 years before Newton published his law of universal gravitation. Kepler had been analysing the motion of the planets in orbit around the Sun, but his laws can be applied to any satellite in orbit around any mass.

FIGURE 6.1.17 Johannes Kepler discovered that the planets travel in elliptical paths, not in circular paths as was widely believed.

Kepler's laws are as follows:

1 The planets move in elliptical orbits, with the Sun at one focus.
2 The line connecting a planet to the Sun sweeps out equal areas in equal intervals of time (Figure 6.1.18).
3 For every planet, the ratio of the cube of its average orbital radius, r, to the square of its period of revolution, T, is the same, i.e. $\frac{r^3}{T^2}$ is constant.

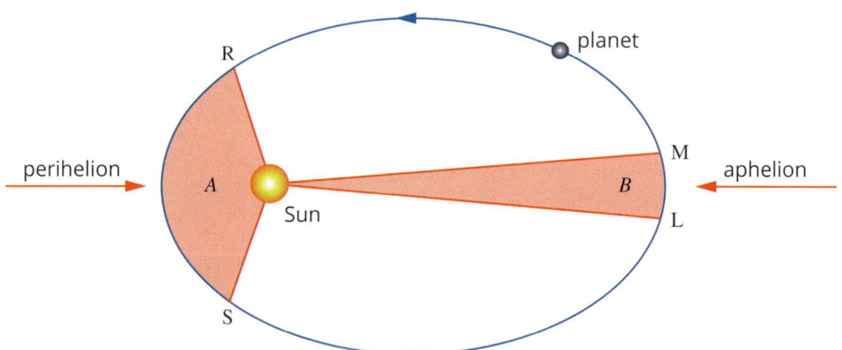

FIGURE 6.1.18 The planets orbit in elliptical paths, with the Sun at one focus. Their speeds vary continually. They are fastest when closest to the Sun. A line joining a planet to the Sun will sweep out equal areas in equal times. Thus if the time it takes for the planet shown to move from R to S is equal to the time it takes to move from L to M, then area A is the same as area B.

From his first law—that planets move in elliptical paths—and the fact that the closer the planet is to the Sun the faster it moves, Kepler could easily derive his second law. But it took Kepler many months of laborious calculations to arrive at his third law.

Newton used Kepler's laws to justify the inverse square relationship. In fact Kepler's third law can be deduced, at least for circular orbits, from Newton's law of universal gravitation.

CALCULATING THE ORBITAL PROPERTIES OF SATELLITES

Although Kepler studied elliptical orbits, which are the most common orbits encountered in the universe, for simplicity we will treat all satellites as having circular orbits. Thus the speed, v, of a satellite will be taken to equal the circumference of its circular orbit, $2\pi r$, divided by the time it takes for it to make one complete revolution, that is, its period, T.

The speed, v, of a satellite in a circular orbit is given by:

$$v = \frac{\text{distance}}{\text{time}} = \frac{2\pi r}{T}$$

where r is the radius of the orbit (m)
T is the period, i.e. the time it takes for one revolution (s)

The centripetal acceleration of a satellite can be determined from the strength of the gravitational field at the satellite's location. Satellites are in free fall. Therefore the only force acting on them is gravity, F_g. The International Space Station (ISS) is in orbit at a distance from Earth where g is $8.8\,\text{N}\,\text{kg}^{-1}$, and so it orbits with a centripetal acceleration of $8.8\,\text{m}\,\text{s}^{-2}$.

The centripetal acceleration, a, of a satellite can also be calculated by considering its circular motion. The equation for speed given above can be substituted into the formula for centripetal acceleration:

$$a = \frac{v^2}{r} \text{ (centripetal acceleration)}$$

$$v = \frac{2\pi r}{T} \text{ (speed)}$$

$$v^2 = \frac{4\pi^2 r^2}{T^2}$$

$$\frac{v^2}{r} = \frac{4\pi^2 r}{T^2} = a$$

Since the centripetal acceleration of the satellite is equal to the strength of the gravitational field acting on it, we can use the equation for gravitational field strength from Chapter 4 to derive the following expression for the centripetal acceleration of a satellite in circular orbit.

$$a = \frac{v^2}{r} = \frac{4\pi^2 r}{T^2} = \frac{GM}{r^2} = g$$

where v is the speed of the satellite ($\text{m}\,\text{s}^{-1}$)
r is the radius of the orbit (m)
T is the period of the orbit (s)
M is the mass of the body around which the satellite orbits (kg)
g is the strength of the gravitational field at r ($\text{N}\,\text{kg}^{-1}$)
G is the gravitational constant, $6.67 \times 10^{-11}\,\text{N}\,\text{m}^2\,\text{kg}^{-2}$

From these relationships certain properties of a satellite's motion can be determined, such as its speed, radius of orbit and period of orbit. As with freely falling objects near the Earth's surface, the mass of the satellite has no effect on any of these properties.

The relationships can also be used to find the mass of the body around which a satellite orbits, i.e. M. The gravitational force, F_g, acting on the satellite can then be found by using Newton's second law.

The gravitational force on a satellite of mass m in a stable circular orbit is given by:

$$F_g = \frac{mv^2}{r} = \frac{4\pi^2 rm}{T^2} = \frac{GMm}{r^2} = mg$$

PHYSICSFILE

Space junk

By mid-2022, about 5500 satellites were in active operation and this number has been increasing every year. There are also more than 3400 satellites that have reached the end of their operational life, or have malfunctioned, but are still in orbit.

In 2007 a Chinese satellite was deliberately destroyed by a missile, creating thousands of pieces of debris. In 2009 a collision between the defunct Russian Cosmos 2251 and the operational US Iridium 33 created even more debris. This debris, and the defunct satellites, are classified as space junk (see figure below).

The presence of this fast-moving space junk puts other satellites, and the International Space Station, at risk from collision. Currently around 27 000 pieces of space junk are being tracked and monitored. There have been a number of occasions where satellites have been moved to avoid collisions with space junk.

The UN has passed a resolution to remove defunct satellites from low-Earth orbits, either by moving them to much higher orbits or by bringing them back towards Earth and allowing them to burn up on re-entry into the atmosphere.

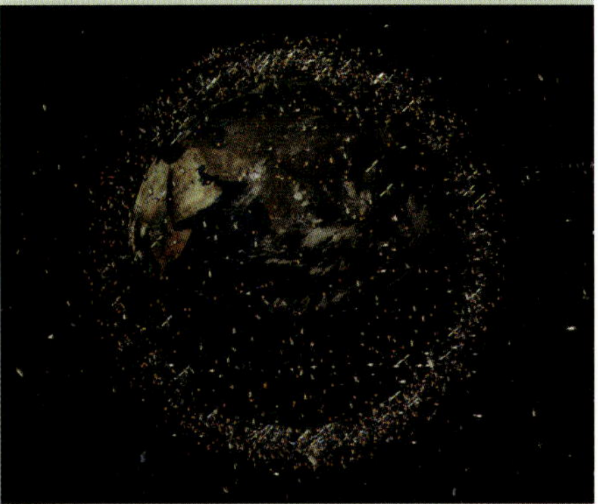

An artist's impression of the space debris and abandoned satellites in near-Earth orbits

PHYSICSFILE

See the International Space Station (ISS) and other satellites

It is easy to see low-orbit satellites if you are away from city lights. The best time to look is just after sunset. Look for any slow-moving objects passing across the background of stars.

There are also websites that allow you to track satellites and even predict their paths. For example, you can use the NASA *Spot the Station* website (https://spotthestation.nasa.gov/) to see when the ISS is passing over your part of the planet. The ISS is so bright that it is easy to see from most locations.

Worked example 6.1.2

WORKING WITH KEPLER'S LAWS

Determine the orbital speed of the Moon assuming it is in a circular orbit of radius 384 000 km around the Earth. Assume that the mass of the Earth is 5.98×10^{24} kg and that G is $6.67 \times 10^{-11}\ \text{N m}^2\ \text{kg}^{-2}$.

Thinking	Working
Convert the radius to standard units.	$r = 384\,000\text{ km} = 3.84 \times 10^8\text{ m}$
Choose the appropriate relationship between the orbital speed, v, and the data that has been provided.	$a = g = \frac{GM}{r^2} = \frac{v^2}{r}$
Make v the subject of the equation.	$v = \sqrt{\frac{GM}{r}}$
Substitute the given values and solve for the orbital speed, v.	$v = \sqrt{\frac{GM}{r}}$ $= \sqrt{\frac{(6.67 \times 10^{-11}) \times (5.98 \times 10^{24})}{3.84 \times 10^8}}$ $= 1.02 \times 10^3\text{ m s}^{-1}$

Worked example: Try yourself 6.1.2

WORKING WITH KEPLER'S LAWS

Determine the orbital speed of a satellite that is in a circular orbit of radius 42 100 km around the Earth. Assume that the mass of the Earth is 5.98×10^{24} kg and that G is $6.67 \times 10^{-11}\ \text{N m}^2\ \text{kg}^{-2}$.

HOW NEWTON DERIVED KEPLER'S THIRD LAW USING ALGEBRA

It took Kepler many months of trial-and-error calculations to arrive at his third law:

$$\frac{r^3}{T^2} = \text{constant}$$

Newton used algebra to derive this same formula from his law of universal gravitation:

$$F_g = \frac{4\pi^2 rm}{T^2} = \frac{GMm}{r^2}$$

$$\therefore \frac{r^3}{T^2} = \frac{GM}{4\pi^2}$$

For any mass, M, around which a satellite orbits in a circular path, $\frac{GM}{4\pi^2}$ is constant. Thus the ratio $\frac{r^3}{T^2}$, being equal to that constant, must also be constant—and for *all* satellites that orbit M (Figure 6.1.19).

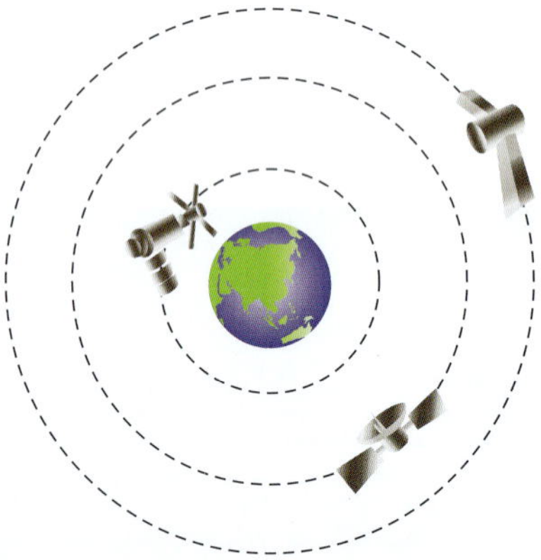

FIGURE 6.1.19 These three satellites are at different distances from Earth and hence, according to Kepler's third law, will have different orbital periods. But for all three, the ratio $\frac{r^3}{t^2}$ will be the same.

So if you know the orbital radius, r, and period, T, of one of the moons of Saturn, for example, you could calculate $\frac{r^3}{T^2}$ and use this as a constant value that applies to every moon of Saturn. If you knew the period of another moon of Saturn, it would be straightforward to calculate its orbital radius.

Worked example 6.1.3

SATELLITES IN ORBIT

Ganymede is the largest of Jupiter's moons. It has a mass of 1.66×10^{23} kg, an orbital radius of 1.07×10^6 km and an orbital period of 6.18×10^5 s (i.e. approximately 7.15 days).

a Use Kepler's third law to calculate the orbital radius (in km) of Europa, another moon of Jupiter, which has an orbital period of 3.55 days.	
Thinking	**Working**
Note the values for the known satellite. You can work in days and km, as this question only requires a ratio.	Ganymede: $r = 1.07 \times 10^6$ km $T = 7.15$ days
For all satellites orbiting the same mass, $\frac{r^3}{T^2}$ is constant. Calculate this ratio for the satellite whose radius and period is known.	$\frac{r^3}{T^2} = \text{constant}$ $= \frac{(1.07 \times 10^6)^3}{7.15^2}$ $= 2.40 \times 10^{16}$
Use this constant value as the ratio to apply to the satellite in question. Make sure T is in days to match the value used in the previous step.	Europa: $T = 3.55$ days, $r = ?$ $\frac{r^3}{T^2} = \text{constant}$ $\frac{r^3}{3.55^2} = 2.40 \times 10^{16}$
Make r^3 the subject of the equation.	$r^3 = 2.40 \times 10^{16} \times 3.55^2$ $= 3.02 \times 10^{17}$
Solve for r. The unit for r is km as the original ratio was calculated using km.	$r = \sqrt[3]{3.02 \times 10^{17}}$ $= 6.71 \times 10^5$ km Note: since Europa has a shorter period than Ganymede, you could conclude that Europa has a smaller orbital radius than Ganymede.

PHYSICSFILE

Ganymede

As of 2023, Jupiter is orbited by 92 known moons, the biggest of which is Ganymede. Ganymede is very large. It is the biggest of all the moons in the solar system and is even bigger than the planet Mercury.

Ganymede, a natural satellite of Jupiter, is the largest moon in the solar system.

b Use the orbital data for Ganymede to calculate the mass of Jupiter.	
Thinking	**Working**
Note the values for the known satellite. You must work in SI units to find the mass in kg.	Ganymede: $r = 1.07 \times 10^9$ m $T = 6.18 \times 10^5$ s $m = 1.66 \times 10^{23}$ kg $G = 6.67 \times 10^{-11}$ N m^2 kg^{-2} Jupiter: $M = ?$
Select the expressions from the equation for centripetal acceleration that best suit your data. $a = \frac{v^2}{r} = \frac{4\pi^2 r}{T^2} = \frac{GM}{r^2} = g$ Use the third and fourth expressions. These use the given variables r and T, and the constant G, so a solution can be found for M.	$\frac{4\pi^2 r}{T^2} = \frac{GM}{r^2}$
Transpose the expressions to make M the subject.	$M = \frac{4\pi^2 r^3}{GT^2}$
Substitute values and solve the equation.	$M = \frac{4\pi^2(1.07\times10^9)^3}{6.67\times10^{-11}\times(6.18\times10^5)^2}$ $= 1.90\times10^{27}$ kg

c Calculate the orbital speed of Ganymede in km s^{-1}.	
Thinking	**Working**
Note the values you will need to use in the equation $v = \frac{2\pi r}{T}$.	Ganymede: $r = 1.07 \times 10^6$ km $T = 6.18 \times 10^5$ s $v = ?$
Substitute values and solve the equation. The answer will be in km s^{-1} if r is expressed in km.	$v = \frac{2\pi r}{T}$ $= \frac{2\pi\times1.07\times10^6}{6.18\times10^5}$ $= 10.9$ km s^{-1}

Worked example: Try yourself 6.1.3

SATELLITES IN ORBIT

Callisto is the second largest of Jupiter's moons. It is about the same size as the planet Mercury. Callisto has a mass of 1.08×10^{23} kg, an orbital radius of 1.88×10^6 km and an orbital period of 1.44×10^6 s (i.e. 16.7 days).

a Use Kepler's third law to calculate the orbital radius (in km) of Europa, another moon of Jupiter, which has an orbital period of 3.55 days.

b Use the orbital data for Callisto to calculate the mass of Jupiter.

c Calculate the orbital speed of Callisto in km s^{-1}.

6.1 Review

OA ✓✓

SUMMARY

- A normal force, F_N, is the force that a surface exerts on an object that is in contact with it. It acts at right angles to the surface and changes as the force exerted on the surface changes.
- Normal force increases or decreases as the surface you are standing on accelerates up or down, respectively.
- Astronauts in orbit experience zero normal force, because they are in free fall around the Earth.
- A satellite is an object that is in a stable orbit around a larger mass.
- The only force acting on a satellite is the gravitational attraction between it and the larger mass.
- Satellites are in continual free fall. They move with a centripetal acceleration that is equal to the gravitational field at the location of their orbit.
- For a satellite in a circular orbit, its speed, v, is given by:

$$v = \frac{2\pi r}{T}$$

- For a satellite in a circular orbit:

$$a = \frac{v^2}{r} = \frac{4\pi^2 r}{T^2} = \frac{GM}{r^2} = g$$

- The gravitational force acting on a satellite in a circular orbit is given by:

$$F_g = \frac{mv^2}{r} = \frac{4\pi^2 rm}{T^2} = \frac{GMm}{r^2} = mg$$

- For any satellite that is orbiting a mass M, $\frac{r^3}{T^2} = \frac{GM}{4\pi^2} = \text{constant}$, so by knowing one satellite's orbital radius and period, you can calculate the orbital radius of another satellite orbiting M if you know its period (or calculate its period if you know its orbital radius).

KEY QUESTIONS

Knowledge and understanding

1 Determine the force due to gravity acting on a 7.20 kg box at the surface of the Earth where $g = 9.8\,\text{m s}^{-2}$ downwards.

2 A box with $F_g = 220\,\text{N}$ sits at rest on the floor. What is the magnitude of the normal force acting on the box?

3 Calculate the normal force acting on a 55.0 kg child standing in a lift that is accelerating upwards at $2.72\,\text{m s}^{-2}$.

4 Calculate the normal force acting on a 55.0 kg child standing in a lift that is moving upwards at a constant speed of $9.00\,\text{m s}^{-1}$.

5 Which of the following objects experiences the greatest normal force?

A a fly flying horizontally
B a fly walking on a table top
C a show-jumping horse mid jump
D the International Space Station

6 Which of the following is correct?

A The Earth orbits Mars.
B The Sun orbits around the Earth.
C The Moon is a satellite of the Sun.
D The Earth is a satellite of the Sun.

Analysis

7 **a** A geostationary satellite is in orbit above Singapore, which is on the equator. Determine the radius from the centre of the Earth at which it orbits. Assume that $M_{Earth} = 5.98 \times 10^{24}\,\text{kg}$.

b A Navstar GPS satellite has a period of 12 hours. Determine the ratio of the Navstar satellite's orbital radius to the orbital radius of the geostationary satellite in part **a**. From your result, calculate the orbital radius of the Navstar satellite.

8 The strength of the gravitational field where the Optus 10 satellite is in stable orbit around the Earth is $0.22\,\text{N kg}^{-1}$. The mass of the satellite is $3.2 \times 10^3\,\text{kg}$.

a Using only the information given, calculate the magnitude of the acceleration of the satellite as it orbits.

b Calculate the net force acting on the satellite as it orbits.

9 Atlas is one of Saturn's moons. It has an orbital radius of $1.37 \times 10^5\,\text{km}$ and a period of 0.60 days. The largest of Saturn's moons is Titan. It has an orbital radius of $1.20 \times 10^6\,\text{km}$. What is the orbital period of Titan in days?

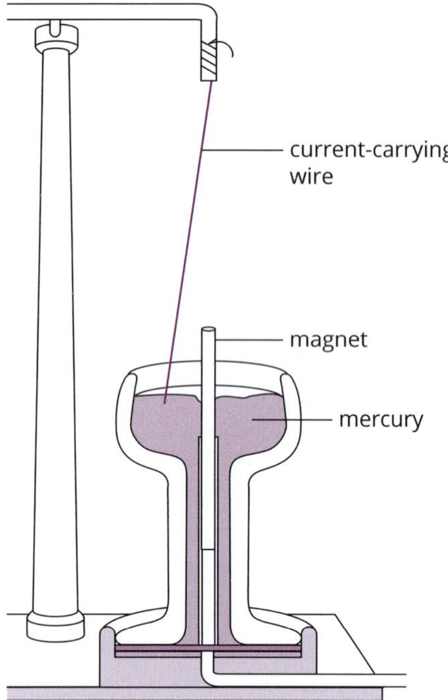

FIGURE 6.2.1 Michael Faraday's electric motor

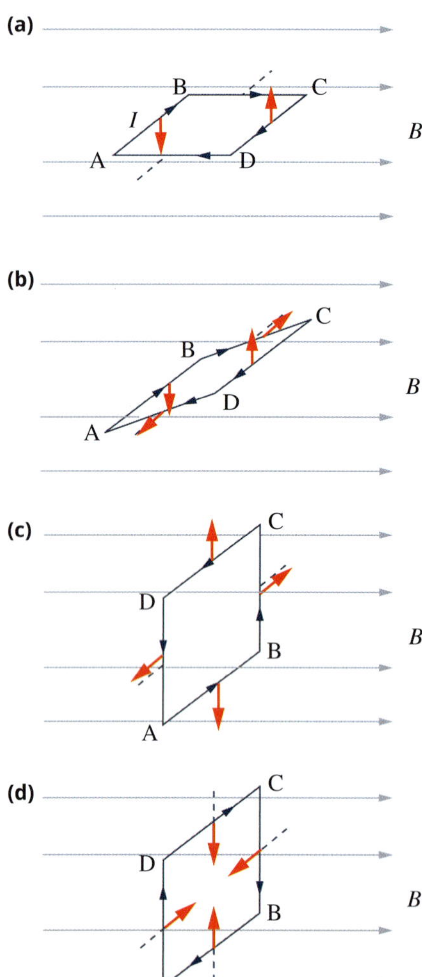

FIGURE 6.2.2 The magnetic force acting on each side of a current-carrying square wire coil in a magnetic field, B, causes the coil to rotate.

6.2 DC motors

Physicists have long been interested in the relationship between electricity and magnetism. It significantly contributes to their understanding of the basic workings of the universe. For the world at large, however, understanding electromagnetism has more practical advantages. It enables, for example, the generation and use of electricity on a large scale. And it led to one of the most useful applications of all time—the electric motor.

PRINCIPLES OF DC MOTORS

The principles on which **direct current** (DC) motors operate have been the same since Michael Faraday built the first one in 1821 (Figure 6.2.1). In Faraday's motor a magnet was mounted vertically in a pool of mercury. A wire carrying a current hung from a support above. (The mercury provided a path for the current.) The magnetic field of the magnet spread outwards from the top of the magnet. Thus there was a component of the field that was perpendicular to the wire. This produced a horizontal force on the wire that kept it rotating around the magnet. Use the right-hand force rule introduced in the previous chapter to convince yourself that if the current flows downwards and the magnetic field points out from the magnet, the wire will rotate clockwise when viewed from above.

In modern DC motors a current-carrying coil of wire in a magnetic field experiences a magnetic force, F, equal to $nIlB$ on two or more of its sides. When any of these variables are increased, the total force will increase. The magnetic field is provided by a number of permanent magnets or by an electromagnet.

Consider a single square-shaped coil of wire, with vertices ABCD, carrying a current, I, in a magnetic field, B (Figure 6.2.2). Initially the wire coil is aligned horizontally in the magnetic field (Figure 6.2.2(a)). Sides AD and BC are parallel to the magnetic field, so no magnetic force will act on them. Sides AB and CD are perpendicular to the field, so both sides will experience a magnetic force. From the right-hand force rule there is a downwards force on AB and an upwards force on CD. These two forces will act together on the coil and, if the coil is free to move, cause it to rotate anticlockwise towards the position shown in Figure 6.2.2(b).

In Figure 6.2.2(b) there will be a magnetic force acting on every side of the coil. Further, the forces acting on sides AD and BC will be equal and opposite in direction. They will tend to stretch the coil outwards but won't affect its rotation. The forces on sides AB and CD will remain vertical and the coil will continue to rotate anticlockwise.

As the coil rotates to the position shown in Figure 6.2.2(c), the forces acting on each side will tend to keep the coil in this position. The force on each side will act outwards from the coil. There are no turning forces at this point, but any further rotation will produce a force in the opposite direction that will cause the coil to rotate clockwise, back to this perpendicular position. For the coil to continue to rotate anticlockwise, the direction of the current needs to be reversed. With the current reversed all the forces are reversed (Figure 6.2.2(d)). Provided that the coil has a little momentum to get it past the perpendicular position shown in Figure 6.2.2(c), it will continue to rotate anticlockwise.

The ability to reverse the direction of the current at the point where the coil is perpendicular to the magnetic field is a key feature of DC motors. This ability is provided by a device called a split ring **commutator**.

Worked example 6.2.1

FORCE ON A COIL

A single square of wire, ABCD, with 5.00 cm sides is free to rotate in a magnetic field, B, of strength 1.00×10^{-4} T. A 1.00 A current is flowing through the coil in the direction indicated by the blue arrows. What is the force (as shown by the red arrows) acting on each side of the coil?

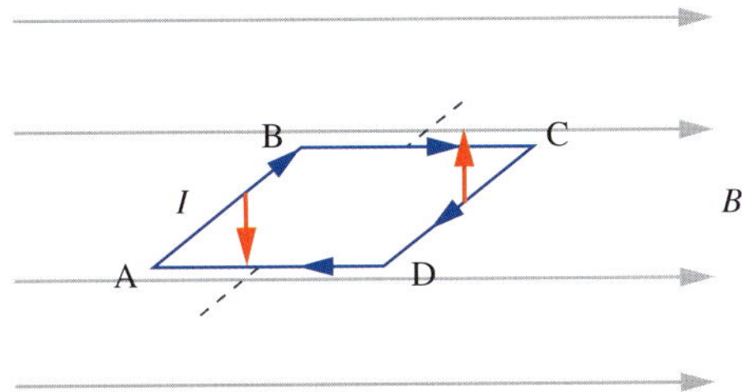

Thinking	Working
Confirm that the coil will experience a force based on the magnetic field and current directions supplied.	The right-hand force rule confirms that a downwards force applies on side AB and an upwards force applies on side CD. The coil will turn anticlockwise. Sides AD and BC lie parallel to the magnetic field and so no force will apply.
Calculate the magnitude of the magnetic force on the sides perpendicular to the field.	For a single wire coil: $F = nIlB$ $= 1 \times 1.00 \times 0.0500 \times 1.00 \times 10^{-4}$ $= 5.00 \times 10^{-6}$ N
State the magnitude and direction of the force on each side.	The force acting on side AB will be equal and opposite to the force acting on side CD. Hence: $F_{AB} = 5.00 \times 10^{-6}$ N downwards $F_{CD} = 5.00 \times 10^{-6}$ N upwards

Worked example: Try yourself 6.2.1

FORCE ON A COIL

A single square of wire with 4.0 cm sides is free to rotate in a magnetic field, B, of strength 1.0×10^{-4} T. A 1.0 A current is flowing through the coil in the direction indicated by the blue arrows. What is the force acting on each side of the coil?

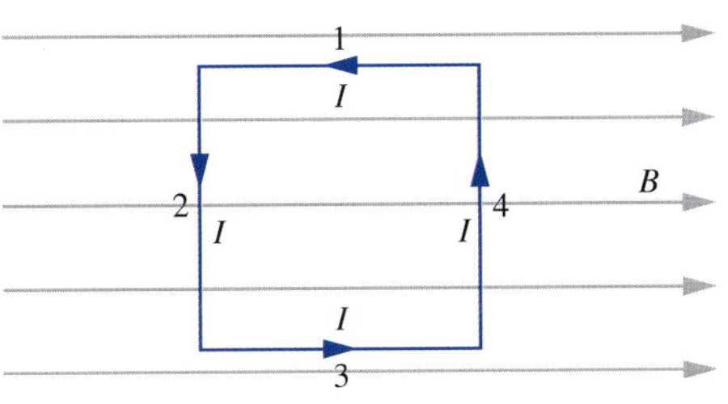

PHYSICSFILE

Michael Faraday

Michael Faraday (1791–1867), shown in the figure below, was an English scientist who worked in the areas of chemistry and physics. He had little formal education. At the age of fourteen he became an apprentice to a London bookbinder. During his apprenticeship he read many of the books that came his way. At the age of 21 he became a laboratory assistant to Sir Humphry Davy, one of the most prominent scientists of the day. Faraday was a gifted experimenter and, after returning from a scientific tour through Europe with Davy, began to be recognised in his own right for the scientific work he was doing. He was admitted to the Royal Society at the age of 32. He is credited with the discoveries of benzene, electromagnetic induction and the basis of the modern electric motor. His contributions to science, and in particular his work in the area of electromagnetism, are recognised through the unit of measurement of capacitance known as the farad. Faraday's work on electromagnetic induction is covered in Chapter 7.

Michael Faraday, from a painting by Thomas Phillips

Torque

The turning force that the coil experiences in an electric motor is referred to as the **torque** on the coil. You may recall from Unit 2 Physics that torque is the turning effect of any force. An example is when you push on a swinging door. To achieve the maximum turning effect the force should be applied at right angles to the door and as far as possible from the side where the door is hinged. This feels easier than when you push close to the hinge, because less force is required to achieve the same torque (Figure 6.2.3).

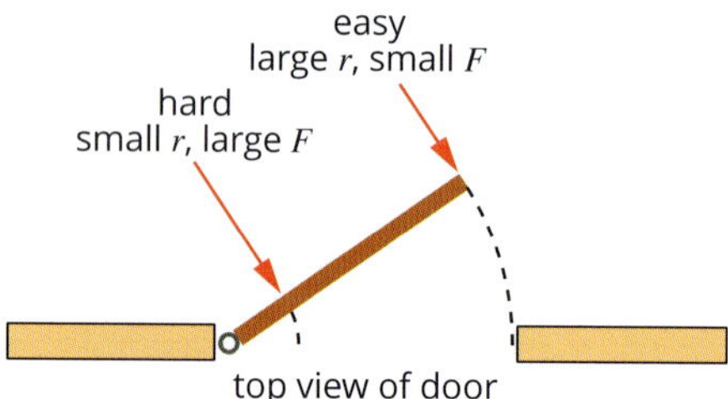

FIGURE 6.2.3 A larger force is required to generate the same torque at a smaller radius. This makes it feel easier to open the door when applying a force further from the hinge.

Torque is defined as follows.

$\tau = r_{\perp} F$

where τ is the torque (N m)

$r_{\perp}$ is the perpendicular distance between the axis of rotation and the point where the force is applied (m)

F is the component of the force perpendicular to the axis of rotation (N)

In the case of a single square or rectangular coil, each of the two sides perpendicular to the magnetic field will experience a force contributing to the total torque of the coil.

MULTI-COIL DC MOTORS

A basic single-coil electric motor with a simple arrangement to reverse the current direction will work, but it won't turn very smoothly. This is because maximum torque will only apply each half turn (i.e. twice for every full rotation). A number of enhancements have been developed to make DC motors the highly practical motive force they are today.

One such enhancement is the split ring commutator. A commutator is usually made from a split ring of copper or another good conducting material on which conducting brushes rub. Each half of the split ring is connected to one end of the coil of wire. (The brushes, which are usually carbon blocks, prevent the wire from becoming tangled as the coil rotates.) The commutator reverses the current at the point where the coil is perpendicular to the magnetic field. This is what keeps the coil rotating (Figure 6.2.4).

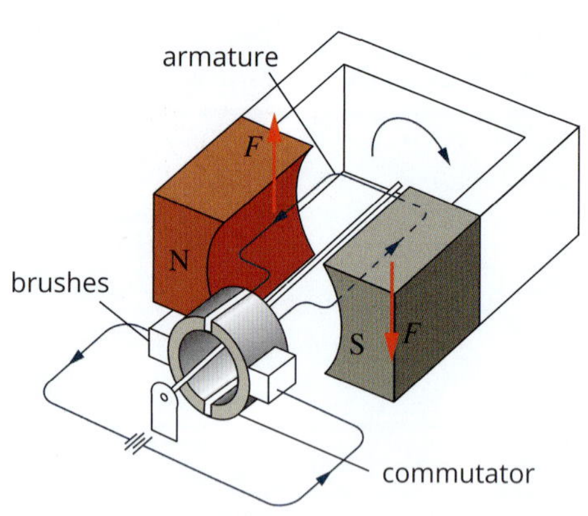

FIGURE 6.2.4 The main parts of a single-coil DC electric motor

Generally motors will have many sets of coils each of many turns, spaced at an angle to each other (Figure 6.2.5). The coils are wound around a soft-iron core to increase the magnetic field that passes through them. The whole arrangement of core and coils is called an **armature** (Figure 6.2.5). In small motors permanent magnets are generally used to provide the magnetic field.

In larger motors electromagnets are used, as they can produce stronger fields. The magnets are usually stationary—unlike the rotating rotor or armature—and are often referred to as the **stator**. The commutator feeds current to the particular coil that is in the best position to provide maximum torque. The total torque will be the sum of the torques on all the individual coils.

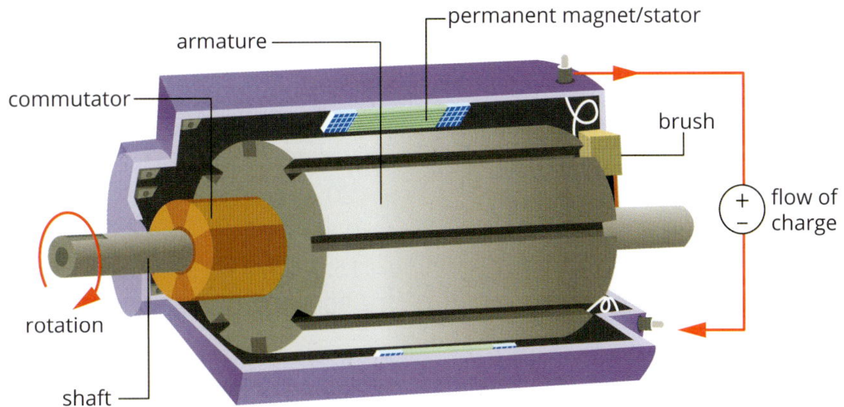

FIGURE 6.2.5 A typical multi-coil DC electric motor. Note that there are many sets of coils wrapped around the shaft offset by an angle from each other. The stator coils produce an electromagnet that provides the magnetic field. The commutator feeds current to the armature coils in the position where maximum torque will be experienced.

Generally speaking, the larger the torque in an electric motor the larger the power produced. Comparing the formulas $\tau = r_{\perp}F$ and $F = nIlB$ we see that the same factors that increase force will also increase torque. (This is in addition to $r_{\perp}$, the perpendicular radius of the coil.) Hence a large torque is achieved by:

- a strong magnetic field
- a large number of turns of wire in each coil
- a high current
- a large area (length and width) of coil.

All these features add to the cost of producing an electric motor, so when designing such a motor, some of these features may need to be compromised to reduce the cost.

Worked example 6.2.2

A SIMPLE DC MOTOR

A DC motor is constructed using a square coil with 20 turns and sides of length 15.00 cm. It is placed in a magnetic field of strength 500 mT. The coil carries a current of 0.25 A in the direction of the blue arrows.

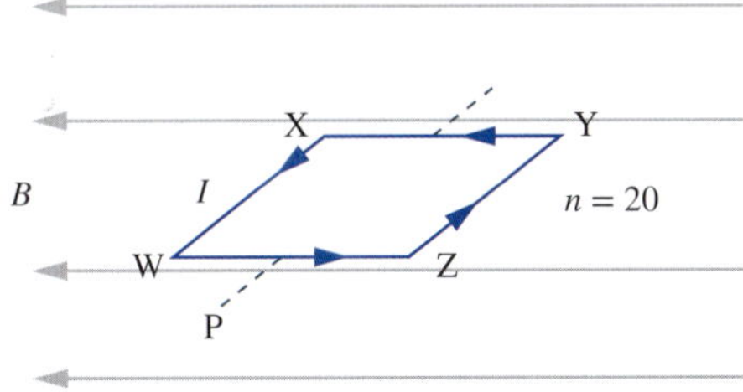

a Viewed from point P, in which direction will the motor rotate: clockwise or anticlockwise?	
Thinking	**Working**
Apply the right-hand force rule to the sides perpendicular to *B* to determine the direction of the magnetic force.	Aligning your thumb with the current's direction on side WX and fingers in the direction of the magnetic field indicates that a downwards force will apply on side WX. Using the right-hand rule similarly on side YZ confirms an upward force here. Viewed from P the coil will rotate anticlockwise.

b Calculate the magnitude of the torque acting on side YZ.	
Thinking	**Working**
Write down the formula for magnetic force and the relevant quantities in SI units.	The magnetic force on a coil is given by $F = nIlB$. $n = 20$ $I = 0.25\,\text{A}$ $l = 15.00\,\text{cm} = 0.1500\,\text{m}$ $B = 500\,\text{mT} = 0.500\,\text{T}$
Substitute the values and calculate the total force.	$F = nIlB$ $= 20 \times 0.25 \times 0.1500 \times 0.500$ $= 0.375\,\text{N}$
Using the side lengths given, compute the torque.	$\tau = r_{\perp}F$ $= 0.15 \times 0.375$ $= 0.056\,\text{Nm}$

c The number of turns in the coil is increased to 40. What will be the new magnitude of the force acting on side YZ?	
Thinking	**Working**
Note that magnetic force depends on the number of coils.	Since $F = nIlB$, if the number of coils, n, is doubled, the total magnetic force also doubles. As a result, the new magnetic force will be $2 \times 0.375 = 0.75\,\text{N}$.

(Note: Quantitative treatment of torque is not required in this course, but is included for completeness in this series of exercises. In practice, exact computation of torque is a very important property of motors.)

Worked example: Try yourself 6.2.2

A SIMPLE DC MOTOR

A DC motor is constructed using a square coil with 25 turns and sides of length 7.50 cm. It is placed in a magnetic field of strength 0.250 T. The coil carries a current of 200 mA in the direction of the blue arrows.

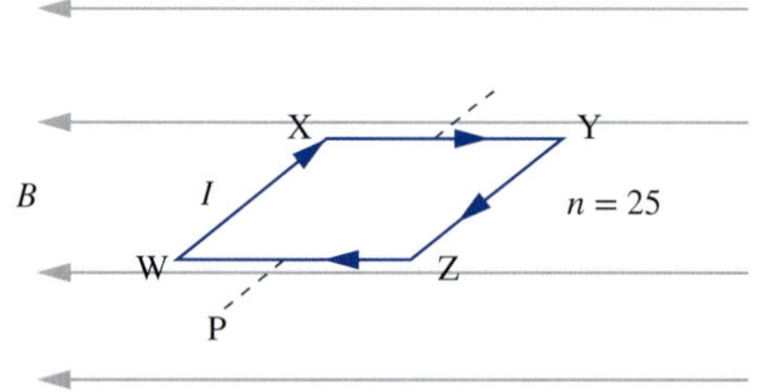

a Viewed from point P, in which direction will the motor rotate: clockwise or anticlockwise?

b Calculate the magnitude of the torque acting on side WX.

c The number of turns in the coil is reduced to 10. What will be the new magnitude of the force acting on side WX?

6.2 Review

SUMMARY

- The magnetic force on a current-carrying wire in a magnetic field is $F = nIlB$.
- There is a torque on a coil of wire carrying a current whenever the current is not parallel to the field. Torque is defined as $\tau = r_{\perp}F$.
- The wire coil of a simple DC motor keeps rotating because the direction of current, and hence torque, is reversed each half-turn by a split ring commutator.
- In the case of a single square or rectangular coil, the total torque applied to the coil will be twice that acting on any one side.
- The armature of a motor consists of a coil or many coils, each of which is fed current by the commutator when it is in a position to deliver maximum torque.
- The total torque will be the sum of the torque on each individual coil.

KEY QUESTIONS

Knowledge and understanding

1 For which of the following situations is torque at a maximum?

A when the force is applied perpendicular to the axis of rotation

B when the force is applied parallel to the axis of rotation

C when the force is applied at a maximum regardless of direction

D when the force applied is zero

2 Explain what would happen to the torque of a square current-carrying wire if the left side and the right side were both placed in a magnetic field of equal but opposite perpendicular direction.

3 Part (a) of the diagram at the right depicts a top view of a single current-carrying coil in an external magnetic field.
Part (b) of the diagram is the corresponding cross-sectional view as seen from point Y. The following data apply:
$B = 0.25\,\text{T}$, PQ = 4.0 cm, PS = QR = 6.0 cm, $I = 3.5\,\text{A}$.

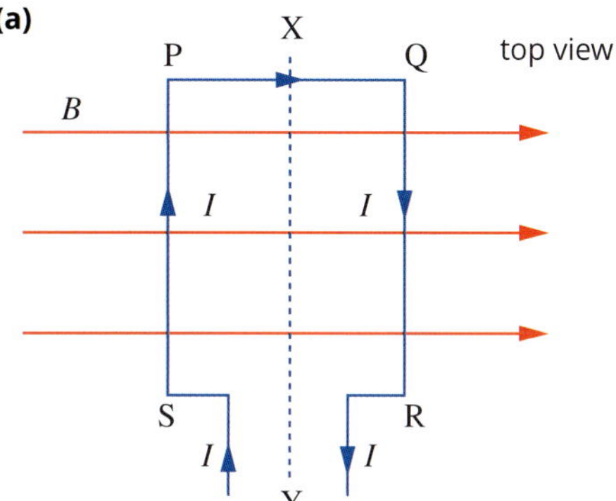

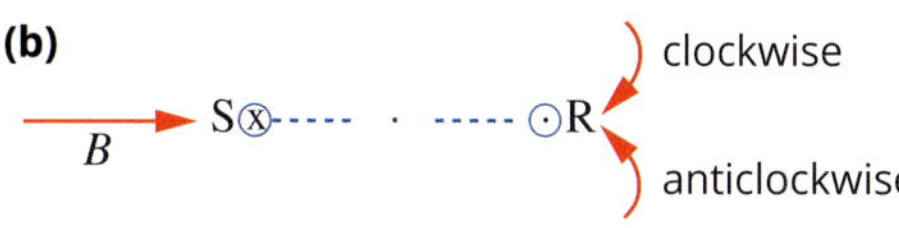

a What is the magnitude and direction of the magnetic force acting on side PS?

b What is the magnitude and direction of the magnetic force acting on side QR?

c What is the magnitude of the force on side PQ?

d The coil is free to rotate about an axis through XY. In what direction, as seen from Y, would the coil rotate?

e Which of the following does not affect the magnitude of the torque acting on the coil?

A the dimensions of the coil

B the magnetic field strength

C the magnitude of the current through the coil

D the direction of the current through the coil

continued over page

6.2 Review *continued*

Analysis

4 A single square coil with sides of length 25 cm has a current of 2.75 A running clockwise through it. The coil is placed in a magnetic field running from left to right so that it is perpendicular to the left and right sides of the coil. However, the magnetic field strength is not uniform: on the left side the magnetic field is 0.75 T and on the right side it is 1.25 T.

- **a** Compute the magnitude of the force acting on each side of the coil.
- **b** Which side of the coil contributes the most to the total net torque?
- **c** What would happen to the net torque if the sides of the coil parallel to the magnetic field were replaced by 50 cm sections?

5 The diagram shows a simplified version of a direct current motor with 15 turns in its coil.

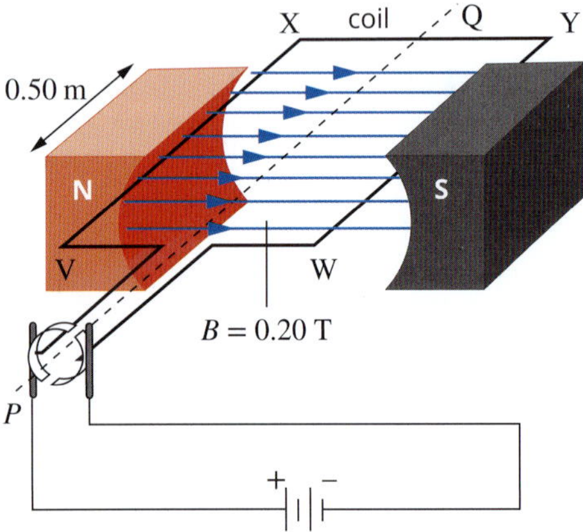

- **a** For the position of the coil shown, calculate the magnitude of the force on side WY when a current of 1.0 A flows through the coil.
- **b** In which direction will the coil begin to rotate? Give your reasoning.
- **c** Explain the role of the split ring commutator in the operation of this DC motor. What would happen to the motor if the commutator were removed and the ends of the coil directly connected to the battery?
- **d** Andrew has an idea to reduce the cost of a DC motor while still maintaining the same force on each side of the coil. He says that if the number of turns in the coil is increased to 30, the length of the wire on each side can be decreased to 0.125 m and the total force will be the same. Is he correct? Explain your reasoning.

6.3 Particle accelerators

Melbourne is the home of the most powerful **synchrotron** in the southern hemisphere (Figure 6.3.1). Shaped like a giant doughnut, it has a circumference of 216 m and produces beams of electromagnetic radiation—from infrared, through visible light, to X-rays.

FIGURE 6.3.1 View of the inside of the Australian Synchrotron

A synchrotron is a type of **particle accelerator**. Electrons and protons are accelerated around its huge evacuated ring to almost the speed of light, reaching energies as high as 3 billion electron volts (3×10^9 eV). The electrons are forced to follow a curved path due to the magnetic field generated by bending magnets. As they accelerate around the ring, the electrons give off bursts of radiation. This synchrotron radiation is channelled along tubes called beamlines and used by researchers in a range of experiments.

This section considers the acceleration of charged particles in uniform electric and magnetic fields, including how electric fields change the speed of particles and magnetic fields change their direction.

A BRIEF HISTORY OF PARTICLE ACCELERATORS

Particle accelerators were originally designed to examine the structure of atoms and molecules. Charged particles—such as electrons, protons or atomic nuclei—were accelerated to speeds often close to the speed of light. The particles travelled through an electric field inside a hollow tube at ultra-high vacuum (where the pressure was comparable to that found in deep space). Strong magnets directed the particles to collide with a target or with another moving particle. Scientists obtained information about the subatomic make-up of the particles fired from the accelerator, or the target samples that were hit, by analysing the results of the collisions.

One of the first particle accelerators was the Van de Graaff accelerator, similar to the Van de Graaff generator (Figure 6.3.2). Developed in the 1930s, it could accelerate charged particles between metal electrodes to energies of about 15 MeV before they collided with a fixed target.

FIGURE 6.3.2 This tandem Van de Graaff accelerator uses two generators to produce beams of charged particles that are accelerated by potential differences of up to 10 million volts.

Currently the world's largest and most powerful particle accelerator is the Large Hadron Collider (LHC). It is located at CERN on the France–Switzerland border. It can produce energies of 13.6 TeV in an underground ring 27 km in circumference. Two sets of particles—usually protons or ions—can be accelerated in opposite directions around its central ring and meet in a collision of massive energy.

ACCELERATING CHARGED PARTICLES

FIGURE 6.3.3 A cathode ray tube

A **cathode ray tube** is a useful type of particle accelerator. A negative terminal, known as a **cathode**, is heated to produce electrons. The electrons are released in a vacuum and accelerate towards a positive terminal, or **anode**. The beam of electrons is collimated (i.e. narrowed) as it passes through a slit, and releases light when it hits a fluorescent screen. The electrons accelerate because there is a potential difference between the cathode and the anode (of approximately 2–3 kV). Older-style televisions (before plasma, LCD and LED screens were invented), certain visual display units and cathode ray oscilloscopes (CROs) all consist of cathode ray tubes (Figure 6.3.3).

A computer monitor, cathode ray oscilloscope and some larger-scale particle accelerators rely on a source of electrons to be accelerated. The device used to provide these electrons is called an **electron gun**.

In an electron gun electrons are boiled off a heated wire filament, or cathode, shown on the left in Figure 6.3.4. They are accelerated from rest across an evacuated chamber towards a positively charged plate, or anode, due to the uniform electric field created between charged plates. Once the electrons pass through a slit in the positive plate, their motion can be further controlled by additional electric and magnetic fields. Focusing magnets are also used to control the width of the beam.

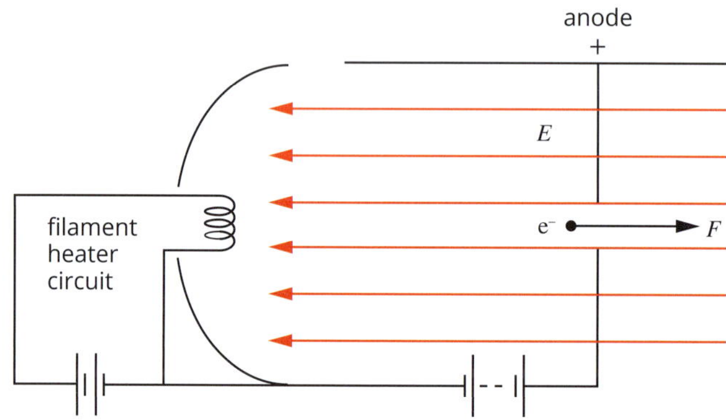

FIGURE 6.3.4 Electron gun assembly

Consider a uniform electric field created between a pair of oppositely charged parallel plates connected to a DC power supply. An electron in the field is attracted to the positive plate and repelled from the negative plate. This electric field is a vector quantity and can be compared in some ways to the Earth's gravitational field. Recall from Chapter 5 that an electric field has units $\mathrm{N\,C^{-1}}$ and its strength is defined as:

$$E = \frac{F}{q}$$

where F is the force (N) experienced by a charged particle due to the field and q is the magnitude of the electric charge of a particle in the field. (In the case of an electron, $q = 1.6 \times 10^{-19}$ C.) Thus a charge will experience a force equal to qE when placed in an electric field.

Recall that the magnitude of an electric field may also be expressed as:

$$E = \frac{V}{d}$$

where d is the distance between the plates (m) and V is the potential difference (V). Combining these two relationships produces an expression for the force on a charge between a pair of parallel charged plates:

$$\frac{F}{q} = \frac{V}{d}$$

$$F = \frac{qV}{d}$$

In addition, calculations of the energy gained by an electron as it is accelerated towards a charged plate by the electric field can be made. The work done in this case is equivalent to:

$$W = qV$$

This equation can be used to calculate the increase in kinetic energy as an electron accelerates from one plate to another:

$$W = \frac{1}{2}mv^2 - \frac{1}{2}mu^2 = qV$$

where v is the final velocity and u is the initial velocity of the electron.

If a charge is accelerated from rest ($u = 0$), then:

$$W = \frac{1}{2}mv^2 = qV$$

$$\frac{1}{2}mv^2 = qV$$

This is often referred to as the electron gun equation.

Worked example 6.3.1

CALCULATING THE SPEED OF ACCELERATED CHARGED PARTICLES

Determine the final speed of a single electron, with a charge of 1.6×10^{-19} C and a mass of 9.1×10^{-31} kg, when it has accelerated from rest across a potential difference of 1.5 kV.

Thinking	Working
Ensure that the variables are in their standard units.	$1.5\,\text{kV} = 1.5 \times 10^3\,\text{V}$
Establish what quantities are known and what is required.	$v = ?$ $q = 1.6 \times 10^{-19}\,\text{C}$ $m = 9.1 \times 10^{-31}\,\text{kg}$ $V = 1.5 \times 10^3\,\text{V}$
Substitute values into the electron gun equation and rearrange it to solve for the speed.	$qV = \frac{1}{2}mv^2$ $1.6 \times 10^{-19} \times 1.5 \times 10^3 = \frac{1}{2} \times 9.1 \times 10^{-31} \times v$ $v = \sqrt{\frac{2 \times 1.6 \times 10^{-19} \times 1.5 \times 10^3}{9.1 \times 10^{-31}}}$ $= 2.3 \times 10^7\,\text{ms}^{-1}$

Worked example: Try yourself 6.3.1

CALCULATING THE SPEED OF ACCELERATED CHARGED PARTICLES

Determine the final speed of a single electron, with a charge of 1.6×10^{-19} C and a mass of 9.1×10^{-31} kg, when it has accelerated from rest across a potential difference of 1.2 kV.

THE EFFECT OF A MAGNETIC FIELD ON A CHARGED PARTICLE

FIGURE 6.3.5 Electron beam being deflected by a magnet

To explore all the forces acting on a beam of charged particles in a particle accelerator (Figure 6.3.5), the effect of a magnetic field on the particles must also be considered. Recall from Chapter 5 that the magnitude of the force, F, on a charge, q, moving with velocity, v, perpendicular to a magnetic field of strength B is given by:

$$F = qvB$$

The direction of the magnetic force exerted on the charge is determined by the right-hand force rule. Note that the direction of current is defined as the direction in which a positive charge would move, so this direction must be reversed to correctly determine the direction of an electron.

If an electron experiences a force of constant magnitude that remains at right angles to its motion, its direction will be changed but not its speed. This is how bending magnets in a particle accelerator alter the path of electrons. As a result the electrons will follow a curved path of radius r (Figure 6.3.6).

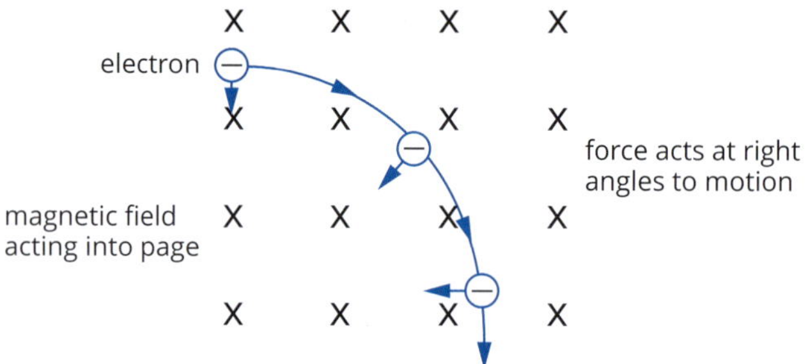

FIGURE 6.3.6 An electron moving in a uniform magnetic field

In this case the net force acting on the charge is:

$$F = ma$$

This is equivalent to the magnetic force on the charge. Thus:

$$qvB = ma$$

The acceleration in this situation is centripetal (i.e. towards the centre of the circular path) and has magnitude:

$$a = \frac{v^2}{r}$$

Substituting this relationship into the previous equation gives:

$$qvB = \frac{mv^2}{r}$$

Rearranging this equation gives an equation for the radius of the path of an electron travelling at right angles to a constant magnetic field.

$$r = \frac{mv}{qB}$$

where r is the radius of the path (m)

m is the mass of the charged particle (9.1×10^{-31} kg for an electron)

v is the speed of the particle (m s^{-1})

q is the charge on the particle (-1.6×10^{-19} C for an electron)

B is the strength of the magnetic field (T)

This relationship can be used to calculate the radius of the path followed by any charged particle travelling at right angles to any uniform magnetic field. The particle could be a low-velocity electron in a cathode ray tube or a high-velocity proton accelerated by the bending magnets inside a powerful particle accelerator.

CASE STUDY

Thomson's $\frac{e}{m}$ experiment

Our knowledge of electrons is relatively recent in science. It was not until 1897 that physicists were able to shed any light on the internal structure of the atom. In that year Joseph John Thomson demonstrated that cathode rays—the rays emanating from a heated cathode—are fundamental constituents of every atom. For the first time the atom was shown to have component particles rather than being indivisible, as was commonly thought at the time. To indicate their importance, cathode rays were renamed electrons.

Thomson's experiment with cathode rays was performed in two stages. At first the forces on a beam of electrons were balanced using an electric field and a magnetic field, as shown by the central dotted line striking the fluorescent screen in Figure 6.3.7. This enabled Thomson to find the speed of the electrons. Then the magnetic field was switched off and the beam was deflected only under the influence of the electric field (see the upper dotted line striking the fluorescent screen in Figure 6.3.7). By measuring the deflection of the beam, Thomson could calculate the charge-to-mass ratio $\left(\frac{e}{m}\right)$ of the cathode rays. (Note that the electron charge here, specifically denoted *e*, is the same as the charge term *q* used throughout this chapter for generic charges.) Thomson repeated the experiment with a variety of cathodes and found that all cathode rays yielded the same ratio: $1 \times 10^{11}\,\text{C kg}^{-1}$. The accepted value today is $1.76 \times 10^{11}\,\text{C kg}^{-1}$.

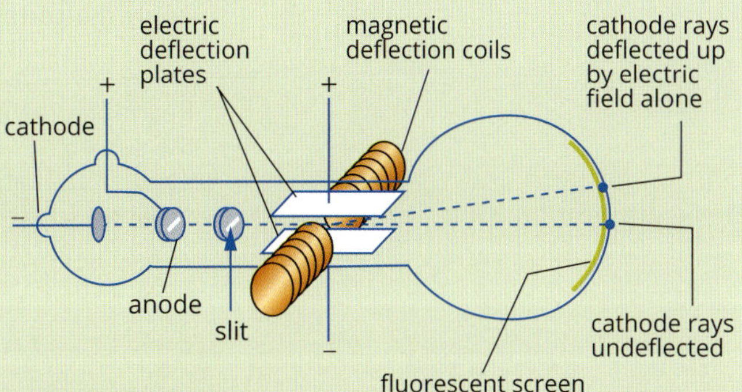

FIGURE 6.3.7 J. J. Thomson's apparatus for finding the charge-to-mass ratio $\left(\frac{e}{m}\right)$ for cathode rays (i.e. electrons)

Worked example 6.3.2

CALCULATING SPEED AND PATH RADIUS OF ACCELERATED CHARGED PARTICLES

An electron gun releases a beam of electrons from its cathode. They are accelerated across a potential difference of 3.2 kV between a pair of charged parallel plates 30 cm apart. Assume that the mass of an electron is 9.1×10^{-31} kg and the magnitude of the charge on an electron is 1.6×10^{-19} C.

a Calculate the strength of the electric field acting on the electron beam.	
Thinking	**Working**
Ensure that the variables are in their standard units.	$3.2\,\text{kV} = 3.2 \times 10^3\,\text{V}$ $30\,\text{cm} = 0.30\,\text{m}$
Apply the correct equation.	$E = \frac{V}{d}$
Solve for *E*.	$E = \frac{3.2 \times 10^3}{0.30}$ $= 1.1 \times 10^4\,\text{V m}^{-1}$

b Calculate the speed of the electrons as they leave the electron gun.	
Thinking	**Working**
Apply the correct equation.	$\frac{1}{2}mv^2 = qV$
Rearrange the equation to make v the subject.	$v = \sqrt{\frac{2qV}{m}}$
Solve for v.	$v = \sqrt{\frac{2 \times 1.6 \times 10^{-19} \times 3.2 \times 10^3}{9.1 \times 10^{-31}}}$ $= 3.4 \times 10^7\,\text{m s}^{-1}$

c The electrons then travel through a uniform magnetic field perpendicular to their motion. If this field is of strength 0.20 T, calculate the radius of the path of the electron beam.	
Thinking	**Working**
Apply the correct equation.	$r = \frac{mv}{qB}$
Solve for r.	$r = \frac{9.1 \times 10^{-31} \times 3.4 \times 10^7}{1.6 \times 10^{-19} \times 0.20}$ $= 9.7 \times 10^{-4}\,\text{m}$

Worked example: Try yourself 6.3.2

CALCULATING SPEED AND PATH RADIUS OF ACCELERATED CHARGED PARTICLES

An electron gun releases a beam of electrons from its cathode. They are accelerated across a potential difference of 2.5 kV between a pair of charged parallel plates 20 cm apart. Assume that the mass of an electron is 9.1×10^{-31} kg and the magnitude of its charge is 1.6×10^{-19} C.

a Calculate the strength of the electric field acting on the electron beam.

b Calculate the speed of the electrons as they leave the electron gun.

c The electrons then travel through a uniform magnetic field perpendicular to their motion. If this field is of strength 0.3 T, calculate the expected radius of the path of the electron beam.

PARTICLE ACCELERATORS AND SCIENTIFIC RESEARCH

To study the basic constituents of matter, physicists accelerate particles—such as electrons and protons—to very high speeds before crashing them into other particles. The by-products from these collisions have revealed a vast array of subatomic particles and led to a better understanding of the fundamental composition of matter. One of the key discoveries of recent particle accelerator experiments is the Higgs boson.

The particles are accelerated by electromagnetic fields, but very long paths are required for the particles to obtain the extremely high speeds needed (speeds very close to the speed of light). To achieve this without the need for tunnels hundreds of kilometres long, particles travel through very strong magnetic fields that cause them to move in a circle. The circumference of the Australian Synchrotron, near Monash University in Melbourne, is 216 m.

CASE STUDY

Accelerating protons vs electrons

Although the focus in this section has been on electrons, all charged particles can be accelerated in an electric field. Advanced modern accelerators, such as the LHC, can accelerate protons to 99.999999% *c*, which is just $3.1\,\text{m s}^{-1}$ slower than the speed of light. Unlike electrons, protons have internal structure. As a result, colliding protons allow physicists to test the predictions made by foundational theories of particle physics. When protons collide at high energies—such as 6.5 TeV at the LHC—they can momentarily generate exotic particles. The behaviour of these exotic particles can then be detected and compared with predictions. In 2012, for example, experiments at the LHC confirmed the existence of the Higgs boson. This particle, first theorised in 1964, is generated by the so-called Higgs field. This is the mechanism that gives all other particles their mass. Since 2012, further properties of the Higgs boson and other particles have been studied at the LHC. However, the accelerator has seen diminishing returns. A new 'high-luminosity' upgrade began in 2018 and is scheduled for completion in 2029.

One of the first major proton accelerators to achieve high energies was the Cosmotron at Brookhaven National Laboratories. In 1953 it accelerated protons to 3.3 GeV.

The mass of protons makes them much harder to keep in an accelerator than electrons. Protons are roughly 1800 times more massive than electrons. Consider the effect of this from the two accelerator equations:

$$\frac{1}{2}mv^2 = qV$$

$$r = \frac{mv}{qB} \text{ (or } qBr = mv\text{)}$$

The magnitude of the charge of the electron and the proton is the same. Hence the first equation implies that an electric field would need to be 1800 times stronger to get a proton to the same speed as an electron. And since the strength of the magnetic field needed to keep a particle travelling at a particular speed around a path of a particular radius is proportional to the mass of that particle (see the second equation), an accelerator would also require magnets that are 1800 times stronger in order to keep the proton in the same size ring as an electron. It therefore requires much more powerful equipment to create a proton accelerator than an electron accelerator. This is necessary for discovering particles, as has been done by the Cosmotron and the LHC. However, for synchrotrons used as light sources for imaging and research—such as the Australian Synchrotron—electron acceleration is far more cost-effective.

The Australian Synchrotron accelerates electrons through an equivalent of 3000 million volts. At this energy, they travel at 99.999 9985% of the speed of light. Because of the relativistic effects that occur at these near-light speeds, their effective mass is about 6000 times that at rest. The electrons are forced onto a circular path by strong magnetic fields. An electron moving through a magnetic field will emit electromagnetic radiation that is proportional to its velocity and the strength of the magnetic field. It is this light that is used for the research projects being conducted with the synchrotron (Figure 6.3.8).

FIGURE 6.3.8 An inside view of part of the Australian Synchrotron

6.3 Review

OA ✓✓

SUMMARY

- Particle accelerators are machines that accelerate charged particles to speeds that can reach close to that of light.
- An electron gun produces a beam of accelerated electrons.
- The work done on a charged particle in an electric field causes a change in the particle's kinetic energy. If the particle is accelerated from rest, the work done is equal to the final kinetic energy:

$$W = qV = \frac{1}{2}mv^2$$

- The magnitude of the force on a charged object in a magnetic field is given by:

$$F = qvB$$

- The right-hand force rule is used to determine the direction of the force on a positive charge moving in a magnetic field. The direction of the force on a negatively charged particle is in the opposite direction.
- The radius of the path of a charged particle travelling at right angles to a uniform magnetic field is given by:

$$r = \frac{mv}{qB}$$

KEY QUESTIONS

Knowledge and understanding

1 How do particle accelerators use electromagnetic fields to change the direction of a charged particle?

A The accelerator is curved around the magnetic field.

B Charged particles are part of the electromagnetic spectrum.

C Charged particles experience a force from the magnetic field that is proportional to the particle's velocity, constantly accelerating the charged particle.

D Charged particles experience a force from the magnetic field that is proportional to the particle's mass, constantly accelerating the charged particle.

2 An electron with a charge of -1.6×10^{-19} C is moving eastwards into a magnetic field of strength 3.2×10^{-4} T acting into the page, as shown below. If the magnitude of the initial velocity is $2.0\,\text{m s}^{-1}$, what is the magnitude and direction of the force it initially experiences as it enters the magnetic field?

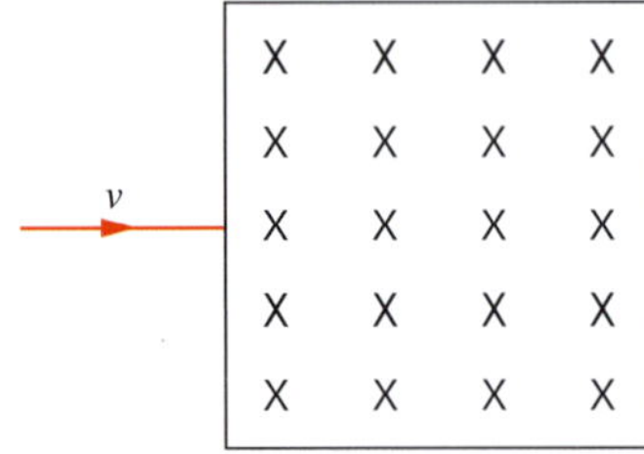

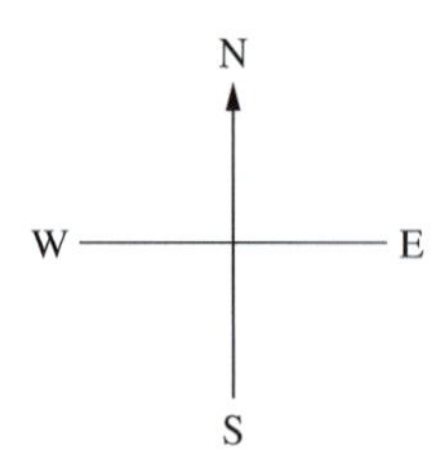

3 Electrons in a cathode ray tube (CRT) are accelerated through a potential difference of 2.5 kV. Calculate the speed at which they hit the screen of the CRT.

4 An electron travelling at $4.1 \times 10^6\,\text{m s}^{-1}$ passes through a magnetic field of strength 9.8×10^{-3} T. The electron moves at right angles to the field.

a Calculate the force exerted on the electron by the magnetic field.

b Given that this force directs the electron in a circular path, calculate the radius of its path.

5 An electron with speed $7.6 \times 10^6\,\text{m s}^{-1}$ travels through a uniform magnetic field and follows a circular path of diameter 9.2×10^{-2} m. Calculate the strength of the magnetic field through which the electron travels.

6 In an experiment similar to that conducted by Thomson for determining the charge-to-mass ratio, $\frac{e}{m}$, of cathode rays, electrons are sent at right angles through a magnetic field of strength 4.0×10^{-4} T. Given that they travel in an arc of radius 10 cm and that $\frac{e}{m} = 1.76 \times 10^{11}$ C kg^{-1}, calculate the speed of the electrons.

7 Modern particle accelerators accelerate protons to very high speeds. Explain, using appropriate theory and relationships, how an accelerator uses magnetic fields to achieve these speeds.

8 An electron beam travelling through a cathode ray tube is subjected simultaneously to electric and magnetic fields. The electrons emerge with no deflection. Given that the potential difference across the two parallel plates is 4.5 kV and the applied magnetic field is 2.3×10^{-3} T, calculate the distance between the plates.

Analysis

9 Sally is designing an electron gun to provide electrons for a small particle accelerator of radius 5.0 m. After the electrons enter the ring they will be subject to a bending magnet of strength 1.0 mT perpendicular to their path. Sally needs to ensure that the electrons are supplied with the correct energy for them to stay in the ring. The electrons are initially accelerated across a pair of charged parallel plates 40 cm apart in an electric field whose strength can be varied.

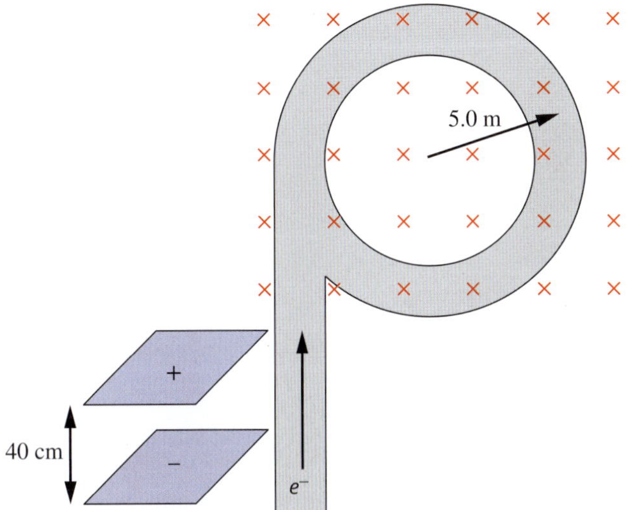

a The electron gun is initially calibrated to accelerate electrons across a potential difference of 45 kV. Calculate the speed of the electrons as they leave the electron gun.

b Sally realises that the electrons are leaving the ring without following the desired path. Is their bending radius too large or too small for the particle accelerator? Should Sally increase or decrease the potential difference of the electron gun in order to compensate?

c Calculate the strength that the electric field must be for the electrons to bend at the required radius.

Chapter review

KEY TERMS

anode
armature
artificial satellite
cathode
cathode ray tube
centripetal acceleration
commutator
direct current
electron gun
free fall
natural satellite
normal force
particle accelerator
satellite
stator
synchrotron
torque

REVIEW QUESTIONS

Knowledge and understanding

1 Calculate the normal force of a 38.0 kg child standing in a lift that is decelerating at $2.95\,m\,s^{-2}$ while travelling upwards.

2 Which description best describes the motion of objects inside a space station orbiting the Earth?

A They float in a reduced gravity environment.
B They fall down very slowly due to the very small gravity.
C They float in a zero gravity environment.
D They fall freely in a reduced gravitational field.

3 Select the statement below that correctly states how a satellite in a stable circular orbit 350 km above the Earth will move.

A It will have an acceleration of $9.8\,m\,s^{-2}$.
B It will have constant velocity.
C It will have zero acceleration.
D It will have acceleration of less than $9.8\,m\,s^{-2}$.

4 Which of the following statements is true about an object that is orbiting the Earth and experiencing no normal force?

A It is in free fall.
B It is in zero gravity.
C It has no mass.
D It is floating.

5 Satellite X, which is in a low-Earth orbit, has an orbital radius of r and period T_X. Satellite Y, which is in a high-Earth orbit, has an orbital radius of $5r$ and period T_Y. In terms of T_X, what is the orbital period of Y?

6 Briefly explain the function of the commutator in an electric motor.

7 Describe the basic set-up of a cathode ray tube and how electrons can be accelerated through it.

8 The diagram below represents an electron being fired at right angles towards a uniform magnetic field acting out of the page.

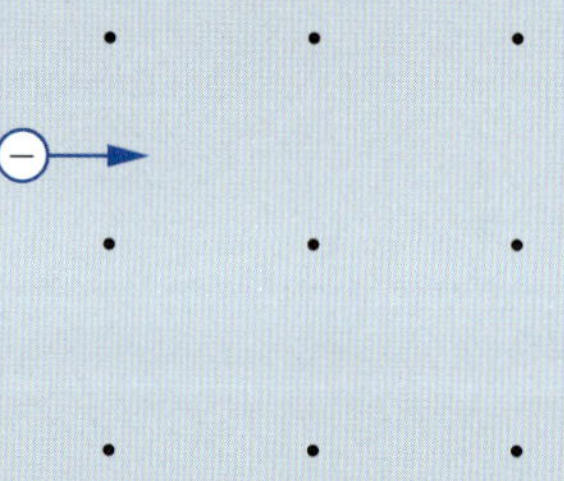

a Copy the diagram and mark on it the path you would expect the electron to follow.
b Which factors would alter the radius of the electron's path?

9 An electron with speed of $4.3 \times 10^6\,m\,s^{-1}$ travels through a uniform magnetic field in a circular path of diameter $8.4 \times 10^{-2}\,m$. Calculate the strength of the magnetic field through which the electron travels.

10 a Calculate the force exerted on an electron travelling at $6.4 \times 10^6\,m\,s^{-1}$ at right angles to a uniform magnetic field of strength $9.1 \times 10^{-3}\,T$.
b Given that this force directs the electron in a circular path, calculate the radius of its path.

Application and analysis

11 Neptune has a mass of $1.02 \times 10^{26}\,kg$. One of its moons, Triton, has a mass of $2.14 \times 10^{22}\,kg$ and an orbital radius of $3.55 \times 10^8\,m$.

a Calculate the orbital acceleration of Triton.
b Calculate the orbital speed of Triton.
c Calculate the orbital period of Triton (in days).

12 Ceres, the first asteroid to be discovered, was found by Giuseppe Piazzi in 1801. It has a mass of $9.38 \times 10^{20}\,kg$ and a radius of 470 km.

a What is the strength of the gravitational field at the surface of Ceres?
b Determine the speed required by a satellite to remain in orbit 10 km above the surface of Ceres.

The following information applies to questions 13–18.

Diagram (a) below shows an end-on view of a current-carrying coil, LM. The coil is free to rotate about a horizontal axis XY. You are looking at the coil from the Y end of the axis. The same coil is seen from the top in diagram (b). Initially sides L and M are horizontal (L_1–M_1). Later they are rotated so that they are vertical (L_2–M_2). The coil is in an external magnetic field of magnitude *B* directed eastwards (at right angles to the axis of the coil). Note the current directions in (a): out of the page in M and into the page in L. Use the up–down, W–E directions in diagram (a) in answering the following questions.

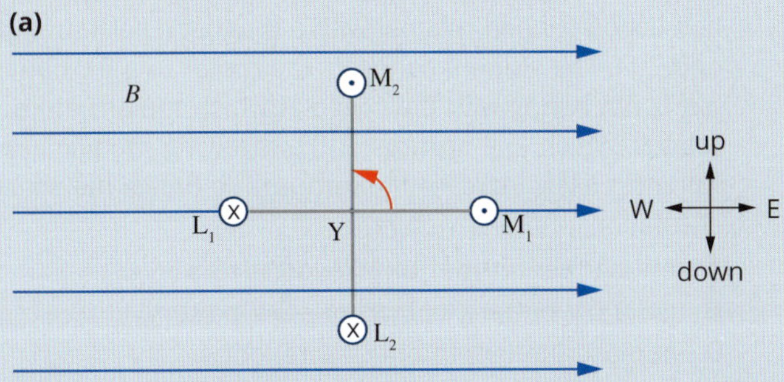

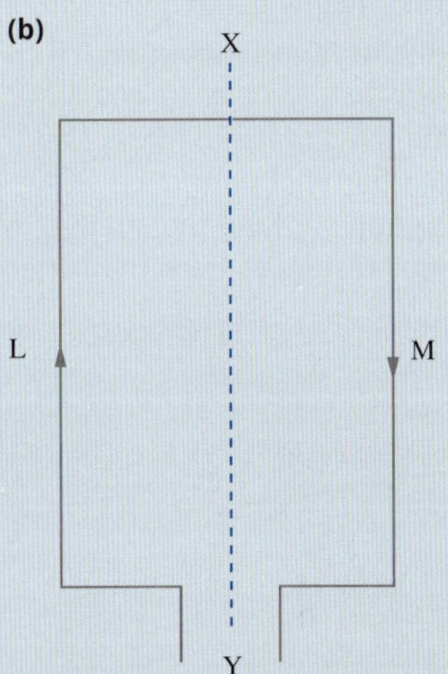

13 When LM is aligned horizontally (L_1–M_1), what is the direction of the magnetic force on the following sides?

a side L

b side M

14 In what direction, as seen from Y, will the coil rotate?

15 Assume that LM is aligned vertically (L_2–M_2).

a What is the direction of the magnetic force on side L?

b What is the direction of the magnetic force on side M?

c What is the magnitude of the torque acting on the coil? Give a reason for your answer.

16 When LM is aligned vertically, which one of the following actions will result in a force acting on the coil that will keep it rotating in an anticlockwise direction? (Assume it still has some momentum when it reaches the vertical position.)

A decrease the current through the loop

B increase the magnetic field strength

C reverse the direction of the current through the loop

17 Seven extra turns are added to the coil, so that there are now eight turns in total. Sides L and M are both 15 cm long. The magnetic field is then fixed at 0.25 T. A current of 30 mA is passed through the coil. When LM is aligned horizontally, what is the magnitude of the magnetic force acting on side M?

18 If the coil were a square, explain how the total torque on the coil depends on the length of the sides.

19 An electron gun emits electrons from a potential difference of 10 kV. Ignore the effects of relativity when answering the following questions.

a Calculate the magnitude of the predicted exit velocity of the electrons.

b After leaving the electron gun, the electrons enter a uniform magnetic field of 1.5 T acting perpendicular to their motion. Calculate the radius of the deflected electron beam.

20 A stream of electrons travels in a straight line through a uniform magnetic field between a pair of charged parallel plates, as shown in the diagram.

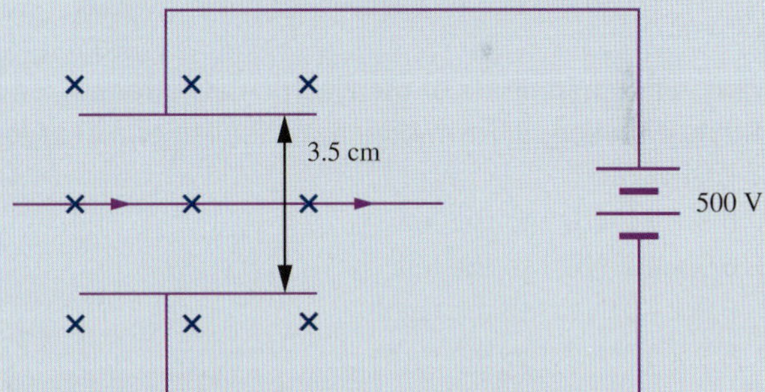

Calculate:

a the strength of the electric field between the plates

b the speed of the electrons, given that the magnetic field is 1.5×10^{-3} T.

21 Electrons in a cathode ray tube are accelerated from a cathode to a fluorescent screen. Calculate the speed at which they hit the screen if the potential difference between the cathode and anode is 4.5 kV.

OA ✓✓

UNIT 3 • Area of Study 2

REVIEW QUESTIONS

How do things move without contact?

Multiple-choice questions

The following information relates to questions 1–4.

Consider an astronaut inside a spacecraft as she travels from launch to a stable orbit. Choose your answers to questions 1 to 4 from the following options:

A normal force
B gravitational force
C neither of the above

1 As the astronaut and spacecraft are launched, which of the above will be greater than usual?

2 As the astronaut and spacecraft are launched, which of the above will remain constant?

3 When the astronaut and spacecraft are in a stable orbit above the Earth, which of the above will apply to the astronaut?

4 If the astronaut and spacecraft ventured into deep space, which of the above would apply to the astronaut?

5 Two charges, one of $+5.9\,\mu C$ and the other of $-7.0\,\mu C$, are placed 40 cm apart in air. What is the magnitude of the force that acts between them?

A 2.0×10^{-4} N
B 7.9×10^{-3} N
C 7.9×10^{-1} N
D 2.0 N

6 What is the magnitude of the electric field strength, and its direction, 3.5 mm directly to the left of a charge of $+9.4\,\mu C$?

A 6.9×10^{3} N C^{-1} to the left
B 6.9×10^{9} N C^{-1} to the left
C 6.9×10^{3} N C^{-1} to the right
D 6.9×10^{9} N C^{-1} to the right

The following information relates to questions 7–9.

The diagram on the left below represents two conductors, both perpendicular to the page and both carrying equal currents into the page (shown by the crosses in the circles). In these questions ignore any contribution from the Earth's magnetic field. Choose the correct options from the arrows A–D and letters E–G.

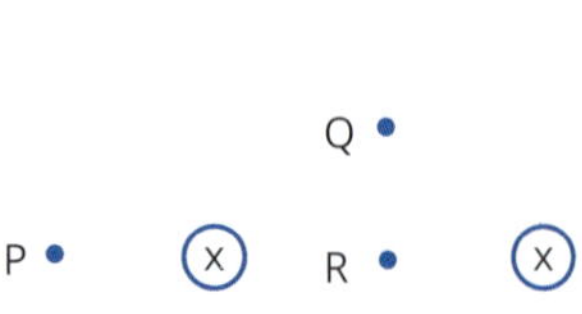

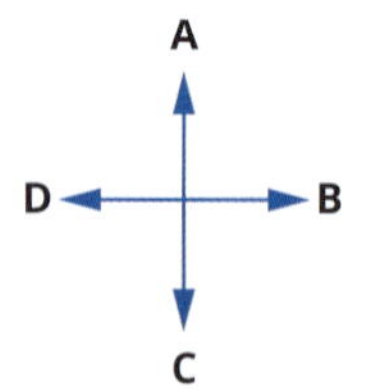

E Out of page
F Into page
G Zero field

7 What is the direction of the magnetic field at point P due to the two currents?

8 What is the direction of the magnetic field at point Q due to the two currents?

9 What is the direction of the magnetic field at point R due to the two currents?

10 A student constructs a simple DC electric motor with an armature consisting of 10 loops of wire and a permanent horseshoe-shaped magnet with a magnetic field strength of 0.20 T. The student connects the motor to a 9 V battery but is not happy with the speed at which the armature rotates. Which one or more of the following modifications will most likely increase the torque, and hence the speed of rotation, of the armature?

A reduce the number of wire loops to 5
B use a 6 V battery instead of a 9 V battery
C increase the length of the wire loops
D use a bar magnet with a magnetic field strength of 0.40 T instead of the horseshoe-shaped magnet

The following information relates to questions 11 and 12.

A small asteroid smashes into the surface of Mars. The crash throws up a lump of Martian rock with mass 20 kg and kinetic energy 40 MJ. A graph showing the strength of the gravitational field at various distances from the surface of Mars is shown below.

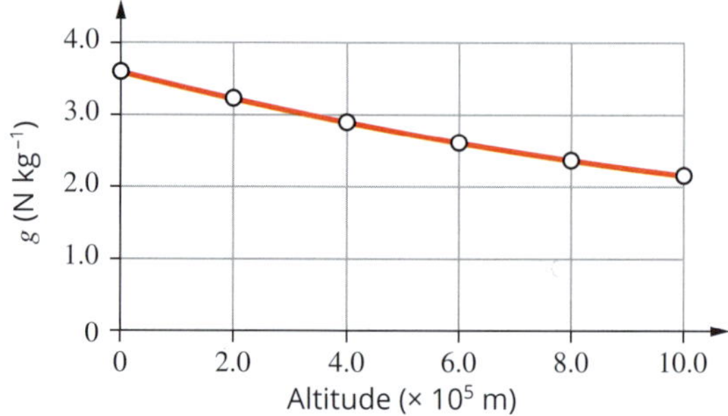

11 What is the gravitational force acting on the Martian rock when it is at an altitude of 300 km?

A 30 N
B 60 N
C 72 N
D 196 N

12 How much kinetic energy (in MJ) does the rock lose as it travels from the surface of Mars to an altitude of 6.0×10^{5} m?

A 1.8 MJ
B 18 MJ
C 36 MJ
D 360 MJ

Short-answer questions

13 Reproduce the diagram below and draw the electric field pattern in the space between and around the two charges.

14 A 10 000 kg spacecraft is drifting directly towards the Earth. When it is at an altitude of 600 km, its speed is 1.5 km s^{-1}. The radius of the Earth is 6400 km. The following graph shows the force due to gravity acting on the spacecraft at various distances from the centre of the Earth.

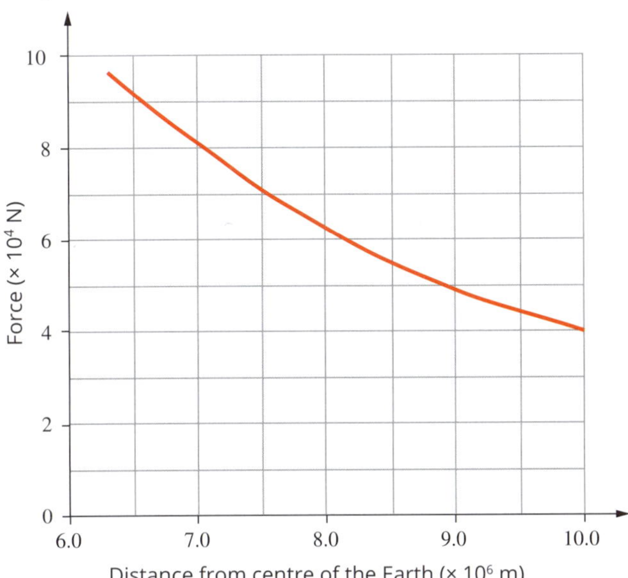

- **a** How much gravitational potential energy does the spacecraft lose as it falls to a distance of 6500 km from the centre of the Earth?
- **b** Determine the speed of the spacecraft at a distance of 6500 km from the centre of the Earth.
- **c** What is the force due to gravity acting on the spacecraft when it is at an altitude of:
  - **i** 3600 km?
  - **ii** 6.0×10^5 m?
- **d** How does the acceleration of the spacecraft change as it moves from an altitude of 600 km to an altitude of 100 km? Include numerical data in your answer.

15 In a Millikan oil drop experiment, students observe that a negatively charged oil drop of mass 1.96×10^{-14} kg is stationary between two horizontal parallel plates. The plates are of 1.6 mm apart with 240 V between them.

- **a** Determine the size and direction of the electric field between the plates.
- **b** Calculate the electric force acting on the oil drop to keep it stationary.
- **c** Calculate the size of the charge that must exist on the oil drop.

16 An electromagnet with a soft-iron core is set up as shown in the diagram below. A small bar magnet with its north end facing towards the electromagnet is placed to the right of it. The switch at S is initially open. The following questions refer to the force between the electromagnet and the bar magnet under various conditions.

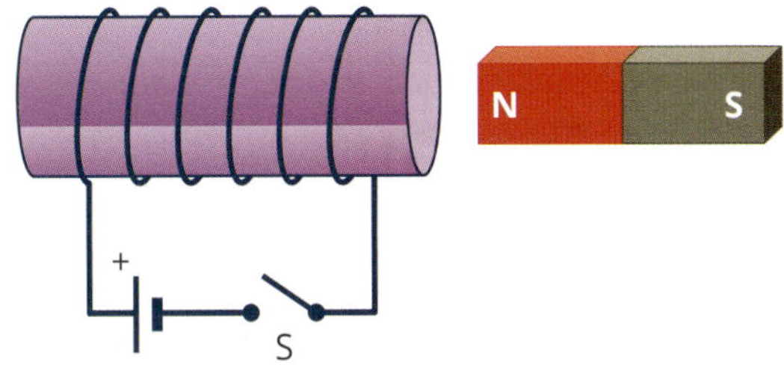

- **a** Describe the force on the bar magnet while the switch is open.
- **b** Describe the force on the bar magnet when the switch is closed and a high current flows.
- **c** The battery is removed and then replaced so that the current flows in the opposite direction. Describe the force on the bar magnet when the switch is closed.

17 The diagram below shows a horizontal east–west electric cable located in a region where the magnetic field of the Earth is horizontal and has a magnitude of 1.0×10^{-5} T. The cable has a mass of 0.05 kg m^{-1}.

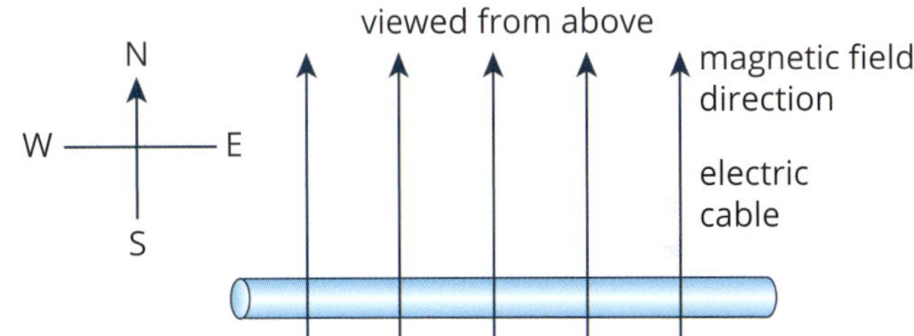

- **a** Calculate the magnitude of the magnetic force on a 1.0 m section of the electric cable if a 100 A current is flowing through it.
- **b** Determine the direction of the current that will produce a force vertically upwards on this cable.
- **c** Magnetic levitation is a phenomenon where objects with mass can be suspended by a magnetic force that is equal in magnitude and opposite in direction to the gravitational force experienced by the object. Calculate the magnitude of current that would be required to produce zero resultant vertical force on a 1.0 m section of this cable.

18 Students set up the electric motor shown below as part of an investigation into DC motors. A square loop of 100 turns and a side length of 20 cm is suspended in a magnetic field, B, of 0.50 T. The direction of the current in the coil is shown by the red arrows (from D anticlockwise to A). Initially, the plane of the loop is parallel to the direction of the field.

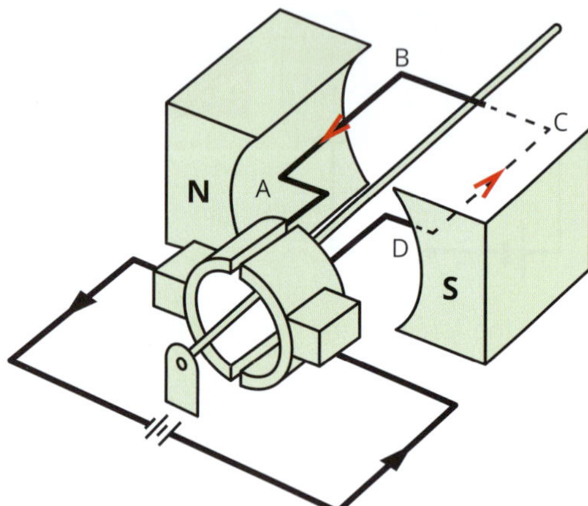

a Determine the direction of the force on sides AB and CD.

b In which position should the coil be in order to experience the greatest amount of torque?

c At one point in the rotation of the coil the torque becomes zero. Explain where this occurs and why the motor continues to rotate.

It is found that there is a force of 40 N on sides AB and CD when they are parallel to the field.

d Calculate the current in each turn of the loop.

e Calculate the torque in sides AB and CD when they are parallel to the field.

19 An electron beam travelling through a cathode ray tube is subjected to an electric field and a magnetic field simultaneously. The electrons emerge with no deflection. The potential difference across the parallel plates in the tube is 3 kV.

a Calculate the final velocity of an electron accelerating from one plate to the other. The mass of an electron is 9.1×10^{-31} kg and its charge is -1.6×10^{-19} C.

The diagram below shows the electron entering a magnetic field of strength 1.2 T.

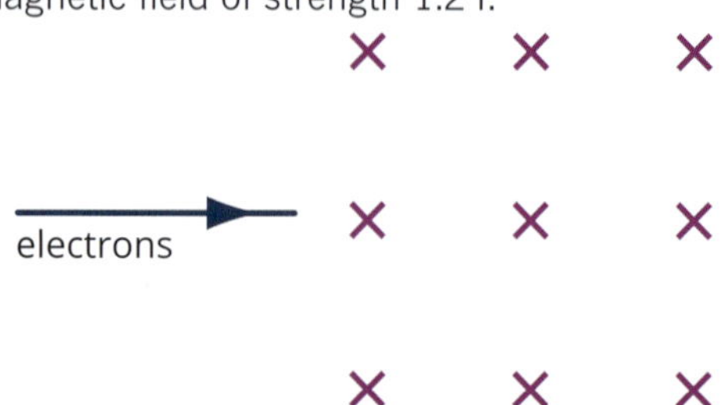

b Reproduce the diagram and show the path the electron will take through the magnetic field.

c Calculate the radius of the path the electron will take through the magnetic field.

20 The Jupiter Icy Moon Explorer (JUICE) is a spaceraft under constrution by the European Space Agency. Its mission is to orbit Jupiter in the same orbital plane as Ganymede, Jupiter's largest moon. JUICE was launched in 2023 for a 2031 arrival at Jupiter.

The following information may be required to answer the questions that follow.

Launch mass of JUICE: 5.3×10^3 kg

Mass of Jupiter: 1.90×10^{27} kg

Orbital radius of Ganymede: 1.07×10^9 m

a Calculate the strength of the gravitational field acting on JUICE when it is in Ganymede's orbit around Jupiter.

b Calculate the gravitational force of Jupiter on JUICE when it is in Ganymede's orbit.

c Calculate the orbital period of JUICE when it is in Ganymede's orbit.

d During JUICE's mission 3000 kg of chemical propellant is used up. What effect will this change in mass have on its orbital period when all the propellant is consumed?

EQ 302

CHAPTER

07 Electromagnetic induction and transmission of electricity

In 1831 Englishman Michael Faraday and American Joseph Henry independently discovered that a changing magnetic flux could induce an electric current in a conductor. This discovery made possible the production of vast quantities of electricity. Today, whether the primary energy source is coal, wind, nuclear fission or falling water, the bulk of the world's production of electrical energy is the result of electromagnetic induction.

In this chapter, electromagnetic induction—the creation of an electric current from a changing magnetic flux—is explored.

Key knowledge

- calculate magnetic flux when the magnetic field is perpendicular to the area, and describe the qualitative effect of differing angles between the area and the field: $\Phi_B = B_{\perp}A$ **7.1**
- investigate and analyse theoretically and practically the generation of electromotive force (EMF) including AC voltage and calculations using induced EMF: $\varepsilon = -N\frac{\Delta\Phi_B}{\Delta t}$, with reference to:
  - rate of change of magnetic flux **7.2**
  - number of loops through which the flux passes **7.2**
  - direction of induced EMF in a coil **7.2**
- explain the production of DC voltage in DC generators and AC voltage in alternators, including the use of split ring commutators and slip rings respectively **7.3**
- describe the production of electricity using photovoltaic cells and the need for an inverter to convert power from DC to AC for use in the home (not including details of semiconductors action or inverter circuitry) **7.4**
- compare sinusoidal AC voltages produced as a result of the uniform rotation of a loop in a constant magnetic field with reference to frequency, period, amplitude, peak-to-peak voltage ($V_{p\text{-}p}$) and peak-to-peak current ($I_{p\text{-}p}$) **7.3**
- compare alternating voltage expressed as the root-mean-square (rms) to a constant DC voltage developing the same power in a resistive component **7.3**
- analyse transformer action with reference to electromagnetic induction for an ideal transformer: $\frac{N_1}{N_2} = \frac{V_1}{V_2} = \frac{I_2}{I_1}$ **7.5**
- analyse the supply of power by considering transmission losses across transmission lines **7.5**

7.1 Inducing an EMF in a magnetic field

PHYSICSFILE

Models and theories

Michael Faraday was not alone in the discovery of electromagnetic induction. The American physicist Joseph Henry (1797–1878) independently discovered the phenomenon a little ahead of Faraday, but Faraday was the first to publish his results. Henry later improved the design of the electromagnet, using a soft iron core wrapped in many turns of wire. He also designed the first reciprocating electric motor. Henry is credited with discovering the phenomenon of self-induction and the unit of inductance is named after him. He also introduced the electric relay, which made it possible to send telegrams. Henry was the first director of the Smithsonian Institution.

While Faraday will be largely referred to throughout this text, it is worth noting that there are often a number of contributors who together build our understanding of key ideas. Joseph Henry's contributions should not be forgotten.

After Oersted's discovery that an electric current produces a magnetic field (see Chapter 5), Michael Faraday was convinced that the reverse should also be true—a magnetic field should be able to produce an electric current.

Faraday wound two coils of wire close together on an iron ring (Figure 7.1.1). The coils were insulated from one another. He then connected a battery to one of the coils, creating a strong electric current through the coil and therefore a strong magnetic field around it. He expected to detect an electric current in the second coil. But no matter how strong he made the magnetic field, he could not detect a current in the other coil.

One day he noticed that a galvanometer (a type of sensitive ammeter) attached to the second coil flickered when he turned on current to the first coil. It gave another flicker, in the opposite direction, when he turned the current off. What he had discovered is that it is not the strength of the magnetic field that could produce an electric current, but a *change* in the magnetic field.

The creation of an electric current in a conductor due to a change in the magnetic field acting on that conductor is now called **electromagnetic induction**.

FIGURE 7.1.1 Michael Faraday's original induction ring. It had two coils of wire—insulated from one another—wound around an iron ring. Passing a current through one coil and turning the current off induced a short-lived voltage in the second coil. The ring is on display in the Faraday Museum at the Royal Institution in London.

CREATING AN ELECTRIC CURRENT

In his attempts to produce an electric current from a magnetic field, Faraday had no success with a constant magnetic field. But he did observe an electric current whenever there was a change in the magnetic field. This current is produced by an induced **EMF**, ε. Although the term EMF is derived from the name 'electromotive force', it is a voltage, or potential difference, rather than a force. Figure 7.1.2 illustrates the induction of EMF, and therefore current, caused by the perpendicular movement of a conducting wire relative to a magnetic field.

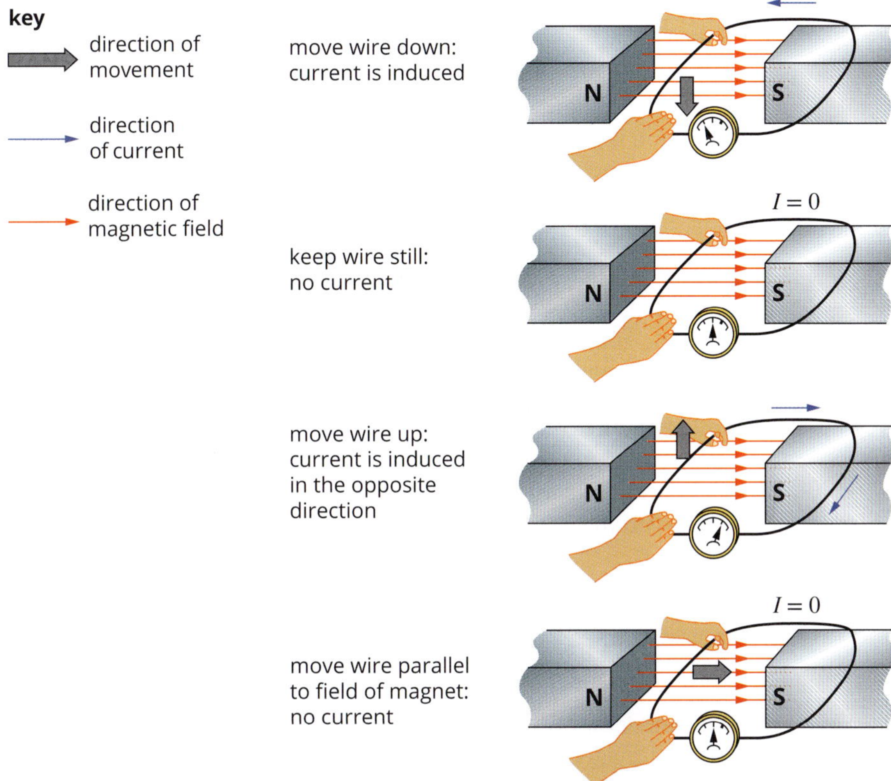

FIGURE 7.1.2 Electromotive force (EMF) is induced in a wire when it moves perpendicular to a magnetic field.

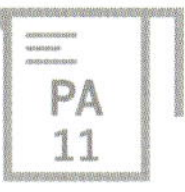

MAGNETIC FLUX

To understand how the change in a magnetic field induces an EMF, which can then induce a current, it is useful to be able to describe the amount of a magnetic field. The amount of a magnetic field is referred to as the **magnetic flux**, a scalar quantity denoted by the symbol Φ_B.

Faraday pictured a magnetic field as consisting of many lines of force. The density of the lines represents the strength of the magnetic field. Magnetic flux can be described as the total number of these lines of force that pass through a particular area. A strong magnetic field acting over a small area can produce the same amount of magnetic flux as a weaker field acting over a larger area (Figure 7.1.3). For this reason, magnetic field strength, B, is also referred to as **magnetic flux density**. In other words, B is proportional to the number of magnetic field lines per unit area perpendicular to the magnetic field. The magnetic flux will be at a maximum when the area examined is perpendicular to the magnetic field, and zero when the area being examined is parallel to the magnetic field.

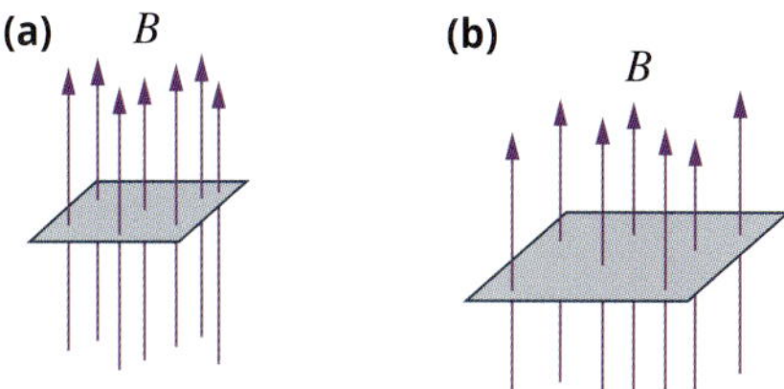

FIGURE 7.1.3 A strong magnetic field acting over a small area (a) can have the same magnetic flux as a weaker magnetic field acting over a larger area (b).

Thus magnetic flux is defined as the product of the strength of the magnetic field, B, and the area of the field perpendicular to the field lines.

$$\Phi_B = B_\perp A$$

where Φ_B is the magnetic flux (Wb)

B is the strength of the magnetic field (T)

A is the area perpendicular to the magnetic field (m^2)

The unit for magnetic flux is the weber (Wb), where $1\,\text{Wb} = 1\,\text{T}\,\text{m}^2$.

The subscript $_\perp$ in the formula indicates that the area referred to is the area that is *perpendicular* to the magnetic field.

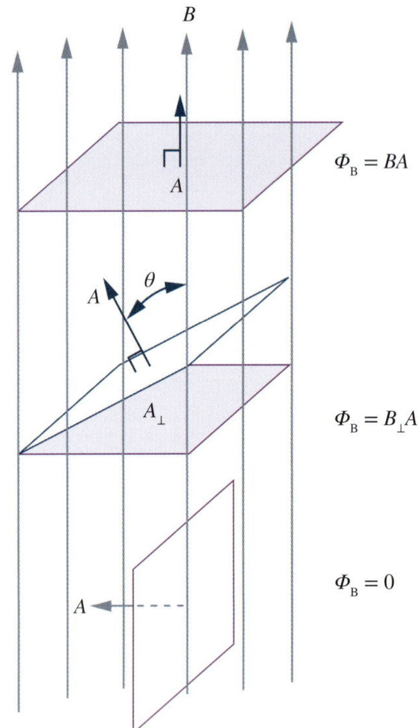

FIGURE 7.1.4 The magnetic flux is the strength of the magnetic field, B, multiplied by the area perpendicular to the magnetic field, given by $A\cos\theta$ and shown as the shaded areas in the diagrams above.

Since it is the area perpendicular to the magnetic field that determines magnetic flux, the angle between the magnetic field and the area through which the field passes affects the amount of magnetic flux. As the angle increases or decreases from 90°, the amount of magnetic flux will decrease. It will reach zero when the area under consideration is parallel to the magnetic field. As Figure 7.1.4 illustrates, the relationship between the amount of magnetic flux and the angle θ to the field can be given as follows.

$$\Phi_B = BA\cos\theta$$

It is important to note that θ is not the angle between the plane of the area and the magnetic field. Rather, it is the angle between a normal to the area and the direction of the magnetic field; hence the use of $\cos\theta$. When the area is at right angles to the magnetic field, the angle θ between the normal and the field is 0° and $\cos 0° = 1$ (top diagram in Figure 7.1.4). When the area is parallel to the magnetic field, the angle θ between the normal and the field is 90° and $\cos 90° = 0$ (bottom diagram in Figure 7.1.4).

Worked example 7.1.1

MAGNETIC FLUX

A student places a horizontal square coil of wire with sides of length 5.0 cm into a uniform vertical magnetic field of 0.10 T. How much magnetic flux passes through the coil?

Thinking	Working
Calculate the area of the coil perpendicular to the magnetic field.	side length: 5.0 cm = 0.05 m area of the square: $0.05^2 = 0.0025\ \text{m}^2$
Calculate the magnetic flux.	$\Phi_B = B_\perp A$ $= 0.1 \times 0.0025$ $= 0.00025\ \text{Wb}$
State the answer in an appropriate form.	$\Phi_B = 2.5 \times 10^{-4}$ Wb or 0.25 mWb

Worked example: Try yourself 7.1.1

MAGNETIC FLUX

A student places a horizontal square coil of wire with sides of length 4.0 cm into a uniform vertical magnetic field of 0.050 T. How much magnetic flux passes through the coil?

Note that in Worked example 7.1.1 an area of 5 cm × 5 cm = 25 cm^2 was considered. This corresponds to 0.0025 m^2 or 25×10^{-4} m^2, so 1 cm^2 = 1×10^{-4} m^2.

You may come across questions asking you to determine how much magnetic flux 'threads' a coil. In this instance, 'threads' means 'passes through'. So 'how much magnetic flux threads the coil' just means 'how much magnetic flux passes through the coil'.

Worked example 7.1.2

MAGNETIC FLUX AT AN ANGLE

A student places a horizontal square coil of wire with sides of length 10.0 cm into a uniform vertical magnetic field of 0.20 T. The square coil is at an angle of 60° to the magnetic field. Would the magnetic flux that passes through the coil be more or less than if it were perpendicular to the magnetic field?

Thinking	Working
All the magnetic field lines pass through the coil when the coil is perpendicular to the magnetic field.	Since the area of the coil is at 60° to the magnetic field, some of the magnetic field lines pass through the coil, but not all.
None of the magnetic field lines pass through the coil when the coil is parallel to the magnetic field.	Therefore the amount of magnetic flux through the coil is less than if it were perpendicular to the magnetic field.

Worked example: Try yourself 7.1.2

MAGNETIC FLUX AT AN ANGLE

A student places a horizontal square coil of wire with sides of length 5.0 cm into a uniform vertical magnetic field of 0.10 T. The plane of the coil is parallel to the magnetic field. How much magnetic flux passes through the coil?

THE INDUCED EMF IN A MOVING CONDUCTOR

When a magnet is moved closer to a conductor, the change in the magnetic field leads to an induced EMF that in turn produces an **induced current** in the conductor. Another way of inducing an EMF is by moving a straight conductor in a magnetic field. It's not hard to understand why this is the case when you realise that charges moving in a magnetic field will experience a force.

Recall that when a charge, q, moves at a velocity, v, perpendicular to a magnetic field, B, the charge experiences a force, F, equal to qvB.

$$F = qvB$$

Consider the direction of positive charges shown in Figure 7.1.5. The force on the positive charges within the conductor when the conductor is moving as shown would be along the conductor and out of the page. The force on the negative charges within the conductor would be along the conductor but into the page.

As the charges in Figure 7.1.5 move apart due to the force they are experiencing from the magnetic field, one end of the conductor will become more positive, the other more negative and a potential difference, ΔV, or EMF will be induced between the ends of the conductor.

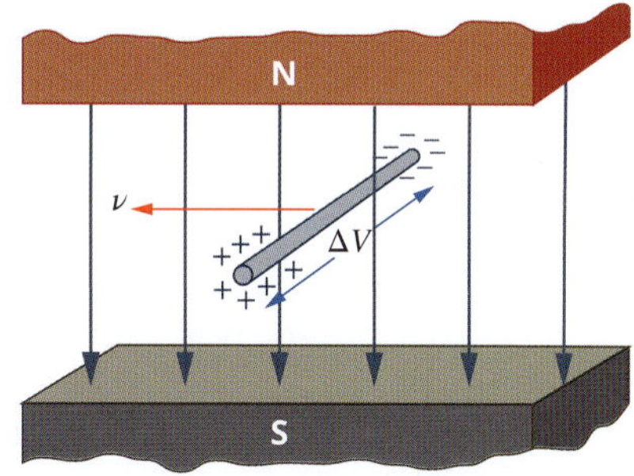

FIGURE 7.1.5 A potential difference, ΔV, will be produced across a straight wire moving perpendicular to a magnetic field. This is due to the force on the charged particles ($F = qvB$). For a straight wire moving to the left in a downwards-pointing magnetic field, the positive charges are moved towards the closer end of the wire, while the negative charges are moved towards the opposite end.

Consider now an electron moving along the conductor. The force from the magnetic field will do work on the electron as it moves along the conductor's length, l. We can calculate the work done as:

$$W = \text{force} \times \text{distance} = qvB \times l$$

Since the EMF is equal to the work done per unit charge:

$$\varepsilon = \frac{W}{q} = \frac{qlvB}{q}$$

which can be simplified as follows.

$\varepsilon = lvB$

where ε is the induced EMF (V)

l is the length of the conductor (m)

v is the speed of the conductor perpendicular to the magnetic field (m s^{-1})

B is the strength of the magnetic field (T)

Worked example 7.1.3

ELECTROMOTIVE FORCE ACROSS AN AIRCRAFT'S WINGS

Can a moving aeroplane develop a dangerous EMF between its wing tips solely from the Earth's magnetic field? Answer by considering an aircraft with a wing span of 64 m flying at a speed of 990 km h^{-1} at right angles to the Earth's magnetic field. Assume that the Earth's magnetic field is 5.0×10^{-5} T.

Thinking	Working
Identify the quantities required in their correct units.	$\varepsilon = ?$ $l = 64\text{ m}$ $B = 5.0 \times 10^{-5}\text{ T}$ $v = 990\text{ km h}^{-1}$ $= \frac{990}{3.6}$ $= 275\text{ m s}^{-1}$
Substitute the values into the appropriate formula and calculate ε.	$\varepsilon = lvB$ $= 64 \times 275 \times 5.0 \times 10^{-5}$ $= 0.88\text{ V}$
State your answer as a response to the question.	$\varepsilon = 0.88\text{ V}$ This is a very small EMF and would not be dangerous.

Worked example: Try yourself 7.1.3

ELECTROMOTIVE FORCE ACROSS AN AIRCRAFT'S WINGS

Consider a fighter jet with a wing span of 25 m, flying at a speed of 2000 km h^{-1} at right angles to the Earth's magnetic field (5.0×10^{-5} T). Will the jet develop a dangerous EMF between its wing tips solely from the Earth's magnetic field?

7.1 Review

OA ✓✓

SUMMARY

- An induced EMF, ε, is produced by a changing magnetic flux in a process called electromagnetic induction.
- Magnetic flux is defined as the product of the strength of the magnetic field, B, and the area of the field perpendicular to the field lines, i.e. $\Phi = B_{\perp}A$.
- The amount of magnetic flux varies with the angle of the field to the area under investigation. It is a maximum when the area is perpendicular (90°) and zero when the area is parallel to the field. The relevant equation is $\Phi_B = BA\cos\theta$.
- The unit for magnetic flux is the weber (Wb). $1\,\text{Wb} = 1\,\text{T m}^2$.
- The induced EMF in a straight conductor moving in a magnetic field, B, is given by $\varepsilon = lvB$ if the conductor is moving perpendicular to the field.

KEY QUESTIONS

Knowledge and understanding

1 Predict whether an EMF will be induced in a long, straight conductor in the following situations.
   a A magnet is brought near the conductor.
   b A conductor is rotated within a magnetic field.
   c A magnet is held stationary alongside the conductor.
   d The conductor is brought into a magnetic field.

2 A student places an 8.0 cm square coil of wire parallel to a uniform vertical magnetic field of 0.025 T. How much magnetic flux passes through the coil?

3 Calculate the magnetic flux through a square coil of wire with sides of length 10 cm perpendicular to a uniform vertical magnetic field of 0.20 T.

4 A rod 15 cm long is moving at a speed of $0.11\,\text{m s}^{-1}$ perpendicular to a magnetic field. If the strength of the magnetic field is 0.60 T, calculate the induced EMF in the rod.

5 Describe two ways to increase the magnetic flux through a coil.

Analysis

6 Suppose that a coil is in a magnetic field and the plane of the coil is perpendicular to the field. The magnetic flux is defined as $\Phi_B = BA\cos\theta$. Plot the magnetic flux as the coil is rotated through one complete revolution.

7 A square loop of wire with sides of length 6.0 cm is in a region of uniform magnetic field of 2.0×10^{-3} T north, as shown below. The loop is free to rotate about the vertical axis XY. When the loop is in its initial position, its plane is perpendicular to the direction of the magnetic field.

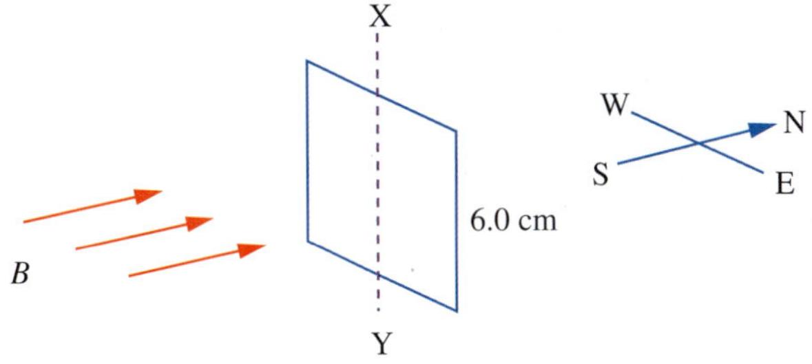

   a Calculate the magnetic flux passing through the loop.
   b Describe what happens to the amount of magnetic flux passing through the loop as the loop is rotated through one complete revolution.

7.2 Induced EMF from a changing magnetic flux

Faraday's early experiments were mostly concerned with investigating electromagnetic induction in coils, or multiple loops, of wire. He found that when a magnet is quickly moved into a coil, an EMF is induced and a current is produced in the coil. When the magnet is removed, current is also produced in the coil but in the opposite direction. Faraday also found that if the magnet is held steady and the coil is moved in such a way that changes the magnetic flux, then again an EMF is induced and an electric current is produced. In other words, it doesn't matter if it is the coil or the magnet that moves. It is the change in magnetic flux, however it is created, that induces the EMF. This discovery led Faraday to his law of induction, which is the subject of this section.

FACTORS AFFECTING INDUCED EMF

Faraday quantitatively investigated the factors affecting the size of the EMF induced in a coil. He noted that an EMF could be induced by a change in the magnetic field. (A simple example of this is to witness the EMF induced when a magnet is brought towards or away from a wire coil.) He discovered that the greater the change in the magnetic field, the greater the EMF induced.

However, he found that it was not only a change in the strength of the magnetic field that induced an EMF. An EMF could also be induced by changing the area perpendicular to the magnetic field through which the magnetic field lines pass. (An example of this is to witness the EMF induced when a wire coil is rotated in the presence of a fixed magnetic field.) Once again, the greater the change, the greater the EMF that is induced.

As we have seen, these discoveries suggested that a necessary condition for inducing EMF was to have a changing magnetic flux, $\Delta\Phi_B$. (Note that Δ is the Greek letter 'delta' and is used to represent change. So $\Delta\Phi_B$ is the *change* in the magnetic flux.)

Faraday also discovered that the faster the change in magnetic flux, the greater the induced EMF.

What Faraday had discovered can be seen in the oscilloscope trace of a magnet falling through a coil (Figure 7.2.1). The magnet is accelerated by gravity as it drops through the coil: its speed as it leaves the coil is greater than its speed as it enters the coil. Hence the peak EMF induced when the magnet first enters the coil is less than the peak EMF induced when the magnet leaves the coil.

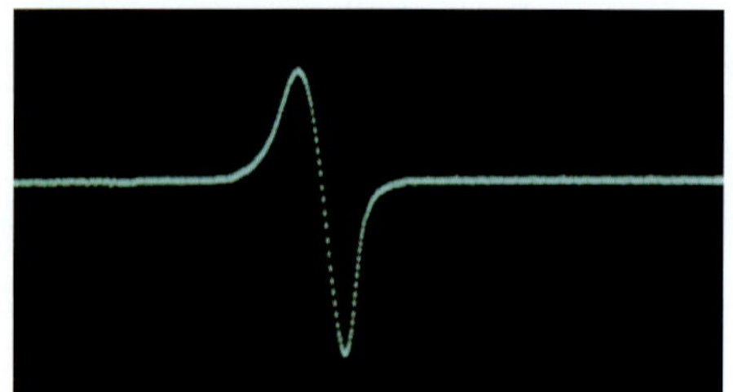

FIGURE 7.2.1 An oscilloscope trace showing the EMF induced as a magnet is dropped through a coil. Notice the spike as the magnet enters the coil and the other spike as the magnet passes out of the coil.

FARADAY'S LAW OF INDUCTION

Faraday's investigations led him to conclude that the average EMF induced in a conducting loop in which there is a changing magnetic flux is proportional to the rate of change of flux. If the magnetic flux through N turns (or loops) of a coil changes from Φ_1 to Φ_2 during time t, then the average induced EMF during this time will be:

$$\varepsilon = -N\frac{(\Phi_2 - \Phi_1)}{t}$$

If the change in magnetic flux, $\Phi_2 - \Phi_1$, is represented as $\Delta\Phi_B$, then the average induced EMF can be written as follows.

$$\varepsilon = -N\frac{\Delta\Phi_B}{\Delta t}$$

where ε is the average induced EMF (V)

N is the number of turns in the coil

Φ_B is the magnetic flux (Wb)

t is the time over which the change in flux occurs (s)

This is now known as **Faraday's law** of induction and it is one of the basic laws of electromagnetism.

The negative sign at the beginning of the equation is a reminder of the direction of the induced EMF (discussed in the next section). Normally you will be concerned only with the *magnitude* of the EMF. Thus you don't need to consider the negative sign or any negative quantities in a calculation.

If the ends of the coil are connected to an external circuit, then a current, I, will be produced. The magnitude of the current is found using Ohm's law as follows.

$$I = \frac{\Delta V}{R}$$

where R is the resistance (Ω)
ΔV is the EMF of the coil (V)

A coil not connected to a circuit will act like a battery not connected to a circuit. There will still be an induced EMF, but no current will be produced.

Worked example 7.2.1

INDUCED EMF IN A COIL

A student winds a coil of area 40 cm^2 with 20 turns. He places it horizontally in a vertical uniform magnetic field of 0.10 T.

a Calculate the magnetic flux perpendicular to the coil.	
Thinking	**Working**
Identify the quantities required to calculate the magnetic flux through the coil. Convert them to SI units where required.	$\Phi_B = B_\perp A$ $B = 0.10\,T$ $A = 40\,cm^2 = 40 \times 10^{-4}\,m^2$
Calculate the magnetic flux. Give your answer in the appropriate units.	$\Phi_B = B_\perp A$ $= 0.10 \times 40 \times 10^{-4}$ $= 4.0 \times 10^{-4}\,Wb$

b Calculate the magnitude of the average induced EMF in the coil when the coil is removed from the magnetic field in 0.5 s.	
Identify the quantities required for determining the induced EMF.	$N = 20$ turns $\Delta\Phi_B = \Phi_2 - \Phi_1$ $= 0 - 4.0 \times 10^{-4}$ $= 4.0 \times 10^{-4}\,Wb$ $\Delta t = 0.5\,s$
Calculate the magnitude of the average induced EMF, ignoring the negative sign that indicates the direction.	$\varepsilon = N\frac{\Delta\Phi_B}{\Delta t}$ $= 20 \times \frac{4.0 \times 10^{-4}}{0.5}$ $= 0.016\,V$

Worked example: Try yourself 7.2.1

INDUCED EMF IN A COIL

A student winds a coil of area 50 cm^2 with 10 turns. She places it horizontally in a vertical uniform magnetic field of 0.10 T.

a Calculate the magnetic flux perpendicular to the coil.

b Calculate the magnitude of the average induced EMF in the coil when the coil is removed from the magnetic field in 1.0 s.

PHYSICSFILE

Inductive-loop traffic detectors

Have you ever wondered why the traffic lights at an intersection can change when vehicles arrive? At some intersections, an induction loop is buried just beneath the road's surface. (You may have noticed a rectangular patch on the road's surface near a traffic light. The induction loop is buried directly beneath the patch.) The loop carries an alternating current and this creates a magnetic field. When a vehicle passes over or stops on the loop, the presence of the vehicle's body changes the magnetic field and a traffic sensor circuit detects the change.

Worked example 7.2.2

NUMBER OF TURNS IN A COIL

A coil of cross-sectional area $1.0 \times 10^{-3}\,m^2$ experiences a change in the strength of a magnetic field from 0 to 0.20 T in 0.50 s. If the magnitude of the average induced EMF is 0.10 V, how many turns must be on the coil?

Thinking	Working
Identify the quantities required to calculate the magnetic flux through the coil when in the presence of the magnetic field.	$\Phi_B = B_{\perp}A$ $B = 0.20\,T$ $A = 1.0 \times 10^{-3}\,m^2$
Calculate the magnetic flux when the coil is in the presence of the magnetic field.	$\Phi_B = B_{\perp}A$ $= 0.20 \times 1.0 \times 10^{-3}$ $= 2.0 \times 10^{-4}\,Wb$
From the question, identify the quantities required by Faraday's law.	$N = ?$ $\Delta\Phi_B = \Phi_2 - \Phi_1$ $= 2.0 \times 10^{-4} - 0$ $= 2.0 \times 10^{-4}\,Wb$ $\Delta t = 0.50\,s$ $\varepsilon = 0.10\,V$
Rearrange Faraday's law and solve for the number of turns in the coil. Ignore the negative sign.	$\varepsilon = N\frac{\Delta\Phi_B}{\Delta t}$ $N = \frac{\varepsilon\,\Delta t}{\Delta\Phi_B}$ $= \frac{0.10 \times 0.50}{2.0 \times 10^{-4}}$ $= 250$ turns

PA 12

Worked example: Try yourself 7.2.2

NUMBER OF TURNS IN A COIL

A coil of cross-sectional area $2.0 \times 10^{-3}\,m^2$ experiences a change in the strength of a magnetic field from 0 to 0.20 T in 1.0 s. If the magnitude of the average induced EMF is 0.40 V, how many turns must be in the coil?

LENZ'S LAW

Lenz's law explains the direction of the current created by the induced EMF resulting from a changing magnetic flux. Understanding how a current is produced by an induced EMF, and being able to determine its direction, has resulted in the invention of a vast array of devices that have transformed modern society, from metal detectors (Figure 7.2.2) to electrical generators.

FIGURE 7.2.2 A diver using a metal detector, first invented by Heinrich Lenz. If the coil of the detector passes over a metal object, an EMF is induced. This creates a detectable change in current.

The direction of an induced EMF

Lenz's law states that an induced EMF always produces a current with a magnetic field that opposes the change in flux that caused the induced EMF.

Figure 7.2.3 illustrates **Lenz's law** by means of a magnet and a single loop of wire. When the north end of a magnet is moved towards the loop from the right (Figure 7.2.3(a)), the magnetic flux from right to left through the loop increases. The induced EMF produces a current that is anticlockwise around the loop when viewed from the right. The magnetic field created by this current, shown by the little circles around the wire, is directed from left to right through the loop. It opposes the magnetic field of the approaching magnet, as Lenz's law states.

When the magnet is moved away from the loop (Figure 7.2.3(b)), the magnetic flux from right to left through the loop decreases. The induced EMF produces a clockwise current when viewed from the right. This creates a magnetic field that is directed from right to left through the loop. This field is in the same direction as the original magnetic field of the retreating magnet. However, note that it is opposing the change in the magnet's flux through the loop by attempting to replace the declining flux.

When the magnet is held stationary (Figure 7.2.3(c)), there is no change in flux to oppose, and so no current is induced.

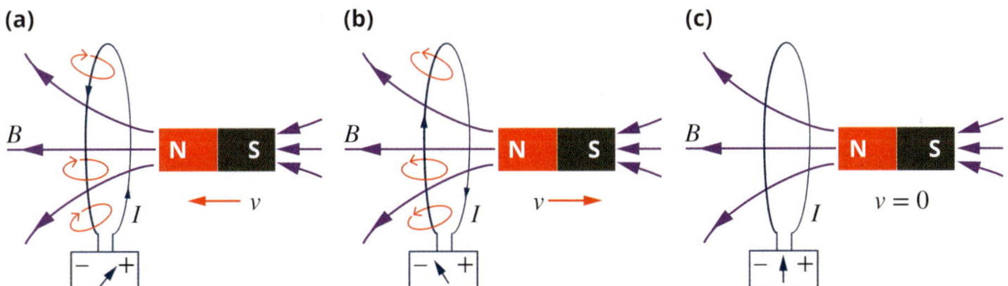

FIGURE 7.2.3 (a) The north end of a magnet is moved towards a loop from right to left, inducing an anticlockwise current. (b) Moving the north end of the magnet away from the loop from left to right induces a current in a clockwise direction. (c) Holding the magnet still creates no change in flux and hence no induced current.

It is worth noting that Lenz's law is a necessary consequence of the law of conservation of energy. If Lenz's law were not true then the new magnetic field created by the changing flux would reinforce the original magnetic force of the magnet. This would add energy to the universe in violation of the law of conservation of energy.

The right-hand grip rule and induced current direction

The right-hand grip rule can be used to find the direction of the induced current. Keep in mind that the current must create a magnetic field that opposes the change in flux due to the relative motion of the magnet and conductor. Point your fingers through the loop in the direction of the field that is opposing the change and your thumb will then indicate the direction of the conventional current (Figure 7.2.4).

To determine the induced current direction according to Lenz's law, the following three questions must be considered.

1. What change in the magnetic flux is occurring?
2. What will oppose the change?
3. In what direction must the current be induced in order to provide this opposition?

These steps will be examined in Worked examples 7.2.3 and 7.2.4 over the page.

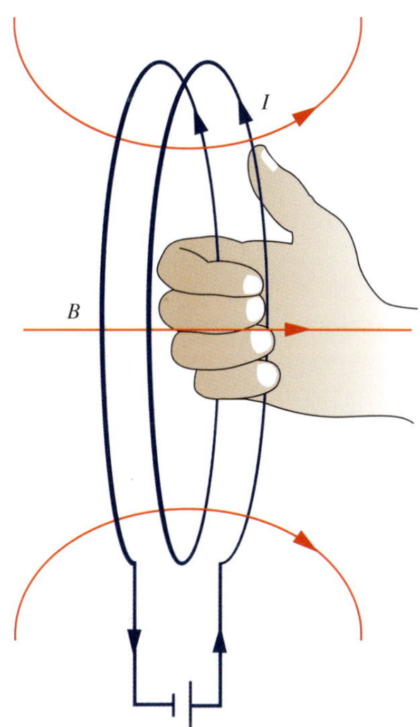

FIGURE 7.2.4 The right-hand grip rule can be used to determine the direction of a magnetic field from a current and vice versa. Your thumb points in the direction of the conventional current in the wire and your curled fingers indicate the direction of the magnetic field through the coil.

Worked example 7.2.3

INDUCED CURRENT IN A COIL FROM A PERMANENT MAGNET

The south pole of a magnet is moved up towards a horizontal coil directly above it. In which direction will the current be induced in the coil?

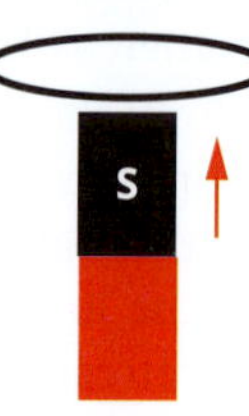

Thinking	Working
Consider the direction of the change in magnetic flux.	The direction of the magnetic field from the magnet will be downwards, towards the south pole. The downwards flux from the magnet will increase as the magnet is brought closer to the coil. So the change in flux is increasing downwards.
What will oppose the change in flux?	The induced magnetic field that opposes the change would act upwards.
Determine the direction of the induced current required to oppose the change.	In order to oppose the change, the current direction would be anticlockwise when viewed from above (as per the right-hand grip rule).

Worked example: Try yourself 7.2.3

INDUCED CURRENT IN A COIL FROM A PERMANENT MAGNET

The south pole of a magnet is moved downwards away from a horizontal coil held directly above it. In which direction will the current be induced in the coil?

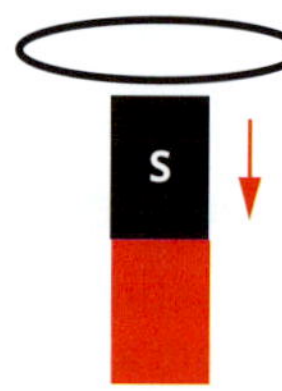

Worked example 7.2.4

INDUCED CURRENT IN A COIL FROM AN ELECTROMAGNET

Instead of using a permanent magnet to change the flux in the loop, an electromagnet could be used (shown on the right in the diagram below). What is the direction of the current induced in the solenoid when the electromagnet is:

a switched on

b left on

c switched off.

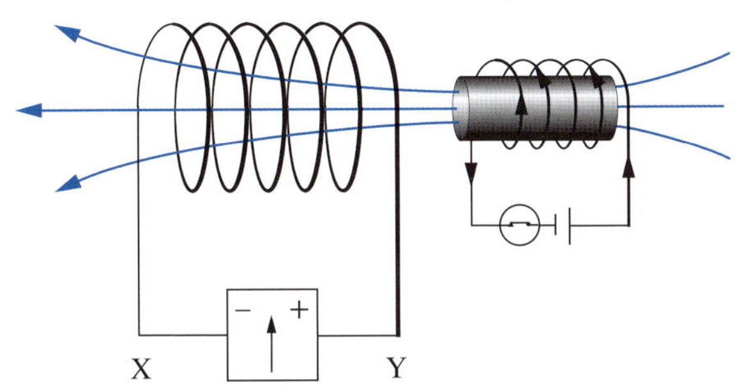

Thinking	Working
Consider the direction of the change in magnetic flux in each case.	**a** Initially there is no magnetic flux through the solenoid. When the electromagnet is switched on, the electromagnet creates a magnetic field directed to the left. So the change in flux through the solenoid is increasing to the left. **b** While the current in the electromagnet is steady, the magnetic flux through the solenoid is constant and thus the flux is not changing. **c** Initially there is a magnetic flux through the solenoid from the electromagnet directed to the left. Then there is no longer a magnetic flux through the solenoid. So the change in flux through the solenoid is decreasing to the left.
What will oppose the change in flux in each case?	**a** The magnetic field that opposes the change in flux through the solenoid is directed to the right. **b** There is no change in flux, no magnetic field created by the solenoid and so no opposition is needed. **c** The magnetic field that opposes the change in flux through the solenoid is directed to the left.
Determine the direction of the induced current required to oppose the change in each case.	**a** In order to oppose the change, the current will be induced in the solenoid in the direction from X to Y (or through the meter from Y to X), as per the right-hand grip rule. **b** There will be no induced EMF or current in the solenoid. **c** In order to oppose the change, the current will be induced in the solenoid in the direction from Y to X (or through the meter from X to Y), as per the right-hand grip rule.

Worked example: Try yourself 7.2.4

INDUCED CURRENT IN A COIL FROM AN ELECTROMAGNET

What is the direction of the current induced in the solenoid shown below when the electromagnet is:

a switched on

b left on

c switched off.

Induced current direction by changing area

It is important to note that an induced EMF is created while there is a change in flux, no matter what causes that change. As magnetic flux $\Phi_B = B_{\perp}A$, a change can be created by any method that causes a relative change in the strength of the magnetic field, B, and/or a change in the area of the coil perpendicular to the magnetic field. So an induced EMF can be created in three ways:

- by changing the strength of the magnetic field
- by changing the area of the coil within the magnetic field
- by changing the orientation of the coil with respect to the direction of the magnetic field.

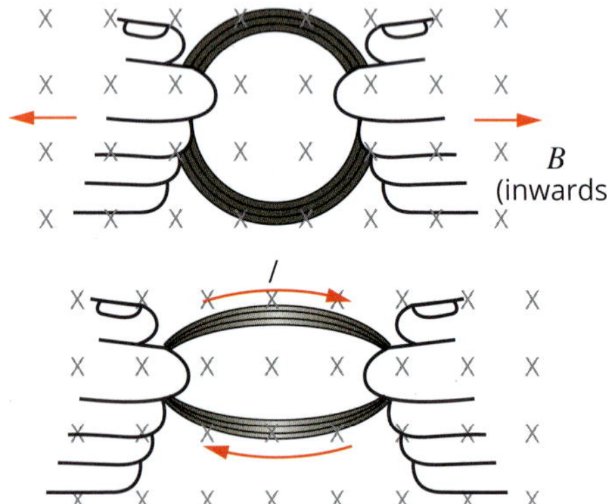

FIGURE 7.2.5 Inducing a current by changing the area of a coil. The amount of flux (the number of field lines) through the coil is reduced and an EMF is therefore induced during the time that the change is taking place. The current is in a direction that creates a field to oppose the reduction in flux into the page.

Figure 7.2.5 illustrates an example of an induced current that results during a decrease in the area of a coil. As the area of the coil decreases due to its changing shape, the flux through the coil (which is directed into the page) also decreases. Applying Lenz's law, the direction of the induced current must oppose this change by acting to increase the magnetic flux through the coil into the page. Using the right-hand grip rule, the current would therefore be in a clockwise direction while the area is changing.

In Figure 7.2.6 a coil is being rotated within a magnetic field. The effect is the same as reducing the area of the coil: the amount of flux flowing through the coil is reduced as the coil moves from being perpendicular to the field to being parallel to the field. An induced EMF is created while the coil is being rotated.

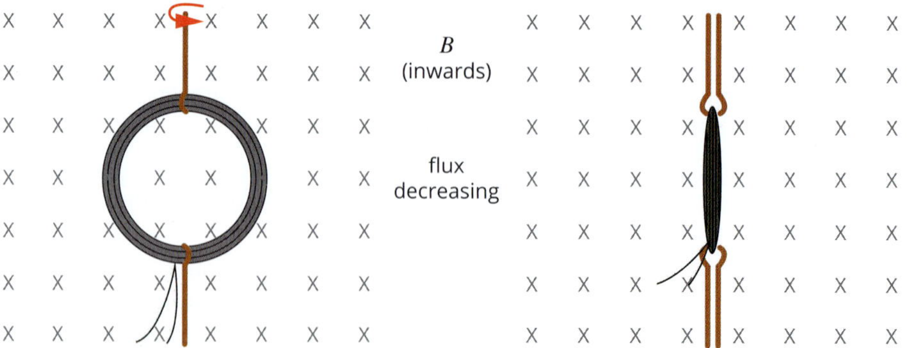

FIGURE 7.2.6 Changing the orientation of a coil within a magnetic field by rotating it reduces the amount of flux through the coil and so induces an EMF in the coil while it is being rotated.

Worked example 7.2.5

FURTHER PRACTICE WITH LENZ'S LAW

The north pole of a magnet is moving towards a coil, into the page. (The south pole is shown at the top, looking down the length of the magnet.) In what direction will the current be induced in the coil while the magnet is moving towards the coil?

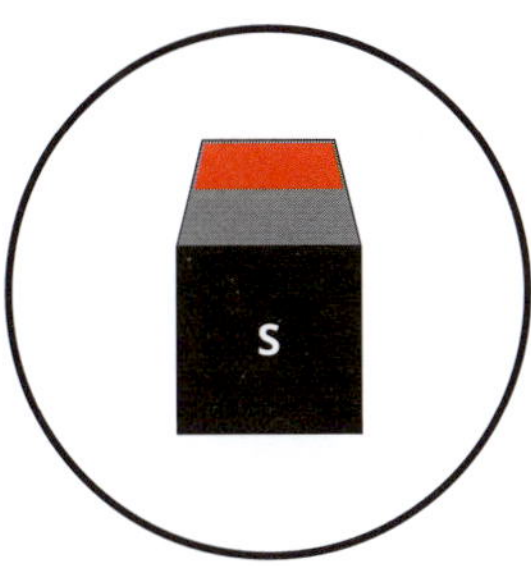

Thinking	Working
Consider the direction of the change in magnetic flux.	The direction of the magnetic field from the magnet will be away from the north pole, into the page. The flux from the magnet will increase as the magnet is brought closer to the coil. So the change in flux is increasing into the page.
What will oppose the change in flux?	The magnetic field that opposes the change will act out of the page.
Determine the direction of the induced current required to oppose the change.	In order to oppose the change, the current direction would be anticlockwise when viewed from above (as per the right-hand grip rule).

Worked example: Try yourself 7.2.5

FURTHER PRACTICE WITH LENZ'S LAW

A coil is moved to the right and out of a magnetic field that is directed out of the page. In what direction will the current be induced in the coil while the coil is moving?

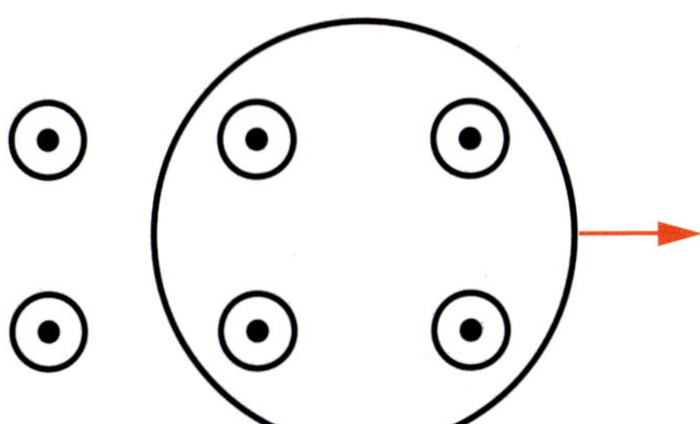

CASE STUDY

Eddy currents

Lenz's law is important for many practical applications, such as metal detectors, induction stoves and regenerative braking systems. These all rely on an eddy current, which is a circular electric current induced in a conductor by a changing magnetic field.

Applying Lenz's law, an eddy current will be in the direction that creates a magnetic field that opposes the change in magnetic flux that created it. Thus eddy currents can be used to apply a force that opposes the source of the motion of an external magnetic field. This is the basis of regenerative braking, where the drag of the opposing magnetic field is used as a braking force.

An eddy current in a conductor with some resistance will also lose energy to the conductor by heating it. This makes eddy currents useful for induction stovetops, where a coil of copper wire is placed within the cooktop. When a pot or a magnetic metal is brought within the field of an induction stovetop, an eddy current is produced in it (Figure 7.2.7). The AC electricity supply produces a changing magnetic field in the coil. This induces an eddy current in the conductive metal pot. The resistance of the metal in the pot, combined with the eddy current, transforms electrical energy into heat and cooks the food.

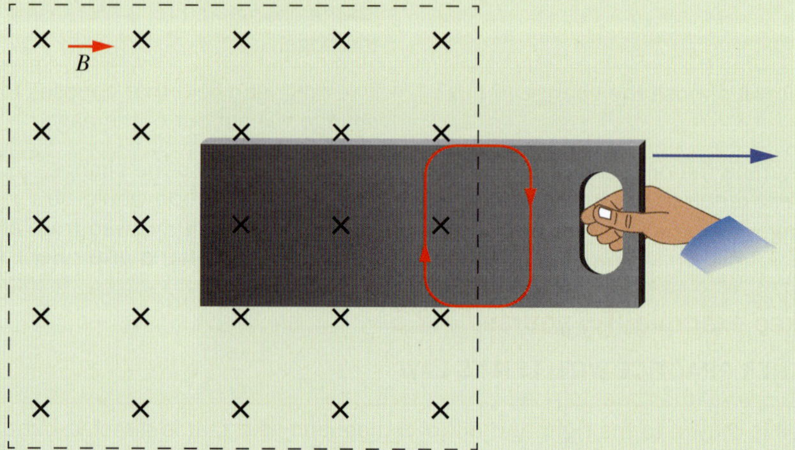

FIGURE 7.2.7 When a magnetic material is brought into the field of an induction stovetop, an eddy current is produced in the material.

7.2 Review

OA ✓✓

SUMMARY

- The EMF induced in a conducting loop in which there is a changing magnetic flux is proportional to the negative rate of change of flux. This can be seen from Faraday's law of induction:

$$\varepsilon = -N\frac{\Delta\Phi_B}{\Delta t}$$

- The negative sign in Faraday's law indicates direction. Where only magnitude is required, the negative sign, and the negative sign on any quantities, can be ignored.
- Lenz's law states that an induced EMF always gives rise to a current whose magnetic field will oppose the change in flux that caused the induced EMF.
- There are three questions that must be considered in determining the direction of the induced current direction according to Lenz's law:
  1. What change in magnetic flux is occurring?
  2. What will oppose the change?
  3. In what direction must the current be to provide this opposition?
- An induced EMF can be created in three ways:
  - by changing the strength of the magnetic field
  - by changing the area of the coil within the magnetic field
  - by changing the orientation of the coil with respect to the direction of the magnetic field.

KEY QUESTIONS

Knowledge and understanding

1 A single rectangular loop of wire is located with its plane perpendicular to a uniform magnetic field of 1.2 mT directed out of the page. The loop is free to rotate about a horizontal axis XY.

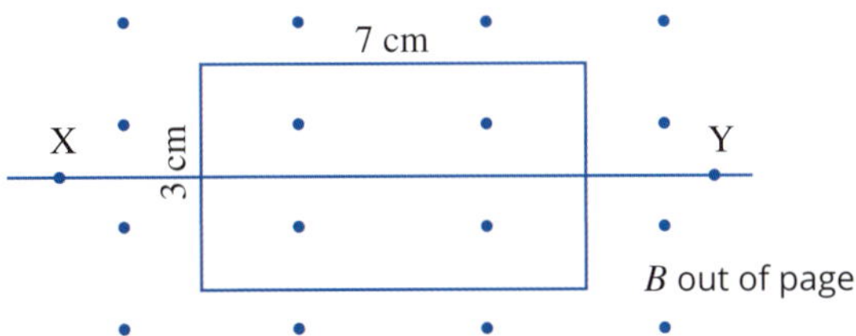

a How much magnetic flux is passing through the loop in this position?

b The loop is rotated about the axis XY through an angle of 90° so that its plane becomes parallel to the magnetic field. How much flux is passing through the loop in this new position?

c If the loop completes one quarter of a rotation in 60 ms, what is the average induced EMF in the loop?

2 A coil of 600 turns each of area 15 cm^2 is wound around a square frame. The plane of the coil is initially parallel to a uniform magnetic field of 65 mT. The coil is then rotated through an angle of 90° so that its plane becomes perpendicular to the field. The rotation takes 25 ms.

a What is the average EMF induced in each turn of wire during the rotation?

b What is the average EMF induced in the coil during the rotation?

3 A student stretches a flexible wire coil of 10 turns and places it in a uniform magnetic field of 10 mT directed into the page. While it is in the field, the student allows the coil to regain its original shape. In doing so the area of the coil changes from 325 cm^2 to 100 cm^2 at a constant rate in 0.25 s. Find the average EMF induced in the coil during this time.

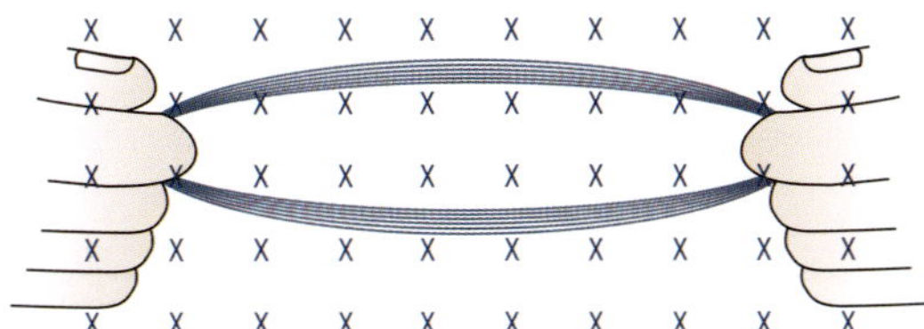

continued over page

7.2 Review *continued*

Analysis

4 When a magnet is dropped through a coil, a voltage sensor will detect an induced voltage in the coil as shown below.

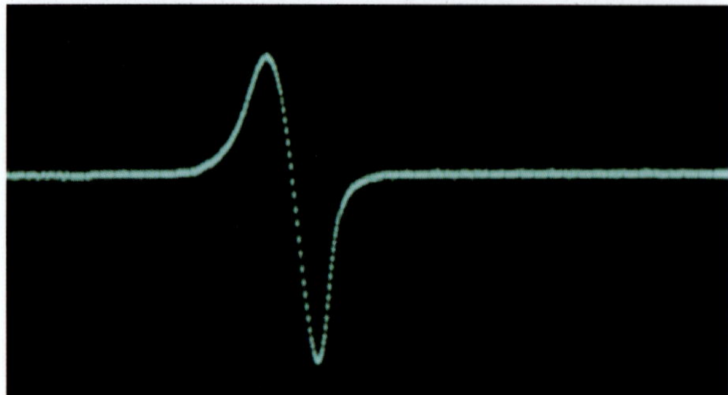

The area of the curve below zero is equal to the area of the curve above zero because:

A The strength of the magnet does not change during the experiment.

B The area of the coil is the same.

C The strength of the magnet and the area of the coil are the same.

D The magnet speeds up as it falls through the coil.

5 A student has a flexible wire coil of variable area with 100 turns. A short distance from the coil is a strong bar magnet that produces a magnetic field of 100 mT. She has been instructed to demonstrate electromagnetic induction by using this equipment to light up an LED rated at 1.0 V. Explain, including appropriate calculations, one method by which she could complete this task.

6 A wire coil consisting of a single turn is placed perpendicular to a magnetic field that experiences a decrease in strength of 0.10 T in 0.050 s. If the EMF induced in the coil is 0.020 V, what is the area of the coil?

7 A wire coil consisting of 100 turns and an area of 50 cm^2 is placed inside a vertical magnetic field of 0.40 T. It is then rotated about a horizontal axis. For each quarter turn, the average EMF induced in the coil is 1600 mV. Calculate the time taken for a quarter turn of the coil.

7.3 Applications of Lenz's law

It is easy to take the supply of electric power to our homes, schools and businesses for granted, and yet it was only in 1881 that the Victorian Electric Light Company demonstrated an electric lamp outside its premises in Swanston Street, Melbourne. (It was lit by a gas-powered generator.)

The electric generator is probably the most important practical application of Faraday's discovery of electromagnetic induction. The principle of electric power generation is the same whether the result is alternating current or direct current—relative motion between a coil and a magnetic field induces an EMF in the coil. A **generator** converts the kinetic energy of a rotating coil into electricity. In small generators, the coil is rotated within a magnetic field. In larger generators, such as those used in power stations and car alternators, the coils are stationary and an electromagnet rotates inside them.

This might all sound quite similar to the way electric motors work. In fact, it is—a generator is basically just the inverse of a motor.

INDUCED EMF IN AN ALTERNATOR OR GENERATOR

A basic electric generator, or **alternator**, consists of many coils of wire wound around an iron core. This is called an armature and it is made to rotate in a magnetic field. The axle is turned by some mechanical means and an EMF is induced in the rotating coil. In other words, mechanical energy is converted to electrical energy.

Consider a single loop of wire in the generator depicted in Figure 7.3.1. The loop is rotated clockwise in a uniform magnetic field. The amount of flux passing through the loop will vary as it rotates. It is this change in flux that induces the EMF.

(a)

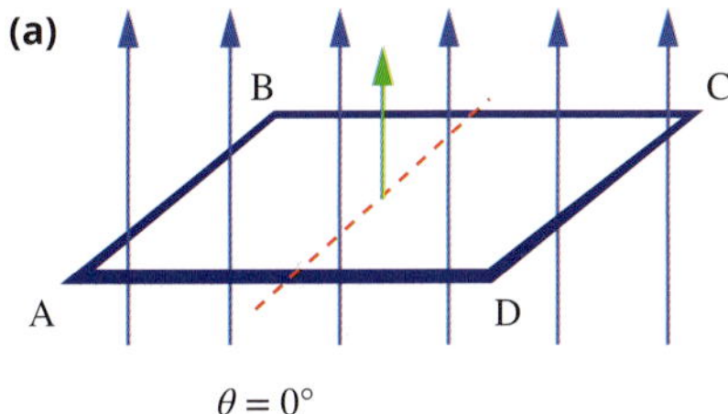

(b)

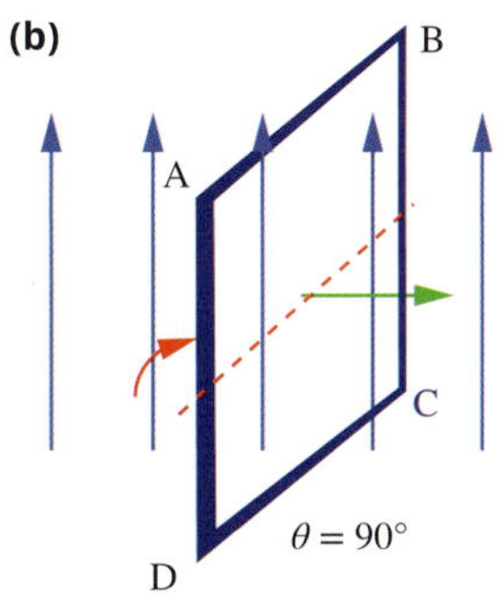

(c)

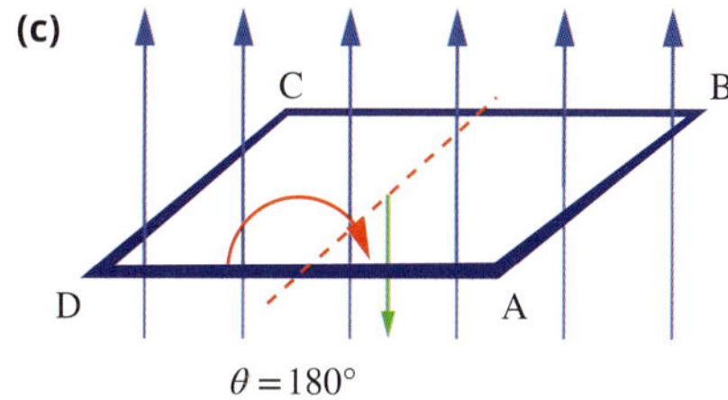

(d)

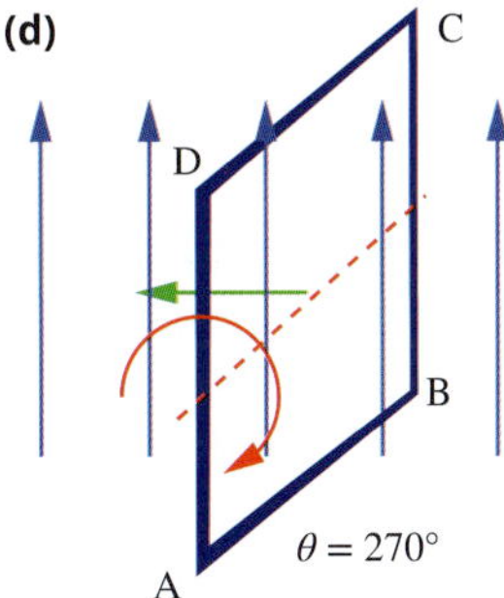

(e)

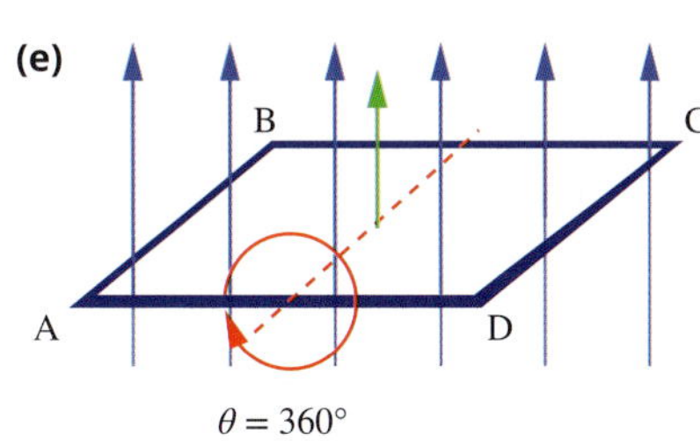

FIGURE 7.3.1 A single loop of a generator rotating in a magnetic field. (a) The plane of the area of the loop is perpendicular to the field, B, and the amount of flux ($\Phi = B_{\perp}A$) is at a maximum. (b) The loop has turned one quarter of a turn and is parallel to the field: $\Phi = 0$. (c) As the loop continues to turn, the flux increases to a maximum but opposite relative to the loop in (a): $\Phi = -B_{\perp}A$. (d) The flux then decreases to zero again, as the loop is parallel to the field, before repeating the cycle again from (e) onwards.

According to Lenz's law, as the flux in the loop decreases from position (a) to (b) in Figure 7.3.1, the induced current generated will flow in whatever direction creates a magnetic field that opposes the change in flux. The right-hand grip rule can then be used to show that the direction of the induced current will be D → C → B → A.

The direction of the induced current will reverse every time the plane of the loop reaches a point perpendicular to the field. The magnitude of the induced EMF will be determined by the rate at which the loop is rotating. It will be a maximum when the rate of change of flux is a maximum. This is when the loop has moved to a position parallel to the magnetic field and the flux through the loop is zero, that is, when the gradient of the flux versus time graph shown in Figure 7.3.2 is a maximum.

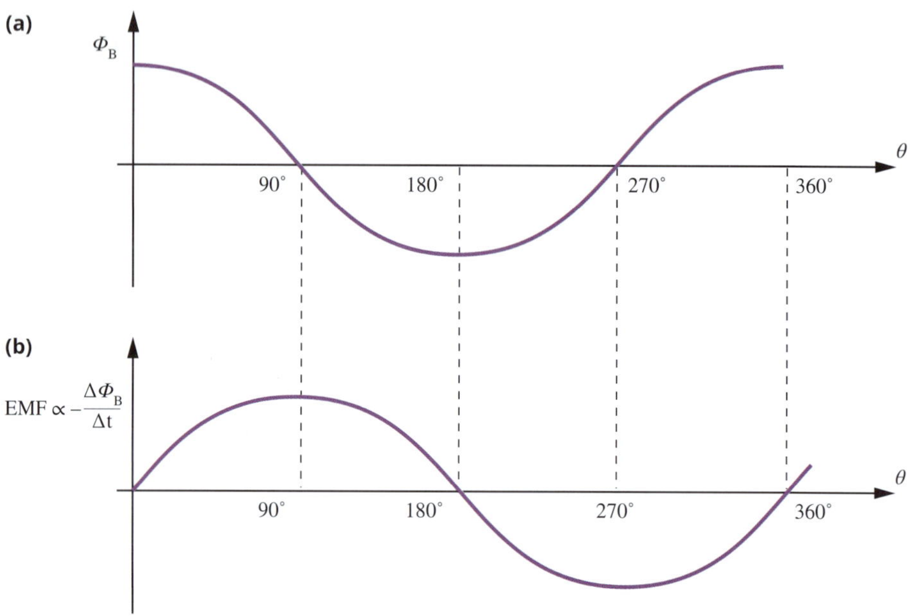

FIGURE 7.3.2 (a) The flux, Φ_B, through the loop shown in Figure 7.3.1 as a function of the angle between the field and the normal to the plane of the area, θ. (b) The rate of change of flux and hence EMF through the loop as a function of the angle between the field and the normal to the plane of the loop, θ. The loop is rotating at a constant speed.

An alternative way to think about how the EMF changes as the loop rotates is to remember that the EMF is actually created as the wires AB and CD cut across the magnetic field lines. Maximum EMF occurs when these wires cut the magnetic field lines perpendicularly, that is, when θ is 90° or 270°. Zero EMF occurs when the motion of these wires is parallel to the field lines, that is, when θ is 0°, 180° or 360°.

AC generators and alternators

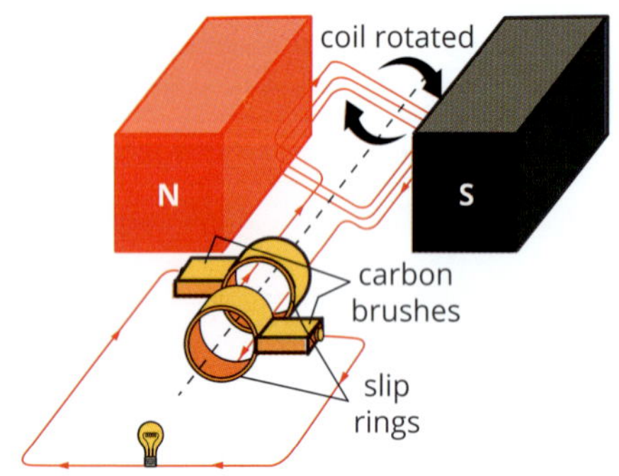

FIGURE 7.3.3 The main components of an AC generator

The construction of an AC generator is basically the same as a motor. The main components are shown in Figure 7.3.3.

Consider again the coil shown in Figure 7.3.1 on page 293. The direction of the resultant EMF alternates, first going above the zero EMF line and then going below it (Figure 7.3.2). This type of EMF produces an **alternating current** (AC) in the coil. How this alternating current is harnessed determines if the device is an AC alternator or a DC generator.

As the coil rotates within the magnetic field, a mechanism is required to keep it in contact with the electric circuit. This is done using **slip rings**. If the output from the coil is transferred to a circuit via continuous slip rings, the alternating current in the coil will be maintained at the output. Carbon **brushes** press against the slip rings to allow a constant output to be transferred to a circuit without a fixed point of connection.

CASE STUDY

Three-phase generators

Many industrial applications require a more constant maximum voltage than is possible from a single coil. These applications require a three-phase power supply. The coils are arranged so that the EMFs vary at the same frequency, but with the peaks and troughs of their waveforms offset to provide three complementary currents with a phase separation of one-third of a cycle, or 120°. The resulting output of all three phases maintains an EMF near the maximum voltage more continuously (Figure 7.3.4). Standard electrical supplies include three phases, but most home applications only require a single phase to be connected. Other advantages of three-phase power systems include less vibrations in motors and generators, and simplified motor design.

A three-phase power supply circuit is also generally more economical than a single-phase power supply circuit. A single-phase power supply circuit uses two conductors: one for power and one neutral (i.e. ground). However, a three-phase circuit has no neutral conductor. So for the cost of an additional conductor, a three-phase circuit can transmit three times as much power (assuming the equivalent voltage and current per phase to the single-phase circuit). This is often why three-phase circuits are used for electricity distribution, particularly over long-distances.

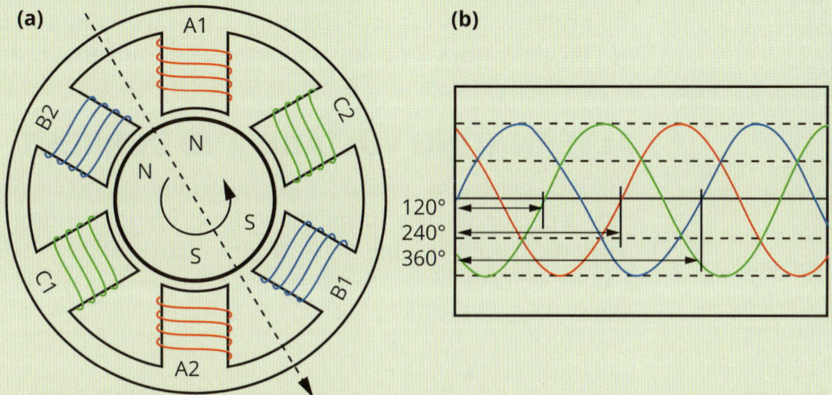

FIGURE 7.3.4 (a) A three-phase power supply has three coils, each producing an output 120° out of phase with the adjoining coil. (b) The resulting output can be combined for a more constant supply voltage.

DC generators

A DC generator is much like an AC generator in basic design. The continuous slip rings are replaced by a **split ring commutator**; that is, the ring picking up output from the coils has two breaks (or splits) in it at opposite sides of the ring. The direction of the output is changed by the commutator every half turn so that the output current is always in the same direction (Figure 7.3.5(a) on page 296).

The output will still vary from zero to a maximum every half cycle, but it can be smoothed by placing a capacitor in parallel with the output. More commonly, the use of multiple armature windings and more splits in the split ring commutator can smooth the output by ensuring that the output is always connected to an armature that is in the position for generating maximum EMF (Figure 7.3.5(b)).

The differences between a slip ring and a split ring commutator are:

- A slip ring is a continuous ring with no breaks or slits.
- A split ring commutator has at least two breaks or slits.
- A slip ring provides a continuous transfer of current and is generally used in AC motors and generators.
- A split ring commutator reverses the polarity of the current and is used in DC motors and generators.

CASE STUDY

Back EMF in DC motors

A DC motor and a generator have a lot in common, and a DC motor could function as a generator and vice versa. In fact, every motor can also be used as a generator. The motors in electric trains, for instance, work as generators when the train is slowing down, converting kinetic energy to electrical energy and putting it back into the electrical supply grid. Regenerative brakes in cars work in a similar way. A DC motor will also generate an EMF when running normally. This is called 'back EMF'.

The back EMF generated in a DC motor is the result of current produced in response to the rotation of the rotor inside the motor in the presence of an external magnetic field. The back EMF, following Lenz's law, opposes the change in magnetic flux that created it, so this induced EMF will be in the opposite direction to the EMF creating it. The net EMF used by the motor is thus always less than the supplied voltage:

$$\varepsilon_{net} = V - \varepsilon_{back}$$

As the motor increases speed, the current induced in it will increase and the back EMF will also increase. When a load is applied to the motor, the speed will generally reduce. This will reduce the back EMF and increase the current in the motor. If the load brings the motor to a sudden halt—say, an electric drill bit getting stuck—the current may be high enough to burn out the motor and the motor windings. To protect the motor, a resistor is placed in series. It is switched out of the circuit when the current drops below a set level and is switched back into the circuit for protection when the current rises above that level.

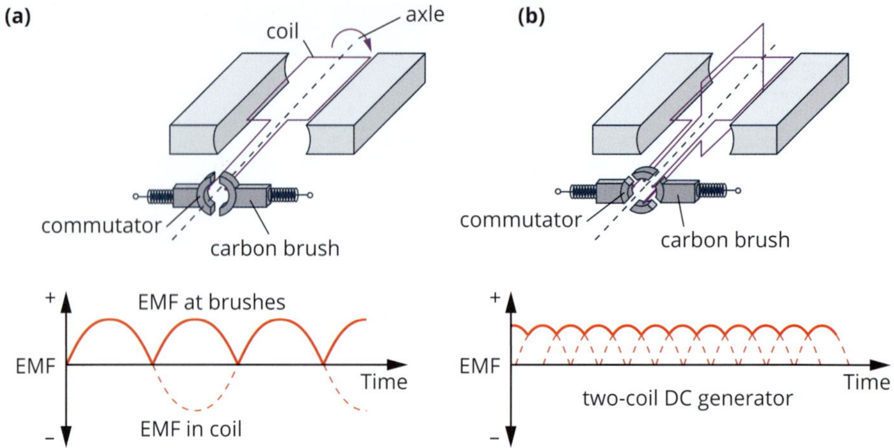

FIGURE 7.3.5 (a) A DC generator has a commutator to reverse the direction of the alternating current every half cycle and so produce a DC output. (b) Multiple armature windings (i.e. loops) can smooth the output.

In the past, cars used DC generators to power ancillary equipment (such as car radios). More common now is the use of AC generators or alternators. These avoid the problems of wear and sparking across the commutator, inherent in the design of DC generators, by using a moving electromagnet inside a set of stationary coils to generate current.

ALTERNATING VOLTAGE AND CURRENT

An AC generator produces an alternating current varying sinusoidally over time with the change in magnetic flux. The maximum EMF is only achieved at particular points in time. In Australia, mains power oscillates at 50 Hz and reaches a peak voltage (V_p) of ±340 V each cycle or a peak-to-peak voltage (V_{p-p}) of 680 V (Figure 7.3.6).

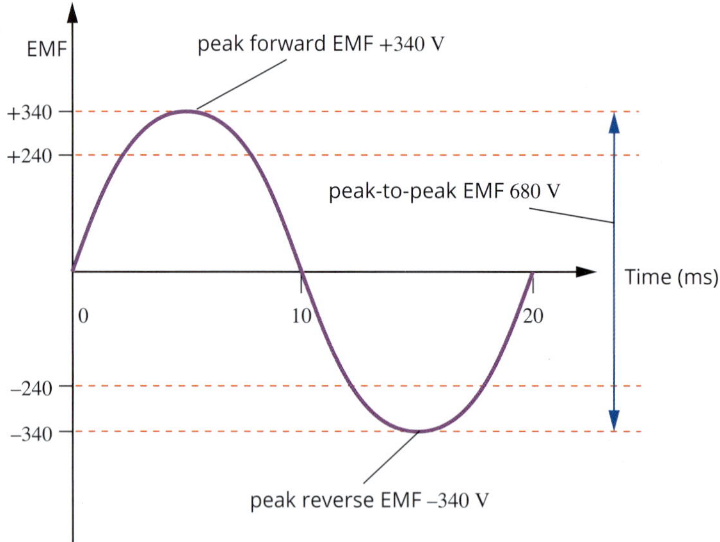

FIGURE 7.3.6 The voltage in Australian power points oscillates between +340 V and –340 V, 50 times each second. The value of a DC supply that would supply the same average power is 240 V.

The graph of current against time also varies sinusoidally. The current reaches a peak (I_p) each cycle with a peak-to-peak current that is double the peak current.

It is often more useful to know the average power produced in a circuit. This can be obtained by using a value for the voltage and current equal to the peak values divided by $\sqrt{2}$. This is referred to as the **root mean square** or rms value.

In effect, the rms values are the values of a DC supply that would be needed to provide the same average power as the AC supply. For example, a 240 V DC power supply would develop the same power in a resistive component that a 240 V rms AC supply would develop. It is the rms value of the voltage $\left(V_{rms} = \frac{340}{\sqrt{2}} = 240\,\text{V}\right)$ that is normally quoted. This is the effective average (mean) value of the voltage and is the value that should be used to find the actual power supplied each cycle by an AC supply.

Worked example 7.3.1

PEAK AND RMS AC CURRENT VALUES

A 60 W light globe is connected to a 240 V AC circuit. What is the peak power use of the light globe?

Thinking	Working
Note that the values given in the question represent rms values. Power is $P = VI$ so both V and I must be known to calculate the power use. The voltage is given, and the current can be calculated from the rms power supplied.	$P_{rms} = V_{rms}I_{rms}$ $I_{rms} = \frac{P_{rms}}{V_{rms}}$ $= \frac{60}{240} = 0.25\,\text{A}$
Substitute the known quantities into the appropriate equation and solve for peak power.	$P_p = \sqrt{2}V_{rms} \times \sqrt{2}I_{rms}$ $= 2V_{rms}I_{rms}$ $= 2 \times 240 \times 0.25$ $= 120\,\text{W}$ $= 1.2 \times 10^2\,\text{W}$

Worked example: Try yourself 7.3.1

PEAK AND RMS AC CURRENT VALUES

A 1000 W kettle is connected to a 240 V AC power outlet. What is the peak power use of the kettle?

PHYSICSFILE

Hybrid vehicles

There has been much attention given in recent years to fully electric vehicles. However, hybrid vehicles have been in widespread commercial use for many years. Hybrid vehicles have both an electric motor and an internal combustion engine (using petrol or diesel). The electric motor in a hybrid vehicle uses energy that is stored in a battery. The batteries in hybrid vehicles are often internally charged by regenerative braking. When the vehicle is braking, the motor acts as a generator. The back EMF produced by the generator is used to charge the battery.

7.3 Review

OA ✓✓

SUMMARY

- The principle of electric power generation is the same whether the result is alternating current or direct current. Relative motion between a coil and a magnetic field induces an EMF in the coil.
- The construction of a generator, or an alternator, is very similar to that of an electric motor.
- A coil rotated in a magnetic field induces an alternating current in the coil. How that current is harnessed determines whether the device is an AC alternator or a DC generator.
- An AC alternator has slip rings that transfer the alternating current in the coil to the output. A DC generator has a split ring commutator to reverse the current direction every half turn so that the output current is always in the same direction.
- The alternating current produced by power stations and supplied to cities in Australia varies sinusoidally at a frequency of 50 Hz. The peak value of the voltage of domestic power (V_p) is ±340 V, and the peak-to-peak voltage (V_{p-p}) is 680 V.
- The root mean square voltage, V_{rms}, is the value of an equivalent steady voltage (DC) supply that would provide the same power:

$$V_{rms} = \frac{V_p}{\sqrt{2}}$$

- The rms value of domestic mains voltage in Australia is 240 V.

KEY QUESTIONS

Knowledge and understanding

1 Describe a benefit of three-phase power generation.

2 The figure below shows a coil rotating through a magnetic field, with positions marked at the following locations: (a) the plane of the area of the loop is perpendicular to the magnetic field, B; (b) the loop has turned one quarter of a turn and is parallel to B; (c) the plane of the area of the loop is perpendicular to B but the flux is opposite relative to the loop in (a); (d) the loop is once again parallel to B; and (e) the coil is in the same position as in (a).
Sketch a graph of the EMF induced in the loop as it rotates in the magnetic field. Indicate the key locations (a) through to (e). You do not need to include values on the axes.

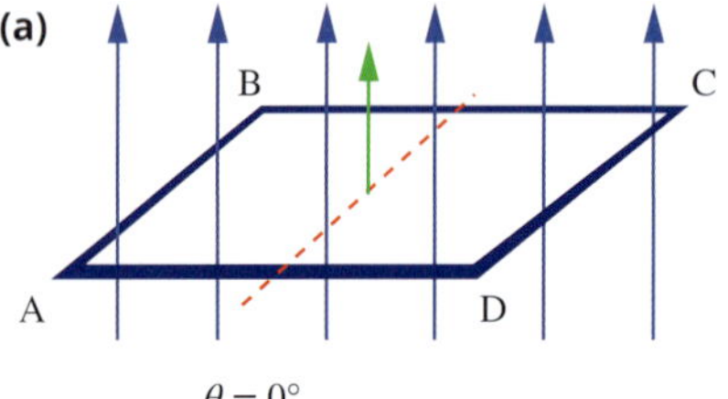

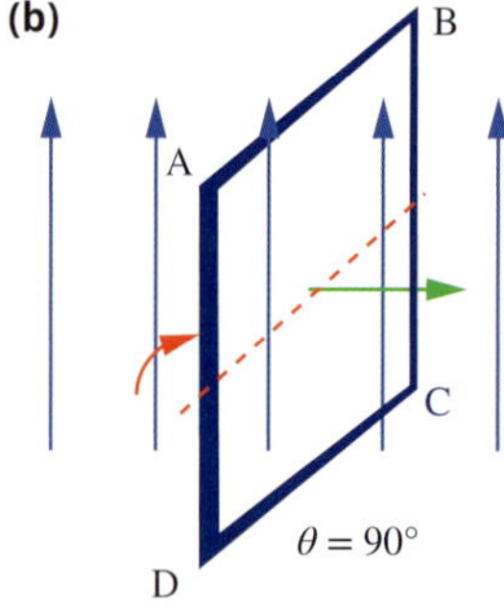

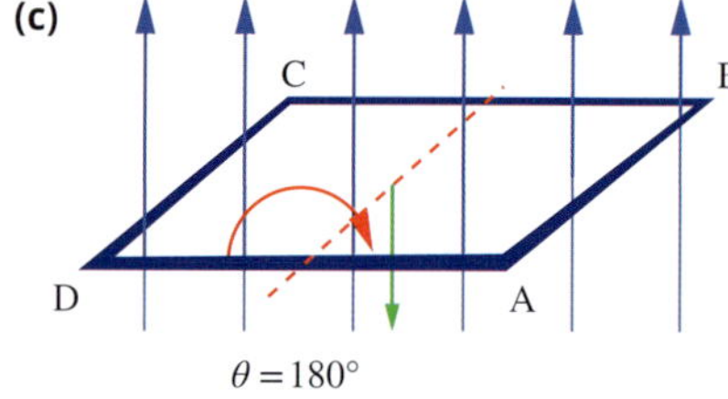

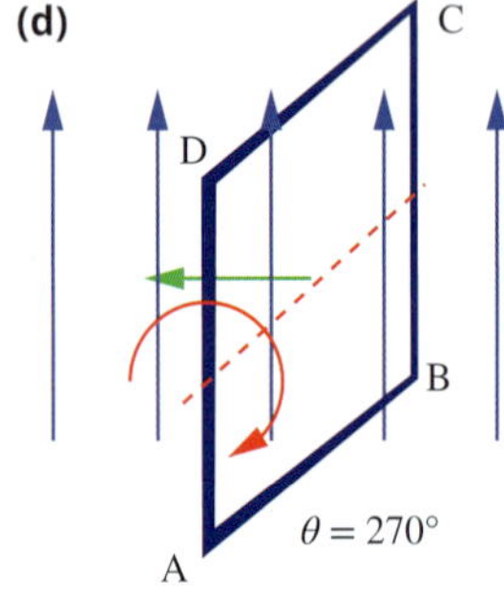

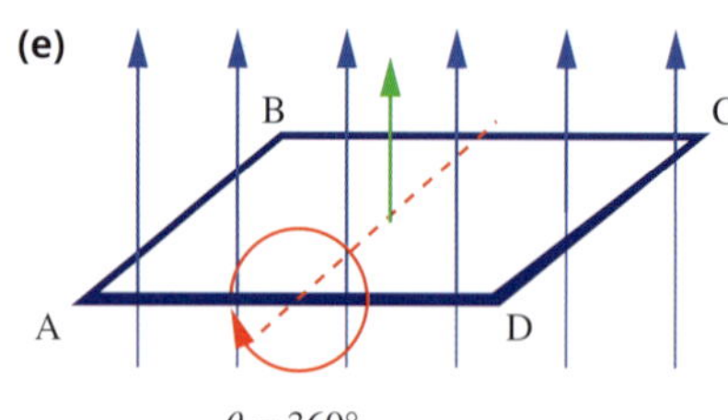

3 Describe the physical and electrical differences between slip rings and a split ring commutator.

4 Sketch the EMF induced in a loop that is connected to an electric circuit using (a) slip rings, and (b) a split ring commutator.

5 An AC supply of frequency 40 Hz is connected to a circuit resulting in an rms current of 2.6 A.

 a Draw a graph to show one full period of the variation of current with time for this circuit.

 b What is the peak and peak-to-peak current in the circuit?

 c What is the period of the AC current supply?

Analysis

6 Consider an anticlockwise rotation of a coil starting at $\theta = 0°$ (i.e. perpendicular) to a constant magnetic field and initially producing a positive current. Which of the graphs best illustrates the variation in the induced current as a function of time for one full revolution of the coil?

A

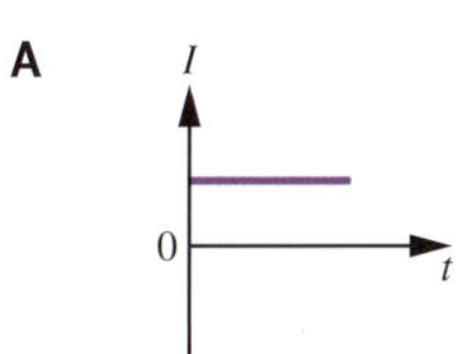

B

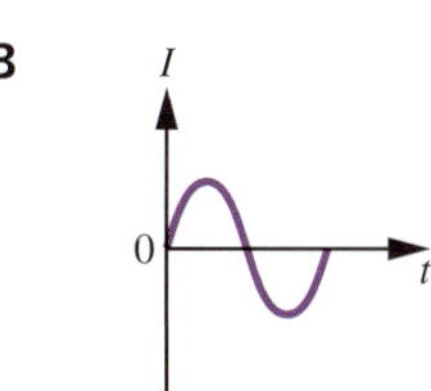

C

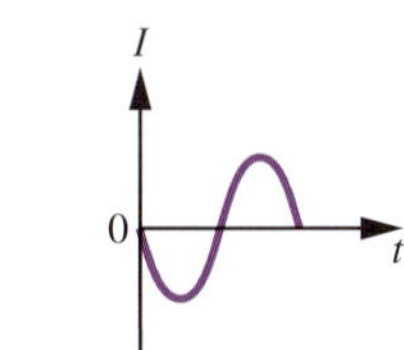

D

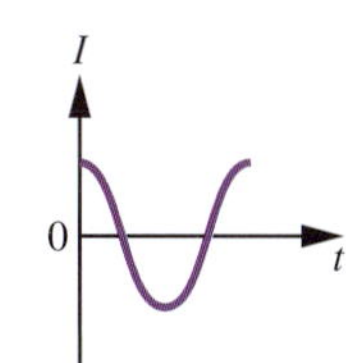

7 A simple generator consists of a coil of 1000 turns, each of radius 10 cm, mounted on an axis in a uniform magnetic field. The following graph shows the voltage output as a function of time when the coil is rotated at a frequency of 50 Hz.

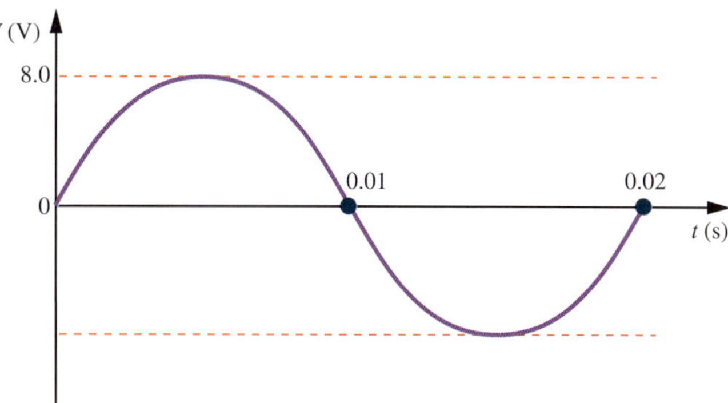

 a Determine the values of V_p, V_{p-p} and V_{rms}.

 b What is the period of the waveform?

 c The generator is modified so that the magnetic field strength is halved and the frequency of rotation is decreased to 25 Hz. The radius of the coil is doubled to 20 cm. Draw a graph representing the new output from the generator.

7.4 Producing electricity—photovoltaic cells

In recent years much attention has been devoted to sources of renewable energy, such as wind, tidal and solar energy. This section explores how solar panels are used to produce electricity—for the home and beyond.

SOLAR PANELS

Solar power generation is based on the principles of the **photovoltaic effect**. This is the effect where a voltage and electric current are generated when certain semiconductor materials are exposed to light. In the case of commercial solar power, the semiconductor materials are embedded in solar panels (Figure 7.4.1).

The semiconductor material in a solar panel is usually made from two forms of silicon. The top layer of the material has excess electrons (n-type silicon) and the bottom layer has a deficit of electrons (p-type silicon). The light from the Sun is in tiny packets of energy called photons. When a solar panel is exposed to sunlight, the photons from the sunlight cause an electron to move away from a silicon atom, setting the electron loose and causing it to move in the semiconductor material. When wires are connected between the top and bottom layers of the semiconductor material, a pathway is created for the electrons to move. Recall that the flow of electrons is an electric current.

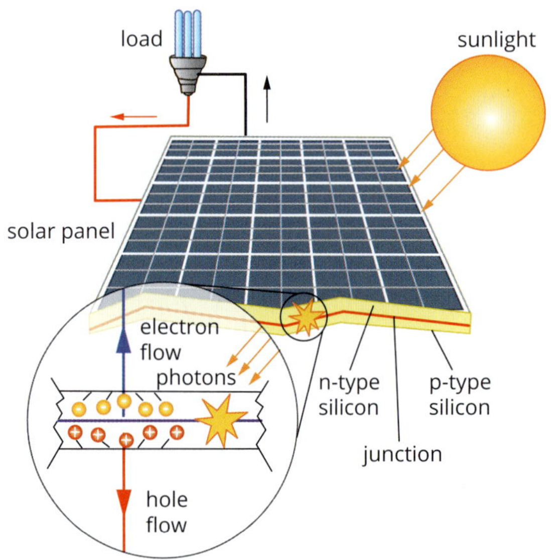

FIGURE 7.4.1 The photovoltaic effect in a solar panel

A solar panel is a collection of many solar cells, mounted on a panel for ease of installation (Figure 7.4.2).

FIGURE 7.4.2 An array of solar panels, with each solar panel made up of solar cells

Solar panels can be mounted in many locations, although the best locations are where they will be exposed to good amounts of sunlight. You will often see solar panels mounted on the rooftops of homes, shopping centres, recreation centres and industrial buildings. They can also be assembled in large quantities to create a solar generation plant, often called a 'solar farm' (Figure 7.4.3).

FIGURE 7.4.3 The Bungala Solar Power Project near Port Augusta, South Australia

The efficiency of solar panels currently in the market is not much greater than 20%. This means that only about 20% of the sunlight the panels absorb is converted into electricity. The rest is converted into other forms of energy, such as heat.

SOLAR POWER GENERATION AND USE

Solar power is an incredibly beneficial energy source. Although solar does not produce electricity all the time, it does during daylight hours when it is often needed the most. Australia is one of the sunniest countries in the world and is a perfect location for generating solar power.

A typical household solar panel can generate 280–300 W of power, is approximately 1.7 m long by 1 m wide and can weigh up to 20 kg. As manufacturing technology improves, the size and mass of panels may decrease. What has certainly decreased over the last decade has been in the cost of purchasing and installation. A household solar production system in 2022 comprising 14 panels and a DC–AC **inverter** (which turns DC into AC) can cost under $5000.

In a solar panel installation, the outputs from each panel are combined to produce the total output power. Thus a 14-panel system in which each panel can produce 280–300W could generate up to 4 kW in total. A typical residential household, with a family of four, may consume 20–30 kW h a day. The generation of solar power will be maximum during the middle of the day, and less during the morning and late afternoon hours. Provided that the household uses most of its electricity during the day, the solar system can meet a significant portion of the household's electricity requirements.

A drawback to the production of solar power is that when the sun is not shining, little or no power is produced. (Most solar panels can produce some power during overcast conditions.) During times when solar generation is inadequate, energy could be taken from the electricity grid. Alternatively, some solar power installations make use of a battery, charging the battery during daylight hours and drawing on the battery when the panels are not generating power.

The positioning of solar panels is key to receiving the full benefit of the Sun. The most effective positioning in Australia is northwards, to capture the morning, midday and afternoon sun. In some installations, the panels are mounted so that they face northwards even if this is not the natural orientation of the mounting platform (e.g. the roof).

PHYSICSFILE

Solar-powered vaccine refrigerators

There are many uses of solar power beyond powering homes and businesses. To meet the need to vaccinate as much of the population as possible during the COVID-19 pandemic, solar-powered refrigerators were used to ensure vaccines remained cool and viable in regions where reliable electricity was not available.

CASE STUDY ANALYSIS

Small-scale solar production

In some rural and developing countries, mains-connected electricity supply may be unreliable or not even exist. Solar production is a practical means of obtaining electricity.

Farmers may have little more than rainfall to rely on for growing and maintaining their crops. But rainfall may be seasonal or, in cases of drought, non-existent. Access to underground water would help, but it may take hours of hard labour to manually pump water to the surface. An alternative is an electrical pump that uses solar power.

Small-scale solar installations can make a world of difference. A 1.65 m^2 solar panel could produce up to 300 W of power. Or a series of smaller panels can be connected to generate the equivalent power. The panels can be mounted on brackets or racks (Figure 7.4.4), or rested on the ground at an angle that maximises the collection of sunlight, making the panels relatively easy to mount and use.

FIGURE 7.4.4 A solar-powered water pump providing an off-grid water supply for small-scale agricultural production

Analysis

A water pump outputs 150 W of hydraulic power. The pump has an efficiency of 60% (that is, it converts 60% of the electrical power it receives into hydraulic power to pump water). How many 400 W solar panels would be required to operate this pump?

Solar panels produce DC current. This is suitable for some equipment (usually small devices such as phone chargers). But to use solar power for most household and industrial applications, the electricity generated needs to be converted from DC to AC. To do this, a DC-to-AC converter (also known as an inverter) is used. An inverter takes a DC current, changes its direction 50 times per second and smooths the output waveform to create a 50 Hz signal. The output voltage is then set to match the needs of the load (e.g. 240 V rms for household use). By the law of conservation of energy, the input power to the inverter will always be greater than or equal to the output power.

Most inverters can connect to the mains electricity grid. This enables excess solar energy to be sent into the grid, and for electricity from the grid to be sent to the house (Figure 7.4.5).

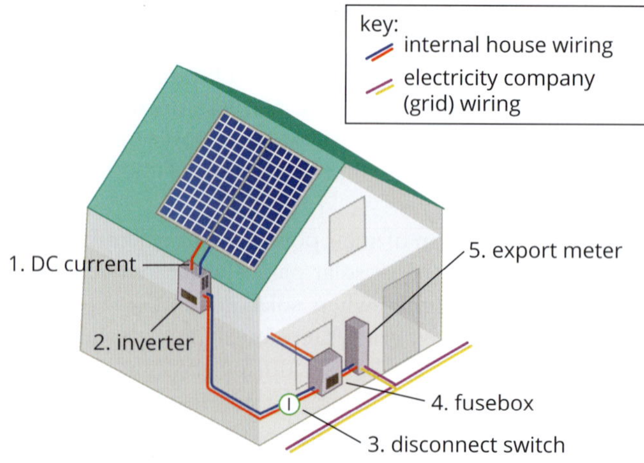

FIGURE 7.4.5 Solar panel installation. The panels are connected to a DC-to-AC inverter, then to the household and also the mains electricity grid.

7.4 Review

OA ✓✓

SUMMARY

- Solar power generation is based on the principles of the photovoltaic effect.
- A drawback to solar power production is that when the sun doesn't shine, little or no power is produced.
- Solar panels produce DC current.
- To use solar power for most household and industrial applications, the generated electricity needs to be converted from DC to AC. To do this, a DC-to-AC converter (also known as an inverter) is used.

KEY QUESTIONS

Knowledge and understanding

1 What is the typical efficiency of a solar cell?

A between 30% and 45%
B under 10%
C over 40%
D approximately 20%

2 Explain what happens when a solar cell is exposed to photons from the Sun.

3 Complete the second sentence below with the appropriate terms.
A solar panel contains a semiconductor material, usually made from silicon. The top layer of the semiconductor has an excess of electrons (_____ silicon) and the bottom layer a deficit of electrons (_____ silicon).

4 Describe the purpose of an inverter in a solar power generation system.

Analysis

5 The power output of a solar power generation system increases from a low in the morning to a peak during the day then tapers off until the sun sets. Knowing this, devise some strategies to maximise the use of solar energy by a household to ensure minimal or no reliance on the external power grid.

7.5 Supplying electricity—transformers and large-scale power distribution

When Faraday discovered electromagnetic induction, he had effectively invented the transformer. A **transformer** is a device for increasing or decreasing an AC voltage. Transformers can be found in just about every electrical device and are an essential part of any electrical distribution system (Figure 7.5.1).

FIGURE 7.5.1 (a) Transformers at an electrical substation take high-voltage electricity from the distribution grid and convert it to a lower voltage. (b) Smaller distribution transformers, found on power poles in every suburban street, lower the voltage further so that it can be used by residential and industrial equipment. See if you can locate one in your street.

THE WORKINGS OF A TRANSFORMER

A transformer takes advantage of the fact that a changing magnetic flux induces an EMF. It consists of two coils, known as the primary coil and the secondary coil. The changing flux originates from the alternating current supplied to the primary coil. The changing magnetic flux is directed to the secondary coil and induces an EMF in that coil (Figure 7.5.2).

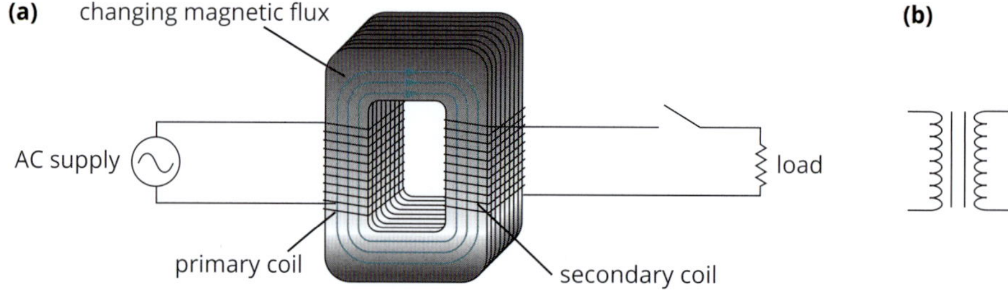

FIGURE 7.5.2 (a) Parts of a transformer; (b) the symbol used in circuit diagrams to indicate an iron-core transformer

The two coils can be interwoven using insulated wire or they can be linked by a soft-iron core that is laminated to minimise loss due to eddy currents.

In an **ideal transformer** all the magnetic flux produced by the primary coil passes to the secondary coil, that is, it is 100% efficient, with no energy loss. However, in a real transformer not all the magnetic flux produced by the primary coil passes to the secondary coil—there is some energy loss. Still, transformers are one of the most efficient devices, with practical efficiencies often better than 99%.

AC VERSUS DC

The general power distribution system distributes alternating current. That may seem odd given that many devices run on direct current. But alternating current has a special advantage: it can be easily transformed from one voltage to another.

PHYSICSFILE

Laminations

Eddy currents produced in the iron core of transformers can generate considerable heat, creating a fire hazard. To reduce eddy current losses, the core is usually made of laminations (see figure below). These are thin plates of iron electrically insulated from each other and stacked so that the insulation reduces eddy currents.

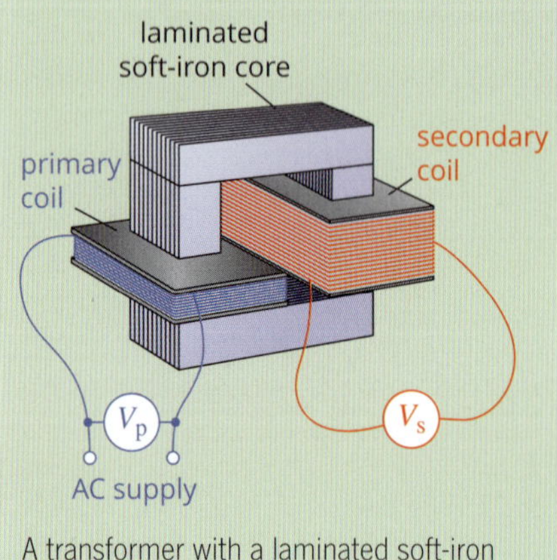

A transformer with a laminated soft-iron core

A transformer works on the basis that a changing current in the primary coil produces a changing magnetic flux which induces a current in the secondary coil. For this to work, the current to the primary coil must be constantly changing, as it does in an AC supply.

A DC voltage has a constant, unchanging current. With no change in the size of the current, no changing magnetic flux will be created by the primary coil and hence no current is induced in the secondary coil. There will be a current induced briefly when a DC supply is turned on (as the input current changes from zero). Current will also be briefly induced when the DC supply is switched off (as the input current returns to zero). But while the DC supply is constant there is no change in magnetic flux to induce a current in the secondary coil.

THE TRANSFORMER EQUATION

When an AC voltage is connected to the primary coil of a transformer, the changing magnetic field induces an AC voltage of the same frequency in the secondary coil. The voltage in the secondary coil depends on the number of turns in each coil.

From Faraday's law, the average voltage in the primary coil, V_1, will affect the rate at which the magnetic flux changes:

$$V_1 = N_1 \frac{\Delta \Phi_B}{\Delta t}$$

Thus:

$$\frac{\Delta \Phi_B}{\Delta t} = \frac{V_1}{N_1}$$

where N_1 is the number of turns in the primary coil. The induced voltage in the secondary coil, V_2, will be:

$$V_2 = N_2 \frac{\Delta \Phi_B}{\Delta t}$$

Thus:

$$\frac{\Delta \Phi_B}{\Delta t} = \frac{V_2}{N_2}$$

where N_2 is the number of turns in the secondary coil.

Assuming that there is little or no loss of flux between the primary and secondary coil, then the flux in each will be the same. It then follows that:

$$\frac{V_1}{N_1} = \frac{V_2}{N_2} \quad \text{or} \quad \frac{V_2}{V_1} = \frac{N_2}{N_1}$$

The transformer equation, relating voltage (V) and the number of turns (N) in each coil, is

$$\frac{N_1}{N_2} = \frac{V_1}{V_2}$$

where the subscript '1' refers to the primary coil and the subscript '2' refers to the secondary coil.

The transformer equation explains how the secondary voltage (output) is related to the primary voltage (input). The equation applies whether rms voltage or peak voltage is used.

In a **step-up transformer** the number of turns in the secondary coil is greater than the number of turns in the primary coil; thus the secondary voltage is greater than the primary voltage. (From the transformer equation, if $N_2 > N_1$ then $V_2 > V_1$.)

In a **step-down transformer** the number of turns in the secondary coil is less than the number of turns in the primary coil; thus the secondary voltage is less than the primary voltage. (From the transformer equation, if $N_2 < N_1$ then $V_2 < V_1$.)

Worked example 7.5.1

TRANSFORMER EQUATION—VOLTAGE

A transformer built into a portable radio reduces the 240 V supply voltage to the 12 V required for the radio to work. If the number of turns in the secondary coil is 100, what is the number of turns in the primary coil?

Thinking	Working
Note the relevant quantities given in the question.	$V_2 = 12\text{ V}$ $V_1 = 240\text{ V}$ $N_2 = 100$ $N_1 = ?$
Substitute the quantities into the transformer equation and solve for N_1.	$\frac{N_1}{N_2} = \frac{V_1}{V_2}$ $\frac{N_1}{100} = \frac{240}{12}$ $N_1 = \frac{100 \times 240}{12}$ $= 2000$ turns

Worked example: Try yourself 7.5.1

TRANSFORMER EQUATION—VOLTAGE

A transformer built into a phone charger reduces the 240 V supply voltage to the required 6 V. If the number of turns in the secondary coil is 100, what is the number of turns in the primary coil?

POWER OUTPUT

Although a transformer very effectively increases or decreases an AC voltage, energy conservation means that the output power cannot be greater than the input power. Since a well-designed transformer with a laminated core can be more than 99% efficient, the power output can be considered equal to the power input, making it an ideal transformer.

Since the power supplied, P, equals VI:

$$V_1 I_1 = V_2 I_2$$

The transformer equation can then be written in terms of current, I.

$$\frac{N_1}{N_2} = \frac{I_2}{I_1}$$

where N is the number of turns
I is the current

Note that the number-of-turns ratio when current is being considered is the inverse of the ratio when voltage is being considered.

Worked example 7.5.2

TRANSFORMER EQUATION—CURRENT

A transformer with 2000 turns in the primary coil and 100 turns in its secondary coil provides a current of 4.0 A. What is the current in the primary coil?

Thinking	Working
Note the relevant quantities given in the question.	$I_2 = 4.0\,\text{A}$ $N_2 = 100$ turns $N_1 = 2000$ turns $I_1 = ?$
Recall the transformer equation written in terms of current. Rearrange the equation, substitute the given quantities and solve for I_1.	$\frac{N_1}{N_2} = \frac{I_2}{I_1}$ $\frac{I_1}{I_2} = \frac{N_2}{N_1}$ $\frac{I_1}{4.0} = \frac{100}{2000}$ $I_1 = \frac{4.0 \times 100}{2000}$ $= 0.20\,\text{A}$

Worked example: Try yourself 7.5.2

TRANSFORMER EQUATION—CURRENT

A phone charger with 4000 turns in its primary coil and 100 turns in its secondary coil provides a current of 0.50 A. What is the current in the primary coil?

Worked example 7.5.3

TRANSFORMERS—POWER

The power drawn from the secondary coil of a transformer by a portable radio is 48 W. What power is drawn from the mains supply if the transformer is an ideal transformer?

Thinking	Working
The energy efficiency of a transformer can be assumed to be 100%. Hence the power in the secondary coil will be the same as that in the primary coil.	The power drawn from the mains supply is the power in the primary coil, which will be the same as the power in the secondary coil: $P = 48\,\text{W}$.

Worked example: Try yourself 7.5.3

TRANSFORMERS—POWER

The power drawn from the secondary coil of the transformer in a phone charger is 3 W. What power is drawn from the mains supply if the transformer is an ideal transformer?

PHYSICSFILE

Standby power

Standby power is a term used for the electrical power consumed by devices that are switched off or not in use. Many things can lead to standby power consumption in a device, including poorly designed or inefficient power supplies, or circuits that are kept active while a device is plugged in (even if the device is not doing any useful work). Theoretically, standby power consumption should be minimal, but it can be up to 10% of power use. This is why devices such as TVs and computers should be switched completely off when not in use. Over the whole community, standby power amounts to megawatts of wasted power and unnecessary greenhouse emissions! Special switches, such as the Ecoswitch shown in the figure below, have been developed. They can be connected between the power outlet and the device to make it easier to remember to turn devices completely off when not in use.

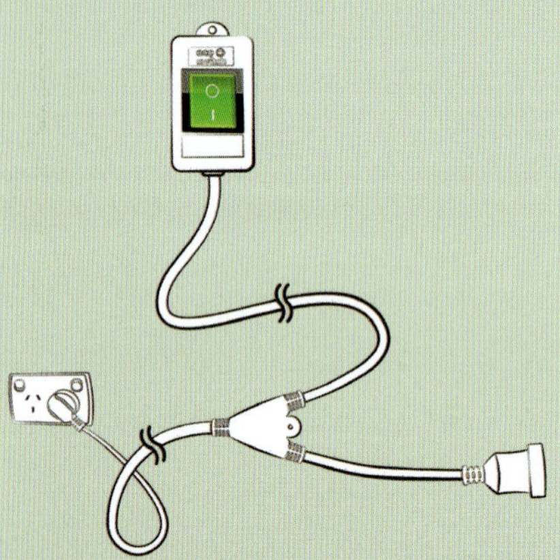

Standby switches such as the Ecoswitch make it easier and more convenient to turn devices completely off when not in use, saving up to 10% on power bills.

POWER FOR CITIES: LARGE-SCALE AC SUPPLY

In your school experiments using electrical circuits, you probably ignored the resistance of the connecting wires because the wires, which are generally made from copper, are good conductors. Thus the resistance is very small over short distances. However, over large distances even relatively good electrical conductors have a significant resistance.

Modern cities use huge amounts of electrical energy, most of which is supplied from power stations built at a considerable distance from metropolitan areas. The efficient transmission of electrical energy—with the least loss of power—is therefore a very important consideration for electrical engineers, particularly given the vast distances between population centres in Australia.

The power lost in an electrical circuit is given by $P_{loss} = \Delta VI$, where ΔV is the voltage drop across the load. By substituting Ohm's law ($\Delta V = IR$) into this equation, power loss can be expressed in terms of either current and load resistance, or voltage drop and load resistance:

$$P_{loss} = \Delta VI = I^2R = \frac{\Delta V^2}{R}$$

Given that $P_{loss} = I^2R$, transmitting large amounts of power using a large current will create very large power losses. If the current in power lines can be reduced, it will significantly reduce the power loss. For example, if the current could be reduced by a factor of 3, power loss will be reduced by a factor of 3^2 or 9.

The challenge, then, is to transmit the large amounts of power being produced at power stations using a very low current. Transformers are the most common solution to this problem. Using a step-up transformer near the power station, the voltage is increased by a certain factor and, importantly, the current is decreased by the same factor.

At this point you might be confused by the alternative equation for power loss: $P_{loss} = \frac{\Delta V^2}{R}$. You might think, from this equation, that increasing the voltage would lead to greater power loss. However, ΔV represents the voltage drop in a circuit. You must be careful not to confuse the voltage being transmitted along the wires with the voltage drop. So even though the voltage being transmitted is increased through the use of a step-up transformer, the voltage drop across the power lines is reduced. This is because $\Delta V = IR$: a smaller current means a smaller voltage drop. And a smaller current also reduces power loss, since $P_{loss} = I^2R$.

AC power from the generator is stepped up by a transformer to between 240 kV and 500 kV prior to transmission. Once the electrical lines reach the city, the voltage is stepped down in stages at electrical substations before being distributed. The power lines in streets will have a voltage of approximately 2400 V before being stepped down further to 240 V by small distribution transformers on power poles in city streets.

Worked example 7.5.4

TRANSMISSION-LINE POWER LOSS

300 MW is to be transmitted from a power station to Melbourne along a transmission line with a total resistance of 1.00 Ω. What would be the total power loss if the initial voltage in the line is 250 kV?

Thinking	Working
Convert the power and voltage to SI units.	$P = 300\text{ MW} = 300 \times 10^6\text{ W}$ $V = 250\text{ kV} = 250 \times 10^3\text{ V}$
Determine the current in the line based on the initial voltage.	$P = VI$ $I = \frac{P}{V}$ $= \frac{300 \times 10^6}{250 \times 10^3}$ $= 1200\text{ A}$
Determine the corresponding power loss.	$P = I^2R$ $= 1200^2 \times 1.00$ $= 1.44 \times 10^6\text{ W}$ or 1.44 MW

Worked example: Try yourself 7.5.4

TRANSMISSION-LINE POWER LOSS

300 MW is to be transmitted from a power station to Melbourne along a transmission line with a total resistance of 1.0 Ω. What would be the total power loss if the voltage along the line is 500 kV?

Note: You can compare the percentage power loss in Worked example 7.5.4 to the power loss in the Try yourself question. In the Worked example, the power loss is $\frac{1.44}{300} \times 100 = 0.480\%$ while in the Try yourself question, the power loss is $\frac{0.36}{300} \times 100 = 0.12\%$. By increasing the voltage by a factor of 2, the current in the line is halved and the power loss decreases by a factor of 4.

Worked example 7.5.5

VOLTAGE DROP ALONG A TRANSMISSION LINE

Power is to be transmitted from a power station to Melbourne along a transmission line with a total resistance of 1.00 Ω. The current in the line is 1200 A. What voltage would be needed at the power station to achieve a supply voltage of 250 kV?

Thinking	Working
Determine the voltage drop along the transmission line.	$\Delta V = IR$ $= 1200 \times 1.00$ $= 1200\text{ V}$
Determine the initial supply voltage.	$V_{\text{initial}} = V_{\text{supplied}} + \Delta V$ $= 250 \times 10^3 + 1200$ $= 251\,200\text{ V}$ or 251 kV

Worked example: Try yourself 7.5.5

VOLTAGE DROP ALONG A TRANSMISSION LINE

Power is to be transmitted from a power station to Melbourne along a transmission line with a total resistance of 1.0 Ω. The current is 600 A. What voltage would be needed at the power station to achieve a supply voltage of 500 kV?

LARGE-SCALE ELECTRICAL DISTRIBUTION SYSTEMS

Large-scale energy transmission is done through an interconnected grid between the power stations and the population centres where the bulk of the electrical energy is used. A wide-area synchronous grid, also known as an interconnection, directly connects a number of generators, delivering AC power with the same relative phase to a large number of consumers.

No matter the source, the path the electrical power takes to the final consumer is very similar (Figure 7.5.3). Step-up transformers in a large substation near the power station raise the voltage from that initially generated to 240 000 V (240 kV) or more. The electrical power is then carried via high-voltage transmission lines to a number of substations near key centres of demand. At the substations, step-down transformers reduce the voltage to a safer level and then distribute it to local power lines. In residential areas, step-down transformers on power poles reduce the voltage to the 240 V AC rms that home and business devices are designed to run on. These suburban step-down transformers each supply 10 to 15 houses with electricity.

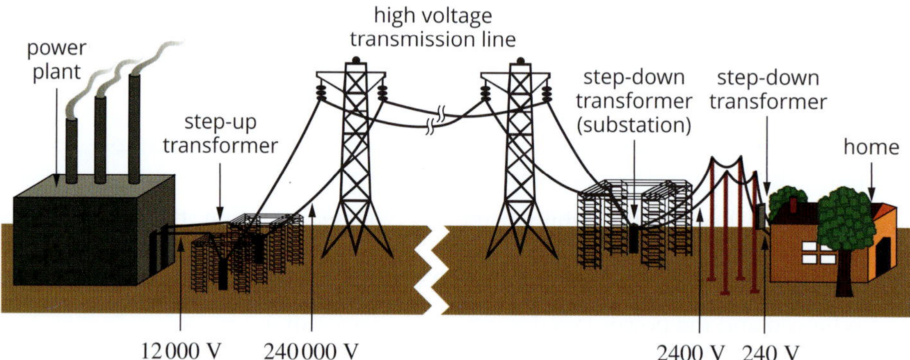

FIGURE 7.5.3 Transmitting electric power from generator to home uses AC power, so transformers are used to minimise power losses through the system.

The use of AC as the standard for distribution enables highly efficient and relatively cheap transformers to convert the initial voltages created at the power station to much higher levels. The same power transmitted at a higher voltage requires less current and therefore less power loss. If it were not for this, the resistance of the transmission wires would need to be significantly reduced, which would require more copper in order to increase their cross-sectional area. This is both expensive and the wires would be heavier. Less metal means that cables can be lighter and thinner, and the supporting towers can be comparatively shorter, cheaper and lighter to build.

ADVANTAGES OF AC VERSUS DC TRANSMISSION

While AC is the primary method of power transmission across the world, there are advantages to DC power transmission. The following is a list of the advantages of AC and DC power distribution, for both short and long-distance transmission.

Advantages of AC distribution

- easier and cheaper to achieve the high voltages required for transmission (which reduces power loss), as DC voltage cannot be stepped up and stepped down easily
- easier and cheaper to maintain AC power distribution systems and substations
- easier and cheaper to control the voltage in AC power distribution systems

Advantages of DC distribution

- power losses are only resistive whereas with AC distribution there are other factors (e.g. non-resistive power losses)
- high-voltage DC (HVDC) can connect AC power networks (including networks with different AC frequencies)
- HVDC can be used for underwater distribution due to lower power losses

Worked example 7.5.6

POWER LOSS THROUGH A TRANSMISSION NETWORK

Consider the power transmission system shown below. The power plant generates AC voltage at 24 kV, which is stepped up to 480 kV using transformer T_1. Transformer T_2 steps the 480 kV voltage down to 48 kV and transformer T_3 steps the 48 kV voltage down to 240 V for the load.

The resistance of the high-voltage transmission line is 5 Ω, the resistance of the medium-voltage distribution line is also 5 Ω and the resistance of the low-voltage distribution line is 10 Ω.

Assume that the transformers are 100% efficient and that the load consumes 1 MW of power.

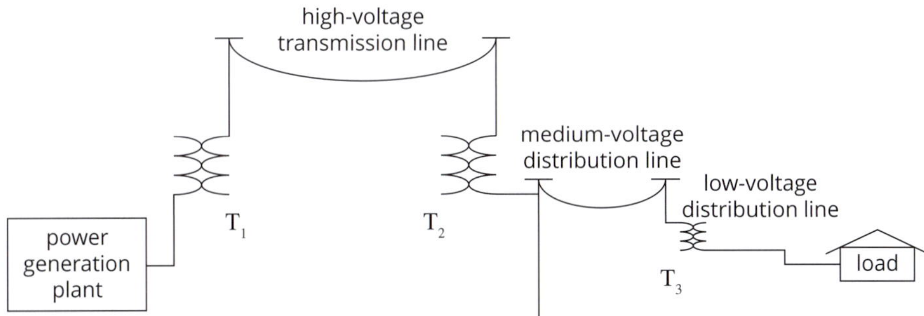

a Calculate the turns ratio of transformers T_1, T_2 and T_3.	
Thinking	**Working**
Rearrange the transformer equation and calculate the turns ratios.	$\frac{N_2}{N_1}=\frac{V_2}{V_1}$ For T_1: $V_1 = 24$ kV and $V_2 = 480$ kV; therefore $\frac{V_2}{V_1}=\frac{480\times10^3}{24\times10^3}=20$ For T_2: $V_1 = 480$ kV and $V_2 = 48$ kV; therefore $\frac{V_2}{V_1}=\frac{48\times10^3}{480\times10^3}=0.10$ For T_3: $V_1 = 48$ kV, $V_2 = 240$; therefore $\frac{V_2}{V_1}=\frac{240}{48\times10^3}=0.0050$

b Determine the power loss through the high-voltage transmission line and through the medium-voltage distribution line.	
Thinking	**Working**
Calculate the current through each part of the system. Recall that the load consumes 1 MW of power. Work back from the load to find the current in the other sections of the line.	At the load: $P = VI$ $I = \frac{P}{V} = \frac{1 \times 10^6}{240}$ $= 4.17\,\text{kA}$ In the high-voltage transmission line: $\frac{I_1}{I_2} = \frac{N_2}{N_1}$ $I_1 = \frac{N_2}{N_1} \times I_2$ $= 0.10 \times 20.85$ $= 2.085\,\text{A}$ In the medium-voltage distribution line: $\frac{I_1}{I_2} = \frac{N_2}{N_1}$ $I_1 = \frac{N_2}{N_1} \times I_2$ $= 0.0050 \times 4.17 \times 10^3$ $= 20.85\,\text{A}$
Calculate the power losses using the currents calculated in the last step. Assume the transformers are ideal.	The power loss through the high-voltage transmission line is: $P = I^2 \times R = (2.085)^2 \times 5$ $= 22\,\text{W}$ The power loss through the medium-voltage distribution line is: $P = I^2 \times R = (20.85)^2 \times 5$ $= 2.2\,\text{kW}$

c Calculate the power loss through the medium-voltage transmission line if the voltage were 96 kV instead of 48 kV.	
Thinking	**Working**
Determine the effect of the change on the turns ratio and therefore on the current ratio. Recall that the power loss is proportional to I^2. If I decreases, so will the power loss.	If the voltage of the medium-voltage transmission line were increased to 96 kV from 48 kV, the turns ratio for that transformer (T_2) would increase by a factor of 2. Therefore the current in the medium-voltage transmission line would decrease by a factor of 2. If the current decreased by a factor of 2, the power loss would decrease by a factor of 4. Therefore, the power-loss through the medium-voltage transmission line would decrease from 2.2 kW to 550 W.

d Describe how the power loss could be minimised throughout the system.	
Thinking	**Working**
Recall that power loss is proportional to I^2 and that voltage and current are inversely proportional in this type of system.	Power loss through the transmission network can be reduced by reducing the effective resistance of the transmission lines and by using higher voltages to transmit power. It is for this reason that power is distributed at such high voltages across the country to major centres and consumers.

Worked example: Try yourself 7.5.6

POWER LOSS THROUGH A TRANSMISSION NETWORK

Consider the power transmission system shown below. The power plant generates AC voltage at 40 kV, which is stepped up to 600 kV using transformer T_1. Transformer T_2 steps the 600 kV voltage down to 60 kV and transformer T_3 steps the 60 kV voltage down to 240 V for the load.

The resistance of the high-voltage transmission line is 2.5 Ω, the resistance of the medium-voltage distribution line is 2.5 Ω and the resistance of the low-voltage distribution line is 7.5 Ω.

Assume that the transformers are 100% efficient and that the load consumes 2 MW of power.

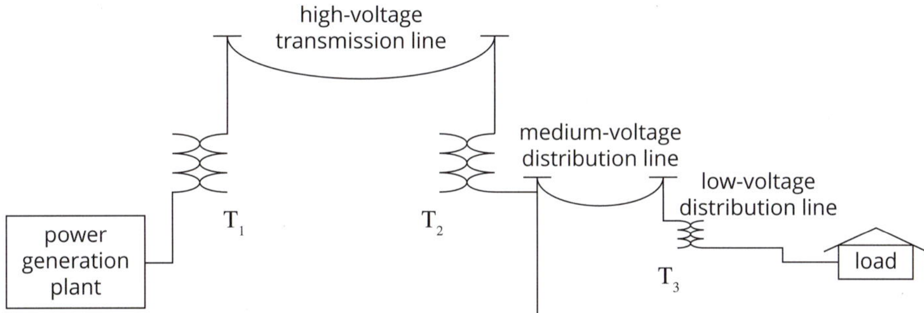

a Calculate the turns ratio of transformers T_1, T_2 and T_3.

b Determine the power loss through the high-voltage transmission and medium-voltage distribution lines.

c Calculate the power loss through the medium-voltage distribution line if it carried 120 kV instead of 60 kV.

d Describe how the power loss could be minimised throughout the system.

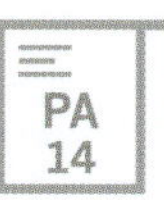

CASE STUDY

The War of Currents

AC and DC power supplies have been in competition for nearly as long as humans have been generating electricity. The heated debates about the benefits and disadvantages of each type of current prompted what has been called the 'War of Currents' in the late 1800s. During this time Thomas Edison, an American inventor and businessman, had started the Edison Electric Light Company and he hoped that his company would supply electricity to large parts of America with his DC generators. Meanwhile Nikola Tesla, a Serbian–American physicist, had invented the AC induction motor and, with financial support from George Westinghouse, hoped AC would become the dominant power supply. Ultimately, the ease with which AC could be stepped up using transformers for long-distance transmission with minimal power loss proved to be the prevailing benefit that led to AC winning the 'war'. However, in his attempt to win the competition, Edison portrayed high voltage AC power as terrifyingly dangerous—and demonstrated his claim by electrocuting elephants. He also invented the AC-powered electric chair for the American government to execute prisoners on death row.

While AC power is now universal in large-scale power distributions, there is a limit to how high the voltage of an AC system can go and still be efficient. Above approximately 100 kV, corona loss (due to the high voltage ionising air molecules) begins to occur, and above 500 kV it no longer becomes feasible to transmit electric power.

To transmit the same power as DC, an AC system needs to operate with a higher peak voltage. During that part of the cycle when the AC is lower than peak voltage, efficiency is compromised, because the higher the voltage the better. Until recently the expense of alternative methods of raising and lowering the voltage at either end of the transmission line outweighed this weakness in AC systems.

High DC voltage levels can now be reached more easily with new technology that employs small, high-frequency switching converters. And some projects—such as the Three Gorges Dam project in China (Figure 7.5.4)—are now transmitting DC at high voltages. There are some other benefits of DC transmission, such as that AC–DC transformers and three-phase industrial power become unnecessary. However, converting the millions of AC devices currently in use would be extraordinarily expensive and there are major issues with safety. For example, safety switches do not work with DC power.

In a way, the competition between Edison and Tesla continues. In 1892, the Edison Electric Company merged with the Thomson-Houston Electric Company to become the General Electric Company (GE). GE exists to this day as one of the largest and most profitable companies in the world, and Westinghouse—whose founder backed Tesla—is still in business as a major manufacturer of home appliances.

FIGURE 7.5.4 The Three Gorges Dam in China transmits DC electricity at higher voltages than is possible with AC, with an aim to reduce transmission losses.

PA 15

7.5 Review

SUMMARY

- A transformer works on the principle that a changing magnetic flux induces an EMF. All transformers consist of two coils, known as the primary coil and the secondary coil.
- Ideal transformers are 100% efficient and suffer no power loss; real transformers suffer some power loss, but it is usually so small (less than 1%) that it can be ignored in calculations.
- The transformer equation can be written in different versions but is based on:

$$\frac{N_1}{N_2} = \frac{V_1}{V_2} = \frac{I_2}{I_1}$$

- In a step-up transformer, the output voltage from the secondary coil is greater than the input voltage sent to the primary coil.
- In a step-down transformer, the output voltage from the secondary coil is less than the input voltage sent to the primary coil.
- Transformers will not work with DC voltage since it has a constant, unchanging current that creates no change in magnetic flux.
- The power supplied in an electrical circuit is given by $P = VI$.
- The power lost in an electrical circuit is given by $P = I^2R$.
- The AC electrical supply from a generator can be stepped up or down by transformers, hence AC is the preferred form of electrical energy in large-scale transmission systems.

KEY QUESTIONS

Knowledge and understanding

1 A non-ideal transformer has a slightly smaller power output from the secondary coil than power input to the primary coil. The voltage and current in the primary coil are denoted V_1 and I_1 respectively. The voltage and current in the secondary coil are denoted V_2 and I_2 respectively. Which of the following expressions describes the power output in the secondary coil?

A V_1I_1

B V_2I_2

C V_1I_2

D $I_2^{\,2}R$

2 Power loss can be expressed by the formula $P = \frac{\Delta V^2}{R} = I^2R$. Knowing this, under what current or voltage conditions is it best to transmit power to minimise losses?

3 A voltage sensor is connected to the output of a transformer. Which of the following graphs represents the most likely output for a steady DC voltage input?

A

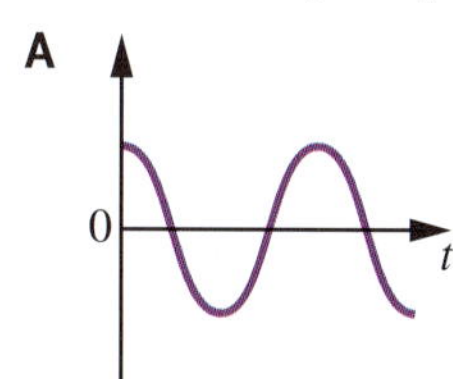

B

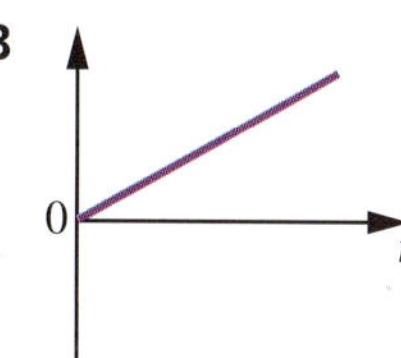

C

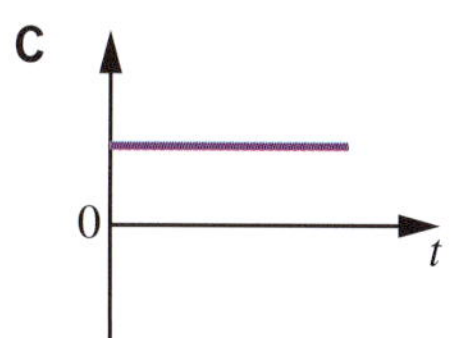

D

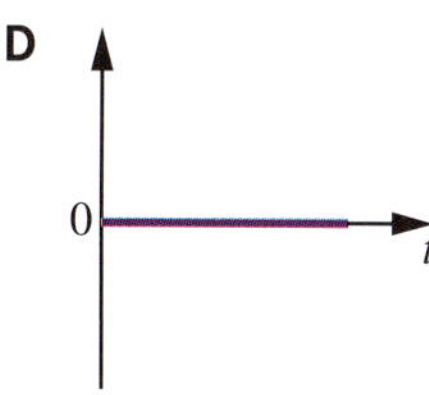

4 A water pump in a garden pond is operated from a mains voltage of 240 V rms through a step-down transformer with 600 turns on the primary coil. The pump operates on an rms voltage of 24 V. How many turns are on the secondary coil?

continued over page

7.5 Review *continued*

5 The primary coil of a transformer has 50 turns and the secondary coil 500 turns. The primary rms voltage input is 10 V and a primary rms current is 5.0 A.

   a What is the rms voltage across the load attached to the secondary coil?

   b What power is being supplied to the load attached to the secondary coil?

   c What is the rms current in the secondary coil?

6 A solar-powered generator produces 9.9 kW of electrical power at 700 V. The power is transmitted to a distant house via twin cables with total resistance of 3.6 Ω. What is the total power loss in the cables?

7 A power station generates 720 MW of power to be used by a town 120 km away. The power lines between the power station and the town have a total resistance of 2.5 Ω.

   a If the power is transmitted at 100 kV, what current would be required?

   b What voltage would be available at the town? Give your answer in kilovolts (kV).

Analysis

8 Suppose that a large-scale solar power generation plant is built approximately 1000 km from Melbourne. The production capacity of the plant is 100 MW at peak capacity. The output from the plant will connect with the rest of the electricity transmission network in Victoria. Compare the advantages of using AC or DC transmission, and recommend which of the two options you would choose.

9 A power station transformer is used to step up an AC voltage from 50 kV to 220 kV. The transformer has 250 turns on the primary coil. When the current in the primary coil is 75 A, the current in the secondary coil is 14.5 A.

   a Calculate the number of turns in the secondary coil.

   b Explain whether this is an ideal transformer.

10 Consider a high voltage electric transmission line that carries a current of 1 kA rms. Assume the equivalent resistance of the line is 0.34 Ω per kilometre. What is the power loss (rms and peak) over a 13 km distance?

Chapter review

KEY TERMS

alternating current
alternator
brushes
electromagnetic induction
EMF
Faraday's law
generator
ideal transformer
induced current
inverter
Lenz's law
magnetic flux
magnetic flux density
photovoltaic effect
root mean square
slip rings
split ring commutator
step-down transformer
step-up transformer
transformer

REVIEW QUESTIONS

Knowledge and understanding

1 Describe the benefits of high voltage DC power transmission over high voltage AC transmission.

2 Describe the purpose of a split ring commutator in a DC motor.

3 A coil is rotated about its vertical axis such that the left-hand side is coming out of the page and the right-hand side is going into it. A magnetic field runs from right to left across the page. In what direction is the induced current in the coil?

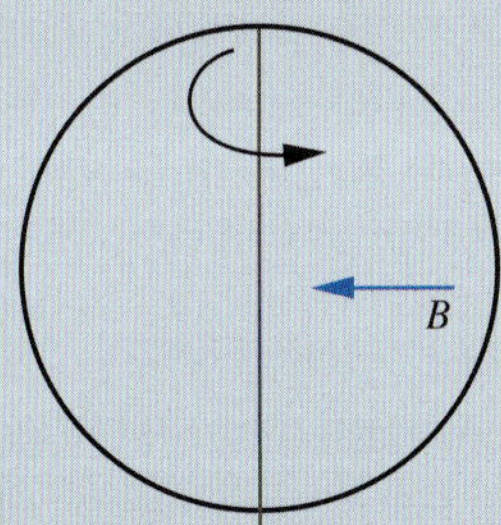

4 A coil in a magnetic field directed into the page is reduced in size. In what direction is the induced current in the coil while the coil is being reduced in size?

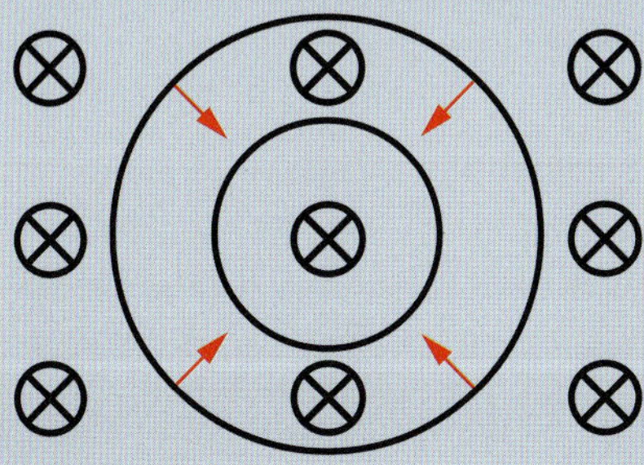

5 A single loop of wire is rotated within a magnetic field, as shown in the diagrams below. While the coil is rotating, an EMF is generated. Which sides of the coil contribute to the generation of the EMF? Give a reason for your answer.

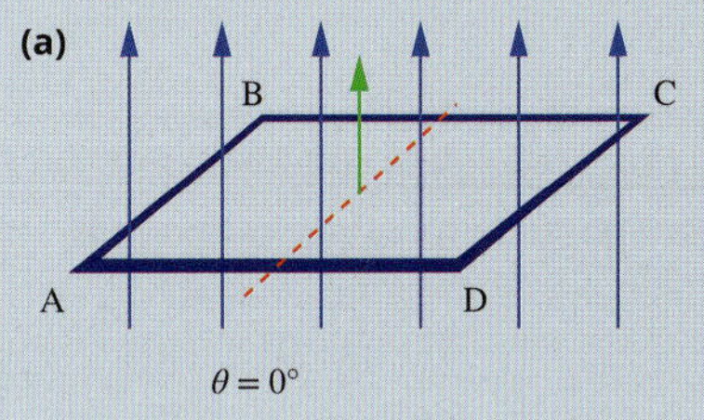

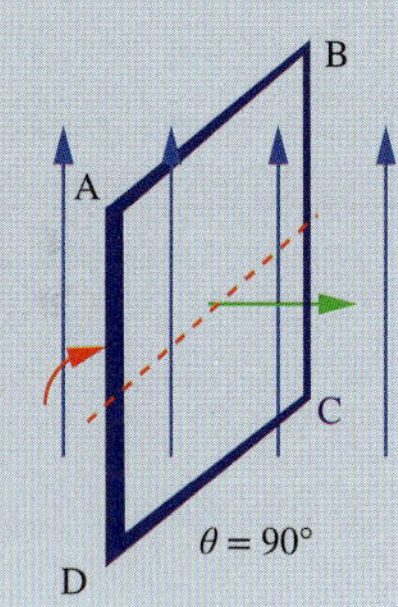

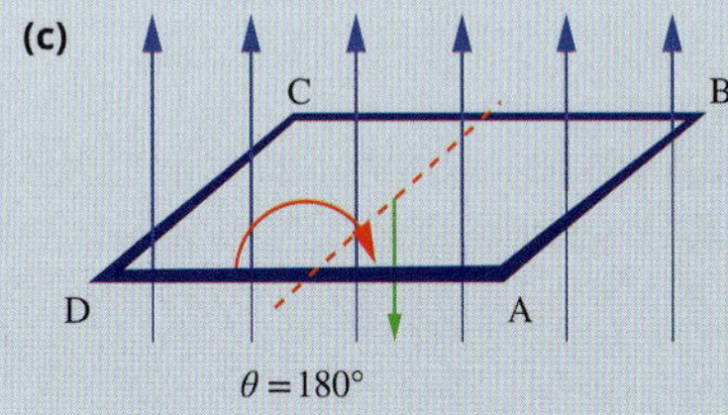

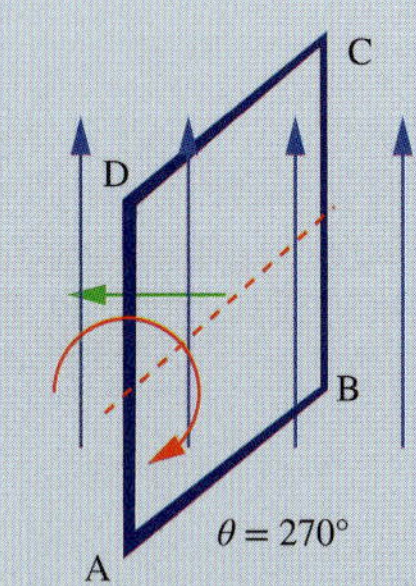

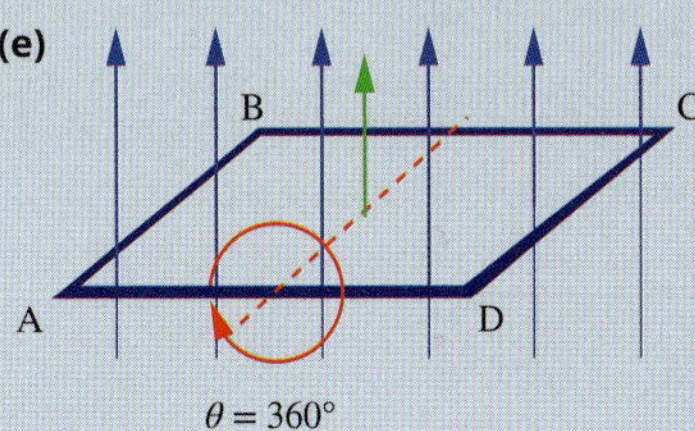

6 A rectangular coil of area 25 cm^2 and resistance 1.0 Ω is in a uniform magnetic field of 8.0×10^{-4} T directed out of the page. The plane of the coil is initially perpendicular to the field, as shown below.

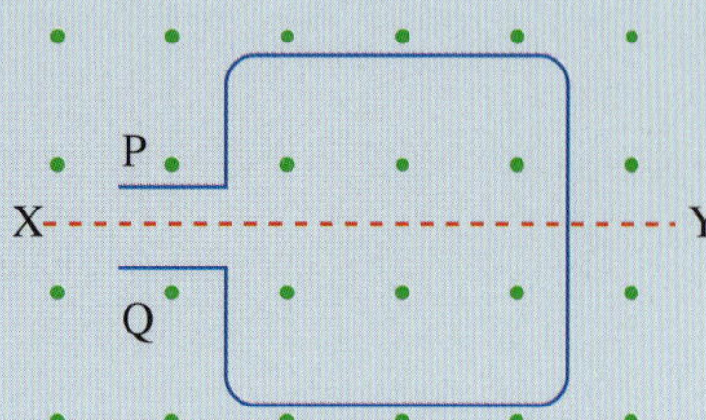

a What is the magnitude of the EMF induced in the coil when the strength of the magnetic field is increased by a factor of four in 3.5 ms?

b What is the direction of the current caused by the induced EMF in the coil when the strength of the magnetic field is halved?

7 During a physics experiment, a student pulls a horizontal circular coil from between the poles of two magnets in 0.15 s. The initial position of the coil is entirely within the field while the final position is completely free of the field. The coil has 65 turns, each of radius 5.0 cm. The field strength between the magnets is 35 mT.

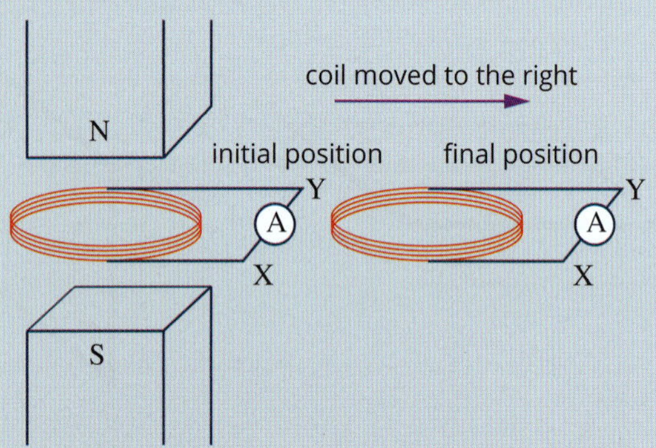

a What is the magnitude of the average EMF induced in the coil as it is moved from its initial position to its final position?

b What is the direction of the current in the coil caused by the induced EMF?

8 A copper rod, XY, of length 45 cm is free to move along a set of parallel conducting rails as shown below. The rails are connected to a switch, S, which completes a circuit when it is closed. A uniform magnetic field of strength 35 mT, directed out of the page, is established perpendicular to the circuit. S is closed and the rod is moved to the right with a constant speed of 2.5 m s^{-1}.

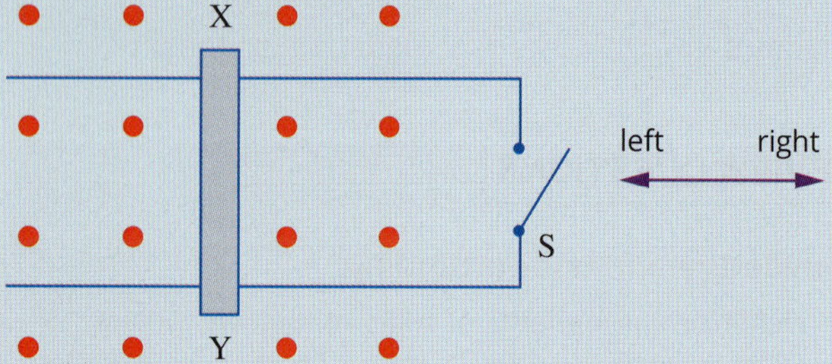

a What is the magnitude of the induced EMF in the rod, in mV?

b What is the direction of the current through the rod caused by the induced EMF?

9 A ship with a vertical steel mast 12.0 m long is travelling due west at 6.5 m s^{-1} in a region where the Earth's magnetic field is horizontal and equal to 7.5×10^{-5} T north. What average EMF, in mV, is induced between the top and bottom of the mast?

10 An ideal transformer is operating with an input voltage of 18 V rms and a primary current of 3.5 A rms. The output voltage is 54 V. There are 45 turns in the secondary winding.

a What is the rms output current?

b How many turns are there in the primary coil?

11 A student builds a simple alternator consisting of a coil containing 750 turns, each of area 10 cm^2, mounted on an axis that can rotate between the poles of a permanent magnet of strength 120 mT. The alternator is rotated at a frequency of 50 Hz.

a Find the average EMF of the alternator.

b Explain what the effect will be on the average EMF if the frequency is doubled to 100 Hz.

12 A wind turbine runs a 250 kW generator with an output voltage of 1000 V. The voltage is increased by transformer T_1 to 10 000 V for transmission to a town 5 km away through power lines with a total resistance of 2 Ω. At the town, another transformer, T_2, reduces the voltage to 250 V. Assume that there is no power loss in the transformers.

a What is the current in the power lines?

b How much power is lost in the power lines?

c It is suggested that some money could be saved by removing the first transformer. Explain, using appropriate calculations, whether this is a good plan.

Application and analysis

13 A coil consisting of 20 loops, each with an area of $0.2\,m^2$, is placed in a uniform magnetic field of 0.05 T so that the plane of the coil is perpendicular to the direction of the field. The magnetic flux through the coil is closest to:

A 0.2 Wb
B 10 mW
C 10×10^{-3} Wb
D 20 mW

14 Two loops of wire move into a uniform magnetic field of 5×10^{-4} T in 0.25 s from point X to point Y, as shown below. The area of the loops is $0.075\,m^2$.

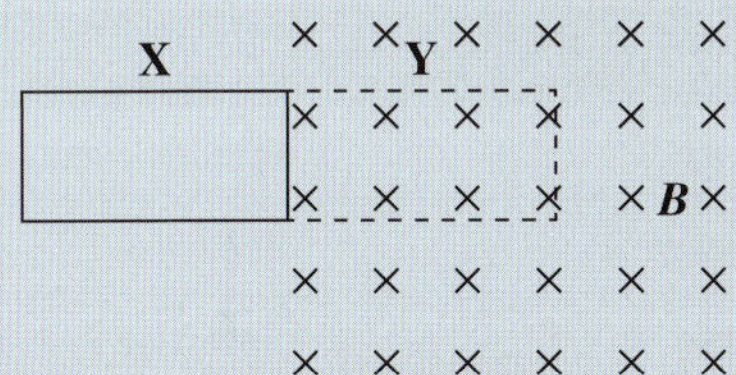

The magnitude of the average induced EMF in the loops is closest to:

A 2×10^{-3} V
B 1.5×10^{-4} V
C 4×10^{-3}
D 3×10^{-4} V

15 An ideal transformer has an input AC voltage of 200 V, 1500 turns in its primary coil and 60 turns in its secondary coil. The output voltage is closest to:

A 8 V
B 5 kV
C 125 mV
D 8 mV

16 Two 100 km AC transmission lines, line A and line B, have the same resistance and can transmit the same amount of power. Transmission line A carries a current of *I* and transmission line B carries a current of 3*I*. What is the ratio of the power lost due to resistance in transmission line A to the power lost due to resistance in transmission line B?

17 Assume that a residential house has solar roof panels that can produce on average 1.75 kW between 9.30 a.m. and 3.15 p.m. The house consumes power as shown in the following table.

Time period	Power use (kW)
10 a.m. to 11 a.m.	1.5
11 a.m. to noon	2.0
noon until 1 p.m.	2.5
1 p.m. to 2 p.m.	2.0
2 p.m. to 3 p.m.	1.5

a On one set of axes, plot graphs to show the household power consumption and solar production over this time. Use this to calculate how much the house draws from or feeds into the grid each hour.

b If the price to take electricity from the grid is $0.25 per kWh, how much does the house pay for electricity between 10 a.m. and 3 p.m.?

c Calculate the cost savings due to the house having solar panels.

18 **a** Referring to the figure below, describe whether this is an AC or DC generator. Justify your answer.

b If the rotor coil were turned clockwise from the position shown, which of the terminals (X or Y) would have the higher electrical potential?

c At what point during the rotor coil's rotation will the induced EMF be zero?

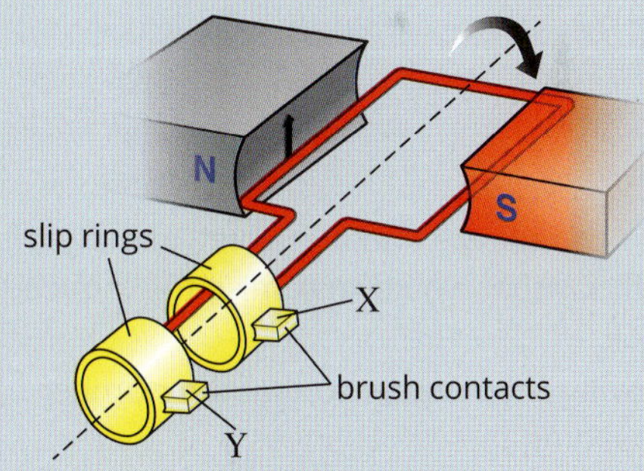

19 When an ideal (100% efficient) generator is used to provide a voltage of 240 V to a circuit, it requires 4.5 kW of power to turn the generator.

a What is the current and the total resistance of the circuit?

b If the total resistance of the circuit were quadrupled, what power would be required to turn the generator to maintain a 240 V potential across the circuit?

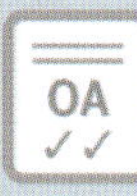

UNIT 3 • Area of Study 3

REVIEW QUESTIONS

How are fields used in electricity generation?

Multiple-choice questions

1 Identify the correct definition of magnetic flux.

A A region of space where a force acts on an object due to its electric charge.

B The force acting per unit current per unit length on a wire placed at right angles to a magnetic field.

C A region in which a moving charge or a current-carrying conductor will experience a force when it is placed in it.

D A measure of the total magnetic field passing through an area, calculated by multiplying the field strength by the perpendicular area.

The following information applies to questions 2 and 3.

A square loop of wire with sides of 5.0 cm is placed in a uniform magnetic field of strength 60 mT.

2 If the loop is aligned so that the field lines are perpendicular to the plane of the loop, which one of the following is closest to the magnetic flux in the loop?

A 3.0 Wb

B 0.15 Wb

C 0.15 mWb

D 0 Wb

3 If the loop is aligned so that the field lines are parallel with the plane of the loop, which one of the following is closest to the magnetic flux in the loop?

A 3.0 Wb

B 0.15 Wb

C 0.15 mWb

D 0 Wb

4 With a coil of wire and a magnet, identify the conditions that would maximise the induced EMF in the coil.

A moving the magnet so that the number of field lines threading the coil changes

B moving the magnet so that the number of field lines threading the coil does not change

C holding the magnet stationary where the field lines thread the coil

D holding the magnet stationary where no field lines thread the coil

5 Which of the following is the best description of how a transformer transfers electrical energy from its primary windings to its secondary windings?

A The current through the primary windings produces a constant electric field in the secondary windings.

B The current through the primary windings produces a steady magnetic field in the secondary windings.

C The current through the primary windings produces a changing magnetic field in the secondary windings.

D The current through the primary windings is connected to the secondary coil through the electrically conductive iron core.

6 A magnet, with its north end facing down, is dropped through a conducting coil, inducing a current. Which of the following options describes the direction of the induced current when viewed from above the coil?

	Magnet entering the coil	Magnet exiting the coil
A	clockwise	anticlockwise
B	anticlockwise	clockwise
C	clockwise	clockwise
D	anticlockwise	anticlockwise

7 Which figure below correctly describes the direction of the induced current in a coil moving out of a magnetic field?

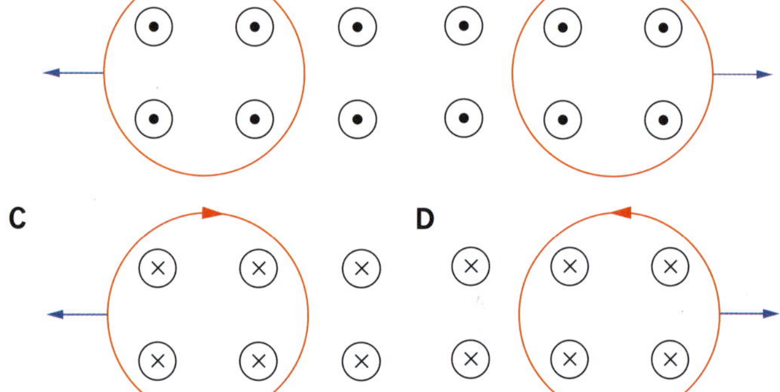

8 A transformer is built to reduce the 240 V supply voltage to 10 V for an appliance. If the number of turns in the secondary coil is 80, what is the number of turns required in the primary coil?

A 3

B 33

C 192

D 1920

9 A household is equipped with solar panels, making use of photovoltaic cells. Why must an inverter be connected before the electricity produced by the cells can be used for household appliances?

A The photovoltaic cells output a positive current, but household appliances need a negative current.

B The inverter converts the DC current output of the photovoltaic cells into AC for use in the home.

C The inverter converts the AC current output of the photovoltaic cells into DC for use in the home.

D The inverter swaps the current and voltage characteristics of the photovoltaic cell output for use in the home.

Short-answer questions

10 A rectangular loop of 100 turns is suspended in a magnetic field where B is 0.50 T. The plane of the loop is parallel to the direction of the field. The dimensions of the loop are 20 cm by 10 cm, with the longer side perpendicular to the field lines and the shorter side parallel to them.

a Calculate the amount of flux that threads the loop in the position described above.

b Describe how the amount of flux threading the loop can be increased.

c Explain how the plane of the loop and the magnetic field direction can be arranged so the maximum possible flux threads the loop.

d Calculate the maximum possible flux threading the loop when it is arranged as described in part **c**.

11 A square conducting loop with 20 cm sides and 0.50 Ω resistance is moving with a constant horizontal velocity of 5.0 cm s^{-1} towards a region of uniform magnetic field of strength 0.40 T directed vertically downwards, as shown in the following diagram. The magnetic field is confined to a cubic region with sides of 30 cm.

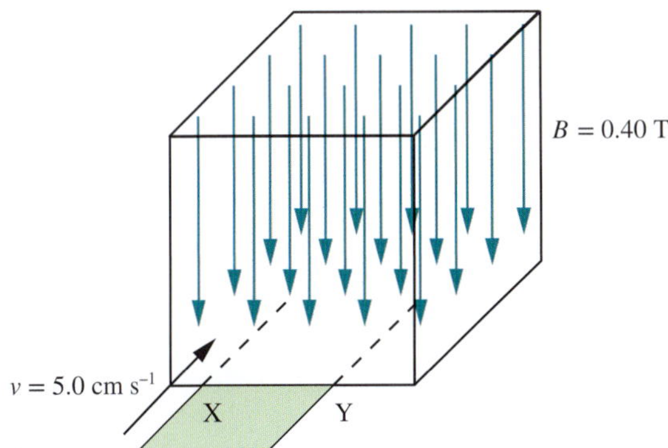

a Describe the direction of the induced current in the loop—from X to Y or Y to X—just as it begins to enter the field. Justify your answer.

b Calculate the average EMF induced in the loop when it is halfway into the field.

c Calculate the current in the loop when it is halfway into the field.

d How much electrical power is consumed in the loop when it is halfway into the field?

e What is the source of this power?

f What is the average EMF induced in the loop 5.0 s after it started to enter the field? Justify your answer.

g What is the direction of the induced current in the side XY just as it begins to emerge from the field? Justify your answer.

12 A 100 mm × 50 mm rectangular conducting loop of 2.0 Ω resistance is located with its plane perpendicular to a uniform magnetic field of strength 1.0 mT.

a Calculate the magnitude of the magnetic flux, Φ_B, threading the loop.

b The loop is rotated through an angle of 90° about an axis so that its plane is now parallel to the magnetic field. Determine the magnetic flux threading the loop in this new position.

c The time interval, Δt, for the rotation is 2.0 ms. Determine the magnitude of the average EMF induced in the loop.

d Determine the value of the average current induced in the loop during the rotation.

e Will the current keep flowing once the rotation is complete and the loop is stationary? Explain your answer.

13 A 5.0 Ω coil of 100 turns and a radius of 3.0 cm is placed between the poles of a magnet so that the flux through it is a maximum. The coil is connected to a sensitive current meter that has an internal resistance of 595 Ω. It is then moved out of the field of the magnet and an average current of 50 μA flows for 2.0 s.

a Had the coil been moved out in only 0.50 s, what would have been the average current?

b What is the strength of the magnetic field?

14 A physics student constructs a simple generator with a coil of 400 turns. The coil is mounted on an axis perpendicular to a uniform magnetic field of strength 50 mT and rotated at a frequency of 100 Hz. It is found that during the rotation, the peak voltage produced is 0.90 V.

a Sketch a graph showing the voltage output of the generator for at least two full rotations of the coil. Include a scale on the time and voltage axes.

b The student now rotates the coil at a frequency of 200 Hz. Describe the changes this would have on the graph.

15 What feature distinguishes an alternator from a DC generator?

16 How does the feature identified in question **15** operate?

17 An ideal transformer is operating with peak input voltage of 600 V and an rms primary current of 2.0 A. The peak output voltage is 3000 V. There are 1000 turns in the secondary windings.

a What is the rms output current?

b Determine the output peak-to-peak voltage.

c Determine the number of turns in the primary windings.

d Determine the rms power consumed in the secondary circuit.

e Calculate the peak power consumed in the secondary circuit.

18 A farmer has installed a wind generator on a hill along with a power line consisting of two cables with a combined total resistance of 2.0 Ω. The output of the generator is given as 250 V AC (rms) with a maximum power of 4000 W. She connects the system and finds that the voltage at the house is indeed 250 V. However, when she turns on various appliances so that the generator runs at its maximum power output (4000 W), she finds that the voltage supplied to the house is lower.

a Explain why the voltage dropped when the farmer turned on appliances in her house.

b Calculate the voltage and power at the house when the appliances are turned on.

She then decides to install ideal transformers at both ends of the same power line so that the voltage transmitted from the generator becomes 5000 V.

c Describe the essential features of the types of transformers that are needed at both ends of the power line.

For parts **d–h**, assume that the generator is operating at full load, i.e. 4000 W.

d Calculate the new current in the power line when the same appliances are turned on.

e Calculate the voltage drop along the power line.

f Determine the power loss in the power line.

g Determine the voltage delivered to the house.

h Determine the power delivered to the house.

i Explain how the power losses in the system without the transformers compare to the power losses in the system with the transformers as a percentage of the power generated.

j Explain why the system operated with much lower power losses when the electricity was transmitted at the higher voltage.

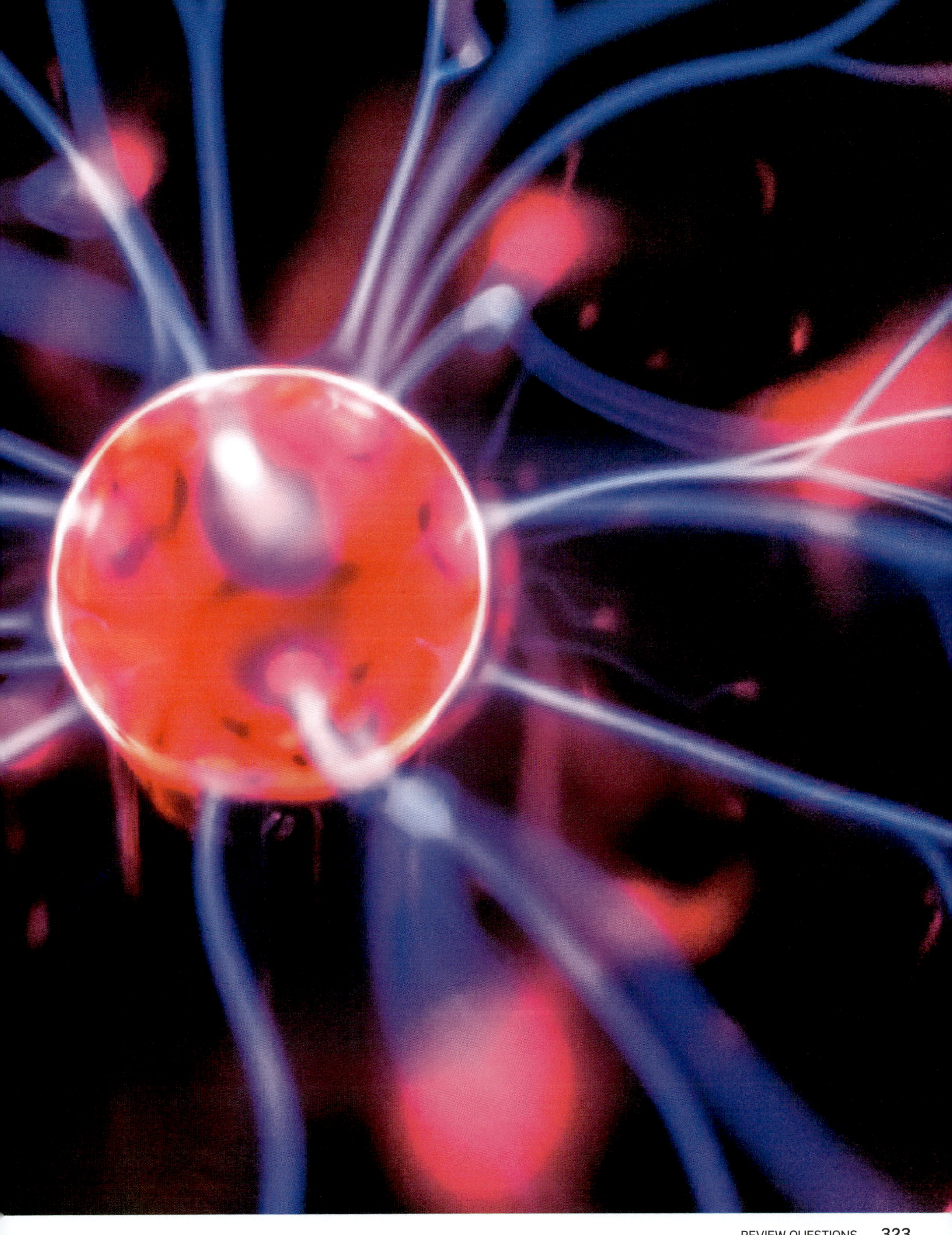

UNIT 4 How have creative ideas and investigation revolutionised thinking in physics?

To achieve the outcomes in Unit 4, you will draw on key knowledge outlined in each area of study and the related key science skills on pages 11 and 12 of the study design. The key science skills are discussed in Chapter 1 of this book.

AREA OF STUDY 1

How has understanding about the physical world changed?

Outcome 1: On completion of this unit the student should be able to analyse and apply models that explain the nature of light and matter, and use special relativity to explain observations made when objects are moving at speeds approaching the speed of light.

AREA OF STUDY 2

How is scientific inquiry used to investigate fields, motion or light?

Outcome 2: On completion of this unit the student should be able to design and conduct a scientific investigation related to fields, motion or light, and present an aim, methodology and method, results, discussion and a conclusion in a scientific poster.

CHAPTER

08 Light as a wave

Discovering the nature of light has been one of the scientific community's greatest challenges. Over the course of history, light has been compared to a geometric ray, a stream of particles and a series of waves. However, these relatively simple models have been found to be limited in their ability to explain all of the properties of light.

The search for a more adequate model has pushed scientists to develop new types of equipment and more sophisticated experiments, and led to a reshaping of our understanding of the nature of matter and energy.

This chapter will describe the evidence for the wavelike nature of light, including resonant effects and Thomas Young's famous double slit experiment. Subsequent chapters will show that light can behave as a wave in some circumstances and as a particle in others.

Key knowledge

- describe light as a transverse electromagnetic wave which is produced by the acceleration of charges, which in turn produces changing electric fields and associated changing magnetic fields **8.3**
- identify that all electromagnetic waves travel at the same speed, c, in a vacuum **8.3**
- explain the formation of a standing wave resulting from the superposition of a travelling wave and its reflection **8.1, 8.2**
- analyse the formation of standing waves (only those with nodes at both ends is required) **8.2**
- investigate and explain theoretically and practically diffraction as the directional spread of various frequencies with reference to different gap width or obstacle size, including the qualitative effect of changing the $\frac{\lambda}{w}$ ratio, and apply this to limitations of imaging using electromagnetic waves **8.3**
- explain the results of Young's double slit experiment with reference to:
  - evidence for the wave-like nature of light **8.3**
  - constructive and destructive interference of coherent waves in terms of path differences: $n\lambda$ and $(n+\frac{1}{2})\lambda$ respectively, where $n = 0, 1, 2,...$ **8.3**
  - effect of wavelength, distance of screen and slit separation on interference patterns: $\Delta x = \frac{\lambda L}{d}$ when $L >> d$ **8.3**

8.1 Wave interactions

In the late seventeenth century, a debate raged among scientists about the nature of light. Isaac Newton considered light to be particles—or 'corpuscles' as he called them—with each colour of the spectrum corresponding to a different type of particle. Robert Hooke (from England) and Christiaan Huygens (from Holland) proposed that light was a type of wave, similar to the water waves observed in the ocean.

A key point of difference between these two theories was that Newton's corpuscular theory suggested that light would speed up as it travelled through a solid material such as glass. In comparison, the wave theory predicted that light would be slower in glass than in air. At that time it was not possible to measure the speed of light accurately, so the question could not be resolved scientifically. Newton's esteemed reputation meant that for many years his corpuscular theory was considered correct.

It was not until the early nineteenth century that experiments convincingly demonstrated the wavelike properties of light. Today our understanding of light draws on aspects of both theories and is more complex than Newton, Hooke or Huygens could ever have imagined.

In this chapter we discuss important properties of all waves, such as superposition and resonance. Thomas Young's double slit experiment, a key experiment that demonstrates the wavelike nature of light, is also discussed.

THE PROPERTIES OF WAVES

In order to understand the wavelike nature of light, it will help if you have a good understanding of **mechanical waves**. Mechanical waves **transmit** energy and they need a **medium** for that energy to travel through. For example, water waves need water molecules to travel through and sound waves need air molecules. The properties of mechanical waves can be used to illustrate superposition, interference and resonance.

Light can be described as a **transverse** wave. Transverse waves were introduced in Unit 1 and can easily be modelled by oscillating a slinky or moving a rope up and down. The vibrations of a transverse wave are at right angles to the direction in which the wave travels.

The displacement vs distance graph in Figure 8.1.1 shows the displacement of particles along the length of a transverse wave at a particular point in time.

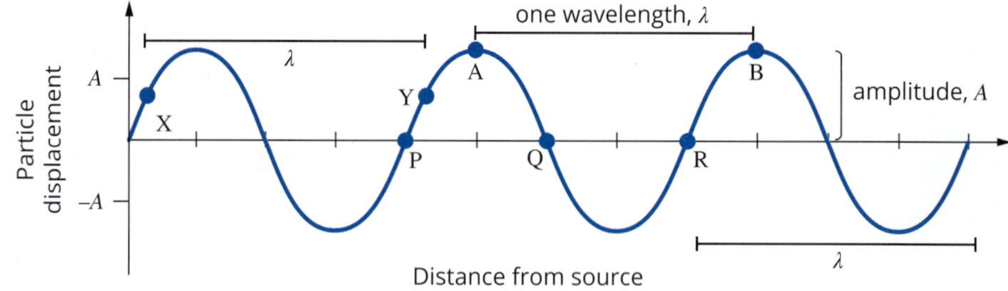

FIGURE 8.1.1 A sine wave representing particle displacements along a wave

The **amplitude** of a wave is the maximum displacement of the particles from their average or rest position. The peak of the displacement in the positive direction is called a **crest** and the peak in the negative direction is called a **trough**.

The **wavelength** (λ) is the length of a complete cycle or oscillation, that is, the distance between any two successive equivalent points on the wave. For example, in Figure 8.1.1 one wavelength is denoted by the distance between points A and B or between points X and Y.

A displacement vs time graph such as the one shown in Figure 8.1.2 traces the position of one point over time as the wave moves through that point. The displacement–time graph looks very similar to a displacement vs distance graph of a transverse wave, such as the one shown in Figure 8.1.1, so be careful to check the label on the horizontal axis.

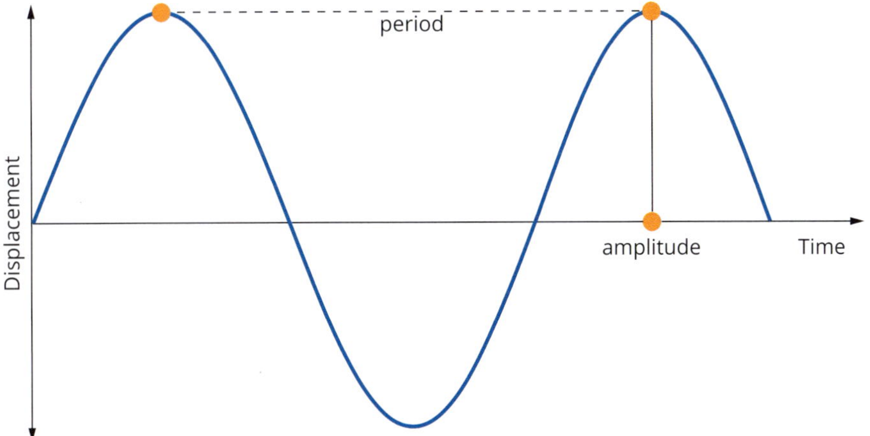

FIGURE 8.1.2 The graph of displacement vs time from the source of a transverse wave shows the movement of a single point on a wave over time as the wave passes through that point.

The **period** of a wave (T), measured in seconds, is the time for one complete oscillation (Figure 8.1.2).

The **frequency** (f), measured in hertz, is the inverse of the period: $f = \frac{1}{T}$.

The velocity of the wave (v), measured in metres per second, is calculated using the relationship $v = f\lambda$.

Note that sound is a **longitudinal** wave, that is, the wave vibration travels in the same direction as the wave. Sound waves have compressions and rarefactions instead of crests and troughs. **Compressions** are areas of increased pressure in a longitudinal wave and **rarefactions** are areas of decreased pressure. However, a longitudinal wave is often represented as a transverse wave, with crests corresponding to compressions and troughs corresponding to rarefactions. Thus the wave properties discussed in the following section apply equally to longitudinal and transverse waves.

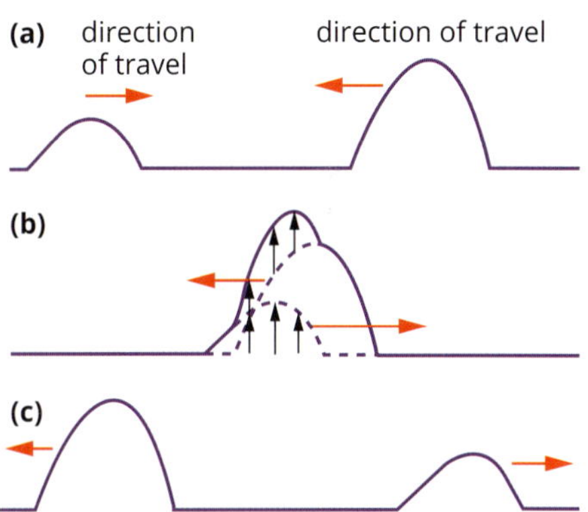

FIGURE 8.1.3 (a) Two wave pulses approach each other. (b) The superposition of the two pulses results in constructive interference. (c) After the interaction, the pulses continue unaltered. The interference does not permanently affect them.

SUPERPOSITION AND INTERFERENCE

Rarely in the real world does one mechanical wave occur in isolation. For example, the sounds produced by musical instruments and the human voice are the result of many waves interacting with each another. Wave interaction can also occur with water waves and in strings.

How waves interact can be explained by the principle of **superposition**. This principle states that when two or more waves interact, the resultant displacement at each point along the wave will be the vector sum of the displacements of the component waves. For example, imagine two transverse mechanical waves both with positive displacement travelling towards each other along a string (Figure 8.1.3(a)). When one wave coincides with another, the resulting displacement of the string is the vector sum of the two individual displacements (Figure 8.1.3(b)). The amplitude at this point is increased and the shape of the string resembles a combination of the two pulses. After they interact, the two pulses continue unaltered (Figure 8.1.3(c)). The principle of superposition can explain many **interference** phenomena. In Figure 8.1.3(b) the increase in amplitude as the two waves are added together is referred to as **constructive interference**.

Now consider the case where a pulse with a positive displacement meets one with a negative displacement (Figure 8.1.4(a)). Again the resulting displacement of the string is the vector sum of the two individual displacements. In this case a negative displacement is added to a positive displacement. The superposition of the pulses produces a pulse of a magnitude that is smaller than the largest of the two individual magnitudes but greater than the smallest of the two individual magnitudes (Figure 8.1.4(b)). This is called **destructive interference**. Once again the pulses emerge from the interaction unaltered (Figure 8.1.4(c)).

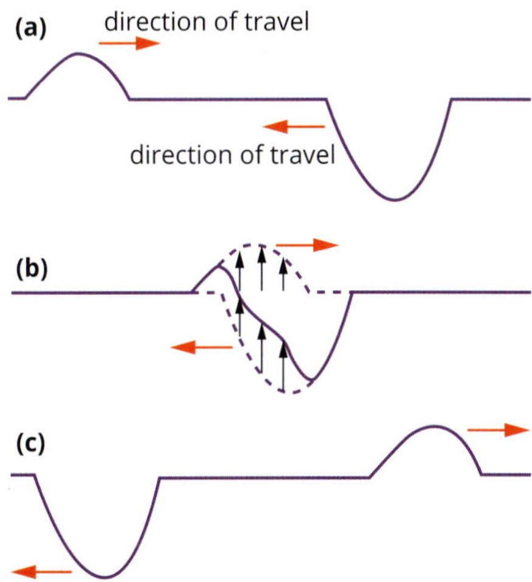

FIGURE 8.1.4 (a) Two wave pulses approach each other. (b) The superposition of the two pulses results in destructive interference. (c) After the interaction, the pulses continue unaltered. The interference does not permanently affect them.

When two waves meet and combine, there will be places where constructive interference occurs and places where destructive interference occurs. Although the wave pulses interact when they meet, passing through each other does not permanently alter the shape, amplitude or speed of either pulse. Longitudinal waves will also be superimposed as they interact, producing constructive, destructive or no interference

A special case of interference is complete constructive interference. This occurs where two waves of the same amplitude are exactly in **phase** (i.e. their peaks and troughs line up). The result is a wave of twice the amplitude (Figure 8.1.5(a)).

In complete destructive interference, two waves of the same amplitude are completely out of phase (i.e. one wave has a positive amplitude when the other has an equivalent negative amplitude). The two waves add together to give zero displacement (Figure 8.1.5(b)).

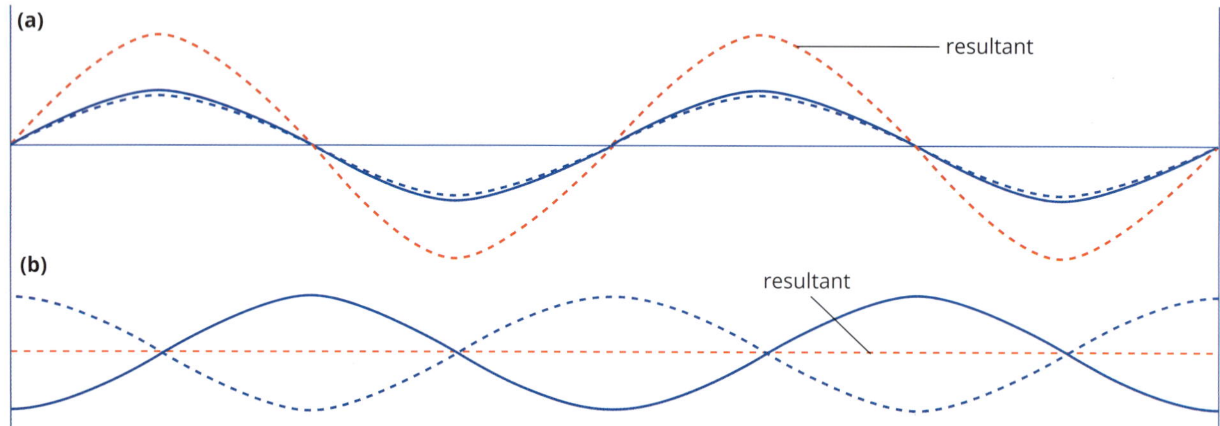

FIGURE 8.1.5 (a) Complete constructive interference occurs when the waves are in phase, i.e. when the peaks and troughs of the waves are perfectly aligned. This doubles the amplitude. (b) Complete destructive interference occurs when the two waves are completely out of phase, i.e. the peak of one wave exactly coincides with a trough of the other wave. The result is zero amplitude.

The effects of interference or superposition of waves can be seen in many everyday phenomena. Consider the ripples in a pond caused by falling raindrops (Figure 8.1.6). Where two ripples meet, a complex wave results from the superposition of the two ripples, after which the ripples continue unaltered. Similarly, all the sounds reaching your ear in a crowded room at one time are superimposed, so that just one complex sound wave arrives at your eardrum.

Superposition is important in understanding the formation of complex sounds. Imagine two single-frequency sound waves, or pure tones, one of which is twice the frequency of the other. When the two waves meet, they are added together to give a more complicated resultant sound wave (Figure 8.1.7).

Note that in Figure 8.1.7, transverse waves are used to represent sound waves. The crests represent compressions (areas of high pressure) and the troughs represent rarefactions (areas of low pressure). Showing longitudinal waves in this way makes certain features, such as period and therefore frequency, easier to see.

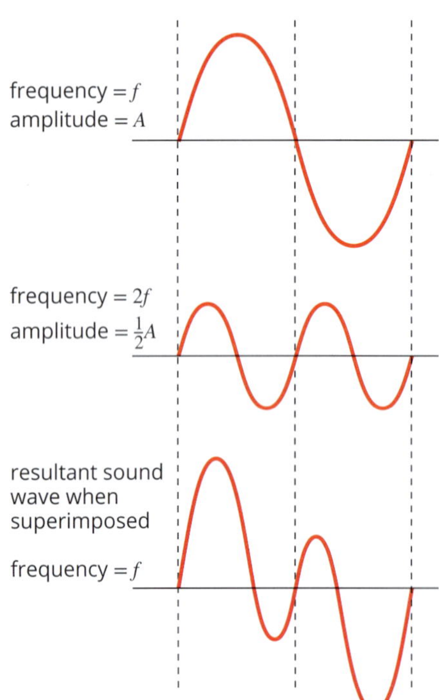

FIGURE 8.1.7 Two sound waves, one twice the frequency of the other, produce a complex wave of varying amplitude when they are superimposed.

FIGURE 8.1.6 The ripples from raindrops striking the surface of a pond behave independently regardless of whether or not they cross each other. Where the ripples meet, a more complex wave will result. After interacting, the ripples continue unaltered.

PHYSICSFILE

Noise cancelling headphones

Noise cancelling headphones use superposition to help reduce unwanted ambient noise. The circuitry in the headphones analyses the sound and, in less time than a human can detect, produces a new sound (i.e. a wave form) that counteracts the unwanted sound. The new sound is an inversion of the unwanted sound—that is, it is out of phase by 180°—so that when the two waves are superimposed, they cancel each other out completely (see figure below).

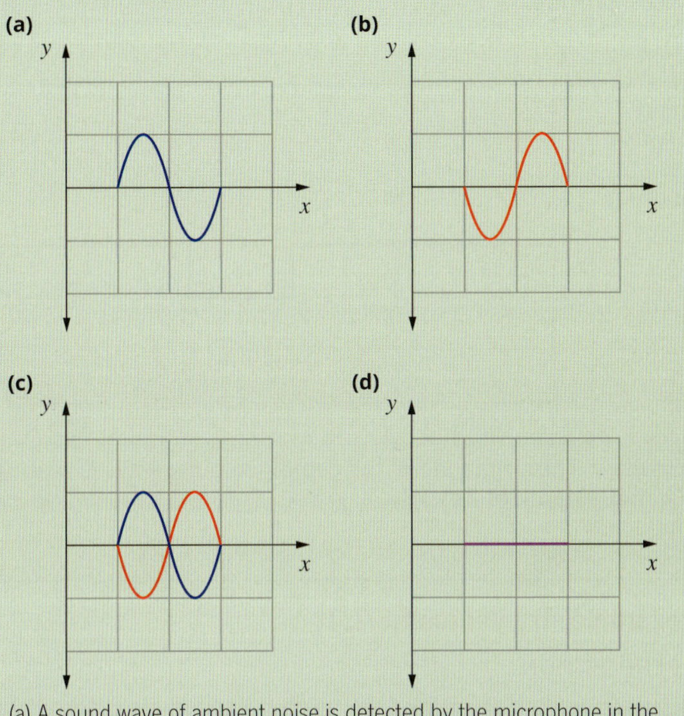

(a) A sound wave of ambient noise is detected by the microphone in the headphones. (b) The circuitry in the headphones produces an inverted waveform of the ambient noise. (c) The two waves are added together. (d) The result is that the ambient noise is cancelled.

PHYSICSFILE

The cocktail party effect

In a crowded room, individual sound waves will interfere with each other repeatedly, but it is still possible to distinguish which person is speaking. If you know the person's voice, then you know that their voice will sound the same. To discern one person's speech amid all the sounds in the room, your brain uses an innate ability to undo the superposition of waves by selecting one person's voice and suppressing all the other noise. This is known as the 'cocktail party effect'. It also explains your ability to hear your own name over the noise of other people talking.

Worked example 8.1.1

WAVE SUPERPOSITION

Two wave pulses are travelling towards each other parallel to the *x*-axis at $1\,\mathrm{m\,s^{-1}}$. The red arrows show the direction of travel. The *x*- and *y*-axes have a scale of 1 m for each box.

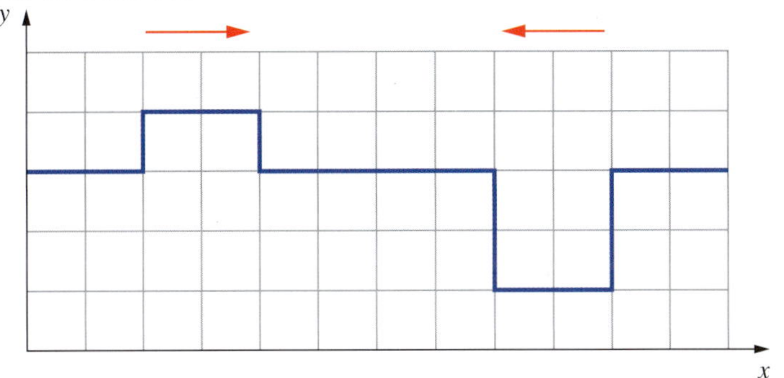

What is the amplitude of the resulting pulse three seconds after the instant shown on the above diagram?

Thinking	Working
Draw a diagram with the position of the two pulses after 3 seconds.	
Draw a new diagram with the pulses superimposed.	
The amplitude is the height of the resultant pulse.	The amplitude of the combined pulse wave is a trough with a depth of 1 m.

Worked example: Try yourself 8.1.1

WAVE SUPERPOSITION

Two wave pulses are travelling towards each other parallel to the *x*-axis at $1\ \mathrm{m\,s^{-1}}$. The directions are shown by the red arrows. The *x*- and *y*-axes have a scale of 1 m for each box.

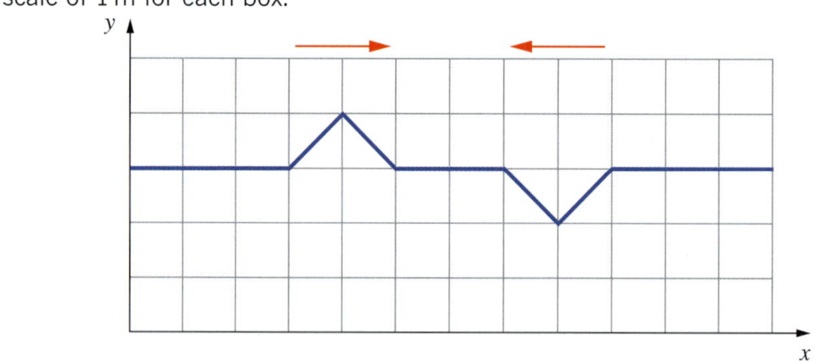

What is the amplitude of the combined pulse when the two pulses interact 2 seconds after the instant shown above?

RESONANCE

You may have heard about singers who can break glass by singing particular notes (Figure 8.1.8). All objects that can vibrate tend to do so at a specific frequency. This is known as their natural or **resonant frequency**. **Resonance** occurs when an object is exposed to vibrations at a frequency equal to its resonant frequency. Typically, a weaker vibration from one object causes a stronger vibration in another object. If the amplitude of the combined vibrations becomes too great, the object can be destroyed.

FIGURE 8.1.8 A glass can be destroyed by the vibrations caused by a singer emitting a sound of the same frequency as the resonant frequency of the glass.

A swing pushed once and left to swing freely back and forward is an example of an object vibrating at its natural frequency. The frequency at which it moves back and forth depends on the design of the swing, mostly on the length of its supporting ropes. In time, the oscillations will fade away as the energy is transferred to the supporting frame and the air.

If you watch a swing in motion, you can determine its natural oscillating frequency. It is then possible to push the swing at exactly the right time so that you match its natural oscillation. The additional energy you add by pushing will increase the amplitude of the swing rather than work against it. Over time, the amplitude will increase and the swing will go higher and higher: this is resonance. The swing can only be pushed at one particular rate to get this increase in amplitude (i.e. to get the swing to resonate). If the rate is faster or slower, the forcing frequency that you are providing will not match the natural frequency of the swing and you will be fighting against the swing rather than assisting it.

Other examples of resonant frequency that you may have encountered are blowing air across the mouthpiece of a flute or drawing a bow across a string of a violin in just the right place (Figure 8.1.9). In each case, a clearly amplified sound is heard when the frequency of the forcing vibration matches the natural resonant frequency of the instrument.

FIGURE 8.1.9 The sound box of a stringed instrument is tuned to resonate for the range of frequencies of the vibrations produced by the strings. When a string is plucked or bowed, the airspace inside the box vibrates in resonance with the natural frequency and the sound is amplified.

Two very significant effects occur when the natural resonant frequency of an object is matched by the forcing frequency.

- The amplitude of the oscillations within the resonating object will increase dramatically.
- The maximum possible energy from the source creating the forced vibration is transferred to the resonating object.

In musical instruments and loudspeakers, resonance is a desired effect. The sounding boards of pianos and the enclosures of loudspeakers are designed to amplify particular frequencies. In other systems, such as car exhaust systems and suspension bridges, resonance is not always desirable, and care must be taken to design the system to prevent resonance.

8.1 Review

OA ✓✓

SUMMARY

- The principle of superposition states that when two or more waves interact, the resultant displacement at each point along the wave will be the vector sum of the displacements of the component waves.
- Interference phenomena, such as constructive and destructive interference, can be explained by the principle of superposition.
- Resonance occurs when the frequency of a forcing vibration equals the natural frequency of the object.
- Two special effects occur with resonance:
  - the amplitude of vibration increases
  - the maximum possible energy from the source is transferred to the resonating object.

KEY QUESTIONS

Knowledge and understanding

1 Which of the following statements about the interaction of wave pulses are true and which are false? For each false statement, rewrite it so that it is true.

  a The displacement of the resultant pulse is equal to the sum of the displacements of the individual pulses.

  b As the pulses pass through each other, the interaction permanently alters the characteristics of each pulse.

  c After the pulses pass through each other, they will have the same characteristics as before the interaction.

2 What conditions are required for resonance to occur?

3 Which of the following statements is true when resonance occurs?

  A The amplitude of the vibration decreases.

  B The amplitude of the vibration increases.

  C The frequency of the vibration increases.

  D The frequency of the vibration decreases.

4 When pushing a child on a swing, explain using the physics explored in this section, what you need to do to make the swing go higher.

5 Two triangular wave pulses are moving towards each other at 1 m s^{-1}. Each pulse is 2 m wide.

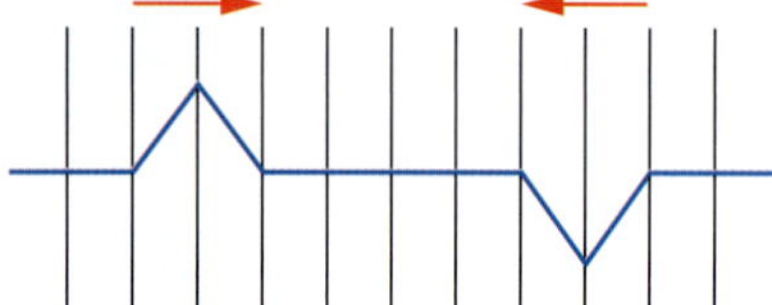

Draw the resultant wave 3 s from the instant shown.

Analysis

6 A footbridge has a natural frequency of oscillation from side to side of approximately 1.5 Hz. When pedestrians cross the bridge at a pace that will produce an oscillation in the bridge close to its natural frequency, resonance will occur. The graph that follows displays relevant data about pedestrians walking or running. A pedestrian completes 1 cycle of their motion every 2 steps. Which activity of the pedestrians is most likely to cause damage to the footbridge over time? Explain your answer.

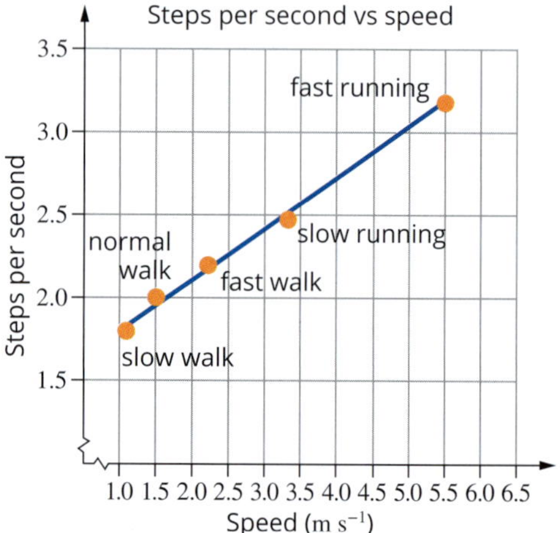

7 The period of a pendulum's oscillation is the time taken for it to swing from one point and then back to the same point. The frequency of a pendulum's oscillation depends on its length. Consider a set of pendulums attached to the same horizontal bar, as shown in the diagram below. If the pendulum on the left is made to oscillate (into and out of the page) it causes the other pendulums to oscillate, as it provides a forcing vibration on the others. Which pendulum, A, B, C or D, would oscillate with the largest amplitude? Explain your answer.

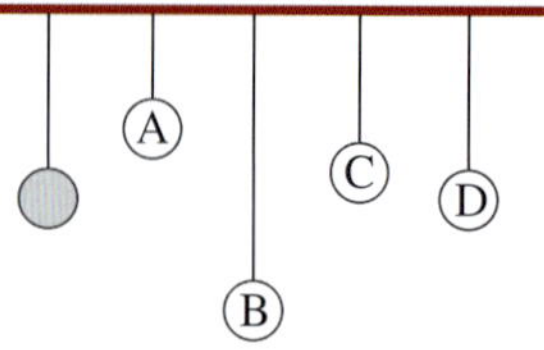

8 While sitting in a stationary truck with the engine idling at 100 Hz, the driver notices that the truck is vibrating strongly. As the driver accelerates to 100 km h^{-1} she notices that the vibration is much less. Explain why this might happen.

8.2 Standing waves in strings

Drawing a bow across a violin string causes the string to vibrate between the fixed bridge of the violin and the finger of the violinist (Figure 8.2.1). The simplest vibration will have maximum amplitude at the centre of the string, halfway between the bridge and the finger. This is a very simple example of a transverse standing wave.

Standing waves are formed from the interference of waves. They occur when two waves of the same amplitude and frequency are travelling in opposite directions towards each other in the same string. Usually, one wave is the reflection of the other. Standing waves are responsible for the wide variety of sounds associated with speech and music.

FIGURE 8.2.1 Transverse standing waves can form along a violin string when the string is bowed by a violinist.

REFLECTION ON A STRING

When a transverse wave pulse reaches a hard surface, such as the fixed end of a rope, the wave is **reflected** (it bounces back). If the end of the rope is fixed, the reflected pulse is inverted; that is, a crest is reflected as a trough and a trough is reflected as a crest (Figure 8.2.2).

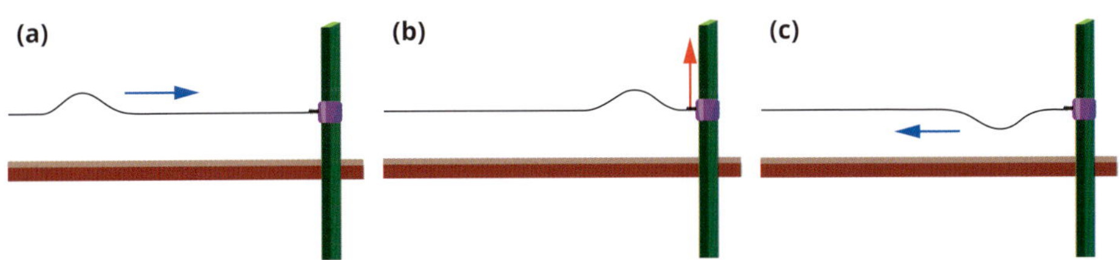

FIGURE 8.2.2 (a) A wave pulse moves along a string to the right and approaches a fixed post. (b) On reaching the post, the string exerts an upwards force on the post. Due to Newton's third law, the post exerts an equal and opposite force on the string. (c) The equal and opposite force inverts the wave pulse and sends its reflection back on the lower side of the string. There is a phase reversal on reflection from the fixed end.

This inversion can also be referred to as a 180° change of phase or, expressed in terms of the wavelength, λ, a shift in phase of $\frac{\lambda}{2}$.

When a wave pulse hits the end of a rope that is free to move (known as a free boundary), the pulse returns with no change of phase, that is, the reflected pulse is the same as the incident pulse. A crest is reflected as a crest and a trough is reflected as a trough (Figure 8.2.3).

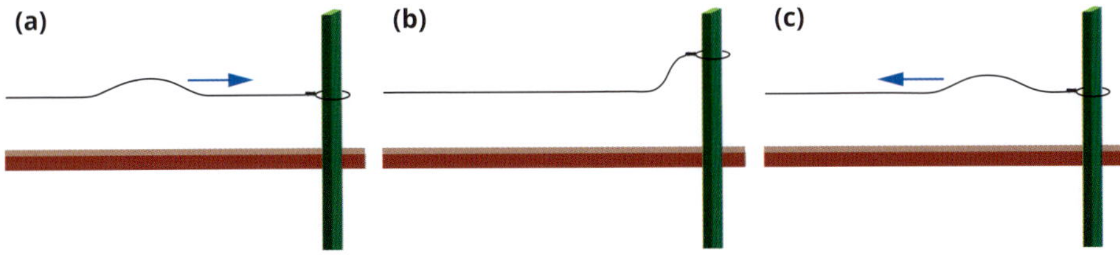

FIGURE 8.2.3 (a) A wave pulse moves along a string to the right and approaches a free end of the string at the post. (b) On reaching the post the free end of the string is free to slide up the post. (c) No inversion happens and the wave pulse is reflected back to the left on the same side of the string, i.e. there is no phase reversal on reflection from a free end.

In reality, when a transverse wave pulse is reflected, the amplitude of the reflected wave isn't quite the same as the original. Part of the energy of the wave is **absorbed** by the post: some will be transformed into heat energy and some will continue to travel through the post. You can see this more clearly by connecting a heavier rope to a lighter rope. The change in density has the same effect as a change in medium (Figure 8.2.4).

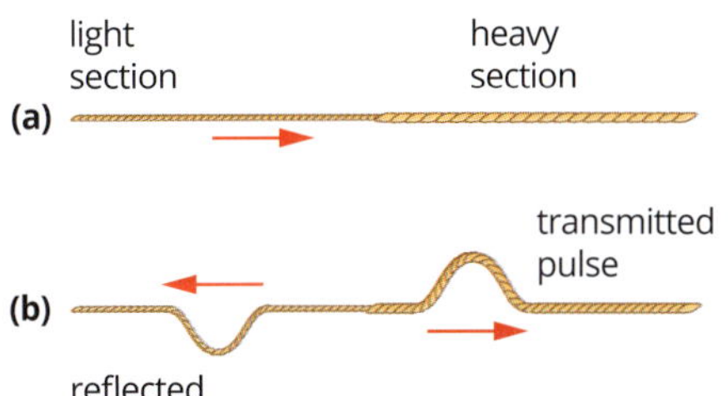

FIGURE 8.2.4 (a) A wave pulse travels along a light rope towards a heavier rope that is connected to it. (b) On encountering a change in density, the wave pulse will be partly reflected and partly transmitted. This is analogous to a change in medium.

When a transverse wave pulse is sent down the rope from the light rope to the heavier rope, part of the wave pulse will be reflected and part of it will be transmitted to the heavier rope. As the second rope is heavier, a smaller proportion of the wave is transmitted to it and a larger proportion of the wave is reflected back.

This example is analogous to a wave pulse striking a wall. The more rigid and/or dense the wall, the more the wave energy will be reflected and the less it will be absorbed. However, there will always be some energy that is absorbed by, or transferred to, the second medium. This explains why sound can travel through walls.

STANDING WAVES IN A STRING

Imagine creating a series of waves in a rope by shaking it vigorously. As the rope continues to be shaken, waves will travel in both directions. The new waves travelling down the rope will interfere with those being reflected back along the rope. This kind of motion will usually create quite a random pattern, with the waves quickly dying away. Shaking the rope at just the right frequency, however, will create a new wave that interferes with the reflection in such a way that the two superimposed waves create a single standing wave of larger amplitude.

This wave is called a standing wave because the wave doesn't appear to be travelling along the rope. The rope just seems to oscillate up and down with a fixed pattern. This situation contrasts with a standard transverse wave where every point on the rope would have a maximum displacement at some time as the wave travels along the rope.

In Figures 8.2.5(a) to (d), two waves (drawn in blue) are shown travelling in opposite directions towards each other along a rope. One wave is a string of pulses (shown as a solid line) and the other is its reflection (shown as a dashed line). The two waves superimpose when they meet and interference results. Since the amplitude and frequency of each wave is the same, the end result, shown in Figure 8.2.5(e), is a standing wave. At the points where destructive interference occurs, the two waves totally cancel each other and the rope will remain still. These are called **nodes**. Where the rope oscillates with maximum amplitude, constructive interference is occurring. These points on the standing wave are called **antinodes**.

In Figure 8.2.5(a), $t = \frac{1}{4}T$. At this time the two waves are completely superimposed (blue lines), and complete constructive interference is occurring, resulting in a wave twice the original amplitude (red dotted line).

In Figure 8.2.5(b), $t = \frac{1}{2}T$. After a further one-quarter of the period, each wave will have moved $\frac{\lambda}{4}$, which means they have moved $\frac{\lambda}{2}$ in relation to each other. The waves are completely out of phase and the resulting displacement is zero. As more time passes, the waves will continue to move past each other.

In Figure. 8.2.5(c), $t = \frac{3}{4}T$. At this point the waves completely superimpose and constructively interfere in the opposite direction.

In Figure 8.2.5(d), $t = T$ and the waves completely cancel each other again.

In Figure 8.2.5(e) the cycles shown in parts (a) to (d) form a standing wave. At an antinode (A) the standing wave swings from maximum positive displacement towards maximum negative displacement, and vice versa. The antinodes lie halfway between the stationary nodes (N). Regardless of the position of the component waves, these nodes stay in the same place, because the displacement at these points is always zero. Successive nodal points lie $\frac{\lambda}{2}$ apart, as do successive antinodal points.

The nodes and antinodes in a standing wave remain in a fixed position for a particular frequency of vibration. Figure 8.2.6 illustrates a series of possible standing waves in a rope with both ends fixed, each corresponding to a different frequency. The lowest frequency of vibration, (a), produces a standing wave with one antinode in the centre of the rope. The ends are fixed so they will always be nodal points. Patterns (b) and (c) are produced at twice and three times the original frequency respectively.

The rope could also vibrate at a frequency four times that of the original and even more. The frequencies at which standing waves are produced are called the resonant frequencies of the rope.

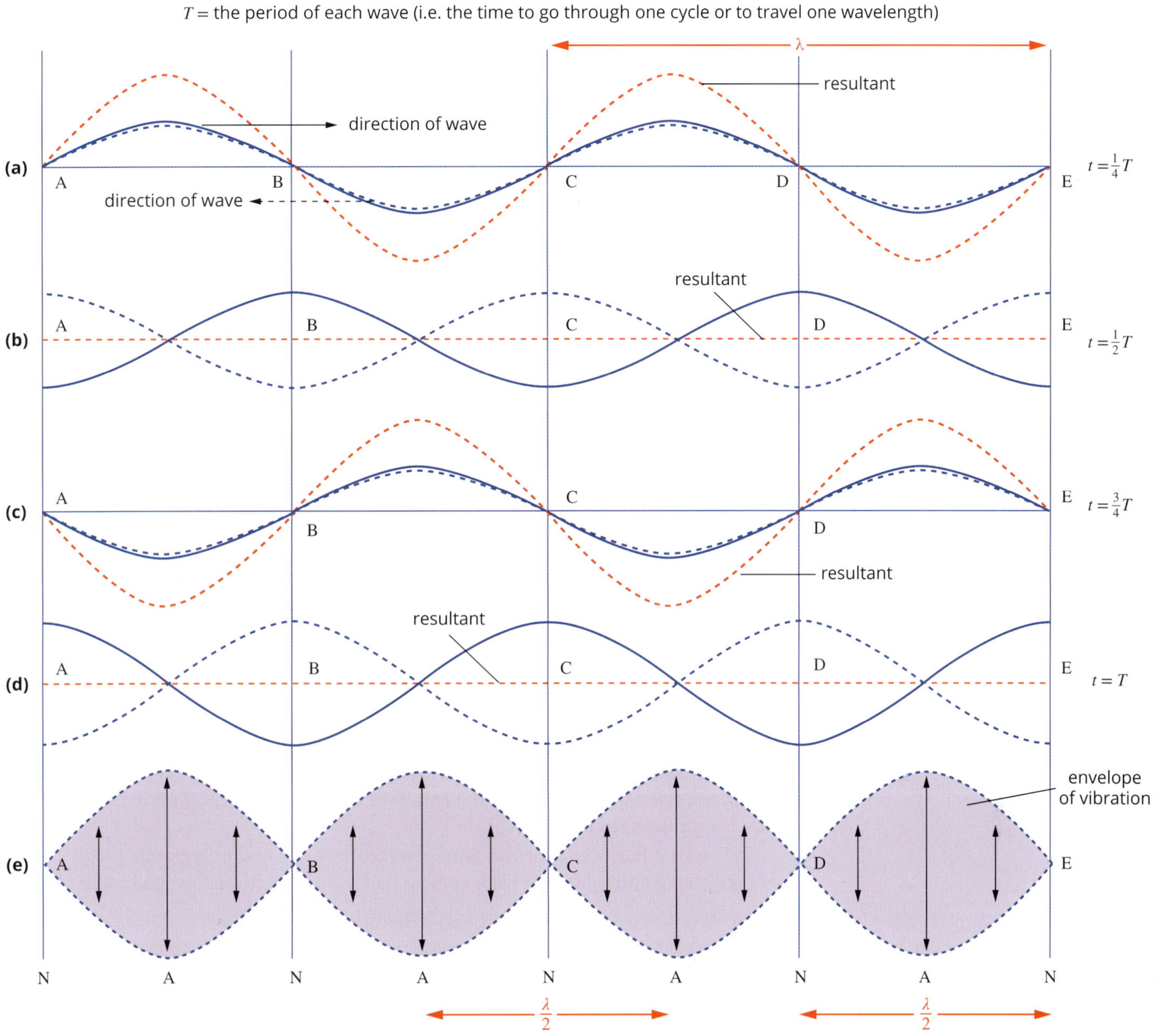

FIGURE 8.2.5 A standing wave created in a rope from two waves travelling in opposite directions, each with the same amplitude and frequency

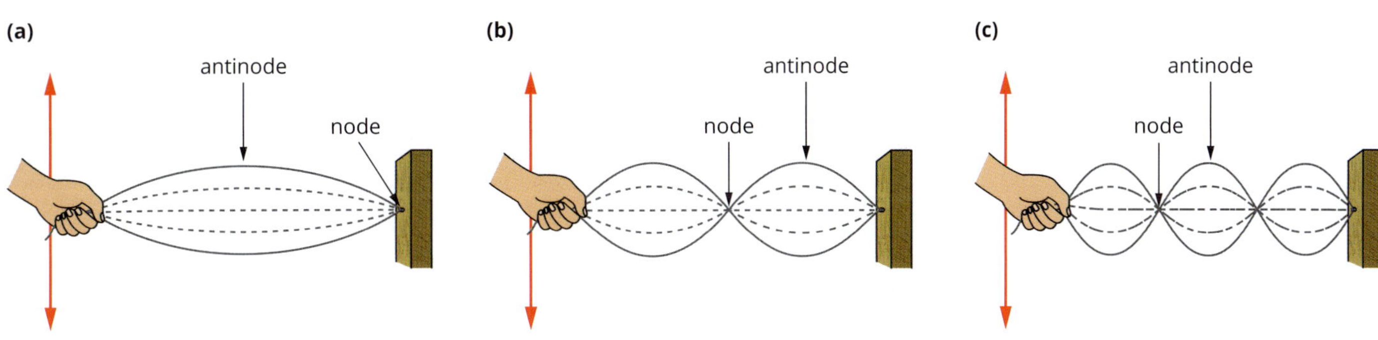

FIGURE 8.2.6 A rope vibrated at three different resonant frequencies, illustrating the standing waves produced at each frequency

Note that the formation of a standing wave does not mean that the rope itself is stationary. It will continue to oscillate as further wave pulses travel up and down the rope. It is the relative position of the nodes and antinodes that remain unchanged. The oscillation between maximum positive amplitude and maximum negative amplitude usually occurs at a rate too fast for the human eye to distinguish, hence a standing wave envelope is observed.

Note, too, that standing waves are not a natural consequence of every wave reflection. Standing waves are only produced by the superposition of two waves travelling in opposite directions that have the same amplitude and frequency.

Standing waves are an example of resonance: the forward-travelling wave resonates with the reflected wave that is travelling back.

This section has discussed standing waves in terms of the everyday world. However, standing waves and resonance are important concepts in understanding quantum effects at the atomic and sub-atomic level. They are also important in understanding the interaction between matter and light. This is covered in detail in Chapter 9.

HARMONICS

The strings of a musical instrument create a large variety of waves. They travel along the string in both directions, reflecting from the fixed ends. Most of these vibrations will interfere in a random fashion and die away. However, those corresponding to the resonant frequencies of the string will form standing waves and remain.

The resonant frequencies produced in this complex vibration of multiple standing waves are termed **harmonics**. The lowest and simplest form of vibration, with one antinode (Figure 8.2.7(a)), is called the **fundamental frequency**. Higher-level harmonics (Figure 8.2.7(b), (c) and (d)) are referred to by musicians as **overtones**.

The fundamental frequency usually has the greatest amplitude, so it has the greatest influence on the sound. The amplitude generally decreases for each subsequent harmonic. Usually all possible harmonics are produced in a string simultaneously, and the instrument and the air around it also vibrate to create the complex mixture of frequencies heard as an instrumental note.

The resonant frequencies or harmonics in a string of length l can be calculated from the relationship between the length of the string and the wavelength, λ, of the corresponding standing wave.

The first harmonic, or fundamental frequency, has one antinode in the centre of the string and a node at each end, so half a wavelength is formed by the string:

$$l = \frac{\lambda_1}{2} \text{ therefore } \lambda_1 = 2l$$

The second harmonic (first overtone) will have two antinodes and three nodes, so there is exactly one wavelength formed by the string:

$$l = 2\frac{\lambda_2}{2} \text{ therefore } \lambda_2 = l$$

The third harmonic (second overtone) will have three antinodes and four nodes, so there are 1½ wavelengths formed by the string:

$$l = 3\frac{\lambda_3}{2} \text{ therefore } \lambda_3 = \frac{2l}{3}$$

In general, any harmonic of a string fixed at both ends can be described as follows.

$$l = n\frac{\lambda_n}{2}$$

$$\therefore \lambda_n = \frac{2l}{n}$$

where λ_n is the wavelength of the nth harmonic (m)
l is the length of the string (m)
n is the number of the harmonic, which is also the number of antinodes: 1, 2, 3, 4 …

The relationship between wavelength, λ, frequency, f, and string length, l, is shown in Figure 8.2.7.

You may recall the wave equation, $v = f\lambda$, from your studies of light in Unit 1 Physics. From the wave equation we can derive the relationship between frequency, velocity and string length.

For the first harmonic, or fundamental frequency:

$$\lambda_1 = 2l \text{ and } v = f_1\lambda_1, \text{ thus } f_1 = \frac{v}{\lambda_1} = \frac{v}{2l}$$

For the second harmonic (first overtone):

$$\lambda_2 = l \text{ and } v = f_2\lambda_2, \text{ thus } f_2 = \frac{v}{\lambda_2} = \frac{v}{l} \text{ and } f_2 = 2f_1$$

For the third harmonic (second overtone):

$$\lambda_3 = \frac{2l}{3} \text{ and } v = f_3\lambda_3, \text{ thus } f_3 = \frac{v}{\lambda_3} = \frac{3v}{2l} \text{ and } f_3 = 3f_1$$

In general, the frequency of any harmonic is given by the following equation.

$$f_n = \frac{nv}{2l} \text{ and } f_n = nf_1$$

where n is the number of the harmonic

f_n is the frequency of the wave for the nth harmonic (Hz)

v is the velocity of the wave ($m\,s^{-1}$)

l is the length of the string (m)

(a) first harmonic (fundamental frequency) $n = 1$

$\lambda_1 = 2l; f_1 = \frac{v}{2l}$

(b) second harmonic (first overtone) $n = 2$

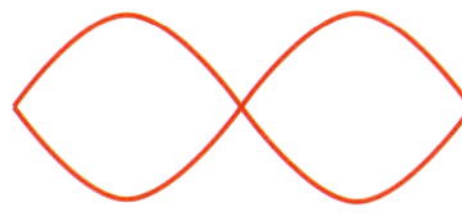

$\lambda_2 = l; f_2 = \frac{v}{l} = 2f_1$

(c) third harmonic (second overtone) $n = 3$

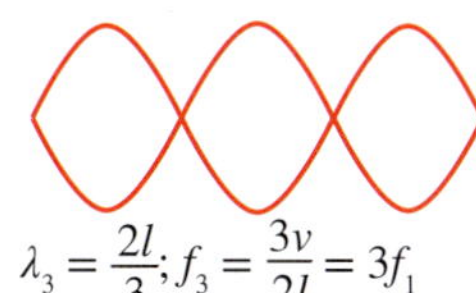

$\lambda_3 = \frac{2l}{3}; f_3 = \frac{3v}{2l} = 3f_1$

(d) fourth harmonic (third overtone) $n = 4$

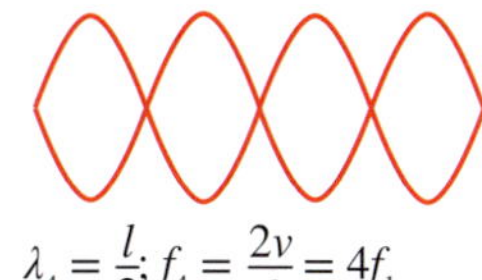

$\lambda_4 = \frac{l}{2}; f_4 = \frac{2v}{l} = 4f_1$

FIGURE 8.2.7 The first four resonant frequencies, or harmonics, in a string fixed at both ends. Being fixed, the ends will always be nodal points.

Worked example 8.2.1

FUNDAMENTAL FREQUENCY

A violin string, fixed at both ends, has a length of 22 cm. It is vibrating at its fundamental frequency.

a What is the wavelength of the fundamental frequency?

Thinking	Working
Note the length of the string (l) in metres and the harmonic number (n).	$l = 22\text{ cm} = 0.22\text{ m}$ $n = 1$
Recall that for any frequency, $\lambda_n = \frac{2l}{n}$. Substitute the given values and solve for λ.	$\lambda_1 = \frac{2l}{n}$ $= \frac{2 \times 0.22}{1}$ $= 0.44\text{ m}$

b What is the wavelength of the second harmonic?

Thinking	Working
Note the length of the string (l) in metres and the harmonic number (n).	$l = 22\text{ cm} = 0.22\text{ m}$ $n = 2$
Recall that for any frequency, $\lambda_n = \frac{2l}{n}$. Substitute the given values into the equation and solve for λ.	$\lambda_n = \frac{2l}{n}$ $= \frac{2 \times 0.22}{2}$ $= 0.22\text{ m}$

PHYSICSFILE

Surface waves

Seismic surface waves travel along the boundary between materials, such as between the Earth's crust and upper mantle. One type of surface wave is called the Rayleigh wave, or ground roll. Rayleigh waves are surface waves that travel as ripples with a motion like that of waves on the surface of water, although the restoring force is elastic rather than gravitational as it is for water waves.

A phenomenon known as the free oscillation of the Earth is the result of the superposition between two such surface waves travelling in opposite directions, thereby creating a surface standing wave.

The first observations of free oscillations of the Earth were made during the 1960 Chile earthquake. Since then thousands of harmonics have been identified.

PHYSICSFILE

Standing waves in air columns

Longitudinal standing waves are also possible in air columns. These create the sounds associated with wind instruments. Blowing over the hole of a flute (see figure below) produces vibrations that correspond to a range of frequencies that create sound waves in the tube.

The compressions and rarefactions of the sound waves, confined within the tube, reflect from the open end. This creates the right conditions for resonance and the formation of standing waves. The length of the pipe will determine the frequency of the sounds that will resonate.

Standing waves are formed in a flute.

Worked example: Try yourself 8.2.1

FUNDAMENTAL FREQUENCY

A standing wave in a string fixed at both ends has a wavelength of 0.50 m for the fundamental frequency of vibration.

a What is the length of the string?

b What is the wavelength of the third harmonic?

Note that the resonant frequencies of a string correspond to a particular tension and mass per unit length. Tightening or loosening the string will change the wavelengths and resonant frequencies for that string. Heavier strings of a particular length will have different resonant frequencies than lighter strings of the same length and tension.

CASE STUDY ANALYSIS

Physics of the guitar

When a guitar string is plucked, standing waves are formed (Figure 8.2.8). The vibrating string creates rarefactions and compressions in the air which move away from the string. These travel towards the listener's ears and are heard as sound. The frequency of the vibrating air molecules is the same as the frequency of the vibrating string. The speed of sound in air depends on the temperature but is around 340 m s^{-1}.

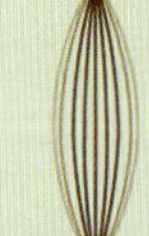

FIGURE 8.2.8 A transverse standing wave is formed on the string of a guitar. The vibration of the string causes vibration of nearby air molecules, which is heard as sound when the vibrating air molecules enter the listener's ears.

A typical guitar has six strings, all of the same length (Figure 8.2.9). The length of a guitar string can vary depending on the model of the guitar, but for the purpose of this case study, consider the length to be 650 mm. The fundamental wavelength ($\lambda_1 = 2l$) will be the same for each string. To change the fundamental frequency without pressing on the frets, the speed of the waves along the string must be changed. This can be done in two ways: by adjusting the tension (T) using the screws at the end of the arm, or by changing the thickness of the string, which changes the mass per unit length (μ). The relationship is described as follows.

$$v = \sqrt{\frac{T}{\mu}}$$

where μ is the mass per unit length (kg m^{-1})
T is the tension of the string (N)
v is the speed of the wave on the string (m s^{-1})

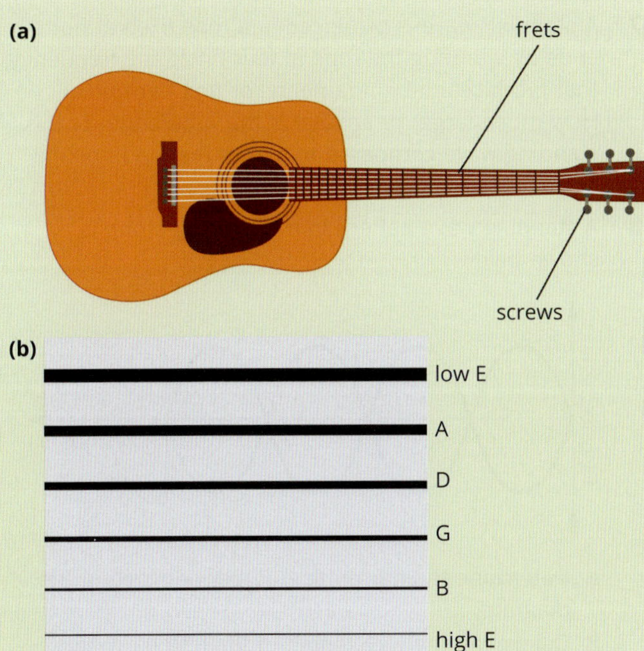

FIGURE 8.2.9 (a) A guitar has six strings of varying thickness and density. The tension in the strings can be adjusted by tightening or loosening the screws at the end of the guitar. (b) The names of the six strings on a standard guitar.

On an acoustic guitar the higher-pitched strings are often made of a lighter material, such as nylon, and the lower-pitched strings are made of a denser material, such as steel. The guitarist tunes a string to the desired note, or frequency, by turning the corresponding screw at the end of the string (which adjusts the string's tension). The standard fundamental frequencies for a tuned guitar are given in Table 8.2.1.

TABLE 8.2.1 The standard fundamental frequencies for a tuned guitar

Name of string	Frequency (Hz)
Low E	82.41
A	110.00
D	146.83
G	196.00
B	246.94
High E	329.63

Once the guitar is tuned, the guitarist can play a variety of notes by pressing the string onto the frets to shorten the length of the string. It should be noted that the dominant frequency will be the fundamental frequency, but the overtones or harmonics can still be heard, which gives the guitar its particular sound.

Analysis

1 Use the information in the text and in Table 8.2.1 to calculate the speed of the wave on the low-E string when a low E is played.

2 Consider the fundamental frequency of each string on the guitar. Describe, using appropriate equations, how the speed of vibration differs in the strings and why the thickness of the strings needs to change from the higher strings to the lower strings.

3 Calculate the wavelength and frequency of the D string when the string is shortened to two-thirds its regular length.

4 Determine how much the low-E string needs to be shortened to match the frequency of the high-E string.

5 Guitarists sometimes use the third harmonic of the low-E string when tuning the B string. Calculate the frequency of the third harmonic. Use resonance and the information in Table 8.2.1 to explain why this works.

8.2 Review

OA ✓✓

SUMMARY

- Standing or stationary waves are formed from the interference of waves and occur as a result of resonance at the natural frequency of vibration.
- Points on a standing wave that remain still are called nodes.
- Points on a standing wave at maximum amplitude are called antinodes.
- The standing wave frequencies are referred to as harmonics. The simplest is referred to as the fundamental frequency.
- Within a string fixed at both ends, the wavelength of the standing waves corresponding to the various harmonics is:

$$\lambda_n = \frac{2l}{n}$$

and the frequency is:

$$f_n = \frac{nv}{2l}$$

All harmonics may be present; that is, n could be any positive integer.

KEY QUESTIONS

Knowledge and understanding

1 Answer true or false to the following statements. Rewrite any incorrect statements.

- **a** The frequency of a wave is reduced when it reflects from a fixed end.
- **b** The speed of a wave is reduced when it reflects from a fixed end.
- **c** There is a phase shift of 180° or $\frac{\lambda}{2}$ when a rope reflects from an end that is free to move.
- **d** The wavelength stays the same when a wave is reflected from a fixed or free end.

2 A transverse standing wave is produced using a rope. Is the standing wave actually standing still? Explain your answer using the principles of superposition (i.e. interference).

3 Describe what happens when a wave on a rope is reflected from a fixed end.

4 What is the wavelength and frequency of the fundamental mode of a standing wave with a speed of 58 m s^{-1} on a string 0.40 m long? Assume that the string is fixed at both ends.

5 Calculate the length of a string fixed at both ends when the wavelength of the fourth harmonic is 0.750 m.

6 A metal string 50.0 cm long is plucked, thereby creating a transverse wave. The speed of the wave created is 300 m s^{-1}. Both ends of the string are fixed.

- **a** Calculate the fundamental frequency.
- **b** Calculate the frequency of the second harmonic.
- **c** Calculate the frequency of the third harmonic.

Analysis

7 A standing wave is produced in a rope fixed at both ends by vibrating the rope at four times the frequency that produces the fundamental or first harmonic. How much larger or smaller is the wavelength of this standing wave compared to that of the fundamental or first harmonic?

8 A standing wave pattern in a 10 m long string is shown below.

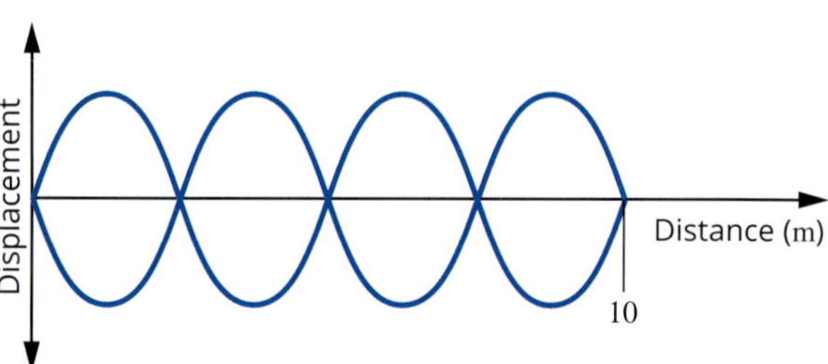

- **a** Determine the harmonic number depicted in the diagram.
- **b** Determine the wavelength of the fifth harmonic.

9 The fundamental frequency of a violin string is 350 Hz and the velocity of the waves along it is 387 m s^{-1}. What is the wavelength of the new fundamental when a finger is pressed on the string to shorten it to two-thirds its original length?

8.3 Evidence for the wave model of light

In many circumstances light behaves like a wave. The wavelike nature of light can be used to explain many phenomena, such as dispersion, the formation of rainbows, polarisation and diffraction. However, unlike mechanical waves, which need a medium to travel through, light can pass through the vacuum of space and thus does not need a medium (Figure 8.3.1). Therefore a more detailed explanation of the wavelike behaviour of light is needed.

FIGURE 8.3.1 Because it can travel through empty space, light cannot be a simple mechanical wave.

ELECTROMAGNETIC WAVES

In 1820 Danish physicist Hans Christian Oersted showed that a wire carrying an electric current generates a magnetic field. (You may recall this from Chapter 5.) In 1830 the English physicist Michael Faraday extended this discovery by showing that a changing magnetic field produces a changing electric field (discussed in Chapter 7). These experiments revealed the relationship between electricity and magnetism. In 1861 Scottish physicist James Clerk Maxwell quantified the relationship through a mathematical study of electric and magnetic effects. His work, known as Maxwell's equations, showed that electric and magnetic fields move at a speed that closely matched the experimental estimates of the speed of light.

The electromagnetic nature of light

Maxwell's study of the relationship between changing electric and magnetic fields led to the development of a comprehensive theory of electromagnetism and **electromagnetic radiation** (EMR). He proposed that an accelerating charged particle moving backwards and forwards would produce a changing electric field, and that the changing electric field would produce a changing magnetic field at right angles to it. The changing magnetic field would, in turn, produce a changing electric field and the cycle would be repeated. (Figure 8.3.2).

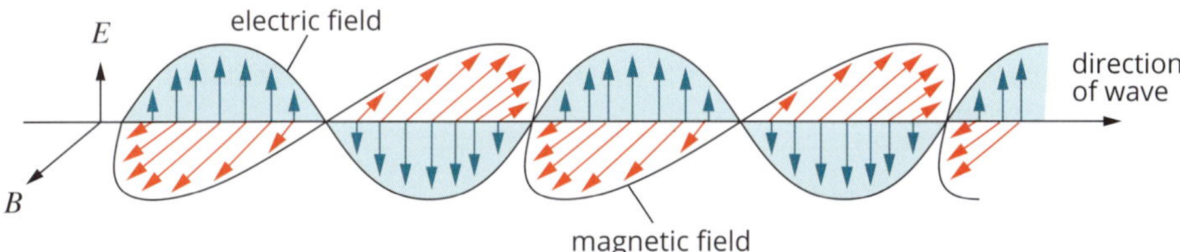

FIGURE 8.3.2 The electric and magnetic fields in electromagnetic radiation are perpendicular to each other and are both perpendicular to the direction of propagation of the radiation.

In effect, this repeated production of changing electric and magnetic fields would generate two mutually propagating fields. Once created, these fields—which constitute electromagnetic radiation—would be self-propagating, no longer dependent on whatever generated it, and thus could extend outwards indefinitely into space. Both the electric and magnetic fields would oscillate at the same frequency, the frequency of the radiation. This theory could be applied to the entire **electromagnetic spectrum**: from gamma rays to visible light to radio waves.

Maxwell's theoretical calculations provided a value for the speed at which electromagnetic radiation should propagate through empty space. This matched the experimental value for the speed of light measured by the French physicist Hippolyte Fizeau in 1849. The accepted value for the speed of light today is $299\,792\,458\,\text{m s}^{-1}$. This is such an important constant that it is designated its own symbol, c. In calculations, the speed of light is usually approximated to $3.0 \times 10^8\,\text{m s}^{-1}$.

PHYSICSFILE

Antennas

When you call someone on your mobile phone, you talk into a microphone. The compressions and rarefactions from the sound waves are converted to a transverse electrical waveform and then sent to the antenna of your phone. The varying electrical signal effectively oscillates the electrons in the metal antenna. The oscillation of these electrons induces a varying magnetic field which in turn induces a varying electric field, thus producing an electromagnetic wave. This electromagnetic wave is received by a mobile phone tower (see figure below) where the electrons are oscillated, producing a new electromagnetic wave. This is relayed to a phone tower near the person you are calling and then relayed to their phone.

Antennas, such as in this mobile-phone tower, are used to generate electromagnetic waves from electrical signals and vice versa.

For light, and other forms of EMR, the familiar wave equation $v = f\lambda$ is rewritten as follows.

$$c = f\lambda$$

where c is the speed of light ($3.0 \times 10^8\,\mathrm{m\,s^{-1}}$)
f is the frequency of the wave (Hz)
λ is the wavelength of the wave (m)

Maxwell's work was pivotal in the history of physics. Not only did he provide an explanation of the nature of light—as a form of electromagnetic radiation—he also brought together a number of formerly distinct areas of study, such as optics (the study of light), electricity and magnetism.

Worked example 8.3.1

USING THE WAVE EQUATION FOR LIGHT

Calculate the frequency of violet light with a wavelength of 400 nm (i.e. 400×10^{-9} m).

Thinking	Working
Recall the wave equation for light.	$c = f\lambda$
Transpose the equation to make frequency the subject.	$f = \frac{c}{\lambda}$
Substitute the appropriate values to determine the frequency.	$f = \frac{3.0 \times 10^8}{400 \times 10^{-9}}$ $= 7.5 \times 10^{14}$ Hz

Worked example: Try yourself 8.3.1

USING THE WAVE EQUATION FOR LIGHT

A particular colour of red light has a wavelength of 600 nm. Calculate the frequency of this colour.

Searching for the aether

One of the characteristics of mechanical waves is that they require a physical medium through which to propagate. For example, sound waves propagate through air and water waves propagate through water. Thinking that all waves need a medium led scientists to propose a medium in which light could travel through space. They called this the 'luminiferous aether' or 'aether'.

The American physicists Albert Michelson and Edward Morley conducted a famous experiment, using what is now known as the Michelson interferometer, to look for evidence that this aether exists. The results they obtained provided no evidence that the aether existed. Thus scientists were forced to conclude that electromagnetic waves are able to propagate through a vacuum without the need for a medium. You will learn more about this famous experiment in Chapter 10.

Resonance in electromagnetic waves

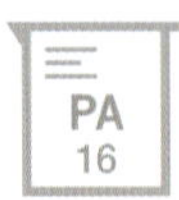

Resonant effects and standing waves also occur with electromagnetic waves. These concepts are important for understanding a range of physical phenomena, including the quantum mechanical description of electron orbitals. An everyday example of the application of standing waves is in heating food using microwave radiation.

CASE STUDY **ANALYSIS**

Heating up food in a microwave

Water molecules naturally vibrate at a frequency of 2.45 GHz. When they are bombarded with microwave radiation of the same frequency, the molecules start to vibrate more quickly. Since this increases their average kinetic energy, the temperature of the water increases. Conduction and convection then transfer this heat to the rest of the food.

Some recipes and heating instructions advise not to use a microwave oven. This is usually because the food in question does not contain much water.

Food sometimes becomes soggy when heated in a microwave oven. This is because the water molecules heat up faster than the food molecules around them.

You can determine the speed of light using a microwave oven. All you need is some food that will melt in a microwave, such as marshmallows. A chocolate bar will also work.

1. Remove the rotating plate from inside a microwave oven.
2. Place some marshmallows on a flat tray and place the tray in the microwave oven (Figure 8.3.3). You may need to add supports to the tray so that it is elevated above the rotor of the microwave.
3. Cook the marshmallows on a low setting until they begin to melt. You will find that the marshmallows will melt at different spots.
4. Measure the distance between the melted spots. This will correspond to half the wavelength of the microwave radiation.
5. Using the known frequency of the microwave and the distance between the melted spots, use the wave equation for light ($c = f\lambda$) to calculate the speed of light.

The reason you get unevenly spaced melted spots on the marshmallows is because most microwave ovens do not cook evenly. The melted spots occur at points where the oven is hottest. (The microwave radiation is emitted from a magnetron into the oven and the oven forms a resonant cavity. The hottest points correspond to antinodes in the microwave radiation.)

FIGURE 8.3.3 Microwave ovens produce electromagnetic radiation with a frequency of 2.45 GHz, which is the resonant frequency of water molecules.

Analysis

1. Using your understanding of resonance, explain why hot spots form in a microwave oven.
2. Express the frequency of the radiation used in microwave ovens in Hz.
3. A student measures the distance between the hot spots to be 6 cm. Calculate the speed of light.
4. Calculate the percentage error between your calculation of the speed of light and the accepted value. Take the accepted value of c to be $2.99792458 \times 10^8 \, m\,s^{-1}$.
5. Explain why there could be a difference between the experimental values and the actual value and list any uncertainties there might be in carrying out this experiment.

DIFFRACTION

FIGURE 8.3.4 Water waves will diffract (bend) as they pass through narrow openings. Sound waves diffract as well, allowing you to hear around corners.

When a plane (straight) wave passes through a narrow opening, it can bend, forming a semicircular pattern (Figure 8.3.4). Waves will also bend as they travel around obstacles. This type of bending is known as **diffraction**.

Diffraction and slit width

In the diffraction of waves, if the wavelength is much smaller than the gap or obstacle, the degree of diffraction is less. This can be seen in Figure 8.3.5, which shows the diffraction of water waves in a ripple tank. In Figure 8.3.5(a), the gap is similar in size to the wavelength, so there is significant diffraction and the waves emerge as circular waves. In Figure 8.3.5(b), the gap is much bigger than the wavelength, so diffraction only occurs at the edges.

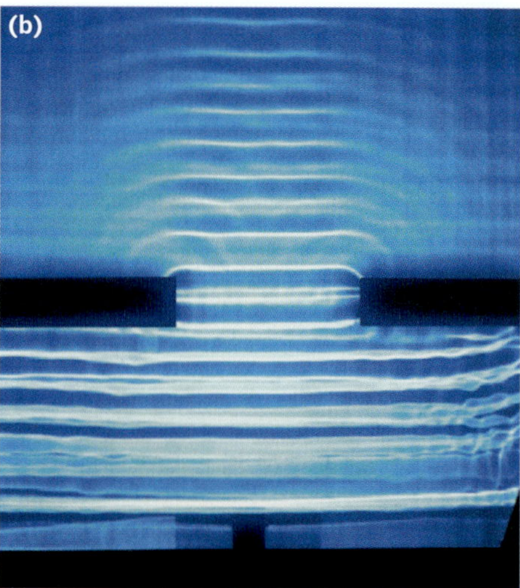

FIGURE 8.3.5 The diffraction of water waves in a ripple tank. (a) Significant diffraction occurs when the wavelength is close to the gap width. (b) As the gap width increases, diffraction becomes less obvious, but it is still present.

There will also be significant diffraction where the wavelength is larger than the width of the gap. In fact, diffraction will occur whenever $\frac{\lambda}{w} \geq 1$, where λ is the wavelength of the wave and w is the width of the gap. The same equation applies to diffraction around objects, where w is the width of the object.

Diffraction and light

Diffraction is significant when the size of the opening or obstacle (w) is similar to or smaller than the wavelength of the wave (λ). Light waves range in wavelength from approximately 700 nm for red light to approximately 400 nm for violet light. (1 nm is equal to 10^{-9} m or a one millionth of a millimetre.) Thus all light waves are less than one thousandth of a millimetre in length. This means that the diffraction of light is difficult to observe because its wavelength is very small.

Diffraction can be observed around natural objects, such as human hair (Figure 8.3.6(a)) or a cotton thread. Diffraction can also occur from artificially constructed materials, such as the surface of CDs and DVDs (Figure 8.3.6(b)), and from commercially produced diffraction gratings.

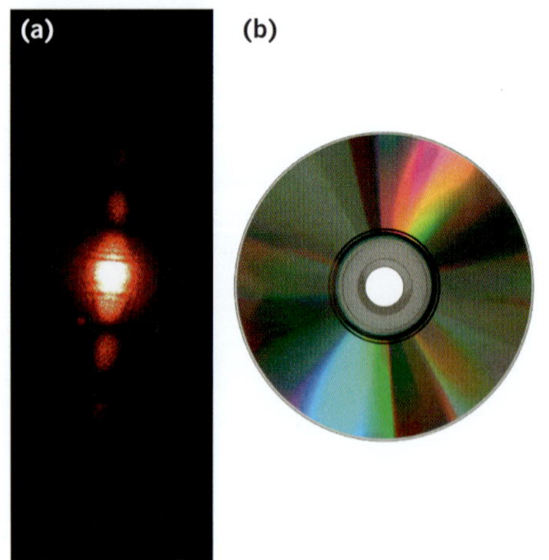

FIGURE 8.3.6 (a) The diffraction pattern of a red laser directed at a human hair. (b) The way information is imprinted on the surface of a CD or DVD creates structures small enough to cause light to diffract.

Diffraction and imaging

Diffraction can be a problem for scientists using microscopes and telescopes: it can result in blurred images. For example, the light from two tiny objects or two distant objects very close together can be diffracted so much that the two objects appear as one blurred object. When this happens we say that the objects are **unresolved**.

The ratio $\frac{\lambda}{w}$ determines the smallest size of object that can be clearly imaged by a particular instrument. As a general rule, optical microscopes cannot yield images of objects that are smaller than the wavelength of the light they use.

While diffraction places a theoretical limit on the resolution of optical telescopes, atmospheric distortion usually has a much larger effect on telescope images than diffraction. However, the Hubble Space Telescope and the James Webb Space Telescope sit above the Earth's atmosphere and thus are not affected by atmospheric distortion. They can resolve images right down to its diffraction limit, i.e. where the separation of the stars is approximately equal to the wavelength of the light being observed.

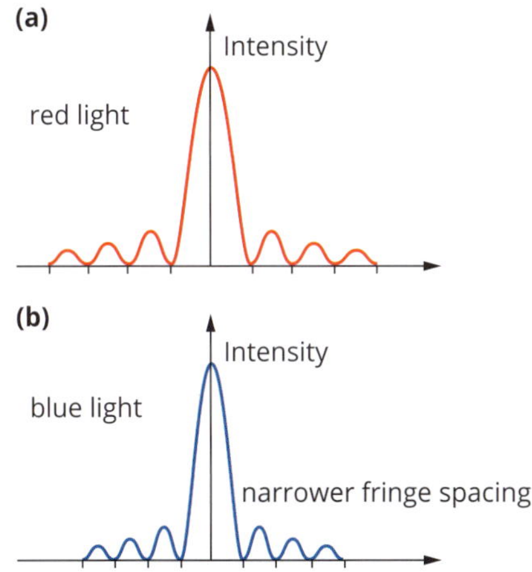

FIGURE 8.3.7 Red light (a) is diffracted to a greater extent than blue light (b). The longer wavelength of red light results in interference fringes that are more widely spaced and have a wider overall pattern.

Diffraction gratings

Light diffracts as it passes through a very small gap. As the light passes through the gap, some of the wavelets making up the wavefront will diffract at the barriers that form the edges of the gap and some will pass through the centre of the gap. As a result the light waves that emerge from the gap will interact (that is, superposition will occur). In some places the interactions will be constructive and in others the interactions will be destructive. When the light waves are made to shine on a screen, the areas of constructive interference will appear as bright bands and the areas of destructive interference will appear as dark bands. This pattern of dark and light bands is called a **diffraction pattern** and the individual bands are called **interference fringes**.

The ratio $\frac{\lambda}{w}$ not only determines the extent of diffraction; it also determines the spacing of dark and light bands in a diffraction pattern. If the wavelength is held constant and the gap made smaller, greater diffraction occurs. If different wavelengths pass through the same gap, those with a smaller wavelength will undergo less diffraction than those with a longer wavelength (Figure 8.3.7). Note that Figure 8.3.7 shows intensity on the y-axis. High intensity is where bright bands will appear on a screen; zero intensity corresponds to dark bands.

Although some diffraction patterns can be observed using natural materials, much clearer diffraction patterns can be generated by passing light through a diffraction grating. A **diffraction grating** is a piece of material that contains a large number of very closely spaced parallel gaps or slits.

A diffraction grating can be thought of as a series of parallel slits all placed side by side. The diffraction pattern from one slit is superimposed on the pattern from an adjacent slit, producing a strong, clear image on the screen.

FIGURE 8.3.8 A diffraction grating disperses white light into its component colours.

Diffraction experiments usually use only **monochromatic** light (i.e. light of only one colour). When white light, which contains a number of different colours, shines through a diffraction grating, each colour is diffracted by a different amount and forms its own set of coloured fringes. This results in the light being dispersed into its component colours (Figure 8.3.8).

YOUNG'S DOUBLE SLIT EXPERIMENT

The start of this chapter briefly describes the historical debate between different models for light. Between the seventeenth and nineteenth centuries, most scientists considered light to be a stream of particles. This idea was based on the corpuscular theory proposed by Sir Isaac Newton. If light behaves purely as a particle, then an explanation was still needed for how interference patterns of light (Figure 8.3.9) could be created.

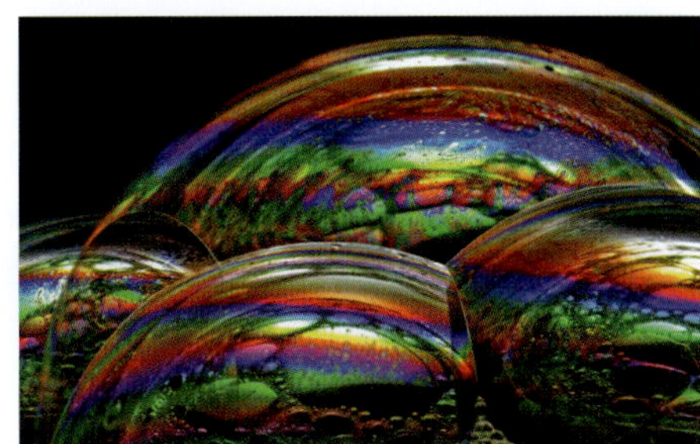

FIGURE 8.3.9 Optical interference can produce spectacular patterns, as shown in these soap bubbles.

Thomas Young's observation of interference patterns in light was pivotal in tipping the scales in the long-running dispute about the nature of light. His work paved the way for a series of discoveries and inventions that would fundamentally change scientists' understanding of energy and matter.

In 1803 Young performed a now-famous experiment in which he shone monochromatic light onto a barrier in which there were two very tiny slits. On the far side of the barrier he placed a screen on which he could observe the pattern produced by light passing through the slits (Figure 8.3.10).

According to the corpuscular theory—where light was considered to be particles—light should have passed directly through the slits and produce two bright lines or bands on the screen (Figure 8.3.10(a)). Instead, Young observed a series of bright and dark bands (Figure 8.3.10(b)).

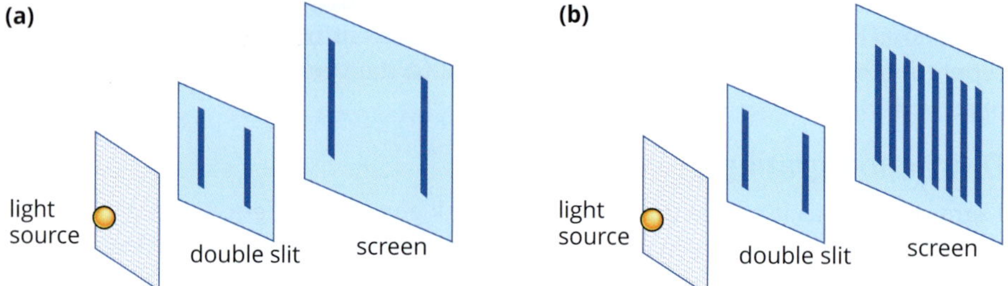

FIGURE 8.3.10 (a) By the particle theory of light, Young's experiment should have produced two bright bands. (b) Instead, he found that a series of bright and dark bands was produced.

Young explained this bright and dark pattern by conceiving of light as a wave. He considered the monochromatic light to be like a series of plane waves. As they passed through the narrow slits, the waves were diffracted into two **coherent** (i.e. of the same frequency) circular waves (Figure 8.3.11). The circular waves would then interact and cause interference. The interference would result in lines of constructive (antinodal) and destructive (nodal) interference that would appear as bright and dark bands on the screen.

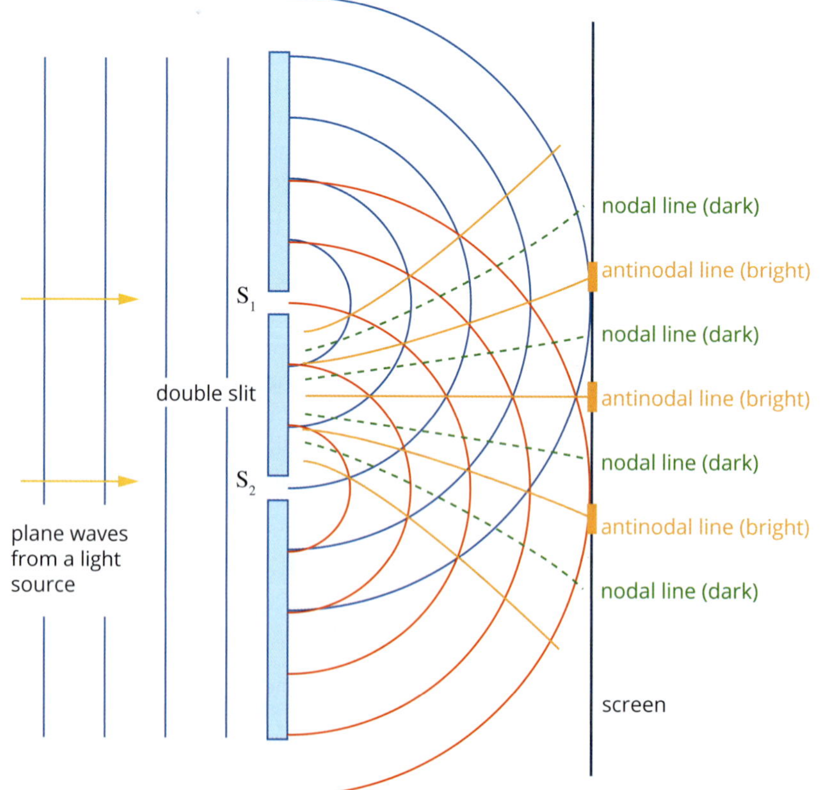

FIGURE 8.3.11 The interaction of two circular waves can produce a pattern of antinodal (constructive interference) and nodal (destructive interference) lines.

Earlier in his scientific career, Young had observed similar interference patterns in water waves (Figure 8.3.12). This may have been the inspiration for his experiment.

Young's results gave greater credibility to the wave model of light proposed by Christiaan Huygens and Robert Hooke many years earlier.

When Young used his data to calculate the wavelength of light, it became clear why no one had ever noticed the wave properties of light before—visible light waves are tiny, with typical wavelengths of less than 1 micrometre (1 µm = 0.001 mm).

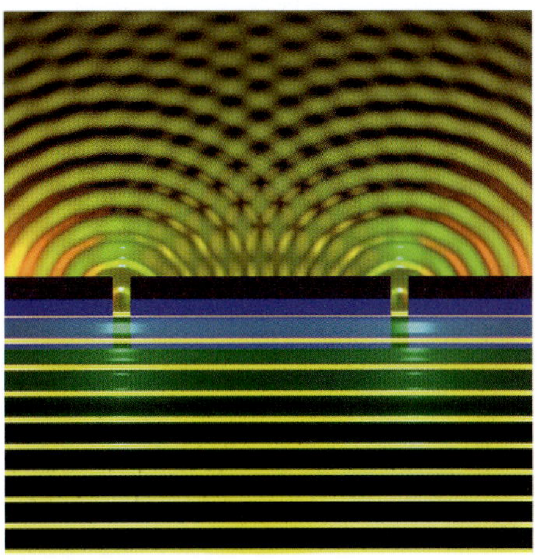

FIGURE 8.3.12 Interference patterns can be observed in water waves (lit here in yellow).

Path difference

To understand Young's experiment more fully, you have to consider how the waves produced by the two slits interact with each other when they hit the screen. Consider Figure 8.3.13 below. At point R on the screen, the wave train from slit S_1 will have travelled a different distance than the wave train from slit S_2. The difference in the distance travelled by each wave train to a point on the screen is called the **path difference**.

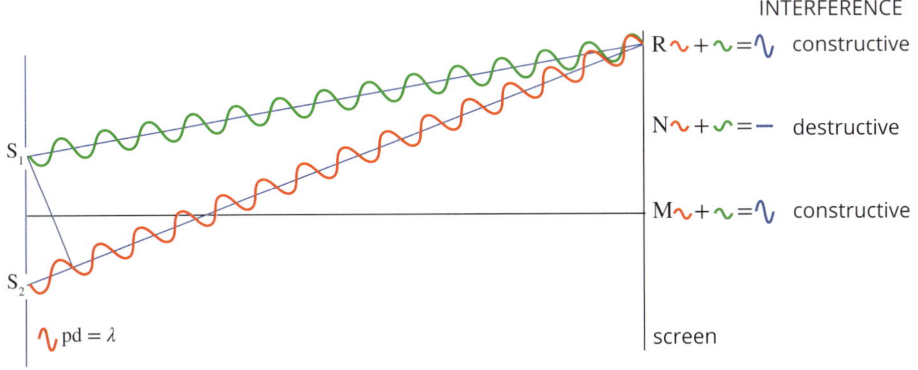

FIGURE 8.3.13 Waves from each slit meeting at R, where the path difference is λ, interact to produce constructive interference. A bright fringe will be seen on the screen.

Path difference can be measured in metres, but it is far more useful to measure it in wavelengths. This will make it easier to determine the light intensity on the screen.

In Figure 8.3.13 point M is equidistant from each slit and so each wave train will have travelled the same distance to reach it. $S_1M = S_2M$, so there is no path difference. The light waves arrive in phase with each other and produce an antinode. A fringe of bright light known as the 'central maximum' is seen on the screen. Recall that the phenomenon that has produced this antinode is called constructive interference.

Complete constructive interference will occur whenever the path difference between two wave trains is zero or differs by a whole number of wavelengths, i.e. pd = $0\lambda, 1\lambda, 2\lambda, 3\lambda$ ... For example, in Figure 8.3.13, the path difference $S_1R - S_2R$ is equal to one λ. The two waves are in phase. The general equation for the path difference where constructive interference occurs is pd = $n\lambda$. In this relationship, $n = 0$ refers to the central maximum band, where there is no path difference, $n = 1$ refers to the first set of bright fringes along from the central maximum, $n = 2$ refers to the second set of bright fringes, and so on.

There will be points on the screen where the path difference is $\frac{\lambda}{2}$ (for example point N in Figure 8.3.13). The two wave trains that meet at these points are completely out of phase. They cancel each other, producing a node. Destructive interference occurs at these points and no light is seen on the screen. This creates the dark fringes that appear between the bright antinodal fringes. Destructive interference occurs wherever the path difference between the waves is $\frac{\lambda}{2}, \frac{3\lambda}{2}, \frac{5\lambda}{2}, \frac{7\lambda}{2}$ ... The general equation for the path difference where destructive interference occurs is pd = $\left(n+\frac{1}{2}\right)\lambda$. In this relationship, $n = 0$ refers to the first dark band adjacent to the central maximum band, where there is path difference of $\frac{\lambda}{2}$.

The path difference for any point, P, between light travelling from wave source S_1 and from wave source S_2, is given by:
pd = $|S_1P - S_2P|$

In summary, constructive and destructive interference occur as follows.

- constructive interference of coherent waves occurs wherever the path difference equals a whole number of wavelengths, that is, pd = $n\lambda$ where n is 0, 1, 2, 3 ...
- destructive interference of coherent waves occurs wherever the path difference equals an odd number of half wavelengths, that is, pd = $(n+\frac{1}{2})\lambda$, where $n = 0, 1, 2, 3$ ...

The sequence of constructive and destructive interference produces a pattern of regularly spaced vertical bands on the screen (the interference fringes). This is represented graphically in Figure 8.3.14. The horizontal axis represents a line drawn across the screen. The centre of the distribution pattern corresponds to the centre of the brightest central fringe, the central maximum.

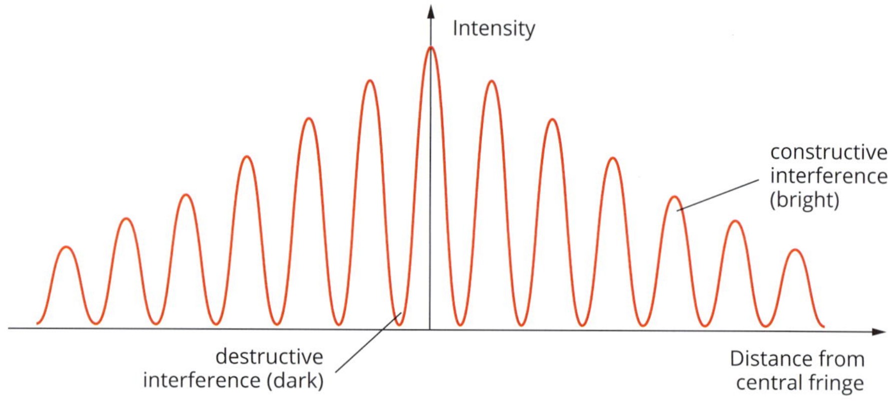

FIGURE 8.3.14 The double slit interference pattern represented by an intensity distribution graph

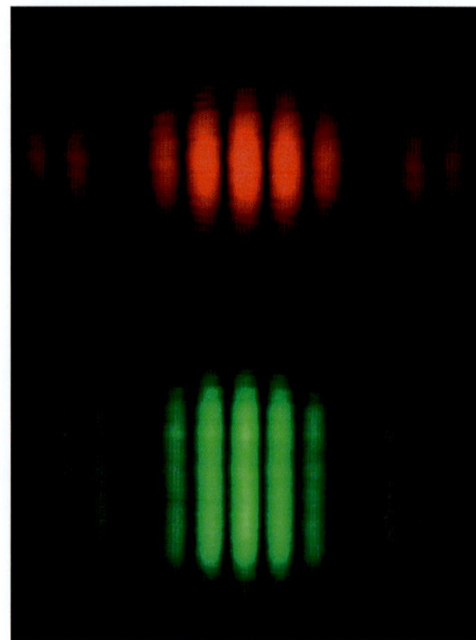

FIGURE 8.3.15 If the distance between the slits and the distance to the screen are unaltered, then the bright fringes produced by red light are further apart than those produced by green light. This is because the wavelength of red light is longer than that of green light.

If you compare the double slit interference pattern in Figure 8.3.14 with the diffraction patterns in Figure 8.3.7 on page 347, you will notice that both have dark and bright bands, but that the diffraction pattern has a much stronger central maximum with the intensity falling off rapidly. This occurs because the observed pattern from a diffraction grating is a combination of a regularly spaced interference pattern and an envelope produced by diffraction effects. In some cases diffraction effects can even lead to the cancellation of some interference fringes, as shown in Figure 8.3.15.

Calculating fringe separation

In a double slit experiment, the distance between adjacent bright bands on the screen is known as the fringe spacing (Δx). This distance depends on the wavelength of light (λ), the separation between the two slits (d) and the distance from the slits to the screen (L), as shown in Figure 8.3.16.

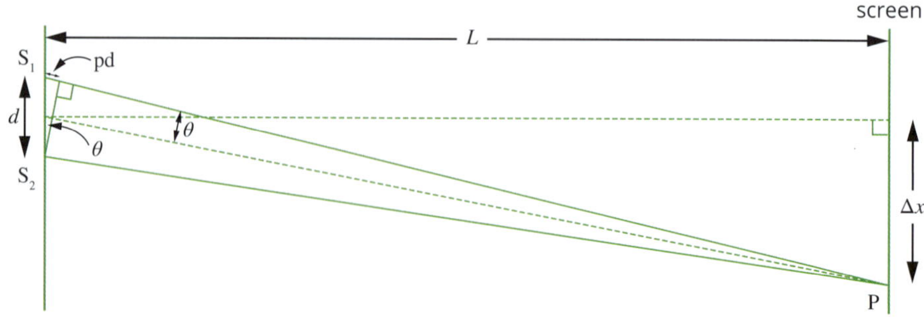

FIGURE 8.3.16 A geometric representation of path difference (pd). To reach point P, the wave train passing through S_1 travels a different distance compared to the wave train passing through S_2.

The distance between the bright fringes can be increased by:

- increasing the distance between the slits and the screen, i.e. $\Delta x \propto L$
- reducing the distance between the two slits, i.e. $\Delta x \propto \frac{1}{d}$
- using light of a longer wavelength, i.e. $\Delta x \propto \lambda$ (Figure 8.3.16).

These relationships can be combined to produce an equation for fringe separation as follows.

$$\Delta x = \frac{\lambda L}{d}$$

where Δx is the fringe separation
λ is the wavelength of the light waves
L is the distance from the slits to the screen
d is the distance between the slits

Worked example 8.3.2

CALCULATING WAVELENGTH FROM FRINGE SEPARATION

Light of an unknown wavelength emitted by a laser is directed through a pair of thin slits 50.0 μm apart. The slits are 2.00 m from a screen on which bright fringes are 2.50 cm apart. Calculate the wavelength of the laser light in nm.

Thinking	Working
Recall the equation for fringe separation.	$\Delta x = \frac{\lambda L}{d}$
Transpose the equation to make λ the subject.	$\lambda = \frac{\Delta x d}{L}$
Substitute the given values and solve for λ. Note: $1\ \mu m = 1 \times 10^{-6}$ m	$\lambda = \frac{0.0250 \times 50.0 \times 10^{-6}}{2.00}$ $= 6.25 \times 10^{-7}$ m
Express your answer in the required units, nm, where $1\ nm = 1 \times 10^{-9}$ m.	The wavelength of the laser light is 625 nm.

Worked example: Try yourself 8.3.2

CALCULATING WAVELENGTH FROM FRINGE SEPARATION

A green laser is directed through a pair of thin slits that are 25.0 μm apart. The slits are 1.50 m from a screen on which bright fringes are 3.30 cm apart. What is the wavelength of the green light in nm?

RESISTANCE TO THE WAVE MODEL

Young's wave explanation for his experimental results was not immediately accepted by the scientific community. Many scientists were reluctant to abandon the corpuscular theory that had been accepted for over a century.

In 1818 the French scientist Augustin-Jean Fresnel provided a mathematical explanation for Young's results based on Huygens' principle. But another French scientist, Simeon Poisson, thought differently. He was a passionate supporter of Newton's particle theory and argued that if the same mathematics were applied to the light shining around a small round disk, then there should be a bright spot in the middle of the shadow created by the disk (Figure 8.3.17). Since nobody had ever observed a bright spot in the middle of a shadow, Poisson believed this proved that the wave model was incorrect.

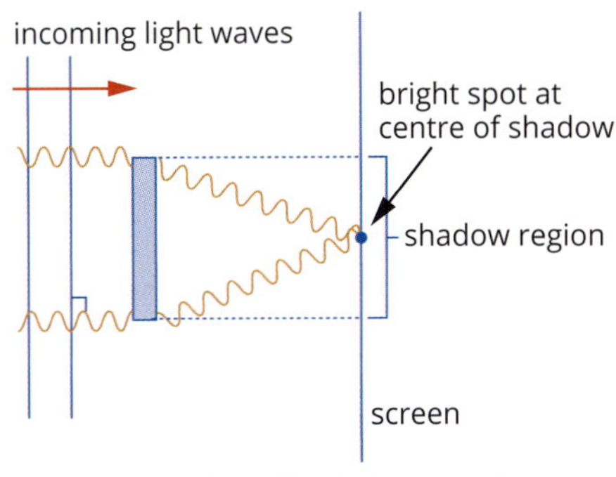

FIGURE 8.3.17 According to the wave theory, light incident on a small solid disk should diffract to give a point of light in the centre of the shadow region. As this effect had not been seen at the time, its absence was considered proof that light is not a wave. The effect has since been observed.

However, one of Poisson's colleagues decided to test these ideas by performing an experiment with a very small bright light source and a round disk. The bright spot predicted by the wave theory was observed (Figure 8.3.18). From then until the end of the 19th century, the wave theory of light became almost universally accepted.

This now famous diffraction pattern has come to be known as the 'Poisson bright spot', which means it is named after the person who predicted that it would not exist!

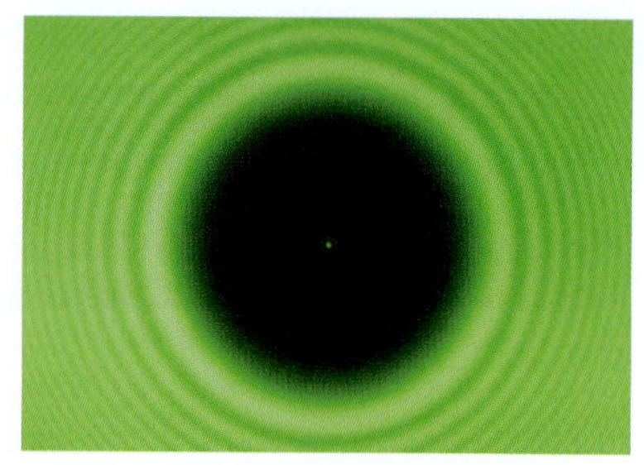

FIGURE 8.3.18 The bright spot inside the shadow region of this image is caused by the diffraction and interference of light waves. The image also shows diffraction and interference patterns surrounding the shadow.

CASE STUDY ANALYSIS

X-ray diffraction

The wavelength of X-rays is similar in size to the distance between atoms in minerals and other crystals. This enables X-ray diffraction to be used to identify the composition of minerals and crystals. The powdered mineral or crystal is placed in the path of the X-ray beam and then rotated. Constructive interference will occur when the path difference ($n\lambda$) of the incoming and reflected X-ray is equal to a multiple of the X-ray wavelength (Figure 8.3.19). This is given by the expression:

$$n\lambda = 2d\sin\theta$$

where n is any whole number, i.e. 0, 1, 2, 3 ...

λ is the wavelength of the X-rays (m)

d is the distance between the atoms in the mineral or crystal (m)

θ, measured from the surface of the crystal, is the angle at which constructive interference occurs

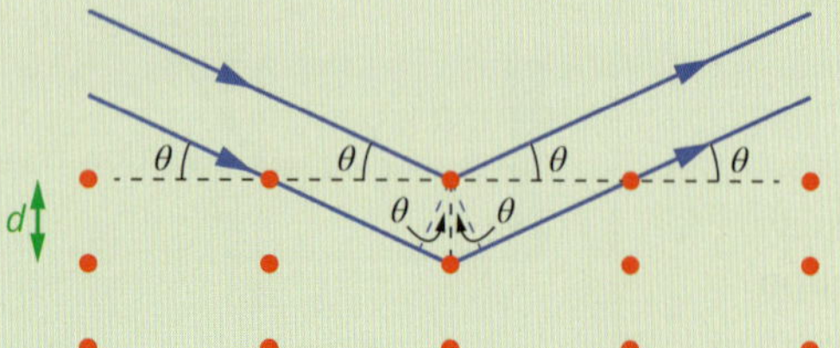

FIGURE 8.3.19 Using X-ray diffraction the angles at which constructive interference occur can be used to identify a mineral or crystal.

By rotating the powdered mineral or crystal through a range of angles—from approximately 5° to 70°—the distance between the atoms can be determined. A typical X-ray diffraction scan is shown in Figure 8.3.20.

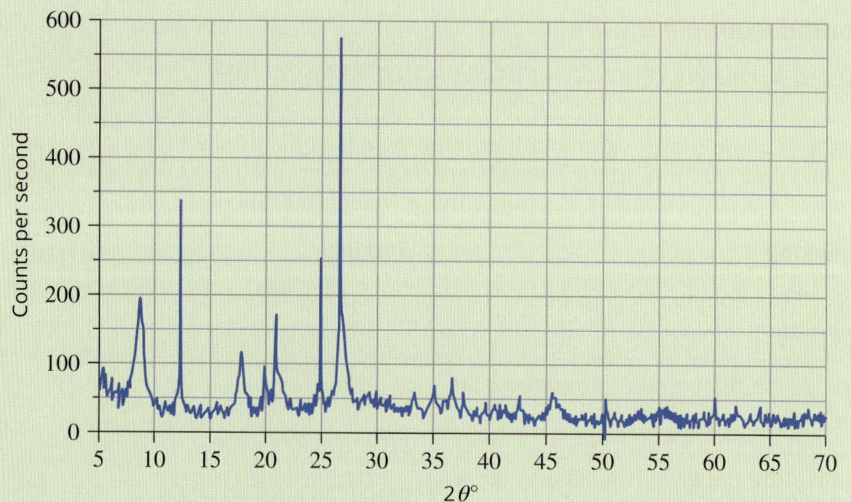

FIGURE 8.3.20 A diffraction image taken when a mineral was bombarded with X-rays at various angles.

The atomic spacing from an unknown mineral or crystal can be compared with the atomic spacing from known materials to determine its composition. This method is used by geologists and geophysicists in many forms of mining.

Analysis

1 Explain why X-rays are used in the analysis of minerals.

2 One characteristic line emitted in X-ray diffraction is at a frequency of 4.23×10^{18} Hz. Workers in X-ray diffraction often express wavelength in angstroms (Å), where 1 angstrom is 10^{-10} m. Calculate the wavelength of the characteristic line and express your answer in angstroms.

3 A typical X-ray diffraction pattern is shown in Figure 8.3.20. Explain why the peaks form.

4 Assuming that the two highest peaks in Figure 8.3.20 correspond to $n = 1$, determine two of the spacings for the crystal being measured and express the values in angstroms. Note that the values on the horizontal axis are 2θ.

8.3 Review

OA ✓✓

SUMMARY

- Electromagnetic radiation is created when charged particles are accelerated. This creates a varying magnetic field that induces a varying electric field which creates another varying magnetic field and so on. The electric fields are at right angles to the magnetic fields.
- When a plane wave (i.e. a straight wave) passes through a narrow opening or meets an obstacle, it undergoes diffraction.
- Significant diffraction occurs when the wavelength of the wave is similar to, or larger than, the size of the diffracting object.
- Young's double slit experiment provided evidence to support the wave model of light.
- Path difference (pd) is the difference in the distance travelled by each wave train from a pair of slits to the same point.
- Constructive interference of coherent waves occurs when the path difference equals an even number of wavelengths, i.e. pd = $n\lambda$, where $n = 0, 1, 2, 3 \ldots$
- Destructive interference of coherent waves occurs when the path difference equals an odd number of half wavelengths, i.e.

$$\text{pd} = \left(n + \frac{1}{2}\right)\lambda, \text{ where } n = 0, 1, 2, 3 \ldots$$

- The distance between the interference fringes produced in a double slit experiment is given by:

$$\Delta x = \frac{\lambda L}{d}$$

KEY QUESTIONS

Knowledge and understanding

1 Light observed from distant stars may have been travelling for billions of years. Explain how it is possible that we can see light from these stars through the vacuum of space.

2 a In order to produce significant diffraction of red light with a wavelength of approximately 700 nm, the slit width in a diffraction experiment would need to be closest to:

A 1 mm
B 0.1 mm
C 0.01 mm
D 0.001 mm

b Justify your answer to part **a**.

3 Explain the features of Young's experiment that convinced scientists that light was a wave.

4 Two students are trying to replicate Young's double slit experiment. One uses torch light and the other uses light from a laser. Which of the following statements explains why the student using the laser light is more likely to obtain the expected interference pattern? (Note: more than one correct answer is possible.) Explain your answer.

A Torch light is monochromatic.
B Torch light is coherent.
C Laser light is monochromatic.
D Laser light is coherent.

5 If Young's double slit experiment were repeated using circular water waves in a ripple tank, which of the following events would correspond to nodal lines? (Note: more than one correct answer is possible.)

A crests meet troughs
B troughs meet troughs
C crests meet crests
D troughs meet crests

6 A version of Young's double slit experiment is set up by directing light from a red laser through a pair of thin slits. An interference pattern appears on the screen behind the slits. The following changes are made to the apparatus. Identify whether the distance between the interference fringes seen on the screen would increase, decrease or stay the same.

a The screen is moved further away from the slits.
b A green laser is used.
c The slits are moved closer together.

Analysis

7 A 580 nm yellow light is directed through a pair of thin slits to produce an interference pattern on a screen. Determine the path difference of the fifth dark fringe from the central maximum.

8 Identify the type of interference (constructive or destructive) that corresponds to the following path differences.

a $\frac{\lambda}{2}$ **b** λ **c** $\frac{3\lambda}{2}$

continued over page

8.3 Review *continued*

9 A 700 nm red light is directed through a pair of thin slits to produce an interference pattern on a screen. Determine the path difference at the second bright fringe from the central maximum.

10 A blue laser is directed through a pair of thin slits that are 40 μm apart. The slits are 3.25 m from a screen on which bright fringes are 3.7 cm apart. What is the wavelength of the light in nm?

11 Consider a wave with a wavelength of 55 cm passing through a gap 50 cm wide. If the width of the gap remains unchanged but the frequency of the wave is increased, what effect would this have on the wavefronts beyond the gap?

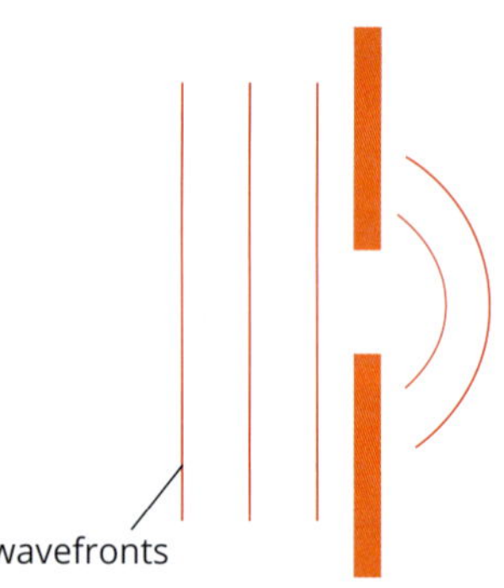

12 The following diagram shows the resulting intensity pattern after light from two slits reaches a screen. Copy the diagram into your workbook and circle the points at which the path difference is equal to 1λ.

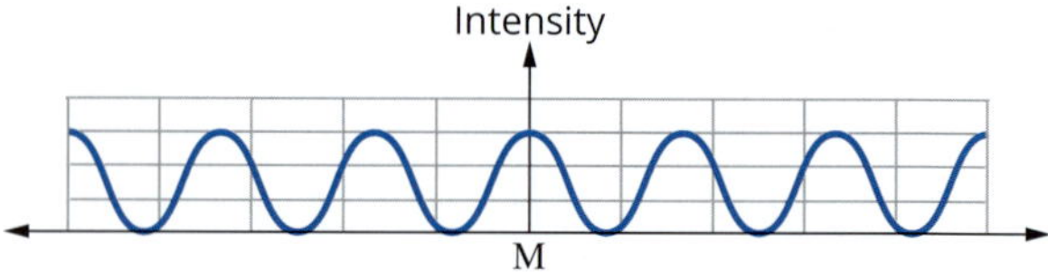

Chapter review

KEY TERMS

absorb
amplitude
antinode
coherent
compression
constructive interference
crest
destructive interference
diffraction
diffraction grating
diffraction pattern
electromagnetic radiation
electromagnetic spectrum
frequency
fundamental frequency
harmonic
interference
interference fringe
longitudinal
mechanical wave
medium
monochromatic
node
overtone
path difference
period
phase
rarefaction
reflect
resonance
resonant frequency
standing wave
superposition
transmit
transverse
trough
unresolved
wavelength

REVIEW QUESTIONS

Knowledge and understanding

1 The following graph shows three waveforms. Two of the wave forms are superimposed to form the third waveform.

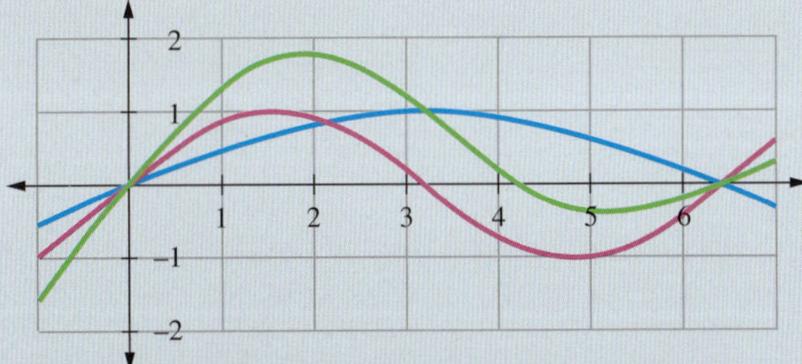

Which waveform is the result of the superposition of the other two?

2 Two wave pulses are travelling along a string towards each other. When they meet, the waves superimpose and cancel each other out completely. What is this called and what characteristics must the two waves have for this to happen?

3 Explain why a car can vibrate more strongly at some speeds than others. Name the phenomenon.

4 **a** Explain what a node in a standing wave is.
b Explain what an antinode in a standing wave is.

5 A string of length 95.0 cm, fixed at both ends, is made to vibrate. The string is kept at constant tension. The speed of the vibrations along the string is 400 $m s^{-1}$.
a What is the lowest frequency of vibration that will produce a standing wave in the string?
b What is the frequency of vibration of the third harmonic?

6 An earthquake causes a footbridge to oscillate up and down with a fundamental frequency of once every 4.0 s. Consider the footbridge to be like a string fixed at both ends. What is the frequency of the second harmonic for this footbridge?

7 The velocity of waves in a particular string at constant tension is 95.0 $m s^{-1}$. The string is fixed at both ends. If a standing wave is formed with a frequency 540 Hz, what is the distance between a node and an antinode that is adjacent to it?

8 A standing wave on a string on a full-size violin has a wavelength of 71.0 cm for the fundamental frequency of vibration.
a Determine the length of the string.
b Determine the wavelength of the fifth harmonic.

9 Light takes 2.537 million years to travel from the Andromeda galaxy to the Earth. Taking the speed of light as 3.0×10^8 $m s^{-1}$, calculate how far away the Andromeda galaxy is in metres.

10 How does light propagate over long distances?

11 Name the phenomenon shown below and explain how you could make the effect stronger.

12 Green light of frequency 5.66×10^{14} Hz is incident on a narrow slit. Determine the slit width in mm that is required for the light to diffract. (The speed of light in air is approximately 3.0×10^{8} m s^{-1}.)

A 0.0010 mm

B 0.005 30 mm

C 0.000 10 mm

D 0.000 530 mm

13 Consider an experiment similar to Young's double slit experiment. Explain how the spacing in the interference pattern would change if a blue laser was replaced with a green laser.

14 Light of an unknown wavelength emitted by a laser is directed through a pair of thin slits separated by 75.0 μm. The slits are 4.00 m from a screen on which bright fringes are 3.10 cm apart.

a Calculate the wavelength of the laser light in nm.

b Using the table below, identify the colour emitted by the laser.

Colour	Approximate wavelength range (nm)
Violet	380–435 nm
Blue	435–500 nm
Cyan	500–520 nm
Green	520–565 nm
Yellow	565–590 nm
Orange	590–625 nm
Red	625–740 nm

15 The following diagram shows the resulting intensity pattern after light from two slits reaches the screen in a double slit experiment. Copy the diagram into your workbook and circle the points at which the path difference is equal to $1\frac{1}{2}\lambda$.

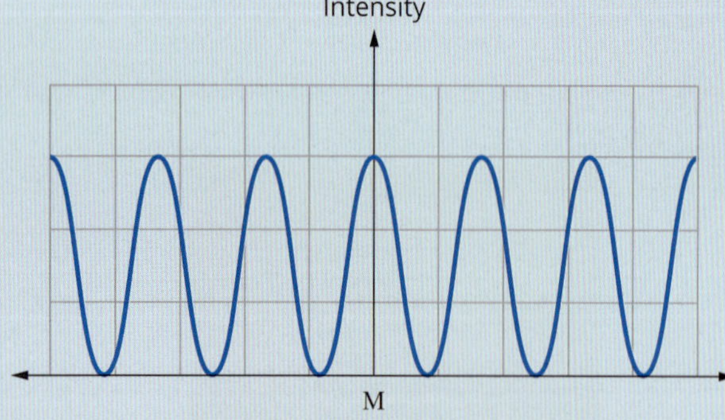

Application and analysis

16 Consider the two wave patterns shown in the diagram below. Use the principle of superposition to draw the resultant wave pattern from combining the two waves.

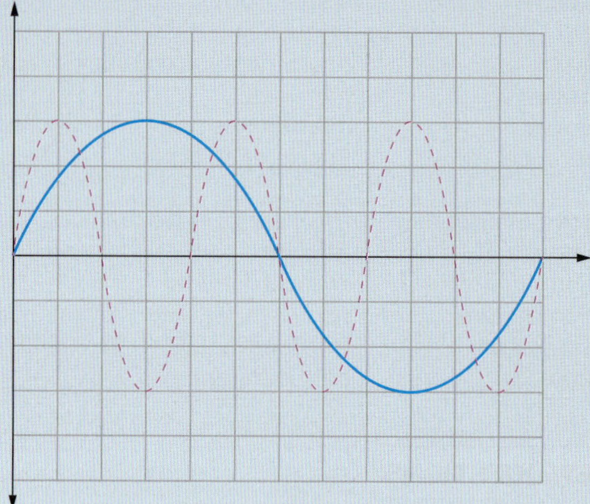

17 Two waves are travelling in opposite directions, as shown in the figure below. Use the principle of superposition to draw the resultant wave pattern during the period shown.

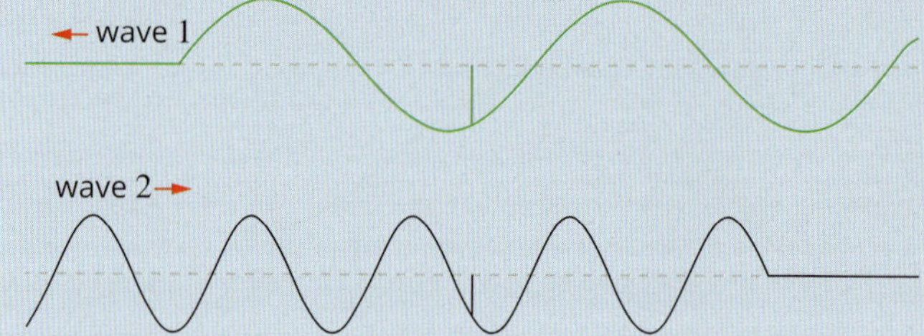

18 In each of the two diagrams below, wavefronts are traveling towards a slit of a particular width. Draw the diffraction wave pattern after the waves have passed through the slit. Explain any differences.

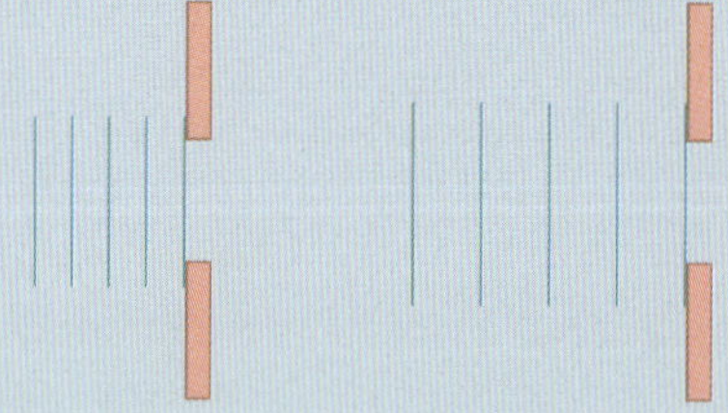

19 At certain frequencies some sounds appear louder to humans than other sounds. These are typically in the region of 2500 Hz. Explain why a 2500 Hz sound from a signal generator might sound louder than other frequencies in the ear canal of humans.

20 The diagram below shows the amplitudes of wave A and wave B at a particular times. The resultant wave pattern shows the combination of the two waves.

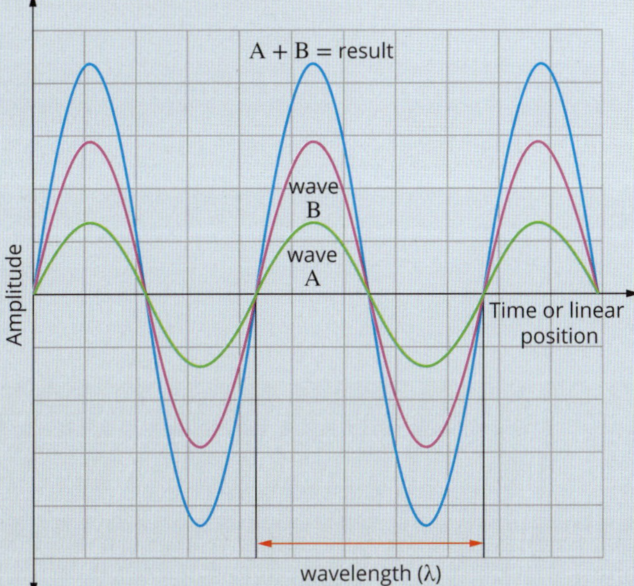

a Name the principle used to construct the resultant interference pattern and comment on the type of interference pattern shown.

b If one of the waves (A or B) is shifted $\frac{\lambda}{2}$ to the left, explain what would happen to the resultant wave pattern. Describe the type of interference pattern produced.

21 A car is travelling along a rough gravel road. The road has ruts on it, as shown below. These are regularly spaced and cause considerable vibration when a car drives over them. The driver adjusts the speed of the car to minimise the vibration. The driver notices that the worst vibration occurs when travelling at a speed of 50 km h^{-1}. If the ruts are 10 cm apart, what is the resonant frequency of the car?

22 When waves pass through narrow openings, diffraction can be observed. Explain why we do not generally observe the diffraction of light in our everyday lives.

23 Explain what is meant by the theoretical limit for the resolution of images viewed with optical instruments.

24 Explain briefly why a microwave oven is tuned to produce electromagnetic waves of a particular frequency.

25 Identify which of the following statements are true and which are false. Rewrite the false statements to make them true.

a Visible light waves have a wavelength between 400 nm and 750 nm and require an opening with a width of about 10^{-2} mm for diffraction to occur.

b Water waves with a frequency of 0.2 Hz and a velocity of 0.5 m s^{-1} diffract around an opening with a width of 2.5 m.

c An acoustic guitar produces a middle C with a frequency of 261.6 Hz. An opening with a width of 1.5 m or larger would cause this sound wave to diffract. (Assume that the wave travels at 400 m s^{-1}.)

d A comparison between light waves and diffraction gaps suggests that red light will diffract through a smaller gap compared to violet light.

e A wave will diffract if the dimensions of the opening, w, and the wavelength of the wave, λ, are identical.

26 A student is trying to determine the slit separation used in a double slit experiment. The student used a green laser with a wavelength of 550 nm and measured the fringe spacing for various distances between the slits and the screen. The student's data is shown in the table below. Use the data to plot a suitable graph. Include the line of best fit. Use the gradient of the line of best fit to determine the slit separation.

Distance to the screen (m)	Fringe spacing (mm)
0.5	4.5
0.6	5.5
0.7	6.5
0.8	7.5
0.9	8.0
1.0	9.0

27 Two sources, S_1 and S_2, are producing waves in phase with identical wavelengths. The wavefronts of both sources are shown in the diagram below and a key is provided for identifying the crests and troughs. The position labelled A is a certain number of wavelengths from both sources. Determine the path difference between the two waves and state whether destructive or constructive interference occurs between the two sources. Show all your working.

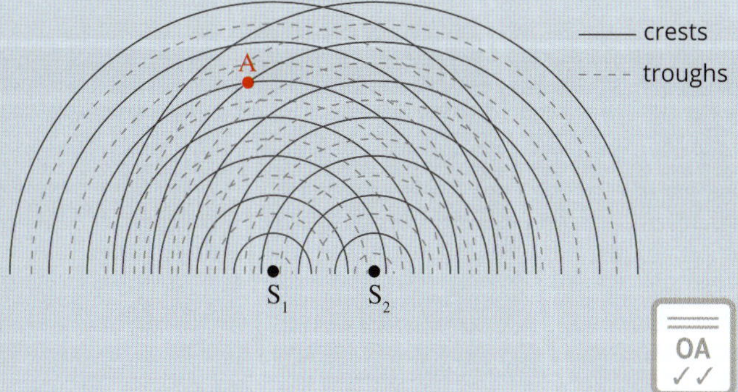

OA ✓✓

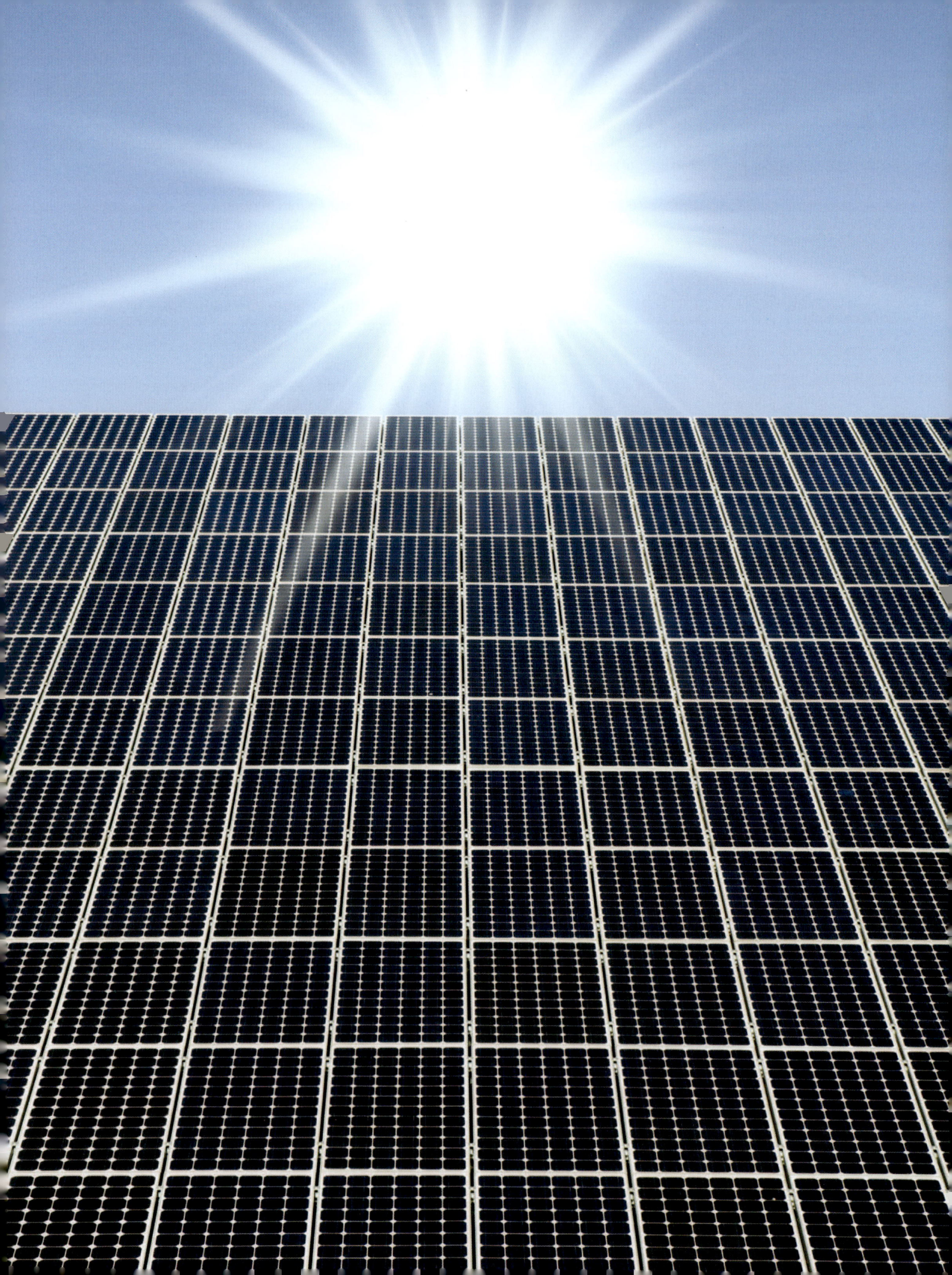

CHAPTER

The dual nature of light and matter

How incredible it would be, if it were possible, to put the giants of physics from throughout history—Galileo, Newton, Maxwell, Heisenberg, Bohr, de Broglie, Einstein and others—together in one room for an hour. The most likely outcome is that the hour would be spent in heated debate. Let's imagine that just one question were posed to them: 'What is light?' None of them would give the same answer. Each would draw on concepts current in their own era. And if another seemingly simple question could be posed—'What is matter?'—the debate would be just as heated!

This chapter will develop your understanding of light and the models that have been built to describe it. It will also introduce the idea that light and matter have much in common—more than you may have thought.

Key knowledge

- apply the quantised energy of photons $E = hf = \frac{hc}{\lambda}$ **9.1**
- analyse the photoelectric effect with reference to:
  - evidence for the particle-like nature of light **9.1**
  - experimental data in the form of graphs of photocurrent versus electrode potential, and of kinetic energy of electrons versus frequency **9.1**
  - kinetic energy of emitted photoelectrons: $E_{k\,max} = hf - \phi$, using energy units of joule and electron-volt **9.1**
  - effects of intensity of incident irradiation on the emission of photoelectrons **9.1**
- describe the limitation of the wave model of light in explaining experimental results related to the photoelectric effect **9.1**
- interpret electron diffraction patterns as evidence for the wave-like nature of matter **9.2**
- distinguish between the diffraction patterns produced by photons and electrons **9.2**
- calculate the de Broglie wavelength of matter: $\lambda = \frac{h}{p}$ **9.2**
- discuss the importance of the idea of quantisation in the development of knowledge about light and in explaining the nature of atoms **9.2**
- compare the momentum of photons and of matter of the same wavelength including calculations using: $p = \frac{h}{\lambda}$ **9.2**
- explain the production of atomic absorption and emission line spectra, including those from metal vapour lamps **9.3**
- interpret spectra and calculate the energy of absorbed or emitted photons: $E = hf$ **9.3**
- analyse the emission or absorption of a photon by an atom in terms of a change in the electron energy state of the atom, with the difference in the states' energies being equal to the photon energy: $E = hf = \frac{hc}{\lambda}$ **9.3**
- interpret the single photon and the electron double slit experiment as evidence for the dual nature of light and matter **9.2**

OA
✓✓

9.1 The photoelectric effect

At the turn of the twentieth century, a number of scientists turned their attention to the features of light that could not be readily explained using Maxwell's electromagnetic wave model. These features required the development of a more sophisticated model of light, and this eventually led to a revolution in the scientific understanding of the nature of energy and matter. One of the scientists who made a significant contribution to this new way of understanding light was Albert Einstein (Figure 9.1.1). Einstein drew on the work of the German physicist Max Planck.

FIGURE 9.1.1 Albert Einstein helped revolutionise our understanding of the nature of light.

PLANCK'S EQUATION

In 1900 the German physicist Max Planck (Figure 9.1.2) was studying the spectrum of light emitted by hot objects. Planck and other scientists had discovered that certain features of this spectrum could not be explained using a wave model of light.

Planck proposed a controversial solution to this problem: light was emitted as discrete packets of energy. He called these packets 'energy quanta' and developed the following equation for the energy, E, of each **quantum**.

$E = hf$
where E is the energy of the quantum (J)
f is the frequency of the light (Hz)
h is the constant 6.63×10^{-34} J s, now known as Planck's constant

FIGURE 9.1.2 Max Planck (1858–1947) proposed that light is emitted as discrete packets.

Because electromagnetic radiation—of which visible light is a form—is more commonly described by its wavelength, scientists often combine Planck's equation with the wave equation for light, $c = f\lambda$, as follows:

$$E = hf \text{ and } f = \frac{c}{\lambda}$$

Therefore:

$$E = \frac{hc}{\lambda}$$

At the time, most scientists disregarded Planck's work. The particle model it suggested was quite at odds with the wave model that had become widely accepted as the correct description of light.

PHYSICSFILE

Max Karl Ernst Ludwig Planck

Max Planck was a German physicist. At the age of 21 he obtained a PhD in physics and, in 1889, was appointed professor at the Friedrich Wilhelm University in Berlin. Planck was the author of numerous works on physics—the quantum theory in particular. On 14 December 1900 he presented the correct version of the Wien formula—which relates the temperature of a black body to the wavelength of light emitted from it—and introduced a new constant: Planck's constant. This date is now recognised as the beginning of the era of quantum mechanics. In 1918 Planck was awarded the Nobel Prize in Physics for the discovery of quantum energy.

Worked example 9.1.1

USING PLANCK'S EQUATION

Calculate the energy in joules of a quantum of ultraviolet light that has a frequency of 2.00×10^{15} Hz.

Thinking	Working
Recall Planck's equation.	$E = hf$
Substitute the appropriate values and solve for E.	$E = 6.63 \times 10^{-34} \times 2.00 \times 10^{15}$ $= 1.33 \times 10^{-18}$ J

Worked example: Try yourself 9.1.1

USING PLANCK'S EQUATION

Calculate the energy in joules of a quantum of infrared radiation that has a frequency of 3.6×10^{14} Hz.

THE ELECTRON VOLT

When studying light, the quantities of energy being considered are usually very small. These are often so small that the joule is no longer a convenient unit. Scientists have adopted a more convenient unit called the **electron volt** (eV). An electron volt is the amount of energy an electron gains when it moves through a potential difference of 1 V. Since the charge on an electron is -1.6×10^{-19} C, then:

$$\begin{aligned} 1\,\text{eV} &= 1e \times 1\,\text{V} \\ &= 1.6 \times 10^{-19}\,\text{C} \times 1\,\text{J C}^{-1} \\ &= 1.6 \times 10^{-19}\,\text{J} \end{aligned}$$

To convert a value expressed in J into eV, divide it by $1.6 \times 10^{-19}\,\text{J eV}^{-1}$

To convert a value expressed in eV into J, multiply it by $1.6 \times 10^{-19}\,\text{J eV}^{-1}$

Worked example 9.1.2

CONVERTING TO ELECTRON VOLTS

A quantum of light has 1.33×10^{-18} J of energy. Convert this energy to electron volts.

Thinking	Working
Recall the rate for converting joules to electron volts and vice versa.	$1\,\text{eV} = 1.6 \times 10^{-19}\,\text{J}$
Divide the value expressed in joules by $1.6 \times 10^{-19}\,\text{J eV}^{-1}$ to convert it to electron volts.	$E = \frac{1.33 \times 10^{-18}}{1.6 \times 10^{-19}}$ $= 8.3\,\text{eV}$

Worked example: Try yourself 9.1.2

CONVERTING TO ELECTRON VOLTS

A quantum of light has 2.4×10^{-19} J of energy. Convert this energy to electron volts.

As worked examples 9.1.1 and 9.1.2 make clear, it is easier to compare the energies of quanta when they are expressed in eV.

Planck's constant can also be expressed in terms of electron volts:

$$\begin{aligned} h &= 6.63 \times 10^{-34}\,\text{J s} \\ &= \frac{6.63 \times 10^{-34}}{1.6 \times 10^{-19}} \\ &= 4.14 \times 10^{-15}\,\text{eV s} \end{aligned}$$

Worked example 9.1.3

CALCULATING QUANTUM ENERGIES IN ELECTRON VOLTS

Calculate the energy in eV of a quantum of ultraviolet light that has a frequency of 2.0×10^{15} Hz. Assume that $h = 4.14 \times 10^{-15}$ eV s.

Thinking	Working
Recall Planck's equation.	$E = hf$
Substitute the given values and solve for E.	$E = 4.14 \times 10^{-15} \times 2.0 \times 10^{15}$ $= 8.3$ eV

Worked example: Try yourself 9.1.3

CALCULATING QUANTUM ENERGIES IN ELECTRON VOLTS

Calculate the energy (in eV) of a quantum of infrared radiation that has a frequency of 3.6×10^{14} Hz. Assume that $h = 4.14 \times 10^{-15}$ eV s.

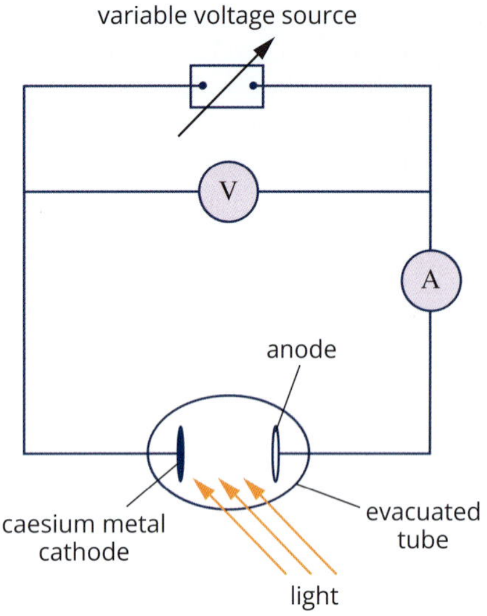

FIGURE 9.1.3 Circuit diagram for investigating the photoelectric effect

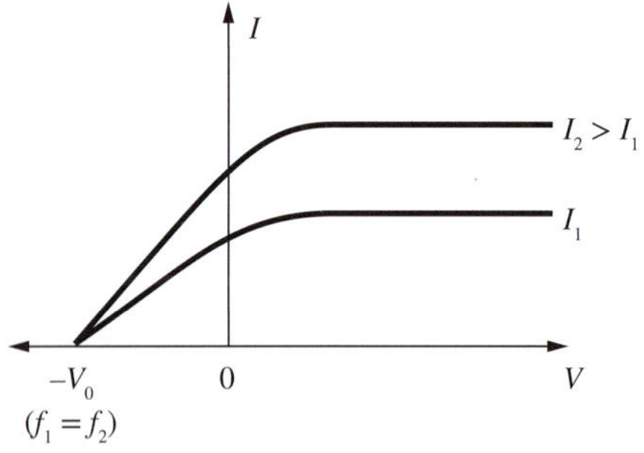

FIGURE 9.1.4 Photocurrent (I) plotted as a function of the voltage (V) applied between the cathode and the anode for different light intensities, I_1 and I_2. Brighter light (i.e. more-intense light) of the same frequency produces a higher photocurrent, but the stopping voltage, V_0, remains the same.

OBSERVING THE PHOTOELECTRIC EFFECT

At the start of the twentieth century, another phenomenon that could not be explained using the wave model of light was being investigated. Scientists noticed that when some types of electromagnetic radiation were projected onto a piece of metal, the metal became positively charged. This positive charge was due to electrons being ejected from the surface of the metal. The electrons became known as **photoelectrons** and the phenomenon was known as the **photoelectric effect**.

Scientists also found that:

- For the photoelectric effect to occur, the electromagnetic radiation projected onto the metal must be above a certain frequency. This is called the **threshold frequency** and it varies from metal to metal.
- If the frequency is kept the same but the intensity of the light is increased, the flow of electrons—the **photocurrent**—also increases.

A common way to investigate the photoelectric effect is to build a circuit like that shown in Figure 9.1.3. The variable voltage supply can make the cathode negative and the anode positive, or vice versa. When the cathode is negative and the anode positive, the electric field created propels the photoelectrons towards the anode. This happens because the electrons will be repelled by the negative potential at the cathode and attracted to the positive potential at the anode. When the cathode is positive and the anode negative, the electric field created repels the photoelectrons and slows them down. As the anode voltage is increased, the photoelectrons are repelled more and more until the photocurrent drops to zero.

Investigating the photoelectric effect in the way illustrated in Figure 9.1.3 reveals some interesting features.

- There is a voltage at which no photoelectrons are emitted from the metal. This is known as the **stopping voltage**. It is indicated by V_0 in Figure 9.1.4. For a particular frequency of light and a particular metal, this stopping voltage is a constant.
- When the light intensity is increased but the frequency remains the same, the photocurrent increases but the stopping voltage remains the same.
- When the applied voltage is positive, photoelectrons are attracted to the anode. A small positive voltage is enough to ensure that every available photoelectron reaches the anode. The current reaches a maximum value and remains there even if the voltage is increased.
- When the applied voltage is negative, the photoelectrons are attracted back towards the cathode and repelled by the anode. The photocurrent is reduced because fewer and fewer photoelectrons have the energy to overcome the opposing electric potential. The photocurrent stops when the applied voltage reaches the stopping voltage.

Recall that the work done on a charged particle by an applied voltage is given by $W = qV$. In the case where the applied voltage prevents electrons from being emitted from the cathode, the voltage is the stopping voltage, V_0. Further, since the charged particle is an electron, $q = q_e$. Hence the work done on the electron to keep it from being emitted is given by $W = q_eV_0$. Since the stopping voltage creates an electric field strong enough to prevent any photoelectrons from being emitted from the cathode, this expression also gives the value of the maximum possible kinetic energy of an emitted photoelectron. For example, if the stopping voltage is 2.5 V, then the maximum kinetic energy of any photoelectron is 2.5 eV.

Figure 9.1.5 shows that when light sources of the same intensity but different frequencies are projected onto the same metal, they produce the same maximum photoelectric current. However, the higher frequency light has a higher stopping voltage.

Finally, provided that the projected light has a frequency above the threshold frequency of the metal, photoelectrons are emitted immediately. This is true regardless of the intensity of the light and cannot be explained if light had only wave properties.

When illuminated with light above the threshold frequency, some photoelectrons are emitted from the first layer of atoms at the surface of the metal and have the maximum kinetic energy possible. Other photoelectrons come from deeper inside the metal and lose some of their kinetic energy due to collisions on their way to the surface. Hence the emitted photoelectrons have a range of kinetic energies from the maximum value downwards.

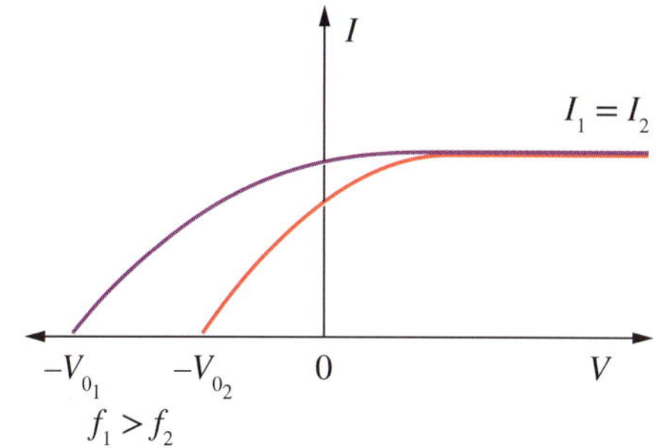

FIGURE 9.1.5 Photocurrent (I) plotted as a function of the voltage (V) applied between the cathode and the anode for different frequencies of projected light with the same intensity.

LIMITATIONS OF THE WAVE MODEL OF LIGHT

The photoelectric effect is an interaction between light and matter that the wave theories of classical physics cannot explain. For example:

- A wave model predicts there should be a time delay between the light striking the metal surface and the emission of photoelectrons. This is because the energy of an electron would have to build up over time while the wave energy is absorbed. What is observed, however, is that there is no time delay: emission is instantaneous.
- A wave is a form of continuous energy transfer, so even low frequency radiation should eventually transfer enough energy to emit photoelectrons if projected onto the surface of the metal for long enough. However, only light above a minimum frequency causes photoelectric emission, and the maximum kinetic energy of the emitted electrons increases with increasing frequencies above this minimum. A further difficulty for the wave model is that this minimum frequency varies between metals.
- Light was considered to have more energy the brighter it is so the brighter the light the more energetic the emitted electrons should be. However, increasing the brightness of the light projected onto metal made no difference to the stopping voltage.

EINSTEIN AND THE PHOTOELECTRIC EFFECT

In 1905 Albert Einstein proposed a solution to this problem. Einstein drew on Planck's earlier work by assuming that light exists as particles, or **photons** (like Planck's quanta), each with an energy of $E = hf$. This assumption made the properties of the photoelectric effect relatively easy to explain.

Einstein's work was a significant extension of Planck's ideas. Although Planck had assumed that light was being emitted in quantised packets, he never questioned the assumption that light was fundamentally a wave phenomenon. Einstein's work went further, challenging scientists' understanding of the nature of light itself.

Einstein proposed that, for a particular metal, the amount of energy required to eject a photoelectron is a constant value that depends on the strength of the bonding within the metal. This energy is called the **work function**, ϕ, of the metal. For example, the work function of lead is 4.14 eV, which means that 4.14 eV of energy is needed to release just one electron from the surface of a piece of lead.

According to Einstein's model, shining light on the surface of a piece of metal is equivalent to bombarding it with photons. When a photon strikes the metal, it can transfer its energy to an electron, that is, a single photon can interact with a single electron, transferring all its energy to the electron at once. What happens next depends on whether the photon has enough energy to overcome the work function of the metal.

If the energy of the photon is less than the work function, then no photoelectrons will be released, as the electrons will not gain enough energy to let them break free of the lead atoms. For example, each photon of violet light with a frequency of 7.50×10^{14} Hz has 3.11 eV of energy. This can be determined by applying the equation $E = hf$:

$$\begin{aligned} E &= hf \\ &= 4.14 \times 10^{-15} \times 7.50 \times 10^{14} \\ &= 3.11 \text{ eV} \end{aligned}$$

This means that violet light shining on lead would not release photoelectrons, since the energy of each photon (3.11 eV) is less than the work function of lead (4.14 eV).

However, each ultraviolet photon with a frequency of 1.20×10^{15} Hz has 4.97 eV of energy:

$$\begin{aligned} E &= hf \\ &= 4.14 \times 10^{-15} \times 1.20 \times 10^{15} \\ &= 4.97 \text{ eV} \end{aligned}$$

Thus ultraviolet light of this frequency would release photoelectrons from the lead, since the energy of each photon (4.97 eV) is greater than the work function of lead (4.14 eV).

Each metal has a threshold frequency, that is, the frequency at which the photons have an energy equal to the work function of the metal, given by the following equation.

$\phi = hf_0$
where ϕ is the work function (J or eV)
h is Planck's constant
f_0 is the threshold frequency for that metal (Hz)

Worked example 9.1.4

CALCULATING THE WORK FUNCTION OF A METAL

Calculate the work function (in J and eV) of aluminium, which has a threshold frequency of 9.8×10^{14} Hz.

Thinking	Working
Recall the formula for work function.	$\phi = hf_0$
Substitute the threshold frequency of the metal and solve for ϕ.	$\phi = 6.63 \times 10^{-34} \times 9.8 \times 10^{14}$ $= 6.5 \times 10^{-19}$ J
Convert this energy from J to eV.	$\phi = \dfrac{6.5 \times 10^{-19}}{1.6 \times 10^{-19}}$ $= 4.1$ eV

Worked example: Try yourself 9.1.4

CALCULATING THE WORK FUNCTION OF A METAL

Calculate the work function (in J and eV) for gold, which has a threshold frequency of 1.2×10^{15} Hz.

THE KINETIC ENERGY OF PHOTOELECTRONS

If the energy of a photon is greater than the work function of a metal, then a photoelectron is released when that photon strikes the metal. The amount of energy in excess of the work function is transformed into the kinetic energy of the photoelectron.

Einstein described this relationship with the photoelectric equation below.

$E_{k\,max} = hf - \phi$

where $E_{k\,max}$ is the maximum kinetic energy of an emitted photoelectron (J or eV)
ϕ is the work function of the metal (J or eV)
h is Planck's constant
f is the frequency of the incident photon (Hz)

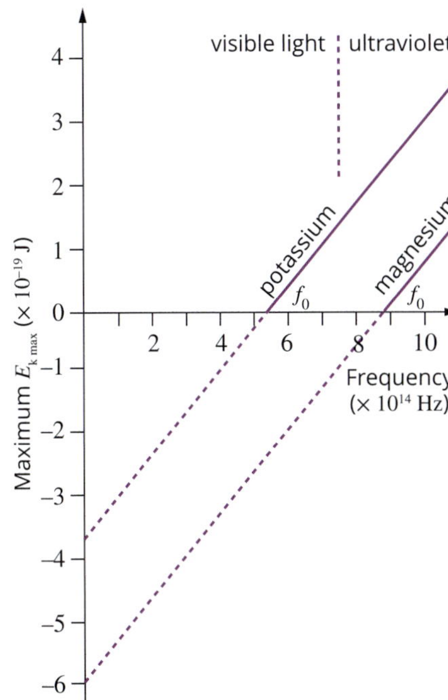

FIGURE 9.1.6 Magnesium has a higher threshold frequency than potassium. (It is in the ultraviolet region whereas potassium's threshold frequency is in the visible region.) The gradient of the graph for each metal is Planck's constant, h. The horizontal intercept gives the threshold frequency, f_0. The vertical intercept gives the work function, ϕ.

Graphing the results when applying Einstein's equation to a metal produces a linear graph like those shown in Figure 9.1.6. Such a graph is useful because it shows key information, such as the threshold frequency of a particular metal. It also shows the work function of that metal. This is found by continuing the graph to the vertical axis. The intercept with the vertical axis gives the magnitude of the work function, ϕ. Further, the gradient of the graph is Planck's constant, h. The graph also shows that as soon as the threshold frequency is exceeded, an electron can be ejected and escape with some kinetic energy. The greater the frequency of the light, the greater the kinetic energy of the photoelectron. At the threshold frequency, electrons are no longer bound to the metal, but they have no kinetic energy. Beyond the threshold frequency, they do have kinetic energy.

Worked example 9.1.5

CALCULATING THE KINETIC ENERGY OF PHOTOELECTRONS

Calculate the kinetic energy in eV of the photoelectrons emitted from lead by ultraviolet light with a frequency of 1.2×10^{15} Hz. The work function of lead is 4.14 eV. Assume that $h = 4.14 \times 10^{-15}$ eV s.

Thinking	Working
Recall Einstein's photoelectric equation.	$E_{k\,max} = hf - \phi$
Substitute the given values and solve for $E_{k\,max}$.	$E_{k\,max} = 4.14 \times 10^{-15} \times 1.2 \times 10^{15} - 4.14$ $= 4.97 - 4.14$ $= 0.83$ eV

Worked example: Try yourself 9.1.5

CALCULATING THE KINETIC ENERGY OF PHOTOELECTRONS

Calculate the kinetic energy (in eV) of the photoelectrons emitted from lead by ultraviolet light with a frequency of 1.50×10^{15} Hz. The work function of lead is 4.14 eV. Assume that $h = 4.14 \times 10^{-15}$ eV s.

CASE STUDY **ANALYSIS**

Lenard's experiment

In the late nineteenth century, a phenomenon that could not be explained by the accepted wave model of light was observed by Heinrich Hertz when conducting research on radio waves. We now call this phenomenon the photoelectric effect.

In 1902 Philipp Eduard Anton von Lenard, an assistant to Hertz, was studying the relationship between the intensity of light projected onto a metal surface and the energy of the electrons emitted. His light source was a high-intensity carbon arc lamp, the most intense light source available at that time. He set up his apparatus similar to that in Figure 9.1.3 on page 362, with the metal he was testing as the cathode and a metal plate as the anode, both housed in an evacuated tube. When the cathode was illuminated by light from the carbon arc lamp, electrons were emitted. These travelled to the anode under the influence of an electric field.

To determine the maximum energy of these electrons, Lenard reversed the polarity of the plates—and hence the direction of the electric field—in order to repel the emitted electrons travelling towards the anode. He found that there was a specific minimum voltage, V_0, between the plates at which the electrons stopped reaching the anode (as measured by an ammeter). He found that when he increased the intensity of the light, the current increased (hence the number of electrons) but the stopping voltage remained the same as for lower-intensity light. The theories of that time suggested that increasing the intensity of the light should increase the energy of the emitted electrons, requiring a higher stopping voltage, but this was clearly not the case. These results are shown in Figure 9.1.4 on page 362.

Lenard decided to repeat his experiment but this time using filters to limit the light source to a particular colour. He found that light with a shorter wavelength (i.e. light with higher energy) caused the emission of photoelectrons with higher energy than light with a longer wavelength (i.e. light with lower energy). He also observed that there was a wavelength above which no electrons were emitted, λ_0, and that this varied for each type of metal illuminated. We now call the corresponding frequency the *threshold frequency*, f_0.

Lenard's observations went against Maxwell's electromagnetic wave theory. Maxwell's theory was the accepted theory of light at that time, and it predicted that the kinetic energy of such electrons was dependent only on the intensity of the light hitting the metal, not its frequency. For his work, Lenard was awarded the 1905 Nobel Prize for Physics.

In 1905 Albert Einstein, using Max Planck's hypothesis that light is carried in discrete quantised energy packages, developed the mathematical equation to explain the photoelectric effect. For this he was awarded the 1921 Nobel Prize for Physics.

Analysis

In an experiment similar to that conducted by Lenard, a potassium surface was bombarded with photons of varying wavelengths.

The following table shows the maximum kinetic energies of the emitted photoelectrons at various wavelengths.

Wavelength (nm)	Maximum kinetic energy (eV)
517	0.14
448	0.53
414	0.75
395	0.86
366	1.09
353	1.25

1. Recreate the table in your workbook or in a spreadsheet program. Add two more columns—one for frequency (Hz) and the other for maximum kinetic energy (J)—and calculate the corresponding values for each wavelength. You should use scientific notation to display your values.
2. Using the data in your table, draw a graph of the maximum kinetic energy (J) of the photoelectrons versus frequency.
3. Using your graph, determine:
   - **a** the gradient of the line of best fit
   - **b** the equation of the line of best fit
   - **c** the *x*-intercept.
4. Analyse your graph to determine:
   - **a** an experimental value for Planck's constant, *h*
   - **b** the threshold frequency of potassium
   - **c** the work function of potassium.

RESISTANCE TO THE QUANTUM MODEL OF LIGHT

This new particle or 'quantum' model of light was not well received by the scientific community. It had already been established that a discrete, particle model for light could not explain many of light's properties, such as polarisation and the interference patterns observed in Young's double-slit experiment.

Most scientists believed instead that wave explanations for the photoelectric effect would eventually be found. However, the quantum model of light eventually became accepted and the Nobel Prize in Physics was awarded to both Planck (1918) and Einstein (1921) for their ground-breaking work in this field.

PHYSICSFILE

Albert Einstein

Although Albert Einstein is most famous for his work on relativity (and its related equation $E = mc^2$), he gained his Nobel Prize 'for his services to Theoretical Physics, and especially for his discovery of the law of the photoelectric effect'. His work on relativity was never formally recognised with a Nobel Prize.

PHYSICSFILE

Solar panels

Each cell in a solar panel (see figure below) functions on the principle of the photovoltaic effect. The photovoltaic effect is based on the same principles as the photoelectric effect. The photons in the sunlight falling on the solar panels have enough energy to cause electrons to be released from atoms in one of the specially treated semiconductor layers in the cells. The result is a net photovoltaic current.

Solar panels use the photoelectric effect to generate electricity.

9.1 Review

OA ✓✓

SUMMARY

- At the atomic level, electromagnetic radiation is emitted or absorbed in discrete packets or quanta called photons.
- A number of phenomena related to the behaviour of light, such as the photoelectric effect, can only be explained using the concept of photons, or light quanta.
- The energy of a photon is proportional to its frequency and inversely proportional to its wavelength:

$$E = hf = \frac{hc}{\lambda}$$

- Planck's constant, h, can be determined experimentally using the photoelectric effect.
- The electron volt is an alternative (non-SI) unit of energy: $1\text{ eV} = 1.6 \times 10^{-19}\text{ J}$.
- The photoelectric effect is the emission of photoelectrons from a metal surface when light is shone onto the surface.
  - If the frequency of the light is less than the threshold frequency, f_0, no electrons are released.
  - If the frequency of light is greater than f_0, the rate of electron release (the photocurrent) is proportional to the intensity of the light and occurs immediately.
- $E_{k\,max} = q_eV_0$, where q_e is the charge on an electron and V_0 is the stopping voltage.
- The work function, ϕ, of a metal is given by $\phi = hf_0$, and is different for each metal. If the frequency of the incident light is greater than the threshold frequency, then a photoelectron will be ejected with some kinetic energy up to a maximum value.
- A graph of $E_{k\,max}$ versus frequency has a gradient equal to Planck's constant, h, and a y-intercept equal to the work function, ϕ.
- The maximum kinetic energy of the photoelectrons emitted from a metal is the energy of the photons minus the metal's work function: $E_{k\,max} = hf - \phi$.
- The wave approach to light could not explain various features of the photoelectric effect, such as:
  - the existence of a threshold frequency
  - the absence of a time delay when using very weak light sources
  - the fact that increasing the intensity of light results in a greater rate of electron release rather than an increase in electron energy.

  The photon model was able to explain all of these features.
- Einstein used Planck's concept of a photon to explain the photoelectric effect, stating that each electron released was due to an interaction with only one photon.

KEY QUESTIONS

Knowledge and understanding

1 When light shines on a metal surface, the metal might become positively charged. Provide an explanation for this and why it does not occur for all metals.

2 Which of the following statements about the photoelectric effect are true and which are false? Rewrite those that are false to make them true.

- **a** When the intensity of light shining on the surface of a metal increases, the photocurrent increases.
- **b** When light sources of the same intensity but different frequencies are used, the higher frequency light has a higher stopping voltage and produces a higher maximum current than the lower frequency light.
- **c** When the applied voltage is positive, photoelectrons are attracted to the anode.

3 Which of the following statements are true and which are false with respect to the value of the stopping voltage obtained when light is projected onto a metal cathode? For those that are false, rewrite them to make them true.

- **a** The stopping voltage indicates how much work must be done to stop the most energetic photoelectrons.
- **b** The stopping voltage is reached when the photocurrent is reduced almost to zero.
- **c** If the intensity of the light projected on the metal is increased, the stopping voltage remains the same.
- **d** For a given metal, the value of the stopping voltage is affected only by the frequency of the light projected on the metal.

4 Light with a frequency of 8.00×10^{14} Hz is shone onto a piece of calcium with a work function of 3.15 eV. Calculate the maximum kinetic energy, in electron volts, of the emitted photoelectrons.

Analysis

5 Calculate the energy (in joules) of the following wavelengths of light. Give your answers to 3 significant figures.

	Colour	Wavelength (nm)
a	red	656
b	yellow	589
c	blue	486
d	violet	397

6 Calculate the work function (in electron volts) of the following metals:

	Metal	Threshold frequency ($\times 10^{15}$ Hz)
a	lead	1.0
b	iron	1.1
c	platinum	1.5

7 In an experiment on the photoelectric effect, different frequencies of light were shone onto a piece of magnesium with a work function of 3.66 eV. By performing the necessary calculations, identify which of the frequencies listed below would be expected to produce photoelectrons.

A 3.0×10^{14} Hz
B 5.0×10^{14} Hz
C 7.0×10^{14} Hz
D 9.0×10^{14} Hz

8 Ultraviolet light with a wavelength of 310 nm is shone onto a piece of cadmium with a work function of 3.77 eV. Calculate the maximum kinetic energy, in electron volts, of the emitted photoelectrons.

9 The metal sodium has a work function of 1.81 eV. Which of the following types of electromagnetic radiation would cause photoelectrons to be emitted from a sodium surface? Justify your choice or choices with appropriate mathematics.

A infrared radiation, $\lambda = 800$ nm
B red light, $\lambda = 700$ nm
C violet light, $\lambda = 400$ nm
D ultraviolet radiation, $\lambda = 300$ nm

10 Green light of wavelength 530 nm is shone onto a metal whose stopping voltage is 0.6 V. Calculate the work function of the metal in electron volts.

11 The light-sensitive cells in the retina of your eye—such as the rods and cones—send signals to your brain when they are stimulated by light. For light of wavelength 535 nm, about 5000 photons must hit a rod at the same time before the rod will respond.

a Calculate the frequency of photons with a wavelength of 535 nm.
b Calculate the energy, in joules, that each photon carries.
c Calculate the smallest amount of light energy, in joules, that can be detected by a single rod.

9.2 The quantum nature of light and matter

In order to explain the photoelectric effect, Einstein used the concept of a photon that Planck had developed. However, like many great discoveries in science, the development of the quantum model of light raised almost as many questions as it answered. It was already well-established that a wave model was needed to explain such phenomena as diffraction and interference. How could these two apparently contradictory models be reconciled to form a comprehensive theory of light?

Answering this question was one of the great scientific achievements of the twentieth century and led to the extension of the quantum model to matter as well as to energy. It led to a fundamental shift in the way the universe is viewed. Some of the great scientists who contributed to these advances are in the historic photograph shown in Figure 9.2.1.

FIGURE 9.2.1 This photo shows some of the participants at the 5th Solvay conference in Brussels in 1927. The conference was attended by many great scientists, including Albert Einstein, Max Planck, Niels Bohr, Marie Curie, Paul Dirac, Erwin Schrödinger and Louis de Broglie. All these scientists contributed to the current knowledge of the universe, the atom and quantum mechanics.

WAVE-PARTICLE DUALITY

In many ways, the wave and particle models of light seem fundamentally incompatible. Waves are continuous and are described in terms of wavelength and frequency. Particles are discrete and are described by physical dimensions such as mass and radius.

In order to understand how these two sets of ideas can be brought together, it is important to remember that scientists describe the universe using models. Models are analogies that are used to illustrate certain aspects of reality that might not be immediately apparent.

Physicists have come to accept that light is not easily compared to any other physical phenomenon. In some situations light has similar properties to a wave; in other situations it behaves more like a particle. This is called the **wave–particle duality** of light. Although this may seem somewhat paradoxical and counter-intuitive, in the century since Einstein did his work establishing quantum theory, many experiments have supported this duality and no scientist has proposed a better explanation.

Experimental evidence for the dual nature of light

In the early years of quantum theory, some scientists believed that the wave properties of light observed in Young's double-slit experiment may have been due to some sort of interaction between photons as they passed through the slits together.

To test this, double slit experiments were done with light sources that were so dim that scientists were confident that only one photon was passing through the slits at a time. In this way, there could be no interactions between photons. After many single photons had passed through the slits, interference patterns formed that were identical to those produced with bright sources (Figure 9.2.2). This clearly demonstrated the dual nature of light.

Interestingly, when a detector is used to measure which slit a photon passes through, the wave pattern disappears and the photon acts like a particle.

De Broglie's wave–particle theory

In 1924 the French physicist Louis de Broglie proposed a ground-breaking theory. He suggested that since light (which had long been considered a wave) sometimes demonstrated particle-like properties, then perhaps matter (which was considered to be made up of particles) might sometimes demonstrate wave-like properties. He quantified this theory by predicting that the wavelength of a particle would be given by the following equation.

$$\lambda = \frac{h}{p}$$

where λ is the wavelength of the particle (m)

p is the momentum of the particle ($kg\,m\,s^{-1}$)

h is Planck's constant (J s)

This is also commonly written as:

$$\lambda = \frac{h}{mv}$$

where m is the mass of the particle (kg)

v is the velocity of the particle ($m\,s^{-1}$)

The wavelength that de Broglie predicted is referred to as the **de Broglie wavelength** of matter.

Worked example 9.2.1

CALCULATING THE DE BROGLIE WAVELENGTH

In a famous experiment conducted by Clinton Davisson and Lester Germer, electrons were fired at a nickel crystal at a speed of $4.0 \times 10^6\, m\,s^{-1}$.

Calculate the de Broglie wavelength of these electrons if the mass of an electron is 9.1×10^{-31} kg.

Thinking	Working
Recall the de Broglie equation.	$\lambda = \frac{h}{mv}$
Substitute appropriate values and solve for λ.	$\lambda = \frac{h}{mv}$ $= \frac{6.63 \times 10^{-34}}{9.1 \times 10^{-31} \times 4.0 \times 10^{6}}$ $= 1.8 \times 10^{-10}$ m or 0.18 nm

Worked example: Try yourself 9.2.1

CALCULATING THE DE BROGLIE WAVELENGTH

Calculate the de Broglie wavelength of a proton travelling at $7.0 \times 10^5\, m\,s^{-1}$. The mass of a proton is 1.67×10^{-27} kg.

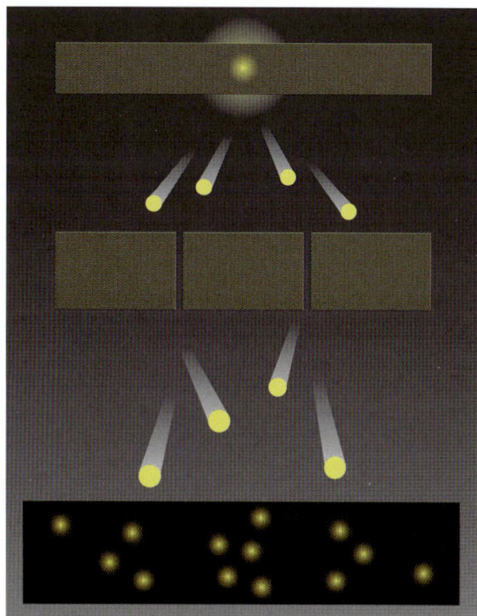

FIGURE 9.2.2 An interference pattern can be built up over time by a series of single photons passing through an apparatus like that used in Young's experiment. This clearly demonstrates the wave–particle duality of light.

PHYSICSFILE

Louis Victor Pierre Raymond de Broglie (1892–1987)

Louis de Broglie was a French physicist. In 1924 he wrote a doctoral thesis entitled *Recherches sur la théorie des quanta* (Research on quantum theory) in which he presented his theory of the wave properties of particles. Building on the work of Einstein and Planck on wave–particle duality, de Broglie proposed what is now known as the de Broglie wave theory. Later he developed his thesis further and formulated the final de Broglie hypothesis. In 1929 he was awarded the Nobel Prize for his research. De Broglie's work has led to a number of practical applications, such as the electron microscope.

Louis de Broglie

Worked example 9.2.2

CALCULATING THE DE BROGLIE WAVELENGTH OF A MACROSCOPIC OBJECT

Calculate the wavelength of a cricket ball of mass 160 g travelling at 150 km h^{-1}.

Thinking	Working
Convert mass and velocity to SI units.	$m = 160\,\text{g} = 0.160\,\text{kg}$ $v = \frac{150}{3.6} = 42\,\text{m s}^{-1}$
Recall de Broglie's equation.	$\lambda = \frac{h}{mv}$
Substitute appropriate values and solve for λ.	$\lambda = \frac{h}{mv}$ $= \frac{6.63 \times 10^{-34}}{0.160 \times 42}$ $= 9.9 \times 10^{-35}\,\text{m}$

Worked example: Try yourself 9.2.2

CALCULATING THE DE BROGLIE WAVELENGTH OF A MACROSCOPIC OBJECT

Calculate the de Broglie wavelength of a 66 kg person running at 36 km h^{-1}.

It can be seen from Worked example 9.2.1 that the de Broglie wavelength of an electron is smaller than that of visible light (although it is still large enough to be measurable). However, the wavelength of an everyday object, such as a cricket ball (Worked example 9.2.2), is extremely small (9.9×10^{-35} m). Hence you will never notice the wave properties of everyday objects. To underscore this point, consider the observable wave behaviour of diffraction. Recall that for diffraction to be noticeable, the size of the wavelength needs to be comparable to the size of the gap or obstacle. Therefore the wavelength from an everyday object, being tiny, would need to pass through a gap much smaller than a fraction of a proton's diameter to produce any noticeable diffraction!

ELECTRON DIFFRACTION PATTERNS

Support for Louis de Broglie's controversial hypothesis that matter could exhibit wave-like behaviour was discovered by accident in 1925 by two American physicists, Clinton J. Davisson and Lester H. Germer. They were investigating the reflection of an electron beam from the surface of crystalline nickel. They used an electron gun to provide a beam of electrons of known speed. A moveable detector could detect electrons scattered from the nickel target in any direction (Figure 9.2.3).

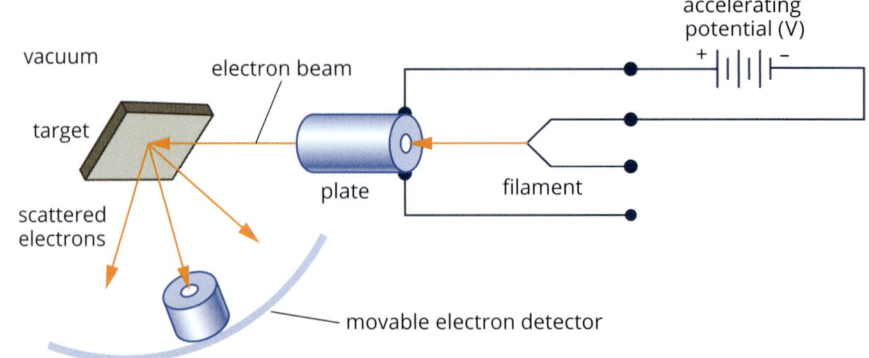

FIGURE 9.2.3 Davisson and Germer used this apparatus to detect electrons scattering from a target.

As a result of an inadvertent recrystallising of one of their nickel samples, Davisson and Germer observed a scattering pattern very different to any seen so far. As they moved the detector through different scattering angles, they encountered a sequence of maximum and minimum electron intensities (Figure 9.2.4). Davisson and Germer did not understand what they had observed until they attended a scientific meeting at Oxford in 1926. There they heard about de Broglie's work and learnt that their results were being used to confirm his hypothesis.

In further experiments, Davisson and Germer showed that the electrons were being scattered by the different layers within the crystal lattice (Figure 9.2.5) and were undergoing interference. When they analysed the diffraction pattern to determine the wavelength of the electron waves, they calculated a value of 0.14 nm. This was consistent with de Broglie's hypothesis and was the first experimental evidence of wave–particle duality. Their work also confirmed de Broglie's prediction that particles have a wavelength that is inversely proportional to their momentum ($\lambda = \frac{h}{mv}$).

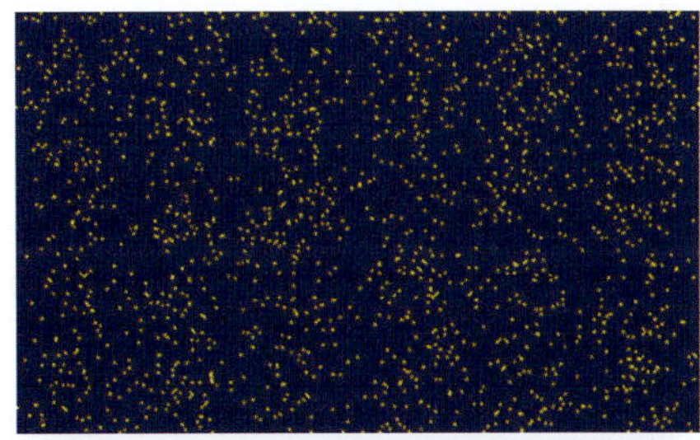

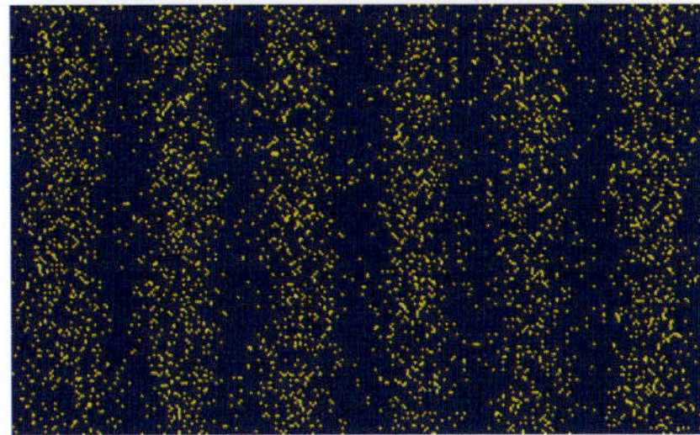

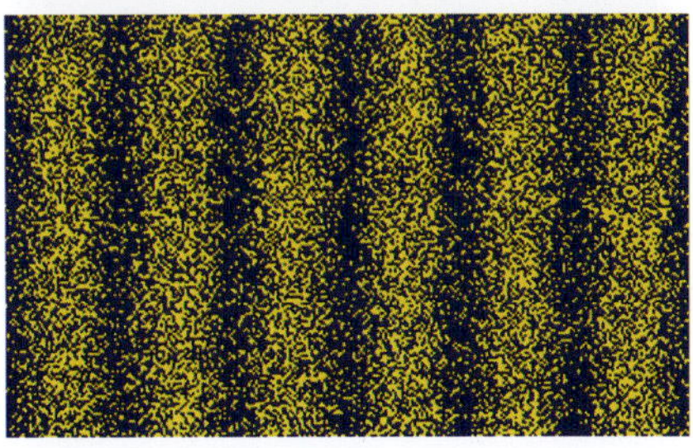

FIGURE 9.2.4 Electron diffraction patterns like those observed by Davisson and Germer are similar to those obtained in experiments with light. This suggests that electrons have a wave property.

Worked example 9.2.3

WAVELENGTH OF ELECTRONS EMITTED FROM AN ELECTRON GUN

Find the de Broglie wavelength of an electron that has been accelerated from rest through a potential difference of 75 V. The mass of an electron is 9.1×10^{-31} kg and the magnitude of the charge on an electron is 1.6×10^{-19} C.

Thinking	Working
Calculate the kinetic energy of the electron from the work done on it by the electric potential. Recall from earlier chapters that $W = qV$.	$W = qV$ $= 1.6 \times 10^{-19} \times 75$ $= 1.2 \times 10^{-17}$ J
Calculate the velocity of the electron.	$E_k = \frac{1}{2}mv^2$ $v = \sqrt{\frac{2E_k}{m}}$ $= \sqrt{\frac{2 \times 1.2 \times 10^{-17}}{9.1 \times 10^{-31}}}$ $= 5.1 \times 10^6 \text{ m s}^{-1}$
Use de Broglie's equation to calculate the wavelength of the electron.	$\lambda = \frac{h}{mv}$ $= \frac{6.63 \times 10^{-34}}{9.1 \times 10^{-31} \times 5.1 \times 10^6}$ $= 1.4 \times 10^{-10}$ m $= 0.14$ nm

Worked example: Try yourself 9.2.3

WAVELENGTH OF ELECTRONS EMITTED FROM AN ELECTRON GUN

Find the de Broglie wavelength of an electron that has been accelerated from rest through a potential difference of 50 V. The mass of an electron is 9.1×10^{-31} kg and the magnitude of the charge on an electron is 1.6×10^{-19} C.

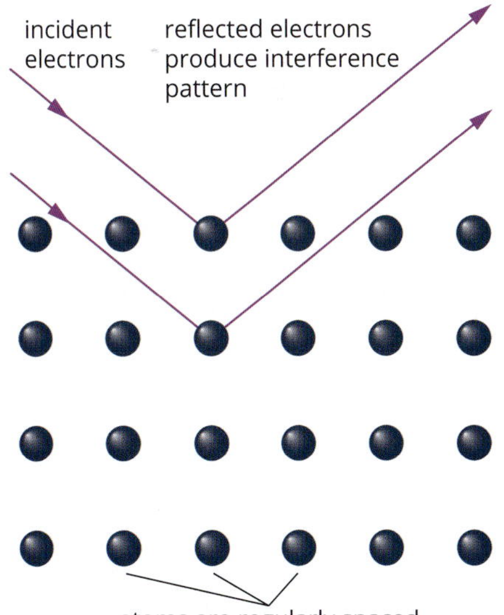

FIGURE 9.2.5 Electrons reflecting from different layers within the crystal structure travel different path distances to reach the observer. The fact they have different path distances can create an interference pattern like those produced by a diffraction grating.

COMPARING THE WAVELENGTHS OF PHOTONS AND ELECTRONS

In the same year that Davisson and Germer conducted their experiment, other supporting evidence for the wave-like nature of matter came from G. P. Thomson (son of J. J. Thomson, discoverer of the electron). Rather than scatter an electron beam from a crystal, Thomson produced a diffraction pattern by passing a beam of electrons through a crystal. Thomson then repeated his experiment using X-rays of the same wavelength instead of electrons. The X-ray diffraction pattern was almost identical to the one made with electrons (Figure 9.2.6).

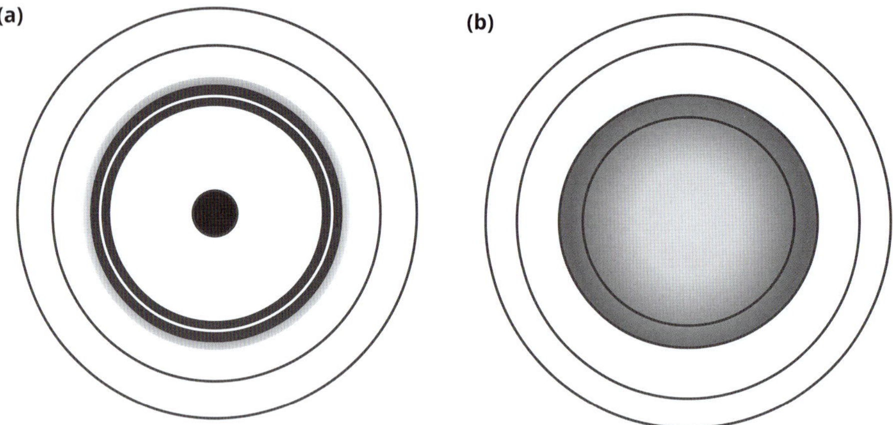

FIGURE 9.2.6 (a) The diffraction pattern formed when X-rays were passed through a crystal. (b) The diffraction pattern formed when electrons were passed through the same crystal. Their similarity suggests that electrons have wave-like behaviour and with a de Broglie wavelength close to the wavelength of X-rays.

As both the X-ray photons and electrons passed through the same gaps in the target crystal and produced a similar diffraction pattern, an important conclusion could be made: the electrons must have a similar wavelength to the wavelength of X-rays. And if their wavelength is similar, their momentum must also be similar (even if their speed is not).

CASE STUDY ANALYSIS

Electron microscope

The discovery of the wave properties of electrons had an important practical application—it led to the development of the electron microscope. In much the same way that an optical microscope makes use of the wave properties of photons to magnify tiny objects, the wave properties of electrons can also be used to create magnified images (Figure 9.2.7).

One of the limitations of an optical microscope is that it can only create a clear image of structures that are larger than the wavelength of the light being used to observe them. This is because the diffraction of the waves at the edges of smaller structures would obscure them. A light microscope is therefore only useful for seeing things down to the lower wavelength of the visible light spectrum—about 390 nm.

In an electron microscope, a stream of high-velocity electrons replaces the beam of visible light. Because the wavelength of the electrons is so small, it allows the observer to see objects as small as several nanometres with a magnification in excess of 10^6. (The maximum useful magnification of a light microscope is around 1000 times.)

There are other differences between light and electron microscopes. For example:

- The electron beam is focused by passing it through a series of electromagnetic coils which cause it to bend (analogous to the way glass lenses bend visible light).
- The image can either be formed as an electron micrograph on a photographic plate or as an image on a computer screen. It cannot be viewed like the image in an optical microscope.

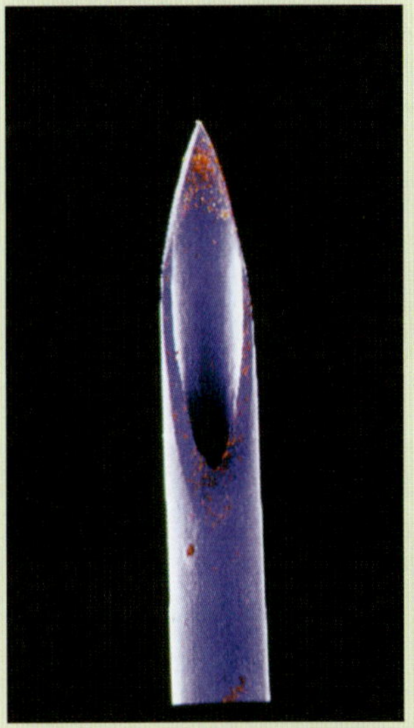

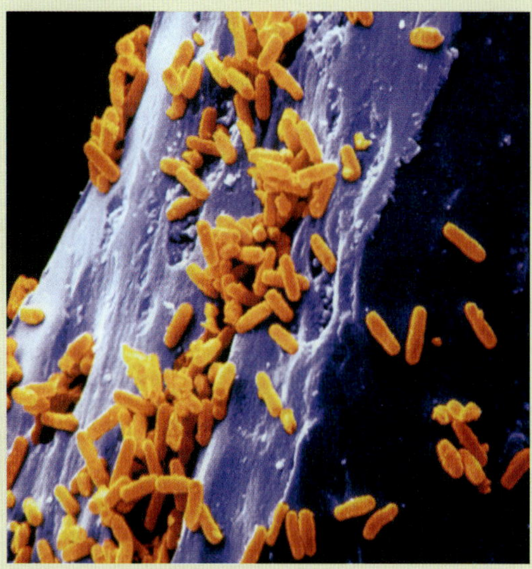

FIGURE 9.2.7 Images formed by an electron microscope of rod-shaped bacteria (coloured orange) clustered on the point of a 35 mm syringe used to administer injections. From left to right, the magnifications are ×9, ×36 and ×560 respectively.

- The preparation of specimens for examination by an electron microscope is much more difficult, as the specimens are usually embedded in a substrate—such as wax—before being cut into very thin slices (about 100 nanometres thick). After preparation they must be placed in a chamber that is at a very high vacuum.

There are several types of electron microscopes, with the most powerful being variations of the transmission electron microscope (TEM). The standard TEM has an electron gun that produces electrons by thermionic emission and accelerates them to a typical energy of 150 keV. The latest development is a high-resolution TEM. It has a magnification above 5×10^7 times and can view structures measuring less than 50 picometres.

Analysis

1. Suggest why the sample to be observed by an electron microscope must be placed in a vacuum and not in a space filled with air.
2. Describe the relationship between the wavelength of an electron and its velocity.
3. What shape does the path of the electron beam take while it traverses the magnetic field of each focusing coil?
4. The strength of the focusing magnetic fields would need to change if the electron beam velocity is changed. Derive the mathematical relationship between the magnetic field strength and the velocity of the electron beam.
5. The typical energy of the electrons leaving the electron gun of an electron microscope is 150 keV. Given that they are accelerated from rest, calculate the velocity of the most energetic electrons leaving the electron gun. The mass of an electron is 9.1×10^{31} kg.

PHOTON MOMENTUM

An interesting corollary of de Broglie's equation, $\lambda = \frac{h}{p}$, is that if a particle has a wavelength, λ, then a photon must have a momentum, p, where $p = \frac{h}{\lambda}$. This is quite counterintuitive, since photons do not have any mass and travel at the speed of light. Nevertheless, de Broglie's equation allows the momentum of a photon to be calculated.

Worked example 9.2.4

CALCULATING PHOTON MOMENTUM

Calculate the momentum of a photon of red light with a wavelength of 650 nm.

Thinking	Working
Recall the formula for the momentum of a photon.	$p = \frac{h}{\lambda}$
Convert 650 nm to m.	$650\,\text{nm} = 650 \times 10^{-9}\,\text{m}$
Substitute the appropriate values and solve for p.	$p = \frac{h}{\lambda}$ $= \frac{6.63 \times 10^{-34}}{650 \times 10^{-9}}$ $= 1.02 \times 10^{-27}\,\text{kg}\,\text{m}\,\text{s}^{-1}$

Worked example: Try yourself 9.2.4

CALCULATING PHOTON MOMENTUM

Calculate the momentum of a photon of blue light with a wavelength of 450 nm.

Clearly, the momentum of a single photon is tiny, which is why you do not feel any physical pressure when light falls on you. However, it is possible to measure the pressure of light using very sensitive equipment.

PHYSICSFILE

Solar sailing

In interplanetary space, where other forces (such as friction) are negligible, light pressure can be used as a form of propulsion. Spacecraft such as Mariner 10 and MESSENGER, which both flew past Mercury and Venus, used solar pressure to decelerate and thus conserve fuel.

In 2010, the Japanese Aerospace Exploration Agency launched IKAROS, the first spacecraft to draw its primary propulsion from a solar sail (see figure at right). IKAROS stands for Interplanetary Kite-craft Accelerated by Radiation Of the Sun. A traditional sail propels a ship using the change of momentum that occurs when air molecules bounce off it. Similarly, a solar sail gains propulsion from changes in photon momentum as light is reflected from it. IKAROS has a $196\,\text{m}^2$ reflective sail which produces a thrust of 1.12 mN.

The last signal received from IKAROS was in 2015, before it went into hibernation mode. The signal indicated that it was approximately 110 million kilometres from Earth.

An artist's impression of IKAROS, the first interplanetary spacecraft to use solar-sail technology.

9.2 Review

OA ✓✓

SUMMARY

- At the atomic level, energy and matter exhibit the characteristics of both waves and particles.
- The wavelength of a particle is given by the de Broglie equation:

$$\lambda = \frac{h}{p} \text{ (i.e. } \lambda = \frac{h}{mv}\text{)}$$

- The results of Young's double-slit experiment are explained with a wave model, but the experiment produces the same interference and diffraction patterns when only one photon at a time (or one electron at a time) is passed through the slits.
- In particle-scattering experiments, beams of particles (electrons usually) are scattered from different layers of atoms in a crystal lattice. Consequently a diffraction pattern is produced which can only be explained if matter has a wave-like nature.
- If photons and matter particles that are being scattered by the same crystal produce the same diffraction pattern, then they must have the same wavelength and momentum.
- All matter, like light, has a dual nature. In our everyday experience matter behaves as a particle, but in some situations it behaves like a wave. This symmetry in nature—the dual nature of light and matter—is referred to as wave–particle duality.
- The momentum of a photon is calculated from:

$$p = \frac{h}{\lambda}$$

KEY QUESTIONS

Knowledge and understanding

1 Explain why a cricketer does not have to consider the wave properties of a cricket ball while batting.

2 Explain why it is possible to observe individual atoms with an electron microscope but not with a light microscope.

3 Which of the following conclusions can be drawn from Louis de Broglie's theory about matter waves?

A All particles exhibit wave behaviour.
B Only moving particles exhibit wave behaviour.
C Only charged particles exhibit wave behaviour.
D Only moving charged particles exhibit wave behaviour.

4 A corollary of de Broglie's work on matter waves is that photons must have momentum. Why can't the momentum of a photon be determined using classical physics?

Analysis

5 A charge q of mass m is accelerated from rest through a potential difference of V. Derive an expression that defines the de Broglie wavelength, λ, of the charge in terms of q, m and V.

6 Consider the technology behind light microscopes and electron microscopes. Explain why images from an electron microscope are produced in greyscale (black and white) and not in their natural colours, while images from a light microscope show the colour of the object itself.

7 Calculate the speed of a positron that has a de Broglie wavelength of 2.5×10^{-9} m. The mass of a positron is 9.1×10^{-31} kg, the same as an electron.

8 Identical diffraction patterns were formed by a beam of electrons and a beam of X-rays when they were fired at a particular crystal. If the frequency of the X-rays was 9.6×10^{17} Hz, calculate:

a the wavelength of the electrons
b the velocity of the electrons.

9 At what speed is a proton travelling if it has the same wavelength as a gamma ray of energy 3.65×10^{-13} J? (The mass of a proton is 1.67×10^{-27} kg.)

9.3 Light and matter

The idea of wave–particle duality is counterintuitive and was not immediately accepted by most scientists, even after the ground-breaking work of Einstein, de Broglie and others.

It was the work of Danish physicist Niels Bohr that finally convinced scientists that the particle model was required as part of a complete understanding of the nature of light. Bohr built on the work of Planck and Einstein to explain the emission and absorption line spectra of hydrogen (Figure 9.3.1). This led to important discoveries in astronomy and eventually a reformulation of our understanding of the nature of energy and matter.

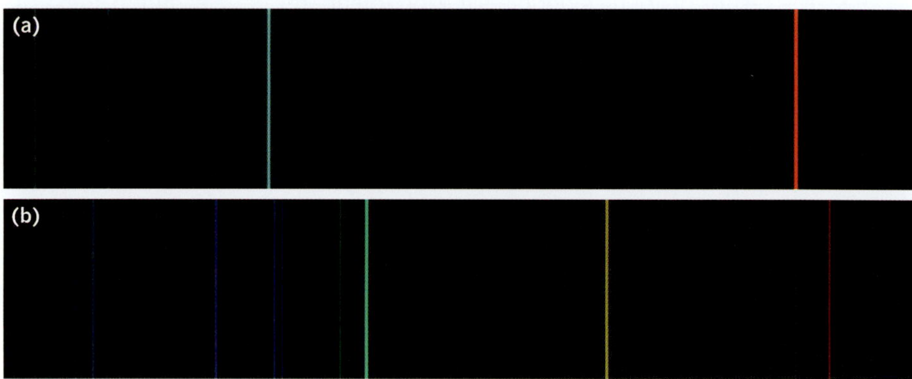

FIGURE 9.3.1 The spectral lines for hydrogen (a) and helium (b). An element's spectral lines are unique to it. They are produced as electrons transition between different energy levels within the atom.

ABSORPTION LINE SPECTRA

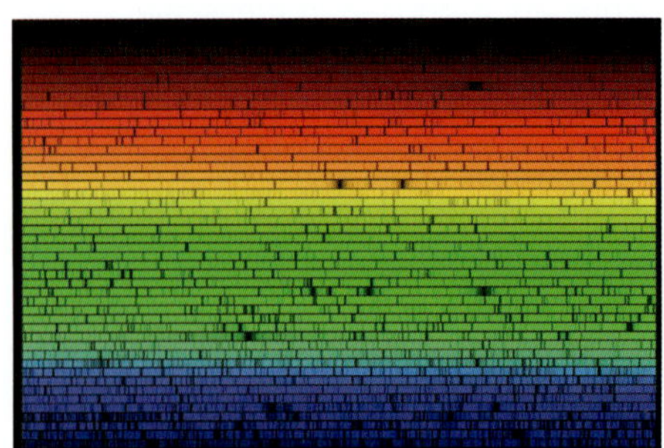

FIGURE 9.3.2 The spectrum of sunlight contains some missing colours, indicated by black lines (known as Fraunhofer lines).

In 1814 the German physicist Joseph von Fraunhofer reported a number of dark lines appearing in the spectrum of sunlight (Figure 9.3.2). You may recall that a spectrum showing all the colour components of white light can be obtained by passing sunlight through a prism. When Fraunhofer did this, he observed that there were some colours missing from the spectrum. The missing colours appeared as black lines at various points along the spectrum. These apparently missing colours came to be known as Fraunhofer lines.

About 50 years later, scientists recognised that some of these lines corresponded to the colours emitted when certain gases were heated to high temperatures. They speculated that the dark lines in the Sun's spectrum were due to the corresponding colours being absorbed by gases as light made its way through the outer atmosphere of the Sun. This **absorption line spectrum** allowed astronomers to determine that the Sun is largely composed of hydrogen, with small quantities of helium and some other heavier elements.

Absorption line spectra are valuable for scientists who wish to know what elements are present in a sample of gas or in a solution, so their use is not limited to astronomy. First, light is directed through a cool sample of a gas or through a solution containing an element. Initially the atoms in the sample are not in an excited state and thus their electrons will be in the ground state. As light passes through, only certain wavelengths (or frequencies) will be absorbed by the elements present in the sample, which means that on viewing the absorption line spectrum, these particular wavelengths will be missing. Because the wavelengths that are absorbed are unique to each type of atom, analysing which wavelengths are missing helps scientists determine exactly what elements are present in the sample.

EMISSION LINE SPECTRA

When elements are heated to high temperatures or have a high-energy electrical current passed through them, they produce light. This happens when electrons within the atoms of the material absorb energy, become excited and move to an energy state higher than the ground state.

Eventually these electrons return to their unexcited or **ground state** and the energy they had absorbed is released as a photon. The colour of this photon will depend on the amount of energy the electron absorbed and the difference in energy of the levels through which it falls on returning to its ground state.

Since atoms usually have a number of different **excited states**, they can produce a number of different colours. The combination of colours produced by a particular element is unique to that element and is known as its **emission line spectrum**. Figure 9.3.1 shows the emission line spectra for hydrogen and helium. The different metals used in fireworks create different colours as their excited electrons return to their ground state (Figure 9.3.3).

FIGURE 9.3.3 The different metals used in fireworks are responsible for the colours in its display. For example, strontium gives red, sodium gives yellow and copper gives green.

SPECTRAL ANALYSIS

In atomic emission spectroscopy, the chemical composition of a material is determined by analysing the light emitted by that material when it is burned, or when an electrical current is passed through it while in a gaseous state. The emitted light can be separated into its component wavelengths using a spectroscope, with the specific wavelengths found being characteristic of each element in the material. (This is much like how a fingerprint or DNA is used to identify an individual person.)

The energy of the emitted photons can be determined by analysing the wavelengths of light in an emission line spectrum.

In his work on the photoelectric effect, Einstein used Planck's equation to determine the energy of an emitted photon, as shown below.

$$E = hf = \frac{hc}{\lambda}$$

where E is the energy of the photon released (J)

h is Planck's constant (6.63×10^{-34} J s or 4.14×10^{-15} eV s)

f is the frequency of the photon (Hz)

c is the speed of light ($3.0 \times 10^{8}\,\text{m s}^{-1}$)

λ is the wavelength of the photon (m)

Worked example 9.3.1 refers to the emission line spectra of **metal vapour lamps**. These lamps emit photons of light as the excited electrons in the vapour return to their ground state. The emitted photons have wavelengths characteristic of the metals whose atoms are being excited in the lamp. A common type of metal vapour lamp is the sodium lamp. These are often used in street lighting and emit a distinctive yellow colour due to the yellow wavelengths in the emission line spectrum of sodium vapour (Figure 9.3.4).

FIGURE 9.3.4 Sodium vapour lamps, commonly used as street lights, emit light when excited sodium atoms release photons on returning to their ground state.

Worked example 9.3.1

SPECTRAL ANALYSIS

The emission line spectrum of light from a sodium vapour lamp shows that most of the light is emitted with a frequency of approximately 5.1×10^{14} Hz. Calculate the energy (in joules) of the photons emitted by the lamp.

Thinking	Working
Recall Planck's equation.	$E = hf$
Substitute appropriate values and solve for E.	$E = hf$ $= 6.63 \times 10^{-34} \times 5.1 \times 10^{14}$ $= 3.4 \times 10^{-19}$ J

Worked example: Try yourself 9.3.1

SPECTRAL ANALYSIS

In the Sun's absorption line spectrum, one of the dark Fraunhofer lines corresponds to a frequency of 6.9×10^{14} Hz. Calculate the energy (in joules) of the photons that corresponds to this line.

HYDROGEN'S ABSORPTION LINE SPECTRUM

In the late nineteenth century, the absorption and emission line spectra of hydrogen were of particular interest to scientists. They had recognised that the lines in the absorption line spectrum of hydrogen matched some of the lines in the Sun's emission line spectrum (Figure 9.3.5).

FIGURE 9.3.5 (a) The absorption line spectrum of hydrogen shows a background of the continuous white light spectrum, with black lines that correspond to absorbed wavelengths. The radiation from the Sun has these same black lines, because these frequencies are absorbed by the hydrogen atoms in the outer surface of the Sun. (b) The emission line spectrum of hydrogen shows a black background with lines in the visible spectrum that correspond to the wavelengths emitted by excited hydrogen atoms as they return to the ground state.

Although scientists had developed an empirical formula (i.e. one based on experimental data) that predicted the wavelength of the lines in the hydrogen spectrum, no one had been able to provide a theoretical explanation using a wave model of light as to why these lines existed.

CASE STUDY

Astronomical spectroscopy

Astronomical spectroscopy is the study of the electromagnetic spectra emitted by stars and is one of the fundamental tools used by astronomers to study the universe. It is especially useful in helping them gain a better understanding of:

- the radiation emitted by stellar bodies such as stars, galaxies, neutron stars and black holes
- the elements these bodies are composed of and the density of those elements
- their temperature
- the nature and strength of their magnetic field
- the speed and direction in which they are moving.

There are three types of spectra: continuous, emission and absorption. Figure 9.3.6 shows examples of these spectra using only the range of radiation visible to humans.

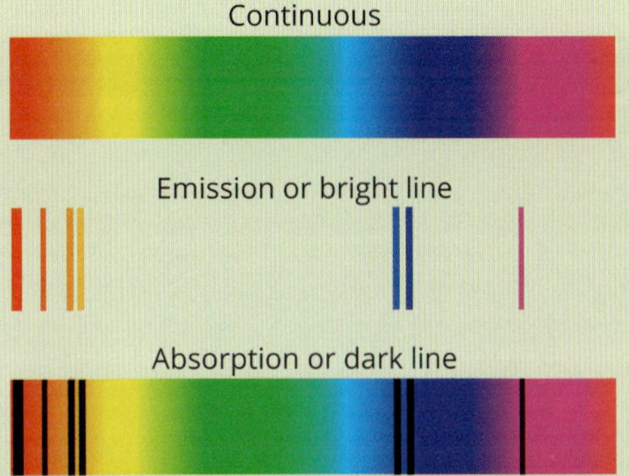

FIGURE 9.3.6 Three types of spectra: continuous, emission and absorption

Spectra with radiation in the visible and near visible regions have most likely resulted from the excitation of electrons in the outer shell of atoms in their gaseous state. The energy levels in atoms are quantised, so when these excited electrons return to their ground state they produce a series of bright lines unique to each element.

The light emitted by stars is the result of the ionisation of elements in the outer region of the star. The spectra from stars can range through all the frequencies of the electromagnetic spectrum. The bulk of it is in the visible spectrum, with the higher energy being emitted by the hotter stars. Astronomers can therefore estimate the temperature of a star from the colours in its spectrum (e.g. blue = hot; red = cool).

The spectral lines from a star usually retain their position in the spectrum. However, if the lines move back and forth it indicates that the star is orbiting another star, being orbited by another star or both. Using this data astronomers can estimate the dimensions and mass of a star system.

The spectrum of a galaxy is a combination of the spectra of the millions of stars in that galaxy. From the analysis of such a spectrum, astronomers can gain information about the type and abundance of each type of star in that galaxy and how fast it is moving away from the centre of the universe.

A non-alternating shift in the position of spectral lines (Figure 9.3.7) is a measure of how fast the universe is expanding. This is called 'cosmological redshift' and is a result of the wavelength increasing as the universe expands. A galaxy that is older has spent longer travelling through this expansion and its spectrum will therefore show a greater redshift.

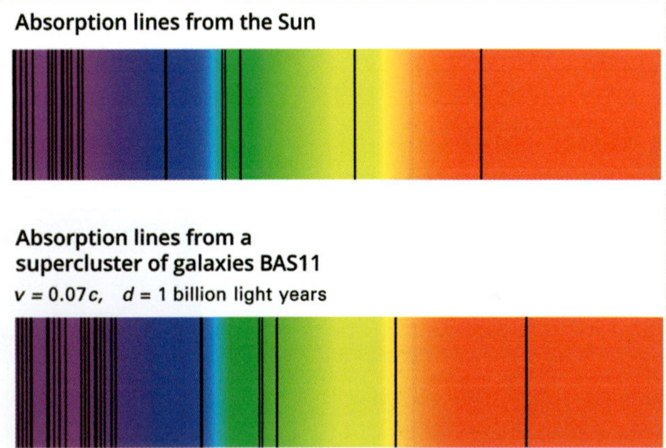

FIGURE 9.3.7 Spectra from the Sun and from a distant galaxy. Notice that the black lines in both spectra are arranged similarly but that the lines in the galaxy's spectrum are further to the right than those in the Sun's spectrum. This is known as redshift.

The Doppler shift is often used as an analogy to explain redshift; however, there is a clear difference between the two. The change in wavelength in the Doppler shift occurs due to the movement of the object producing the sound relative to the observer. In cosmological redshift the wavelength of light originally emitted is increased as space itself expands (Figure 9.3.8), not as a result of the motion of the galaxy through space.

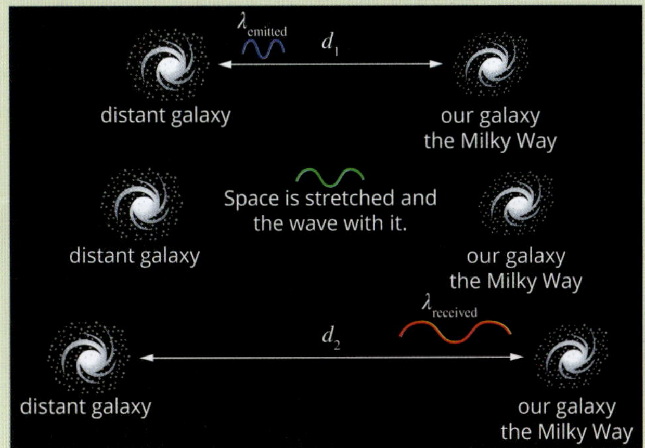

FIGURE 9.3.8 As the universe expands, so too does the wavelength of light travelling through it.

BOHR MODEL OF THE ATOM

In 1913 the Danish physicist Niels Bohr proposed an explanation for the emission line spectrum of hydrogen that drew on the quantum ideas proposed by Planck and Einstein. Bohr proposed that:

- The absorption line spectrum of hydrogen showed that the hydrogen atom was only capable of absorbing a small number of different frequencies of light and therefore only specific energies—in other words, the absorbed energy was *quantised.*
- The emission line spectrum of hydrogen showed that hydrogen atoms were also capable of emitting quanta of the exact energy value that it was able to absorb.
- If the frequency, and hence energy, of light passing through hydrogen gas was below a certain value, the light would pass straight through the gas without any absorption occurring.
- Hydrogen atoms have an ionisation energy of 13.6 eV. Light of this energy or greater can remove an electron from a hydrogen atom, creating a positive ion.
- Photons of light with energies above the ionisation value for hydrogen are continuously absorbed by the hydrogen.

Bohr's explanation relied on a significant refinement of Rutherford's planetary model of the atom. He devised a sophisticated model of electron energy levels (Figure 9.3.9), a development for which he was awarded the Nobel Prize in Physics.

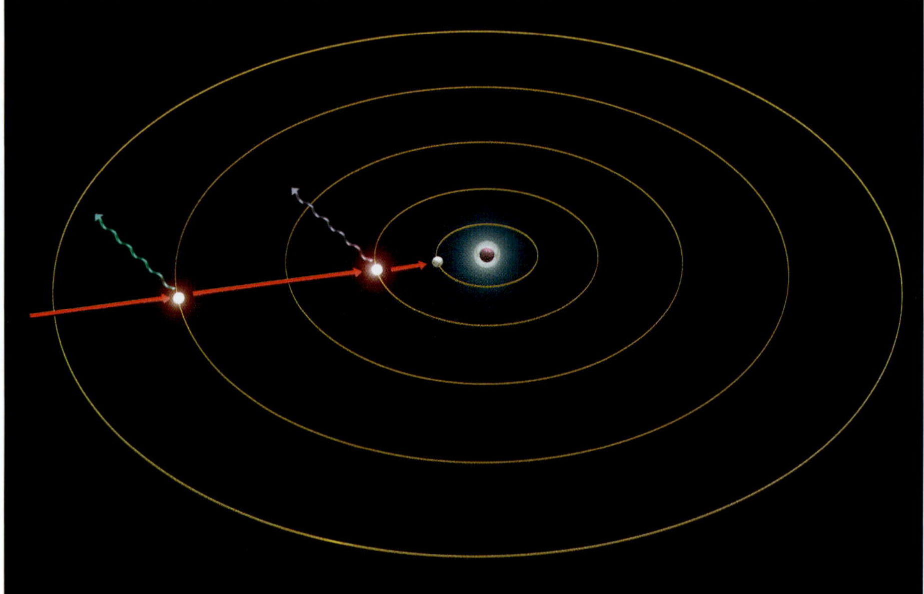

FIGURE 9.3.9 A diagram showing hydrogen spectrum emission levels based on the Bohr model of the atom. Electrons can only be in orbits of specific energy, shown by the concentric circles. They absorb energy and move to a higher orbit. They emit light when they move to a lower orbit. The light emitted is of specific wavelengths unique to hydrogen.

In Bohr's model of the atom:

- Electrons move in circular orbits around the nucleus of the atom.
- The centripetal force keeping an electron in a circular orbit is the electrostatic force of attraction between the positive nucleus and the negative electron.
- A number of allowable orbits of different radii exist for each atom (labelled $n = 1, 2, 3 \ldots$ and known as the 'principal quantum number'). Electrons can only occupy these orbits.
- An electron usually occupies the orbit of the lowest energy (i.e. the ground state).
- An electron does not radiate energy while it is in its ground state.
- Electromagnetic radiation (in the form of photons) can be absorbed by an atom when the photon's energy is equal to the difference in energies between an orbit occupied by an electron and a higher unoccupied orbit.

- Electromagnetic radiation is emitted by an excited atom when an electron returns from an orbit of higher energy to an orbit of lower energy. The photon's energy will be exactly equal to the energy difference between the electron's initial and final orbits.

Bohr labelled the possible electron orbits of the hydrogen atom with a quantum number (n), and he was able to calculate the energy associated with each quantum number. Using these energy levels, he could theoretically predict the wavelengths of all the lines in the hydrogen emission line spectrum using Planck's equation:

$$E = \frac{hc}{\lambda}$$

where E is the difference in energy between two energy levels, i.e. $E_{initial} - E_{final}$.

Figure 9.3.10 shows the energy levels for the hydrogen atom. These energies indicate how strongly the electron is bound to the nucleus. The ground state ($n = 1$) represents the orbit that is closest to the nucleus, that is, the unexcited state. An electron in this orbit has an energy of −13.6 eV, which means that it would need to gain 13.6 eV of energy to escape the atom. Higher energy levels represent orbits that are further from the nucleus.

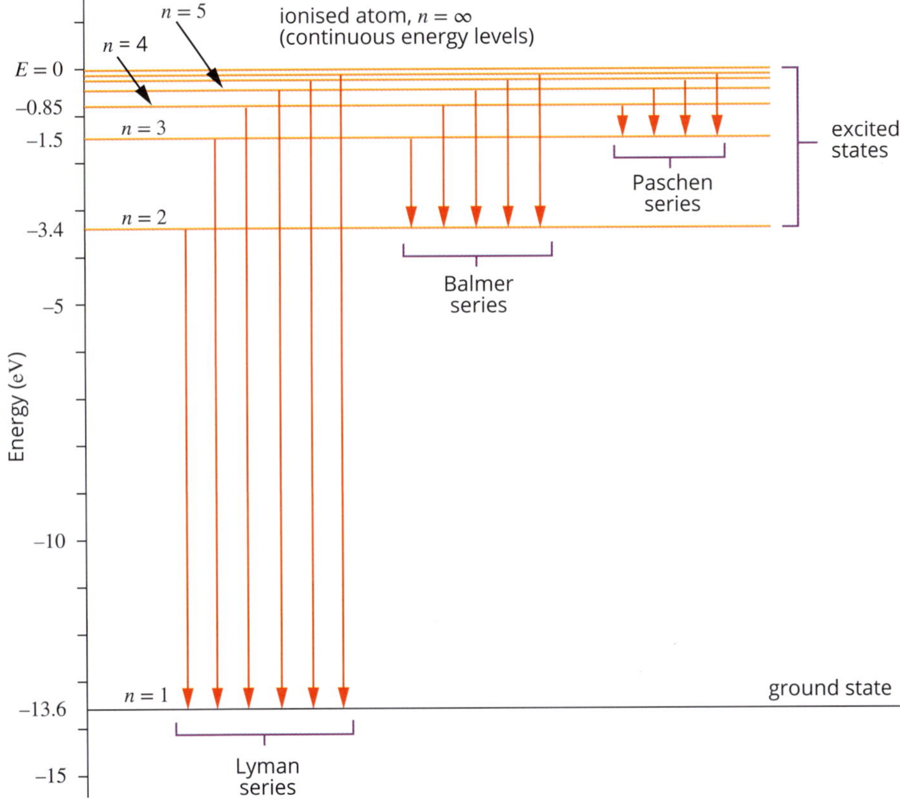

FIGURE 9.3.10 An energy level diagram for hydrogen. An electron in the ground state ($n = 1$) has an energy of 13.6 eV. For higher energy levels ($n > 1$), the energy levels are closer together.

When a hydrogen atom gains energy—either by being heated or from an electric current—its electron moves from the ground state to one of the higher energy levels. The atom is now described as being in an excited state. If the atom gains even more energy, the electron could move to an even higher energy level. Eventually the electron will drop from a higher energy level to a lower level and will emit a photon with an energy equal to the difference in energy between the levels.

You can see from Figure 9.3.10 that energy levels are all negative in value. A free electron (at $n = \infty$) must have zero potential energy, as it has escaped the electrostatic attraction of the proton in the nucleus. To raise an electron from one energy level to another, the appropriate amount of energy must be delivered to it. As an electron falls back to its previous energy level, its energy value decreases (it becomes a larger negative number).

Figure 9.3.10 also shows that the spectral lines of hydrogen can be explained in terms of electron transitions. The different series shown on the diagram (Lyman, Balmer and Paschen) represent specific transitions. The Balmer series, for example, shows transitions back to $n = 2$ from various higher energy levels. These transitions represent the wavelengths of the visible lines in the hydrogen emission line spectrum.

Worked example 9.3.2

USING THE BOHR MODEL OF THE HYDROGEN ATOM

Calculate the wavelength (in nm) of the photon produced when an electron drops from the $n = 4$ energy level of a hydrogen atom to the $n = 2$ energy level. Identify the spectral series to which the corresponding spectral line belongs.

Use Figure 9.3.10 on page 383 to calculate your answer.

Thinking	Working
Note the energy of the relevant energy levels of the hydrogen atom.	$E_4 = -0.85\,\text{eV}$ $E_2 = -3.4\,\text{eV}$
Calculate the difference in energy. You do not need to include the negative sign in your answer because energy is a scalar quantity. It has only magnitude.	$\Delta E = E_4 - E_2$ $= -0.85 - (-3.4)$ $= 2.55\ \text{eV}$
Calculate the wavelength of a photon with this amount of energy.	$E = \frac{hc}{\lambda}$ $\lambda = \frac{hc}{E}$ $= \frac{4.14 \times 10^{-15} \times 3.0 \times 10^{8}}{2.55}$ $= 4.87 \times 10^{-7}\,\text{m}$ $= 487\,\text{nm}$
Identify the spectral series.	The electron falls to the $n = 2$ energy level. Therefore the spectral line belongs to the Balmer series.

Worked example: Try yourself 9.3.2

USING THE BOHR MODEL OF THE HYDROGEN ATOM

Calculate the wavelength (in nm) of the photon produced when an electron drops from the $n = 3$ energy level of the hydrogen atom to the $n = 1$ energy level. Identify the spectral series to which the corresponding spectral line belongs.

Use Figure 9.3.10 to calculate your answer.

ABSORPTION OF PHOTONS

The Bohr model also explains the absorption line spectrum of hydrogen (Figure 9.3.5(a) on page 380). The missing lines in the spectrum correspond to the energies of light that a given atom is capable of absorbing. This is due to the energy differences between the electron's allowable orbits. Only light carrying just the right amount of energy to raise an electron to an allowed level can be absorbed.

An electron ordinarily occupies the lowest energy orbit. If light does not have enough energy to raise the electron from the lowest energy level to the next level, it cannot be absorbed by the atom. It would simply pass straight through. If the light has greater energy than the ionisation energy of an atom, the excess energy is translated into extra kinetic energy for the released electron. (Recall the photoelectric effect from Section 9.1 of this chapter.) When this happens, the atom is said to be ionised. (For a hydrogen atom to become ionised, it needs to absorb a photon with 13.6 eV or more, as Figure 9.3.10 on page 383 shows.)

PHYSICSFILE

The special case of hydrogen

The hydrogen atom was a relatively simple place to begin exploring the field that would come to be known as *quantum mechanics*. The hydrogen atom contains just two charged particles: the positively charged nucleus (which usually contains a single proton) and the negatively charged electron. This means that only one electrical interaction needs to be considered.

In more complex atoms, such as helium, electrical interactions between the electrons are also significant. This makes the construction of mathematical models of these atoms vastly more complicated than for hydrogen.

The hydrogen atom contains only two charged particles: the proton in the nucleus and the electron.

Worked example 9.3.3

ABSORPTION OF PHOTONS

Some of the energy levels for atomic mercury are shown in the diagram below.

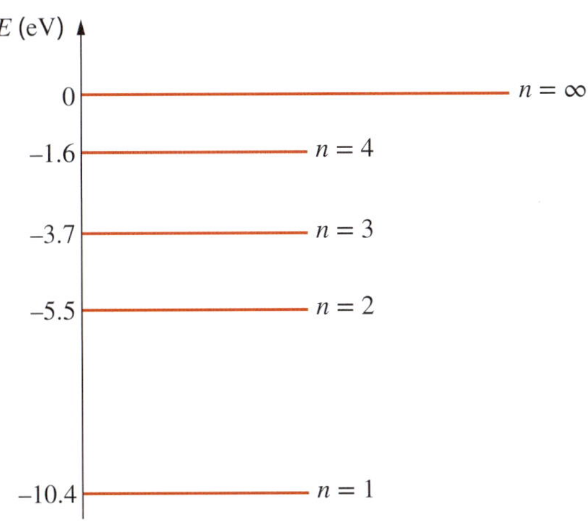

Ultraviolet light with photon energies 4.9 eV, 5.0 eV and 10.50 eV passes through some mercury gas. What could happen as a result?

Thinking	Working
Calculate the difference in energy between each level.	E (eV); 0, −1.6, −3.7, −5.5, −10.4; $n = \infty$, $n = 4$, $n = 3$, $n = 2$, $n = 1$; 4.9 eV, 6.7 eV, 8.8 eV, 10.4 eV, 1.8 eV, 3.9 eV, 5.5 eV, 2.1 eV, 3.7 eV, 1.6 eV
Check whether a given photon energy corresponds with any energy difference.	A photon of 4.9 eV corresponds to the energy required to promote an electron from the ground state to the first excited state ($n = 1$ to $n = 2$). The photon will be absorbed. A photon of 5.0 eV cannot be absorbed since there is no difference between any energy levels of exactly 5.0 eV. A photon of 10.5 eV will ionise the mercury atom. The ejected electron will leave the atom with 0.1 eV of kinetic energy.

Worked example: Try yourself 9.3.3

ABSORPTION OF PHOTONS

Some of the energy levels for atomic mercury are shown in the diagram below.

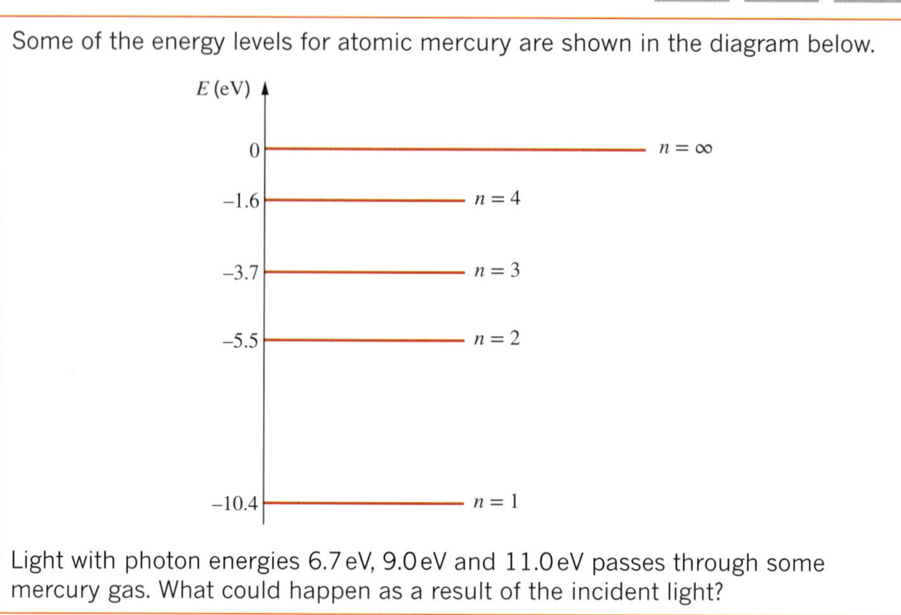

Light with photon energies 6.7 eV, 9.0 eV and 11.0 eV passes through some mercury gas. What could happen as a result of the incident light?

Problems with Bohr's model

Bohr's model of the hydrogen atom applied a quantum approach to the energy levels of atoms to explain a set of important, previously unexplained phenomena—the lines in the emission and absorption spectra of hydrogen. In principle, Bohr's work on the hydrogen atom could be extended to other atoms and, in 1914, the German scientists James Franck and Gustav Hertz demonstrated that mercury atoms had energy levels similar to hydrogen atoms. Bohr's model signified an important conceptual breakthrough.

However, Bohr's model is limited in its application. It could only be accurately applied to single-electron atoms, such as hydrogen and ionised helium. It modelled inner-shell electrons well but could not predict the higher-energy orbits of multi-electron atoms. Nor could it explain the discovery of the continuous spectrum emitted by compounds. Further studies showed that the model failed to fully explain the emission line spectrum of hydrogen. Some of the observed emission lines could be resolved into two very close spectral lines, and Bohr's model could not explain this. A more complex quantum approach was required.

STANDING WAVES AND THE DUAL NATURE OF MATTER

While this content is not explicitly included in the study design and will not be examinable, it is included to extend your understanding of the dual nature of matter.

In earlier sections the idea that matter has the characteristics of a wave was explored. This is especially relevant in considering small objects travelling at great speeds. For example, wave behaviour could be used to indicate the probability of the path of an electron in some semiconductor applications. If particles can be thought of as matter waves, then these matter waves must be able to maintain steady energy values if the particles are to be considered stable.

Louis de Broglie, the scientist who proposed the idea of matter having wavelengths, applied his idea to Bohr's model of the hydrogen atom. He viewed the electron orbiting the nucleus as a matter wave and suggested that the electron could only maintain a steady energy level if it established a standing wave. (Standing waves were introduced in Chapter 8 in relation to waves in strings.)

De Broglie reasoned that if an electron of mass m were moving with speed v in an orbit with radius r, the orbit would only be stable if it matched the condition:

$$mvr = n\frac{h}{2\pi} \text{ where } n \text{ is an integer}$$

This can be rearranged to:

$$2\pi r = n\frac{h}{mv}$$

Note that $2\pi r$ is the circumference, C, of a circle. Also note that, by the de Broglie equation, $\frac{h}{mv} = \lambda$. Combining these two yields $C = n\lambda$.

The stable orbits of the hydrogen atom are those where the circumference is exactly equal to a whole number of electron wavelengths: $C = n\lambda$.

This can be visualised by imagining a conventional standing wave joined end-to-end in a continuous loop in three dimensions (Figure 9.3.11).

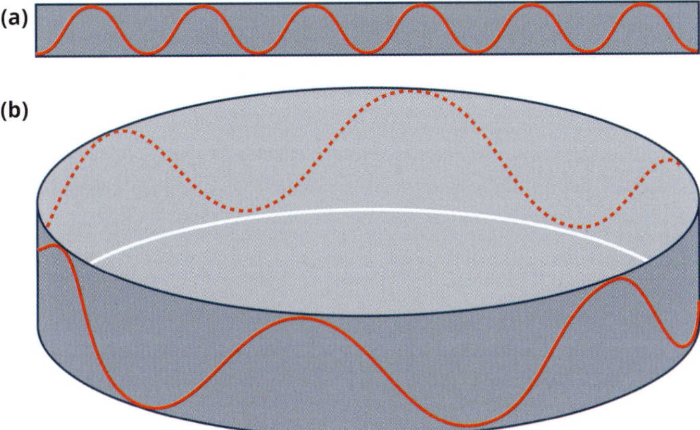

FIGURE 9.3.11 (a) A standing wave. (b) A standing wave joined end-to-end in a continuous loop, which is only possible if the circumference of the circle is equal to a whole number of wavelengths

If the circumference of the circle is not equal to a whole number of wavelengths, destructive interference will occur. In this case a standing wave pattern cannot be established and the orbit cannot represent a stable energy level (Figure 9.3.12).

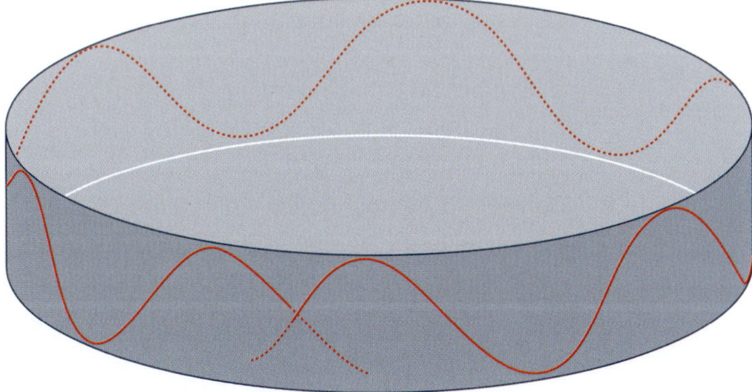

FIGURE 9.3.12 A stable circular standing wave cannot be established if the circumference of the circle is not equal to a whole number of wavelengths.

9.3 Review

OA ✓✓

SUMMARY

- The features of spectra suggest a particular internal structure in atoms. An emission line spectrum is produced by excited atoms losing energy. An absorption line spectrum is produced when light passes through a gas and the gas absorbs some of the light's energy.
- An element's spectrum is unique to that element.
- Bohr suggested that electrons in atoms orbit at discrete energy levels. No radiation is emitted or absorbed unless the electron can jump from one energy level to another. Electron energies are said to be quantised, since only certain values are allowed.
- The frequency of a photon emitted or absorbed by an atom can be calculated from the difference between the energy levels involved:

$$E = E_{\text{initial}} - E_{\text{final}} = hf = \frac{hc}{\lambda}$$

- Although the Bohr model of the atom is limited in its application, it was a significant development at the time. It took a quantum approach to the energy levels of atoms and explained the quantum nature of electromagnetic radiation.
- De Broglie viewed electrons as matter waves. He suggested that the only way that an electron could maintain a steady energy level was if it formed a standing wave. This standing-wave model could explain why electrons can only occupy particular energy levels in atoms.

KEY QUESTIONS

Knowledge and understanding

1 In the context of an electron having several energy levels in an atom, explain what the following terms mean.
  - a quantisation
  - b ground state
  - c excited state
  - d ionisation energy

2 State the conditions necessary for the production of an emission line spectrum by an element such as mercury.

3 Bohr's quantised model of the atom was revolutionary in developing our understanding of atomic structure but was limited in its depth. State the major limitations of his model.

4 When we observe both the emission line spectrum of an atom and its absorption line spectrum, we see more lines on the emission spectrum. Explain why this is the case.

5 An emission line corresponding to a frequency 5.3×10^{14} Hz is observed in the emission line spectrum of a particular elemental gas. What is the energy, in joules, of photons corresponding to this frequency?

6 Photons of energy 1.84 eV are emitted by a particular atom as it returns from an excited state to its ground state. What is the corresponding wavelength of these photons?

Analysis

7 The first four energy levels above the ground state for atomic lithium are −30.6 eV, −13.6 eV, −7.65 eV and −4.9 eV. A narrow beam of a continuous visible spectrum was passed through a sample of lithium gas in a transparent container. The emergent beam had a series of thin dark lines on it at several wavelengths, one of which corresponded to 452 nm.
  - a Explain how these dark lines were produced.
  - b Determine, with calculations, why there is a dark line corresponding to 452 nm.

8 Consider a hydrogen atom in its ground state.
  - a Calculate the energy of the photon required to move the electron from its ground state to the $n = 3$ energy level. Refer to Figure 9.3.10 on page 383.
  - b Calculate the wavelength of the photon.

9 When an electron in a hydrogen atom drops from the $n = 5$ energy level to the $n = 2$ energy level, a 410 nm photon is released. The $n = 2$ level has an energy of −3.4 eV. What is the energy of the $n = 5$ level?

Chapter review

KEY TERMS

absorption line spectrum
de Broglie wavelength
electron volt
emission line spectrum
excited state
ground state
metal vapour lamp
photocurrent
photoelectric effect
photoelectron
photon
quantum
stopping voltage
threshold frequency
wave–particle duality
work function

REVIEW QUESTIONS

Knowledge and understanding

1 Davisson and Germer conducted an experiment in which electrons were scattered after being fired at a target.

a What was observed by a detector moving through the scattering angles?

b What was the implication of this observation?

2 A particular atom has four energy levels. What does it mean to say that the levels are quantised?

3 Explain why the development of the Bohr model of the hydrogen atom was significant in the development of a comprehensive understanding of the nature of light.

4 Describe the relationship between the colours seen in the emission line and absorption line spectra of sodium.

5 Using your knowledge of spectra, explain how scientists know that the Sun's atmosphere contains hydrogen and helium.

6 The dual nature of light and matter has been accepted since the work of de Broglie and Einstein in the early 1900s. State two pieces of evidence for this duality.

7 Is it possible for a particle and a photon to have the same momentum? Justify your answer.

8 What name is given to the electrons released from the surface of a metal due to the photoelectric effect?

9 What is the energy, in electron volts, of light with a frequency of 6.0×10^{14} Hz?

10 What is the approximate value of the energy, in joules, of a quantum of light with energy of 5.0 eV?

11 If the work function of nickel is 5.0 eV, what is nickel's threshold frequency?

12 Platinum has a threshold frequency of 1.5×10^{15} Hz. Calculate the maximum kinetic energy, in electron volts, of the emitted photoelectrons when ultraviolet light with a frequency of 2.2×10^{15} Hz shines on platinum.

13 The stopping voltage obtained using a particular photocell is 1.95 V. Determine the maximum kinetic energy, in electron volts, of the photoelectrons.

Application and analysis

14 From the graph below, determine the value of the work function for each of the metals.

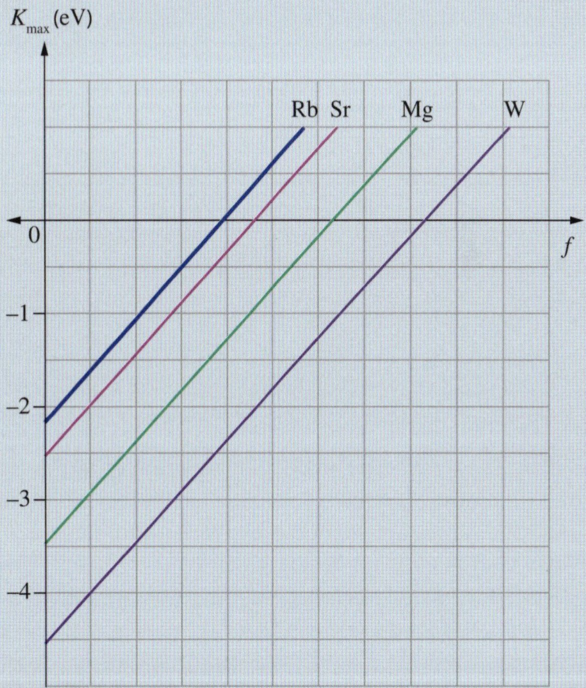

15 The cathode of a photocell shown below is coated with rubidium. Light of varying frequencies is directed onto the cathode and the maximum kinetic energy of the photoelectrons is logged. The results are summarised in the following table.

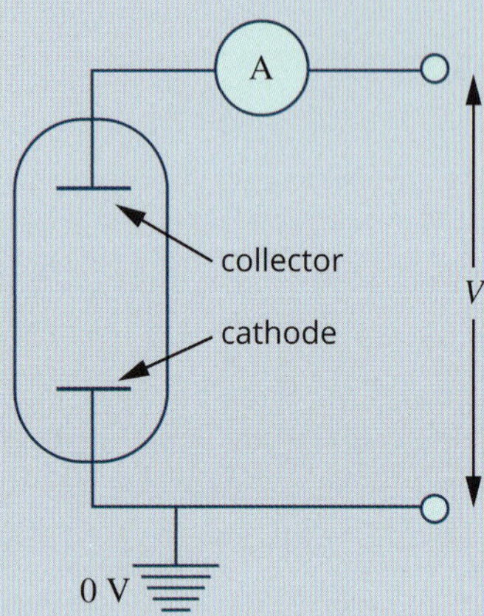

Frequency (Hz) × 10^{14}	$E_{k\ max}$ (eV)
5.20	0.080
5.40	0.163
5.60	0.246
5.80	0.328
6.00	0.411
6.20	0.494

a Plot the points from the table onto a graph.
b Calculate the gradient of the graph.
c Extrapolate your graph to estimate a value for the threshold frequency of rubidium.
d Will red light of wavelength 680 nm cause photoelectrons to be emitted from the rubidium coating? Justify your answer.

16 In an X-ray diffraction experiment a beam of X-rays from a synchrotron is directed onto a sheet of thin aluminium foil. The X-rays are scattered by the foil and detected by a charge coupled device (CCD) behind the foil. The device forms a digital image of the resulting pattern.

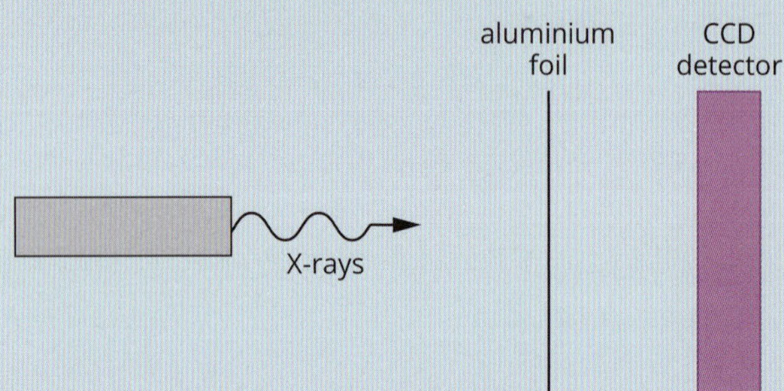

a If the wavelength of the X-rays is 260 pm (260×10^{-12} m), what is the energy of the X-rays? Give your answer in keV.
b The CCD device displays an image of the diffraction pattern formed on a computer screen (part (a) in figure below). A beam of accelerated electrons is then substituted for the X-rays. A very similar diffraction pattern is observed (part (b) in figure below). Why do the electrons produce a diffraction pattern with similar spacing to that of the X-rays?

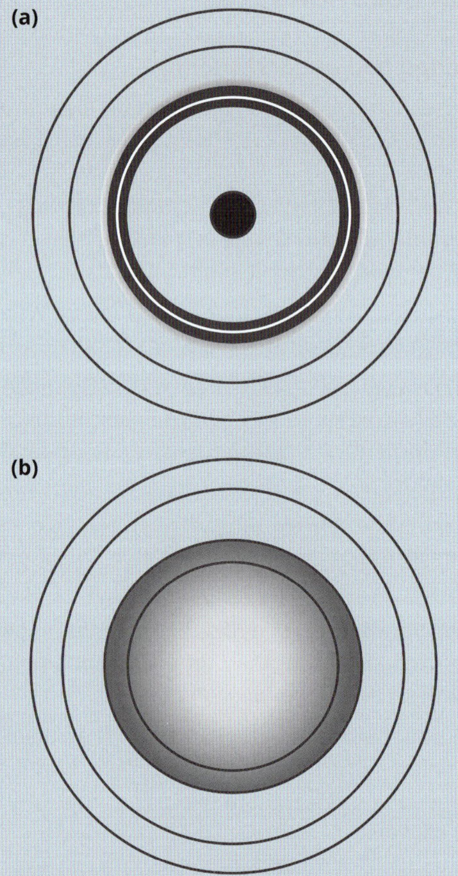

c Based on the assumption that the two diffraction patterns have the same radius, what is the momentum of the electrons?

17 Which of the following would have the longest wavelength? Justify your answer with appropriate calculations.
A electron: $m = 9.1 \times 10^{-31}$ kg, $v = 7.5 \times 10^{6}$ m s^{-1}
B blue light: $\lambda = 470$ nm
C X-ray: $f = 5 \times 10^{17}$ Hz
D proton: momentum = 1.7×10^{-21} kg m s^{-1}

18 **a** What is the de Broglie wavelength of a 40 g bullet travelling at 1.0×10^3 m s^{-1}?

b Would diffraction be noticeable for the bullet? Explain your answer.

19 An electron in an excited hydrogen atom drops from the $n = 4$ level to its ground state, $n = 1$. With reference to the hydrogen spectrum in Figure 9.3.10 on page 383, determine:

a the number of possible spectral lines that will be present as a result of this transition

b the frequency of the highest energy line

c the region of the electromagnetic spectrum in which the highest energy line is situated.

20 A beam of 3.72 eV electrons is passed through a tube containing sodium vapour. The kinetic energy of the electrons emerging from the tube was measured as 3.72 eV, 1.61 eV, 0.52 eV and 0.10 eV.

a Calculate the velocity of the electrons in the beam directed at the vapour.

b Sketch a representation of the energy levels in the sodium atom that can be determined from the given data.

c State the energies, in eV, corresponding to the spectral lines that would be present in the emission spectrum due to the excitation of electrons from the ground state to the highest of the given energy levels.

d The brightest lines on the sodium spectrum are two yellow lines very close together, with wavelengths of 589 nm and 589.6 nm (referred to as a doublet). Determine the electron transition that results in the production of the 589 nm line.

e Using the given data, determine the energy level transition that produces the spectral line with the lowest frequency.

OA ✓✓

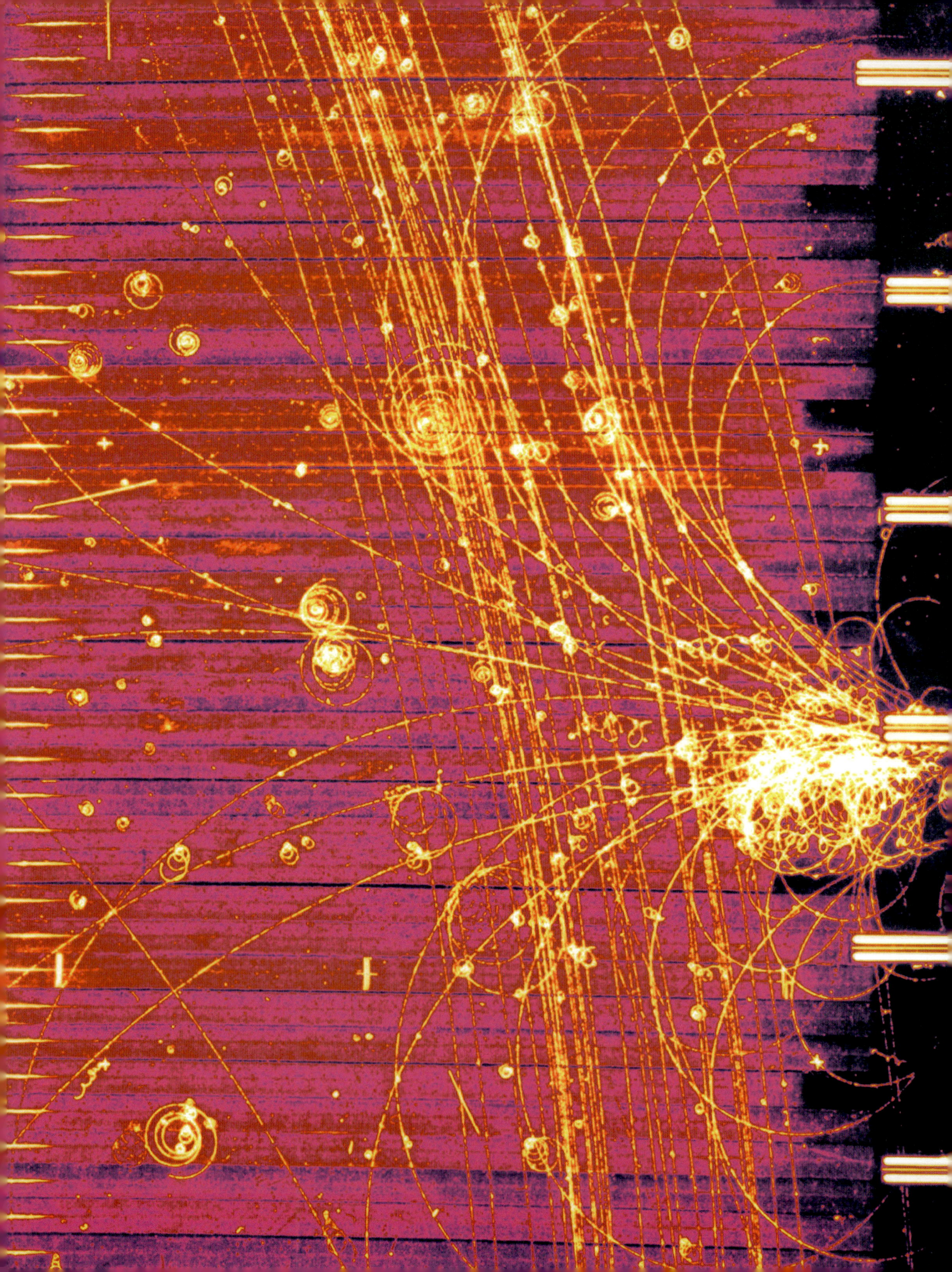

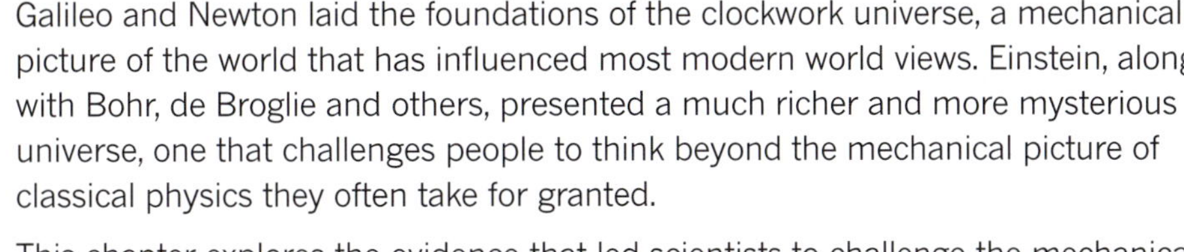

CHAPTER 10

Einstein's special theory of relativity

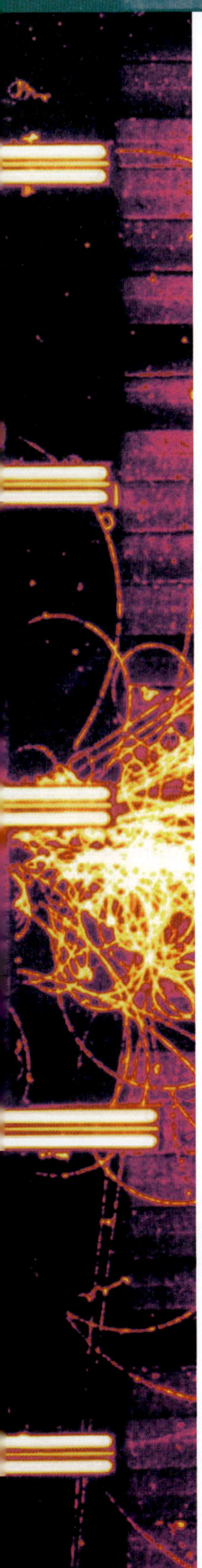

Galileo and Newton laid the foundations of the clockwork universe, a mechanical picture of the world that has influenced most modern world views. Einstein, along with Bohr, de Broglie and others, presented a much richer and more mysterious universe, one that challenges people to think beyond the mechanical picture of classical physics they often take for granted.

This chapter explores the evidence that led scientists to challenge the mechanical picture of the universe presented in the classical physics of Galileo and Newton. In particular, it looks at how Einstein's theory of special relativity solved a particular problem of classical physics: the behaviour of objects approaching the speed of light.

Key knowledge

- describe the limitation of classical mechanics when considering motion approaching the speed of light **10.1, 10.2**
- describe Einstein's two postulates for his special theory of relativity that:
  - the laws of physics are the same in all inertial (non-accelerated) frames of reference **10.1**
  - the speed of light has a constant value for all observers regardless of their motion or the motion of the source **10.1**
- interpret the null result of the Michelson–Morley experiment as evidence in support of Einstein's special theory of relativity **10.1**
- compare Einstein's special theory of relativity with the principles of classical physics **10.1, 10.2, 10.5**
- describe proper time (t_0) as the time interval between two events in a reference frame where the two events occur at the same point in space **10.3**
- describe proper length (L_0) as the length that is measured in the frame of reference in which objects are at rest **10.4**
- model mathematically time dilation and length contraction at speeds approaching c using the equations: $t = \gamma t_0$ and $L = \frac{L_0}{\gamma}$ where $\gamma = \frac{1}{\sqrt{1 - \frac{v^2}{c^2}}}$ **10.3, 10.4**
- explain and analyse examples of special relativity including that:
  - muons can reach Earth even though their half-lives would suggest that they should decay in the upper atmosphere **10.2, 10.3, 10.4**
  - particle accelerator lengths must be designed to take the effects of special relativity into account **10.4**
  - time signals from GPS satellites must be corrected for the effects of special relativity due to their orbital velocity **10.3**
- interpret Einstein's prediction by showing that the total 'mass-energy' of an object is given by: $E_{tot} = E_k + E_0 = \gamma mc^2$ where $E_0 = mc^2$, and where kinetic energy can be calculated by: $E_k = (\gamma - 1)mc^2$ **10.5**
- apply the energy–mass relationship to mass conversion in the Sun, to positron-electron annihilation and to nuclear transformations in particle accelerators (details of the particular nuclear processes are not required) **10.5**

10.1 Einstein's special relativity

Galileo and Newton developed theories of motion. These theories mathematically modelled the motion of objects travelling at relatively low speeds. This section presents observations that challenged Galilean relativity and Newtonian physics, eferred to as **classical physics**, and explains the key principles that led to the new physics described by Albert Einstein. This is the physics known as the special theory of relativity.

EINSTEIN'S BRILLIANT THEORY

FIGURE 10.1.1 Einstein as a teenager

When Albert Einstein was just 5 years old, his father gave him a compass. He was fascinated by the fact that it was responding to some invisible field that enveloped the Earth. His curiosity was aroused and, fortunately for physics, he never lost it. In his teens he turned his attention to the question of light (Figure 10.1.1).

Perhaps it was lucky that in his early twenties Einstein was not part of the physics establishment. He was working as a patent clerk in the Swiss Patent Office. It was an interesting enough job, but it didn't engage him fully: it left him time to think about light and its relationship to the Galilean principle of relativity.

Galileo was particularly interested in relative motion. One of his famous experiments involved dropping a cannon ball from the top of the mast of a moving ship. Galileo found that the motion of the cannon ball was not affected by the motion of the ship: the cannon ball landed next to the base of the mast. His principle of relativity was this: you cannot tell if you are moving or not without looking outside your own frame of reference.

> A **frame of reference** is a coordinate system that is based on the position of an object or observer.

Based on the work of Galileo, Newton established detailed models of the motion of objects, such as planets, moons and comets—even falling apples. According to his equations, the velocity of an object can be calculated relative to any frame of reference as long as the velocity of that frame of reference is known. The principle that the velocities of objects and frames of reference can be added together to determine the velocity of the object in another frame of reference is common throughout Newton's equations and laws.

Consider an object moving in frame of reference A. This frame of reference is moving in another frame of reference, B. The velocity of the object in B is given by:

$$v_{\text{object in B}} = v_{\text{object in A}} + v_{\text{A in B}}$$

An example of this is when a person on a train runs towards the front of a carriage. Here the train is frame of reference A and the track beside the railway line is frame of reference B. Suppose that the person on the train is running at $5.0\,\text{m s}^{-1}$ and the train is travelling at $20.0\,\text{m s}^{-1}$. The velocity of the person relative to B, on the track beside the railway line, is:

$$\begin{aligned} v_{\text{person on track}} &= v_{\text{person on train}} + v_{\text{train}} \\ &= 5.0 + 20.0 \\ &= 25.0\,\text{m s}^{-1} \end{aligned}$$

That is, the person is moving with a velocity of $25.0\,\text{m s}^{-1}$ towards the front of the carriage when measured from the track beside the railway line.

> Inertial frames of reference are frames that are moving with a constant velocity relative to each other. In these frames the law of inertia applies, that is, objects will continue with their velocity if no external, unbalanced force is applied. Non-inertial frames of reference occur when one frame is accelerating relative to another.

Einstein was a typical theoretician. The only significant experiments he ever did were thought experiments, or ***Gedanken*** experiments as they are called in German. Many of his *Gedanken* experiments involved thinking of situations that involved two frames of reference moving with a steady relative velocity. These are situations in which the principles of Galilean relativity applied. Newton referred to these as **inertial frames of reference**, as the law of inertia applied within them.

Einstein and Galilean relativity

Einstein thought that the elegance of the Galilean principle of relativity meant that it simply had to be true. Nature did not appear to have a special frame of reference, and Einstein could see no reason to believe that there was one waiting to be discovered. In other words, there is no such thing as an absolute velocity. It is not possible to have a velocity relative to space itself, only to other objects within space.

So the velocity of any object can always be stated as relative to some other object. In the case of the person running in the train, their velocity can be stated as either 5.0 m s^{-1} relative to the train or 25.0 m s^{-1} relative to the track beside the railway line.

Einstein expanded the Galilean principle to state that all inertial frames of reference must be equally valid, and that the laws of physics must apply equally in any frame of reference that is moving at a constant velocity. It follows that there is no physics experiment you can do entirely within a single frame of reference that will tell you that you are moving. In other words, as you speed along in your train with the blinds down, you cannot measure your speed. You can tell easily enough if you are accelerating: just hang a pendulum from the ceiling. However, the pendulum will hang straight down whether you are travelling steadily at 100 km h^{-1} or are stopped at a station. Figure 10.1.2 illustrates these points.

FIGURE 10.1.2 There is no observation that could show the difference between two inertial frames of reference. For example, parts (a) and (b) are identical, but one illustrates a stationary train and the other a train travelling smoothly at 100 km h^{-1}. There is no observation that will tell which one is which. In parts (c) and (d), the direction of the handles hanging from the ceiling of the train indicate that these trains are not moving at a constant speed.

Einstein decided that the relativity principle could not be abandoned. Recall that Einstein was, at the time, thinking about the relationship between light and relativity. Whatever the explanation for the strange behaviour of light, it could not be based on a flaw in the principle of Galilean relativity.

In the middle of the nineteenth century, the Scottish physicist James Clerk Maxwell gained a key insight into the nature of light waves. In his mathematical study of electric and magnetic effects, he realised that some of the constants in the equations he developed could be combined in a way that yielded a value very close to the speed of light. Not only did Maxwell's equations predict the speed of light; they also suggested that the speed of light was constant. Maxwell went on to develop a comprehensive theory of electromagnetism in which light is described as a form of electromagnetic radiation (EMR).

Einstein's fascination with the nature of light led him to a deep understanding of Maxwell's work on the electromagnetic nature of light waves. He was convinced of the elegance of Maxwell's equations and of their prediction that the speed of light was constant. Most physicists at the time believed that the constant speed predicted by Maxwell's equations referred to the speed of light relative to a **medium** (the substance it travelled through). It was thought that the speed predicted would be the speed in the medium in which light travelled, and that any measured speed would have to be adjusted for one's own speed through that medium.

This thought arose from the assumption that light waves are mechanical waves and mechanical waves need a medium in which to travel. Ripples in a pond need water to travel through, just as sound waves need the particles in a solid, liquid or gas. Because light travels as a wave between the Sun and the Earth, clearly some kind of medium must exist between the two bodies. Physicists gave it the name '**aether**', as it seemed to be an ethereal substance. It was thought, following Maxwell's work, that the aether must be some sort of massless, rigid medium that somehow carried electric and magnetic fields.

This was a real problem for Einstein. If the speed of light was fixed in free space, this would be in direct conflict with the principle of Galilean relativity, which Einstein was reluctant to abandon.

Resolving the problem of the aether

As in any conflict, the resolution is usually found by people who are prepared to look at it in new ways. This was the essence of Einstein's genius. Instead of looking for faults in what appeared to be two perfectly good principles of physics, he decided to see what happened if they were both accepted, despite their apparent contradiction.

Einstein's approach was to dismiss the notion of the aether outright, arguing that it was simply unnecessary. It had been invented only to be a medium for light waves, and no one had found any evidence for its existence. Electromagnetic waves, he said, could travel through empty space without a medium. Doing away with the aether, however, did not solve the basic conflict between the constant speed of light and the principle of relativity.

THE MICHELSON–MORLEY EXPERIMENT

The existence of an aether would be a serious blow for the principle of relativity. It would imply that a frame of reference was attached to space itself, and if this was the case, an absolute zero velocity was possible.

Scientists therefore needed to look for evidence of the aether. Since the Earth is in orbit around the Sun, an aether wind should, presumably, be blowing past the Earth as it moved in its orbit. This suggested to American physicist Albert Michelson that it should be possible to measure the speed at which the Earth is moving through the aether. This could be done by measuring the small changes in the speed of light as the Earth changed its direction of travel. For example, if light was travelling through the aether in the same direction as the Earth is moving, its apparent speed should be slower than usual: $c - v$, where v is the velocity of the Earth (Figure 10.1.3). Light would be travelling against an aether wind created by the motion of the Earth passing through it. If the light was travelling against the Earth's motion, its apparent speed should be faster as it would be travelling with the wind at $c + v$. The differences would be tiny, less than 0.01%, but Michelson was confident that he could measure them.

In the 1880s Michelson, and his collaborator Edward Morley, set up a device known as an interferometer (Figure 10.1.4). The device could not measure the speed of light but it could detect changes in the speed of light that might have been due to the aether wind. It would do so by measuring the time taken for light to travel in two perpendicular directions. Michelson and Morley could rotate the whole apparatus and hoped to detect the small difference that should result from the fact that one of the directions was the same as that in which the Earth was travelling and the other was at right angles to it. However, they found no difference. Perhaps, then, the Earth at that time was stationary with respect to the aether. Six months later, however, when the Earth would be travelling in the opposite direction relative to the aether, there was still no difference in the measured speeds. Other scientists performed similar experiments and got the same result. In whatever direction the Earth was moving it seemed to be at rest in the aether. Or perhaps there was no aether at all.

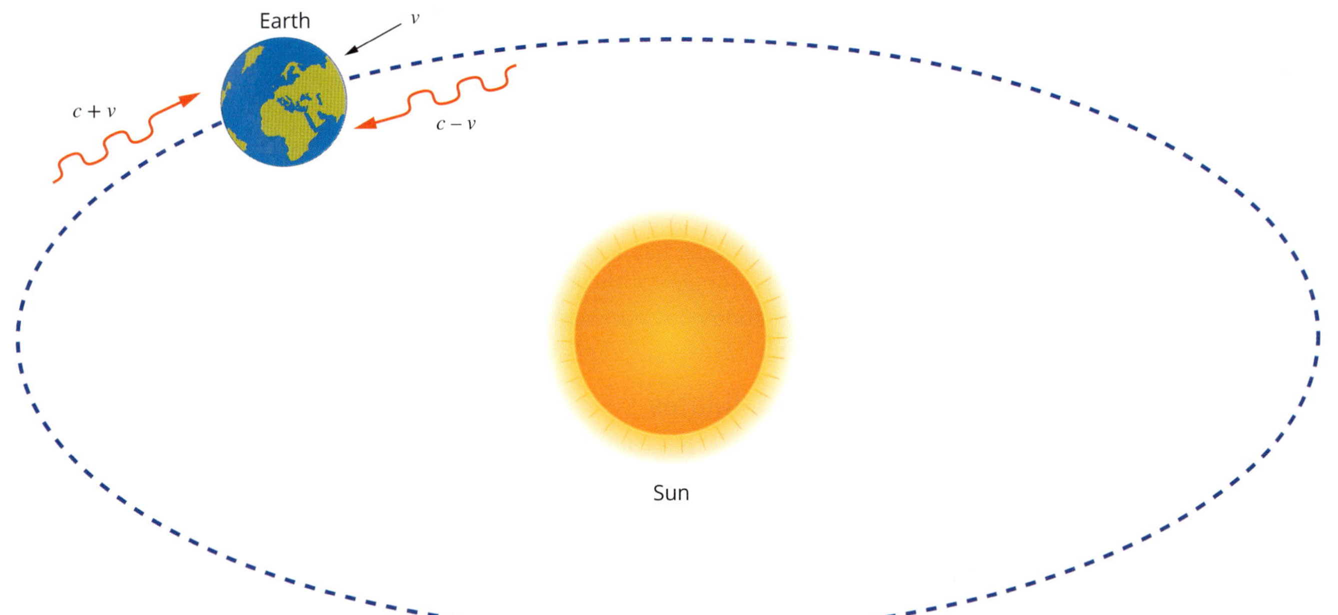

FIGURE 10.1.3 The basic principle behind the Michelson–Morley experiment. Assume that the aether is fixed relative to the Sun. If light is travelling at c relative to the aether and in the same direction as the Earth travelling at v, light's apparent speed should be less than c, i.e. $c - v$. If the light was travelling in the opposite direction to the Earth, it should appear faster than c, i.e. $c + v$.

Michelson and Morley's results were consistent with Maxwell's prediction that the speed of light would always appear to be the same for any observer. However, most physicists found this difficult to accept and believed that some flaw in the theory behind the experiment, or in its implementation, would soon be discovered. Einstein, however, wondered about the consequences of accepting Maxwell's prediction about the speed of light while at the same time holding on to the relativity principle. Ultimately, the null result of the Michelson–Morley experiment would be accepted as strong evidence to support the special theory of relativity.

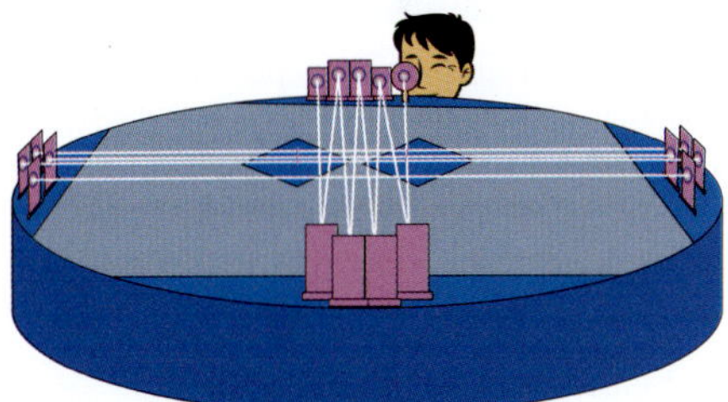

FIGURE 10.1.4 Michelson and Morley used an interferometer to compare the paths of light waves travelling in perpendicular directions.

Michelson and Morley relied on the interference of light waves to look for changes in the speed of light. Their interferometer used a partially silvered mirror to split a single beam of light into two perpendicular paths. Using mirrors, the light was reflected back and forth many times to increase the distance it had to travel, before being recombined into a single beam. They then directed the single beam to a detector that measured the intensity of the light. If the two beams combined together in phase, the intensity of light would be highest. However, if one beam had travelled faster than the other, they would be slightly out of phase and the intensity would be less than its maximum value. The whole apparatus needed to be very stable, so Michelson and Morley floated it on a bath of liquid mercury mounted on a large block of granite.

EINSTEIN CHALLENGES NEWTON'S ASSUMPTIONS

Einstein accepted the theories of both Galileo and Maxwell despite the apparent contradiction between the two, but this still left the question: How could two observers travelling at different speeds see the same light beam travelling at the same speed? The answer, Einstein said, was in the very nature of space and time.

In 1905 Einstein sent a paper to the respected physics journal *Annalen der Physik* entitled 'On the electrodynamics of moving bodies'. In the paper he put forward two simple **postulates** (statements assumed to be true) and followed them to their logical conclusion. It was his conclusion that was so astounding.

Einstein's two postulates are:

- The laws of physics are the same in all inertial (i.e. non-accelerated) frames of reference.
- The speed of light has a constant value for all observers regardless of their motion or the motion of the source.

The first postulate means that there is no preferred frame of reference and so is sometimes stated as: *no law of physics can identify a state of absolute rest.*

Einstein's postulates

FIGURE 10.1.5 Einstein's two postulates are seemingly contradictory. His first postulate implies that the speed of the laser light measured by Binh, who is travelling away from the space station at speed v, should be $c - v$ whereas his second postulate states it should be c. Einstein revisited Newton's assumptions to resolve this problem.

The first postulate is basically that of Newton, but Einstein extended it to include the laws of electromagnetism, so elegantly expressed by Maxwell. The second postulate simply takes Maxwell's prediction about the speed of electromagnetic waves in a vacuum at face value.

These two postulates sound simple enough. The only problem was that, according to Newtonian physics, they were contradictory.

Consider the situation illustrated in Figure 10.1.5. Binh is in his spaceship travelling away from Clare at speed v. Clare turns on a laser beam to send a signal to Binh. The first postulate seems to imply that the speed of the laser light, as measured by Binh, should be $c - v$, where c is the speed of light in Binh's frame of reference. This is what you would expect if, for example, you were to measure the speed of sound as you travel away from its source: as your velocity gets closer to the speed of sound, the slower the sound waves appear to be travelling.

The second postulate, however, says that when Binh measures the speed of Clare's laser light, he will find it to be c. So at first glance these two postulates appear to be mutually exclusive. To resolve this problem Einstein went back to the assumptions on which Newton based his theories.

Newton's assumptions

In 1687 Isaac Newton published his famous *Principia*. At the start of this incredible work, which was the basis for all physics in the following two centuries, he noted the following assumptions.

The following two statements are assumed to be evident and true:

- *Absolute, true, and mathematical time, of itself, and from its own nature, flows equably without relation to anything external.*
- *Absolute space, in its own nature, without relation to anything external, remains always similar and immovable.*

Newton based all his laws on these two assumptions: that space and time are constant, uniform and straight. According to Newton, space is like a big set of x-y-z axes that always have the same scale and in which distances can be calculated exactly according to Pythagoras's theorem. You expect a metre rule to be the same length whether it is held vertically or horizontally, north–south or east–west, in your classroom or on the International Space Station.

In this space, time flows at a constant rate, which is the same everywhere. One second in Perth is the same as one second in Melbourne, and one second on the ground is the same as one second up in the air.

Einstein realised that the assumptions that Newton made may not be valid, at least not on scales involving huge distances and speeds approaching the speed of light. The only way in which Einstein's two postulates can both be true is if both space and time are not fixed and unchangeable, but are somehow related.

> Einstein's special theory of relativity is a theory based on the relationship between space and time. The theory is based on the two postulates outlined in this section.

10.1 Review

OA ✓✓

SUMMARY

- The Michelson–Morley experiment sought to prove the existence of an aether; however, the results failed to support the hypothesis that the aether existed. This was taken as evidence in support of the special theory of relativity.
- Einstein decided that Galileo's principle of relativity was so elegant it simply had to be true. He was also convinced that Maxwell's electromagnetic equations, and their predictions, were sound.
- Einstein's two postulates of special relativity can be abbreviated to:
  i The laws of physics are the same in all inertial frames of reference.
  ii The speed of light is the same to all observers.
- Einstein realised that accepting both postulates implied that space and time were not absolute and independent, but were related in some way.

KEY QUESTIONS

Knowledge and understanding

1 Why did the physicists of the late nineteenth century feel the need to propose the idea of the aether?

2 Two spaceships are travelling for a while at a constant relative velocity. One then begins to accelerate. A passenger with a laser-based velocity measurer sees the relative velocity increase. Suggest a way that this passenger could tell whether it was their own or the other ship that had begun to accelerate.

3 Tom, who is in the centre of a train carriage moving at constant velocity, rolls a ball towards the front of the train. At the same time he blows a whistle and shines a laser towards the front of the train. What will Jana, who is on the ground beside the train, observe compared to Tom regarding the speed of the ball, the sound and the light?

4 Give two examples of objects within an inertial frame of reference.

5 Give two examples of objects within a non-inertial frame of reference.

6 Based on your answers to questions **4** and **5**, give an example of how Einstein's first postulate would provide proof that the frames of reference were either inertial or non-inertial.

7 Max is behind the safety rail holding a ball while watching his little sister go around on a merry-go-round. He sees her come around the bend and throws the ball straight at her, hoping she will catch it. Max's throw misses his sister. Later, his sister wants to know how Max made the ball curve in the air, but Max says he threw it straight. Discuss why they saw the ball's movement differently.

Analysis

8 The speed of sound in air is $340\,m\,s^{-1}$. At what speed would the sound from a fire truck siren appear to you to be travelling if:

a you are driving towards the stationary fire truck at $30\,m\,s^{-1}$

b you are driving away from the stationary truck at $40\,m\,s^{-1}$

c you are stationary and the fire truck is heading towards you at $20\,m\,s^{-1}$

d you are driving at $30\,m\,s^{-1}$ and approaching the rear of the fire truck, which is travelling at $20\,m\,s^{-1}$ in the same direction.

9 Alex is at the front of a ferry that is moving at $6.00\,m\,s^{-1}$ south. She kicks a soccer ball towards her brother Bill who is 7.00 m away at the northern end of the deck. Bill traps it with his foot 1.59 s after it was kicked. Carla is watching all this from a stationary fishing boat.

a At what velocity does Carla see the soccer ball travelling?

b How far, in Carla's frame of reference, did the ball move while rolling between Alex and Bill?

c For how long was the ball rolling in Carla's frame of reference?

10 Imagine that the speed of light has suddenly slowed down to $70.0\,m\,s^{-1}$. Alex (still at the front of the ferry moving at $6.00\,m\,s^{-1}$ south as in question **9**) sends a flash of light from her phone camera towards Bill.

a From Alex's point of view, how long does it take the light flash to reach Bill?

b How fast was the light travelling in Bill's frame of reference?

c In Carla's frame of reference, how far did the ferry travel in 0.100 s?

d How fast was the light travelling in Carla's frame of reference?

10.2 Einstein's *Gedanken* train

An unusual consequence of special relativity arises when two events that are seen as occurring at the same instant by one observer are seen to occur at different times by another observer. This seems to defy logic: both observers are looking at the same two events yet for one observer the two events are **simultaneous** while for another observer they are not. Einstein proposed that this lack of simultaneity occurs when the same two events are seen by observers in different frames of reference.

SIMULTANEITY

FIGURE 10.2.1 Amaya and Binh measure the light taking the same time, $\frac{l}{c}$ seconds, to reach the front and back walls of the carriage.

FIGURE 10.2.2 Clare sees the light reach the back wall first and then the front wall.

To illustrate the consequences of accepting the two postulates he put forward, Einstein offered a simple thought experiment. It involves a train moving at a constant velocity.

Amaya and Binh are on Einstein's train and Clare is watching the train pass by (Figure 10.2.1). The train has a flashing light bulb attached to the ceiling right in the centre of the carriage. Amaya and Binh are watching the flashes of light and notice that the flashes reach the front and back walls of the carriage at the same time. In other words, the two events occur simultaneously. Outside, Clare sees the same flashes of light. Einstein was interested in when Clare would see the flashes reach the end walls.

To appreciate Einstein's ideas, you need to contrast them with what you might ordinarily expect. Consider a situation in which Amaya and Binh are rolling balls towards opposite ends of the carriage. It is important to appreciate that, while Clare, the outside observer, sees the ball's velocity differently from Amaya and Binh, the balls are seen by all three as hitting the walls at the same time.

The same is true of sound. If a pulse of sound waves is sent from the centre of the carriage, Clare would agree with Amaya and Binh that the time taken for the sound waves to reach the end walls is the same. But what about light?

Einstein's second postulate tells you that all observers will observe light to be traveling at the same speed. Amaya, Binh and Clare will all observe the light travelling at $3.0 \times 10^8 \, m\,s^{-1}$. They do not add or subtract the velocity of the train.

Although Clare measures the light to be travelling at the same speed whether it is directed at the front wall or the back wall of the carriage, she will observe the light reaching the back wall first (Figure 10.2.2). This is because that wall is moving towards the light, whereas the front wall is moving away from the light, and so the light will take longer to catch up to it. Amaya and Binh note that the light flashes reach the ends of the carriage at the same time; Clare notes them reaching the walls at different times.

The idea that two events that are simultaneous for one set of observers but are not simultaneous for another may at first seem nonsensical. It goes against the principles of Newtonian physics.

Simultaneity and spacetime

The big difference between the situation for light and that for balls or sound is the strange notion that both sets of observers measure the speed of light to be exactly the same. The velocity of a thrown ball or the velocity of sound in Amaya and Binh's frame of reference will always be different from that in Clare's frame of reference by exactly the velocity of the train. For light, however, there is no difference. As a result, events that are simultaneous for one set of observers are not simultaneous for the others. This very strange situation is referred to as a lack of simultaneity.

Although Einstein's experiments were purely hypothetical, other experiments based on these ideas are well within the capacity of modern experimental physics. In all cases they confirm Einstein's ideas to a high degree of accuracy.

Einstein said that the only reasonable explanation for how two events that were simultaneous to one set of observers were not simultaneous to another is that time itself is not behaving as we commonly expect. The amount of time that has elapsed in one frame of reference is not the same as that which has elapsed in another.

PHYSICSFILE

Measurement in a thought experiment

The people in the train example would need extremely good measuring devices, such as an atomic clock, and amazingly quick reflexes to take their measurements.

Under normal circumstances there is no chance of detecting the lack of simultaneity of light beams hitting the front and back walls of a train. This is because the difference in time is about a millionth of a microsecond, well beyond the capacity of even the best stopwatches. The reflexes required to see the light reach the back wall, then see the light encounter the front wall, would also be beyond human ability.

FIGURE 10.2.3 The famous clock tower in Bern, Switzerland, near Einstein's apartment, which is thought to have inspired his special theory of relativity. Its hands move at one minute per minute, but only in the same frame of reference as the clock.

In the example shown in Figure 10.2.1, Amaya and Binh measured the light flashes and found that they took the same time to reach the walls. In Clare's frame of reference, the times were different. Time, which has one dimension, seems to depend on the frame of reference in which it is measured, and a frame of reference is just a way of defining three-dimensional space. Clearly time and space are somehow interrelated. This four-dimensional relationship, which includes the three dimensions of space and the one dimension of time, is called **spacetime**. Special relativity is all about spacetime.

This was a profound shock to the physicists of Einstein's time. Many of them refused to believe that time was not the constant and unchanging quantity they had assumed it had always been. To think that it might flow at a different rate in a moving frame of reference was beyond the comprehension of many. It could mean that if you went on a train trip, your clock would go slower, and you would come back having aged slightly less than those who stayed behind.

Einstein's idea was that time and distance are relative. They can have different values when measured by different observers. Simultaneous events in one frame of reference are not necessarily simultaneous when observed from another frame of reference. An observer might see light travelling through a distance d in a time t at a speed c. A different observer might see the same light travelling through a different distance, d', in a different time, t', but still at the same speed, c. Our basic understanding of time and distance (and perhaps mass too) needs adjustment, especially when considering objects travelling at close to the speed of light.

Probably because of the tiny differences in time involved, and the highly abstract nature of the work, many physicists simply disregarded Einstein's concepts and got on with their work. They thought it could never have any practical implications and, besides, no clock in existence could measure what could only be incredibly small differences in time.

ATOMIC CLOCKS

Measuring time is an exercise in precision, replicating an interval of one second over and over again—86 400 times for just one day. There have been many mechanical solutions to keeping track of time, mostly using cogs and levers, weights and dials. The accuracy of these devices varied, with some gaining or losing seconds or minutes every day.

Historically, to correct your clock you needed to align it with some recognised standard clock. To help in this recalibration, radio stations would regularly broadcast a time signal, usually a series of five beeps counting down to each hour. You could also phone a number to hear such messages as 'at the tone it will be six o'clock ... beep'.

For scientists, standard clocks were limited in their usefulness. They were reliable for measuring events to one or two decimal places, which is fine if all you wanted to do was measure relatively slow motion.

Before 1967 the standard of one second was based on a fraction of the time it took for the Earth to orbit the Sun—a far from ideal standard. In 1967 the basis for the unit of time was changed: it became a certain number of transitions of the outermost electron of a caesium-133 isotope. One second became defined as 9 192 631 770 oscillations of the 6s electron of the Cs-133 isotope. The remarkable precision of this oscillation resulted in the invention of atomic clocks (Figure 10.2.4) with an accuracy of 1 second in 1.4 million years, and the ability to measure time to an incredible number of decimal places. It is at these levels of measurement that the predictions of Newton's laws of motion vary from the values measured.

FIGURE 10.2.4 The first atomic clock was developed in 1955. From 1967, the atomic clock was used to set the standard of one second.

OBSERVATIONS THAT NEWTON'S LAWS CANNOT EXPLAIN

With the invention of more-accurate measuring devices for time and distance, it became evident that some actual measurements didn't match predictions based on Newton's laws acting in a framework of Galilean relativity. An example is the lifetimes of unstable particles.

Unstable particles typically have very short lifetimes. For example, a stationary positive pion, π^+, has a mean lifetime of just 0.000 000 026 033 s (26.033 ns) before it decays into a muon and a muon neutrino. Hence a sample of stationary positive pions has a lifetime of approximately 26.033 ns. However, when a positive pion is accelerated to 99% of the speed of light, its mean lifetime is 1845.4 ns. This means that the faster moving positive pions exist seven times longer than stationary pions. Classical Newtonian physics cannot explain how or why this occurs.

In the Earth's atmosphere, high-energy cosmic rays interact with the nuclei of oxygen atoms about 15 km above the surface of the Earth and create a cascade of high-velocity subatomic particles. One of these particles is the muon, and it too is unstable. The mean lifetime of a stationary muon is 2.2 μs. However, the muons created by cosmic radiation typically travel at 99.97% of the speed of light. At this speed Newtonian physics would predict that the longest-living muons would travel about 660 m during their lifetime:

$$\begin{aligned} s &= v\Delta t \\ &= 0.9997 \times 3.0 \times 10^{8} \times 2.2 \times 10^{-6} \\ &= 659.8\,\text{m} \end{aligned}$$

After forming 15 km above the Earth, and decaying by the time they have travelled 0.66 km, you would expect that no muons would be detected reaching the surface of the Earth. However, muons created by cosmic radiation *are* detected at the surface of the Earth. This means that the fast-moving muons have existed for longer than they should have. Muons that strike the surface of the Earth must have existed at least 22 times longer than they would have had they been stationary. Once again, Newtonian physics and Galilean relativity cannot explain this observation. Section 10.3 explains why this happens.

10.2 Review

OA ✓✓

SUMMARY

- Simultaneous events in one frame of reference are not necessarily simultaneous when observed from another frame of reference.
- Time depends on the frame of reference in which it is measured.
- A four-dimensional relationship exists between the three dimensions of space and the one dimension of time. This is called spacetime.
- Observations of the lifetimes of subatomic particles that are accelerated to high speeds indicate that they exist for longer than when they are stationary.
- According to Newtonian physics, high-speed muons created in the Earth's upper atmosphere should not last long enough to reach the Earth's surface, but they do. The moving muons have longer lifetimes than stationary muons.

KEY QUESTIONS

Knowledge and understanding

1 Recall why Einstein said that we must use four-dimensional spacetime to describe events that occur in situations where high speeds are involved.

2 Why did Einstein say that the effects of special relativity can only be considered by using *Gedanken* experiments?

3 An observer on a train moving at a constant velocity rolls a ball from the centre of a carriage to the front wall and from the centre of the carriage to the back wall at the same time. An observer on the train sees both balls strike the walls at the same time. Why would an observer outside the train also see the balls strike the wall at the same time?

4 Imagine that Amaya is at the front of a train carriage that is moving at $10.0\,m\,s^{-1}$. She shines a torch towards Binh, who is at the other end of the carriage. Clare is watching this from the side of the track. Determine the velocity at which the light is travelling as measured by Clare.

5 What fundamental property causes the lack of simultaneity in events involving light travelling from the centre of a train carriage to the front and back walls when viewed by observers inside and outside of the carriage?

6 Complete the following sentences by selecting the correct term in bold.

Muons have **very short/long** lives. They are created approximately 15 km up in the atmosphere. As they travel down through the atmosphere the muon's speed is **about a tenth of/very similar to** the speed of light. According to Newtonian laws, muons **should/should not** reach the Earth's surface. However, **many do/do not**.

Analysis

7 Barry is standing still at an equal distance between two stationary light sources. The sources each emit an instantaneous flash. The flashes reach him at the same time, so Barry concludes that the flashes were emitted at the same time. However, Stephen, who is moving towards the left light source disagrees with this statement. Why might Stephen disagree?

8 Reflect on the term 'simultaneity' and illustrate its meaning with an analogy.

9 Why was the development of atomic clocks important to the advancement of Einstein's special theory of relativity?

10 A particle with a very short half-life is the hydrogen-7 isotope. It has a half-life of 23 yoctoseconds ($23 \times 10^{-24}\,s$). After 10 half-lives it is expected that all the particles in a sample of hydrogen-7 particles will have decayed.

- **a** Estimate the maximum lifetime of a hydrogen-7 atom in that sample.
- **b** A sample of hydrogen-7 atoms was ejected from a source at a speed of 0.993 times the speed of light. Using Newtonian physics, calculate the distance from the source where there would be no hydrogen-7 particles left.
- **c** Experiments have shown that hydrogen-7 particles can be found over 8 times the distance away from their source than that calculated in part **b**. Suggest why this might be possible.

10.3 Time dilation

FIGURE 10.3.1 Extremely precise atomic clocks enable scientists to determine the life span and speed of subatomic particles.

Extremely precise atomic clocks (Figure 10.3.1) enabled very short-lived events to be measured to a large number of decimal places. At this level of precision, some unusual observations have been made regarding the life spans of certain high-speed subatomic particles when compared to their life spans when they are at rest. This section explores the concept of **time dilation** as an explanation for these observations.

TIME IN DIFFERENT FRAMES OF REFERENCE

The consequences of Einstein's two postulates have been discussed in the context of simple *Gedanken* situations, such as in a moving train. Observers inside the train see two simultaneous events while those outside see the same two events occurring at different times. The differences are extremely small and would not be noticeable by an observer in any actual train, unless they had an atomic clock and no reflex time. For aircraft flying at supersonic speeds, the differences, while still very small, become measurable by the most precise clocks. For subatomic particles, such as pions in accelerators, the differences in time become more significant. In situations like this, where speeds approach the speed of light, it is important that calculations take Einstein's theory into account.

The light clock

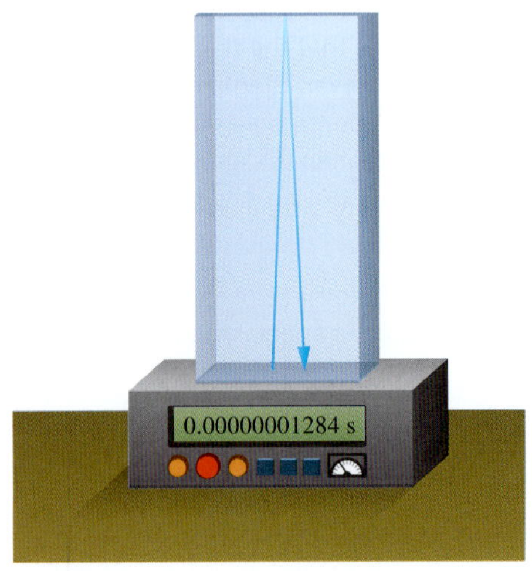

FIGURE 10.3.2 The light clock 'ticks' each time the light pulse reflects off the bottom mirror.

If you want to observe time dilation in a moving train or among moving subatomic particles, you need to watch a very precise clock in a moving reference frame to see if time is actually going slower. The term 'dilation' in this context means 'the slowing of time'.

Consider Amaya and Binh travelling in a spaceship close to the speed of light. Clare is watching them from a space station which, according to Clare, is a stationary frame of reference. Amaya and Binh have taken along a clock which (it is assumed) Clare can read, even from a long distance away. Like any clock, this clock is governed by a regular oscillation that defines a period of time.

Amaya's clock has a light pulse that bounces back and forth between two mirrors. One mirror is on the floor and the other on the ceiling (Figure 10.3.2). A light pulse oscillating from one mirror to the other and back can be considered one unit of time. Clare has an identical clock in her space station which she can compare to Amaya's clock.

The advantage of this type of clock is that it can be used to predict how motion will affect it by using Pythagoras's theorem and some simple algebra. The clock has been set up so that the light pulses oscillate up and down a distance d that is at right angles to the direction of travel. The distance d is shown by a black arrow in the centre position of the moving spacecraft in Figure 10.3.3. As the spaceship speeds along, the light will trace out a zigzag path, as shown by the red dotted line in Figure 10.3.3.

Only one oscillation of the light pulse needs to be considered, as all the other oscillations will have the same geometry. One unit of time is the time taken for the light pulse to oscillate once. In the frame of reference of the spaceship, Amaya and Binh measure a unit of time as t_a. Clare, from her frame of reference, will measure a different time, t_c.

Amaya and Binh measure the light pulse travelling at the speed of light, c, along a distance of $2d$ (from the bottom mirror to the top and back again) in time t_a. So the distance that the light pulse travels is:

$$2d = c \times t_a \qquad \text{(Equation 10.3.1)}$$

FIGURE 10.3.3 Clare sees that the light pulses travel a zigzag path between the mirrors in one unit of time.

But Clare measures the light travelling a longer path (the red dotted line in Figure 10.3.3).

The spaceship moves with a velocity, v, and so in one unit of time as measured by Clare, t_c, the spaceship has travelled a distance of $2 \times d_s$. This is equal to the velocity multiplied by the time taken for her to see one oscillation:

$$2d_s = v \times t_c$$

Consider just half of the light oscillation. The light pulse not only travels the vertical distance d between the mirrors in the clock; it also travels forward as the spaceship moves through the distance d_s, making the combined distance d_c. Therefore, by Pythagoras's theorem:

$$\begin{aligned} d_c^2 &= d^2 + d_s^2 \\ &= d^2 + \left(\frac{vt_c}{2}\right)^2 \\ d_c &= \sqrt{d^2 + \left(\frac{vt_c}{2}\right)^2} \end{aligned} \qquad \text{(Equation 10.3.2)}$$

Clare measures this light pulse travelling twice the combined distance at the speed of light, c, during time t_c as measured on her clock. So:

$$2d_c = c \times t_c \qquad \text{(Equation 10.3.3)}$$

Rearranging Equation 10.3.3 and substituting d_c from Equation 10.3.2 gives:

$$\begin{aligned} \frac{c \times t_c}{2} &= \sqrt{\left(d^2 + \left(\frac{vt_c}{2}\right)^2\right)} \\ c \times t_c &= 2 \times \sqrt{\left(d^2 + \left(\frac{vt_c}{2}\right)^2\right)} \\ c \times t_c &= \sqrt{\left(4d^2 + 4 \times \left(\frac{vt_c}{2}\right)^2\right)} \\ t_c &= \frac{\sqrt{4d^2 + (vt_c)^2}}{c} \end{aligned} \qquad \text{(Equation 10.3.4)}$$

From Amaya and Binh's frame of reference—where they measure the light pulse travelling a distance $2d$ at speed c in a time t_a—we noted earlier (in Equation 10.3.1 on page 404) that:

$$2d = c \times t_a \text{ or } d = \frac{c \times t_a}{2}$$

Note that you have used the same value for c in both frames of reference, something you would never do in classical physics (the physics of Galileo and Newton), but something Einstein insisted that you must.

Substituting Equation 10.3.1 into Equation 10.3.4 gives:

$$t_c = \frac{\sqrt{4 \times \left(\frac{ct_a}{2}\right)^2 + (vt_c)^2}}{c}$$

Now square both sides and simplify to make t_c^2 the subject:

$$t_c^2 = \frac{\frac{4(ct_a)^2}{4} + (vt_c)^2}{c^2}$$

$$t_c^2 = \frac{c^2 t_a^2 + v^2 t_c^2}{c^2}$$

$$t_c^2 = \frac{c^2 t_a^2}{c^2} + \frac{v^2 t_c^2}{c^2}$$

$$t_c^2 = t_a^2 + \frac{v^2 t_c^2}{c^2}$$

Group the terms with t_c^2 together and factorise:

$$t_c^2 - \frac{v^2 t_c^2}{c^2} = t_a^2$$

$$t_c^2 \left(1 - \frac{v^2}{c^2}\right) = t_a^2$$

Take the square root of both sides and make t_c the subject:

$$t_c \sqrt{1 - \frac{v^2}{c^2}} = t_a$$

$$t_c = \frac{t_a}{\sqrt{1 - \frac{v^2}{c^2}}}$$

(Equation 10.3.5)

As v can never be larger than c, the denominator in Equation 10.3.5 must always be less than one. Since any number divided by a number less than one must result in a larger number, it follows then that $t_c > t_a$. Hence the time that Clare measures is greater than the time that Amaya and Binh measure *for the same event.*

When expressed in a general form (see page 407), Equation 10.3.5 is Einstein's equation for time dilation.

CALCULATING TIME DILATION

In Einstein's equation for time dilation, the t represents the time that a stationary observer (Clare, in the previous example) measures using a stationary clock of an event that the observer sees occurring in a moving frame of reference. The symbol t_0 is the time that passes on the moving clock, which is also known as the **proper time**.

Generalising Equation 10.3.5 opposite, Einstein's equation for time dilation can be written as follows:

$$t = \frac{t_0}{\sqrt{\left(1 - \frac{v^2}{c^2}\right)}}$$

The factor that the proper time (t_0) is multiplied by is usually given the symbol gamma, γ, so the equation for time dilation can be written as follows.

$$t = \gamma t_0$$

where $\gamma = \frac{1}{\sqrt{\left(1 - \frac{v^2}{c^2}\right)}}$

v is the speed of the moving frame of reference
c is the speed of light in a vacuum ($3.0 \times 10^8\,\text{m s}^{-1}$)
t is the time observed in a stationary frame
t_0 is the time observed in the moving frame (i.e. the proper time)

The Dutch physicist H. A. Lorentz first introduced the factor γ in an attempt to explain the results of the Michelson–Morley experiment, so it is often known as the **Lorentz factor**.

Table 10.3.1 and Figure 10.3.4 show the effect of varying the value of v (as a proportion of the speed of light) on the value of γ.

TABLE 10.3.1 The value of the Lorentz factor at various speeds

v (m s^{-1})	$\frac{v}{c}$	γ
3.00×10^2	0.000001	1.000000000
3.00×10^5	0.00100	1.0000005
3.00×10^7	0.100	1.005
1.50×10^8	0.500	1.155
2.60×10^8	0.867	2.00
2.70×10^8	0.900	2.29
2.97×10^8	0.990	7.09
2.997×10^8	0.999	22.4

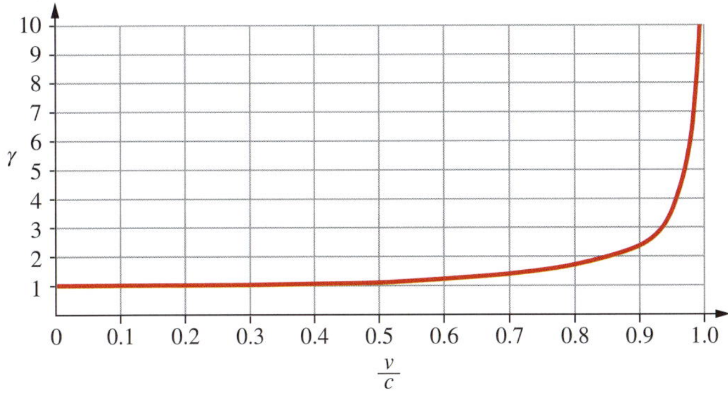

FIGURE 10.3.4 A graph of the Lorentz factor versus $\frac{v}{c}$

Sometimes it is useful to make v the subject of the equation for the Lorentz factor:

$$v = c\sqrt{\left(1 - \frac{1}{\gamma^2}\right)}$$

The data in Table 10.3.1 on page 407 show that a velocity of $300\,\text{m}\,\text{s}^{-1}$ has a Lorentz factor that is essentially 1. So for relatively low speeds, a stationary observer (such as Clare in her space station) measures the oscillation of light in the light clock on Amaya and Binh's slow-moving spaceship to be the same as they measure using their own stationary light clock. This implies that time is passing at essentially the same rate in both frames of reference.

When the spaceship is travelling at 0.990c, a stationary observer will measure a single oscillation of light in the spaceship's light clock as taking the equivalent of seven oscillations of their own stationary light clock. From their perspective, time for the objects and people in the moving frame of reference has slowed down to one-seventh of what they will consider as normal time.

As the speed approaches the speed of light, time in the moving frame, as viewed from the stationary frame, slows down even more. If you were able to see the clock travelling on a light wave, the clock would not be ticking at all—time would be seen to stand still.

It is important to realise that Amaya and Binh do not perceive their time slowing down at all. To them, their clock keeps ticking away at the usual rate and events in their frame of reference take the same time as they normally would. It is the series of events that a stationary observer (such as Clare) measures in Amaya and Binh's frame of reference that go slowly. Amaya and Binh are moving in slow motion because, according to a stationary observer's observations, time for Amaya and Binh has slowed down (Figure 10.3.5).

FIGURE 10.3.5 As Clare watches Amaya and Binh play space squash in their spaceship travelling at near the speed of light, the players and the ball seem to be moving much more slowly than would if she were playing a game of squash in her own frame of reference.

Worked example 10.3.1

TIME DILATION

A stationary observer on the Earth sees a very fast car passing by, travelling at $2.50 \times 10^8\,m\,s^{-1}$. In the car is a clock on which the stationary observer measures 3.00 s passing. Calculate how many seconds pass by on the stationary observer's own clock during this observation. Assume that $c = 3.00 \times 10^8\,m\,s^{-1}$.

Thinking	Working
Identify the variables: the time for the stationary observer is t, the proper time for the moving clock is t_0 and the velocities are v and c.	$t = ?$ $t_0 = 3.00\,s$ $v = 2.50 \times 10^8\,m\,s^{-1}$ $c = 3.00 \times 10^8\,m\,s^{-1}$
Use Einstein's time dilation formula with the Lorentz factor.	$t = \gamma t_0 = \dfrac{t_0}{\sqrt{1 - \dfrac{v^2}{c^2}}}$
Substitute the appropriate values and solve for t.	$t = \dfrac{3.00}{\sqrt{1 - \dfrac{(2.50 \times 10^8)^2}{(3.00 \times 10^8)^2}}}$ $= \dfrac{3.00}{0.55277}$ $= 5.43\ s$

Worked example: Try yourself 10.3.1

TIME DILATION

A stationary observer on the Earth sees a very fast scooter passing by, travelling at $2.98 \times 10^8\,m\,s^{-1}$. On the wrist of the rider is a watch on which the stationary observer measures 60.0 s passing. Calculate how many seconds pass by on the stationary observer's clock during this observation. Assume that $c = 3.00 \times 10^8\,m\,s^{-1}$.

PROPER TIME

In Einstein's time dilation equation, t_0 is referred to as the proper time. The proper time is the time between two events that are measured within the same frame of reference. For example, when a light bulb in the train flashes and Amaya (who is also inside the train) measures the time for the flash to reflect off a mirror and return to her, she has measured proper time. This is because the stopwatch remained stationary at the point in space inside the frame of reference where the light started and where it finished. Proper time is illustrated in Figure. 10.3.6

It is important that the clock isn't moved from one frame of reference to another if you want to measure proper time. This is because, as soon as the clock is changing its frame of reference, the time for that clock slows slightly.

Looking back to the stationary observer

Galileo said that all inertial frames of reference are equivalent. Thus Amaya and Binh, as they look out the window at Clare in her space station receding from them (Figure 10.3.5), will think it is they who are at rest and it is Clare who is moving away at a velocity near the speed of light. This follows from Galileo's principle of relativity and Einstein's first postulate.

If Amaya and Binh watch the light clock in Clare's space station, they see that time has slowed down for her, for they would observe Clare's light clock oscillations taking longer than the oscillations of their stationary light clock. This raises the question: Whose time actually runs slowly?

FIGURE 10.3.6 A clock measuring proper time. The clock is positioned at the place where the event started (the light starting out) and is at the same place when the event ends (the light returning).

The answer is that they are both right. The whole point of relativity is that you can only measure quantities relative to some particular frame of reference, not in any absolute sense. Certainly Amaya and Binh see events in Clare's frame of reference in slow motion, and Clare sees events in Amaya and Binh's frame of reference in slow motion. Remember that there is no absolute frame of reference, so there is no absolute clock ticking away the absolutely right time. All that you can be sure of is that time in your own inertial frame of reference is ticking away at a rate of one second per second.

The twin paradox

If Clare measures time for Amaya and Binh running slowly, then Amaya and Binh will age slowly. But if Amaya and Binh measure time for Clare as running slowly, then Clare will age more slowly. So what happens when Amaya and Binh decide to turn their spaceship around and come home?

This seems to be a paradox (i.e. a situation that appears to have contradictory elements). To solve this paradox Einstein described a thought experiment in which one of a set of twins goes on a long space journey and the other twin stays on the Earth.

The travelling twin finds that when she returns to the Earth, her twin has become quite elderly (Figure 10.3.7). While each twin is in constant motion relative to the other, they both see the other twin ageing more slowly. So why did the twin on the spaceship age more slowly than the twin on the Earth?

FIGURE 10.3.7 In the twin paradox, one twin ages less quickly than the other after travelling in a non-inertial frame of reference.

Here we need to point out that Einstein's 1905 theory of relativity deals only with frames of reference that are in constant motion, that is, inertial frames of reference. For this reason it is called the special theory of relativity. Special relativity does not deal with accelerated frames of reference. Ten years later, Einstein put forward the general theory of relativity, which does deal with situations in which acceleration occurs, that is, in non-inertial frames of reference. As part of this theory, he showed that in an accelerated frame of reference, time also slows down.

If we apply the twin paradox situation to Amaya, Binh and Clare, the general theory of relativity tells us that Clare's view of Amaya and Binh in their non-inertial frame shows them ageing very slowly. During this time, Amaya and Binh see Clare's time passing quickly. As a result they will see Clare age quite rapidly while they were accelerating, and then slowly when travelling at constant velocity. Clare sees Amaya and Binh aging very slowly as they change speed, then aging slowly as they travelled at a constant speed. Amaya and Binh never age rapidly.

But how do we know that it is Amaya and Binh who have accelerated and not Clare, because that is what it would look like for Amaya and Binh looking out of their window at Clare? For the answer to this we need to ask Amaya and Binh if they noticed anything unusual in their frame of reference. For example, did the surface of the water in their bottles tilt at an angle to the horizontal, or did the handles hanging down from the ceiling lean forwards or backwards. If we asked Clare these questions she would say no, while Amaya and Binh would say yes. Therefore it was Amaya and Binh that accelerated and not Clare.

Although it is often called a paradox, there is actually nothing impossible or illogical about this story. Einstein himself pointed out that, due to the Earth's rotation, and therefore its centripetal acceleration, a clock on the Earth's equator would run a little more slowly than one at the poles. This has now been found to be the case. In fact, in 1971 accurate atomic clocks were flown around the world on commercial flights. When compared with clocks left behind, the difference of about a quarter of a microsecond was just what Einstein had predicted. Now there are many satellites in orbit around the Earth, so the theory has been well and truly tested many times. Indeed, global positioning systems (GPS) must take relativistic corrections into account to ensure their accuracy.

At the speed at which satellites travel, special relativity effects due to their speed account for time slowing down by 7 microseconds a day. General relativistic affects due to the decreased gravitational fields at the altitude of the satellites account for time speeding up by 45 microseconds a day. Overall there is a 38-microsecond difference in time experienced by the satellite clocks that must be corrected if the satellites are to maintain the same time as the clocks on the Earth.

HIGH-ALTITUDE MUONS

Section 10.2 discussed the surprising observation that high-speed muons originating 15.0 km above the Earth can reach the Earth's surface. This can only be explained if the mean lifetime of these short-lived particles is somehow extended far beyond their normal mean lifetime. Time dilation provides an explanation for this unusual observation.

The normal mean lifetime of a muon is about 2.20 μs. However, this is the mean lifetime when measured in a stationary frame of reference. Muons travel very fast. In fact a speed as great as $0.999c$ is possible. At this speed an observer on the Earth will measure the lifetime of a sample of muons as far greater:

$$
\begin{aligned}
t &= \gamma t_0 \\
&= \frac{t_0}{\sqrt{1-\frac{v^2}{c^2}}} \\
&= \frac{2.20\times10^{-6}}{\sqrt{1-\frac{(0.999c)^2}{c^2}}} \\
&= \frac{2.20\times10^{-6}}{\sqrt{1-0.999^2}} \\
&= 49.2\,\mu\text{s}
\end{aligned}
$$

This is more than 22 times as long as in a stationary frame of reference. An observer on the Earth measures the muon's time running much slower. The slower time means that many muons live long enough to reach the Earth's surface.

PHYSICSFILE

Gold and relativity

A combination of relativistic and quantum mechanical affects explains the unique colour of gold.

Gold has a large number of nucleons in its nucleus, with 79 protons and 118 neutrons making up the very stable gold-197 isotope. With 79 positively charged protons in its nucleus, the inner electrons are subject to immense electrostatic charge. To avoid collapsing into the nucleus, the electrons in the closest energy level (1s orbital) need to be circling the nucleus at a speed of approximately $0.533c$. At this speed there are relativistic effects that cause the radius of the 1s-orbital to decrease, bringing it slightly closer to the nucleus.

There are also quantum mechanical effects that result in s-orbitals that are further away from the nucleus, shrinking their orbital radius as well, while the radii of the other orbitals (p, d, f and g) remain largely unaffected.

The colour of metals is mainly due to the transition of electrons between outer d and s orbitals. With most silvery metals the transition between the 4d and 5s orbitals absorbs photons in the UV range. This results in no loss of colour in the photons being reflected, so they appear shiny silver. In the case of gold, the transition between the 6d and 5s orbitals corresponds to lower energy blue light, as the levels are closer together. This means that blue light is absorbed and scattered from the incoming light, resulting in less blue light being reflected. The resulting light is predominantly red, orange, yellow and green, which combine to produce the yellowish colour we perceive as gold.

10.3 Review

OA ✓✓

SUMMARY

- The pulses in a light clock in a moving frame of reference have to travel further when observed from a stationary frame.
- Because of the constancy of the speed of light, this means that time appears to a stationary observer to have slowed in a moving frame.
- Time in a moving frame of reference seems to flow more slowly according to the equation $t = \gamma t_0$, where t_0 is the time in the moving frame of reference (the proper time), t is the time observed from the stationary frame of reference and γ is the Lorentz factor:

$$\gamma = \frac{1}{\sqrt{1 - \frac{v^2}{c^2}}}$$

- Observers in relative motion both see time slowing in the other frame of reference; that is, each sees the other ageing more slowly.
- Time dilation explains how short-lived, fast-moving muons can reach the surface of the Earth when, by classical physics, they should not be able to.

KEY QUESTIONS

For the following questions, assume that $c = 3.00 \times 10^8\,\text{m s}^{-1}$.

Knowledge and understanding

1 Complete the following sentences by selecting the correct term from those in bold.
In a device called a **light/mechanical/digital** clock, the **speed/oscillation/wavelength** of light is used to measure **time/mass**, as the speed of light is **variable/constant** no matter from which inertial frame of reference it is viewed.

2 An observer on one of the moons of Jupiter looks at their watch and measures the time for a deep-space probe, moving at 0.975*c*, to pass by. From this observer's frame of reference, what time has the observer measured: t or t_0? Explain.

3 An observer with a stopwatch is standing on a platform as a very fast train passes at a speed of $1.75 \times 10^8\,\text{m s}^{-1}$. The observer notices the proper time (t_0) on a passenger's phone as the passenger drops the phone to the floor. According to the clock on the phone, it takes 1.05 s to hit the floor. Calculate how much time has passed on the observer's stopwatch during this time.

4 An observer standing on a comet is watching as a satellite approaches at $2.30 \times 10^8\,\text{m s}^{-1}$. The observer notes that the solar panels on the satellite unfold in 75.0 s. Calculate the proper time (t_0) that would have passed on the satellite's clock as the panels unfold.

5 A student standing by the side of a road sees a very fast sports car driving past. The driver notes from his car's clock that it takes 5.50 s for the student to pick up her bag. If the car is moving at $2.75 \times 10^8\,\text{m s}^{-1}$, calculate the proper time (t_0) that the driver measures on the student's watch as she picks up her bag.

6 If Anna sees Ben fly past at 0.50*c*, how long, in her frame of reference, would it take Ben's clock to tick 1 second?

Analysis

7 Nala's light clock has a height of 1.0 m between its mirrors and, relative to Gemma, her spaceship is travelling at 0.90*c*. One tick is the time for light to go from one mirror to the other.

a How far does the light travel in Nala's frame of reference in one tick?

b What is the tick time for the clock in Nala's frame of reference?

c As the light takes a zigzag path in her frame of reference, Gemma measures the clock ticking at a slower rate, t_G. In terms of c and t_G, what is the length of the zigzag path that the flash travels in one tick in Gemma's frame of reference?

d What is the tick time of the clock in Gemma's frame of reference?

e What is the ratio of Gemma's tick to Nala's tick?

10.3 Review *continued*

8 A muon created at an altitude of 15.0 km above the Earth is moving at a speed of 0.9992c. The mean lifetime of a muon at rest is 2.20×10^{-6} s.

  a Calculate the mean lifetime of moving muons as timed by a stationary observer.

  b Using equations from classical physics and the result from part **a**, calculate the non-relativistic distance and the relativistic distance travelled by the moving muons during their lifetime.

9 In a LINAC particle accelerator, a sample of radioactive particles is accelerated to $2.93 \times 10^{8}\,\text{m s}^{-1}$. A researcher analysing the images of the path of the particles notes that the particles produce a trail 15.5 cm long before they all decayed.

  a Calculate the lifetime of the moving particles.

  b Calculate the lifetime of the same radioactive particles if they were at rest relative to the researcher and their timer.

10 Briefly explain why Einstein said that a clock at the Earth's equator should run slightly slower than one at the Earth's poles. Why do we not find this to be a problem?

10.4 Length contraction

The previous section described how time can only be measured relative to some particular frame of reference and not in any absolute sense. Because of the constancy of the speed of light, time appears to have slowed in a moving frame relative to the frame of reference of a stationary observer. Einstein described how space and time are interrelated, so it follows that space, and therefore length, is not absolute (Figure 10.4.1). This section explores the effect on the length of an object due to its motion in an inertial frame of reference.

FIGURE 10.4.1 Length is relative to the frame of reference and the direction of motion.

LENGTH IN DIFFERENT INERTIAL FRAMES

FIGURE 10.4.2 Einstein showed that the length of a moving object is reduced by the Lorentz factor, γ. However, the height and width of the carriage remain unchanged.

You may already have an inkling that length depends on who is doing the measuring and the frame of reference in which they make the measurement.

The light clock analysis in the previous section enabled us to compare the proper time on a clock in a moving frame of reference (observed by Clare) with the time measured on a clock in the stationary frame of reference (with Clare). The light clock was used as it only depends on light, not on some complicated mechanical arrangement that may have been altered by relative motion. There was, however, one important condition in the clock analysis—that both Amaya and Clare would agree on the distance, *d*, between the mirrors. This enabled the two expressions for *d* to be equated and a relationship found between proper time, t_0, and time, t.

The clock was set up in the spaceship so that the light path, of distance *d*, was perpendicular to the velocity. Distances in this perpendicular direction are unaffected by motion. However, Einstein showed that while perpendicular distances are unaffected, relative motion affects length in the direction of travel (Figure 10.4.2).

Calculating length contraction

Consider the situation in which Clare is standing on a train platform while Amaya and Binh pass by at speed v. Both Clare and Binh want to measure the length of the platform on which Clare is standing. Using a measuring tape, Clare measures the length of the platform (which is at rest from her perspective) as L_0, and concludes that Amaya and Binh cover this distance in a time equal to:

$$t = \frac{L_0}{v} \qquad \text{(Equation 10.4.1)}$$

Amaya and Binh observe the platform passing in time t_0. The relationship between the time in Amaya and Binh's frame of reference and the time that Clare measures is:

$$t_0 = \frac{t}{\gamma}$$

$$t_0 = t\sqrt{1-\frac{v^2}{c^2}} \qquad \text{(Equation 10.4.2)}$$

Substituting Equation 10.4.1 into Equation 10.4.2 gives:

$$t_0 = \frac{L_0}{v}\sqrt{1-\frac{v^2}{c^2}} \qquad \text{(Equation 10.4.3)}$$

Amaya and Binh measure the platform moving at a velocity of v relative to them, so they can say that the distance from the start to the end of the platform is:

$$L = vt_0 \qquad \text{(Equation 10.4.4)}$$

Substituting Equation 10.4.3 into Equation 10.4.4 gives:

$$L = v \times \frac{L_0}{v}\sqrt{1-\frac{v^2}{c^2}}$$

This simplifies to:

$$L = L_0\sqrt{1-\frac{v^2}{c^2}}$$

This is Einstein's equation for **length contraction**, incorporating the Lorentz factor. It shows that an object with a **proper length** of L_0 when measured at rest will, when in motion, have a shorter length, L, parallel to the direction of its moving frame of reference when measured by an observer who is in a stationary frame of reference. The proper length is contracted by a factor of $\frac{1}{\gamma}$. Given that length contraction can be represented as:

$$L = L_0\sqrt{\left(1-\frac{v^2}{c^2}\right)}$$

it follows that it can also be expressed in terms of the Lorentz factor, γ.

$$L = \frac{L_0}{\gamma}$$

where $\gamma = \dfrac{1}{\sqrt{1-\frac{v^2}{c^2}}}$

L_0 is the proper length, that is the length measured at rest in the stationary frame of reference

L is the length in the moving frame as measured by an observer in a stationary frame of reference

The effects of time dilation and length contraction cannot be seen in real life because our everyday speeds are well below the speed of light.

PHYSICSFILE

The strange behaviour of observed length

In the *Gedanken* experiment of a relativistic train, the stationary observer sees that the train appears shorter due to length contraction. However, an observer would actually see something much stranger, because of the finite speed of light.

At low speeds there is not much difference between the time light takes to reach us from the front and the back of the train. However, if the train were travelling near the speed of light, this difference would become significant—the train would travel a significant distance in the time it took for light to reach the observer. This would cause the train to appear stretched. However, the train would also appear length contracted. For small objects these effects cancel out, so the train would appear the same length but would appear to be rotated.

For Year 12 Physics, we assume an observer measures the object's contracted length without this distortion. However, these effects must be considered in more-advanced physics.

Worked example 10.4.1

LENGTH CONTRACTION

A stationary observer on the Earth sees a very fast car travelling by at $2.50 \times 10^8 \text{ m s}^{-1}$. When stationary, the car is 3.00 m long. Calculate the length of the car as measured by the stationary observer. Assume that $c = 3.00 \times 10^8 \text{ m s}^{-1}$.

Thinking	Working
Identify the relevant variables: the length measured by the stationary observer is L, the proper length of the car is L_0 and the velocities are v and c.	$L = ?$ $L_0 = 3.00 \text{ m}$ $v = 2.50 \times 10^8 \text{ m s}^{-1}$ $c = 3.00 \times 10^8 \text{ m s}^{-1}$
Use Einstein's length contraction equation and the Lorentz factor.	$L = \frac{L_0}{\gamma}$ $= L_0\sqrt{1-\frac{v^2}{c^2}}$
Substitute the appropriate values and solve for L.	$L = 3.00 \times \sqrt{1-\frac{(2.50 \times 10^8)^2}{(3.00 \times 10^8)^2}}$ $= 3.00 \times 0.553$ $= 1.66 \text{ m}$

Worked example: Try yourself 10.4.1

LENGTH CONTRACTION

A stationary observer on the Earth sees a very fast scooter travelling by at $2.98 \times 10^8 \text{ m s}^{-1}$. The stationary observer measures the scooter's length as 22.0 cm. Calculate the proper length of the scooter, that is the length measured when the scooter is at rest. Assume that $c = 3.00 \times 10^8 \text{ m s}^{-1}$.

Looking out the window

The situations considered so far involve an object in a moving frame of reference. The object is measured as being shorter in the direction of the motion according to an observer that is in a stationary frame of reference. Length contraction can also be applied to the *distance* that a moving object covers as it travels at very high speed.

Recall that no inertial frame of reference is special. Consider Amaya and Binh in their spacecraft travelling from the Earth to the Moon. According to them they are stationary; it is space that rushes by at high speed. As space rushes by, they are in fact travelling a proper distance of 384 400 km—the distance from the Earth to the Moon. This is the *proper length*, as it is measured by a device that is in the same frame of reference as the Earth and the Moon. As Amaya looks out of the window, she measures a much shorter distance to travel.

Worked example 10.4.2

LENGTH CONTRACTION FOR DISTANCE TRAVELLED

The pilot of a spaceship travelling at 0.997c is travelling from the Earth to the Moon. The proper distance between the Earth and the Moon is 384 400 km. When the pilot looks out the window, the distance between the Earth and the Moon looks much less than that. Calculate the distance that the pilot sees.

Thinking	Working
Identify the relevant variables: the length seen by the pilot is L, the proper length of the distance is L_0 and the velocity is v.	$L = ?$ $L_0 = 384\,400\,\text{km}$ $v = 0.997c$
Use Einstein's length contraction formula and the Lorentz factor.	$L = \frac{L_0}{\gamma}$ $= L_0\sqrt{1-\frac{v^2}{c^2}}$
Substitute the appropriate values and solve for L.	$L = 384\,400 \times \sqrt{1-\frac{(0.997c)^2}{c^2}}$ $= 384\,400 \times \sqrt{1-(0.997)^2}$ $= 384\,400 \times 0.0774$ $= 29\,800$ $= 2.98 \times 10^4\,\text{km}$

Worked example: Try yourself 10.4.2

LENGTH CONTRACTION FOR DISTANCE TRAVELLED

A stationary observer on the Earth sees a very fast train approaching a tunnel at a speed of 0.986c. The stationary observer measures the tunnel's length as 123 m. Calculate the length of the tunnel as seen by the train's driver.

The result from Worked example: Try yourself 10.4.2 leads to an interesting observation. If the proper length of the train is 100 m, then the driver could park the train in the 123 m tunnel with 11.5 m of tunnel extending beyond each end of the train. But when the train is moving at 0.986c, then, according to the train driver, the train will not fit entirely in the tunnel. There could be up to approximately 39.8 m of train extending past each end of the tunnel (Figure 10.4.3).

Similarly, a train that is longer than the tunnel will fit completely inside the tunnel if its length was measured by a stationary observer as it was moving past very quickly. In this scenario, the length of the train would be contracted according to the stationary observer (Figure 10.4.4).

(a)

Both train and tunnel are stationary

(b)

The tunnel is moving towards the observer in the train

FIGURE 10.4.3 The train both fits in the tunnel and doesn't fit in the tunnel, depending on your frame of reference.

(a)

The stationary train does not fit in the tunnel

(b)

The train is contracted

FIGURE 10.4.4 The train both doesn't fit in the tunnel and also does fit in the tunnel, depending on your frame of reference.

PROPER LENGTH

The proper length is the distance between two points whose positions are measured by an observer at rest with respect to the two points.

Recall the example of Amaya and Binh on a train and Clare on the platform observing them passing. Clare measures the length of the platform as the proper length, as her measuring tape is at rest on the platform, while Amaya and Binh measure the platform as a shorter length. If Amaya and Binh measure the length of the carriage they are in, this will be the proper length. If Claire measures the length of the carriage as it passes by, she will measure a shorter length of carriage.

Remember that length contraction occurs only in the direction of travel, not in a perpendicular direction. To Clare, the length of the carriage passing by will appear shortened, but its width and height (the dimensions of the train perpendicular to the direction of travel) will appear unaltered.

An example of length contraction with a tennis ball is shown in Figure 10.4.5. The dimension in the direction of the motion is contracted, but not the other dimensions.

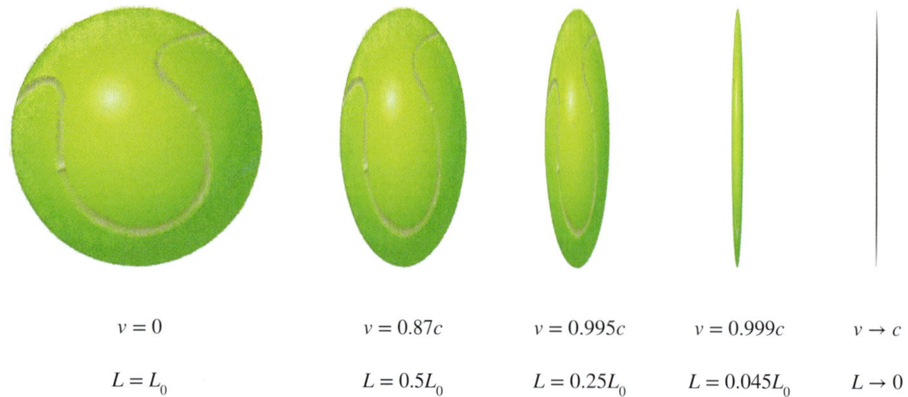

FIGURE 10.4.5 As the tennis ball moves faster to the right, its dimension in the left–right direction is contracted, but its other dimensions remain the same.

ALTERNATIVE EXPLANATION FOR MUONS REACHING THE EARTH

In Section 10.3 an explanation was provided for the strange observation of high-speed, short-lived muons that originated 15 km above the Earth but still reached the surface of the Earth, when classical physics says that none would. Time dilation explained how the muons could reach the Earth: the mean lifetime of the short-lived particles was extended far beyond their normal mean lifetime.

The observation can also be explained in terms of length contraction. The proper distance that the muons must travel to strike the Earth's surface is 15 km. However, observers sitting on the fast-moving muons would measure the distance that the muons need to travel as much less than 15 km:

$$\begin{aligned} L &= \frac{L_0}{\gamma} \\ &= L_0\sqrt{1-\frac{v^2}{c^2}} \\ &= 15\sqrt{1-\frac{(0.999c)^2}{c^2}} \\ &= 15\sqrt{1-0.999^2} \\ &= 0.67\,\text{km} \end{aligned}$$

WS 37 WS 38

This is more than 22 times shorter than the proper length. The muon could easily travel this contracted distance during their 2.20 μs lifetime.

CASE STUDY ANALYSIS

How length contraction affects linear particle accelerators

One of the principal tools of particle physicists is the particle accelerator. With these devices physicists can collide particles to investigate how matter behaves under extreme conditions. One such accelerator was the Stanford Linear Collider (SLC) at the SLAC National Accelerator Laboratory in California. The SLC was 3.2 km long, which made it the longest linear particle accelerator in the world. It would accelerate electrons and positrons to 50 GeV, or 99.999 999 994 77% of the speed of light. Once accelerated, the electrons travelled through a transfer line to a collision point 1.4 km away—or more accurately, 1.4 km *in the reference frame of a stationary observer*. So how long is the 1.4 km transfer line in the reference frame of a 50 GeV electron?

FIGURE 10.4.6 The Australian Synchrotron's 100 MeV linac

Using the length contraction equation, we can calculate that the 1.4 km length of the transfer line would be only 1.43 cm long from the frame of reference of a 50 GeV electron.

At the Australian Synchrotron facility, a linear accelerator, or linac (Figure 10.4.6), is used for the initial stage of electron acceleration. The linear accelerator provides the electrons with 100 MeV of energy, resulting in velocities of $2.99788 \times 10^8\,\text{m s}^{-1}$ or 99.9987% of the speed of light. After they are accelerated, the electrons travel through a 16 m transfer line. At this speed, the 16 m transfer line would be just 8.13 cm in the electrons' reference frame.

Analysis

1 With advances in technology, the accelerating potential of the Australian Synchrotron's linear facility could be boosted to 150 MeV. This would result in electrons travelling at $2.9979073 \times 10^8\,\text{m s}^{-1}$ or 99.999 423 66% of the speed of light. Calculate the length of the transfer line from the frame of reference of the electron if the transfer line is 16.0 m long in the frame of reference of the laboratory.

2 The storage ring at the Australian Synchrotron facility is a circular particle accelerator with a circumference of 216 m. When the electrons reach the storage ring they have about 3000 MeV of energy and are travelling at 99.999 998 55 the speed of light. Calculate the circumference of the storage ring as seen from the frame of reference of a 3000 MeV electron.

3 Determine the radius of the storage ring for an observer that is stationary in the frame of reference of the laboratory and compare it to the radius that would be seen from the frame of reference of a 3000 MeV electron in the storage ring.

10.4 Review

OA ✓✓

SUMMARY

- The special theory of relativity implies that time and space are related. Motion affects space in the direction of travel.
- A moving object will appear shorter, or appear to travel less distance, by the inverse of the Lorentz factor, γ. Einstein's length contraction equation is given by:

$$L = \frac{L_0}{\gamma}$$

where L_0 is the proper length in the stationary frame of reference, L is the contracted length as seen in a moving frame of reference and γ is the Lorentz factor:

$$\gamma = \frac{1}{\sqrt{1 - \frac{v^2}{c^2}}}$$

- The proper length, L_0, is the length measured by an observer at rest with respect to the object being measured.
- Length contraction for high-speed muons can also explain why muons are detected on the surface of the Earth when classical physics predicts that they can't reach the Earth.

KEY QUESTIONS

For the following questions, assume that $c = 3.00 \times 10^8\,\text{m s}^{-1}$.

Knowledge and understanding

1 Under what conditions are you able to measure the proper length, L_0, of an object?

2 If you are standing on the Moon and observe a speeding spaceship moving in the same direction as its length, what would you notice about its length, height and width?

3 An observer is standing on a train platform as a very fast train passes at a speed of $1.75 \times 10^8\,\text{m s}^{-1}$. The observer notices that a passenger is holding a metre rule in the direction that the train is moving. Calculate the length, L, of the metre rule that the stationary observer measures.

4 An observer standing on a comet is watching a satellite approach at $2.30 \times 10^8\,\text{m s}^{-1}$. The observer knows that the proper length of the satellite in the direction of its motion is 5.25 m. Calculate the length of the satellite that the observer measures as it passes.

5 A builder makes a mistake and builds a garage too short for the owner's car to fit in. The proper length of the garage is 1.50 m and the proper length of the car is 3.50 m. The builder suggests that if the owner drives fast enough, the builder could stand by the garage and see that the car would fit.

  a Calculate the speed that the car would need to travel to just fit in the garage when observed by the builder.

  b Explain why the car owner would not be happy about the builder's suggestion by calculating the length of the garage as seen by the driver.

Analysis

6 Corey is standing by the side of an athletics track watching Emily run in an 800.0 m race.

  a At what speed must Emily run in order for the race to be only 400.0 m long in her frame of reference?

  b Corey notices that Emily is thinner than she normally is but just as wide and just as tall. Calculate the fraction of Emily's thickness while she is running to her thickness when she is standing still.

7 A drone flies at $754\,\text{m s}^{-1}$ past its operator who is standing on the ground. If the length of the drone is 35.0 cm when stationary on the ground at the operator's feet, calculate the length that the operator would measure as the drone flies past.

8 A wealthy entrepreneur wanting to go fishing in the Sea of Tranquillity flies their rocket towards the Moon at a speed of 0.934c. On the way the entrepreneur decides to rig up a 2.55 m fishing rod and hold it in the direction in which the rocket is moving.

  a Determine the length of the fishing rod as observed by an astronaut in the International Space Station.

  b What is the length of the fishing rod as observed by the entrepreneur in the rocket?

9 An observer on the Earth notices a variety of differently shaped spacecraft flying past at velocities very close to the speed of light. Determine the shape of each spacecraft that the observer would see if the stationary spacecraft were:

a a sphere with a radius of 125 m

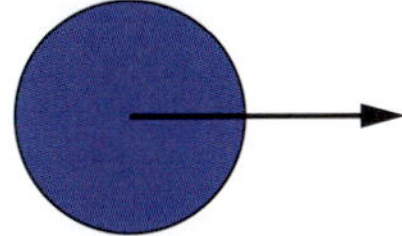

b a cube with sides of 141 m moving with four sides parallel to the direction of motion

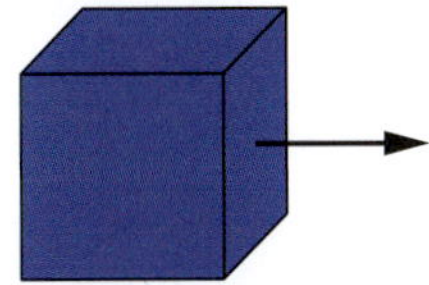

c a 175 m long hollow cylinder with a radius of 30.0 m moving with the side of the cylinder parallel to the direction of travel

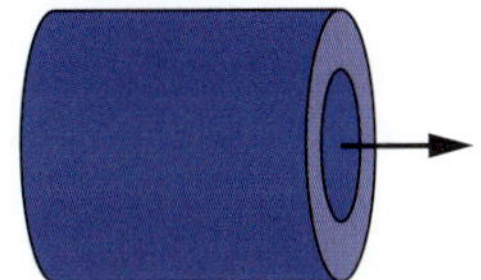

d a square-bottomed pyramid with base sides of 50.0 m and an apex 80.0 m above the base moving with the base parallel to the direction of travel.

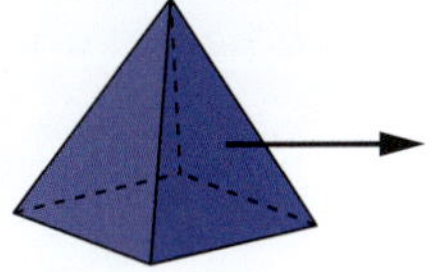

10.5 Einstein's mass-energy relationship

This section explains how Einstein's relativistic principles led to his most famous equation: $E = mc^2$. It begins by looking at what happens to the momentum of an object as its speed approaches the speed of light.

APPROACHING THE SPEED OF LIGHT

Recall the Lorentz factor that was introduced earlier in this chapter:

$$\gamma = \frac{1}{\sqrt{1 - \frac{v^2}{c^2}}}$$

At low speeds γ is so close to 1 that the effects of special relativity can be ignored. However, γ rapidly increases as the speed, v, comes closer to the speed of light, c. In fact, γ rapidly increases towards infinity.

It is interesting to wonder what would happen at the speed of light. According to Einstein's equations, the length of an object shrinks to zero and, to an observer, time inside its frame of reference appears to stop altogether. Einstein took this to mean that it is not possible for any real object to reach the speed of light. However, vanishing time and length were not the only reasons Einstein came to this conclusion.

Relativistic momentum

If a rocket ship (Figure 10.5.1) is travelling at $0.99c$, why can't it simply turn on its rockets and accelerate up to c or more? An answer to this question was not given in Einstein's original 1905 paper on relativity. Some years later he showed that as the speed of a spaceship approaches c, its momentum increases, but this does not lead to a corresponding increase in speed.

FIGURE 10.5.1 This rocket ship is moving at $0.99c$ and accelerating, and yet it can never reach c.

Imagine a spaceship accelerating from rest to a high speed as viewed by an observer in a stationary frame of reference. The change in momentum of the ship is given by:

$$F\Delta t_0 = m\Delta v \qquad \text{Equation 10.5.1}$$

where t_0 is the time in the ship's frame of reference
$m\Delta v$ is just the classical Newtonian change in momentum.

In the stationary observer's frame of reference, the time is dilated:

$$\Delta t = \gamma \Delta t_0$$

$$\Delta t_0 = \frac{\Delta t}{\gamma}$$

Substituting Δt_0 into Equation 10.5.1 gives:

$$F\frac{\Delta t}{\gamma} = m\Delta v$$

$$F\Delta t = \gamma m \Delta v$$

That is, the impulse as seen by the stationary observer is equal to the product of the Lorentz factor, γ, and the change in Newtonian momentum. This means that as the spaceship approaches the speed of light, the impulse is multiplied by a factor that grows very rapidly. Thus the change in momentum in the stationary observer's frame of reference is equal to:

$$\Delta p = \gamma m \Delta v$$

$$= \gamma \Delta p_0$$

If we assume that the spaceship starts at zero velocity, the final relativistic momentum can be written as follows.

$$p = \gamma m v$$

$$= \gamma p_0$$

where p_0 is momentum as defined in classical mechanics
p is the relativistic momentum

The momentum increases very rapidly as a spaceship approaches the speed of light. You might argue that this is expected—after all, momentum is a function of velocity. If you graph the relativistic momentum, p, against the velocity, v, and on the same graph show the classical momentum, you can see that the relativistic momentum increases at a rate far greater than it would if it were due to the increase in velocity alone (Figure 10.5.2).

This result can be interpreted by thinking of the mass as a quantity that also increases with speed. Thus there is a relationship between the rest mass, m, which is the mass measured while the object is at rest in the frame of reference, and the relativistic mass, γm, which is the mass measured as the object is moving relative to the observer. As the Lorentz factor increases with an increase in velocity, the relativistic mass must also increase. (Einstein was never happy with the term 'relativistic mass', preferring the term 'relativistic momentum of an object'.)

Notice too how the classical treatment allows the object to have a speed greater than the speed of light, but the relativistic treatment sees the mass become so large that the speed of light is never actually reached.

Returning to the example of the rocket ship attempting to increase its velocity to the speed of light: with an increase in its relativistic mass, it becomes harder for its engines to cause a change in velocity. The closer the rocket ship approaches c, the greater the amount of impulse that is required to accelerate the ship to the speed of light. In fact, as the velocity approaches c, the relativistic mass, γm, approaches infinity. Thus the rocket ship can never reach the speed of light.

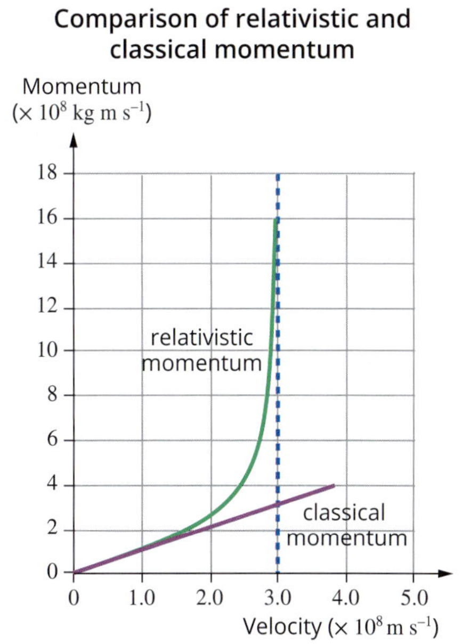

FIGURE 10.5.2 The relationship between classical momentum and velocity, and between relativistic momentum and velocity, for a 1 kg mass

CASE STUDY

Particles gaining mass

If an object moves in a circular path it is because of a centripetal force that acts towards the centre of the path. Centripetal force therefore acts continuously at right angles to the velocity of the object. Centripetal force can arise in a number of ways, such as from the tension in a string or from a gravitational force (such as the gravitational force of the Earth on the Moon).

Another source of circular motion is the force on a charged particle that is moving at right angles to a magnetic field. The equation that represents the relationship between the magnetic force (F_B) and the centripetal force (F_C) is:

$$F_B = F_C$$

$$qvB = \frac{mv^2}{r}$$

$$r = \frac{mv}{qB}$$

$$r = \frac{m}{qB} \times v$$

The final relationship implies that if the mass (m), charge (q) and magnetic field strength (B) are all kept constant, the radius of the circular path is directly proportional to the velocity of the charged particle. So if the velocity increases by a factor of 2, you might expect the radius to also increase by a factor of 2. However, this is not the case.

Consider a circular particle accelerator, such as the one at the Australian Synchrotron facility. As the velocity of a charged particle increases, the radius of its path also increases—but to a much greater degree than expected. According to the relationship derived above, if the charge (q) and the magnetic field strength (B) are kept constant, then the only explanation for the extra increase in radius over that expected is that the mass of the particle (m) must have increased.

In fact the mass of an electron travelling at 99.999 99% of the speed of light increases to 6000 times the mass of an electron at rest. There is no explanation for this phenomenon in Galilean relativity or Newtonian physics. This will all be explained later in the chapter using Einstein's theories.

EINSTEIN'S FAMOUS EQUATION

As the momentum of an object increases so does its kinetic energy. The classical relationship between the two can be written as:

$$E_k = \frac{1}{2}mv^2$$

$$= \frac{1}{2}(mv)v$$

$$= \frac{1}{2}pv$$

This form of the equation shows that the kinetic energy of an object is related to the object's momentum as well as its velocity.

Einstein showed, however, that the classical expression for kinetic energy was not correct at high speeds. The mathematics involved is beyond the scope of this course, but Einstein, working from the expression for relativistic momentum and the usual assumptions about work, forces and energy, showed that the kinetic energy of an object was given by the expression:

$$E_k = (\gamma - 1)mc^2$$

Although it is not obvious from this expression, if the velocity (which is hidden in the γ term) is small, the expression approaches the classical equation for E_k, that is, $\frac{1}{2}mv^2$. Even at speeds of $0.10c$ the classical expression is accurate to better than $\pm 1\%$.

Einstein's expression can be expanded to:

$$E_k = \gamma mc^2 - mc^2$$

This equation can be rearranged as:

$$\gamma mc^2 = E_k + mc^2$$

Einstein interpreted the left side of this expression as being an expression for the total energy of the object:

$$E_{tot} = \gamma mc^2$$

The right side of the expression ($E_k + mc^2$) implies that there are two parts to the total energy: the kinetic energy, E_k, and another term that only depended on the rest mass, m. Einstein referred to the second term, mc^2, as the rest energy of the object, E_0, as it does not depend on the object's speed. This appeared to imply that somehow there was energy associated with mass. This would be an astounding proposition to a classical physicist, but as you have seen, in relativity, the mass of an object increases as you add kinetic energy to it.

The conservation of energy relationship can therefore be stated as:

$$E_{tot} = E_k + E_0$$

$E_{tot} = E_k + E_0 = \gamma mc^2$
where $E_k = (\gamma - 1)mc^2$
$E_0 = mc^2$

Worked example 10.5.1

TOTAL ENERGY OF AN OBJECT

Calculate the total energy of a very fast vintage Vespa scooter if its rest mass is 210 kg and it is travelling at a speed of $2.55 \times 10^8\,\text{m s}^{-1}$. Assume that $c = 3.00 \times 10^8\,\text{m s}^{-1}$ and that *Gedanken* conditions apply for this question.

Thinking	Working
Identify the relevant variables. Given the question asks for the total energy, E_{tot}, the relevant variables are the mass of the scooter, m, the speed, v, and the speed of light, c.	$E_{tot} = ?$ $m = 210\,\text{kg}$ $v = 2.55 \times 10^8\,\text{m s}^{-1}$ $c = 3.00 \times 10^8\,\text{m s}^{-1}$
Use Einstein's total energy formula and the Lorentz factor.	$E_{tot} = \gamma mc^2$ $= \dfrac{mc^2}{\sqrt{1-\dfrac{v^2}{c^2}}}$
Substitute the appropriate values and solve for E_{tot}.	$E_{tot} = \dfrac{mc^2}{\sqrt{1-\dfrac{v^2}{c^2}}}$ $= \dfrac{(210)(3.00\times10^8)^2}{\sqrt{1-\dfrac{(2.55\times10^8)^2}{(3.00\times10^8)^2}}}$ $= \dfrac{(1.89000\times10^{19})}{(0.52678)}$ $= 3.59\times10^{19}\,\text{J}$

Worked example: Try yourself 10.5.1

TOTAL ENERGY OF AN OBJECT

Calculate the total energy of an electron speeding through the Australian Synchrotron if its rest mass is 9.11×10^{-31} kg and it is travelling at a speed of $2.9979 \times 10^8\,\text{m s}^{-1}$. Assume that $c = 3.00 \times 10^8\,\text{m s}^{-1}$ and that *Gedanken* conditions apply for this question.)

CONVERTING MASS TO ENERGY

You have, no doubt, seen part of Einstein's equation for total energy before:

$$E = mc^2$$

This equation tells you that mass and energy are interrelated. The relationship is known as the **mass–energy equivalence**. In a sense you can say that mass has energy and energy has mass. Recall that in Unit 1 this relationship was used to calculate the energy absorbed and released during fission and fusion nuclear reactions.

Nuclear reactions involve vastly more energy per atom than chemical reactions (Figure 10.5.3). When a uranium atom splits into two fission fragments, about 200 million electron volts of energy are released. By comparison, most chemical reactions involve just a few electron volts.

FIGURE 10.5.3 In a nuclear bomb a few grams of mass is converted into energy. As the uranium undergoes fission, it releases hundreds of gigajoules (10^{12} J) of energy. Millions of tonnes of a chemical explosive (e.g. TNT) would be needed to produce this much explosive energy.

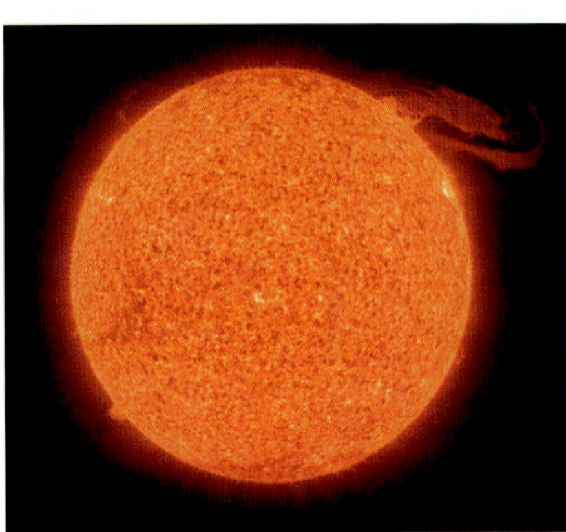

FIGURE 10.5.4 In nuclear fusion in the Sun, about 4 million tonnes of mass is converted into energy every second.

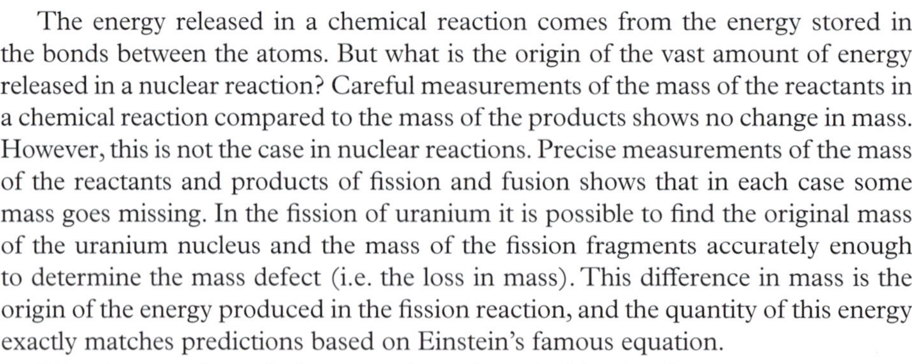

The energy released in a chemical reaction comes from the energy stored in the bonds between the atoms. But what is the origin of the vast amount of energy released in a nuclear reaction? Careful measurements of the mass of the reactants in a chemical reaction compared to the mass of the products shows no change in mass. However, this is not the case in nuclear reactions. Precise measurements of the mass of the reactants and products of fission and fusion shows that in each case some mass goes missing. In the fission of uranium it is possible to find the original mass of the uranium nucleus and the mass of the fission fragments accurately enough to determine the mass defect (i.e. the loss in mass). This difference in mass is the origin of the energy produced in the fission reaction, and the quantity of this energy exactly matches predictions based on Einstein's famous equation.

Likewise, **nuclear fusion** reactions deep inside the Sun release the huge amounts of energy that stream from it. This energy comes from the conversion of about 4 million tonnes of mass into energy every second (Figure 10.5.4).

NUCLEAR FUSION

Nuclear fusion occurs when two light nuclei combine to form a larger nucleus (Figure 10.5.5).

As with radioactive decay and nuclear fission, the mass of the reactants is slightly greater than the mass of the products when the nuclei combine during fusion. The energy created by this missing mass can be determined as follows.

$$\Delta E = \Delta mc^2$$

where ΔE is the energy (J)

Δm is the mass defect (kg)

c is the speed of light

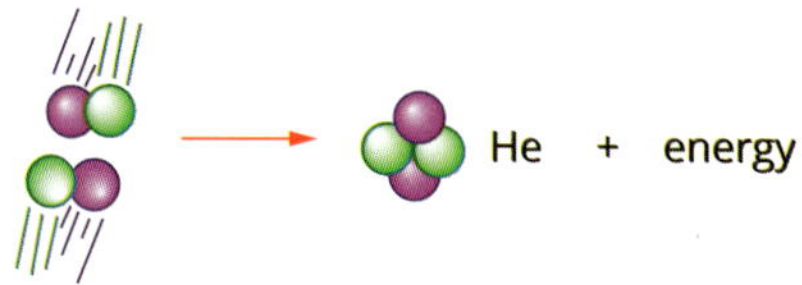

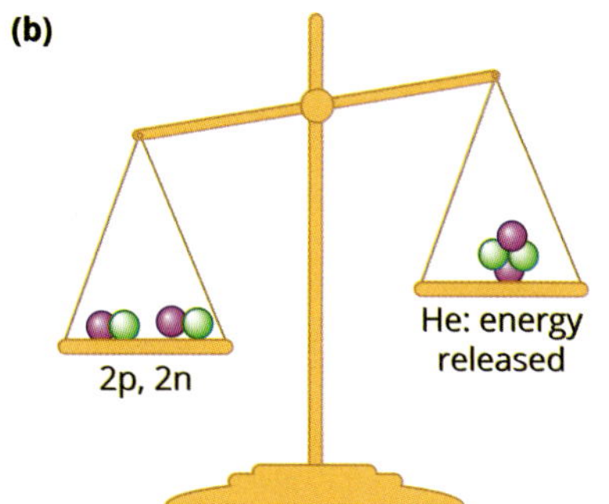

FIGURE 10.5.5 (a) When two isotopes of hydrogen fuse to form a helium nucleus, energy is released. (b) The binding energy of the nucleus appears as a loss in mass, Δm, which can be calculated using $\Delta E = \Delta mc^2$.

Nuclear fusion is a very difficult process to create in a laboratory. The main problem is that nuclei are positively charged and thus repel one another. Slow-moving nuclei with relatively small amounts of kinetic energy will not be able to get close enough for the strong nuclear force to come into effect, and so fusion will not happen. Only if nuclei have enough kinetic energy to overcome the repulsive force can they get close enough for the strong nuclear force to start acting. If this happens, fusion occurs (Figure 10.5.6).

Typically, temperatures in the order of hundreds of millions of degrees are required for fusion to occur. Such temperatures are present inside the Sun.

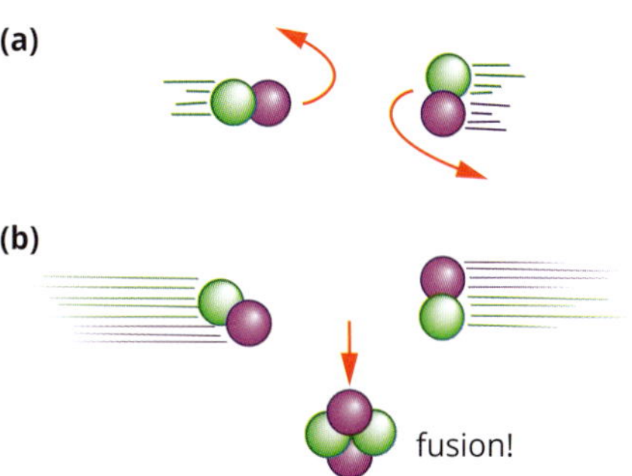

FIGURE 10.5.6 (a) Slow-moving nuclei do not have enough energy to fuse together. Electrostatic forces cause them to be repelled from each other. (b) If the nuclei have sufficient kinetic energy, they will overcome the repulsive forces and move close enough for the strong nuclear force to come into effect. At this point fusion occurs and energy is released.

FUSION IN THE SUN AND SIMILAR STARS

In the Sun, many different fusion reactions take place. The main reaction is the fusion of hydrogen nuclei to form helium. Each second about 657 million tonnes of hydrogen and hydrogen isotopes fuse to form about 653 million tonnes of helium. Each second a mass defect in the order of 4 million tonnes results from these fusion reactions. The amount of energy released is enormous and can be found from the equation $\Delta E = \Delta mc^2$. A tiny proportion of this energy reaches the Earth and sustains life as we know it.

The sequence of fusion reactions shown in Figure 10.5.7 has been occurring inside the Sun for the past 5 billion years and is expected to continue for another 5 billion years. Hydrogen nuclei are fused together and, after several steps, a helium nucleus is formed. This process releases about 25 MeV of energy.

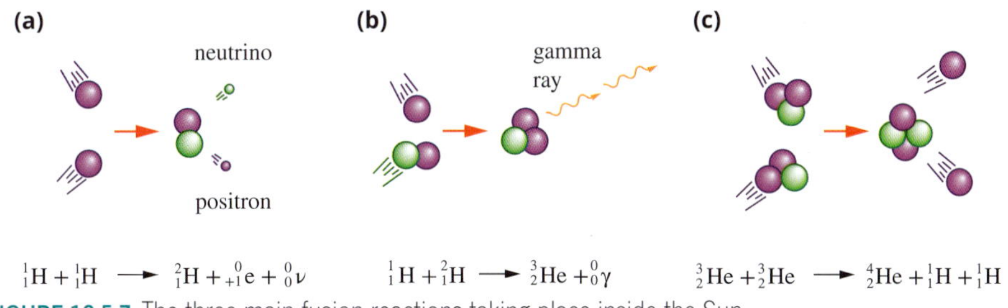

FIGURE 10.5.7 The three main fusion reactions taking place inside the Sun

The Sun is a second- or third-generation star. It was formed from the remnants of other stars that exploded much earlier in the history of the galaxy. As this giant gas cloud contracted under the effect of its own gravity, the pressure and temperature at the core reached extreme values, sufficient to sustain these fusion reactions.

Worked example 10.5.2

FUSION

Consider the fusion reaction shown below. A proton fuses with a deuterium nucleus (i.e. a hydrogen nucleus with one neutron) in the Sun. A helium nuclide is formed and a γ-ray released. A total of 20.0 MeV of energy is released during this process.

$$^{1}_{1}p + ^{2}_{1}H \rightarrow ^{x}_{2}He + \gamma$$

a What is the value of the unknown mass number, *x*?	
Thinking	**Working**
Analyse the mass numbers. The gamma ray has atomic and mass numbers of zero.	$1 + 2 = x + 0$ $x = 3$ A helium-3 nucleus is formed.

b How much energy is released in joules?	
Thinking	**Working**
Recall that $1\ eV = 1.60 \times 10^{-19}\ J$	$20.0\ MeV = 20.0 \times 10^{6} \times 1.60 \times 10^{-19}$ $= 3.20 \times 10^{-12}\ J$

c Calculate the mass defect for this reaction.	
Thinking	**Working**
Use $\Delta E = \Delta mc^2$.	$\Delta m = \frac{\Delta E}{c^2}$ $= \frac{(3.20 \times 10^{-12})}{(3.00 \times 10^{8})^2}$ $= 3.55555 \times 10^{-29}$ $= 3.56 \times 10^{-29}\ kg$

Worked example: Try yourself 10.5.2

FUSION

A fusion reaction in the Sun fuses two helium nuclides. A helium nucleus and two protons are formed, and 30.0 MeV of energy is released.

$$^{3}_{2}He + ^{3}_{2}He \rightarrow ^{x}_{2}He + 2^{1}_{1}H$$

a What is the value of the unknown mass number *x*?

b How much energy is released in joules?

c Calculate the mass defect for this reaction.

POSITRON-ELECTRON ANNIHILATION

When matter was first created from energy, moments after the Big Bang, every particle created was accompanied by an antimatter counterpart. We know that energy has no charge, so for every positive particle created an equivalent negative particle was also created. There are other properties of antimatter that are opposite to its matter equivalent, such as spin. However, the property of mass is always identical.

The antimatter particles of the proton, neutron and electron are the anti-proton, anti-neutron and the positron. As these matter–antimatter pairs were created from pure energy, when matter meets its antimatter equivalent, **annihilation** results. When this occurs all the mass of the particles is converted into energy in the form of two identical photons that head in opposite directions.

As the mass of an electron is identical to the mass of the positron, Einstein's equation relating mass and energy can be used to calculate the energy of the two photons created in the annihilation process:

$$\begin{aligned} E &= mc^2 \\ &= (2\times 9.11\times 10^{-31})(3.00\times 10^{8})^2 \\ &= 1.63980\times 10^{-13}\,\text{J} \\ &= 1.02488\times 10^{6}\,\text{eV} \\ 2\times E_{\text{photon}} &= 1.02488\,\text{MeV} \\ E_{\text{photon}} &= 0.512\,\text{MeV per photon} \end{aligned}$$

The energy of each photon is also the energy equivalent of the mass of an electron or positron. A photon with an energy equivalent to twice that value can undergo the opposite process of **pair production**, in which a photon creates an electron–positron pair.

TRANSFORMATIONS IN PARTICLE COLLIDERS

When two particles with significant kinetic energy collide, the kinetic energy is transformed into matter and new particles can be created. The Large Hadron Collider (LHC) can cause collisions between protons that have been accelerated to energies up to 6.50 TeV. Since the LHC began operating, 59 new particles have been discovered. The energy available to create matter comes from the protons' combined kinetic energies (2×6.50 TeV) for a total of 13.0 TeV. The most notable particle created by the LHC is the Higgs boson, which has a mass–energy equivalence of 0.125 TeV.

The search continues for more massive particles, which requires particles with even higher kinetic energies to collide.

As in any interaction between matter and energy, there are rules of conservation that must be obeyed. For example, the particles produced must comply with the rules of:

- conservation of energy
- conservation of charge.

By the conservation of energy, both the kinetic energy and the energy–mass equivalent must be accounted for. Any residual energy remaining after particles are produced becomes the kinetic energy shared between those particles.

The sum of the charges of the colliding particles must also equal the sum of the charges of the particles that emerge from the interaction.

Worked example 10.5.3

TRANSFORMATIONS AND CONSERVATIONS

When two protons each with 95.0 MeV of kinetic energy collide, they produce three particles. One is a proton and another is a positive pion (π^+) with a mass–energy equivalence of 139.57039 MeV. The total kinetic energy available to all three particles after the collision is 49.1362784 MeV. Assume that $c = 3.00 \times 10^8$ m s^{-1} and use the data in the table below to answer the following questions.

Particle	Mass (kg)
proton/antiproton	1.672622×10^{-27}
neutron/antineutron	1.674928×10^{-27}
electron/positron	9.109384×10^{-31}

a Calculate the total energy of the reactants before the collision.

Thinking	Working
Use Einstein's equation to calculate the total mass–energy equivalence of the two protons before the collision (in joules).	$E = mc^2$ $= 2(1.672622 \times 10^{-27})(3.00 \times 10^8)^2$ $= 3.010720 \times 10^{-10}$ J
Convert the energy to MeV.	$E = \frac{(3.010720 \times 10^{-10})}{(1.60 \times 10^{-19})}$ $= 1.881700 \times 10^9$ eV $= 1.881700 \times 10^3$ MeV
Add the kinetic energy of the colliding protons to get the total energy before the collision.	$E_{initial} = (1.881700 \times 10^3) + 2(95.0)$ $= 2.071700 \times 10^3$ $= 2.07 \times 10^3$ MeV

b Determine the total energy of the known particles present after the collision and hence identify the unknown particle.

Thinking	Working
Use Einstein's equation to calculate the total mass–energy of the proton after the collision (in joules).	$E = mc^2$ $= (1.672622 \times 10^{-27})(3.00 \times 10^8)^2$ $= 1.505360 \times 10^{-10}$ J
Convert the energy to MeV.	$E = \frac{(1.505360 \times 10^{-10})}{(1.60 \times 10^{-19})}$ $= 9.408499 \times 10^8$ eV $= 9.408499 \times 10^2$ MeV
Add the mass–energy equivalents of the known particles to the kinetic energy of each product.	$E = (9.408499 \times 10^2) + 139.57039 + 49.136278$ $= 1.129557 \times 10^3$ MeV
Subtract the total energy of the known products in eV from the total energy of the reactants to find the mass–energy of the unknown particle.	$E_{unknown} = (2.071700 \times 10^3) - (1.129557 \times 10^3)$ $= 9.421435 \times 10^2$ MeV $= 9.421435 \times 10^8$ eV

Convert the mass–energy equivalent of the unknown particle to mass in kilograms. To do this ensure that the energy value is in joules. Identify the particle from the table provided.	$E = mc^2$ $m = \frac{(9.421435 \times 10^8)(1.60 \times 10^{-19})}{(3.00 \times 10^8)^2}$ $= 1.674922 \times 10^{-27}$ kg This is close to the mass of a neutron, so the unknown particle is a neutron.

c Show that charge is conserved during the collision.	
Thinking	**Working**
Write the reaction for the collision showing the charge on each particle.	$p^+ + p^+ \longrightarrow p^+ + n^0 + \pi^+$
Determine the sum of the charges on each side of the collision.	$(2 \times +) \longrightarrow (2 \times +) + (1 \times 0)$ $2+ \longrightarrow 2+$ Hence charge is conserved.

Worked example: Try yourself 10.5.3

TRANSFORMATIONS AND CONSERVATIONS

Two protons each with 82.0 MeV of kinetic energy collide to produce a neutral pion (π^0) with a mass–energy equivalence of 134.9768 MeV. One of the two protons remains intact after the collision, while the fate of the other proton is unknown. The total kinetic energy available to the particles after the collision is 29.02290 MeV. Assume that $c = 3.00 \times 10^8\,\text{m}\,\text{s}^{-1}$ and use the data in the table below to answer the following questions.

Particle	Mass (kg)
proton/antiproton	1.672622×10^{-27}
neutron/antineutron	1.674922×10^{-27}
electron/positron	9.109384×10^{-31}

a Calculate the total energy of the reactants before the collision.

b Determine the total energy of the known particles present after the collision and hence identify the particle that one of the protons transforms into.

c Show that charge is conserved during the collision.

10.5 Review

OA ✓✓

SUMMARY

- Relativistic momentum includes the Lorentz factor, γ. Hence as more impulse is added, the mass increases towards infinity. The speed gets closer to, but never equals, c. The relativistic momentum equation is $p = \gamma mv = \gamma p_0$.
- The relativistic mass, γm, can be used to determine the mass of a moving object.
- The rest energy (i.e. the energy associated with the rest mass of an object) is given by $E_0 = mc^2$.
- The relativistic kinetic energy (i.e. the energy associated with a moving object) is given by $E_k = (\gamma - 1)mc^2$.
- Mass and energy are different forms of the same thing. This means that mass can be converted into energy, and energy can be converted into mass.
- Nuclear fission and fusion reactions result in a mass defect (i.e. a loss of mass). It is this difference in mass that is converted to the energy released in nuclear reactions. This mass is related to the energy produced according to $\Delta E = \Delta mc^2$.
- Nuclear fusion is the combining of light nuclei to form heavier nuclei. Extremely high temperatures are required for fusion to occur. This is the process occurring in stars.
- Hydrogen nuclei fuse to form deuterium. Further fusions result in the formation of isotopes of helium.
- Particle colliders convert the kinetic energy of colliding particles into matter according to $\Delta E = \Delta mc^2$, but only in compliance with the laws of conservation of energy and conservation of charge.

KEY QUESTIONS

Knowledge and understanding

1 What happens to the mass of a particle as it approaches the speed of light?

2 The equation for the fusion of two isotopes of hydrogen (deuterium and tritium) is shown below.

$$^{2}_{1}\text{H} + {}^{3}_{1}\text{H} \longrightarrow {}^{4}_{2}\text{He} + {}^{1}_{0}\text{H}$$

Explain why energy is released during this process.

3 Often particle physics describes the mass of an electron as 0.510 MeV. Explain why this is possible when the mass of an electrons is 9.11×10^{-31} kg.

Analysis

4 The Rosetta spacecraft as observed by scientists at the European Space Agency has a rest mass of 1230 kg and a speed of 775 m s^{-1}.

- **a** Calculate the relativistic momentum of Rosetta.
- **b** How does the relativistic momentum calculated in part **a** compare to the momentum calculated by the equations of classical physics?

5 Calculate the relativistic momentum of a carbon-12 nucleus in a linear accelerator if its rest mass is $1.99264824 \times 10^{-26}$ kg and it is travelling at 0.950c.

6 A very fast arrow has a rest mass of 15.1 g and a speed of 0.840c.

- **a** Calculate the relativistic kinetic energy of the arrow.
- **b** Calculate the kinetic energy of the arrow according to classical physics.
- **c** What accounts for the difference in the answers to parts **a** and **b**?

7 Calculate the energy in joules produced by the Sun in one day if 4.00 million tonnes of matter are converted into energy every second.

8 One of the fusion reactions occurring in the Sun fuses two protons to form a deuterium nucleus, a positron and an electron neutrino.

$$^{1}_{1}\text{H} + {}^{1}_{1}\text{H} \longrightarrow {}^{2}_{1}\text{H} + {}^{0}_{1}\text{e}^{+} + {}^{0}_{0}\nu$$

Calculate the energy in MeV of one fusion reaction of this type given the data in the following table.

Particle	Mass (kg)
proton	1.672622×10^{-27}
deuterium	3.34358×10^{-27}
positron	9.10938×10^{-31}
electron neutrino	2.14000×10^{-37}

9 A proton and antiproton each with 105.0 MeV of kinetic energy collide and are completely annihilated. The result is two gamma rays of equal energy. In answering the following questions, take the mass of a proton and an antiproton to each be 1.672622×10^{-27} kg.

- **a** Calculate the total energy of the particles before the collision.
- **b** Calculate the energy of each gamma ray after the collision.
- **c** Show that charge is conserved during the collision.

10 Two neutrons each with 145.0000 MeV of kinetic energy collide and produce two negative pions (π^-), with a mass–energy equivalence of 139.57039 MeV each, and two other identical particles. The four particles produced each continue with the same kinetic energy of 2.072649 MeV. Use the data in the table below to answer the following questions.

Particle	Mass (kg)
proton/antiproton	1.672622×10^{-27}
neutron/antineutron	1.674922×10^{-27}
electron/positron	9.109384×10^{-31}

- **a** Calculate the total energy of the particles in MeV before the collision.
- **b** Determine the total energy of the known particles present after the collision and hence identify the particles that the two neutrons are transformed into.
- **c** Show that charge is conserved during the collision.

Chapter review

10

KEY TERMS

aether
annihilation
classical physics
frame of reference
Gedanken
inertial frame of reference
length contraction
Lorentz factor
mass–energy equivalence
medium
nuclear fusion
pair production
postulate
proper length
proper time
simultaneous
spacetime
time dilation

REVIEW QUESTIONS

Knowledge and understanding

1 Which of the following are inertial frames of reference? More than one answer is possible.
- **A** an aircraft in steady flight
- **B** an aircraft taking off
- **C** a car turning a corner
- **D** a car driving up a hill of constant slope at a steady velocity

2 Where on the Earth's surface are we closest to an inertial frame of reference?

3 A passenger in a train carriage travelling at a constant velocity sees a ray of light go from a globe located in the centre of the carriage towards the front wall of the carriage. At the same time the passenger observes a ray of light go from the globe towards the back wall of the carriage. The passenger sees the two rays of light strike the walls at the same time. Describe what an observer on a track beside the railway line would see. Explain why they would see what they do.

4 In order to resolve an apparent conflict resulting from his two postulates, Einstein rejected some of Newton's assumptions. Which of the following statements is a consequence of this rejection?
- **A** Time is not constant in all frames of reference.
- **B** Absolute, true and mathematical time, of itself, and from its own nature, flows equably without relation to anything external.
- **C** One second in any inertial frame of reference is the same as one second in any other inertial frame of reference.
- **D** Space and time are independent of each other.

5 In 1905 Einstein put forward two postulates. Which two of the following best summarises them?
- **A** All observers will find the speed of light to be the same.
- **B** In the absence of a force, motion continues with constant velocity.
- **C** There is no way to detect an absolute zero velocity.
- **D** Absolute velocity can only be measured relative to the aether.

6 Which of the following is closest to Einstein's first postulate?
- **A** Light always travels at approximately $3.0 \times 10^8\,m\,s^{-1}$.
- **B** There is no way to tell how fast you are going unless you can see what is around you.
- **C** Velocity can only be measured relative to something else.
- **D** Absolute velocity is that measured with respect to the Sun.

7 Which one or more of the following conditions is sufficient to ensure that we will measure the proper time between two events?
- **A** We must be in the same frame of reference as the events.
- **B** We must be in a frame of reference which is travelling at the same velocity.
- **C** We must be stationary.
- **D** We must not be accelerating with respect to the frame of reference of the two events.

8 If you were riding in a very smooth, quiet train with the blinds drawn, how could you tell the difference between the train (i) being stopped at the station, (ii) accelerating away from a station and (iii) travelling at a constant speed?

9 Explain the term 'four-dimensional spacetime'.

10 Spaceships A and B leave the Earth and travel towards Vega, both at a speed of 0.9*c*. An observer on the Earth sees the crews of A and B moving in slow motion. Describe how the crew of A see the crew of B moving, and how they see people on the Earth moving.

11 You are in a spaceship travelling at a very high speed past a new colony on Mars. Do you notice time going slowly for you (e.g. do you find your heart rate is slower than normal)? Do the people on Mars appear to be moving normally? Explain your answers.

12 Imagine that Amaya is at the front of a train carriage that is moving at $32.0\,m\,s^{-1}$. She shines a torch towards Binh who is at the other end of the carriage. Clare is watching this from the side of the track. At what velocity does Clare see the light travelling?

Application and analysis

13 An observer sitting in a very fast *Gedanken* jet plane is looking out the window at a clock placed on top of a mountain. The passenger, using the mountain's clock, notes that it takes a goat 20.0s to run along a rocky ridge. If the plane is flying at $2.00 \times 10^8\,m\,s^{-1}$, calculate how much time has passed on the passenger's clock during the goat's run.

14 A spectator is standing next to a pool watching the clock on the wall as a swimmer swims at $2.12 \times 10^8 \text{ m s}^{-1}$. The spectator notices that on the pool clock the swimmer completes one length of the pool in 21.5 s.

a Calculate how much time the spectator sees pass on the swimmer's wristwatch.

b The swimmer keeps swimming laps until her own wristwatch shows that 21.5 s have passed. How much time does the swimmer see has passed on the pool clock at this moment?

15 a At what speed, in terms of c, would a rocket ship be going if it is observed to be half its normal length?

b The rocket ship is then observed to accelerate until its length is halved again (i.e. it is now one-quarter of its original length). To what speed did it accelerate relative to c?

16 Amaya and Binh are playing a game of ten-pin bowling in their spaceship as they fly by Clare in her space station at a relative speed of 238000 km s^{-1}. Binh uses his watch to observe that it takes 4.00 s for the ball to leave his hand and strike the pins. Their bowling alley is aligned in the direction of their spaceship's motion and is 18.3 m long and 1.07 m wide.

a Calculate the time it takes to bowl the ball and hit the pins as measured by Clare.

b Calculate the length and width of the bowling alley as measured by Clare.

17 Barnards Star is 5.96 light-years from the Earth. Space adventurer Amelia heads from the Earth towards Barnards Star at 0.900c.

a For those watching from the Earth, how many years will it take Amelia to reach Barnards Star?

b From Amelia's perspective how long will it take her to reach Barnards Star?

c Explain why it is that, although Amelia knows that Barnards Star is 5.96 light-years from the Earth and that she travelled at 0.900c, it took much less time than she might have expected to reach the star.

18 A lunar orbiter is orbiting the Moon at a speed of 0.885c. As it travels from west to east it takes a photograph of the Longomontanus crater, which is 145 km wide and 4.50 km deep in the frame of reference of the Moon. Because of its speed, the camera will see everything on the Moon slightly contracted.

a How much less than 145 km wide will Longomontanus appear to be in the photograph?

b Will the depth of Longomontanus be less as well? Explain your answer.

19 From our position on the Earth we watch a deep-space explorer travelling at 99.7% of the speed of light on the way to Ross 128, a dim red dwarf star. We see that the explorer's clock has slowed down.

a By what factor has the explorer's clock slowed down?

b Does this mean that the explorer would experience this slowing down of time?

c Ross 128 is 10.89 light-years from the Earth. Thus in our frame of reference on the Earth it takes light from the star 10.89 years to reach us, and vice versa. How long will it take the explorer to reach Ross 128?

d How long will the explorer observe the journey to Ross 128 to take based on an on-board clock?

e Does your answer to part **d** imply that the explorer is able to get to Ross 128 in less time than it does for light? Explain your answer.

20 Muons are high-speed particles that are created 15.0 km above the Earth's surface. Classical physics dictates that, due to their short life spans, muons should not reach the Earth's surface even though they travel at very high speeds (approximately 0.992c). However, they do reach the Earth's surface. Explain how this is possible, referring to both the frame of reference of an observer on the Earth and the frame of reference of a muon.

21 A *Gedanken* sparrow has a rest mass of 30.0 g and a speed of 0.925c.

a Calculate the relativistic kinetic energy of the sparrow.

b Calculate the kinetic energy of the sparrow according to classical physics.

22 One of the fusion reactions occurring in the Sun fuses two helium-3 nuclei to form helium-4 and two protons.

$$^{3}_{2}\text{He} + ^{3}_{2}\text{He} \rightarrow ^{4}_{2}\text{He} + ^{1}_{1}\text{p}^{+} + ^{1}_{1}\text{p}^{+}$$

Calculate the energy in MeV of one fusion reaction of this type given the following data.

Particle	Mass (kg)
proton	1.672622×10^{-27}
helium-3	5.00823×10^{-27}
helium-4	6.64648×10^{-27}

23 Two neutrons each with 155.0 MeV of kinetic energy collide to produce a negative pion (π^-) with a mass–energy equivalence of 139.6 MeV. One of the two neutrons remains intact after the collision, while the other neutron is transformed into a proton.

a Calculate the kinetic energy available to the particles after the collision.

b Show that charge is conserved during the collision.

24 An electron and a positron each with 42.0000 MeV of kinetic energy collide and are completely annihilated, producing two gamma rays of equal energy. In answering the following questions, assume that the mass of an electron is the same as the mass of a positron: $9.10938356 \times 10^{-31}$ kg.

a Calculate the total energy of the particles before the collision.

b Calculate the energy of each gamma ray after the annihilation.

c Show that charge is conserved during the collision.

OA ✓✓

UNIT 4 • Area of Study 1

REVIEW QUESTIONS

How has understanding about the physical world changed?

Multiple-choice questions

1 A gamma ray is composed of electric and magnetic fields which are:

A perpendicular to each other and to the direction of motion.

B perpendicular to each other and parallel to the direction of motion.

C parallel to each other and perpendicular to the direction of motion.

D parallel to each other and to the direction of motion.

2 Which of the following images show emission line spectra? Select all that apply.

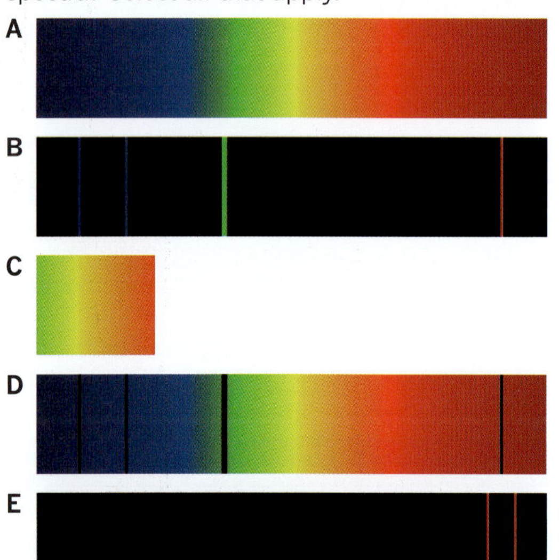

3 For an electron and a proton to have the same wavelength:

A the electron must have the same energy as the proton.

B the electron must have the same speed as the proton.

C the electron must have the same momentum as the proton.

D is impossible.

4 Which of the following has the longest wavelength?

A an electron with a mass of 9.1×10^{-31} kg travelling at 7.5×10^{6} m s^{-1}

B UV light of wavelength 150 nm

C X-rays of frequency 3.0×10^{17} Hz

D a proton with momentum 1.7×10^{-21} kg m s^{-1}

The following information relates to questions 5–8.

In an experiment on the photoelectric effect, light passing through a yellow filter is incident on a metal plate, as shown in diagram (a) below. The potential difference can be altered using the variable voltage source. At the stopping voltage, V_0, the photocurrent is zero. The current in the circuit is plotted as a function of the applied voltage in diagram (b).

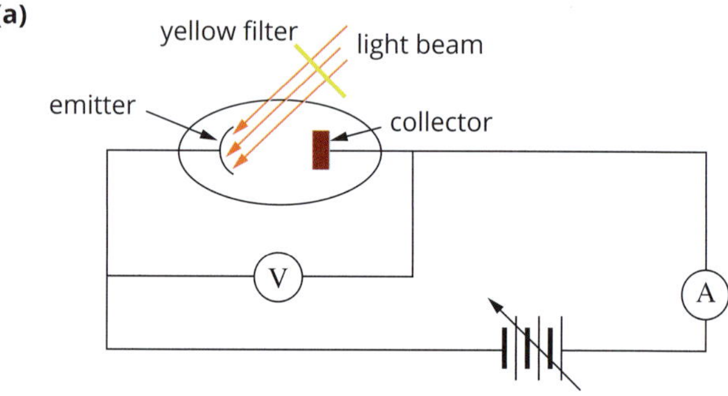

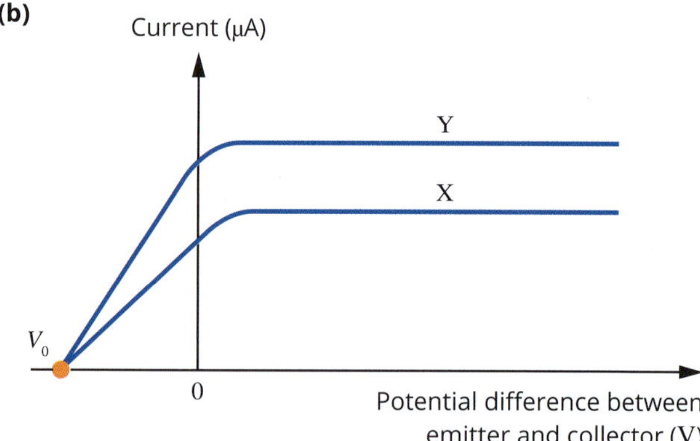

5 Which of the following changes would result in an increase in V_0?

A replacing the yellow filter with a red filter

B replacing the yellow filter with a blue filter

C increasing the intensity of the yellow light

D decreasing the intensity of the yellow light

6 Which one of the following options best describes why there is zero current in the circuit when the applied voltage equals the stopping voltage?
   A The threshold frequency of the emitter increases to a value higher than the frequency of yellow light.
   B The work function of the emitter is increased to a value higher than the energy of a photon of yellow light.
   C The emitted photoelectrons do not have enough kinetic energy to reach the collector.
   D There is a gap in the circuit.

7 Which of the following descriptions of graphs X and Y in diagram (b) are correct?
   A Both graphs are produced by yellow light of different intensities.
   B Both graphs have the same intensity but different stopping voltages.
   C Graph X is produced by yellow light and graph Y is produced by blue light.
   D Each graph is produced by light of a different colour and different intensity.

8 The emitter of the photocell is coated with nickel. The filter is removed and light of wavelength 200 nm is directed onto the emitter. The stopping voltage, V_0, that will result in zero current in the circuit is 1.21 V. What is the work function of nickel?
   A 0 eV
   B 1.21 eV
   C 5.0 eV
   D 6.2 eV

9 In 1905 Einstein put forward two postulates of special relativity. Which two of the following best summarises them?
   A All observers will find the speed of light to be the same.
   B In the absence of a force, motion continues with constant velocity.
   C There is no way to detect an absolute zero of velocity.
   D Absolute velocity can only be measured relative to the aether.

10 From the options below, select the one that best describes inertial frames of reference.
   A Inertial frames move at a constant velocity.
   B Inertial frames accelerate at a constant rate.
   C Inertial frames change velocity at a constant rate.
   D Inertial frames do not move relative to one another.

11 One of the fastest objects made on the Earth is the Galileo probe. In 1995, under the influence of Jupiter's huge gravity, it entered Jupiter's atmosphere at nearly 50 000 m s^{-1}. Which of the following is the best estimate of the Lorentz factor for the probe?
   A Less than 1
   B 1.000 000 00
   C 1.000 000 01
   D 1.1

12 A sphere is approaching an observer at a speed of 0.8c.

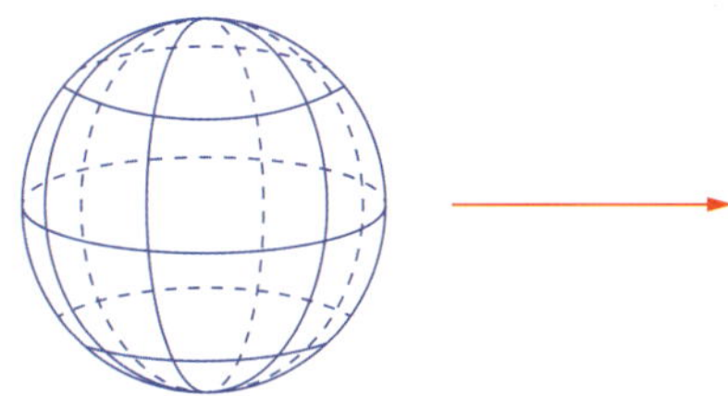

From the figures below, identify which shape would be seen by the observer.

A

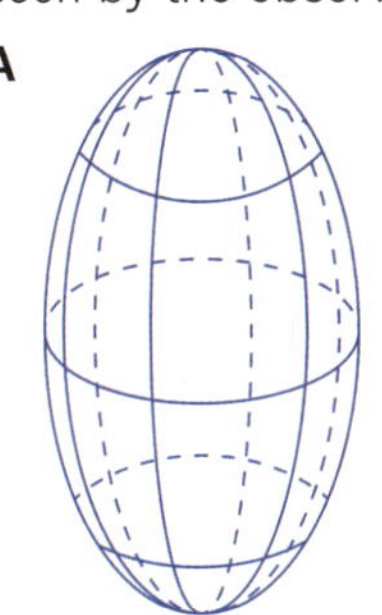

B

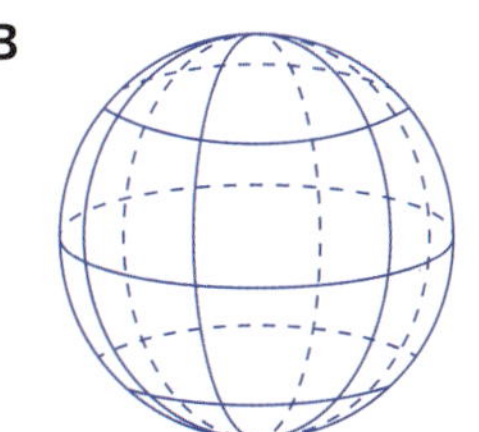

C

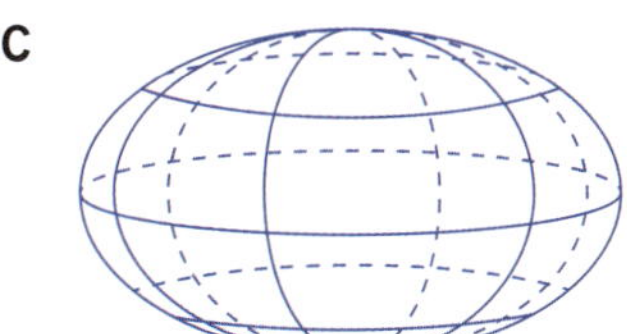

D

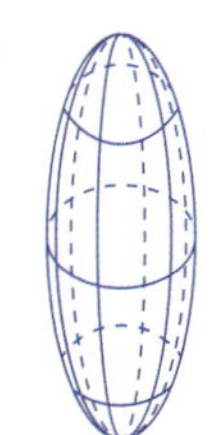

13 A star is losing mass at the rate of 4.0×10^9 kg s^{-1}. How much energy in radiation is being produced per second?

A 3.6×10^{26} J
B 3.7×10^{26} J
C 3.8×10^{26} J
D 3.9×10^{26} J

14 Which of the following is not one of the findings of the Michelson–Morley experiment?

A There is no evidence of an aether.
B The speed of light is constant independent of the motion of the Earth.
C The speed of light depends on the direction of the Earth's travel through the aether.
D There is support for Einstein's theory of special relativity.

15 Assume you are travelling at half the speed of light (0.5c) and approaching a beacon that is emitting light. What would you measure the speed of light from the beacon to be?

A 1.0c
B 1.5c
C 0.50c
D 0.75c

Short-answer questions

16 Describe what happens when two or more waves meet while travelling through the same medium. Comment on what happens to the shape and speed of the waves at the instant they meet, and then describe what happens after the interaction has passed.

17 The diagram below shows the standing wave pattern made by a rope when it is oscillated back and forth. In this case, each end of the rope can be considered to be a fixed end. The tension in the rope is constant.

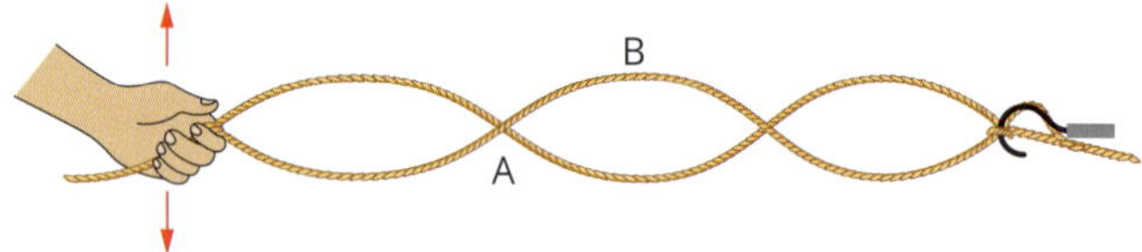

a Identify the names given to points A and B.
b State why there are two images of the rope.
c Identify the harmonic shown in the diagram.
d Explain how a standing wave is formed.
e If the wave has a speed of 2.0 m s^{-1} and the distance from the hand to the hook is 80.0 cm, determine the frequency of the wave shown.

18 Young's double slit experiment is performed using red light and an interference pattern is observed. Explain the effect on the interference pattern of:

a halving the separation of the slits
b doubling the distance between the slits and the screen
c doubling the brightness of the light
d swapping the red light for blue light
e covering one of the slits.

19 At the time when Young carried out his famous double slit experiment, there were two competing models for explaining the nature of light.

a State the names of the two competing models.
b Explain how Young's experiment supported one of these models and not the other.

20 The image below shows the wave fronts from two sources of waves. The red lines represent wave crests.

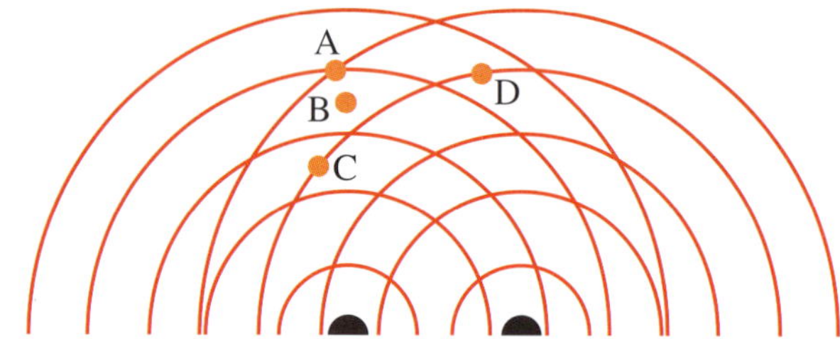

At which of the positions marked A, B, C and D does maximum constructive interference occur? Explain your answer.

21 Physicists replicating Young's double slit experiment observe that adjacent dark fringes on the interference pattern (for example, the third and fourth fringe) differ in distance from one of the slits by only 500 nm. Determine the wavelength of the monochromatic light being used.

22 Describe three experimental results associated with the photoelectric effect that cannot be explained by the wave model of light.

23 In a double slit experiment, an electron beam travels through two slits 20 mm apart in a piece of copper foil. A pattern is detected photographically at a distance of 2.0 m. The speed of the electrons is 0.1% of the speed of light.

a State what you expect to see on the photographic plate. What is this type of pattern called?
b State what the pattern on the photographic plate suggests about the behaviour of the electrons.
c Calculate the de Broglie wavelength of the electrons used in the experiment.
d If the experiment were repeated using neutrons, which have a mass of 1.67×10^{-27} kg, calculate:
  i the speed a neutron needs to travel to have the same de Broglie wavelength as the electrons in the original experiment
  ii the speed a neutron needs to travel to produce the same pattern on the photographic plate as that produced by the electrons in the original experiment.

24 The energy levels for atomic mercury are as follows.

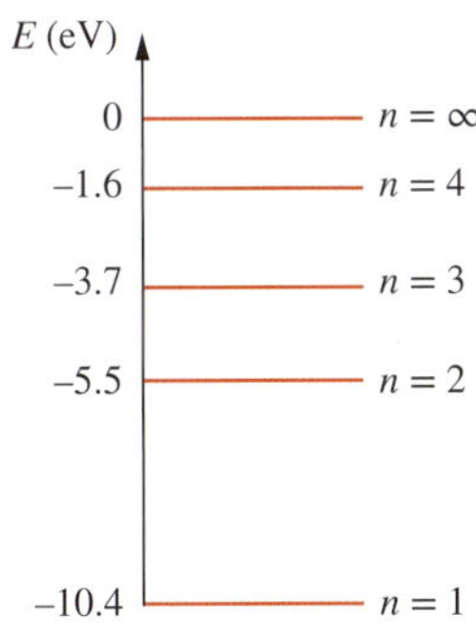

a Determine the energy and frequency of the light emitted when an atom of mercury makes the following transitions:

i $n = 4$ to $n = 1$

ii $n = 2$ to $n = 1$

iii $n = 4$ to $n = 3$

b Copy the diagram above and draw on it the transition that would emit a photon with the highest frequency.

c Copy the diagram above and draw on it the transition that would emit a photon with the largest wavelength.

An electron beam with energy 7.0 eV passes through mercury vapour in the ground state and excites the electrons in the vapour to the $n = 3$ energy level.

d List all the possible photon energies that would be present in the emission spectrum when the electrons return from the $n = 3$ energy level to the ground state.

25 An electron is accelerated across a potential difference of 65 V.

a Determine the maximum speed the electron will reach.

b What is the de Broglie wavelength of the electron?

26 How did Bohr explain that when the frequency of light is below a certain value, it passes straight through a sample of hydrogen gas without any absorption occurring?

27 Physicists can investigate the spacing of atoms in a sample of powdered crystal using electron diffraction. This involves accelerating electrons to known speeds using an accelerating voltage. In a particular experiment, electrons of mass 9.1×10^{-31} kg are accelerated to a speed of 1.75×10^{7} m s^{-1}. The electrons pass through a powdered crystal sample and the diffraction pattern is observed on a fluorescent screen.

a Calculate the de Broglie wavelength (in nm) of the accelerated electrons.

b Describe and explain the main features of the diffraction pattern.

c If the accelerating voltage is increased, what difference would you expect to see in the diffraction pattern produced? Explain your answer.

28 Discuss and compare the wave–particle duality of light and of matter.

29 When conducting an experiment to investigate the photoelectric effect, a student correctly observes that the energy of emitted electrons depends only on the frequency of the incident light and is independent of the intensity.

a Explain how the particle model accounts for this observation.

b Explain why the wave model cannot account for this observation.

30 In the photoelectric effect, the relationship between the maximum kinetic energy of emitted photoelectrons and the frequency of the light incident on the metal plate is $E_{k\ max} = hf - \phi$.

a Explain the meaning of the terms $E_{k\ max}$, f and ϕ in this equation.

b If the intensity of the light striking the metal is increased but the frequency remains the same, what effect does this have on the value of $E_{k\ max}$?

c If the intensity of the light striking the metal is increased but the frequency remains the same, what effect does this have on the value of the photocurrent?

31 The image below shows diffraction images that have been obtained by scattering X-rays and electrons off the same sample, which is made up of many tiny crystals with random orientation. The X-rays have a frequency of 8.3×10^{18} Hz.

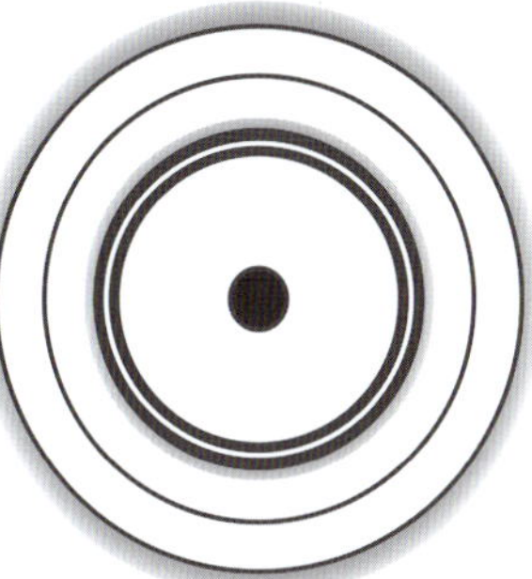

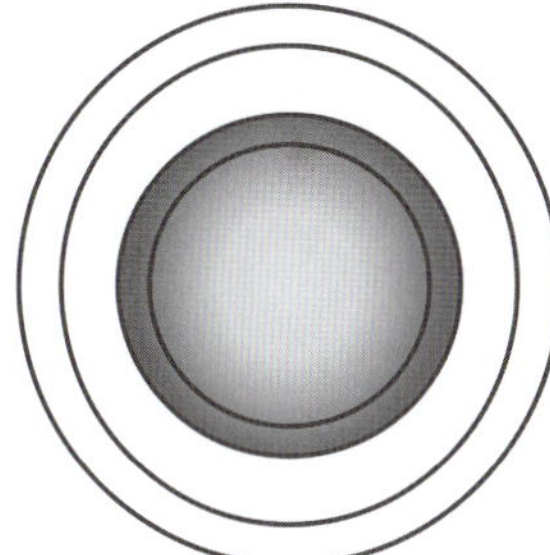

a The X-rays and electrons have produced the same diffraction pattern. Explain why this is the case, and what this says about their respective wavelengths.

b Calculate the momentum of the electrons.

c Do the X-rays and the electrons have the same energy? Explain your answer.

d With regards to diffraction, explain why an electron microscope is preferable to an optical microscope for observing extremely small structures.

32 Xquar is a star 5 light-years from the Earth. Space adventurer Raqu heads from the Earth towards Xquar at a speed of 0.9*c*.

- **a** For those watching from the Earth, how long does it take Raqu to reach Xquar?
- **b** From Raqu's point of view, how long does it take her to reach Xquar?
- **c** Explain why it is that, although Raqu knew that Xquar was 5 light-years from the Earth and she was travelling at 0.9*c*, it took her much less time than she might have expected to reach Xquar.

33 The fusion reaction that powers the Sun combines four protons (each with a rest mass of 1.673×10^{-27} kg) to form a helium nucleus of two protons and two neutrons (with a total rest mass of 6.645×10^{-27} kg). The total power output of the Sun is 3.9×10^{26} W.

- **a** How much energy is released by each fusion of four protons to form a helium nucleus?
- **b** How many helium nuclei are being formed every second in the Sun?
- **c** How much mass is the Sun losing every day?

Answers

The answers to questions that involve calculations are given to the least number of significant figures as given in the question.

See page 24 in Chapter 1 for more details.

Chapter 1 Scientific investigation

1.1 Planning scientific investigations

1 a testable prediction that proposes an answer to a research question

2 the values manipulated, observed or kept constant during research

3 Quantitative data involves numeric variables that can be measured; qualitative data can be described, categorised or counted but not measured.

4 (I) state the research question to be investigated, (II) form a hypothesis, (III) plan the experiment and equipment, (IV) perform the experiment, (V) collect the results, (VI) question whether the results support the hypothesis, (VII) draw conclusions

5 grip of the bat

6 dependent variable: flight displacement (range); independent variable: release angle; controlled variables: release velocity, release height, landing height and air resistance

7 **a** A **b** D **c** B

8 **a** independent: surface area of the container; dependent: time to cool to room temperature
b independent: thickness of foam bumper; dependent: force
c independent: number of coils; dependent: torque
d independent: applied force; dependent: acceleration
e independent: position of diver; dependent: speed of rotation

9 qualitative

10 **a** dependent **b** controlled **c** independent

11 A, because it specifies what is being measured, how it is measured and how the dependent variable is expected to relate to the independent variable.

12 C, because it tests only one independent variable and it predicts the relationship between the independent and dependent variables.

1.2 Conducting investigations

1 **a** systematic error **b** random error

2 Accuracy is the ability to obtain the correct measurement. Precision refers to how close two or more measurements are to each other.

3 The choice of equipment and instruments will influence the repeatability of the experiment.

4 Student answers will vary. One way to ensure validity is to test only one variable at a time. One way to ensure repeatability is to take repeat readings or measurements.

5 **a** mistake **b** random error
c systematic error **d** systematic error
e mistake **f** random error

6 elimination, substitution, engineering controls, administrative controls, personal protective equipment

7 Student answers will vary. For example, an SDS provides important safety and first-aid information.

8 Student answers will vary. For example, as the chemical is corrosive, any skin that has come into contact with it should be thoroughly washed.

9 Student answers will vary. An example is provided in the fully worked solutions.

1.3 Data collection and quality

1 Raw data: data you record in your logbook; processed data: raw data that has been organised, altered or analysed to produce meaningful information.

2 **a** 22 (2 significant figures) **b** 19
c 21

3 **a** if the voltage is measured in units of the number of batteries
b if the voltage is measured with a voltmeter

4 2500 μm

5 two

6 **a** 2.55×10^5 **b** 4.32×10^{-7}
c It allows very large and very small numbers to be written quickly and compactly. It can also aid with providing results to the appropriate number of significant figures.

7 **a** 6.63×10^{-34} J s **b** 1.78×10^{-3} MeV

8 Student answers will vary. Examples are provided in the fully worked solutions.

1.4 Data analysis and presentation

TY 1.4.1 9.9 m s^{-2}

TY 1.4.2 0.04 W

CSA: The Hubble constant

1

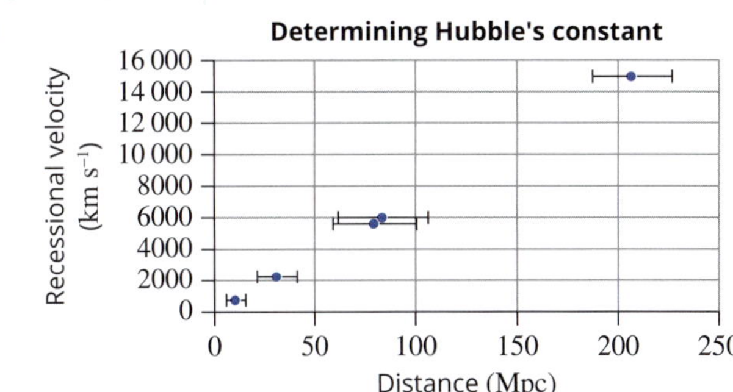

2

Determining Hubble's constant

Recessional velocity (km s^{-1}); Distance (Mpc)

$y = 72.563x - 73.211$

$y = 82.965x - 514.48$

$y = 64.279x + 408.6$

maximum gradient; line of best fit; minimum gradient

3 72 km s^{-1} Mpc^{-1} (to 2 significant figures)

4 9.3 km s^{-1} Mpc^{-1} (to 2 significant figures)

Key questions

1 **a** linear **b** parabolic or square
c hyperbolic or inverse

2 linear

3 form of systematic error resulting from the researcher's personal preferences or motivations (e.g. poor definitions of concepts or variables, incorrect assumptions, and errors in methodology)

4 a data point that significantly differs from the trend

5

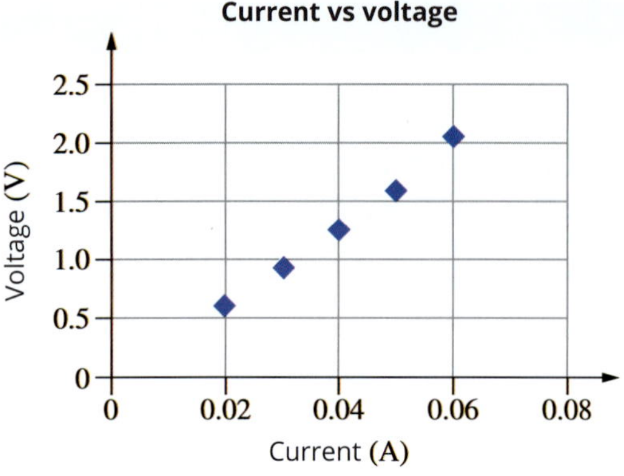

6 extension = 15 × mass

7 a Student answers will vary. One example is that the heading is non-descriptive. Other examples are given in the fully worked solutions.

b Student answers will vary. See the fully worked solutions for examples.

8 a Formula 1: P on the y-axis and T^4 on the x-axis; formula 2: a on the y-axis and T^{-2} on the x-axis will produce a straight-line graph

b Formula 1: a y-intercept of zero and gradient of $A\sigma$; formula 2: a y-intercept of zero and gradient of $4\pi r$.

9 a

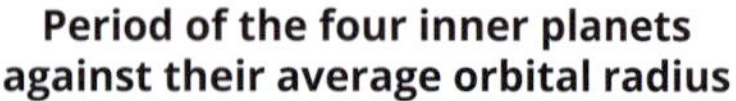

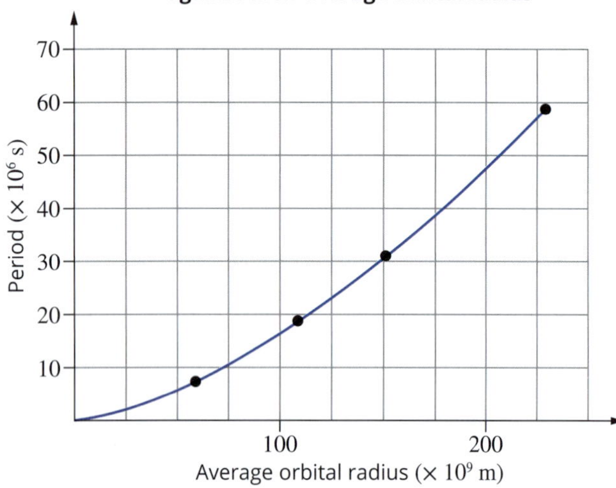

b Make the new y-axis T^2 and the new x-axis r^3.

c

Period squared against average orbital radius cubed for the four inner planets

Period squared (× 10^{12} s²): 0, 500, 1000, 1500, 2000, 2500, 3000, 3500, 4000

$y = 0.2978x - 3.2194$

Average orbital radius cubed (× 10^{30} m³): 0, 4000, 8000, 12 000

d $2.978 \times 10^{-19}\,s^2\,m^{-3}$

e 1.99×10^{30} kg

1.5 Conclusion and evaluation

1 An increase in current from 0.03 A to 0.05 A produced an increase of 0.88 V across the resistor.

2 B

3 The results show an inverse relationship: as the mass increases, the acceleration decreases. This supports the hypothesis.

4 The results show different drop times for objects of different mass or surface area when they are dropped from the same height. The experiment did not find a correlation between mass and time taken to fall.

5 Repeating an experiment means that you can average the results, thereby minimising random errors.

6 an evaluation of the method; an analysis of the data; an explanation of the link between the findings and relevant concepts in physics

7 D. A is incorrect as it is a generalisation to all springs. B is incorrect as it is not a conclusion. C is incorrect as it is an extrapolation.

1.6 Reporting investigations

1 B 2 D

3 results

4 to give credit to others where it is due, avoid plagiarism and direct the reader to further information

5 bibliography: all sources consulted during the research; reference list: only those sources cited in the report

6 a repeatability (i.e. reliability)

b validity and repeatability

c accuracy

d precision

Chapter 2 Newtonian theories of motion

2.1 Newton's laws of motion

TY 2.1.1 a 0 N

b 6.0 N 30° above the horizontal

c 6.0 N 30° below the horizontal.

TY 2.1.2 a 1.1×10^4 N in the direction of motion

b 2.2×10^3 N

TY 2.1.3 a 2.9×10^2 N

b 7.8×10^2 N

c $3.4\,m\,s^{-1}$ down the slope

Key questions

1 No. Phil's inertia made him remain stationary as the tram moved forward. This made it look like Phil was thrown backwards relative to the tram.

2 12 N

3 (a)

Force exerted on racquet by ball

(b)

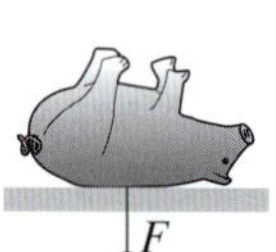

Force exerted on ground by pig

(c)

Force exerted on ground by wardrobe

(d)

Gravitational force of attraction that seal exerts on Earth

4 9.8×10^{-2} N upwards

5 a 45.0 N b 165 N

6 a 1.5 m s^{-2}
b 1.2×10^2 N forwards c 60 N

7 a 0 b 66 N
c 66 N on Matt

8 9.2 N

9 a 1840 N b 580 N

10 a A b C
c 490 N up the hill d 4.9 m s^{-2}
e If there is no friction, acceleration is not affected by mass.

2.2 Circular motion in a horizontal plane

TY 2.2.1 7.5 km h^{-1}

TY 2.2.2 a 521 m s^{-2}
b 3.6×10^3 N

TY 2.2.3 a 1.53 m
b

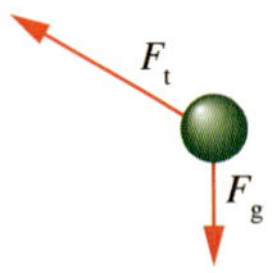

c 2.34 N towards the centre of the circular path
d 3.05 N

Key questions

1 A and D

2 a 8.0 m s^{-1} b 8.0 m s^{-1} south
c 7.0 m s^{-2} west

3 8.4×10^3 N west

4 a 8.0 m s^{-1} north b east

5 The force needed to give the car a larger centripetal acceleration could eventually exceed the maximum frictional force acting between the tyres and the road surface. At this time, the car would skid out of its circular path.

6 a 0.90 m s^{-2}
b The skater has an acceleration, so forces are unbalanced.
c 67.5 N

7 a 0.40 s b 19 m s^{-1}
c 3.0×10^2 m s^{-2} d 4.5×10^2 N

8 a 1.20 m
b the force due to gravity acting vertically downwards and the tension in the rope acting along the rope towards the top of the maypole
c towards B
d 170 N towards B
e 2.61 m s^{-1}

9 a 3.1×10^4 N
b The friction between the road surface and the car's tyres provides the centripetal force.
c The inertia of the passengers causes them to continue to move in a straight path while the car makes the turn. The passengers exert a force outwards on the side of the car as it moves around the turn.
d If there is any ice or oil on the road, the friction between the road and the car's tyres will be reduced. This will reduce the centripetal force on the car. As a result, the car's inertia will cause the car to continue to move in a straight line at a tangent to the circle from the point where the car hits the oil or ice on the road.

2.3 Circular motion on banked tracks

TY 2.3.1 a 5.9×10^2 N towards the centre of the circle
b 17 m s^{-1}

TY 2.3.2 7.3°

Key questions

1 horizontally towards the centre of the circle

2

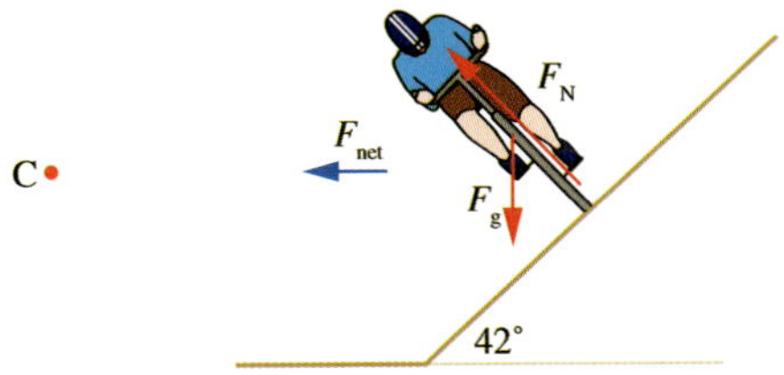

3 45°

4 a 46 km h^{-1} b 8.1×10^2 N
c 735 N, which is less than the normal force

5 48° 6 8.1°

7 increase the banking angle or increase the track's radius

2.4 Circular motion in a vertical plane

TY 2.4.1 a 4.9 m s^{-1}
b 16 N up
c 3.7 m s^{-1}
d 6.74 N down

Key questions

1 a constant b at the bottom
c at the top d at the bottom
e 3.5 m s^{-1}

2 a force due to gravity; the normal force from the road
b 1.3×10^3 N
c Yes. When the driver is driving over a hump, the normal force is less than their force due to gravity (*mg*). Their apparent force due to gravity is given by the normal force that is acting, and so the driver feels lighter at this point.
d 36 km h^{-1}

3 a 37 m s^{-1} b 27 m s^{-1}
c 2.3×10^4 N down d 13 m s^{-1}

4 2.0×10^2 N down

5 a 16 m s^{-2} up b 1.9×10^3 N up

6 188 m s^{-1}

7 a 9.8 m s^{-2} down b 2.0 m s^{-1}

8 a 9.6×10^3 N down b 5.1×10^3 N up
c 9.9 m s^{-1}

2.5 Projectiles launched horizontally

TY 2.5.1 a 2.5 s
b 49.4 m
c 31 m s^{-1} at 50° below the horizontal

Key questions

1 a 0.59 s b 3.2 m
c 9.8 m s^{-2} down

2 a 0.71 s b 0.71 s
c 3.2 m

3 a 76 m s^{-1} b 23°

4 a 6.5 m s^{-1} forwards b 4.4 m s^{-1} down
c 7.8 m s^{-1} at 34° below the horizontal
d 0.45 s
e 2.9 m
f

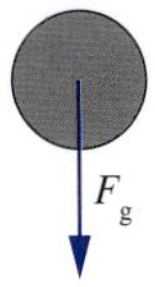

5 26 m s^{-1}

2.6 Projectiles launched obliquely

TY 2.6.1 **a** 6.11 $m s^{-1}$ horizontally to the right
b 0.25 m
c 0.45 s

CSA: The physics of shot putting

1 6.5 $m s^{-1}$ **2** 3.8 $m s^{-1}$
3 0.38 s **4** 2.3 m
5 6.5 $m s^{-1}$ **6** 7.0 m

Key questions

1 The horizontal velocity remains constant throughout the javelin's flight (ignoring the effect of air resistance).

2 Ollie. The optimal launch angle to give the greatest range for any ideal projectile is 45° (if air resistance is ignored, and if the start and end point are at the same height).

3 19 $m s^{-1}$

4 **a** 22 $m s^{-1}$ **b** 13 $m s^{-1}$
c 9.8 $m s^{-2}$ down **d** 22 $m s^{-1}$

5 **a** 11 $m s^{-1}$ **b** 5.1 $m s^{-1}$
c 0.52 s **d** 3.1 m
e 11 $m s^{-1}$ **f** 14 m

6 **a** **i** 21.7 $m s^{-1}$ **ii** 21.7 $m s^{-1}$
iii 21.7 $m s^{-1}$
b **i** 3.82 $m s^{-1}$ **ii** 1.37 $m s^{-1}$
iii 1.08 $m s^{-1}$ down
c 21.7 $m s^{-1}$
d 21.7 $m s^{-1}$
e 16.9 m
f Air resistance is a force that acts in the opposite direction to the velocity of the ball, thereby producing a horizontal and vertical deceleration of the ball during its flight. Hence the maximum height of the ball is less than it would be with no air resistance and the range of flight is shorter.

Chapter 2 Review

1 The bowling ball is increasing in speed at a constant rate, i.e. with constant acceleration.

2 B

3 **a** 6.9 $m s^{-2}$ **b** $0.71F_g$

4 Emma. The forces acting on the water when the bucket is directly overhead are the force of gravity and the normal force from the base of the bucket on the water. Both of these forces act downwards. There is no separate outwards or centrifugal force.

5 **a** 6.5 $m s^{-2}$ **b** 33 N

6 **a** perpendicular = 8.5×10^2 N; parallel = 2.3×10^2 N
b 852 N
c 2.5 $m s^{-2}$ down the slope

7 **a** 5.0×10^4 N in the direction of motion
b 1.2×10^4 N

8 4.5×10^3 N

9 **a** 6.3 $m s^{-2}$ **b** 6.6 $m s^{-1}$

10 **a** 2.5×10^2 N
b 8.9 $m s^{-2}$ down the ramp
c 5.3×10^2 N down the ramp
d 9.4 $m s^{-1}$
e 5.3×10^2 N up the ramp

11 **a** 3.70 $m s^{-1}$
b 17.1 $m s^{-2}$ towards the centre of the circle
c 0.428 N
d

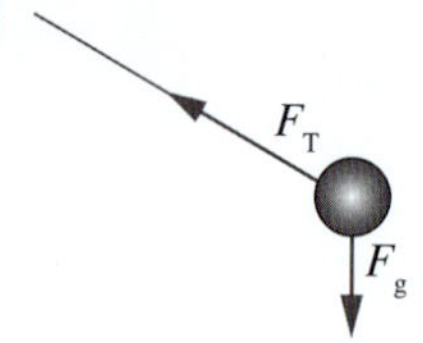

e 0.49 N

12 **a** 3.8 $m s^{-2}$ towards the centre of the circle
b The centripetal force is created by the friction between the tyres and the ground.

13 16 $m s^{-1}$

14 **a** **i** 365 N up
ii 615 N up
b D

15 **a** 5.6 m **b** 9.8 $m s^{-2}$ downwards

16 **a** 15.0 $m s^{-1}$ **b** 4.99 $m s^{-1}$
c 15.8 $m s^{-1}$

17 **a** 12.9 $m s^{-1}$ **b** 13.3 $m s^{-1}$
c 10.7 m

18 **a** 9.8 $m s^{-2}$
b

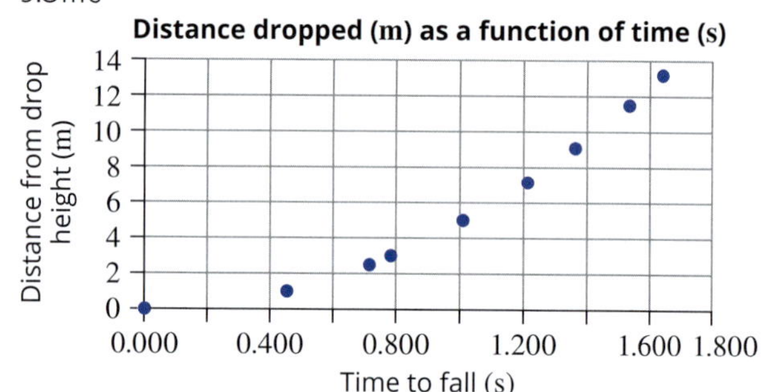

c 1.75 s

19 **a**

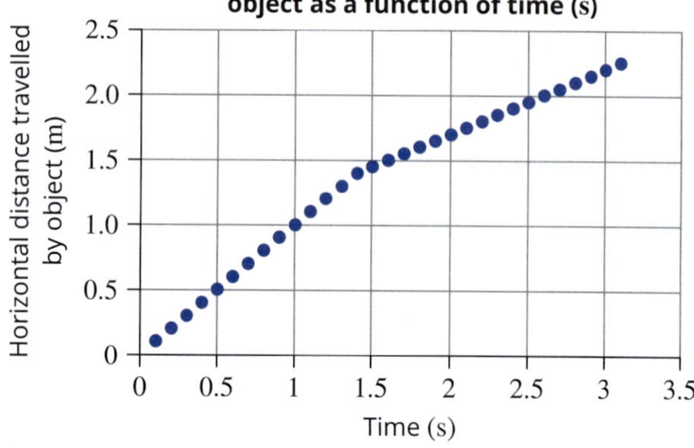

b See the 'Horizontal velocity' column in the corresponding table in the fully worked solutions.

c

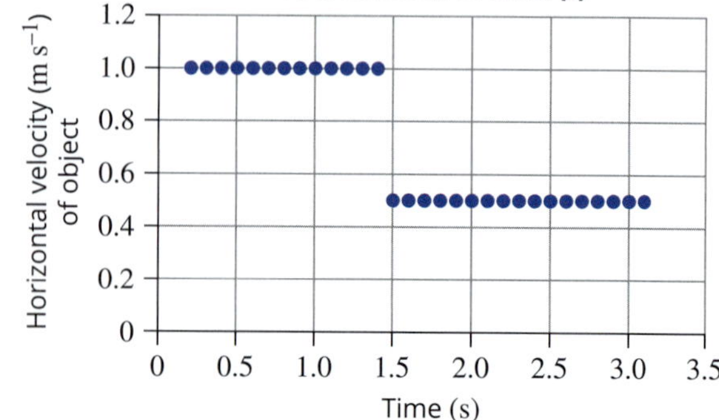

The velocity is constant at 1 m s^{-1} from initial launch of the object to the point where the object reaches its maximum height. The velocity is constant at 0.5 m s^{-1} from the object's maximum height to the point where it reaches the ground.

d See the 'Horizontal acceleration' column in the corresponding table in the fully worked solutions.

e

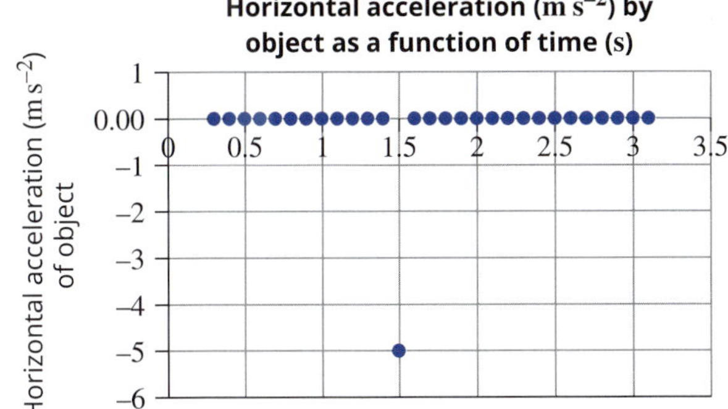

There is an outlier in the data and the effect of air resistance on the horizontal component is negligible as the acceleration is zero.

Chapter 3 The relationship between force, energy and mass

3.1 Conservation of momentum

TY 3.1.1 **a** 9.59 m s^{-1} west
b 3.79×10^4 kg m s^{-1} west
c 3.79×10^4 kg m s^{-1} east
d $\Delta p_c + \Delta p_b = (-3.79 \times 10^4) + (3.79 \times 10^4) = 0$

TY 3.1.2 **a** 1.06 10^{-3} kg m s^{-1} south
b 0.253 m s^{-1} north
c 1.12×10^{-3} kg m s^{-1} north
d 1.12×10^{-3} kg m s^{-1} south

TY 3.1.3 **a** 0 kg m s^{-1}
b 1.39 m s^{-1} south
c 136 kg m s^{-1} north
d 136 kg m s^{-1} south

Key questions

1 Before you step forward to get off the board, the sum of the initial momentum of you and the board is zero. Therefore the total momentum after you step forward must also zero. Hence your forward momentum must equal the board's backward momentum. But as the board's mass is probably less than yours, it will move rapidly backwards while you move slowly forward. However, a ferry's mass is much larger than yours, so its backwards velocity will be extremely small compared to your forward velocity.

2 As the two toy cars are moving in opposite directions, one of the cars will have a positive momentum and the other car will have a negative momentum. Therefore the sum of their initial momenta adds to zero to match their overall final momentum of zero.

3 2.83 m s^{-1} west **4** 5.13 m s^{-1} north

5 **a** **i** 1.00×10^4 kg m s^{-1} east
ii 1.00×10^4 kg m s^{-1} west
iii 0 kg m s^{-1}
b **i** 0 m s^{-1}
ii The momentum of each vehicle before the collision has combined and is conserved. The sum of the momentum before the collision becomes the sum of the momentum after the collision.
iii 1.00×10^4 kg m s^{-1} west
iv 1.00×10^4 kg m s^{-1} east

6 9.39 m s^{-1} to the right **7** 1.93 m s^{-1} west
8 0.762 m s^{-1} south **9** 5.05 m s^{-1} west
10 6.11 m s^{-1} away from the spaceship

3.2 Impulse

TY 3.2.1 3.93×10^4 N s south-west
TY 3.2.2 101 N s
TY 3.2.3 5.04×10^4 N

CSA: Car safety and crumple zones

1 1.50×10^3 kg m s^{-1} north
2 1.50×10^3 kg m s^{-1} south
3 The impulse of the car is the change in its momentum. There is no change if the time over which it stops is extended, as momentum is independent of time.
4 The impulses experienced by both drivers will be the same, as they have the same mass, initial velocity and final velocity.
5 1.52×10^3 N south **6** 1.84×10^4 N south
7 inverse relationship

Key questions

1 7.10 N s **2** 38.0 N **3** 2.14×10^3 N
4 **a** 1.39×10^6 kg m s^{-1}
b 1.39×10^6 N s
5 A, C, B
6 Student answers will vary. Any factor that increases the time of deceleration will decrease the force. Similarly, any factor that decreases the impulse will decrease the force. One example: wearing good runners will reduce the force, while being barefoot will increase the force. Other examples are provided in the fully worked solutions.
7 $22.4 \times F_A$
8 **a** 200 N s **b** 1.00×10^3 N
9 13 N

3.3 Work done

TY 3.3.1 **a** 106 J
b 46.1 J
c 60.0 J
TY 3.3.2 90 J

Key questions

1 Student answers will vary. An example: someone standing on a horizontal travelator holding a suitcase above the ground.
2 Work is only done if there is force in the same direction as the displacement. The force of gravity is at right angles to the motion of the Earth. Therefore the force and any displacement are perpendicular.
3 **a** 36.0 J **b** 24.0 J **c** 12.0 J
4 0.27 J **5** 1.8×10^3 J
6 510 J
7 **a**

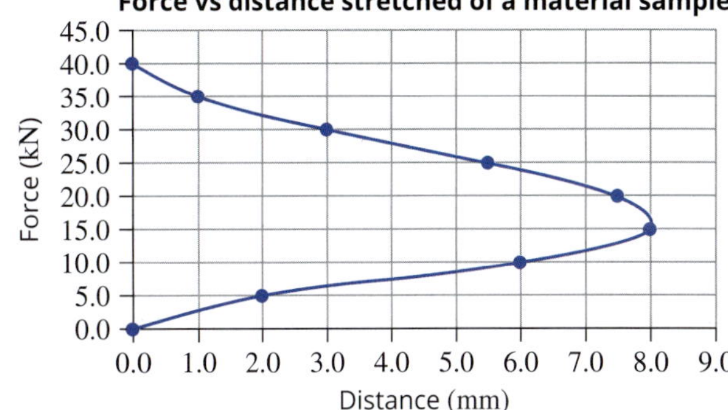

b 68 J
8 3.8×10^2 J

3.4 Elastic potential energy

TY 3.4.1 **a** 2.0×10^4 N m^{-1}
b 2.3 J
c 14.5 J

Key questions

1 C, B, A
2 **a** high **b** medium **c** low
3 **a** stiff = 200 N m^{-1}; weak = 50 N m^{-1}
b 3.0 J
4 0.0800 m
5 **a** 32.0 N **b** 4.00 J
6 **a**

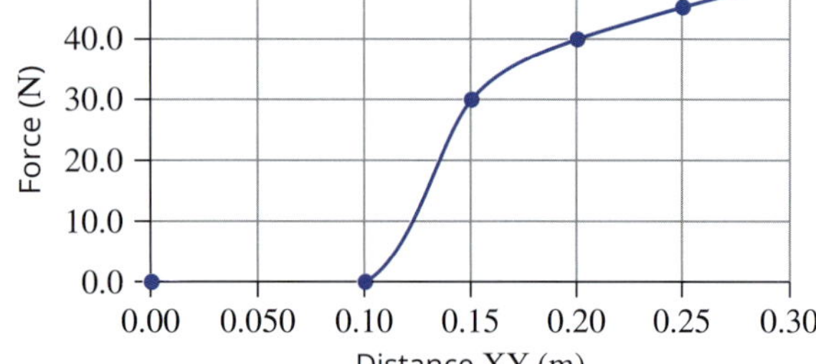

b 7 J
c 7 J
d No. Hooke's law is not obeyed, as the force vs distance graph is not a straight line between 10.0 cm and 30.0 cm.
e distance = 0.15 m; force = 30 N

3.5 Kinetic and potential energy

TY 3.5.1 inelastic
TY 3.5.2 4.5×10^8 J
TY 3.5.3 **a** 1.4×10^{10} J (approx.)
b 6.2 km s^{-1}

Key questions

1 A and E 2 A, C and D 3 2.5×10^6 J
4 12.8 J 5 1.3×10^3 J 6 B and D
7 **a** 1.6×10^7 J **b** 5.3×10^8 J **c** 1.5×10^3 m s^{-1}
8 2.5×10^{11} J

3.6 Conservation of energy

TY 3.6.1 50 m
TY 3.6.2 76 m s^{-1}

Key questions

1 D
2 **a** Both sticks land at the same time.
b brown stick
c Both sticks land with the same speed.
d brown stick
3 1.1×10^2 m 4 14 m s^{-1} 5 6.4 m s^{-1}
6 **a** 46.7 J **b** 46.7 J **c** 24 m
7 12 m s^{-1}
8 **a** 1.8 m s^{-1} **b** 1.0 m
c 4.5 m s^{-1} **d** 4.5 m s^{-1}
9 **a** 1.47×10^3 J **b** 1.47×10^3 J
c 2.0 m

Chapter 3 Review

1 B, C, A, D 2 C
3 345 kg m s^{-1} north 4 B and D
5 Impulse is the change in momentum of an object. Airbags increase the duration of a collision, which decreases the force on the driver.
6 The wall undergoes no displacement, so no work is done.
7 When an object has access to energy it has the *capacity* to do work, whereas work occurs when energy is transferred or transformed.
8 The gradient of an F–x graph is $\frac{\Delta F}{\Delta x}$ with unit $\frac{\text{N}}{\text{m}}$ = N m^{-1} = the unit for the spring constant. The area under the graph = $\Delta F \times \Delta x$ with unit N m = kg m s^{-2} = the unit for joule. Therefore the area represents the elastic potential energy.
9 D
10 No work is done on the backpack as it did not rise in the gravitational field to increase its gravitational potential energy. Further, it did not increase its kinetic energy as its speed was constant. As no energy was transferred to the backpack, no work was done on it.
11 45.5 J
12 The ball's kinetic energy is now stored in the gravitational field.
13 0.354 m s^{-1} in the opposite direction to the rower
14 3.00 m s^{-1} in the opposite direction to the gas
15 5.68 kg m s^{-1} 16 480 N
17 **a** 3.87×10^4 J **b** 13.1 m s^{-1}
18 3×10^3 J 19 0.032 J
20 5.73×10^4 J 21 3.81×10^5 J
22 **a** 4.6 J **b** 2.3 J
23 2.4×10^6 J
24 **a** yes
b inelastic. Hooke's law is not obeyed, as the force vs distance graph is not a straight line between 10.0 cm and 30.0 cm.
c 2.00 m s^{-1}
25 1.2×10^3 N m^{-1} 26 $4 \times h$
27 A
28 **a** 7.2×10^{10} J **b** 2.58×10^{11} J
29 6.2×10^{13} J
30 **a** 0.258 m s^{-1} **b** 0.0141 J **c** 0.0138 J
d **i** more than **ii** is not
iii is **iv** is
v not, kinetic energy

Unit 3 Area of Study 1 Review

1 C 2 C 3 C
4 D 5 B 6 A
7 D 8 D 9 D
10 **a** C **b** A **c** B and D
11 D 12 D
13 **a** unbalanced, balanced
b 12.6 m s^{-1}
c 31.8 m s^{-2}
d 1.91×10^3 N
14 **a**

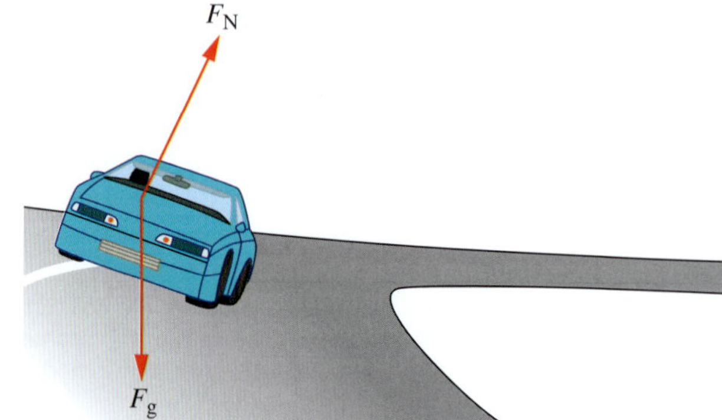

b 47°

15 **a** 4.9 m **b** $9.8\,ms^{-2}$ down
c $11\,ms^{-1}$

16 **a** $9.8\,ms^{-2}$ down **b** $2.2\,ms^{-1}$
c $4.9\,ms^{-1}$

17 **a** 1.4×10^3 J **b** 1.4×10^3 J
c $3.7\,ms^{-1}$ **d** 24 J

18 **a** $4.4\,ms^{-1}$
b **i** $1.9 \times 10^2\,kg\,ms^{-1}$ **ii** $1.9 \times 10^2\,kg\,ms^{-1}$
c 1.6×10^3 N
d inelastic

19 **a**

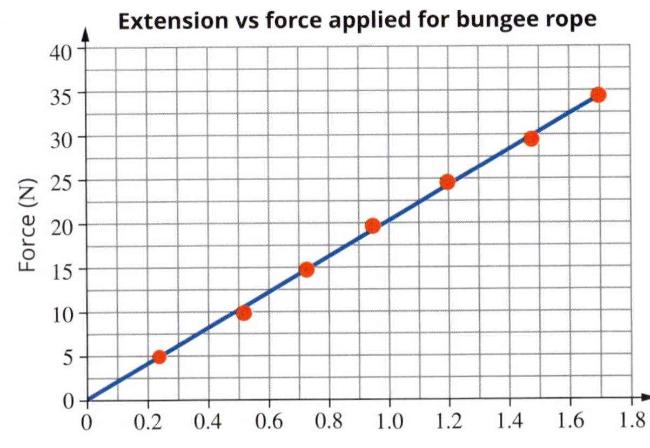

b $20\,N\,m^{-1}$ (approx.)
c 2.3×10^3 J
d $8.7\,ms^{-1}$

20 Aristotle's ideas agree with our everyday observations. We experience objects as slowing or stopping without an external force to keep them going, and we cannot see that there are actually forces acting to slow them down. In a space station we would often experience objects moving with constant velocity as they floated around the ship in freefall without friction from surfaces to slow their motion.

Chapter 4 Gravity

4.1 Newton's law of universal gravitation

CSA: Measuring the gravitational constant, *G*

1 1.45×10^{-7} N
2 1.20%

TY 4.1.1 7.1×10^{-9} N
TY 4.1.2 3.53×10^{22} N
TY 4.1.3 $\frac{a_{Earth}}{a_{Sun}} = \frac{5.85 \times 10^{-3}}{1.76 \times 10^{-8}} = 3.33 \times 10^5$

CSA: Extrasolar planets

1 A hot Jupiter is an exoplanet that is at least as large as Jupiter and orbits its host star much closer than Mercury orbits the Sun. Because of their large mass and close proximity to their host star, hot Jupiters exert a large gravitational pull on their host star. The resultant wobble of the host star is detectable by astronomers.

2 **a** 1.07×10^{26} N **b** $2.96 \times 10^{-5}\,ms^{-2}$

TY 4.1.4 The force is the same: 9.8 N (to 2 significant figures).
TY 4.1.5 8.1×10^2 N. The person will feel lighter than when they are standing on the ground.

Key questions

1 when objects have a small mass and/or are far apart
2 *r* is the distance between the centres of the two objects, in metres.
3 **a** doubles
b becomes one quarter of the original force
c decreases by a factor of 16

4 **a** 5.4×10^{22} N **b** $1.1 \times 10^{-2}\,ms^{-2}$
5 Deimos has a smaller mass than Mars and therefore experiences a larger acceleration from the same gravitational force.
6 **a** 1.50×10^{-6} N
b $1.0 \times 10^{-8}\,ms^{-2}$ towards each other
7 295 N
8 **a** 6.7×10^2 N **b** 5.9×10^2 N
9 **a** 1.34×10^{20} N **b** 4.17×10^{23} N
c The mass of the Sun is much greater than the mass of Jupiter. This difference accounts for the difference in the forces calculated.
d Comparison of forces: $(4.17 \times 10^{23}) \div (1.34 \times 10^{20}) = 3112$. The force between the Sun and Jupiter is 3112 times that between Jupiter and Saturn.
Comparison of masses: $(1.99 \times 10^{30}) \div (5.68 \times 10^{26}) = 3504$. The Sun's mass is 3504 times the mass of Saturn.
The force between the Sun and Jupiter is approximately 3000 times more than the force between Jupiter and Saturn. The Sun–Saturn mass ratio is also approximately 3000. The discrepancy (3112 versus 3504) can be attributed to the difference in the two distances.
10 **a** 8.1×10^{16} N **b** 5.03×10^{15} N **c** 6.2%
11 **a** $g_{Mercury} = G\frac{m_{Mercury}}{(r_{Mercury})^2}$
b greater gravitational acceleration at the surface
c less gravitational acceleration at the surface
d $3.69\,ms^{-2}$ towards the surface of Mercury
e 2.8×10^2 N towards the surface of Mercury

4.2 Gravitational fields

TY 4.2.1 **a**

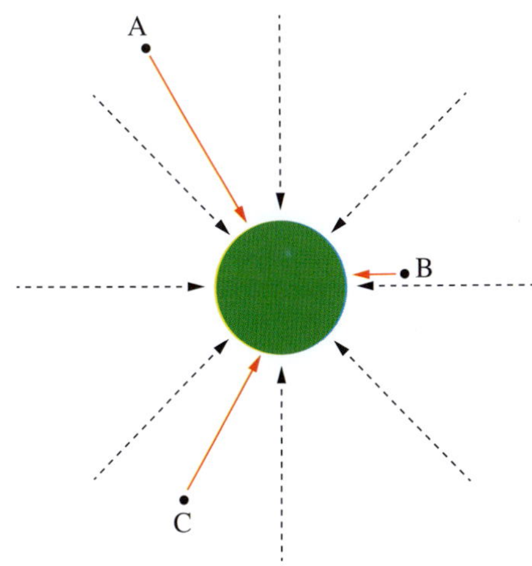

b in increasing strength: A, C, B

TY 4.2.2 $9.6\,N\,kg^{-1}$
TY 4.2.3 $9.79\,N\,kg^{-1}$
TY 4.2.4 $3.73\,N\,kg^{-1}$. At their surfaces, the Earth's gravitational field strength is approximately 2.5 times that of Mars.

Key questions

1 **a** $9.8\,N\,kg^{-1}$. An average value for *g* is used because there is variation in the Earth's gravitation field strength depending on location, rock density and altitude.
b Because the Earth is so large, small distances can be considered to be flat. Therefore gravitational field arrows are approximately parallel at the surface over a small area.
2 $7.5\,N\,kg^{-1}$
3 One-sixteenth
4 $8.70\,N\,kg^{-1}$
5 $6.7 \times 10^{-4}\,N\,kg^{-1}$
6 **a** $1.36\,N\,kg^{-1}$ **b** $0.233\,N\,kg^{-1}$
c $0.146\,N\,kg^{-1}$

7 a 2.0×10^{12} N kg^{-1}
b The gravitational field strength of a neutron star is 200 billion times that at the surface of the Earth.
c 4.5×10^{9} m

8 a g at the equator is approximately 0.7 that at the poles, i.e. 5.6 N kg^{-1}.
b 5.6 N kg^{-1}

9 3.4×10^{8} m 10 10 Earth radii

4.3 Work in a gravitational field

TY 4.3.1 3.1×10^{3} MJ
TY 4.3.2 5.2 m s^{-1}
TY 4.3.3 5.2×10^{8} J
TY 4.3.4 5.4×10^{9} J

Key questions

1 2.1×10^{5} J 2 1.4×10^{12} J
3 4.0 m s^{-1}
4 a The field strength increases.
b The acceleration increases.
c i increase ii decrease
iii stay the same
5 a 2.3×10^{2} m s^{-1} b 1822.5 m (approx. 1.8 km)
6 a between 9 N and 9.2 N b 2.6×10^{6} m
c 8.0×10^{6} J d 1.9×10^{7} J
e 2.7×10^{7} J f 7.3 km s^{-1}
7 a 1.7×10^{9} J b 1.6×10^{6} m
8 3.5×10^{11} J

Chapter 4 Review

1 983 N 2 3.78×10^{8} m 3 1.6 m
4 a The forces are equal.
b The acceleration of Jupiter caused by the Sun is greater than the acceleration of the Sun caused by Jupiter.
5 2.1×10^{-7} m s^{-2}
6 a 2.48×10^{4} N b 2.48×10^{4} N
c 24.8 m s^{-2} d 1.31×10^{-23} m s^{-2}
7 3.7 m s^{-2}
8 a 4.6×10^{2} N b 4.9×10^{2} N
9 a 3.2×10^{3} N b 0 N
c 6.6×10^{2} N
10 a

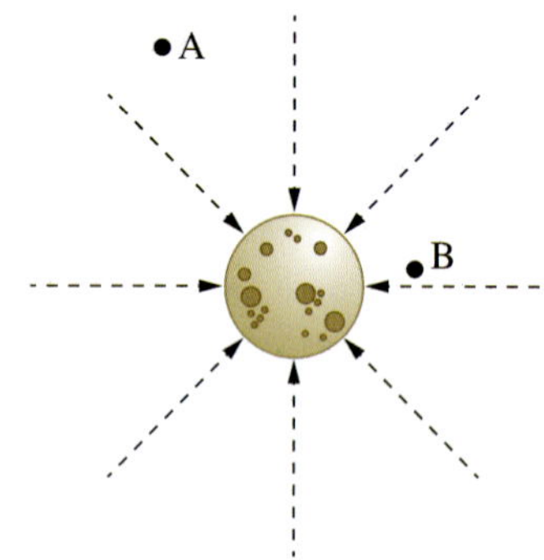

b B, because the field lines are closer at B than at A.
11 a 500 g b 500 g
12 9.80 N kg^{-1}
13 a 11.58 N kg^{-1} b 102.07% (2.07% more)
14 a 11.1 N kg^{-1} b 11.1 m s^{-2}
15 3.7×10^{9} m
16 100 N
17 a 3×10^{7} J b 4×10^{7} J
c 2000 m s^{-1} (2.0 km s^{-1}) d 3.5 N kg^{-1}
18 a 5×10^{6} m or 5000 km (approx.)
b Strength at ground level is 10 N kg^{-1} (approx.); strength at an altitude of 6500 km is 2.5 N kg^{-1} (approx.). Altitude = the Earth's radius (or 2*r*). By the inverse square law, strength at that altitude should be a quarter of the strength at ground level. This is confirmed by the graph.
19 16
20 a $\frac{9}{4}$ times that at $2r$ b $\frac{9}{4}$ times that at $2r$
21 a 8.5 N kg^{-1} (approx.)
b gain in kinetic energy per kilogram of the satellite (which is also the loss of gravitational potential energy)
c 1.08×10^{10} J
d No. Air resistance affects the satellite as it re-enters the Earth's atmosphere.

Chapter 5 Electric and magnetic fields

5.1 Electric fields

TY 5.1.1 5.6×10^{-4} N C^{-1}

CSA: Gravitational force and electric force

1.3×10^{-15} g
TY 5.1.2 2.2×10^{-18} J; work done on the field

Key questions

1 A drawing similar to the one below:

a positive and a negative charge

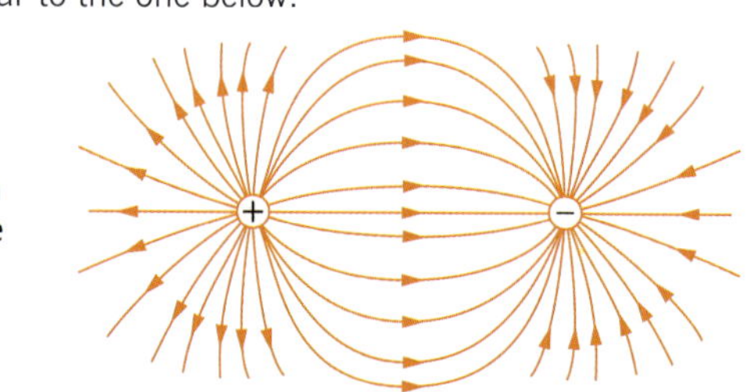

2 A drawing similar to the one below:

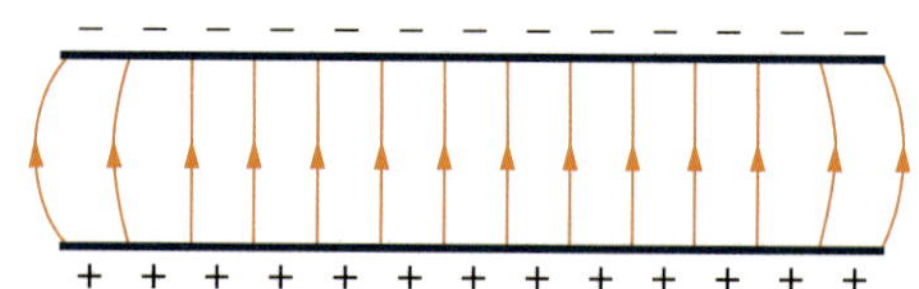

3 a true b false c false
d true e true f false
g false
4 a by the field b no work is done
c on the field d no work is done
e by the field
5 1.25×10^{-2} N 6 1.39 mC
7 5.7×10^{11} m s^{-2} 8 1200 V
9 a 1.09×10^{-19} J b on the field
10 20

5.2 Coulomb's law

TY 5.2.1 $q_1 = +6.35 \times 10^{-10}$ C; $q_2 = -6.35 \times 10^{-10}$ C
TY 5.2.2 8.0×10^{5} N C^{-1} to the right

Key questions

1 a double, repel b quadruple, repel
c double, attract d quadruple, repel
2 2.4×10^{4} N C^{-1} 3 8.99×10^{3} N
4 37 N 5 1.43 m
6 2.0×10^{13}
7 Student answers will vary. See the fully worked solutions for suggestions.

5.3 The magnetic field

TY 5.3.1 perpendicular to the wire

Key questions

1 Each new piece will be a separate magnet with two poles.

2 towards the magnetic North Pole

3 The force decreases.

4 B

5 **a** A = to the east; C = to the west
b zero

6 A = east, B = south, C = west, D = north

5.4 Forces on charged objects due to magnetic fields

TY 5.4.1 4.8×10^{-22} N

TY 5.4.2 outwards from the screen

TY 5.4.3 2.5×10^{-3} N per metre of power line

TY 5.4.4 vertically downwards

TY 5.4.5 **a** 0 N
b 1.0×10^{-3} N out from Santa's house

Key questions

1 **a** south **b** C
c remains constant **d** path A
e particles with no charge, e.g. neutrons

2 towards you

3 If doubled, the force doubles; if quadrupled, the force quadruples

4 0 N

5 out of the page

6 1.0 N downwards

7 4.8×10^{-24} N south

8 $2F$ north

9 2.0×10^{-4} N north

10 **a** 0.18 N downwards
b Same as before. The change in height has no effect on the perpendicular components of the magnetic field (south–north) and the wire's direction.

5.5 Comparing fields—a summary

1 B

2 **a** monopoles **b** both **c** dipoles

3 radial, static, non-uniform.

4 negative

5 D

6 resultant, individual

7 The direction of a field line at any point is the resultant field vector. At either end, the field outside the plates is less than between the plates. The horizontal component of the resultant field vector would point outwards.

8 8.3×10^{-18} m

9 **a** 8.2×10^{-8} N **b** 3.6×10^{-47} N
c The gravitational force of attraction is significantly less than the electrical force of attraction between the two particles. The electrical force of attraction is approximately 2×10^{39} times greater than the gravitational force of attraction.

Chapter 5 Review

1 *Electrical potential* is the work done per unit charge to move a charge from infinity to a point in the electric field. When you have two points in an electric field separated by a distance that is parallel to the field, the *potential difference* is the change in the electrical potential between these two points.

2 **a** done by the field
b done on the charged particle

3 The electric field strength is the same at all points between the plates.

4 **a** B into the page **b** $5B$ into the page

5 **a** decreases by a factor of 9 **b** increases by a factor of 16

6 The magnitude of the magnetic force on a conductor aligned so that the current is running parallel to a magnetic field is zero.

7 The palm is the direction of the force applied by the magnetic field, the fingers are pointing in the direction of the magnetic field and the thumb is pointing in the direction of the conventional current.

8 0.0225 N

9 D

10 25 V

11 1.63×10^{-17} J

12 2.0×10^{-14} N

13 8.43 mN

14 3.3 A

15 **a** 5.0×10^{-9} N towards you **b** 2.0×10^{-3} N away from you

16 9.6×10^{-15} N

17 The east–west line would experience the greater magnetic force as it runs perpendicular to the Earth's magnetic field.

18 7.0×10^{4} m s^{-1}

19 59.8 m

20 $+1.63 \times 10^{-4}$ C

21 **a** 7.9×10^{-6} N **b** 1.9×10^{-48} N
c The gravitational force of attraction is significantly less than the electrical force of attraction between the two particles. The electrical force of attraction is approximately 4×10^{42} times greater than the gravitational force of attraction.

Chapter 6 Application of field concepts

6.1 Satellite motion

TY 6.1.1 **a** 5.6×10^{2} N
b 6.7×10^{2} N
c 9.2×10^{2} N

CSA: Four satellites

1 A geostationary orbit is where the satellite has an orbital period of 24 hours, so it will appear as though the satellite is stationary with respect to the Earth's surface. This is ideal for weather satellites because it permits them to study the weather and climate of a single region in depth for an extended period.

2 0.224 m s^{-2}, 8.21 m s^{-2}, and 7.69 m s^{-2} respectively

3 2.65×10^{8} m, 4.38×10^{7} m and 4.52×10^{7} m respectively

4 86 400 s, 5796 s and 6084 s respectively

5 3.07×10^{3} m s^{-1}, 7.56×10^{3} m s^{-1} and 7.45×10^{3} m s^{-1} respectively

TY 6.1.2 3.08×10^{3} m s^{-1}

TY 6.1.3 **a** 6.70×10^{5} km
b 1.90×10^{27} kg
c 8.20 km s^{-1}

Key questions

1 71 N

2 220 N

3 6.9×10^{2} N

4 5.4×10^{2} N

5 B

6 D

7 **a** 4.23×10^{7} m **b** 2.66×10^{7} m

8 **a** 0.22 m s^{-2} **b** 7.0×10^{2} N

9 15.6 days

6.2 DC motors

TY 6.2.1 4.0×10^{-6} N out of the page and 4.0×10^{-6} N into the page

TY 6.2.2 **a** clockwise
b 0.0070 N m
c 0.0375 N

Key questions

1 A

2 The force acting on each side of the coil will be equal in magnitude, but the resulting torques from each side will be opposite in direction. Since the total torque is the sum of the individual torques, the net torque will be zero and hence the coil will not turn.

3 a 5.3×10^{-2} N into the page b 5.3×10^{-2} N out of the page
c 0 N d anticlockwise
e D

4 a left: 0.52 N; right: 0.86 N b right side
c The net torque will double.

5 a 1.5 N
b Current flows into brush P and around the coil from V to X to Y to W. The force on side VX is down and the force on side YW is up, so the rotation is anticlockwise.
c The commutator keeps the motor rotating in the same direction by reversing the polarity of the current after each half rotation. The net effect is that the left and right sides of the coil always have their direction of current pointing either upwards or downwards, regardless of whether the coil has turned. If the coil were directly connected to the battery, it would rotate half a revolution before experiencing a force in the other direction. It would continue to oscillate backwards and forwards rather than rotating smoothly.
d Since $F = nIlB$, if we increase the number of turns in the coil to 30, the force will double. However, by decreasing the side length from 0.50 m to 0.125 m, the force will decrease to one quarter. Overall the coil will experience half the amount of force.

6.3 Particle accelerators

TY 6.3.1 2.1×10^{7} m s^{-1}

TY 6.3.2 a 1.3×10^{4} V m^{-1}
b 3.0×10^{7} m s^{-1}
c 5.7×10^{-4} m

Key questions

1 C 2 1.0×10^{-22} N south

3 3.0×10^{7} m s^{-1}

4 a 6.4×10^{-15} N b 2.4×10^{-3} m

5 9.4×10^{-4} T 6 7.0×10^{6} m s^{-1}

7 A charged particle in a magnetic field will experience a force ($F = qvB$). As the force is proportional to velocity, it will increase as the velocity increases. This will continue while the charge remains in the magnetic field, continuously accelerating the charge.

8 4.9×10^{-2} m

9 a 1.3×10^{8} m s^{-1}
b Since the bending radius (0.72 m) is less than 5 m, it is too small for the particle accelerator. Sally should increase the potential difference of the electron gun.
c 5.5×10^{6} V m^{-1}

Chapter 6 Review

1 2.6×10^{2} N 2 D 3 D

4 A 5 $11.2T_x$

6 The commutator's function is to reverse the direction of the current in the coil after every half turn. This keeps the coil rotating in the same direction.

7 Electrons are released from a negative terminal (or hot cathode) of the evacuated tube and accelerate towards a positively charged anode. They can be detected as they hit a fluorescent screen at the rear of the tube. The electrons are accelerated by a large potential difference between the cathode and anode.

8 a

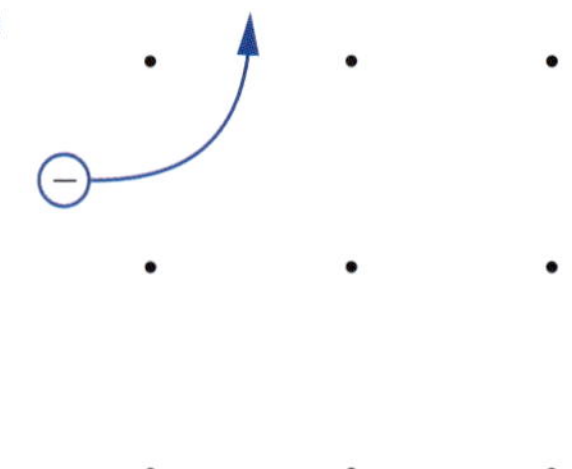

b The electron's velocity and the magnitude of the magnetic field that is acting on it.

9 5.8×10^{-4} T

10 a 9.3×10^{-15} N b 4.0×10^{-3} m

11 a 0.0540 m s^{-2} b 4.38×10^{3} m s^{-1}
c 5.90 days

12 a 0.283 N kg^{-1} b 361 m s^{-1}

13 a down the page b up the page

14 anticlockwise

15 a down the page b up the page
c There is zero torque, as the forces are trying to pull the coil apart rather than turn it. The force is parallel to the coil rather than perpendicular to it.

16 C 17 9.0×10^{-3} N

18 Each torque depends linearly on the perpendicular radius. So if we double the side lengths, each torque will also double. Since the total torque is equal to the sum of all the torques, the total torque is also linearly dependent on the length of the sides of the coil.

19 a 5.9×10^{7} m s^{-1} b 2.2×10^{-4} m

20 a 1.4×10^{4} V m^{-1} b 9.3×10^{6} m s^{-1}

21 4.0×10^{7} m s^{-1}

Unit 3 Area of Study 2 Review

1	A	2	B	3	B
4	C	5	D	6	B
7	A	8	B	9	G
10	C and D	11	B	12	C

13

14 a 4.3×10^{10} J
b 3.3×10^{3} m s^{-1}
c i 4.0×10^{4} N ii 8.1×10^{4} N
d acceleration increases from 8.1 m s^{-2} to 9.2 m s^{-2}

15 a 1.5×10^{5} N C^{-1} (or V m^{-1}) downwards
b 1.9×10^{-13} N upwards
c 1.3×10^{-18} C

16 a to the left b more strongly to the left
c to the right

17 a 1.0×10^{-3} N b west to east
c 4.9×10^{4} A

18 a AB is upwards; CD is downwards
b horizontal
c When the coil is in the vertical position. It continues to rotate because (i) of its momentum and (ii) at that point the commutator reverses the direction of the current, which reverses the forces and makes the coil continue to rotate for another half turn, at which point the current reverses again and the rotation continues.
d 4.0 A
e 8.0 N m

19 a 3.2×10^{7} m s^{-1}
b

electrons

c 1.5×10^{-4} m

20 **a** 1.11×10^{-1} N kg^{-1} **b** 5.9×10^{2} N
c 6.18×10^{5} s **d** no effect

Chapter 7 Electromagnetic induction and transmission of electricity

7.1 Inducing an EMF in a magnetic field

TY 7.1.1 8.0×10^{-5} Wb

TY 7.1.2 no magnetic flux

TY 7.1.3 An EMF of 0.70 V is too small to be dangerous.

Key questions

1 **a** An EMF is induced. **b** An EMF is induced.
c No EMF is induced. **d** An EMF is induced.

2 0 Wb **3** 2.0×10^{-3} Wb

4 9.9×10^{-3} V

5 increasing the magnetic field strength over the surface area, increasing the area of the loop and/or adjusting the angle so that the maximum magnetic field passed through the loop

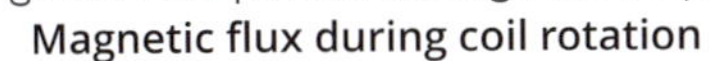

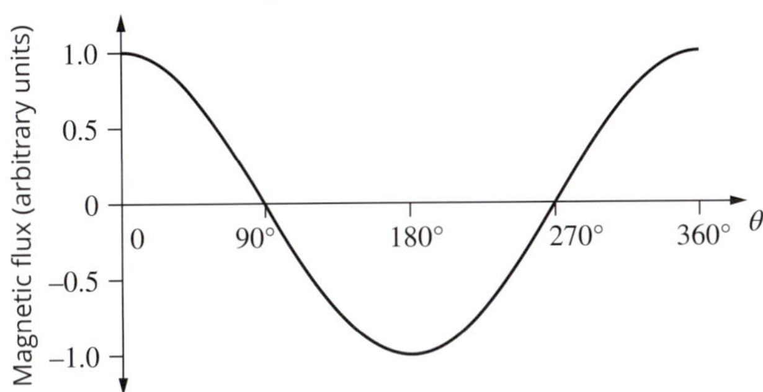

7 **a** 7.2×10^{-6} Wb

b The magnetic flux decreases from 7.2×10^{-6} Wb to 0 after one-quarter of a turn. Then it increases again to 7.2×10^{-6} Wb through the opposite side of the loop after half a turn. Then it decreases to 0 again after three-quarters of a turn. After a full turn it is back to 7.2×10^{-6} Wb.

7.2 Induced EMF from a changing magnetic flux

TY 7.2.1 **a** 5.0×10^{-4} Wb
b 5.0×10^{-3} V

TY 7.2.2 1000

TY 7.2.3 clockwise

TY 7.2.4 **a** from Y to X
b current is steady
c from X to Y

TY 7.2.5 anticlockwise

Key questions

1 **a** 2.5×10^{-6} Wb **b** no flux **c** 4.2×10^{-5} V

2 **a** 3.9×10^{-3} V **b** 2.3 V

3 9.0×10^{-3} V

4 C

5 The student must induce an EMF of 1.0 V in the wire by changing the magnetic flux through the coil at an appropriate rate. A change in flux can be achieved by changing the strength of the magnetic field or by changing the area of the coil. The magnetic field can be changed by changing the position of the magnet relative to the coil. The area can be changed by changing the shape of the coil or by rotating the coil relative to the magnetic field. (Relevant calculations are shown in the fully worked solutions.)

6 0.010 m^2 **7** 0.13 s

7.3 Applications of Lenz's law

TY 7.3.1 2.00×10^{3} W

Key questions

1 The resulting output of all three phases maintains an EMF near the maximum voltage more continuously than output from a single coil.

2

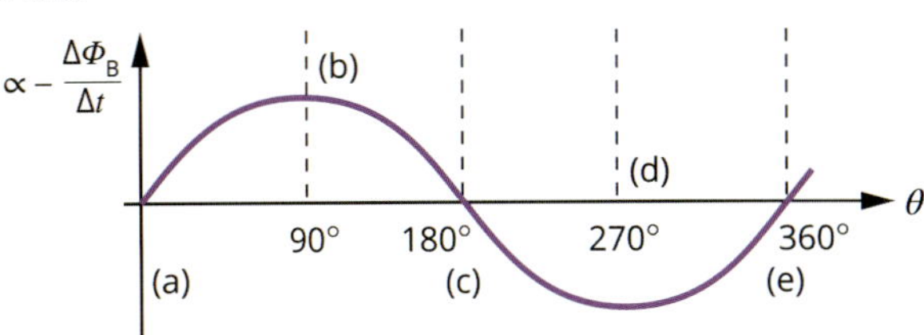

3 A slip ring is a continuous ring with no breaks or cuts, providing a continuous transfer of current. A split-ring commutator has at least two breaks and reverses the polarity of the current.

4

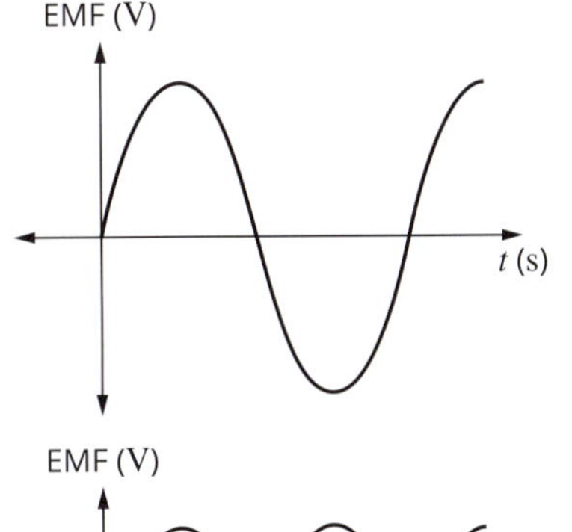

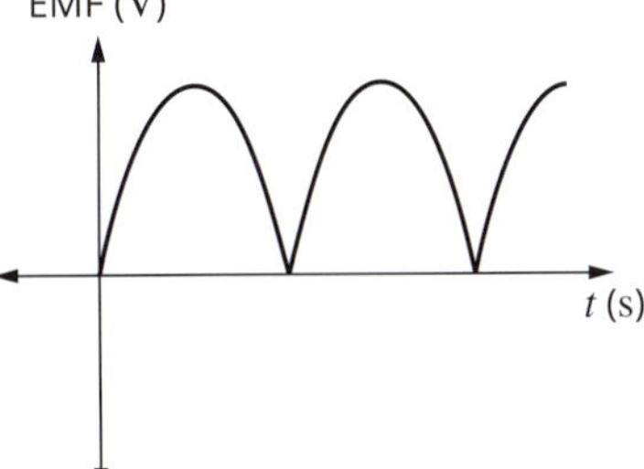

5 **a**

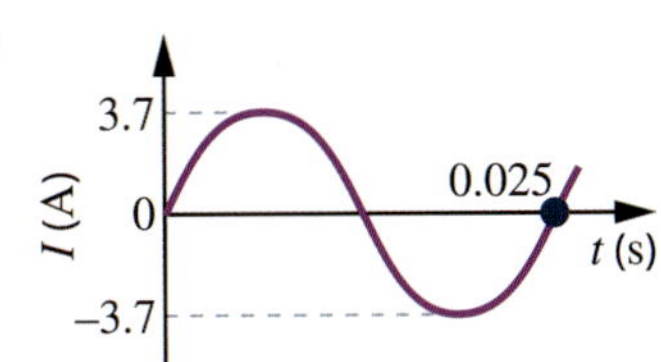

b 7.4 A

c 0.025 s

6 B

7 **a** $V_p = 8.0$ V, $V_{p-p} = 16$ V, $V_{rms} = 5.7$ V

b 0.02 s

c

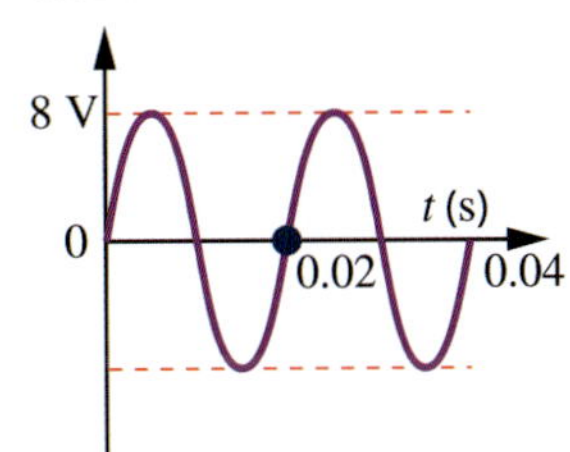

7.4 Producing electricity—photovoltaic cells

CSA: Small-scale solar production

One 400 W panel will be enough to operate the pump.

Key questions

1 D

2 The photons from sunlight cause an electron to move away from a silicon atom, setting the electron loose and causing it to move in the semiconductor material.

3 The top layer of the semiconductor has an excess of electrons (*n-type* silicon) and the bottom layer a deficit of electrons (*p-type* silicon).

4 An inverter converts the DC current generated by the solar cell into AC current that can be used by household and industrial appliances (or fed into the electricity transmission grid).

5 To maximise the use of solar energy, it would be best to run appliances that consume large amounts of power during peak daylight hours even if they are not required at that time. Appliances that require less power can be run during the morning and late afternoon.

7.5 Supplying electricity—transformers and large-scale power distribution

TY 7.5.1 4000

TY 7.5.2 0.013 A

TY 7.5.3 3 W

TY 7.5.4 0.36 MW

TY 7.5.5 5.0×10^5 V

TY 7.5.6 **a** $T_1 = 15$, $T_2 = 0.10$, $T_3 = 4.0 \times 10^{-3}$
b high = 28 kW, medium = 2.8 kW
c 675 W
d reducing the effective resistance of the transmission lines or using higher voltages to transmit power

Key questions

1 B

2 at high voltage and low current

3 D **4** 60

5 **a** 100 V **b** 50 W **c** 0.50 A

6 7.2×10^2 W

7 **a** 7.20×10^3 A **b** 82 kV

8 Student answers will vary. For suggestions, see the fully worked solutions.

9 **a** 1100
b Since the output power is less than the input power, it is not an ideal transformer.

10 8.8 MW

Chapter 7 Review

1 Student answers will vary. One example: only two conductors (live and return) are required. For more examples, see the fully worked solutions.

2 To reverse the connection between the rotor coil and power supply each half-cycle, thus ensuring that the current in the armature produces a torque in one direction as the rotor coil rotates.

3 anticlockwise **4** clockwise

5 AB and CD. Both sides cut across lines of flux as the coil rotates.

6 **a** 1.7 mV **b** anticlockwise

7 **a** 0.12 V **b** from Y to X

8 **a** 39 mV **b** from X to Y

9 5.9 mV

10 **a** 1.2 A **b** 15

11 **a** 18 V
b Doubling the frequency halves Δt, which doubles the average EMF to 36 V.

12 **a** Without the first transformer, the current is 250 A. When the voltage is stepped up to 10 000 V, the current is reduced to 25 A.
b 1 kW
c If the voltage was not stepped up, the current in the transmission line would be 250 A. The power lost in the transmission line would be 125 kW. Power supplied to the load would then be 125 kW. This is a 50% power loss.

13 C **14** D **15** A

16 The power loss in transmission line B would be 9 times as much as in transmission line A.

17 **a**

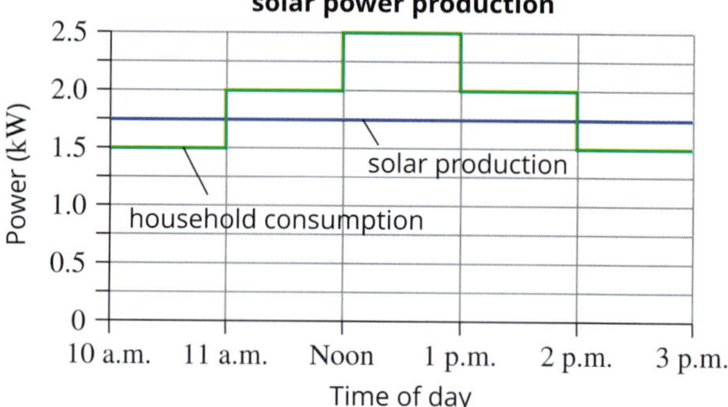

The energy for each one-hour period is as follows:

Time period	Energy to/from grid (kW h)
10 a.m. to 11 a.m.	0.25 to grid
11 a.m. to noon	0.25 from grid
noon until 1 p.m.	0.75 from grid
1 p.m. to 2 p.m.	0.25 from grid
2 p.m. to 3 p.m.	0.25 to grid

b \$0.31 **c** \$2.07

18 **a** It is an AC generator, as it has slip rings.
b Y
c when the rotor coil is perpendicular to the magnetic field

19 **a** 19 A, 13 Ω **b** 1.1 kW

Unit 3 Area of Study 3 Review

1 D **2** C **3** D

4 A **5** C **6** B

7 C **8** D **9** B

10 **a** no flux
b Rotate the loop or the magnetic field so that they are no longer parallel.
c when the plane of the loop and the direction of the magnetic field are perpendicular to each other
d 1.0×10^{-2} Wb

11 **a** As the loop enters the magnetic field there is a flux increasing down through the loop. Lenz's law states that the induced current in the loop will oppose the change in flux that causes it. Therefore there will be an induced field (or flux) up through the loop. Using the right-hand grip rule, align your fingers so that they are pointing up on the inside of the loop. Your thumb will point in the direction of the induced current, that is, from Y to X.
b 4.0×10^{-3} V
c 8.0×10^{-3} A
d 3.2×10^{-5} W
e the external force moving the loop into the magnetic field

f After 5.0 seconds the loop has moved 25 cm and has been totally within the magnetic field for 1.0 second. Since there is now no flux change there will be no EMF induced in the loop at this moment.

g As the loop emerges from the magnetic field there is flux decreasing down through the loop. Lenz's law states that the induced current in the loop will oppose the change in flux that causes it. Therefore there will be an induced field (or flux) down through the loop. Using the right-hand grip rule, align your fingers so that they are pointing downwards on the inside of the loop. Your thumb will point in the direction of the induced current, that is, from X to Y.

12 **a** 5.0×10^{-6} Wb
b no flux
c 2.5×10^{-3} V
d 1.3×10^{-3} A
e No. Once the loop is stationary there is no change in flux. Therefore no EMF is generated and no current flows in the loop.

13 **a** 200 μA **b** 0.21 T

14 **a**

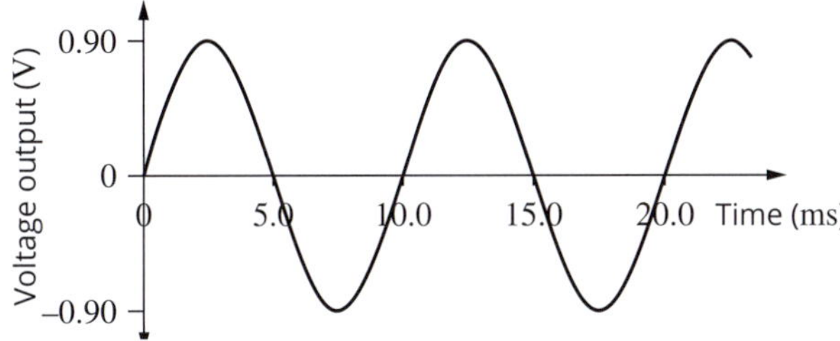

b The output graph would have half the period and twice the amplitude.

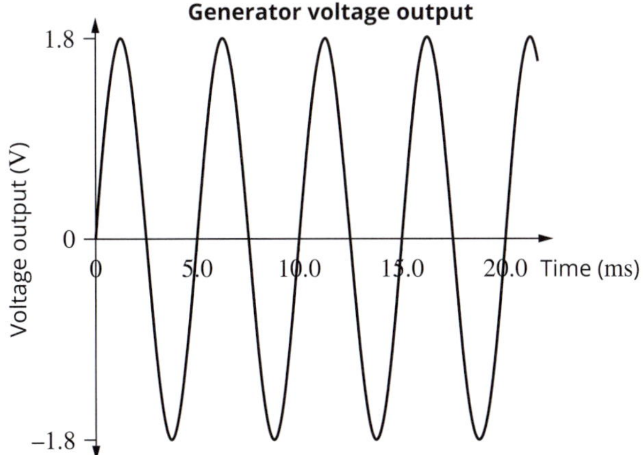

15 An alternator has a pair of slip rings instead of a split ring commutator.

16 AC is generated in the coils of an alternator. Each slip ring connects to each end of the coil. The slip rings are continuous and so maintain the AC generated in the coil at the output. Carbon brushes press against the slip rings to allow a constant output to the circuit without a fixed point of connection.

17 **a** 0.40 A **b** 6000 V **c** 200
d 8.5×10^2 W **e** 1.7×10^3 W

18 **a** With little or no current in the power line there was almost no voltage drop. When the house appliances were turned on, there was a higher current in the power line and hence a voltage drop along the line, leaving a lower voltage at the house.
b voltage = 2.2×10^2 V, power = 3.5×10^3 W
c At the generator end a 1:20 step-up transformer is required. At the house end a 20:1 step-down transformer is required.
d 0.8000 A
e 1.6 V
f 1.3 W
g 2.5×10^2 V
h 4.0×10^3 W
i power loss before transformers = 512 W (about 12.8%); power loss with transformers = 1.3 W (about 0.03%)
j Power loss depends on the square of the current. Since the current was reduced by a factor of 20 and the resistance remained constant, the power loss decreased by a factor of 20^2 or 400.

Chapter 8 Light as a wave

8.1 Wave interactions

TY 8.1.1 zero

Key questions

1 **a** true **b** false **c** true

2 the applied frequency is equal to the object's natural frequency of vibration

3 B

4 To push a child on the swing you need to work out the frequency of the swing and then push at exactly that rate. This maximises the transfer of energy from the person pushing to the swing, thereby increasing the amplitude of the swing.

5 The waves will cancel.

6 As the vibration of the truck is noticeably larger when the truck is stationary, the natural frequency of vibration of the body of the truck must be close to 100 Hz, the frequency of vibration of the motor. As the truck accelerates to a higher speed, the frequency of vibration of the motor increases and is no longer at the resonant frequency of the truck body. Therefore the amplitude of vibration becomes much less.

7 The bridge resonates at 1.5 Hz. A pedestrian moving at a frequency of 1.5 Hz would be completing 1.5 cycles per second. Since one cycle is 2 steps, that pedestrian would be taking 3 steps per second. This corresponds closely to fast running on the graph. Thus a pedestrian running fast may cause an increase in the amplitude of the oscillation of the bridge which, over time, could damage it.

8 D. It has the same length as the first pendulum. Therefore it's natural frequency of vibration is the same as the first pendulum. The frequency of vibration of the first pendulum becomes the applied frequency for pendulum D; thus there is maximum energy transfer.

8.2 Standing waves in strings

TY 8.2.1 **a** 0.25 m
b 0.17 m

CSA: Physics of the guitar

1 107 m s^{-1}

2 The wavelength of the fundamental frequency for each string remains the same. The fundamental frequency decreases for each string as the guitarist moves from the higher strings to the lower strings. Therefore, by the equation $v = f_1\lambda_1$, the speed of the wave along the string must also decrease. Thus from the relationship $v = \sqrt{\frac{T}{\mu}}$,the mass per unit length must increase (as it is an inverse relationship) or the tension must decrease. The mass per unit length can increase by using a low-density material such as nylon for the higher strings and denser steel for the lower strings.

3 $\lambda = 0.866$ m; $f = 220$ Hz

4 0.163 m

5 247 Hz. This is almost the same as the fundamental frequency of the B string. Thus the $n = 3$ vibration of the low-E string should resonate with the B string and cause it to vibrate.

Key questions

1 a false b false c false d true

2 No. A transverse wave moving along a rope is reflected from a fixed end. The interference that occurs during the superposition of the reflected wave and the original wave creates a standing wave. This standing wave consists of locations called nodes (where the movement of the rope is cancelled out) and antinodes (where maximum movement of the rope occurs).

3 There is a 180° change in phase (and thus a crest is reflected as a trough and vice versa).

4 $\lambda = 0.80$ m; $f = 73$ Hz 5 1.50 m

6 a 300 Hz b 600 Hz c 900 Hz

7 one quarter

8 a 4 b 4.0 m

9 0.737 m

8.3 Evidence for the wave model of light

TY 8.3.1 5.0×10^{14} Hz

CSA: Heating up food in a microwave

1 The microwave oven is a resonant cavity. Its electromagnetic radiation is reflected from the cavity walls and forms a standing wave. The hot spots are where the antinodes occur.

2 2.45×10^{9} Hz 3 2.94×10^{8} m s^{-1}

4 2%

5 marshmallows may have melted unevenly; uncertainty in measuring the position of the antinodes

TY 8.3.2 550 nm

CSA: X-ray diffraction

1 The wavelength of X-rays is similar in size to the lattice spacing of the crystal, i.e. the spacing between the atoms.

2 0.71 Å

3 Peaks occur where the path difference is equal to one wavelength or multiples of one wavelength (which is where constructive interference occurs).

4 left peak, 3.4 Å; right peak, 1.5 Å

Key questions

1 Light does not require a medium to travel in. Accelerating charges produce varying magnetic fields. The varying magnetic field produces a varying electric field, which produces a varying magnetic field and so on. Therefore light is self-propagating and we can see light from far away.

2 a D

b Significant diffraction occurs when $\frac{\lambda}{w}$ is approximately 1 or greater. Red light has a wavelength of 700 nm, which is approximately 10^{-6} m, and a diffraction width of 0.001 mm is 10^{-6} m.

3 If light were a particle it would be expected to create two bright bands on the screen behind the slits. However, an interference pattern with alternating bright and dark lines was seen, which is characteristic of wave behaviour.

4 C and D 5 A and D

6 a Increase b decrease c increase

7 2.61×10^{-6} m

8 a destructive b constructive c destructive

9 1400 nm 10 455 nm

11 The diffraction effects would be less significant.

12

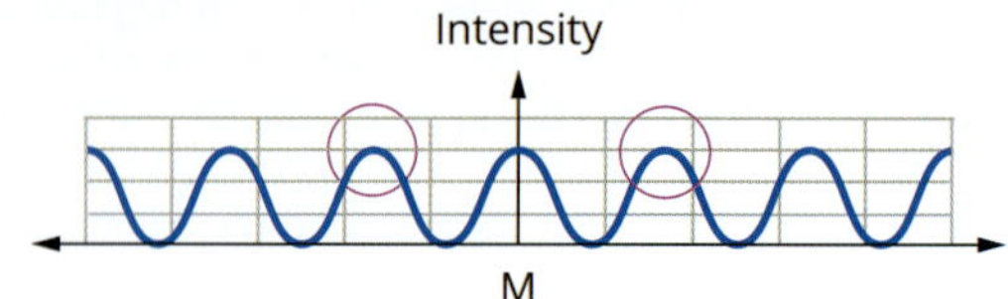

Chapter 8 Review

1 green wave

2 Destructive interference: the two wave pulses must have the same frequency (or wavelength), the same amplitude and be out of phase by 180°.

3 The motor produces different frequencies of vibrations depending on its speed. At speeds where the car is vibrating more strongly, the motor frequency can be assumed to be at a similar frequency to the natural frequency of the car. This is known as resonance.

4 a where the amplitude of the standing wave is always zero

b where the amplitude of the standing wave varies from maximum amplitude to zero to minimum

5 a 211 Hz b 633 Hz

6 0.50 Hz 7 0.044 m

8 a 0.355 m b 0.142 m

9 2.537 million light-years

10 A varying magnetic field is produced by accelerating oscillating charged particles. The varying magnetic field induces a varying electric field which in turn produces a magnetic field. This sequence repeats indefinitely.

11 diffraction. Narrowing the gap would make the effect stronger.

12 D

13 Since spacing is proportional to wavelength, it will increase in changing from blue light ($\lambda = 460$ nm) to green light ($\lambda = 525$ nm).

14 a 581 nm b yellow

15

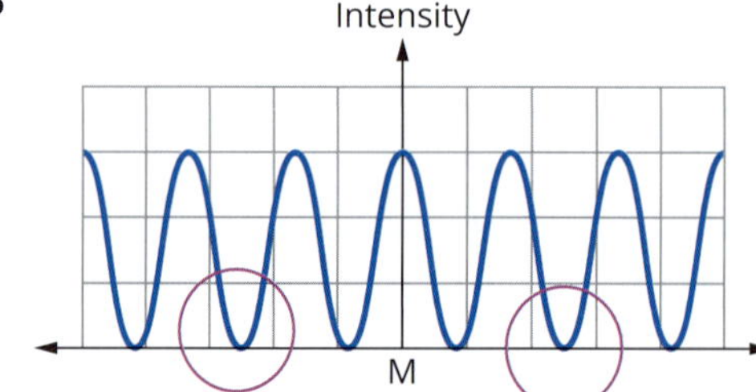

16

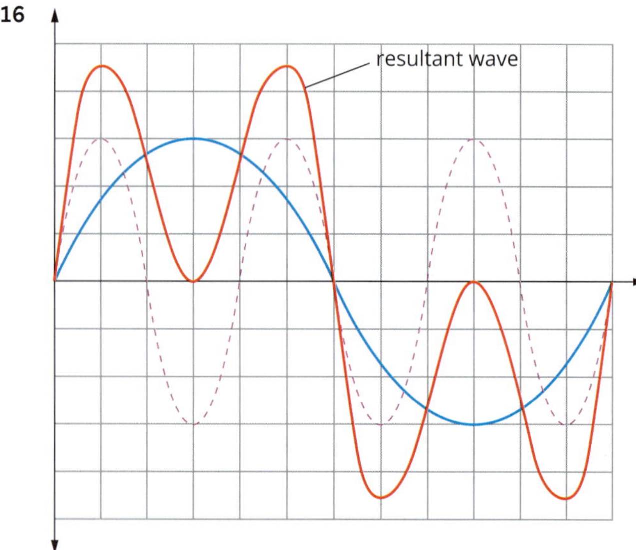

17

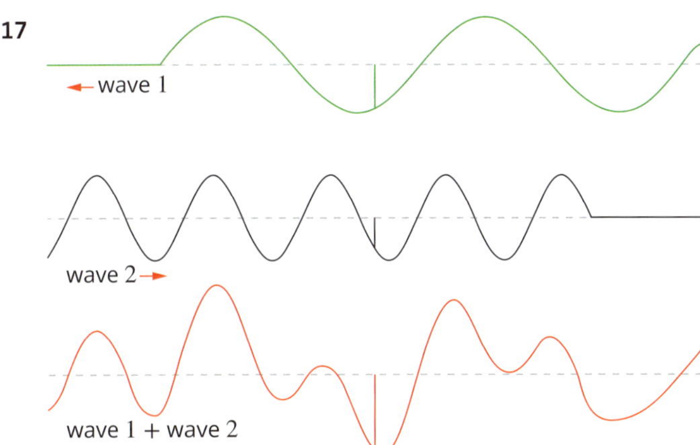

18

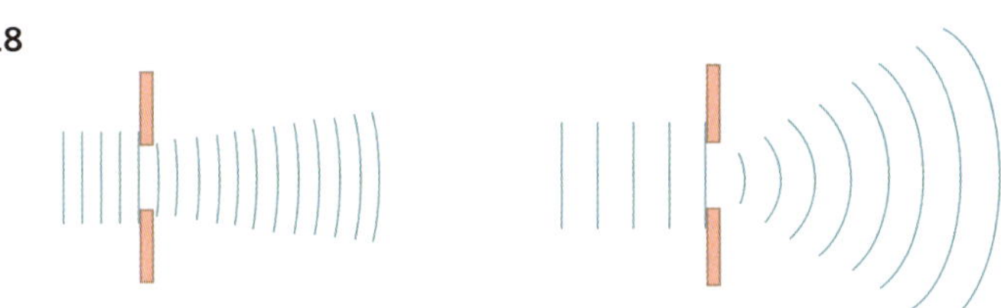

The shorter wavelength shows a smaller diffraction effect (because the width of the gap is much greater than the wavelength) and the larger wavelength shows significant diffraction (because the width of the gap is similar to the wavelength).

19 This occurs if the natural frequency of the ear canal is 2500 Hz. If the applied frequency of the signal generator is the same as the natural frequency of the ear canal, there would be maximum transfer of energy from the generator to the ear, which is perceived as a louder sound.

20 a principle of superposition
b Its amplitude will be less than the amplitudes of wave A and wave B. In this case it would be complete destructive interference and the resultant amplitude would be zero.

21 139 Hz

22 Significant diffraction only occurs when $\frac{\lambda}{w} \geq 1$. The wavelength of light is too small ($\approx 10^{-7}$ m) to diffract around most everyday objects.

23 The visible light used in optical instruments varies in wavelength from 400 nm to 750 nm. For objects with spacings of 400 nm or less, diffraction effects will limit the ability of the microscope or telescope to resolve the image.

24 A microwave oven is tuned to produce electromagnetic waves with a frequency of 245 GHz. This is the resonant frequency of water molecules. When food is bombarded with radiation at this frequency, the water molecules in the food start to vibrate. The energy of the water molecules is transferred to the rest of the food, heating it up.

25 a false. Visible light waves have a wavelength range from 400 nm to 750 nm and require an opening with a width of 400 nm to 750 nm for diffraction to occur.
b true
c true
d false. Red light has a longer wavelength than violet light, therefore red light will diffract through a larger gap than violet light.
e true

26

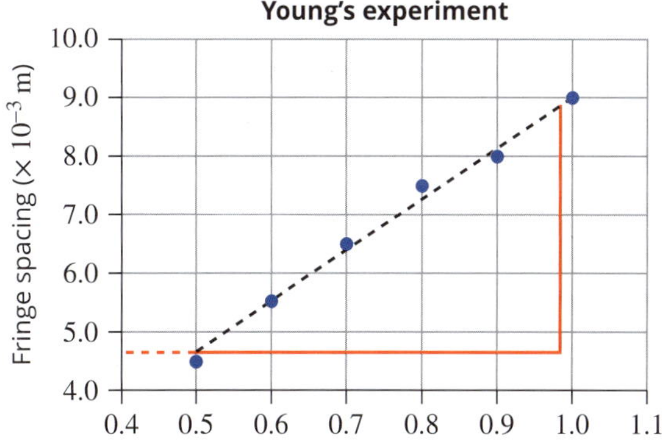

Answers for the slit separation of between 60 and 65 μm are acceptable.

27 Count the number of wavelengths between S_1 and position A:
d_1 = five wavelengths = 5λ
Count the number of wavelengths between S_2 and position A:
d_2 = six wavelengths = 6λ
Path difference = $6\lambda - 5\lambda = \lambda$
Thus constructive interference occurs at position A. This is called an antinodal point.

Chapter 9 The dual nature of light and matter

9.1 The photoelectric effect

TY 9.1.1 2.4×10^{-19} J
TY 9.1.2 1.5 eV
TY 9.1.3 1.5 e
TY 9.1.4 5.0 eV
TY 9.1.5 2.07 eV

CSA: Lenard's experiment

1

λ (nm)	Maximum kinetic energy (eV)	f (× 10^{14} Hz)	Maximum kinetic energy (× 10^{-19} J)
517	0.14	5.8	0.22
448	0.53	6.7	0.85
414	0.75	7.2	1.20
395	0.86	7.6	1.38
366	1.09	8.2	1.74
353	1.25	8.5	2.00

2

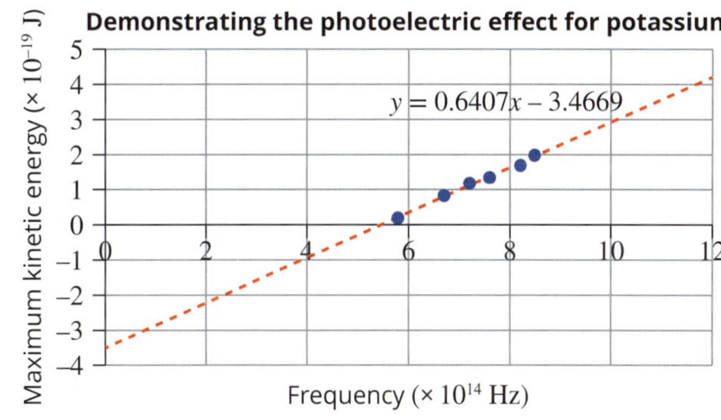

3 **a** 6.4×10^{-34} J s **b** $y = (0.64 \times 10^{-33})x - 3.5$
c 5.4×10^{14} Hz

4 **a** 6.4×10^{-34} J s **b** 5.4×10^{14} Hz
c 3.5×10^{-19} J

Key questions

1 If the light has a frequency above the threshold frequency of the metal, electrons are released. This leaves electron-deficient atoms, hence the positive charge. If the incident light is below the threshold frequency, no electrons are released.

2 **a** true
b false. When light sources of the same intensity but different frequencies are used, the higher frequency light has a higher stopping voltage, but it produces the same maximum current as the light of the lower frequency.
c true

3 **a** true
b false. The stopping voltage is reached when the photocurrent is reduced to zero.
c true
d true

4 0.162 eV

5 **a** 3.0×10^{-19} J **b** 3.4×10^{-19} J
c 4.1×10^{-19} J **d** 5.1×10^{-19} J

6 **a** 4.1 eV **b** 4.6 eV **c** 6.2 eV

7 D

8 0.24 eV

9 C and D. According to the equations $\phi = hf_0 = \frac{hc}{\lambda_0}$, a work function of 1.81 eV corresponds to a wavelength of 686 nm. Photons with wavelengths shorter than this (i.e. violet light and ultraviolet radiation) will cause photoelectrons to be emitted.

10 1.74 eV

11 **a** 5.61×10^{14} Hz **b** 3.72×10^{-19} J
c 1.86×10^{-15} J

9.2 The quantum nature of light and matter

TY 9.2.1 5.7×10^{-13} m

TY 9.2.2 1.0×10^{-36} m

TY 9.2.3 0.17 nm

CSA: Electron microscope

1 High-energy electrons colliding with gas atoms will produce anomalies in the resulting images.

2 By de Broglie's equation, the wavelength is inversely proportional to the velocity.

3 While an electron is in any part of a magnetic field its path will be circular.

4 If the velocity is reduced, the magnetic field must be reduced by the same factor in order to maintain the same force.

5 2.3×10^{8} m s^{-1}

TY 9.2.4 1.47×10^{-27} kg m s^{-1}

Key questions

1 The wavelength of a cricket ball is so small that its wave-like behaviour could not be seen by a cricket player.

2 The wavelength of light is larger than the radius of the atom so it cannot be reflected from the atom. An electron microscope can observe individual atoms because the wavelength of the electron is very small and comparable to the radius of the atom.

3 B

4 Classical physics requires an object to have a rest mass in order to calculate its momentum, but a photon has no rest mass.

5 The image from a light microscope is white light reflected from the object and white light is composed of many colours. The image from an electron microscope is produced by electrons with wavelengths outside the visible spectrum. Where electrons pass easily through parts of the object, white regions result; where they can't, shades of grey to black result.

6 $\lambda = \frac{h}{\sqrt{2qVm}}$

7 2.9×10^{5} m s^{-1}

8 **a** 3.1×10^{-10} m **b** 2.4×10^{6} m s^{-1}

9 7.3×10^{5} m s^{-1}

9.3 Light and matter

TY 9.3.1 4.6×10^{-19} J

TY 9.3.2 103 nm, Lyman series

TY 9.3.3 Photons of 6.7 eV may be absorbed, photons of 9.0 eV cannot be absorbed and photons of 11.0 eV may ionise the atom (with the ejected electron having 0.6 eV of kinetic energy).

Key questions

1 **a** Energy levels are restricted to certain discrete values.
b When the electrons in an atom are not in an excited state.
c Electrons can be excited by the addition of energy.
d The minimum energy needed to overcome the forces keeping the electron in the atom.

2 The electrons in an element in the gaseous state become excited to a higher energy level when the gas is heated or an electric current flows through it. On returning to their ground state, the electrons emit the energy gained as a photon of light.

3 Bohr's model could not explain the spectra of multi-electron atoms, the continuous emission spectrum of compounds and the two close spectral lines in hydrogen that are revealed at high resolution.

4 Absorption is observed when electrons move mainly from the ground state, while emission includes all possible downward transitions from a higher excited state to a lower excited state.

5 3.5×10^{-19} J

6 675 nm

7 **a** The energies of incident photons that correspond to the exact differences in energy levels in the lithium atom are absorbed if they collide with an electron in a gas atom and excite that electron to a higher energy level. The wavelengths that correspond to these energies are removed from the incident beam, leaving a dark line on the spectrum of the emergent beam.
b 2.75 eV

8 **a** 12.1 eV **b** 103 nm

9 −0.37 eV

Chapter 9 Review

1 **a** The detector observed a sequence of maximum and minimum intensities.
b As the electron beam is diffracted, the electrons are exhibiting wave-like behaviour. Electrons are not light but, like light, a beam of electrons can be diffracted.

2 Energy levels in an atom cannot be any value within a continuous range but are restricted to certain discrete values, i.e. the levels are quantised.

3 Bohr's work on the hydrogen atom and his idea of electrons revolving around the nucleus in orbits with specific energies convinced many scientists that a particle model was needed to explain the way light behaves in certain situations. It built significantly on the work of Planck and Einstein.

4 The emission line spectrum of sodium appears as a series of coloured lines. The absorption line spectrum of sodium appears as a full visible spectrum with a number of dark lines. The colours that are missing in the absorption line spectrum match the colours that are visible in the emission line spectrum.

5 The Sun's spectrum is an absorption line spectrum with the dark lines being the same as those in the emission line spectra of hydrogen and helium. This indicates that the two gases are present in the Sun's atmosphere.

6 The photoelectric effect is explained by the particle nature of light; electron diffraction patterns are explained by the wave nature of particles. The interference pattern in Young's double-slit experiment is used to justify the wave nature of light, so an electron diffraction pattern produced in a similar way suggests electrons have a wave property.

7 From de Broglie's equation, momentum $= \frac{h}{\lambda}$. It is possible for particles and photons to have the same momentum provided both the particle and photon have the same wavelength.

8 photoelectrons

9 2.5 eV

10 8.0×10^{-19} J

11 1.2×10^{15} Hz

12 2.9 eV

13 1.95 eV

14 Rb = 2.1 eV, Sr = 2.5 eV, Mg = 3.4 eV, W = 4.5 eV

15 a

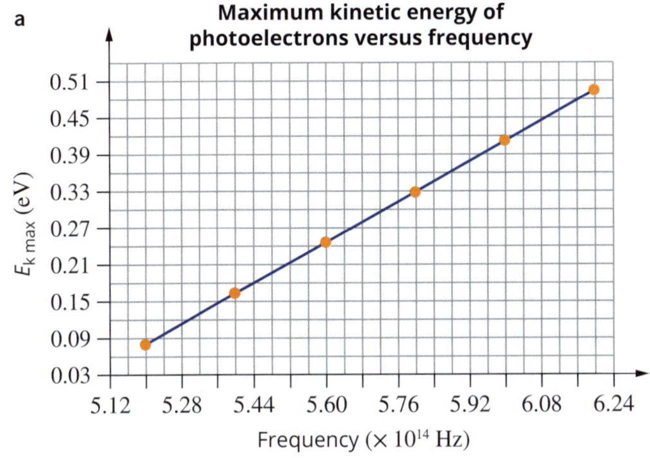

b 4.14×10^{-15} eV s

c $\approx 5.0 \times 10^{14}$ Hz

d No. The frequency of red light (4.41×10^{14} Hz) is below the threshold frequency for rubidium.

16 a 4.78 keV

b The electrons have a de Broglie wavelength that is similar to the wavelength of the X-rays.

c 2.55×10^{-24} kg m s^{-1}

17 B. Electron: $\lambda = 9.7 \times 10^{-11}$ m; blue light: $\lambda = 4.7 \times 10^{-7}$ m; X-ray: $\lambda = 6 \times 10^{-10}$ m; proton: $\lambda = 3.9 \times 10^{-13}$ m

18 a 1.7×10^{-35} m

b No. The wavelength is much smaller than the size of everyday objects (even smaller than the radius of an atom). Significant diffraction only occurs when the wavelength is equal to or greater than the size of the object.

19 a 6 spectral lines

b 3.08×10^{15} Hz

c ultraviolet region

20 a 1.14×10^{6} m s^{-1}

b
———— 3.62 eV
———— 3.20 eV
———— 2.11 eV
———— 0 eV

c 3.62 eV, 3.62 – 2.11 eV, 3.62 – 3.20 eV, 3.20 eV, 3.20 – 2.11 eV, and 2.11 eV

d 2.11 eV

e $n = 4$ to $n = 3$

Chapter 10 Einstein's special theory of relativity

10.1 Einstein's special relativity

Key questions

1 They believed that all waves needed to travel in a medium, so just as air is the medium for sound, they invented the aether to be the medium for light.

2 Place a hanging pendulum in both spaceships. The pendulum in the accelerating spaceship will move from its normal vertical position when the spaceship accelerates.

3 The speed of the ball is greater for Jana. The speed of the sound in the carriage is greater on the approach than it is on departure for Jana, while for Tom it is the same both forwards and backwards. The speed of light is the same for Jana and Tom.

4 Student answers will vary. Two examples are a person standing on a train that has a constant velocity and a passenger sitting on a plane that is travelling with constant velocity.

5 Student answers will vary. Two examples are a person sitting on a Ferris wheel that is rotating at a constant speed and a driver in a racing car accelerating away from the start line.

6 Student answers will vary. Einstein's first postulate is that the laws of physics are the same in all inertial frames of reference. For example, if an object is dropped from a plane moving at constant velocity, it would hit the ground vertically below the point from which it was dropped. This applies in all inertial frames of reference. However, if a driver were to drop a ball inside an accelerating racing car, it would hit the floor at a point not vertically below the point from which it is dropped. This shows that in non-inertial frames of reference, the laws of physics are inconsistent with the laws in inertial frames of reference.

7 Max's sister is in circular motion, so she is in a non-inertial frame of reference. Thus she sees the path of the ball curve. Max sees the ball move in a straight line, as the ball is in his inertial frame of reference. He should have thrown the ball slightly in front of his sister for her to catch it.

8 a 370 m s^{-1} b 300 m s^{-1}

c 360 m s^{-1} d 350 m s^{-1}

9 a 1.60 m s^{-1} south b 2.54 m south

c 1.59 s

10 a 0.100 s b 70.0 m s^{-1}

c 0.600 m south d 70.0 m s^{-1}

10.2 Einstein's *Gedanken* train

Key questions

1 The three dimensions of space and the fourth dimension of time are interdependent, i.e. the passage of time can be affected by motion in space. It is most noticeable when the velocity involved is very fast. Similarly, distances in space are also affected by the motion of an object through space at high speeds.

2 The effects of relativity occur in situations that we cannot normally observe. In a similar way, observations of simultaneous events would not be possible using our senses. Thus we must use thought experiments.

3 Galilean relativity applies to objects moving in inertial frames of reference. The outside observer would see the balls moving at different speeds in their frame of reference. So the ball travelling forwards is going faster by the amount required to cover the extra distance that the front wall moves away in the same time that the ball moving backwards is slowed to cover the shorter distance as the back wall moves forward.

4 3.0×10^8 m s^{-1}

5 The constancy of the speed of light viewed from any frame of reference.

6 very short; very similar to; should not; many do

7 Stephen is in a different frame of reference to Barry. Stephen will be moving towards the light source on the left and away from the light source on the right. Thus Stephen sees the light arrive from the left first and the light from the right arrive some time later. He then deduces that the flashes of light were sent at different times.

8 Student answers will vary. An example: a person sitting in a plane travelling at high speed and sending flashes of light simultaneously towards the front and back of the plane. If a stationary observer on the ground could see this light, they will see the flash of light strike the back of the plane first, whereas the person on the plane will see the flashes of light strike the front and back at the same time.

9 They enable extremely short durations to be timed to many decimal places. Differences in time for the same event to occur, when measured by observers in different inertial frames of reference, indicate that time is not uniform between the two inertial frames. The ability to time these very small differences enables researchers to gather evidence to support Einstein's special theory of relativity.

10 a 2.3×10^{-22} s
b 6.9×10^{-14} m
c To still exist over 8 times the distance from their source, the hydrogen-7 atoms must be living for a longer time. By existing for a longer time, they can travel further.

10.3 Time dilation

TY 10.3.1 5.20×10^2 s

Key questions

1 light; oscillation; time; constant

2 Proper time, t_0. From their frame of reference, the observer can hold their stopwatch stationary in one location, start it when the front of the space probe is in line with the stopwatch and stop it when the back of the probe is in line with the stopwatch.

3 1.29 s 4 48.2 s

5 2.20 s 6 1.2 s

7 a 1.0 m b 3.3×10^{-9} s
c $d = ct_G$ d 7.6×10^{-9} s
e 2.3

8 a $5.501\,10 \times 10^{-5}$ s (55.0 μs)
b Non-relativistic: 660 m; relativistic: 1.649×10^4 m

9 a 5.29×10^{-10} s b 1.14×10^{-10} s

10 The equator clock is moving faster relative to the poles, so special relativity effects mean that the clock will run slower relative to an observer at the poles. It is also accelerating and hence will run slower due to general relativity effects. The combined effect is well below what we can detect with any clock, as the speed of the equator is only about 460 m s^{-1}, which is about 1.5 millionth the speed of light.

10.4 Length contraction

TY 10.4.1 1.91 m

TY 10.4.2 20.5 m

CSA: How length contraction affects linear particle accelerators

1 0.0544 m 2 0.0368 m 3 34.4 m

Key questions

1 The object must be at rest relative to the observer.

2 Width and height are not affected as they are at right angles to the direction of motion, but the stationary observer will see the spaceship with a contracted length.

3 0.812 m 4 3.37 m

5 a 2.71×10^8 m s^{-1} b 0.643 m

6 a 2.598×10^8 m s^{-1} b 0.5000

7 35.0 cm

8 a 0.911 m b 2.55 m

9 a an almost flat circular disc with a radius of 125 m
b an almost flat square with sides of 141 m
c an almost flat circular ring 30.0 m in radius
d an almost flat triangle with a base of 50.0 m and an apex 80.0 m above the base

10.5 Einstein's mass–energy relationship

TY 10.5.1 2.19×10^{-12} J

TY 10.5.2 a 4
b 4.80×10^{-12} J
c 5.33×10^{-29} kg

TY 10.5.3 a 2.05×10^3 MeV
b $1.672\,623 \times 10^{-27}$ kg
c $(2\times+) \longrightarrow (2\times+)+(1\times0)$
$2+ \longrightarrow 2+$

Key questions

1 approaches infinity

2 Mass goes missing when the two particles fuse and create the two products. This missing mass is converted into energy according to Einstein's equation $E = mc^2$.

3 Einstein's equation suggests that matter and energy are interchangeable. The mass in kilograms of the electron can be completely converted into the equivalent amount of energy. For an electron, 0.510 MeV is the energy-equivalent of its mass.

4 a 9.53×10^5 kg m s^{-1}
b Given that the value of γ is essentially 1, at this speed relativistic effects would not be noticed. Thus the value for the classical momentum would be identical to the relativistic momentum to many decimal places.

5 1.82×10^{-17} kg m s^{-1}

6 a 1.15×10^{15} J
b 4.79×10^{14} J
c The momentum of the arrow at relativistic speeds increases much more rapidly than classical theory predicts.

7 3.11×10^{31} J

8 0.424 MeV

9 a 2.09 MeV
b 1.05×10^3 MeV
c $p^+ + p^- \longrightarrow \gamma^0 + \gamma^0$
$(1\times+)+(1\times-) \longrightarrow (2\times0)$
$0 \longrightarrow 0$

10 a $2.174\,287 \times 10^3$ MeV
b $1.677\,205 \times 10^{-27}$ kg. This is close to the mass of a proton, so the two particles produced are protons.
c $n^0 + n^0 \longrightarrow p^+ + p^+ + \pi^- + \pi^-$
$(2\times0) \longrightarrow (2\times+)+(2\times-)$
$0 \longrightarrow 0$

Chapter 10 Review

1 A and D

2 The observer would see the ray of light strike the back wall before the ray of light strikes the front wall. This is because (a) the rays of light travel at the same speed forwards and backwards, (b) the front of the carriage is moving forward, thus extending the distance that the forward-moving ray must travel before it hits the front wall and (c) the back wall has moved forward, thus decreasing the distance that the rear-moving ray must travel before it hits the back wall. The ray that travels the shorter distance takes less time to hit the wall than the ray that travels the longer distance.

3 A

4 A and C

5 C

6 at the poles

7 A and B

8 You could not tell the difference between (i) and (iii), but in (ii) you could see whether an object, such a pendulum, hangs straight down.

9 Space and time are interdependent: motion in space alters an object's passage through time as measured by a stationary observer. Thus the three dimensions of physical space must be considered relative to the one dimension of time, as they are not independent.

10 Crews A and B will see each other normally as there is no relative velocity between them. They will both see people on the Earth moving in slow motion, as the Earth has a high relative velocity.

11 In your frame of reference time proceeds normally. Your heart rate would appear normal. As Mars is moving at a high speed relative to you, people on Mars appear to be in slow motion, as time for them—as seen by you—will be dilated.

12 3.0×10^8 m s^{-1}

13 26.8 s

14 a 15.2 s b 15.2 s

15 a 0.866c b 0.968c

16 a 6.57 s b 11.1 m

17 a 6.62 years b 2.89 years
c 2.60 light-years

18 a 77.5 km
b No. Since the motion is perpendicular to the depth, this dimension is not affected.

19 a 12.9196 or ≈13 times
b No, they do not experience any difference in the rate at which time passes in their own frame of reference.
c 10.92 years
d 0.845 years
e No. The moving explorer travels a contracted distance between the Earth and Ross 128, which is about 13 times shorter than we see on the Earth. The speed of the spaceship is constant in both frames of reference, so the explorer travels only 0.845 light-years to the star.

20 The observer on the Earth will not measure the proper time of the muon's lifetime. Instead they will see that the muon's lifetime is slow according to the equation $t = t_0\gamma$, where t_0 is the lifetime of a muon at rest. The result is that the observer sees the muon live a much longer time, t, and therefore makes it to the surface. The muon will see the Earth approach at a very high speed (approx. 0.992c) and will see the distance contracted. It will not be 15.0 km. Instead it will travel a much shorter distance according to the equation $L = \frac{L_0}{\gamma}$. As the distance the muon travels is shorter, it can make the journey to the Earth's surface during its proper lifetime.

21 a 4.41×10^{15} J b 1.16×10^{15} J

22 13.9 MeV

23 a 1.72×10^2 MeV
b $n^0 + n^0 \longrightarrow n^0 + p^+ + \pi^-$

$(2 \times 0) \longrightarrow (1 \times 0) + (1 \times +) + (1 \times -)$

$0 \longrightarrow 0$

24 a 85.0248 MeV
b 42.5124 MeV
c $e^+ + e^- \longrightarrow \gamma^0 + \gamma^0$

$(1 \times +) + (1 \times -) \longrightarrow (2 \times 0)$

$0 \longrightarrow 0$

Unit 4 Area of Study 1 Review

1 A
2 B and E
3 C
4 B
5 B
6 C
7 A
8 C
9 A and C
10 A
11 C
12 D
13 A
14 C
15 A

16 The resulting waveform will be the vector addition of the displacement of the individual particles in the waves due to the principle of superposition. Although there is a different displacement as the waves are superimposed, passing through each other does not permanently alter the shape, amplitude or speed of the individual waves.

17 a A = node, B = antinode
b The two images show the maximum and minimum positions of the rope as it oscillates.
c third harmonic
d The oscillating hand makes the incident waves. Those waves approach the hook. As the hook is a fixed end, the wave is inverted and reflected back towards the hand. The continuous incident waves superimpose with the reflected waves, forming a series of nodes and antinodes in the standing wave.
e 3.8 Hz

18 a Δx doubles b Δx doubles
c no effect d Δx decreases
e a wider central band

19 a wave model and particle model
b The bright and dark bands seen on the screen can only be due to the interference of waves. The particle model could not explain these bands.

20 A and B. At A, two crests are arriving in phase. At B, two troughs are arriving in phase. At C and D, a trough and a crest meet 180° out of phase, resulting in destructive interference.

21 500 nm

22 Student answers will vary. One answer is that only certain frequencies of light will emit photoelectrons. See the fully worked solutions for more.

23 a a series of bright fringes (indicating that electrons were detected) and dark fringes (indicating that no electrons were detected)
b The electrons are exhibiting wave-like behaviour.
c 2.4×10^{-9} m
d i 1.6×10^2 m s^{-1}
ii 1.6×10^2 m s^{-1}

24 a i 2.1×10^{15} Hz
ii 1.2×10^{15} Hz
iii 5.1×10^{14} Hz
b

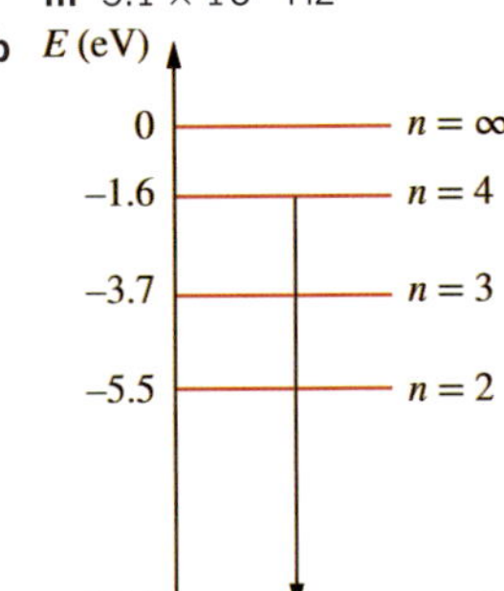

c 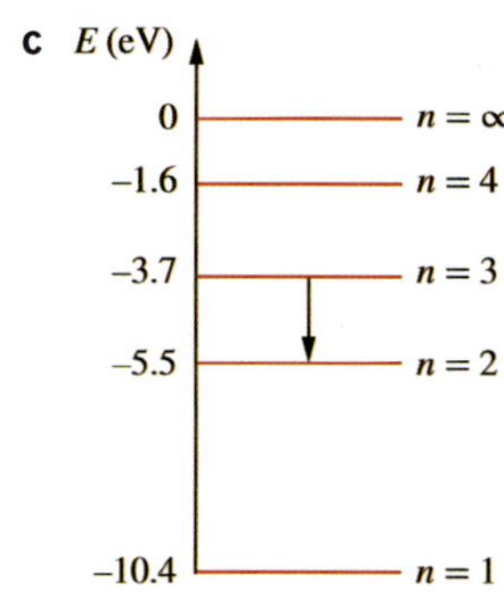

d 1.8 eV, 4.9 eV, 6.7 eV

25 a 4.8×10^{6} m s^{-1}

b 1.5×10^{-10} m

26 If incident light has energy less than the minimum energy difference between the lowest and next orbital levels within the hydrogen atom, the light would not cause any orbital changes. Therefore the light would not be absorbed by the atom.

27 a 0.0416 nm

b The electrons were diffracted as they passed through the gaps between the atoms in the crystal, thus creating a diffraction pattern. This pattern would be circular fringes of specific spacing around a common central point. Dark bands are due to destructive interference; bright bands are due to constructive interference.

c As the accelerating voltage is increased, the speed of the electrons would increase and thus their momentum would increase. This reduces their wavelength. Since the amount of diffraction is proportional to wavelength, less diffraction occurs. Less diffraction means the overall pattern is smaller, i.e. the circular bands are more closely spaced.

28 Electrons have a de Broglie wavelength and their orbit must fit an integral number of wavelengths for a standing wave to be formed. Only energy levels corresponding to these wavelengths exist, which means that atoms have quantised energy levels.

29 After discovering that light displays both particle and wave properties, physicists discovered that matter also has wave properties (when moving very fast) as well as particle properties. Both matter and light display wave–particle duality, but matter waves are not experienced in our macroscopic, low-speed world.

30 a In the particle model, the energy of the incident photons is set by their frequency. Each photon interacts with only one electron, therefore the energy of an emitted electron will depend only on the frequency of the incident light. Electron energy is not altered by altering the intensity because this only varies the number of photons, not their energy. Therefore the energy of the emitted electrons is not affected, only the number emitted.

b The wave model predicts that altering the intensity of light corresponds to waves of greater amplitude. Hence the wavefronts should deliver more energy to the electrons and the emerging electrons should thus have higher energy (which is not observed).

31 a $E_{k\,max}$ = the maximum kinetic energy with which electrons are emitted; f is the frequency of the light incident on the metal plate; ϕ is the work function, which is the minimum energy required to eject an electron

b no change

c a greater photocurrent

32 3

33 a The amount of diffraction is proportional to $\frac{\lambda}{w}$. Given that the X-rays and electrons are both diffracting through the same sample of crystal, they are diffracting through the same width, w. Therefore they must have equivalent wavelengths.

b 1.8×10^{-23} kg m s^{-1}

c No. The energy of the X-rays, being a form of electromagnetic radiation that travels at c, is given by $E = \frac{hc}{\lambda}$ and the energy of the electrons is given by $\Delta E_k = \frac{1}{2}mv^2$, as electrons have mass and do not travel at c.

d As optical microscopes use visible light, they will encounter much diffraction around objects with widths similar to the wavelength of light, resulting in blurriness. As electrons have wavelengths smaller than visible light, less diffraction will occur with widths similar to the wavelength of light, resulting in less blurry images.

34 a 5.6 years

b 2.4 years

c Relative to Raqu, the distance appeared to be shortened by the factor γ, thus the distance she travelled was much less than 5 light-years.

35 a 4.2×10^{-12} J

b 9.3×10^{37}

c 3.8×10^{14} kg

Glossary

RED ENTRIES VCE Physics Study Design extracts © VCAA (2022); reproduced by permission

A

absorb To take in (energy).

absorption line spectrum A spectrum containing dark lines in the positions of the wavelengths that are absorbed by a gas as light passes through it. See also *emission line spectrum*.

acceleration due to gravity The rate at which a falling object will accelerate in a gravitational field. Equivalent to the gravitational field strength. It is measured in m s^{-2} and its symbol is g.

accuracy A measurement value is considered to be accurate if it is judged to be close to the true value of the quantity being measured. Accuracy is a qualitative term; a measurement value or measurement result may be described, for example, as being 'less accurate' or 'more accurate' when compared with a true value.

aether An invisible, massless, rigid substance that was proposed as the medium in which light waves propagate. There is no experimental evidence for the existence of the aether.

aim A statement describing what will be investigated in an experiment.

air resistance A retarding force that acts in the opposite direction to the motion of an object as it passes through air.

alternating current An electric current that periodically reverses direction and with a magnitude that varies over time. Abbreviated to AC. Household power supplies usually operate at 240V AC.

alternator An electric generator that produces alternating current (AC).

altitude The height above an object's surface (such as a planet or moon).

amplitude The absolute value of the maximum displacement from zero during one period of one oscillation.

annihilation The complete conversion of matter into energy when matter meets its antimatter equivalent. This is not a chemical process, where matter in one form is converted to matter of another form, as occurs in burning.

anode A positively charged plate in an electron gun. Electrons are accelerated towards the anode.

antinode The region of maximum amplitude between two adjacent nodes in a standing wave or interference pattern. See also *node*.

armature A revolving structure in an electric motor or generator that is wound with a coil or coils that carry current. It rotates within a magnetic field to induce an EMF.

artificial satellite A body that is made by humans and placed in orbit around a planet or the Moon, e.g. Sputnik, the Hubble Space Telescope and NOAA-19.

B

banked track A track inclined at an angle to the horizontal. This enables vehicles to corner at higher speeds than if the track were horizontal.

bias A form of systematic error resulting from the researcher's personal preferences or motivations.

breaking point The point at which a material cannot be extended any further without it breaking.

brushes Devices that transfer the current in the rotating coil to a stationary external circuit by pressing against the split ring commutator or against the slip rings.

C

cathode The filament in a cathode ray tube which, when heated, produces electrons.

cathode ray tube A vacuum tube in which a hot cathode emits a beam of electrons that travel towards a high-voltage anode. The beam can be focused or deflected by electrodes that generate an electric field.

centripetal acceleration The acceleration directed towards the centre of the circle when an object moves with constant speed in a circular path.

centripetal force The force that causes an object to travel in a circular path, e.g. gravity, tension, normal force and friction.

classical physics The physics of Galileo and Newton, in which the addition of velocities has no limit, and which treats length and time as fixed properties across all frames of reference. Accurately describes the macroscopic world but becomes inaccurate for things that are very small or moving very fast.

coherent Describes waves that have the same frequency and waveform and are either in phase or have a constant phase difference.

commutator The part of an electric motor that allows current to be reversed. This enables the coil to keep rotating due to a magnetic force.

compression An area of increased pressure within a longitudinal wave.

conclusion An evidence-based statement that is developed from an analysis of experimental results.

conserved Not created or destroyed, but remaining constant.

constructive interference The process by which two or more waves combine to reinforce each other. The amplitude of the resulting wave is equal to the sum of the amplitudes of the superimposed waves.

continuous variable A variable that can have any numerical value within a given range.

controlled variable A variable which is kept constant in order to reliably find the effect of changing the independent variable.

crest The maximum or highest point in a cycle of a transverse wave.

D

de Broglie wavelength The wavelength of a particle due to its motion.

deformation The change in shape of an object as a result of the application of a force. This is often used to describe a permanent change, as when work is done to change the structure of the object.

dependent variable The variable that is to be measured and that is expected to change in response to changes in the independent variable.

design speed The speed at which a vehicle experiences no sideways force as it travels around a banked track. It is dependent on the banking angle.

destructive interference The result observed when waves interact and the amplitude of the resulting wave is equal to the difference between the amplitudes of the superimposed waves.

diffraction The bending of waves around an obstacle or through gaps in a barrier.

diffraction grating A piece of material with numerous closely spaced parallel slits used to create diffraction patterns.

diffraction pattern The pattern of dark and light bands seen when light passes through a single small gap. Areas of constructive interference appear as bright bands and areas of destructive interference appear as dark bands.

dipole A pair of electric charges or magnetic poles that have equal magnitudes but opposite signs, usually separated by a small distance.

direct current A continuous electric current that flows in one direction only and without substantial variation in magnitude. Batteries are a source of direct current. Abbreviated to DC.

discrete variable A variable that can be counted or measured but which can only have certain values, e.g. the number of times an experiment is repeated or the number of protons in a nucleus.

E

elastic A material that returns to its original shape after the application of a force.

elastic collision A collision in which kinetic energy is conserved.

elastic limit The maximum force that can be applied to a material before permanent deformation occurs.

elastic potential energy The energy stored in a material when it is stretched or compressed. If the material is elastic, the energy can be returned to the system. If the material is inelastic, a permanent change occurs.

electric field A region of space where charged objects experience a force due to the field created by another charged object.

electric field strength A measure of the force per unit charge on a charged object within an electric field, with units N C^{-1}. Electric field strength can also be a measure of the difference in electrical potential per unit distance, with units V m^{-1}.

electrical potential The work required per unit charge to move a charged object from infinity to a point in the electric field. The unit in J C^{-1}.

electromagnet An iron or steel core wound with a coil of wire through which a current can be passed. The core becomes magnetised when current is flowing.

electromagnetic induction The creation of an electric current, or an EMF, in a loop of wire as the result of a changing magnetic flux through the loop.

electromagnetic radiation Energy emitted in continuous waves with two transverse, mutually perpendicular components. It is comprised of a varying magnetic field and a varying electric field.

electromagnetic spectrum The range of all possible frequencies of electromagnetic radiation. The visible spectrum is just one small part of the electromagnetic spectrum.

electron gun A device where a heated cathode produces a beam of electrons which is accelerated by a series of charged plates.

electron volt An amount of energy equal to the charge of an electron multiplied by 1 volt, i.e. $1\,\text{eV} = 1.6 \times 10^{-19} \times 1 = 1.6 \times 10^{-19}\,\text{J}$. It is an alternative to the joule as a unit for measuring energy.

EMF A source of energy that can cause a current to flow in an electrical circuit or device.

emission line spectrum A spectrum of coloured lines in the positions of the wavelengths of light emitted when a gas is heated or has an electric current passed through it. See also *absorption line spectrum*.

excited state An energy state of an atom that is above its ground state (i.e. $n > 1$).

F

Faraday's law The average EMF generated in a coil is proportional to the rate of change of magnetic flux and the number of turns in the coil.

field A region of space around an object where a force can be felt by other objects.

field lines A 2-dimensional representation of a field where arrows are used to indicate the direction of the field. The closer the field lines, the stronger the field.

frame of reference A coordinate system that is based on the position of an object or observer.

free fall A body's motion where gravity is the only force acting on it.

frequency The number of waves passing a given point in one second or the number of repeats of a cycle each second. Measured in hertz (Hz).

fundamental frequency The harmonic with the lowest frequency. It is the simplest form of vibration, having only one antinode.

G

Gedanken German word for 'thought'. Einstein used this term to describe his theoretical 'experiments' on relativity.

generator An electrical device that converts kinetic energy into direct current (DC) electricity. It usually has a rotating coil that cuts across a magnetic field.

gravimeter A sensitive instrument used by geologists to detect small variations in gravitational field strength.

gravitational constant A universal constant of value $6.67 \times 10^{-11}\,\text{N}\,\text{m}^2\,\text{kg}^{-2}$.

gravitational field The region around an object where other objects will experience a gravitational force.

gravitational field strength The strength of gravity, usually measured at the surface of a planet or moon. It is equivalent to the acceleration due to gravity, g, and measured in newtons per kilogram ($\text{N}\,\text{kg}^{-1}$).

gravitational force The force of attraction acting between two objects that have mass.

gravitational potential energy The energy that a body possesses due to its position in a gravitational field. It is a scalar quantity measured in joules (J).

ground state The lowest energy state of an atom (i.e. where $n = 1$).

H

harmonic A frequency that is a whole-number multiple of the fundamental frequency.

hypothesis A proposed explanation for an observed phenomenon. It is a prediction based on scientific reasoning that can be tested experimentally.

I

ideal transformer A transformer where the output power is the same as the input power (i.e. one that is 100% efficient). Real transformers come close to this value.

impulse The change in the momentum of an object. It can be calculated as the difference between the final momentum and the initial momentum. For collisions, it can also be calculated by multiplying the average force by the duration of the interaction or by finding the area under a force–time graph.

inclined plane A sloping surface; a ramp.

independent variable The variable that the experimenter varies.

induced current The electric current produced by changing a magnetic flux in the region of a conductor, or by moving the conductor in a magnetic field.

inelastic collision A collision in which kinetic energy is not conserved.

inertial frame of reference A frame of reference that is either moving with constant velocity or is stationary (i.e. it is not accelerating).

interference The process in which two or more waves combine to reinforce or weaken each other. The amplitude of the resulting wave is equal to the vector sum of the amplitudes of the superimposed waves. See also *constructive interference* and *destructive interference*.

interference fringe A bright or dark band in an interference pattern. The bright bands are caused by constructive interference and the dark bands are caused by destructive interference.

inverse square law The relationship between two variables where one is proportional to the reciprocal of the square of the other. An example is Newton's law of universal gravitation.

inverter A device that converts DC current to AC. They are commonly used to convert current from solar panels so that it can be used in the home.

isolated system A system where only internal forces act between the objects in it, i.e. there are no interactions with objects outside the system.

K

kinetic energy The energy of a moving body, measured in joules (J). Kinetic energy is a scalar quantity.

L

law of conservation of energy Energy cannot be created or destroyed, but only changed or transformed from one form to another.

law of conservation of momentum In any collision or interaction between two or more objects in an isolated system, the total momentum of the system remains constant, i.e. the total initial momentum is equal to the total final momentum.

length contraction An object's length when measured in a moving frame of reference is less when measured by a stationary observer.

Lenz's law The direction of the induced current in a conductor is such that its associated magnetic field opposes the change in flux that caused the current.

line of best fit A straight line that best represents scatterplot data. It does not necessarily pass through every or any data point. Also known as a trend line.

linear relationship A relationship where the variables are in direct proportion to each other. It produces a straight trend line.

longitudinal Extending in the direction of the length of something rather than its width or height. The vibrations of a longitudinal wave are in the same direction as—that is, parallel to—the direction in which the wave travels.

Lorentz factor Describes how much length and time change for a moving object: $\gamma = \dfrac{1}{\sqrt{1-\dfrac{v^2}{c^2}}}$

Lorentz force The force experienced by a point charge moving along a wire that is in a magnetic field. The force is at right angles to both the current and the magnetic field.

M

magnetic field A region influenced by a magnet.

magnetic flux The strength of a field in a given area expressed as the product of the area and the component of the field strength at right angles to the area (i.e. $\Phi = B_{\perp}A$).

magnetic flux density The amount of magnetic flux per unit area, i.e. the closeness of magnetic field lines. Same as magnetic field strength.

magnetic pole One of the two limited regions of a magnet where the magnetic field is most intense.

magnitude The size of a quantity without regard for its direction.

mass–energy equivalence The relationship between mass and energy. A particle at rest has energy equal to its mass multiplied by the square of the speed of light.

mean The average of the numeric data in a set.

mechanical energy The energy that a body possesses due to its position or motion. Kinetic energy, gravitational energy and elastic potential energy are all forms of mechanical energy.

mechanical wave A wave that propagates as an oscillation of matter and therefore transfers energy through a medium.

median The middle number in an ordered set of data.

medium A physical substance, such as air or water, through which a mechanical wave is propagated.

metal vapour lamp A lamp that contains a gas that, when excited, emits photons with the colour that is characteristic of the main element in the gas, e.g. a sodium vapour lamp.

method The specific steps that are taken to collect data during an investigation.

methodology The approach taken to investigate a research question or hypothesis, e.g. a controlled experiment, fieldwork, literature review, modelling or a simulation.

mistake An error made by an experimenter that could have been avoided.

mode The value that appears the most often in a data set.

momentum The product of the mass and the velocity of an object. Momentum is a vector and measured in kg m s^{-1}.

monochromatic Of a single colour, e.g. red light.

monopole A single mass or point electric charge. A mass is considered to be a monopole at its centre of mass. Magnetic poles only exist, as far as we know, as dipoles. See also *dipole*.

N

natural satellite A body not made by humans that is in orbit around another body, e.g. the Moon or a planet.

Newton's law of universal gravitation The attractive force between two masses due to gravity is directly proportional to the product of their masses and inversely proportional to the square of the distance between them.

node A point at which the amplitude of two or more superimposed waves has a zero or minimum value. See also *antinode*.

normal force The force with which a surface pushes back on an object. It is always at right angles to the surface. Same as the apparent weight of an object. It is measured in newtons and its symbol is F_N or N.

nuclear fusion A reaction in which nucleons are joined to form a new species of nucleus and, in the process, release energy.

O

observation The gathering of information in a variety of ways, such as using your senses, employing instruments or through laboratory techniques.

outlier A reading that lies a long way from other results. Repeating readings may be useful in further examining an outlier.

overtone Any harmonic other than the fundamental frequency.

P

pair production The creation of a particle and its antiparticle. This is commonly the result of two photons interacting or a photon interacting with an atomic nucleus.

particle accelerator A machine that can accelerate a charged particle—such as a proton or electron—or an atomic nucleus to very high speeds (in some cases approaching the speed of light).

path difference The difference in the lengths of the paths from each slit to the screen in a double-slit experiment.

period The time interval taken to complete one cycle of a regularly repeating phenomenon, such as a rotating object or in a sound wave. The SI unit is seconds (s).

personal protective equipment (PPE) Clothing that is worn to minimise personal risk in an investigation, e.g. safety glasses, protective gloves, ear muffs and a laboratory coat.

phase The fraction of a cycle of a wave that has been completed at a specific point in time, usually expressed as an angle. A particular stage in a periodic process such as a wave.

photocurrent Current caused by the flow of photoelectrons during the photoelectric effect.

photoelectric effect The spontaneous emission of electrons from a metal surface when it is illuminated by light of particular frequencies and energies.

photoelectron An electron released from an atom due to the photoelectric effect.

photon A packet or bundle of electromagnetic radiation such as light.

photovoltaic effect The generation of current when certain semiconductor materials are exposed to sunlight (as occurs in solar panels).

point charge An ideal situation in which all the charge on an object can be considered to be concentrated at a single point. The point size is negligible in considering the distance between it and another point charge

postulate A suggestion that is put forward as a fact to form a basis for further discussion or reasoning.

potential difference The work required per unit charge to move a charged object between two points in an electric field. Its units are J C^{-1} or volts (V).

precision A measure of the repeatability or reproducibility of scientific measurements and refers to how close two or more measurements are to each other. A set of precise measurements will have values very close to the mean value of the measurements. Precision gives no indication of how close the measurements are to the true value and is therefore a separate consideration to accuracy.

primary data The data you have collected yourself.

primary source Information created by a person directly involved in an investigation.

processed data Raw data that has been organised, altered or analysed to produce meaningful information.

projectile An object moving freely through the air without an engine or power source driving it.

proper length The length of an object as measured in the frame of reference in which the object is stationary.

proper time The duration as measured by a clock that doesn't move relative to the observer measuring the duration.

Q

qualitative data Non-numeric data from categorical variables that can be counted but not measured, e.g. colour and brightness.

qualitative variable Variables that can be observed but not measured, e.g. colour and brightness.

quantitative data Numeric data from variables that can be measured, e.g. wavelength and temperature.

quantitative variable Variables that can be measured, e.g. wavelength and temperature.

quantum (Plural: quanta) A discrete packet of electromagnetic radiation emitted from objects. Each quantum has energy proportional to its frequency, as per the equation $E = hf$.

R

random errors Unpredictable variations in the measurement process. They are present in all measurements except those involving counting. Random errors affect the precision of a measurement. They can be reduced by making more or repeated measurements and calculating a new mean and/or by refining the measurement method or technique.

range The difference between the highest and lowest values in a numeric data set.

rarefaction An area of decreased pressure within a longitudinal wave.

raw data Data that has not been processed or analysed.

reflect To cause something—such as light, sound, particles or waves—to bounce back after reaching a boundary or surface.

repeatability The closeness of the agreement between the results of successive measurements of the same quantity being measured in an experiment, carried out under the same conditions of measurement.

reproducibility The closeness of the agreement between the results of measurements of the same quantity being measured, carried out under changed conditions of measurement.

research question A question that defines the focus of an investigation.

resonance The state of a system in which an abnormally large vibration is produced in response to an external vibration. Resonance occurs when the frequency of the external vibration is the same, or nearly the same, as the natural vibration frequency of the system.

resonant frequency The natural frequency of an object's vibration.

right-hand force rule Tells us the direction of the force (palm) on a current (thumb) in a magnetic field (straight fingers).

right-hand grip rule Tells us the direction of the magnetic field (curled fingers) around a current (thumb).

risk assessment An activity to identify, assess and control hazards.

root mean square The square root of the arithmetic mean of the squares of the numbers in a given set. For alternating power the root mean square value, P_{rms}, is $\frac{P_{peak}}{2}$. Similarly, it is the effective mean (average) power rating value of an AC supply.

S

safety data sheet (SDS) Important information about the possible hazards in using a substance and how it should be handled and stored.

satellite An object in a stable orbit around a central body. A satellite could be natural, such as a moon, or artificial, such as a communications satellite.

scientific method The process scientists use to construct and test theories that explain observations.

secondary data Data you have not collected yourself.

secondary source A synthesis, review or interpretation of primary sources. For example, textbooks and news articles are secondary sources.

significant figures The number of digits used in a calculation. For example, 5.1 has two significant figures, whereas 5 has just one. Some numbers may have an ambiguous number of significant figures. For example, 100 could have three or one significant figure. In VCE physics, trailing zeros are significant, thus 100 can be assumed to have three significant figures.

simultaneous Occurring at exactly the same time.

slip rings Components of alternators (i.e. AC generators) that allow a constant electrical connection to be made between the rotating armature and the static external circuit through which the alternating current that is generated flows.

solenoid A coil of wire that acts as an electromagnet when electric current is passed through it due to the magnetic field created by the current. Solenoids are often used to control the motion of metal objects, such as the switch of a relay.

spacetime A term used to describe the combination of the three-dimensional space coordinate system with the one-dimensional time system.

split ring commutator A component of DC generators and motors that resembles a ring that has been cut into two equal parts. Each part has a fixed connection to the ends of the coil while also making contact with stationary brushes. This means that the connection between the rotating coil and the static circuit is reversed every half turn, which ensures that the direction of current in the circuit is constant (in the case of the generator) or the direction of rotation is constant (in the case of the motor).

spring constant A quantity that describes the stiffness of a particular spring or other elastic material. A larger spring constant indicates a stiffer material.

standing wave The periodic disturbance in a medium resulting from the combination of two waves of equal frequency and intensity travelling in opposite directions. Also called a stationary wave.

stator A portion of a machine that remains stationary with respect to rotating parts, especially the collection of stationary parts in the magnetic circuits of a motor or generator.

step-down transformer A device that outputs a lower voltage than the voltage of the input.

step-up transformer A device that outputs a higher voltage than the voltage of the input.

stopping voltage The applied voltage required to stop all photoelectrons from leaving a particular metal. For a particular frequency of incident light on a particular metal, the stopping voltage is a constant.

superposition Relating to the interaction of two or more waves. The principle of superposition states that the resulting wave at any moment is the vector sum of the displacements of each of the interacting waves.

synchrotron Large particle accelerator that accelerates electrons in a circular path, producing a very intense, very narrow beam of electromagnetic radiation called synchrotron radiation.

systematic errors Errors that cause readings to differ from the true value by a consistent amount each time a measurement is made, i.e. all readings are shifted in one direction from the true value. Systematic errors affect the accuracy of a measurement and the accuracy cannot be improved by repeating those measurements.

T

tangential Describes a direction that is a tangent to a curve.

threshold frequency The minimum frequency of electromagnetic radiation for which the photoelectric effect can occur for a given material.

time dilation When an observer watches events in a frame of reference that is moving very fast relative to them, time in that frame of reference will appear to go more slowly. People in the moving frame do not experience any difference in the rate at which time passes. This effect is one of the results of Einstein's theory of special relativity.

torque Any force or system of forces that causes or tends to cause rotation. A turning or twisting force effect.

transform To change form, e.g. the stored energy in a compressed spring changes to kinetic energy as the spring is released.

transformer A device that transfers an alternating current from one circuit to one or more other circuits, usually with an increase (step-up transformer) or decrease (step-down transformer) in voltage. The input goes to a primary coil and the output is taken from a secondary coil (or from windings linked by induction to the primary coil).

transmit To allow light, heat, or sound etc. to pass through.

transverse Lying or extending across something. The vibrations of a transverse wave are at right angles to the direction of travel of the wave.

trough The minimum or lowest point in a cycle of a transverse wave.

true value The value that would be found if the quantity could be measured perfectly.

U

uncertainty The true value is expected to lie in the interval expressed as mean ± uncertainty. For example, for the measurement result 20 ± 1 mm, the uncertainty is 1 mm and the interval is between 19 to 21 mm.

uncertainty bars Graphical representations on graphs to show the uncertainty in a measurement and thus indicate the variability of data.

uniform Constant, unvarying.

unresolved Being unable to distinguish between multiple objects in an image (such as from a telescope or microscope).

V

validity A valid experiment investigates what it sets out and/or claims to investigate. Both experimental design and the implementation should be considered when evaluating validity. An experiment and its associated data may not be valid, for example, if the investigation is flawed and controlled variables have been allowed to change. Data may not be valid, for example, if there is observer bias.

variable A factor or condition that can change.

W

wavelength The distance between two successive points on a wave during the same phase.

wave–particle duality The theory that light and matter sometimes behave like waves and sometimes like particles.

work The transfer or transformation of energy. Work is done when a force causes a displacement in the direction of the force.

work function The energy required to remove an electron from an atom. It is measured in joules or electron volts.

Index

G

H

I

J

K

L

M

N

O

P

Q

R

S

T

U

V

W

X

Y

Attributions

Heinemann Physics 12 5th edition student book attributions

Front and back covers: Science Photo Library: Parker, David.

123RF: 1xpert, p. 179b; Dimitrova, Diyana, p. 367; Guseva, Alexandra, p. 138; imagesource, p. 102l; jewhyte, p. 70t; Kuvaiev, Denys, p. 132r; lightpoet, p. 144; lightpoet, p. 414t; Machin, Rafael Angel Irusta, p. 71bl; manganganath, p. 401; Prudek, Daniel, p. 186; Sutandio, Alexius, p. 201 (shark); tristan3d, p. 251.

AAP: Davy, Adam/PA Wire, p. 81; Ugeda, Yukihiro/AP, p. 118.

Air Liquide: p. 19.

agefotostock: Nunuk, David, p. 379b.

Alamy: ABACAPRESS, p. 243t; Actionplus, p. 121l; Agencja Fotograficzna Caro, p. 123r; Alchemy, p. 213br; Andrew Lambert Photography/Science Photo Library, pp. 262t, 282, 292; Barnes, Stephen/Sport, p. 94b; Bond, Martin/Science Photo Library, p. 220t; Brookes, Andrew, National Physical Laboratory/ Science Photo Library, p. 404t; CAMERIQUE/ClassicStock, p. 426t; Chillmaid, Martyn F./Science Photo Library, p. 13b; cisfo/mauritius images GmbH, p. 333t; Clariva, Carlos/Science Photo Library, p. 382; Cooper, Ashley/Science Photo Library, p. 302t; corleve/Alamy Live News, p. 120 (tennis player); Dalhoff, Henning/Science Photo Library, p. 376; Dorling Kindersley/ UIG/Science Photo Library, p. 165; Dr. Brain, Tony/Science Photo Library, pp. 375c, 375l, 375r; Dr. Hweeler, Keith/Science Photo Library, p. 347b; Dr. Settles, Gary/Science Photo Library, p. 152; Dunkley, Dan/Science Photo Library, p. 49; European Space Agency/AOES Medialab/Science Photo Library, p. 245b; European Space Agency/Science Photo Library, pp. 185, 249; Fowler, Lincoln, p. 301; Frantic, p. 4; Garlick, Mark/Science Photo Library, p. 245t; Gl0ck, p. 300b; Goode, Tim/PA Images, p. 80t; Griffiths, Tom/ZUMA Wire/ZUMAPRESS.com, p. 121c; H.S. Photos, pp. 378rt, 387br; Hebditch, Paul, p. 83; imageBROKER, p. 284; Kightley, Russell/Science Photo Library, pp. 326–7, 349r; King-Holmes, James, p. 236–7; koene, ton, p. 17; Marmaduke St. John, p. 218lb; NASA/Science Photo Library, p. 62; NOAA/ Science Photo Library, p. 29; Parker, David/Science Photo Library, p. 324–5, 436–40; Photo12 Collection, p. 370; Popular Click Photography, p. 75tl; Prisma Archivo, p. 248tr; Royal Astronomical Society/Science Photo Library, p. 175br; Royal Institution of Great Britain/Science Photo Library, p. 276; Scherl/Süddeutsche Zeitung Photo, pp. 360b, 371b; Science History Images, p. 392–93; Science Photo Library, p. 418; sciencephotos, p. 346tl; Scientific Picture Research/Science Photo Library, p. 18lb; Sirulnikoff, Alan/Science Photo Library, p. 18lt; Trevor Clifford Photography/Science Photo Library, pp. 215b, 224bl; VintageCorner, p. 164l.

Australian Nuclear Science and Technology Organisation (ANSTO): pp. 419, 261t, 267.

Binte Mohamed Sapi'ee, Amirah Fatin: p. 345.

Brookhaven National Laboratory: p. 261b.

Burrows, Keith: p. 212cr.

Carbon Reduction Industries Pty Ltd: Image courtesy of EcoSwitch ® A Division of Carbon Reduction Industries Pty Ltd, p. 308.

Chau, Audrey: p. 340l.

Cross, Malcolm: pp. 346cl, 346cr.

Crouch, Vanessa: Courtesy of Jonathan Horner, p. 169.

Fotolia: Ezume Images, p. 8; Ireland, Michael, p. 200; Jewell, Craig, p. 199t; Mr Twister, p. 330br; rueangwit, p. 333b.

Getty Images: Bettmann, p. 360t; cruphoto, p. 358–9; Dwyer, Philip A./Bellingham Herald/Tribune News Service via Getty Images, p. 179t; Ellis, Paul/AFP, p. 121r; fStop Images/Caspar Benson, p. 122t; Isakovic, Andrej/AFP, p. 98l; MacNicol, Ian, p. 60–1; moodboard, p. 244l; Pierse, Ryan, p. 126; Thornton, Jayme, p. 109; TR/AFP/ p. 314.

Len, Patrick M: Cuesta College Physical Sciences Division instructor Dr. Patrick M. Len. Licensed under CC BY 2.0, link to licence: https://creativecommons.org/licenses/by/2.0/, p. 346bl.

Mann, Dale: p. 201 (grass seeds).

NASA: Courtesy of nasaimages.org, pp. 175bl, 182b, 187r, 426b; ESA/Hubble Telescope, p. 246; ESA, CSA, and STScI: p. 1–3; Rohr, Dixon, p. 213t; ISS Expedition 12 Crew, p. 248l; JPL, pp. 162–3, 175t.

National Physical Laboratory: p. 402.

Oakley, Terry: pp. 373tt, 373tc, 373tb.

PASCO Scientific: pp. 22tl, 22tr, 22bl, 22br.

Pearson Education Ltd: p. 224t; HL Studios, p. 277.

Pearson Education Asia Ltd: Leung, Terry, p. 347c; Yuen, Coleman, p. 214cr.

Phillips, Thomas (1842): p. 255br.

Science Photo Library: pp. 94t, 212lt, 230t, 319; Andrew Lambert Photography, p. 355r; N.A.SHARP, NOAO/NSO/KITT PEAK FTS/AURA/NSF, p. 378lb; NASA/Walter Myers, p. 422; Winters, Charles D., pp. 220b, 264t.

Science Source: GIPhotoStock, pp. 350l, 352t.

Shutterstock: p. 346br; Assmy, Gunar p. 100; Dedyukhin, Dmitry, p. 123l; Fleet, Paul, p. 238t; Frazao, Filipe, p. 166; FreshPaint, p. 110; Graiki, Mauricio, p. 199b; Habich, Arina P. p. 87t; HTU, pp. 58–9, 158–61, 272–74, 320–23; Iryna1, p. 164r; JRC-Stop Motion, p. 148; Kinney, Brian, p. 85l; koya979, p. 384; KPhrom, p. 344; Miss Kanithar Aiumla-Or, p. 304tl; MNI, p. 210; muzsy, p. 70bl; Pichitchai, p. 304tr; pistolseven, p. 335 (violinist); Publicover, Keith, p. 167; pukach, p. 208 (balloons); RemarkEliza, p. 341lt; Santa Maria., Carlos E., p. 379t; solarseven, p. 343t; Stetson, Brian L. p. 196–98; Tristan3D, p. 182t; Troke, Matthew, p. 275.

United Nations Commission for Europe (UNECE): p. 19.

Unknown via Wikipedia: p. 394.

Victorian Curriculum and Assessment Authority (VCAA): Selected extracts from the VCE Physics Study Design (2023–2027) are copyright Victorian Curriculum and Assessment Authority (VCAA), reproduced by permission. VCE® is a registered trademark of the VCAA. The VCAA does not endorse this product and makes no warranties regarding the correctness or accuracy of its content. To the extent permitted by law, the VCAA excludes all liability for any loss or damage suffered or incurred as a result of accessing, using or relying on the content. Current VCE Study Designs and related content can be accessed directly at www.vcaa.vic.edu.au, pp. 1–3, 59, 61, 109, 163, 197–8, 237, 275, 325, 327, 359, 393.